Geotechnologies and the Environment

Volume 21

Series Editors

Jay D. Gatrell, *Department of Geology & Geography, Eastern Illinois University, Charleston, IL, USA*

Ryan R. Jensen, *Department of Geography, Brigham Young University, Provo, UT, USA*

The *Geotechnologies and the Environment* series is intended to provide specialists in the geotechnologies and academics who utilize these technologies, with an opportunity to share novel approaches, present interesting (sometimes counterintuitive) case studies, and most importantly to situate GIS, remote sensing, GPS, the internet, new technologies, and methodological advances in a real world context. In doing so, the books in the series will be inherently applied and reflect the rich variety of research performed by geographers and allied professionals.

Beyond the applied nature of many of the papers and individual contributions, the series interrogates the dynamic relationship between nature and society. For this reason, many contributors focus on human-environment interactions. The series are not limited to an interpretation of the environment as nature per se. Rather, the series "places" people and social forces in context and thus explore the many sociospatial environments humans construct for themselves as they settle the landscape. Consequently, contributions will use geotechnologies to examine both urban and rural landscapes.

More information about this series at http://www.springer.com/series/8088

G. P. Obi Reddy • S. K. Singh
Editors

Geospatial Technologies in Land Resources Mapping, Monitoring and Management

Springer

Editors
G. P. Obi Reddy
Principal Scientist, Division of Remote Sensing Applications
ICAR-National Bureau of Soil Survey and Land Use Planning
Nagpur, India

S. K. Singh
Director, ICAR-National Bureau of Soil Survey and Land Use Planning
Nagpur, India

ISSN 2365-0575 ISSN 2365-0583 (electronic)
Geotechnologies and the Environment
ISBN 978-3-319-78710-7 ISBN 978-3-319-78711-4 (eBook)
https://doi.org/10.1007/978-3-319-78711-4

Library of Congress Control Number: 2018950388

Printed on acid-free paper

This Springer imprint is published by the registered company Springer Nature Switzerland AG
The registered company address is: Gewerbestrasse 11, 6330 Cham, Switzerland

Foreword

Land resources are central to sustainable agriculture and are intrinsically linked to global challenges of food insecurity, poverty, climate change adaptation and mitigation. With increasing population pressure on land resources and the need for increased agricultural production, necessity arises for improved management of the land resources, especially soils and water. Precise information on the nature, extent and spatial distribution of land resources is pre-requisite to address various issues of land degradation and climate change which affect the livelihoods of millions of rural people across the world. The inventory of land resources provides information about their nature, extent, evolution, their potentials and limitations to various uses.

Geospatial technologies encompasses the advanced tools such as remote sensing, geographic information system (GIS) and global positioning system (GPS), which play an important role in mapping, monitoring and management of land resources over the conventional methods. The rapidly advancing geospatial technologies have immense potential in land resource mapping and generation of spatial databases on a regular basis for monitoring and management of various action plans more precisely and efficiently at different levels. These technologies are therefore being effectively used for precise mapping and judicious management of land resources. The combination of these technologies provides a cost effective means of acquiring high resolution real-time data through remote sensing, data management and analysis through GIS and geo-referencing the ground truth data with GPS, putting all data in an information system and utilisation of the information for a specific purpose.

The major objective of this publication is thus to build awareness on various aspects of geospatial technologies and their principles and applications in land resource mapping, monitoring and management. The eminent authors from leading institutes like ICAR-NBSS&LUP, IIT-B, ICRISAT, and NRSC shared their valuable experiences and insights on principles of geospatial technologies and their applications in land resources mapping, monitoring and management.

It gives me an immense pleasure to note that the Editors have brought out this excellent publication on ***Geospatial Technologies in Land Resources Mapping, Monitoring and Management.*** It is a comprehensive and valuable reference material for the scientific and teaching community, extension scientists at research institutes, students at agricultural universities and colleges and those involved in planning and managing land resources for sustainable agriculture and livelihood security.

I complement the Editors for their sincere efforts in bringing out this valuable publication.

Trilochan Mahapatra

Secretary, DARE, Govt. of India &
Director General, ICAR, Ministry of
Agriculture and Farmers Welfare
New Delhi, India
18th July, 2018

Preface

The rapid advancing geospatial technologies, which include remote sensing, photogrammetry, cartography, geographic information systems (GIS), global positioning systems (GPS), and information technologies (IT), have revolutionized the field of land resources mapping and generation of location-specific spatial databases for effectively and efficiently monitoring and management at different scales. Chapter 1 in the volume provides an overview on applications of geospatial technologies in land resources mapping, monitoring, and management. The remaining 28 chapters in the volume have been organized under four sections namely Geospatial Technologies – Principles and Applications, Geospatial Technologies in Land Resources Mapping, Geospatial Technologies in Land Resources Monitoring, and Geospatial Technologies in Land Resources Management. Each chapter is authored by an expert(s) in the respective field of research. Each chapter begins with abstract and ends with conclusions and supported by suitable diagrams to help the readers to understand the complex geospatial technologies and their applications in land resources mapping, monitoring, and management.

Part I, "Geospatial Technologies – Principles and Applications," consists of seven chapters. As many are new to modern geospatial technologies, it is important to introduce the readers to the principles, applications, and analytical tools of geospatial technologies. Chapter 2 illustrates satellite remote sensing sensors – principles and concepts of different satellite remote sensing sensors and their resolutions. Chapter 3 describes principles and concepts of GIS, philosophy and components of GIS, geographic data representation, principal functions of GIS, and data formats of spatial features. Chapter 4 gives details on history of GPS, different segments of GPS, differential GPS, and precise satellite tracking. Chapter 5 provides details on components of Geo-ICDT architecture, sensing platforms, data processing, data management, dissemination, and multimodal platforms in providing advisory services. Chapter 6 provides the details on digital image processing techniques, image restoration, enhancement, and classification algorithms. Chapter 7 gives detail note on database generation in GIS, spatial data models, database query, overlay analysis, and modeling of spatial problems. Chapter 8 describes non-geostatistical and geostatistical interpolators, sampling strategy, error estimation, and use of geostatistics in spatial mapping of soil properties.

Part II, "Geospatial Technologies in Land Resources Mapping," consists of nine chapters. Chapter 9 provides an overview on land resources management for agricultural land use planning – challenges, conceptual model for land resources inventory on 1:10000 scale, characterization of landscape ecological units, and agricultural land use planning at 1:10000 scale with a case study. Chapter 10 provide insights on open source satellite datasets, GIS software and their applications in land resources management. Chapter 11 gives a note on remote sensing and GIS in digital terrain modeling; it includes digital elevation models – sources and resolutions, GIS and remote sensing in digital terrain analysis, and impact of scales in deriving host of terrain attributes. Chapter 12 describe various aspects of remote sensing and GIS applications in geomorphological mapping in different environments like fluvial, arid (aeolian), glacial, karst, and coastal. Chapter 13 gives a brief note on geospatial technologies for generation of semi-automated baseline database in large-scale land resources inventory with case studies. Chapter 14 describes importance of pedogeomorphic analysis in land resources inventory and remote sensing and GIS applications in land resources inventory with a case study. Chapter 15 is devoted to describe land resources inventory and mapping – tools and techniques, types of soil surveys, and remote sensing and GIS applications in detailed land resources survey with a case study. Chapter 16 provides a thorough theoretical background on applications of remote sensing in land resources inventory, mapping and assessment of soil properties using soil reflectance spectroscopy. Chapter 17 gives a note on land resources mapping for village level planning and soil suitability evaluation with a case study.

Part III, "Geospatial Technologies in Land Resources Monitoring," consists of five chapters. Chapter 18 provides an overview on applications of various remote sensing datasets for land resources monitoring and management and parametric methods of land evaluation. Chapter 19 devoted to remote sensing and GIS applications in land resources monitoring, crop condition, and drought assessment. Chapter 20 provides a note on remote sensing and GIS applications in mapping, monitoring, and assessment of land degradation. Chapter 21 describes on monitoring of spatiotemporal dynamics of *rabi* rice fallows in South Asia using temporal MODIS data. Chapter 22 describes the development of land resource information system and BHOOMI Geoportal for land resources monitoring and management.

Part IV, "Geospatial Technologies in Land Resources Management," consists of seven chapters. Chapter 23 gives details on land use planning – basics and approaches in land use planning. Chapter 24 devoted to explain various aspects of integrated remote sensing, GIS, and GPS applications in agricultural land use planning. Chapters 25 and 26 provide trends in land resource management, land use planning, and impacts of soil-based land use planning on livelihood security. Chapter 27 gives details on application of geospatial technologies in integrated watershed management. Chapter 28 illustrates geospatial applications in water resource management with special reference to climate change. Chapter 29 elaborates on applications of remote sensing and GIS in micro-level planning, land and water resources development, and monitoring at micro level.

We place on record our sincere regards to the Secretary, Department of Agricultural Research and Education (DARE), Government of India and Director General, Indian Council of Agricultural Research (ICAR), Ministry of Agriculture and Farmers Welfare, New Delhi, Deputy Director General (NRM) and Assistant Director General (S&WM), NRM Division, ICAR, New Delhi for proving valuable guidance, necessary facilities and encouragement in brought out this volume. We also sincerely thank to the Head, NRDMS (DST) for valuable support, guidance, and encouragement. We are also indebted for all the contributors for their untiring efforts to make this endeavour a success and sharing their experiences in applications of geospatial technologies in their respective field of research. In fact, without their support and timely inputs, this volume could not be a reality. We sincerely thank to the Heads of the Divisions, colleagues from Division of Remote Sensing Applications and PME Cell, ICAR-NBSS&LUP, Nagpur for their support and cooperation. We sincerely thanks to the editorial team of Springer especially J.D. Gatrell and R.R. Jensen, Series editors of 'Geotechnologies and the Environment', Mr. Marielle Klijn, Mr. R. Rameshbabu, Ms. Deepthi Vasudevan and Mrs. Rathika Ramkumar for their valuable support and co-operation in brought out the volume.

The book is extremely valuable for scientists, academicians, planners, policy makers, and students, who are interested to explore the applicability of modern geospatial techniques in addressing the issues of mapping, monitoring, and management of land resources. We strongly believe that this book is highly beneficial to the readers in better understanding the geospatial technologies and thier applications in land resources mapping, monitoring, and management to formulate future research and development programs in optimizing the land resources for sustainable agricultural land use planning and livelihood security.

ICAR-NBSS&LUP, Nagpur, India G. P. Obi Reddy
S. K. Singh

Abbreviations

AER	Agro-ecological Region
AESR	Agro-ecological Sub-region
AFPS	Air-Filled Pore Space
AFSIS	Africa Soil Information Service
AGPS	Assisted Global Positioning System
AHP	Analytical Hierarchical Process
AIS&LUS	All India Soil and Land Use Survey
ALES	Automated Land Evaluation System
ALOS	Advanced Land Observation Satellite
AOI	Area of Interest
AOS/CS	Agricultural Ontology Service Concept Server
APAR	Absorbed Photosynthetically Active Radiation
API	Application Programming Interface
ASAR	Advanced Synthetic Aperture Radar
ASRIS	Australian Soil Resource Information System
ASSOD	Soil Degradation in South and Southeast Asia
ASTER	Advanced Space-Borne Thermal Emission and Reflection Radiometer
AVHRR	Advanced Very High Resolution Radiometer
AVIRIS	Visible Infrared Imaging Spectrometer
AWC	Available Water Content
AWiFS	Advanced Wide Field Sensor
BCM	Billion Cubic Meters
BD	Bulk Density
BDS-R	Beidou Navigation Satellite System-Reflectometry
BI	Brightness Index
BIL	Band Interleaved by Line
BIP	Band Interleaved by Pixel
BK	Block Kriging
BRDF	Bidirectional Distribution Function
BSD	Berkeley Software Distribution

CAF	Crop Area Fraction
CanSIS	Canadian Soil Information Service
CAPE	Crop Acreage and Production Estimation
CASA	Carnegie-Ames-Stanford Approach
CAZRI	Central Arid Zone Research Institute
CC	Crop Cultivation
CEC	Cation Exchange Capacity
CGIS	Canada Geographic Information System
CI	Crust Index
CMU	Composite Mapping Units
CPI	Crop Productivity Index
CR	Consistency Ratio
CRP	Conservation Reserve Program
CSI	Composite Soil Index
CSM	Crop Simulation Model
CTI	Compound Topographic Index
CV	Coefficient of Variation
CWR	Crop Water Requirement
DBMS	Database Management System
DC	Double Crop
DEM	Digital Elevation Model
DGPS	Differential Global Positioning System
DH	Dryland Horticulture
DIARES-IPM	Diagnostic Advisory Rule-Based Expert System for Integrated Pest Management
DIFF	Relative Gas Diffusion Coefficient
DIP	Digital Image Processing
DLG	Digital Line Graph
DMSP	Defense Meteorological Satellite Program
DN	Digital Number
DOD	Department of Defense
DOS	Department of Space
DOT	Department of Transportation
DPR	Detail Project Report
DRG	Digital Raster Graph
DSM	Digital Soil Mapping
DSM	Digital Surface Model
DSS	Decision Support System
DSSAT	Decision Support System for Agro-technology Transfer
DTM	Digital Terrain Model
DXF	Digital Exchange Format
EC	Electrical Conductivity
EC	European Commission
ECEF	Earth-Centered Earth-Fixed

EDSS	Environmental Decision Support System
EEA	European Environment Agency
EEZRP	Energy Efficient Zone-Based Routing Protocol
EMR	Electromagnetic Radiation
EMS	Electromagnetic Spectrum
ENSO	El Niño/Southern Oscillation
EO	Earth Observation
EPL	Eclipse Public License
EQ	Socio-economic Qualities
ERS	European Remote Sensing
ERTS	Earth Resources Technology Satellite
ESA	European Space Agency
ESP	Estimation of Scale Parameter
ESRI	Environmental Systems Research Institute
ET	Evapotranspiration
ETM+	Enhanced Thematic Mapper Plus
EUSIS	European Soil Information System
EVI	Enhanced Vegetation Index
FACE	Free Air Carbon Enrichment
FAO	Food and Agricultural Organization
FASAL	Forecasting Agricultural Output using Space, Agro-meteorology and Land-Based Observations
FC	Field Capacity
FCC	False Color Composite
FCC	Fertility Capability Classification
FFRAS	Indian Forest Fire Response and Assessment System
FIRS	Automated Feature Information Retrieval System
FLDs	Front Line Demonstrations
FOSS	Free/Open Source Software
FPA	Full Pixel Areas
FPR	Farmer Participatory Research
FSI	Forest Survey of India
GAC	Global Area Coverage
GAGAN	GPS Aided Geo Augmented Navigation
GCPs	Ground Control Points
GCS	Geographic Coordinate System
GDEM	Global Digital Elevation Model
GEO GLAM	GEO Global Agricultural Monitoring Initiative
GEO	Group on Earth Observations
GEOBIA	Geospatial Object-Based Image Analysis
Geo-ICDT	Geographical-Information, Communication and Dissemination Technology
GeoIOT	Geo-Internet-of-Things
GEOSS	Global Earth Observation System of Systems

GGCMs	Global Gridded Crop Model
GIB	Great Indian Bustard
GIS	Geographic Information System
GLASOD	Global Assessment of Soil Degradation
GLOPEM	Global Production Efficiency Model
GNSS	Global Navigation Satellite System
GoI	Ministry of Agriculture
GPDP	Gram Panchayat Development Plan
GPL	General Public License
GPP	Gross Primary Productivity
GPS	Global Positioning System
GRACE	Gravity Recovery and Climate Exchange
GRASS	Geographic Resources Analysis Support System
GSI	Geological Survey of India
GSI	Top Soil Grain Size Index
GSP	Global Soil Partnership
GT	Grass Land with Trees
GUI	Graphical User Interface
GW	Ground Water
GWC	Gravimetric Water Content
GWR	Weighted Regression Method
GWRK	Geographically Weighted Regression Kriging Method
HI	Hypsometric Integral
HPC	High Performance Computing
HRS	Hyperspectral Remote Sensing
HRSI	High Resolution Satellite Image
HRV	High Resolution Visible
IARI	Indian Agricultural Research Institute
IAS	Image Analysis System
IBIN	Indian Bioresource Information Network
ICDTs	nformation Collection and Dissemination System
ICT	Information and Communication Technologies
IDS	Inverse Distance Squared
IDW	Inverse Distance Weighing
IFOV	Instantaneous Field of View
IFS	Integrated Farming System
IGU	International Geographical Union
IHS	Intensity, Hue, and Saturation
IIRS	Indian Institute of Remote Sensing
ILWIS	Integrated Land and Water Information System
IMD	India Meteorological Department
INARIS	Integrated National Agricultural Resource Information System
IoT	Internet of Things
IRRI	International Rice Research Institute

IRS	Indian Remote Sensing Satellite
ISO	International Organization for Standardization
ISRIC	International Soil Reference and Information Centre
ISRO	Indian Space Research Organisation
IT	Information Technology
ITC	International Institute for Aerial Survey and Earth Sciences
IUSS	International Union of Soil Science
IVLP	Institution Village Linkage Programme
IWM	Integrated Watershed Management
JAXA	Japan Aerospace Exploration Agency
JERS	Japanese Earth Resources Satellite
JMF	Joint Membership Function
JPEG	Joint Photograph Experts Group
LAC	Local Area Coverage
LACIE	Large Area Crop Inventory Experiment
LADA	Land Degradation Assessment in Drylands
LAI	Leaf Area Index
LBS	Location-Based Services
LC	Land Characteristics
LCC	Lambert Conformal Conic
LCS	Land Change Science
LEU	Land Ecological Unit
LGP	Length of Growing Period
LGPL	Lesser General Public License
LiDAR	Light Detection and Ranging
LIS	Land Information System
LISS	Linear Imaging Self Scanning
LMU	Land Management Unit
LOOCV	Leave One Out Cross Validation
LP DAAC	Land Processes Distributed Active Archive Center
LQ	Land Qualities
LRDP	Land Resource Development Plan
LRI	Land Resource Inventory
LRIS	Land Resource Information System
LRS	Land Remote Sensing
LS	Slope Length
LSWI	Land Surface Water Index
LUE	Light Use Efficiency
LULC	Land Use/Land Cover
LUR	Land Use Requirements
LUT	Land Utilization Type
LWCI	Leaf Water Content Index
M&E	Monitoring and Evaluation
M2M	Machine-to-Machine

MAE	Mean Absolute Error
MCDA	Multi-criteria Decision Analysis
MCS	Matter Control Station
ME	Mean Error
METI	Ministry of Economy, Trade and Industry
MFAFF	Ministry for Food, Agriculture, Forestry and Fisheries
MIR	Mid-Infrared
MLC	Maximum Likelihood Classifier
MLP	Micro-level Planning
MMA	Monitoring, Management and Adaptation
MNCFC	Mahalanobis National Crop Forecast Centre
MOA	Ministry of Agriculture
MODIS	Moderate Resolution Imaging Spectroradiometer
MPL	Mozilla Public License
MQ	Management Qualities
MRS	Microwave Remote Sensing
MRT	MODIS Reprojection Tool
MSE	Mean Squared Error
MSI	ultispectral Instrument
MSL	Mean Sea Level
MSS	Multi-spectral Scanner
MVC	Maximum Value Composites
NACIS	North American Cartographic Information Society
NADAMS	National Agricultural Drought Assessment and Monitoring System
NASA	National Aeronautics and Space Administration
NATP	National Agricultural Technology Project
NBSS&LUP	National Bureau of Soil Survey and Land Use Planning
NCSS	National Cooperative Soil Survey
NDSDI	Normalized Difference Sand Dune Index
NDSI	Normalized Difference Salinity Index
NDVI	Normalized Difference Vegetation Index
NDWI	Normalized Difference Water Index
NDWI	Normalized Difference Wetness Index
NIR	Near-Infrared
NN	Nearest Neighbors
NOAA	National Oceanic and Atmospheric Administration
NPP	Net Primary Production
NRCS	Natural Resources Conservation Service
NRM	Natural Resources Management
NRSC	National Remote Sensing Centre
NSDB	National Soil Database
NWDB	National Wasteland Development Board
OBIA	Object-Based Image Analysis
OC	Organic Carbon

ODbL	Data Commons Open Database License
OGC	Open Geospatial Consortium
OK	Ordinary Kriging
OLI	Operational Land Imager
OLS	Operational Linescan System
OS	Open Source
OSGeo	Open Source Geospatial Foundation
OSGS	Open Source GIS Software
OSM	Open Street Map
OSMF	Open Street Map Foundation
PA T	Polygon Attribute Table
PA	Precision Agriculture
PA	Producer's Accuracy
PAN	Panchromatic
PAR	Photosynthetically Active Radiation
PCA	Principal Components Analysis
PCMs	Pairwise Comparison Matrices
PFT	Plant Functional Type
PLU	Physiography and Land Use
PLUP	Participatory Land Use Planning
PPS	Precise Positioning Service
PRN	Pseudo Random Noise
PSP	Participatory Sensing
PWP	Permanent Wilting Point
QGIS	Quantum GIS
RADAR	Radio Detection and Ranging
RBPS	Radio-Based Positioning System
RDBMS	Relational Database Management System
RFID	Radio-Frequency Identification
RGB	Red, Green and Blue
RISAT	Radar Imaging Satellite
RMSE	Root Mean Square Error
RS	Remote Sensing
RTK	Real Time Kinematic
RTLS	Real-Time Locating System
RUSLE	Revised Universal Soil Loss Equation
RVPS	River Valley Projects
SAC	Space Application Center
SAGA	System for Automated Geoscientific Analyses
SAR	Synthetic Aperture Radar
SASI	Shortwave Angle Slope Index
SAU	State Agricultural University
SAVI	Soil Adjusted Vegetation Index
SBAS	Satellite-Based Augmentation Services

SBI	Soil Brightness Index
SDI	Spatial Data Infrastructure
SDSS	Spatial Decision Support System
SDTS	Spatial Data Transfer Standard
SHc	Saturated Hydraulic Conductivity
SI	Salinity Index
SIR	Storie Index Rating
SK	Simple Kriging
SLM	Sustainable Land Management
SLUP	Soil-Based Land Use Planning
SLUSI	Soil and Land Use Survey of India
SMAF	Soil Management Assessment Framework
SMT	Spectral Matching Techniques
SOC	Soil Organic Carbon
SOM	Soil Organic Matter
SONAR	Sound Navigation and Ranging
SOTER	Soil and Terrain Database
SPA	Sub-pixel Area
SPCS	State Plane Coordinate System
SPI	Soil Productivity Index
SPI	Stream Power Index
SPS	Standard Positioning Service
SQ	Soil Qualities
SQI	Soil Quality Index
SRM	Soil Resource Mapping
SRS	Satellite Remote Sensing
SRTM	Shuttle Radar Topographic Mission
SSURGO	Data and Soil Survey Geographic
STATSGO	State Soil Geographic
STI	Sediment Transport Index
SVG	Scalable Vector Graphics
SW	Surface Water
SWAP	Soil-Water-Atmosphere-Plant
SWAT	Soil and Water Assessment Tool
SWE	Sensor Web Enablement
SWIR	Short Wave Infrared
TCM	Technical Cooperation Mission
TD	Transformed Divergence
TGA	Total Geographical Area
TIFF	Tagged Image File Format
TIGER	Topologically Integrated Geographic Encoding and Referencing
TIN	Triangulated Irregular Network
TIRS	Thermal Infrared Sensor
TIROS-1	Television Infrared Observation Satellite

TM	Thematic Mapper
TORT	Pore Tortuosity Factor
TPI	Topographic Position Index
TPS	Total Pore-Space
TQ	Topography Qualities
TWI	Topographic Wetness Index
UA	User's Accuracy
UAV	Unmanned Aerial Vehicles
UNCED	United Nations Conference on Environment and Development
UNEP	United Nations Environmental Programme
USBR	United States Bureau of Reclamation
USDA	United States Department of Agriculture
USDM	United States Drought Monitoring
USGS	United States Geological Survey
USLE	Universal Soil Loss Equation
UTM	Universal Transverse Mercator
VAC	Volumetric Air Content
VI	Vegetation Index
VLIS	Village Level Information System
VLSI	Very Large-Scale Integration
VLUP	Village Level Land Use Planning
VNIR	Visible and Near-Infrared
VPM	Vegetation Production Model
VRA	Variable Rate Application
VRT	Variable Rate Technology
VWC	Volumetric Water Content
WAAS	Wide Area Augmentation Service
WFPS	Water-Filled Pore Space
WFS	Web Future Services
WGS84	World Geodetic System of 1984
WI	Wetness Index
WiFS	Wide Field Sensor
WISE	World Inventory of Soil Emission Potential
WMS	Web Map Services
WSA	Water Spread Area
WSN	Wireless Sensor Network
XML	Extensible Markup Language

Contents

Contributors

R. K. Bajpai Department of Soil Science & Agricultural Chemistry, IGKV, Raipur, India

B. P. Bhaskar ICAR-National Bureau of Soil Survey & Land Use Planning, Nagpur, India

S. Chattaraj ICAR-National Bureau of Soil Survey & Land Use Planning, Nagpur, India

S. Chatterji ICAR-National Bureau of Soil Survey & Land Use Planning, Nagpur, India

A. Daripa ICAR-National Bureau of Soil Survey & Land Use Planning, Nagpur, India

S. K. Gangopadhyay ICAR-National Bureau of Soil Survey & Land Use Planning, Regional Centre, Kolkata, India

Murali Krishna Gumma International Crops Research Institute for the Semi-Arid Tropics, Patancheru, Hyderabad, India

Adinarayana Jagarlapudi Centre of Studies in Resources Engineering, Indian Institute of Technology Bombay, Powai, Mumbai, India

Soumyashree Kar Centre of Studies in Resources Engineering, Indian Institute of Technology Bombay, Powai, Mumbai, India

Y. B. Katpatal Visvesvaraya National Institute of Technology, Nagpur, India

Y. V. N. Krishna Murthy National Remote Sensing Centre (ISRO), Hyderabad, India

Nirmal Kumar ICAR-National Bureau of Soil Survey & Land Use Planning, Nagpur, India

Suresh Kumar ISRO-Indian Institute of Remote Sensing, Dehardun, India

Narayan Lal ICAR-National Research Centre on Litchi, Muzaffarpur, India

Jugal Kishore Mani Regional Remote Sensing Centre-Central, National Remote Sensing Centre, ISRO, Nagpur, India

V. N. Mishra Department of Soil Science & Agricultural Chemistry, IGKV, Raipur, India

Rohit Nandan Centre of Studies in Resources Engineering, Indian Institute of Technology Bombay, Powai, Mumbai, India

D. C. Nayak ICAR-National Bureau of Soil Survey & Land Use Planning, Regional Centre, Kolkata, India

G. P. Obi Reddy ICAR-National Bureau of Soil Survey & Land Use Planning, Nagpur, India

Rahul Raj Centre of Studies in Resources Engineering, Indian Institute of Technology Bombay, Powai, Mumbai, India

V. Ramamurthy ICAR-National Bureau of Soil Survey & Land Use Planning, Regional Centre, Bengaluru, India

S. K. Ray ICAR-National Bureau of Soil Survey & Land Use Planning, Regional Centre, Jorhat, India

Nisha Sahu ICAR-National Bureau of Soil Survey & Land Use Planning, Nagpur, India

T. K. Sen ICAR-National Bureau of Soil Survey & Land Use Planning, Nagpur, India

S. K. Singh ICAR-National Bureau of Soil Survey & Land Use Planning, Nagpur, India

N. K. Sinha ICAR- Indian Institute of Soil Science, Bhopal, India

G. Sreenivasan Regional Remote Sensing Centre-North, National Remote Sensing Centre (ISRO), New Delhi, India

Rajeev Srivastava ICAR-National Bureau of Soil Survey & Land Use Planning, Nagpur, India

Saurabh Suradhaniwar Centre of Studies in Resources Engineering, Indian Institute of Technology Bombay, Powai, Mumbai, India

Pardhasaradhi Teluguntla U.S. Geological Survey (USGS), Western Geographic Science Center, Flagstaff, AZ, USA

Prasad S. Thenkabail U.S. Geological Survey (USGS), Western Geographic Science Center, Flagstaff, AZ, USA

A. O. Varghese Regional Remote Sensing Centre-Central, National Remote Sensing Centre, ISRO, Nagpur, India

Anthony M. Whitbread International Crops Research Institute for the Semi-Arid Tropics, Patancheru, Hyderabad, India

Chapter 1
Geospatial Technologies in Land Resources Mapping, Monitoring, and Management: An Overview

G. P. Obi Reddy

Abstract Geospatial technologies broadly includes remote sensing, photogrammetry, cartography, geographic information system (GIS), global positioning system (GPS), and information technology (IT). Geospatial technologies deal with the acquisition, storage, processing, production, presentation, and dissemination of geoinformation. Remote sensing technology allows us to observe the earth features from the space, and there are several techniques to differentiate the information collected from remote sensing on land, vegetation, water, etc. GIS is a computer-based system to capture, organise, store, edit, analyse, display, and plot geographically referenced data. GPS allow the surveys to collect the precise locational information and increase the accuracy in mapping and monitoring over the conventional surveying techniques. IT is the use of any computers; storage, networking, and other physical devices; and infrastructure to create, process, store, secure, and exchange all forms of electronic data. The information generated through conventional methods, remote sensing, and GPS techniques could be used effectively to create database in GIS and perform variety of spatial analysis in sustainable management of land resources and generate environment-friendly action plans. Some of the applications of geospatial technologies are digital terrain modeling, geomorphological mapping, soil resource inventory and mapping, soil-landscape modeling, land use/land cover mapping, croplands mapping and monitoring, assessment and monitoring of droughts, soil erosion assessment, mapping and assessment of land degradation, water resource management, watershed management, agricultural land use planning, spatial decision support systems, etc., which have a far-reaching impact on mapping, monitoring, and management of land resources on sustainable basis.

Keywords Geographic information system · Global positioning system · Information technology · Land resources · Remote sensing

---

G. P. O. Reddy (✉)
ICAR-National Bureau of Soil Survey & Land Use Planning, Nagpur, India

G. P. O. Reddy, S. K. Singh (eds.), *Geospatial Technologies in Land Resources Mapping, Monitoring and Management*, Geotechnologies and the Environment 21,
https://doi.org/10.1007/978-3-319-78711-4_1

1.1 Introduction

Land is a delineable area of the earth's terrestrial surface, encompassing all attributes of the biosphere immediately above or below this surface, including those of the near-surface climate, the soil and terrain forms, the surface hydrology, the near-surface sedimentary layers and associated groundwater reserve, the plant and animal populations, the human settlement pattern, and physical results of past and present human activity (FAO 1995). Millions of hectares of land per year are being degraded in all climatic regions in the world (Yagi et al. 2015). The decline in land quality caused by human activities has been a major global issue since the twentieth century and will remain high on the international agenda in the twenty-first century (Eswaran et al. 2001). The availability of accurate and timely information on nature, extent and spatial distribution of land resources enables and rationalizes the decision-making processes of planners, policy makers and managers in a cost-effective and time-efficient manner. Mapping, monitoring, and management of land resources, evaluation of their potentials and limitations for wide range of land use options, and formulation of sustainable land use plans assumed a greater significance in the twenty-first century to meet the ever-increasing demand of food, fodder, and fuel (Reddy et al. 2013a, 2017a, b). Information on the nature, extent, and spatial distribution of land resources is prerequisite for their mapping, monitoring, and management on sustainable basis. The rapid advancing geospatial technologies have revolutionized the land resources mapping and generation of spatial databases on a regular basis for monitoring, management, and implementation of various plans more precisely and efficiently at different levels (Maji et al. 2001, Reddy et al. 2016a, b, 2017a, b).

Geospatial technologies are a combination of mapping and surveying technologies, which include remote sensing, photogrammetry, cartography, geographic information system (GIS), global positioning system (GPS), and information technology (IT). Geospatial technology is a multidisciplinary science that integrates the technologies and principles of digital cartography, remote sensing, photogrammetry, surveying, GPS, GIS, and automated data capture systems using high-resolution geo-referenced spatial information from aerial and space based remote sensing platforms. It is fundamental to all the disciplines, which use data identified by their locations. Thus, geospatial technologies provide tools that allow for acquisition, processing, manipulation, and analysis of spatial data into information to make decisions, about portions of the earth and environmental problems. Geospatial technologies deal with spatial and non-spatial data, their methods of acquisition, management, analysis, display, and dissemination. The emerging field of geospatial technologies enables us to take advantage of the unprecedented amount of digital data and computing power available through electronic networks.

Geospatial technologies play a key role in realizing the sustainable management of land resources through their core potential, which includes *mapping* through collection of thematic and quantitative baseline resources data, *measuring* through more rigorous mapping process by quantifying and documenting the historical in geographic format, and *monitoring* through regular assessment of the conditions by recording the changes in natural phenomena and human-induced activities. The key catalysts that have transformed the geospatial technologies include the widespread acceptance of GPS technologies, smart phones, and mapping services offered by a variety of mapping portals such as Google Earth/Maps. GPS provides positionally accurate coordinates, thus useful to establish geographic location and acquire cost-effective *in situ* data. GPS is a promising tool for regular monitoring of dynamic features through acquiring location-based data repetitively about earth features and phenomena. It is estimated that over 2 billion people on this planet are users of remotely sensed data and geospatial datasets, which include data from a variety of geospatial technologies such as GPS, GIS, and remote sensing (Digital Globe 2015). The core thrust of geospatial technologies consists of thematic, spatial, and temporal data and design of systems for data analysis and computing. The tools used are those of spatial database design, GIS application development for spatial analysis, and general computer programming (Burrough and McDonell 1998). Harnessing full potential of latest geospatial technologies in capture, collate, analyze, and disseminate data is of immense help to manage the land resources on sustainable basis.

1.2 Geospatial Technologies: Definition

Geospatial technologies, commonly known as geomatics, refer to technology used for visualization, measurement, and analysis of features or phenomena that occur on the earth. Geospatial technologies encompass four important technologies such as remote sensing, GPS, GIS, and information technology, which are directly related to mapping, monitoring, and management of features on the surface of the earth. Geospatial technologies are art, science, and technologies, which deal with acquisition, storage, processing, production, presentation, and dissemination of geoinformation. Geoinformatics has also been described as "the science and technology dealing with the structure and character of spatial information, its capture, its classification and qualification, its storage, processing, portrayal and dissemination, including the infrastructure necessary to secure optimal use of this information" (Groot 1989). Ehlers and Amer (1991) define it as "the art, science or technology dealing with the acquisition, storage, processing production, presentation and dissemination of geoinformation."

1.3 Geospatial Technologies: Concepts and Principles

Although technology for geoinformation management is changing rapidly that it is difficult to keep pace, the user interfaces of these systems are a fairly stable technology based on storage techniques and manipulation requirements. The core thrust in geoinformation management consists of thematic, spatial, and temporal data and design of systems for data manipulation and computing. The tools used are those of spatial database design, GIS application development for spatial analysis, and general computer programming. It also includes higher-dimensional topology, fuzzy logic, temporal and spatial data, object-oriented application development, spatial database design, and Internet programming techniques. Integration of geospatial technologies with other analytical tools and techniques is often desirable and essential to produce better information thereby enhancing our understanding for better management of land resources.

1.4 Geospatial Technologies: Applications in Land Resource Mapping, Monitoring, and Management

Remote sensing, GIS, GPS, and information technologies have become indispensable in land resource mapping, monitoring, and management. The integration of these technologies has taken land resource management domain to the new digital age. Efficiency in use of land resources can be augmented through their effective mapping, monitoring, and management using geospatial technologies. The generation of precise database on land resources can ensure greater reliability of estimates and forecasting that helps in the process of planning and policy making. Efforts are needed to harness the full potential of latest remote sensing and information technologies to capture, collate, analyze to develop action plans, and disseminate information to appropriate destinations for implementation in land resource management. Geospatial technologies individually as well as jointly play a significant role in mapping, monitoring, and management of land resources. Important applications of geospatial technologies in mapping, monitoring, and management of land resources are discussed below.

1.4.1 Digital Terrain Modeling

Geospatial technologies play a significant role in digital terrain modeling and replacing the traditional qualitative approaches in characterization of topography using digital elevation models (DEMs) available at various spatial resolutions. Application of geospatial technologies in digital terrain models has become an integral part of digital processing of spatial geographical information. Remote sensing, GIS, and GPS technologies can be effectively used at a wide range of scales in digital terrain analysis, topographical mapping, generation of digital terrain database, and integrated analysis. It has many comparative advantages in quantitative measurement of elevation, derive terrain attributes quantitatively, visualize topography in more realistic way than ever before, and enables to store, update, proliferate, and manipulate topographic data digitally (Li et al. 2005; Moore et al.

1993; Wilson and Gallant 2000a, Chattaraj et al. 2017). It further provides the possibility of deriving indices, which can be used as indicators for environmental processes (Pike 1988; Wilson and Gallant 2000b). In GIS, modeling, analyzing, and representation of topography and relief can be performed. Reddy et al. (2012) carried out digital terrain analysis using Shuttle Radar Topographic Mission (SRTM) DEM (90 m) data and generated various geomorphometric parameters like slope, hillshade, relief, aspect, contours, plan curvature, profile curvature, total curvature, flow direction, flow accumulation, drainage, and topographic wetness index (TWI) at India and state level.

1.4.2 Geomorphological Mapping

Remote sensing and GIS technologies have a great potential in geomorphological mapping and generating geomorphic maps at different scales. The precise geomorphological mapping helps to understand the landscape features, evolution, and process of landforms. In earlier days, geomorphometric properties were originally measured by calculating the geometry of the landscape manually (Horton 1945; Miller 1953; Coates 1958). In the 1960s and early 1970s, the general availability of computers made possible more complex, statistically based methods to identify landscape features (Chorley 1972; Evans 1972). The use of DEM is important to derive landscape attributes that are utilized in landform characterization (Burrough 1986; Dobos et al. 2000). DTM provides greater functionalities than the qualitative and nominal characterization of topography. DEM can be manipulated to provide many kinds of data that can assist the soil surveyor in mapping and giving a quantitative description of landforms and their variability. DEM can be effectively used to generate slopes, aspects, rate of change of slope, drainage network on catchments areas, etc. (Burrough 1986; Brabyn 1997). Availability of many digital elevation datasets in public domain, advances in processing capabilities of computers, and spatial analytical methods have reoriented the field of geomorphological mapping (Pike 1999; Hengl and Reuter 2009; Reddy et al. 2012). These advancements contributed significantly in development of computer-based algorithms for discriminating geomorphometric properties of terrain and generate landform maps. Landforms are acceptable as integrated classifiers of the landscape and can be used to divide the landscape into discrete segments (Bocco et al. 2001).

1.4.3 Soil Resource Inventory and Mapping

Remote sensing, GIS, and GPS have immense potential in time-efficient and cost-effective soil resource inventory and mapping. The use of multi-temporal remote sensing data complemented with terrain information was found to be very useful for deriving reliable soil classification categories (Dobos et al. 2000; McBratney et al.

2000). In interpretation of satellite image for soil resource mapping, proper identification of land type, drainage pattern, vegetation, land use, slope, and relief is essential (Dwivedi 2001). Several authors have reported on the potential of satellite remote sensing and GIS as promising tools in soil resource mapping (Srivastava and Saxena 2004; Reddy et al. 2008, 2013). GIS applications in soil survey and mapping can be divided into two parts. Firstly, GIS applications can be used in pre-survey and mapping for database generation and development of base maps. A detailed landform map could be generated through analysis of topographical sheets, satellite data, and DEMs, which can form as base map in soil resource inventory and mapping. The detailed landform maps depict spatial variations of terrain features, which in turn immensely help in inventory and mapping of soils and finalize the soil mapping units in the region (Reddy et al. 2013). Secondly, GIS can be effectively used in the post-survey and mapping applications. Soil mapping units, forming the skeleton of the soil data base, are then labeled through appropriate GIS functions to tabular data base containing essential soil characteristics, to form an integrated soil coverage, comprising both spatial and descriptive information of soil mapping unit. The resulting coverage can be analyzed using the analytical capabilities of GIS and tabular database; numerous thematic and derivative maps could be developed to use in various planning processes of land resource management.

1.4.4 Soil-Landscape Modeling

Terrain characterization and landform analysis of a terrain are essential to develop landscape-soil relationship for precise soil resource mapping, land degradation mapping, and other environmental applications (Reddy et al. 1999a, b; 2002a, b). Many of the attributes of landscape have potential use for spatial analysis of the soil properties. Primary terrain attributes like areal measures such as specific catchment area and point measures such as slope, aspect, and profile curvature can be directly calculated from elevation data. Secondary attributes involve combinations of the primary attributes, which quantify the contextual nature of points or characterize the spatial variability of the specific processes occurring in the landscape or both. The terrain features, like slope or aspect, which are recognized as leading forces of the soil formation within a relatively small area, show significant relationship with soil attributes, but often represent low predictive value when used individually. However, when these terrain variables are combined in one model, the predictive value can raise relatively high. The use of digital terrain parameters as soil predictors is certainly not only the way of organizing our soil-landscape knowledge but one of the most powerful ways certainly. The GIS-based multi-criteria overlay analysis of various landscape parameters like elevation, slope, aspect, morphometry, hydrological parameters, and land use/land cover allows us to establish broad inter-relationship between landscape and soil attributes in any given landscape. The analysis of landform-soil relationship in association with drainage and elevation properties in GIS immensely helps to understand spatial patterns of soil attributes in similar climatic and geological conditions (Bell et al. 1994; Gessler et al. 1995; Reddy and Maji 2003; Reddy et al. 2013).

1.4.5 Land Use/Land Cover Mapping

Land use/land cover inventories form essential components in land resources evaluation, erosion, and environmental studies. The pattern, extent, and transformation of land use systems are more influenced on the status of productivity of any given ecosystem. Apart from slope, soil, topography, and water availability, the physiography and structural characteristics of the vegetative cover and their spatial distribution are considered to be important parameters (Sudhakar et al. 2000). High spatial resolution remote sensing data provides valuable information on location and extent of given land use/land cover class. With the increase in spatial, spectral, and temporal resolutions of the satellite sensors, the satellite data becomes a more effective source for accurate land use/land cover mapping and monitoring. The satellite remote sensing data with their repetitive nature have proved to be quite useful in mapping land use/land cover and change with time. The capability of data storage and retrieval and analysis of GIS play an important role in land use system studies and analysis. Recent advances in remote sensing and GIS technologies have emerged as promising tools to meet ever-increasing demand for generation of precise and timely information on land use/land cover and their dynamics (Kar et al. 2018).

1.4.6 Cropland Mapping and Monitoring

The role of remote sensing and GIS in crop studies can be broadly categorized into two groups, i.e., inventorying/mapping and management. Remote sensing data is being widely used for inventorying purposes like crop acreage estimation, crop condition assessment, crop yield forecasting, etc. and management like cropping system analysis, precision farming, etc. Use of remote sensing data in cropland mapping and monitoring has been investigated in India (Reddy et al. 2009; Dheeravath et al. 2010) and many other parts of the world (Thenkabail et al. 2009; Biradar et al. 2009). Identification and discrimination of crop types require quantitative use of subtle differences in their spectral data and hence rely mostly on digital image processing techniques. The acreage estimation procedure broadly consists of identifying representative sites of various crop classes on the image based on the ground truth collected, generation of signatures for different training sites, and classifying the image, using training statistics. The available high-resolution temporal satellite data for the growing season and ground-based observations help immensely in mapping, monitoring, and crop acreage estimation. The multitemporal data available from satellite sensors like Landsat, SPOT-VGT, Moderate

Resolution Imaging Spectroradiometer (MODIS), Indian Remote sensing Satellite (IRS)-P6- Advanced Wide Field Sensor (AWiFS), Linear Imaging Self Scanning Sensor (LISS)-III, LISS-IV, Sentinel-2, etc. have immense potential in deriving time-series vegetation indices at different resolutions for discrimination and monitoring of various crops on regular basis. Depending on the objectives of application, satellite imageries ranging from low to high spatial resolution could be used for studying agricultural landscapes (Vinciková et al. 2010).

1.4.7 Assessment and Monitoring of Droughts

In agriculture sector, robust monitoring, warning, and insurance systems, as well as finding ways to reduce its environmental impacts without compromising food security and rural development, assume greater importance (FAO 2010a). In agriculture-based countries with a dependence on subsistence systems, the priority is to increase productivity to achieve food security (FAO 2010b). Advanced research/methodologies of geospatial technologies have immense scope in assessment and monitoring of droughts to cope with climate change/variability/extreme weather events. Monitoring the crop conditions at regular intervals during the crop growth cycle is essential to take appropriate curative measures and to assess the probable loss in production. Vegetation index (VI) derived from spaceborne data is sensitive to moisture stress in crops and serves as surrogate measures to assess agricultural drought. Inadequacy of ground-based data prompted the use of indices like normalized difference vegetation index (NDVI) and normalized difference water index (NDWI) derived from remote sensing data to provide certain weather-based evidence (Razali and Nuruddin 2011; Rojas et al. 2011). In India, a nationwide project on National Agricultural Drought Assessment and Monitoring System (NADAMS) is being implemented since 1987 to monitor the drought during *kharif* (southwest monsoon) season by generating NDVI from temporal NOAA-AVHRR data. NDVI profile of the current season is compared with the normal year to infer stress conditions. This is complemented by *in situ* observations on rainfall and other agricultural practices with availability of AWiFS data. With the availability of microwave satellite data, the accuracy in drought assessment has been improved to a great extent. NADAMS project gives fortnightly information during monsoon season at district level using satellite-derived NDVI information as input. It is also increasingly considered a viable alternative to the index-based crop insurance, due to unreliable weather station-based data sources (Kamble et al. 2013). There is great potential for index-based insurance in the future because majority of smallholder farmers perceive climate change as a threat to their livelihoods (Morton 2007; Bunce et al. 2010).

1.4.8 Soil Erosion Assessment

Soil erosion can be assessed at different levels using Universal Soil Loss Equation (USLE) (Wischmeier and Smith 1978). GIS integrated with Image Analysis Systems (IAS) and database management systems (DBMS) are vital tool in the generation of spatial data layers and their integration to estimate soil loss of a given area. These techniques have been used extensively in quantification of USLE parameters to estimate soil erosion (Jain and Kothari 2000; Dabral et al. 2008). The factors of rainfall erosivity (R), soil erodibility (K), slope length (LS), cover (C), and management (P) are the components of USLE and can be computed in GIS using information on rainfall, slope, soil, and land use/land cover generated from remotely sensed data and field surveys. GIS helps in generation and integrated analysis of USLE parameters in quantification of potential and actual soil loss (Srinivas et al. 2002; Reddy et al. 2004a, b, 2016a, b). Revised Universal Soil Loss Equation (RUSLE) (Renard et al. 1991) is also widely used to estimate the status of soil loss. Using advanced remote sensing and GIS techniques and modeling, many authors developed methods for assessment of erosion hazards, such as integrated and systematic approaches (Tian et al. 2008), geo-statistical multivariate approaches (Conoscenti et al. 2008), sensitivity analysis approaches (Mendicino, 1999), soft computing methods (Gournellos et al. 2004), and analytical risk evaluation methods (Wu and Wang 2007).

1.4.9 Mapping and Assessment of Land Degradation

Land degradation affects about 2.6 billion people in more than a hundred countries and over 33% of the earth's land surface (Adams and Eswaran 2000). Around 73% of rangelands in drylands are currently being degraded, together with 47% of marginal rainfed croplands and a significant percentage of irrigated croplands (CSD 2000). The decline in land quality caused by human activities has been a major global issue since the twentieth century and will remain high on the international agenda in the twenty-first century (Eswaran et al. 2001). Combating land degradation is critical for ensuring sustainability of agriculture to support current and future demands in crop and livestock production. Sustaining productivity of existing agricultural and grazing land is, therefore, essential to meet current and future aspirations for increasing food production without compromising ecosystem goods and services. Land degradation is, therefore, a major factor in the fight against poverty, hunger, food insecurity, and natural resource conflicts throughout the developing world. Human-induced land degradation most likely alters the vegetation cover and ecosystem functions, for example, increasing the extent of soil erosion or changing the local climate through positive feedbacks (Charney et al. 1977; Xue and Fennessy 2002). Human interventions, those causing land degradation, are deforestation, overgrazing by livestock, mismanagement of agricultural land, overexploitation of vegetative cover, and (bio)industrial activities.

In India, recent estimates show that about 120.72 Mha is under different categories of land degradation (Maji et al. 2010). The causes of land degradation are inappropriate land use practices, which lead to degradation of soil, water, and vegetative cover and loss of both soil and vegetative biological diversity, affecting ecosystem structure, functions, and services. Degraded lands are more susceptible to the adverse effects of climatic change such as increased temperature and more severe droughts. Remote sensing has been successfully utilized in land degradation assessment and monitoring over a range of spatial and temporal scales (Jafari et al. 2008; Reddy et al. 2002a, b). Ostir et al. (2003) pointed out that remote sensing has developed as an important tool for assessment and monitoring of vegetation and erosion. It can provide calibrated, quantitative, repeatable, and cost-effective information for large areas and can be related to the field data (Pickup 1989; Tueller 1987; Jafari et al. 2008). In selected areas of India, high-resolution remote sensing data has been used to map and assess the land degradation and understand the causes and consequences to develop the location-specific database and mitigation strategies using the latest geospatial technologies (Ahmad and Pandey 2018).

1.4.10 Water Resource Management

Water is the most important renewable and finite natural resources since it is required for agriculture, industry, and domestic purposes. In the field of water resources, there are several areas where geospatial technologies are being used for scientific planning, monitoring, and management (Reddy et al. 2001). In the important study like hydrological studies, geospatial technologies are being used in rainfall estimation, forecasting and monitoring, hydrological and water balance modeling. As far as groundwater is concern, there are a number of factors, which control the groundwater regime in a given area such as lithology, stratigraphy, structure, landforms, recharge conditions, slope, soil, and land use/land cover (Reddy and Rao 1994; Krishnamurthy et al. 1996). Remote sensing and GIS have become widely accepted as effective tools in the field of hydrogeological investigations and groundwater exploration, due to the robust integration of spatial features and tabular data, spatial analysis capability, intelligent maps, and a geographic user interface to the databases.

Analysis of remotely sensed data on drainage, geological, geomorphological and lineaments characteristics of terrain in an integrated way facilitates to effectively delineate groundwater potential zones (Saraf and Choudhuray 1998; Reddy et al. 1999a, b, 2001). Remote sensing data provides reliable information on available water resources and its utilization. Remote sensing inputs have been significantly contributing in water management, both in its conservation and control aspects. The integration of remotely acquired spectral information with other data like physiographical, geological, geophysical, hydrogeological, geochemical, and geometrical

GIS provides a holistic approach to understand hydrogeological characteristics of the area. Remote sensing and GIS are widely used in delineation of groundwater potential zones (Chi and Lee 1994; Krishnamurthy et al. 1996) and vulnerability assessment (Rundquist et al. 1991; Laurent et al. 1998) and groundwater modeling (Watkins et al. 1996) and management (Hendrix and Buckley 1989).

1.4.11 Watershed Management

Watershed is a geo-hydrological unit area that drains to a common point and is considered as an appropriate physical unit for natural resource evaluation, planning, and management. Watershed management implies the rational utilization of land and water resources for optimum production with minimum hazard to natural environment (Bhat 1989). The concept of watershed management involves accurate and timely monitoring of land resources using available latest technologies. This approach aim at proper utilization of land and water resources, protecting land against all forms of deterioration, and monitoring and maintaining soil health, on-site conservation of water, proper management of local drainage and sediment reduction, and increasing productivity from all land uses through suitable land use planning. Since watershed management involves decision-making for efficient utilization of natural resources, a multidisciplinary approach is essential. Watershed management is the process of formulating and carrying out a course of action involving modification of the natural system of watershed to achieve specified objectives befitting to the local needs. These include optimum production with minimum hazard to natural resources and assured economic benefits. It also helps to restore the fragile geo-ecological balance at microlevel through adoption of various conservation measures.

The various parameters of the watershed, i.e., stream network (drainage), physiography, and land use/land cover, can be mapped and monitored using remote sensing data. Geospatial technologies could be effectively used in strategic planning adopted for watershed development comprised likely of delineation and codification of watershed, prioritization of watersheds, soil resource inventory, treatment of very high and high priority watersheds, and evaluation and monitoring of the impact of the treatment. In recent years, high-resolution satellite remote sensing techniques proved to be of immense value for evaluation of land resources and land use systems at watershed level preparing land management strategies and monitoring changes at regular intervals of time (Reddy et al. 2008). Analysis of satellite data for drainage, lithology, geomorphology, soil, and land use/land cover aspects in conjunction with collateral data facilitates effective evaluation of status of soil loss at watershed level (Reddy and Sarkar 2012). Along with the remote sensing, GIS is widely accepted as a standard tool in land resource management and environmental planning at watershed level. The datasets in the core of GIS provide an excellent means of storing, retrieving, and analyzing geo-referenced information. This is essential for the conservation and management of land and water resources of watersheds for optimum productivity. Geospatial technologies have immense potential in watershed characterization, including landforms, soil and land use/land cover mapping, erosion intensity mapping and identification of erosion prone areas, development of water conservation strategies, selection of sites for construction of water harvesting structures, and monitoring the impact of interventions in watershed management (Reddy et al. 2004a).

1.4.12 Agricultural Land Use Planning

Crop land suitability analysis is a prerequisite to achieve optimum utilization of available land for agricultural production in a sustainable manner. In order to practice it, one has to grow the crops where they suit best, for which the first and foremost requirement is to carry out land suitability analysis (Sys 1985; Sys et al. 1991; NBSS&LUP 1994; Naidu et al. 2006). Hence, proper evaluation of land resources is essential for sustainable agriculture land use planning. Various approaches of land evaluation have been developed, and each one has a specific methodological procedure (Storie 1978; Riquier et al. 1970; FAO 1976). There is an increasing tendency in the adoption of geospatial technologies in quantitative land suitability evaluation models for delineation of suitable areas for different crops (Vishakha et al. 2016; Kumar et al. 2017). In India, by using remote sensing and GIS-based applications, agricultural land use plans have been developed at village (Ramamurthy et al. 2006), watershed (Ramamurthy and Sarkar 2009), and district level (Ramamurthy et al. 2015).

1.4.13 Spatial Decision Support System (SDSS)

Decision support systems (DSS) are defined as computer-based information systems designed to support decision-makers interactively in thinking and taking decisions about relatively unstructured problems. Traditionally, DSS have three major components, a database, a model base, and a user interface. An extension of the DSS concepts Spatial Decision Support Systems (SDSS), which are the integration of DSS and GIS, was initiated by Densham and Goodchild (1988). A significant capability of the SDSS is the ability to use spatial analysis and display tools with the sectorial models that would form the model base of SDSSs. The modeling capability of GIS allows the user of the SDSS to stimulate changes in objects and attributes. The database components of the SDSS can supply input data for the models. After the models are run, the resulting output can be returned to the database for later display through user interface. For planning purpose, this ability to dynamically change information and forecast on perform sensitivity analysis are essential. Both GIS and DSS have been widely used in natural resource management. This

GIS-based SDSS helps in modeling capability, which can integrate with simulation techniques to solve complex natural resource management problems.

1.5 Conclusions

Modern technologies such as high-resolution satellite data, GPS, GIS, and information technology can be effectively used in land resource inventory, monitoring, and management. The satellite data provides integrated information on landforms, geological structures, soil types, erosion, land use/land cover, surface water bodies, and qualitative assessment of groundwater potential at different scales. The integrated approach of remote sensing, GPS and GIS is highly useful in digital terrain analysis, geomorphological mapping, soil resource inventory and mapping, soil-landscape modeling, land use/land cover mapping, cropland mapping and monitoring, assessment and monitoring of droughts, soil erosion assessment, mapping and monitoring of land degradation, identification and mapping of perspective groundwater potential zones, watershed management and development of agricultural land use plans at various scales. In the current and new generation of high-resolution satellite imageries, GPS and GIS ease the task of land resource planning and management on sustainable basis. The generated accurate spatial database in GIS on natural/physical resources, socioeconomic and contemporary technologies, and integrated analysis provides a sophisticated mechanism to generate alternative action plans in land resource development. The evaluation and management of land resources based on remotely sensed data, GPS, and GIS can be effectively utilized in assessment of various types of degraded lands, viz., water logging, soil erosion, salinity/alkalinity, and shifting cultivation areas for reclamation/conservation, which results in bringing additional areas into cultivation and also to improve productivity level in marginal lands to meet ever-increasing food, fodder, and fuel demand.

References

Adams CR, Eswaran H (2000) Global land resources in the context of food and environmental security. In: Gawande SP, Bali JS, Das DC, Sarkar TK, Das DK, Narayanaswamy G (eds) Advances in land resources management for the 21st century. Soil Conservation Society of India, New Delhi, pp 35–50

Ahmad N, Pandey P (2018) Assessment and monitoring of land degradation using geospatial technology in Bathinda district, Punjab, India. Solid Earth 9(1):75–90

Bell JC, Cunningham RL, Havens MW (1994) Soil drainage class probability mapping using a soil-landscape model. Soil Sci Soc Am J 58:464–470

Bhat LS (1989) Status report on land system analysis for evaluation of resources at micro level. Discussion paper for the fifth review meeting for NRDMS Project. DST, Govt. of India

Biradar CM, Thenkabail PS, Noojipady P, Li Y, Dheeravath V, Turral H, Velpuri M, Gumma MK, Reddy GPO, Cai XL, Xiao X, Schull MA, Alankara AD, Gunasinghe S, Mohideen S (2009) A global map of rainfed cropland areas (GMRCA) at the end of last millennium using remote sensing. Int J Appl Earth Obs Geoinf 11:114–129

Bocco G, Mendoza M, Velazquez A (2001) Remote sensing and GIS-based regional geomorphological mapping-a tool for land use planning in developing countries. Geomorphology 39:211–219

Brabyn L (1997) Classification of macro landforms using GIS. ITC J 97(1):26–40

Bunce M, Rosendo S, Brown K (2010) Perceptions of climate change, multiple stressors and livelihoods on marginal African coasts. Environ Dev Sustain 12(3):407–440

Burrough PA (1986) Principles of geographical information system for land resource assessment. Monograph on Soil and Land Resource Survey No. 12, Oxford University

Burrough PA, McDonell RA (1998) Principles of geographical information systems. Oxford University Press, New York, p 190

Charney J, Quirk WJ, Chow SH, Kornfield J (1977) A comparative study of the effects of albedo change on drought in semi-arid regions. J Atmos Sci 34:1366–1385

Chattaraj S, Srivastava R, Barthwal AK, Giri JD, Mohekar DS, Reddy GPO, Daripa A, Chatterji S, Singh SK (2017) Semi-automated object-based landform classification modelling in a part of the Deccan Plateau of Central India. Int J Remote Sens 38(17):4855–4867

Chi KH, Lee BJ (1994) Extracting potential groundwater area using remotely sensed data and GIS techniques. In: Proceedings of the regional seminar on integrated applications of remote sensing and GIS for Land and Water Resources Management, Bangkok (ES-CAPE), pp 64–69

Chorley RJ (1972) Chapter 1: Spatial analysis in geomorphology. In: Chorley RJ (ed) Spatial analysis in geomorphology. Harper and Row Publishers, New York, pp 3–16

Coates DR (1958) Quantitative geomorphology of small drainage basins in Southern Indiana, 1st edn. Columbia University, New York

Conoscenti C, Maggio CD, Rotigliano E (2008) Soil erosion susceptibility assessment and validation using a geostatistical multivariate approach: a test in Southern Sicily. Nat Hazards 46:287–305

CSD (2000) Secretary General's Report on Land Chapter of Agenda 21 to Commission on Sustainable Development (CSD8, UN, New York 2000). UNCED Agenda 21, Rio de Janeiro, 1992 and UNCCD, Paris, 1994

Dabral PP, Baithuri N, Pandey A (2008) Soil erosion assessment in a hilly catchment of North Eastern India using USLE, GIS and remote sensing. Water Resour Manag 22:1783–1798

Densham PJ, Goodchild MF (1988) Spatial decision support systems. A research agency. GIS/LIS' 88 proceeding value 2 Bethesda, Maryland: American Congress on surveying and mapping, November–December 1988

Dheeravath V, Thenkabail PS, Chandrakantha G, Noojipady P, Reddy GPO, Biradar CM, Gumma MK, Velpuri M (2010) Irrigated areas of India derived using MODIS 500m time series for the years 2001–2003. ISPRS J Photogramm Remote Sens 65(1):42–59

Digital Globe (2015) Remote sensing technology trends and agriculture. White paper, Digital Globe, p 17

Dobos E, Micheli E, Baumgardner MF, Biehl L, Helt T (2000) Use of combined digital elevation model and satellite radiometric data for regional soil mapping. Geoderma 97:367–391

Dwivedi RS (2001) Soil resource mapping: a remote sensing perspective. Remote Sens Rev 20:89–122

Ehlers M, Amer S (1991) Geoinformatics: an integrated approach to acquisition, processing and production of geo-data. In: Proceedings, EGIS'91, Brussels, Belgium, pp 306–312

Eswaran H, Lal R, Reich PF (2001) Land degradation: an overview. In: Bridges EM, Hannam ID, Oldeman LR, Pening de Vries FWT, Scherr SJ, Sompatpanit S (eds) Responses to land degradation. Proceedings of the 2nd international conference on land degradation and desertification, Khon Kaen, Thailand. Oxford Press, New Delhi, India

Evans IS (1972) General geomorphometry, derivatives of altitude and descriptive statistics. In: Chorley RJ (ed) Spatial analysis in geomorphology. Methuen, London, pp 17–90

FAO (1976) A framework for land evaluation, FAO soils bulletin 32. FAO, Rome

FAO (1995) Land and water bulletin 2, planning for sustainable use of land resources: towards a new approach. FAO, Rome

FAO (2010a) Climate-smart agriculture: Policies, practices and financing for food security, adaptation, and mitigation. Rome: FAO

FAO (2010b). Climate-smart agriculture. Press release. 28 October 2010. Rome

Gessler PE, Morre ID, McKenzie NJ, Ryan PJ (1995) Soil-landscape modelling and spatial pridiction of soil attributes. Int J Geogr Inf Syst 9:421–432

Gournellos T, Evelpidou N, Vassilopoulos A (2004) Developing an erosion risk map using soft computing methods (case study at Sifnos Island). Nat Hazards 31:63–83

Groot R (1989) Meeting educational requirements in Geomatics. ITC J 1:1–4

Hendrix WG, Buckley DJA (1989). Geographic information system technology as a tool for groundwater management. In: Proceedings of the annual convention of ACSM-ASPRS, Falls Church, Virginia, pp 230–239

Hengl T, Reuter HI (eds) (2009). Geomorphometry: concept, software, applications. Developments in Soil Science, 33. Elsevier Science

Horton RE (1945) Erosional development of streams and their drainage basins: hydrophysical approach to quantitative morphology. Geol Soc Am Bull 56:275–370

Jafari R, Lewis MM, Ostendorf B (2008) An image-based diversity for assessing land degradation in an arid environment in South Australia. J Arid Environ 72:1282–1293

Jain MK, Kothari UC (2000) Estimation of soil erosion and sediment yield using GIS. Hydrol Sci J 45(5):771–786

Kamble B, Irmak A, Hubbard K (2013) Estimating crop coefficients using remote sensing based vegetation index. Remote Sens 5(4):1588–1602

Kar R, Reddy GPO, Kumar N, Singh SK (2018) Monitoring spatio-temporal dynamics of urban and peri-urban landscape using remote sensing and GIS – a case study from Central India. Egypt J Remote Sens Space Sci. https://doi.org/10.1016/j.ejrs.2017.12.006

Krishnamurthy J, Kumar Venkates N, Jayaraman V, Manivel M (1996) An approach to demarcate groundwater potential zones through remote sensing and a geographic information system. Int J Remote Sens 17(10):1867–1884

Kumar N, Reddy GPO, Chatterji S, Srivastava R, Singh SK (2017) Soil suitability evaluation for soybean using temporal satellite data and field data – a case study from semi-arid region of Central India. In: Reddy GPO, Patil NG, Chaturvedi A (eds) Sustainable management of land resources – an Indian perspective. Apple Academic Press, Oakville/Waretown, pp 387–410

Laurent R, Anker W, Graillot D (1998) Spatial modeling with geographic information system for determination of water resources vulnerability application to an area in Massif Central (France). J Am Water Resour Assoc 34(1):123–134

Li Z, Zhu Q, Gold C (2005) Digital terrain modeling: principles and methodology. CRC Press, Boca Raton

Maji AK, Nayak DC, Krishna NDR, Srinivas CV, Kamble K, Reddy GPO, Velayutham M (2001) Soil information system of Arunachal Pradesh in a GIS environment for land use planning. Int J Appl Earth Obs Geoinf 3:69–77

Maji AK, Reddy GPO, Sarkar D (2010) Degraded and wastelands of India, status and spatial distribution. ICAR and NAAS Publication, pp 1–158

McBratney AB, Odeh IOA, Bishop TFA, Dundar MS, Shatar TM (2000) An overview of pedometric techniques for use in soil survey. Geoderma 97:293–327

Mendicino G (1999) Sensitivity analysis on GIS procedures for the estimate of soil erosion risk. Nat Hazards 20:231–253

Miller JC (1953) A quantitative geomorphic study of drainage basin characteristics in the cinch mountain area. Virginia and Tennessee Technical Report 3

Moore ID, Grayson RB, Ladson AR (1993) Digital terrain modeling: a review of hydrological, geomorphological, and biological applications. In: Beven KJ, Moore ID (eds) Terrain analysis and distributed modeling in hydrology, advances in hydrological processes, Wiley, Chichester

Morton JF (2007) The impact of climate change on smallholder and subsistence agriculture. PNAS 104:19680–19685

Naidu LGK, Ramamurthy V, Challa O, Hegde R, Krishnan P (2006) Manual on soil site suitability criteria for major crops. NBBS Publication No. 129, NBSS and LUP, Nagpur, pp 1–3

NBSS&LUP (1994) Proceedings of the National meet on soil-site suitability criteria for different crops, February 7–8, held at NBSS&LUP (ICAR), Nagpur

Ostir K, Veljanovski T, Podobnikar T, Stancic Z (2003) Application of satellite remote sensing in natural hazard management: the Mount Mangart landslide case study. Int J Remote Sens 24(20):3983–4002

Pickup G (1989) New land degradation survey techniques for arid Australia: problems and prospects. Aust Rangeland J 11:74–82

Pike RR (1988) The geometric signature – quantifying landslide terrain types from digital elevation models. Math Geol 20(5):491–511

Pike RJ (1999) A bibliography of geomorphometry, the quantitative representation of topography-supplement 3 (Open-file report 99-140). US Geological Survey

Ramamurthy V, Sarkar D (2009) Soil based agro technologies for livelihood improvement. NBSS Publ. No. 144, NBSS&LUP, Nagpur, 44 p

Ramamurthy V, Gajbhiye KS, Lal S, Maji AK (2006) Land resource management through technology assessment and refinement. NBSS Publ. No.133, NBSS&LUP, Nagpur, 106 p

Ramamurthy V, Naidu LGK, Nair KM, Ramesh Kumar SC, Srinivas S, Thayalan S, Sarkar D, Chaturvedi A, Singh SK (2015) District land use planning, Mysore, Karnataka. NBSS Publ. No.169, 95 pp

Razali SM, Nuruddin AA (2011) Assessment of water content using remote sensing normalized difference water index: preliminary study. In: Proceedings of the 2011 I.E. international conference on space science and communication (IconSpace), pp 265–268

Reddy GPO, Maji AK (2003) Delineation and characterization of geomorphological features in a part of lower Maharahstra metamorphic plateau, using IRS-ID LISS-III data. J Indian Soc Remote Sens 31(4):241–250

Reddy GPO, Rao MS (1994) Hydro-geology and hydro-geomorphological conditions of Anantapur district (AP), India. Indian Geogr J 69(2):128–135

Reddy GPO, Sarkar D (2012) Assessment of soil loss for prioritization of subwatersheds – a remote sensing and GIS approach. NBSS Publ. No.137. NBSS&LUP, Nagpur, p 55

Reddy GPO, Chandramouli K, Srivastav SK, Maji AK, Srinivas C (1999a) Evaluation of groundwater potential zones using remote sensing data – a case study. J Indian Soc Remote Sens 28(1):19–32

Reddy GPO, Shekinah DE, Maurya UK, Thayalan S, Prasad J, Ray SK, Bhaskar BP (1999b) Landscape-soil relationship in part of Bazargaon plateau, Maharashtra. Geogr Rev 63(3):280–291

Reddy GPO, Maji AK, Krishna NDR, Srinivas CV, Velayutham M (2001) Integrated remote and GIS approach for delineation of groundwater potential zones and identification of sites for artificial recharge – a case study. In: Muralikrishna IV (ed) Spatial information technology (remote sensing and geographical information systems), vol I. BSP Publications, Hyderabad, pp 649–658

Reddy GPO, Maji AK, Gajbhiye KS (2002a) GIS for morphometric analysis of drainage basins. GIS India 11(4):9–14

Reddy GPO, Maji AK, Srinivas CV, Velayutham M (2002b) Geomorphological analysis for inventory of degraded lands in a river basin of basaltic terrain, using remote sensing data and geographical information systems. J Indian Soc Remote Sens 30:15–31

Reddy GPO, Maji AK, Gajbhiye KS (2004a) Drainage morphometry and its influence on landform characteristics in a basaltic terrain, Central India – a remote sensing and GIS approach. Int J Appl Earth Obs Geoinf 6:1–16

Reddy GPO, Maji AK, Chary GR, Srinivas CV, Tiwary P, Gajbhiye KS (2004b) GIS and remote sensing applications in prioritization of river sub basins using morphometric and USLE parameters – a case study. Asian J Geoinf 4(4):35–49

Reddy GPO, Maji AK, Nagaraju MSS, Thayalan S, Ramamurthy V (2008) Ecological evaluation of land resources and land use systems for sustainable development at watershed level in different agro-ecological zones of Vidarbha region, Maharashtra using remote sensing and GIS techniques. Project Report, NBSS & LUP, Nagpur, pp 270

Reddy GPO, Dheeravath V, Thenkabail PS, Chandrakantha G, Biradar CM, Noojipady P, Velpuri M, Maji AK (2009) Irrigated areas of India derived from satellite sensors and national statistics – a way forward from GIAM experience. In: Thenkabail PS, Lyon JG, Hugh T, Biradar CM (eds) Remote sensing of global croplands for food security. CRC Press, London, pp 139–176

Reddy GPO, Maji AK, Das SN, Srivastava R (2012) Development of GIS based seamless mosaic of SRTM elevation data of India to analyze and characterize the selected geomorphic parameters. Project Report, NBSS&LUP, Nagpur, 54p

Reddy GPO, Nagaraju MSS, Ramteke IK, Sarkar D (2013) Terrain characterization for soil resource mapping in part of semi-tract of Central India using high resolution satellite data and GIS. J Indian Soc Remote Sens 41(2):331–343

Reddy GPO, Sarkar D, Prasad J, Ramamurthy V (2013a) Geospatial modeling in assessment of biophysical resources for sustainable land resource management. Trop Ecol 54(2):227–238

Reddy GPO, Kurothe RS, Sena DR, Harindranath CS, Niranjana KV, Naidu LGK, Singh SK, Sarkar D, Mishra PK, Sharda VN (2016a) Assessment of soil erosion in tropical ecosystem of Goa, India using universal soil loss equation, geostatistics and GIS. Indian J Soil Conserv 44(1):1–7

Reddy GPO, Sarkar D, Mandal C, Srivastava R, Bhattacharyya T, Naidu LGK, Sidhu GS, Baruah U, Singh SK, Singh RS, Nair KM, Sen TK, Chandran P, Sahoo AK, Srinivas S, Kumar N, Chavan S (2016b) Digital soil resource database and information system. In: Roy PS, Dwivedi RS (eds) Geospatial technology for integrated natural resources management. Yes Dee Publishing Pvt Ltd, Chennai, pp 370–406

Reddy GPO, Patil NG, Chaturvedi A (eds) (2017a) Sustainable management of land resources – an Indian perspective. Apple Academic Press, Oakvile/Waretown, pp 769

Reddy GPO, Singh SK, Kumar N, Mondal C, Srivastava R, Maji AK, Sarkar D (2017b) Geospatial technologies in development of soil information system and prototype geoportal. In: Reddy GPO, Patil NG, Chaturvedi A (eds) Sustainable management of land resources – an Indian perspective. Apple Academic Press, Oakvile/Waretown, pp 411–442

Renard KG, Foster GR, Weesies GA, Porter JP (1991) RUSLE, revised universal soil loss equation. J Soil Water Conserv 46(1):30–33

Riquier J, Cornet JP, Braniao DL (1970) A new system of soil appraisal in terms of actual and potential productivity. 1st Approx. World Soil Res. FAO, pp 44

Rojas O, Vrieling A, Rembold F (2011) Assessing drought probability for agricultural areas in Africa with coarse resolution remote sensing imagery. Remote Sens Environ 115:343–352

Rundquist DC, Peters AJ, Di L, Rodekohr DA, Ehram RL, Murray G (1991) Statewide groundwater vulnerability assessment in Nebraska using the DRASTIC/GIS model. Geocarto Int 6(2):51–57

Saraf AK, Choudhuray PR (1998) Integrated remote sensing and GIS for groundwater exploration and identification of artificial recharge sites. Int J Remote Sens 19(10):1825–1841

Srinivas CV, Maji AK, Reddy GPO, Chary GR (2002) Assessment of soil erosion using remote sensing and GIS in Nagpur District, Maharashtra, for prioritization and delineation of conservation units. J Indian Soc Remote Sens 30(4):197–211

Srivastava R, Saxena RK (2004) Techniques of large-scale soil mapping in basaltic terrain using satellite remote sensing data. Int J Remote Sens 25(4):679–688

Storie RE (1978) Storie index soil rating (revised). Special publication division of agricultural science. University of California, Berkeley

Sudhakar S, Sridevi G, Ramana IV, Rao VV, Raha AK (2000) Techniques of classification for land use/land cover with special reference to forest type mapping in Jaldapara Wildlife Sanctuary, Jalpaiguri district, West Bengal – a case study. J Indian Soc Remote Sens 27(4):217–224

Sys C (1985) Land evaluation. Algemeen Bestuur vande Ontwikkelingss. International Training Centre for Post-Graduate Soil Scientists. State University of Ghent, Ghent

Sys C, Van Ranst E, Debaveye J (1991) Land evaluation. Part I: Principles in land evaluation and crop production calculations. General Administration for Development Cooperation. Agric. Publ. No. 7

Thenkabail PS, Dheeravath V, Biradar CM, Reddy GPO, Noojipady P, Chandrakantha G, Velpuri M, Gumma MK, Li Y (2009) Irrigated area maps and statistics of India using remote sensing and national statistics. Remote Sens 1(2):50–67

Tian YC, Zhou YM, Wu BF, Zhou WF (2008) Risk assessment of water soil erosion in upper basin of Miyun Reservoir, Beijing, China. Environ Geol 57(4):937–942

Tueller PT (1987) Remote sensing science applications in arid environment. Remote Sens Environ 23:143–154

Vinciková H, Hais M, Brom J, Procházka J, Pecharová E (2010) Use of remote sensing methods in studying agricultural landscapes – a review. J Lands Stud 3:53–63

Vishakha D, Maji AK, Reddy GPO, Ramteke IK (2016) Land suitability evaluation for rice (***Oryza sativa*** *L.*) in Tirora tehsil of Gondia District, Maharashtra – a GIS approach. Agropedology 26(01):69–78

Watkins DW, Mckinney DC, Maidment DR, Lin MD (1996) Use of geographic information system in groundwater flow modeling. J Water Resour Plan Manag 122(2):88–96

Wilson DJ, Gallant JC (2000a) Digital terrain analysis. In: Wilson DJ, Gallant JC (eds) Terrain analysis: principles and applications. Willey, New York, pp 1–27

Wilson PJ, Gallant JC (2000b) Secondary topographic attributes. In: Wilson PJ, Gallant JC (eds) Terrain analysis: principles and applications. Willey, New York, pp 87–131

Wischmeier WH, Smith DD (1978) Predicting rainfall erosion losses – a guide to conservation planning. USDA Agricultural Handbook No. 587

Wu Q, Wang M (2007) A framework for risk assessment on soil erosion by water using an integrated and systematic approach. J Hydrol 337:11–21

Xue Y, Fennessy MD (2002) Under what conditions does land cover change impact regional climate? In: Reynolds JF, Smith DMS (eds) Under what conditions does land cover change impact regional climate? Dahlem University Press, Berlin, pp 59–74

Yagi K, Agus F, Arao T, Aulakh MS, Bai Z, Carating R, Jung K, Kadono A, Kawahigashi M, Lee SH, Ma L, Reddy GPO, Sidhu GS, Takata Y, Tien TM, Xu R, Yan X, Yokoyama K, Zhang F, Zhou D (2015) In: Regional assessment of soil changes in Asia. Status of the World's Soil Resources–Main Report. Food and Agriculture Organization of the United Nations and Intergovernmental Technical Panel on Soils, Rome, Italy, pp 287–329

Part I
Geospatial Technologies – Principles and Applications

Chapter 2
Satellite Remote Sensing Sensors: Principles and Applications

G. P. Obi Reddy

Abstract Remote sensing (RS) refers to the science of identification of Earth surface features and estimation of their geo-biophysical properties using electromagnetic radiation as a medium of interaction. Satellite remote sensing, with its synoptic view of the Earth's features, regular repetitive coverage over large areas, and digital mode of data capture, offers an effective means of mapping, monitoring, and management of land resources and environmental impacts near real time, providing a historical profile for monitoring the Earth's features. Satellite sensors record information about the Earth's surface by measuring the transmission of energy from the surface in different portions of the electromagnetic spectrum (EMS). The optical remote sensing devices operate in the visible, near-infrared, middle-infrared, and shortwave infrared portion of the electromagnetic spectrum. These devices are sensitive to the wavelengths ranging from 300 nm to 3000 nm. The sensors, which operate in thermal range of electromagnetic spectrum, record the energy emitted from the Earth features in the wavelength ranges of 3000–5000 nm and 8000 nm to 14,000 nm. A microwave remote sensor records the backscattered microwaves in the wavelength range of 1 mm–1 m of electromagnetic spectrum. A wide variety of satellite remote sensing data from Moderate Resolution Imaging Spectroradiometer (MODIS), Landsat, IRS-IC, IRS-ID, IRS-P6, Cartosat-1, Cartosat-2, QuickBird, and Google are available to the Earth scientists for generation of spatial database on natural resources for various applications. Analysis of multiple-date satellite imagery provides information in mapping, monitoring, and management of land resources on periodical basis.

Keywords Electromagnetic radiation · Electromagnetic spectrum · Remote sensing · Remote sensing sensors

---

G. P. O. Reddy (✉)
ICAR-National Bureau of Soil Survey & Land Use Planning, Nagpur, India

G. P. O. Reddy, S. K. Singh (eds.), *Geospatial Technologies in Land Resources Mapping, Monitoring and Management*, Geotechnologies and the Environment 21,
https://doi.org/10.1007/978-3-319-78711-4_2

2.1 Introduction

Remote sensing (RS) can be defined as the acquisition of information about an object without being in physical contact with it (Elachi 1987). Remote sensing technology allows us to observe the Earth feature from the space fundamentally, and there are several techniques to differentiate the information collected from remote sensing like extraction of information on forest, water, agriculture, etc. (Sabins and Floyd 1978; Sabins 1997). Detection and discrimination of objects on the Earth surface mean detecting and recording of radiant energy reflected or emitted by objects or surface material. The advances in remote sensing technology provide data for detail inventory, mapping, and monitoring of land resources in large scale (Reddy 2012). A wide variety of satellite remote sensing data from MODIS, Landsat, IRS-IC, IRS-ID, IRS-P6, Cartosat-1, Cartosat-2, QuickBird, and Google are available to the Earth scientists for generation of spatial database on land resources for various applications. The temporal remote sensing data support us to analyze the real Earth situation of the resources to know the type, stock, and spatial extent of the resources over a period of time. Remotely sensed data from variety of satellites are being extensively used in various applications like soil resource mapping, assessment of land degradation, land use/land cover changes, evaluation of water resources, environmental impact assessment, etc. However, remotely sensed data coupled with field observations and contemporary technology provides more realistic datasets as compared to image interpretation alone. Besides database derived from satellite data, socioeconomic profile of the people serves as vital inputs for land resource management on sustainable basis. With the advancements in the field of computer and geographic information systems (GIS), the resource management issues can be dealt more efficiently and effectively by applying these technologies rather than the traditional methods of management.

2.2 History of Remote Sensing

The concept and practice of remote sensing were first developed in the 1840s, when it was realized that a different and perhaps more revealing view of a particular landscape could be obtained by taking a photograph from a vantage point. Subsequently, to map topographic features, efforts were made to look down at the Earth's surface by taking pictures with the aid of cameras fitted with the balloons. By the First World War, cameras were mounted on airplanes, which provided aerial views of fairly large surface areas, and were used as a method of data and information acquisition in military reconnaissance. At the end of the Second World War, radar had come into civilian use, thermal sensors were being developed, and the concept of multispectral photography was finding applications.

In early 1946, efforts involved in installing automated sensors and cameras on captured German V-2 rockets to took the first high-altitude aerial images. Till the early 1960s, aerial photography remained the single standard tool for depicting the surface from a vertical or oblique perspective. In the late 1960s, with the emergence of the space program, cosmonauts and astronauts started taking photographs out of the window of their spacecraft in which they were orbiting the Earth. This was followed by the development of meteorological satellites in the 1960s, the first of which was the Television Infrared Observation Satellite (TIROS-1). Further refinements in imaging sensors revealed the surface of the Earth in a new light, and the modern era of multispectral imaging had begun. The value of the multispectral aspects of imagery from space was tested by using multiband imagery from manned aircraft, where it is still widely used. This early phase led to the birth of the Earth Resources Technology Satellite (ERTS, later renamed Landsat) that was launched in 1972. Seven other Landsat satellites followed, each of which had several visible light and infrared bands. The success of such missions led to a remote sensing industry and to more satellites (e.g., SPOT, JERS, IRS) each with improved capabilities.

Today, remote sensing is carried out using airborne and satellite technology, not only utilizing film photography but also digital camera, scanner and video, as well as radar and thermal sensors. Unlike in the past, when remote sensing was restricted to only the visual part of the electromagnetic spectrum, i.e., what could be seen with the naked eye, today through the use of special filters, photographic films, and other types of sensors, the parts of the spectrum, which cannot be seen with the naked human eye, can also be utilized. In the microwave region of the spectrum, the active sensor like radar provides its own source of electromagnetic radiation (EMR), transmitted to the Earth's surface, and EMR reflected (backscattered) from the surface is recorded and analyzed. Further, hyperspectral remote sensing is emerged as promising technology in remote sensing for studying the Earth surface materials in detail both spectrally and spatially. Thus, today remote sensing is largely utilized in land resource management, which frequently requires rapid, accurate, and up-to-date data collection and dissemination.

2.3 Principles and Concept of Remote Sensing

Remote sensing (RS) refers to the science of identification of Earth surface features and estimation of their geo-biophysical properties using electromagnetic radiation as a medium of interaction.

2.3.1 Characteristics of Electromagnetic Spectrum and Radiation

Electromagnetic (EM) spectrum represents the continuum of EMR arranged on the basis of wavelengths or frequency. EMS ranges from shorter wavelengths (gamma rays to X rays) to the longer wavelengths (microwave and radio waves). Most common remote sensing systems operate in one or several of the visible, infrared, and microwave portions of the electromagnetic spectrum. Within the infrared portion of the spectrum, only thermal infrared energy is directly related to the sensation of heat but not the near- and mid-infrared ones. EMR travels in the form of waves, and those EMR waves reflected by or emitted or backscattered from objects give us information about the objects themselves. Visible and infrared remote sensing normally uses wavelength to denote the range of plane waves, but work in microwave remote sensing commonly employs both wavelength and frequency. Optical remote sensing is a term used to describe remote sensing at visible (400–700 nm) and near-infrared wavelengths; radar remote sensing operates at microwave frequencies.

Most sensors record information about the Earth's surface by measuring the transmission of energy from the surface in different portions of the electromagnetic (EM) spectrum (Fig. 2.1). Since Earth's surface varies in nature, the transmitted energy also varies, and this variation in energy allows images of the surface to be created. Human eyes see this variation in energy in the visible portion of the EM spectrum. Sensors detect variations in energy in both the visible and non-visible areas of the spectrum.

EMR is energy that propagates in wave form at a velocity of $C = 3 \times 10^{10}$ cm/sec. The parameters that characterize a wave motion are wavelength (λ), frequency (ν), and velocity (C). The relationship between these parameters is $C = \nu\lambda$. Figure 2.1 illustrates spectral bands used in remote sensing.

The areas of the EM spectrum that are absorbed by atmospheric gases such as water vapor, carbon dioxide, and ozone are known as absorption bands. In Fig. 2.2,

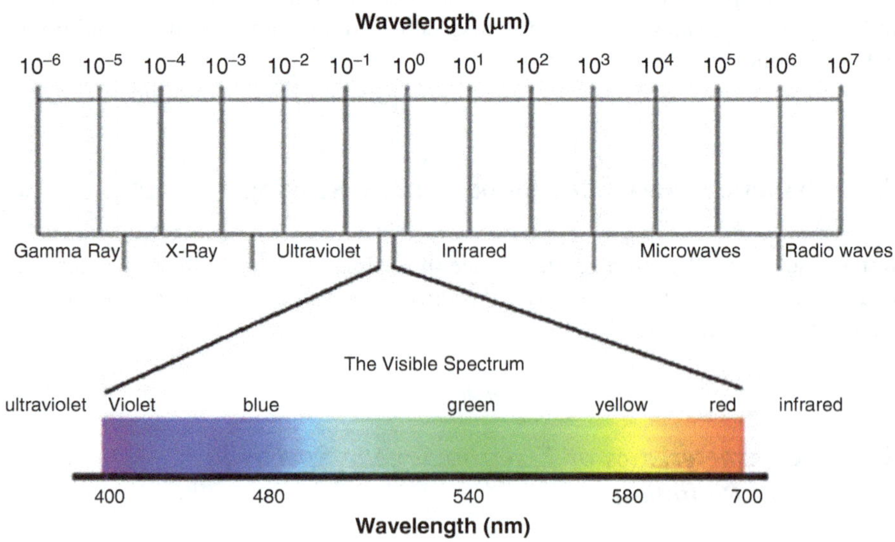

Fig. 2.1 Wavelength, frequency, and energy. Note that wavelength has been reversed (shorter wavelengths to the right)

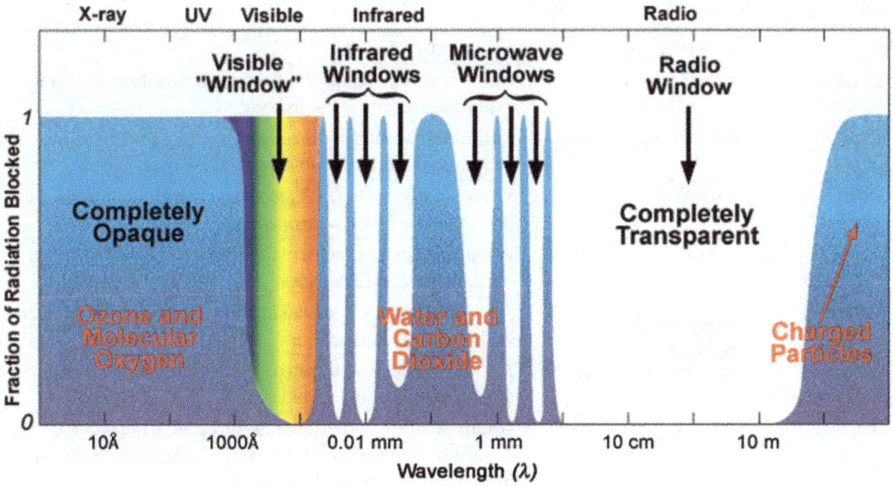

Fig. 2.2 Atmospheric windows (Source: Campbell and Wynne 2011)

absorption bands (shown in brown) are represented by a low transmission value that is associated with a specific range of wavelengths. Trying to obtain remotely sensed imagery in the absorption bands is nearly impossible; thus, sensors are generally designed not to record information in these portions of the spectrum.

In contrast to the absorption bands, there are areas of the EM spectrum (Fig. 2.2 and Table 2.1), where the atmosphere is transparent (little or no absorption of energy) to specific wavelengths. The spectral bands for which the atmosphere is relatively transparent are known as atmospheric windows. Atmospheric windows are present in the visible part (0.4–0.76 μm) and the infrared regions of the EM spectrum. In the visible part, transmission is mainly affected by ozone absorption and by molecular scattering. The atmosphere is transparent again beyond about $\lambda = 1$ mm, the region used for microwave remote sensing. It is in these windows that sensors are used to gather information about Earth phenomena.

Satellite images are generally captured using sensors and digital technology. A sensor often records simultaneously in several different regions of the spectrum creating multi-images taken at the same time. The regions of the spectrum scanned are called "bands." A band is identified in nanometers (nm). For example, an image or band scanned in the visible light region of the spectrum would be identified as 400–700 nm. With multibands, various color composite images can be created.

A single band image shows features in various gray tones, but if several images are combined, they can form a color composite. A true color composite is based on using the red, green, and blue portions of the visible region of the EM spectrum. It relates to what the human eye would see if a person was on a satellite or aircraft looking down at the Earth. On the other band, combinations form false color composites (FCC). A false color composite generally enhances certain features on an image, features that might not be as apparent on a true color composite. Using

Table 2.1 Major regions of the electromagnetic spectrum

Region name	Wavelength	Comments
Gamma ray	< 0.03 nanometers	Entirely absorbed by the Earth's atmosphere and not available for remote sensing
X-ray	0.03–30 nanometers	Entirely absorbed by the Earth's atmosphere and not available for remote sensing
Ultraviolet	0.03–0.4 micrometers	Wavelengths from 0.03 to 0.3 micrometers absorbed by ozone in the Earth's atmosphere
Photographic ultraviolet	0.3–0.4 micrometers	Available for remote sensing the earth. Can be imaged with cameras and sensors
Visible	0.4–0.7 micrometers	Available for remote sensing the earth. Can be imaged with cameras and sensors
Near- and mid-infrared	0.7–3.0 micrometers	Available for remote sensing the earth. Can be imaged with cameras and sensors.
Thermal infrared	3.0–14 micrometers	Available for remote sensing the earth. This wavelength cannot be captured by film cameras. Sensors are used to image this wavelength band
Microwave or radar	0.1–100 centimeters	Longer wavelengths of this band can pass through clouds, fog, and rain. Images using this band can be made with sensors that actively emit microwaves
Radio	> 100 centimeters	Not normally used for remote sensing the earth

different color composites is one way that a remote sensing specialist detects features on the Earth. The FCC image generated from Landsat-8 OLI Band 3 (Green), Band 4 (Red), and Band 5 (Infrared) is shown in Fig. 2.3.

Table 2.2 gives the wavelengths and, where appropriate, frequencies commonly found in remote sensing. Note that the most commonly used units for optical remote sensing are micrometers (10^{-6} m).

2.4 Types of Remote Sensing

2.4.1 Passive Remote Sensing

The sun provides a very convenient source of energy for remote sensing. The sun's energy is either reflected, as it is for visible wavelength, or absorbed and then reemitted (for thermal infrared wavelength). Remote sensing systems, which measure this naturally available energy, are called *passive sensors*. The sensor then records the reflected energy as it has been altered by the target. This can only take place, when the sun is illuminating the Earth. There is no reflected energy available from the sun at night. Energy that is naturally emitted can be detected day and night provided that the amount of energy is large enough to be recorded.

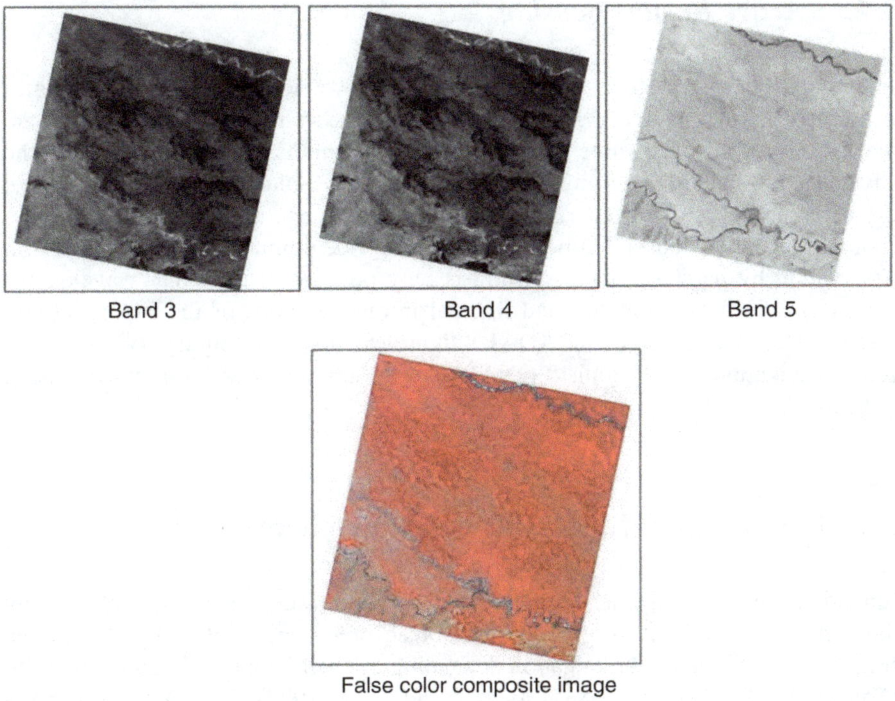

Fig. 2.3 FCC image generated from Landsat-8 OLI Band 3 (green), Band 4 (red), and Band 5 (infrared)

Table 2.2 Wavelengths and frequencies used in satellite remote sensing EM region

EM region	Wavelength	Frequency
Ultraviolet	<0.4 μm	
Visible	0.4–0.7 μm	
Near-infrared	0.7–1.5 μm	
Very near-infrared	*0.7–1.5 μm*	
Shortwave infrared	*1.5–2.5 μm*	
Middle-infrared	2.5–8.0 μm	
Thermal infrared	8.0 -14 μm	
Microwave	10–300 mm	1–12.5 GHz
X band	24–38 mm	8–12.5 GHz
C band	38–75 mm	4–8 GHz 4
L band	150–300 mm	1–2 GHz

Note: 1 μm = 10^{-6} m (1 micrometer = one millionth of a meter) 1 μm = 1000 nm (1 micrometer = one thousand nanometers)

2.4.2 *Active Remote Sensing*

The remote sensing systems, which provide their own source of energy for illumination, are known as *active sensors*. Most active systems for remote sensing are radar systems, which operate at microwave wavelengths. These sensors have the advantage of obtaining data any time of day or season. Solar energy and radiant heat are examples of passive energy sources. Synthetic aperture radar (SAR) is an example of active sensor. Atmospheric windows determine that successful Earth observation by passive remote sensing is only possible at particular wavelengths within visible, near-, middle-, and thermal infrared regions of the EM spectrum. Radio detection and ranging (RADAR), light detection and ranging (LIDAR), and sound navigation and ranging (SONAR) are the important active remote sensing systems.

2.5 Interaction of EMR with the Atmosphere

Radiation from the sun, when incident upon the Earth's surface, is either reflected by the surface, transmitted into the surface, or absorbed and emitted by the surface (Fig. 2.4). Objects emit the radiation depending on their temperature and emissivity (Campbell 1996; Colwell 1983; Liliesand and Kiefer 1994). The EMR, on interaction, experiences a number of changes in magnitude, direction, wavelength, polarization, and phase. These changes are detected by the remote sensor and enable the interpreter to obtain useful information about the object of interest. The remotely sensed data contain both spatial information (size, shape, and orientation) and spectral information (tone, color, and spectral signature).

The EMR interacts with the atmosphere while traveling from the source to Earth features and from Earth features to the sensor. During this whole path, the EMR changes its properties due to loss of energy and alteration in wavelength, which

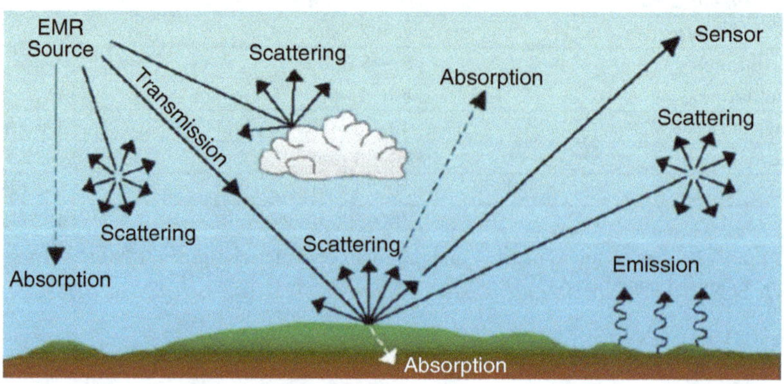

Fig. 2.4 Interaction of energy with the Earth's surface. (Source: Liliesand and Kiefer 1993)

ultimately affects the sensing of the EMR by the sensor. This interaction often leads to *atmospheric noise.*

From the viewpoint of interaction mechanisms, with the object-visible and infrared wavelengths from 0.3 μm to 16 μm can be divided into three regions. The spectral band from 0.3 μm to 3 μm is known as the reflective region. In this band, the radiation sensed by the sensor is that due to the sun, reflected by the Earth's surface. The band corresponding to the atmospheric window between 8 μm and 14 μm is known as the thermal infrared band. The energy available in this band for remote sensing is due to thermal emission from the Earth's surface. Both reflection and self-emission are important in the intermediate band from 3 μm to 5.5 μm. In the microwave region of the spectrum, the sensor is radar, which is an active sensor, as it provides its own source of EMR. The EMR produced by the radar is transmitted to the Earth's surface, and the EMR reflected (back scattered) from the surface is recorded and analyzed. The microwave region can also be monitored with passive sensors, called microwave radiometers, which record the radiation emitted by the terrain in microwave region.

Reflection Of all the interactions in the reflective region, surface reflections are the most useful and revealing in remote sensing applications. Reflection occurs when a ray of light is redirected as it strikes a nontransparent surface. The reflection intensity depends on the surface refractive index, absorption coefficient, and the angles of incidence and reflection.

Transmission Transmission of radiation occurs when radiation passes through a substance without significant attenuation. For a given thickness, or depth of a substance, the ability of a medium to transmit energy is measured as transmittance (T).

$$T = \frac{Transmitted\ radiation}{Incidence\ radiation}$$

2.6 Spectral Signature

Spectral reflectance, [ρ(λ)], is the ratio of reflected energy to incident energy as a function of wavelength. Various materials of the Earth's surface have different spectral reflectance characteristics. Spectral reflectance is responsible for the color or tone in a photographic image of an object. Trees appear green because they reflect more of the green wavelength. The values of the spectral reflectance of objects averaged over different, well-defined wavelength intervals comprise the spectral signature of the objects or features by which they can be distinguished. To obtain the necessary ground truth for the interpretation of multispectral imagery, the spectral characteristics of various natural objects have been extensively measured and recorded. The spectral reflectance is dependent on wavelength; it has different values at different wavelengths for a given terrain feature. The reflectance characteristics of the Earth's surface features are expressed by spectral reflectance, which is given by:

$$\rho(\lambda) = [ER(\lambda)/EI\ (\lambda)] \times 100$$

where:

$\rho(\lambda)$ = spectral reflectance (reflectivity) at a particular wavelength
ER (λ) = energy of wavelength reflected from object
EI (λ) = energy of wavelength incident upon the object

The plot between $\rho(\lambda)$ and λ is called a spectral reflectance curve. This varies with the variation in the chemical composition and physical conditions of the feature, which results in a range of values. The spectral response patterns are averaged to get a generalized form, which is called generalized spectral response pattern for the object concerned. Spectral signature is a term used for unique spectral response pattern, which is characteristic of a terrain feature. Figure 2.5 shows a typical reflectance curves for three basic types of Earth surface features, healthy vegetation, dry bare soil (gray-brown and loamy), and clear lake water.

The spectral characteristics of the three main Earth surface features are as follows:

Vegetation The spectral characteristics of vegetation vary with wavelength. Plant pigment in leaves called chlorophyll strongly absorbs radiation in the red and blue wavelengths but reflects green wavelength (Curran 1985). The internal structure of healthy leaves acts as diffuse reflector of near-infrared wavelengths. Measuring and

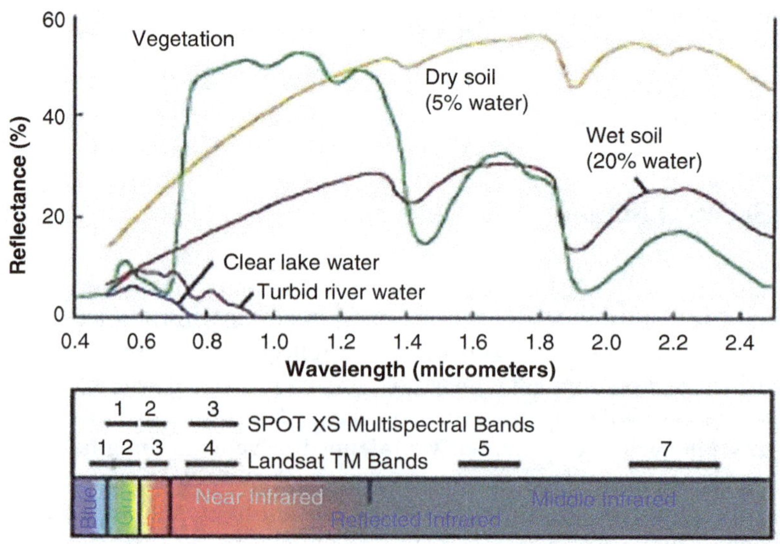

Fig. 2.5 Typical spectral reflectance curves for vegetation, soil, and water. (Source: Liliesand and Kiefer 1994)

monitoring the near-infrared reflectance are ways that scientists determine how healthy particular vegetation may be.

Water Majority of the radiation incident upon water is not reflected but is either absorbed or transmitted. Longer visible wavelengths and near-infrared radiation is absorbed more by water than by the visible wavelengths. Thus water looks blue or blue green due to stronger reflectance at these shorter wavelengths and darker if viewed at red or near-infrared wavelengths. The factors that affect the variability in reflectance of a water body are depth of water, materials within water, and surface roughness of water.

Soil The majority of radiation incident on a soil surface is either reflected or absorbed, and little is transmitted. The characteristics of soil that determine its reflectance properties are its moisture content, organic matter content, texture, structure, and iron oxide content. The soil curve shows less peak and valley variations. The presence of moisture in soil decreases its reflectance. By measuring the energy that is reflected by targets on Earth's surface over a variety of different wavelengths, we can build up a spectral signature for that object. And by comparing the response pattern of different features, we may be able to distinguish between them, which we may not be able to do if we only compare them at one wavelength. For example, water and vegetation reflect somewhat similarly in the visible wavelength but not in the infrared.

Signature of any object and/or its condition comprises a set of observable characteristics, which directly or indirectly lead to the identification of an object and/or its condition. There are four principal characteristics of signatures to identify an object.

- Spectral variations: Changes in the reflectance or emittance as a function of wavelength
- Spatial variations: Variations in the reflectance/emittance determined by the shape, size, and texture of the target
- Temporal variations: Diurnal and/or seasonal changes in reflectance or emittance
- Polarization variations: Changes in the polarization of the radiation reflected or emitted by an object

2.7 Energy Interaction with Atmosphere

An interaction of the direct solar radiation and reflected radiation from the target with the atmospheric constituents interferes with the process of remote sensing and is called as "atmospheric effects." The atmospheric constituents scatter and absorb the radiation modulating the radiation reflected from the target by attenuating it, changing its spatial distribution, and introducing into field of view radiation from sunlight scattered in the atmosphere and some of the energy reflected from nearby ground area. Both scattering and absorption vary in their effect from one part of the spectrum to the other. The solar energy is subjected to modification by several physical processes as it passes the atmosphere, viz., (1) scattering, (2) absorption, and (3) refraction.

2.7.1 Atmospheric Scattering

Scattering is the redirection of EMR by particles suspended in the atmosphere or by large molecules of atmospheric gases. Scattering not only reduces the image contrast but also changes the spectral signature of ground objects as seen by the sensor. The amount of scattering depends upon the size of the particles, their abundance, the wavelength of radiation, depth of the atmosphere through which the energy is traveling, and the concentration of the particles. The concentration of particulate matter varies both in time and over season. Thus the effects of scattering will be uneven spatially and will vary from time to time.

Theoretically scattering can be divided into three categories depending upon the wavelength of radiation being scattered and the size of the particles causing the scattering. The three different types of scattering from particles of different sizes are summarized below:

Scattering process	Wavelength	Approximate dependence particle size	Kind of particles
Selective			
(i) Rayleigh scattering	λ^{-4}	< 0.1 μm	Air molecules
(ii) Mie scattering	λ^{0} to λ^{-4}	0.1 to 10 μm	Smoke, haze
Nonselective	λ^{0}	> 10 μm	Dust, fog, clouds

Rayleigh Scattering Rayleigh scattering appears when the radiation wavelength is much larger than the particle size, for example, scattering of visible light (0.4–0.76 μm) by pure gas molecules (10^{-4} λm) in a clear atmosphere. These could be particles such as small specks of dust or nitrogen and oxygen molecules. Rayleigh scattering causes shorter wavelengths of energy to be scattered much more than longer wavelengths. Rayleigh scattering is the dominant scattering mechanism in the upper atmosphere. Rayleigh scattering causes the sky to appear blue. The scattering coefficient is proportional to the inverse fourth power of wavelength. Radiation in shorter blue wavelengths is scattered toward the ground much more strongly than radiation in the red wavelength region. Due to Rayleigh scattering, multispectral data from the blue portion of the spectrum is of relatively limited usefulness.

Mie Scattering Mie scattering occurs when radiation wavelength is comparable to the size of the scattering particles. In remote sensing, Mie scattering usually manifests itself as a general deterioration of multispectral images across optical spectrum under conditions of heavy atmospheric haze. Depending upon the particle size relative to the wavelength, Mie scattering may fall anywhere between λ^{-4} and λ^{-0} . The incident light is scattered mainly in the forward direction. Dust, pollen, smoke, and water vapor are common causes of Mie scattering, which tends to affect wavelengths longer than those affected by Rayleigh scattering. Mie scattering occurs mostly in the lower portions of the atmosphere where larger particles are more abundant and dominates when cloud conditions are overcast.

Nonselective Scattering Selective scattering usually occurs when the particle size is much larger than the radiation wavelength. Scattering does not depend on the wavelength of radiation. This type of scattering usually occurs when the atmosphere is heavily dust laden and results in a severe attenuation of the received data. There is a uniform attenuation at all wavelength. The whitish appearance of the sky under haze condition is due to nonselective scattering. Occurrence of this scattering mechanism gives a clue to the existence of large particulate matter in the atmosphere above the scene of interest, which itself is a useful data.

2.7.2 Atmospheric Absorption

Absorption is the other main mechanism at work when electromagnetic radiation interacts with the atmosphere. In contrast to scattering, this phenomenon causes molecules in the atmosphere to absorb energy at various wavelengths. Ozone, carbon dioxide, and water vapor are the three main atmospheric constituents, which absorb radiation. Ozone absorbs the high-energy, short wavelength portions of the ultraviolet spectrum ($\lambda < 0.24$ μm), thereby preventing the transmission of this radiation to the lower atmosphere. Carbon dioxide is important in remote sensing as it effectively absorbs the radiation in mid- and far-infrared regions of the spectrum. It strongly absorbs in the region from about 13–17.5 μm, whereas two most important regions of water vapor absorption are in bands 5.5–7.0 μm and above 27 μm. Absorption relatively reduces the amount of light that reaches our eye making the scene looks relatively duller. The atmosphere selectively transmits energy of certain wavelengths. The spectral bands for which the atmosphere is relatively transparent are known as atmospheric windows. Atmospheric windows are present in the visible part (0.4–0.76 μm) and the infrared regions of the EM spectrum. In the visible part, transmission is mainly affected by ozone absorption and by molecular scattering. The atmosphere is transparent again beyond about $\lambda = 1$ mm, the region used for microwave remote sensing.

Refraction The phenomenon of refraction that is bending of light at the contact between two media also occurs in the atmosphere as the light passes through the atmospheric layers of varied clarity, humidity, and temperature. These variations influence the density of atmospheric layers, which in turn, causes the bending of light rays as they pass from one layer to another. The most common phenomena are the mirage-like apparitions sometimes visible in the distance on hot summer days.

2.8 Satellite Remote Sensing Sensors and Their Resolutions

There are four types of *resolutions* that are defined for the remote sensing systems, which are spatial, spectral, radiometric, and temporal resolutions.

Spatial Resolution Spatial resolution is defined as "a measure of the smallest angular or linear separation between two objects that can be resolved by the remote sensing system" (Jenson 2007). It represents area covered by a pixel on the ground. It is nothing but the smallest object that can be detected and distinguished from a point. Mostly, it is measured in meters. The most frequently used measure, based upon the geometric properties of an imaging system, is the instantaneous field of view (IFOV). The IFOV is the area on the surface that is theoretically viewed by the instrument from a given altitude at a given time. The greater the sensor's resolution, the greater the data volume and smaller the area covered. The area coverage and resolution are interdependent, and these two factors determine the scale of imagery. For example, Cartosat-1 sensor has a spatial resolution of 2.5 × 2.5 m, IRS-P6 Linear Imaging Self-Scanning Sensor (LISS)-IV has a spatial resolution of 5.6x5.6 m for its multispectral bands, and LISS-III has spatial resolution of 23.5 × 23.5 m in its first three bands. The smaller the spatial resolution, the greater the resolving power of the sensor system. That's why one can detect even a car in the satellite image acquired by IKONOS (spatial resolution 1 × 1 m) but can see hardly even a village in a satellite image acquired by *Advanced Very High Resolution Radiometer* (AVHRR) (spatial resolution 1.1 × 1.1 km). Different spatial resolution images of IRS-P6 LISS-IV (5.8 m), Sentinel 2 (10 m), and Landsat-8 OLI (30 m) are shown in Fig. 2.6.

Spectral Resolution Spectral resolution refers to the specific wavelength intervals in the electromagnetic spectrum for which a satellite sensor can record the data. It can also be defined as the number and dimension of specific wavelength intervals in the electromagnetic spectrum to which a remote sensing instrument is sensitive. Multi-spectral imaging is the viewing of a given area in several narrow bands to obtain better identification and classification of objects. For example, band 1 of the Landsat

Fig. 2.6 IRS-P6 LISS-IV (5.8 m), Sentinel 2 (10 m), and Landsat-8 OLI (30 m) spatial resolution data

TM sensor records energy between 0.45 μm and 0.52 μm in the visible part of the spectrum. The spectral channels containing wide intervals in the electromagnetic spectrum are referred to as coarse spectral resolution and narrow intervals are referred to as fine spectral resolution. For instance, the SPOT panchromatic sensor is considered to have coarse spectral resolution because it records EMR between 0.51 μm and 0.73 μm. On the other hand, band 2 of the Advanced Spaceborne Thermal Emission and Reflection Radiometer (ASTER) sensor has fine spectral resolution, and it records EMR between 0.63 μm and 0.69 μm.

Radiometric Resolution Radiometric resolution is defined as the sensitivity of a remote sensing detector to differentiate in signal strength as it records the radiant flux reflected or emitted from the terrain. It refers to the dynamic range, or number of possible data file values in each band. This is referred to by the number of bits into which the recorded energy is divided. The ability to distinguish the finer variation of the reflected or emitted radiation from different objects is characterized by the radiometric resolution. This depends on the number of quantization levels within the spectral band. It is the number of bits of digital data in the spectral band or the number of gray level values that will decide the sensitivity of the sensor. An image with more number of gray level values increases the capability for the extraction of information about different features. For instance, ASTER records data in 8-bit for its first nine bands; it means the data file values range from 0 to 255 for each pixel, while the radiometric resolution of LISS-III is 7-bit; here the data file values for each pixel range from 0 to 128 (Table 2.3).

Temporal Resolution By a suitable selection of spacecraft altitude and the inclination angle of the orbit, the spacecraft can be made to cover the same area on the Earth at regular intervals. The temporal resolution of a satellite system refers to how frequently it records imagery of a particular area. For example, Cartosat-1 can acquire images of the same area of the globe every 5 days, while LISS-III does it every 24 days. The temporal resolution of a satellite sensor is very much helpful in change detection. For instance, agricultural crops have unique crop calendars in each geographic region. To measure specific agricultural variables, it is necessary to acquire remotely sensed data at critical dates in the phenological cycle. Analysis of multiple-date imagery provides information on how the variables are changing through time. Multi-date satellite images are also used to detect change in forest cover.

Table 2.3 Some of satellite sensors and their radiometric resolutions

Sensor	Radiometric resolution	Color
LISS-III	7 bits/128 level	0–127 colors
PAN	7 bits/128 level	0–127 colors
WiFS	6 bits/64 level	0–63 colors
AWiFS	10 bits/1024 level	0–1023 colors
Cartosat-1	10 bits/1024 level	0–1023 colors

2.9 Based on Range of Electromagnetic Spectrum

2.9.1 Optical Remote Sensing Sensors

The optical remote sensing sensors operate in the visible, near-infrared, middle-infrared, and shortwave infrared portion of the electromagnetic spectrum (Joseph 1996). These sensors are sensitive to the wavelengths ranging from 300 nm to 3000 nm. Most of the remote sensors record the EMR in this range, e.g., bands of IRS-P6 LISS-IV sensor are in optical range of EMR.

2.9.1.1 Panchromatic Imaging System

Imaging is performed in a single spectral band, corresponding to the visible part of the electromagnetic spectrum. The panchromatic band in SPOT 1, 2 HRV covers 0.51–0.73 μm. For SPOT 4 HRVIR, the panchromatic band has a narrower bandwidth centered at the red band (0.61–0.68 μm). The panchromatic mode of the SPOT 4 HRVIR is named as the monospectral (M) mode, to differentiate it from the panchromatic mode of the SPOT 1, 2 HRV. The single channel imaging mode (P or M mode) supplies only black and white images with a pixel width of 10 m. This band is intended primarily for applications calling for fine geometrical detail. Indian remote sensing satellites like IRS-1C/1D and Cartosat-1, 2, 2A, and 2B have PAN imaging capabilities.

2.9.1.2 Multispectral Imaging Sensors

Landsat First Landsat satellite was launched in 1972, and subsequently series of more sophisticated satellites with multispectral imaging sensors, named TM-Thematic Mapper, have been added ranging from Landsats 4 (1982), 5 (1984), 6 (1993, launch failed) to 7 (1999) (Enhanced Thematic Mapper Plus, ETM+) were launched. Landsat carries two important multispectral sensors, i.e., the first is the multispectral scanner (MSS), which acquires imagery in four spectral bands: blue, green, red, and near-infrared. The second is the Thematic Mapper (TM), which collects seven bands: blue, green, red, near-infrared, two mid-infrared, and one thermal infrared. The MSS has a spatial resolution of 80 m, while that of the TM is 30 m. Both the sensors image has 185-km-wide swath with the repetivity of 16 days. Landsat 7, support for TM imagery with the addition of a co-registered 15 m panchromatic band. Landsat-8 Operational Land Imager (OLI) has nine spectral bands in visible, near infrared and shortwave infrared regions with a spatial resolution of 30 m for bands 1 to 7 and 9. The resolution of band 8 (panchromatic) is 15m.

SPOT The images acquired by SPOT Earth observation satellites are useful for mapping, monitoring, and managing land resources and human-induced activities.

So far five SPOT satellites, i.e., SPOT 1 to SPOT 5, have been launched in the year of 1986, 1990, 1993, 1998, and 2002, respectively. SPOT imagery comes in a full range of resolutions from 1 km global scale (SPOT vegetation imagery) down to 2.5 m local scale. SPOT satellites carry two high-resolution visible (HRV) push broom sensors 7, which operate in multispectral or panchromatic mode. The multispectral images have 20 m spatial resolution, while the panchromatic images have 10 m resolution. SPOT satellites 1–3 provide three multispectral bands: green, red, and infrared. SPOT 4, launched in 1998, provides the same three bands plus a shortwave infrared band. The panchromatic band for SPOT 1–3 is 0.51–0.73 m while that of SPOT 4 is 0.61–0.68 m. All SPOT images cover a swath 60 km wide. The SPOT sensor may be pointed to image along adjacent paths. This allows the instrument to acquire repeat imagery of any area 12 times during its 26-day orbital period.

IRS-1C and IRS-1D IRS-1C and IRS-1D are identical and were launched in December 1995 and September 1997, respectively. They carry three cameras: panchromatic camera (PAN), Linear Imaging Self-Scanning (LISS-III), and Wide Field Sensor (WiFS). The PAN camera provides data with a spatial resolution of 5.8 m and a ground swath of 70 km at nadir view. This camera can be steered up to +/− 26 degrees, which can be used to acquire stereopairs, and this also improves the revisit capability to 5 days. LISS-III camera provides multispectral data in four bands. The spatial resolution for visible (two bands) and near-infrared (one band) is 23.5 m with a ground swath of 141 km. The fourth band (shortwave infrared band) has a spatial resolution of 70.5 m with a ground swath of 148 km. The repetivity of LISS-III is 24 days. WiFS camera collects data in two spectral bands with a spatial resolution of 188 m and a ground swath of 810 km. By virtue of its wide swath, there is huge side lap between adjacent paths. A repetivity of 3 days can be achieved by suitably combining paths. The detail specifications of PAN, LISS-III, and WiFs sensors are given in Table 2.4.

Resourcesat-1 (IRS-P6) Resourcesat-1 carries three sensors that deliver an array of spectral bands and resolutions ranging from 5.8 m to 60 m. Data products derived from Resourcesat-1 can be used for advanced applications in vegetation dynamics, crop yield estimates, disaster management support, etc. In addition, Resourcesat-1

Table 2.4 Specification of PAN, LISS-III, and WiFS sensor of IRS-1C/1D

		LISS-III		
Sensor parameters	PAN	VNIR	SWIR	WiFs
Spatial resolution (m)	5.8	23.5	70.5	188
Swath (km)	70	142	148	810
Spectral band (microns)	0.5–0.75	0.52–0.59 0.62–0.68 0.77–0.86	1.55–1.7	0.62–0.68
Quantization (bit)	6	7	7	7
Revisit	5 days by tilting camera	24 days	24 days	3 days

Table 2.5 Sensor parameters of IRS-P6 (Resourcesat-1)

Sensor parameters	Bands	LISS-IV		LISS-III	AWiFS
		Mono mode	MX mode		
Spatial resolution	Band 2 (green)	5.8 m	5.8 m	23.5 m	60 m . . . 70 m
	Band 3 (red)		5.8 m	23.5 m	60 m . . . 70 m
	Band 4 (NIR)		5.8 m	23.5 m	60 m . . . 70 m
	Band 5 (SWIR)			23.5 m	60 m . . . 70 m
Swath width	All bands	70 km	23.9 km	140 km	700 km
Radiometric resolution, quantization	All bands	10 bit	10 bit	7 bit (VNIR)	10 bit
		7 bit transmission	7 bit transmission	10 bit (SWIR)	
Spectral coverage	Band 2 (green)	620–680 nm	520–590 nm	520–590 nm	520–590 nm
	Band 3 (red)		620–680 nm	620–680 nm	620–680 nm
	Band 4 (NIR)		770–860 nm	770–860 nm	770–860 nm
	Band 5 (SWIR)			1550–1700 nm	1550–1700 nm

has 120 gigabits of onboard memory that allows for out-of-contact imaging. IRS-P6 has several improved features over its predecessors. These include availability of 5.8 m spatial resolution in three bands from LISS-IV camera and improved LISS-III with MIR band information at 23.5 m resolution as other visible and NIR bands. In addition, the AWiFS provides data in the same spectral channels as LISS-III at about 56 m resolution with 10-bit radiometry, 5-day revisit, and scene coverage of 740 km for regional studies. Unique to IRS-P6 is the availability of simultaneous multispectral data at three spatial resolutions from the same platform with scene coverage varying from 576 sq. km to 1,9600 sq. km to 5,42,000 sq. km (Table 2.5).

Resourcesat-2 Resourcesat-2 satellite was launched by Indian Space Research Organization (ISRO) in the year 2011, and it is intended to continue the remote sensing data services to the users and to provide data with enhanced multispectral and spatial coverage. The payloads are similar to Resourcesat-1 with enhanced radiometric resolution. The swath of the LISS-IV multispectral sensor has been enhanced from 23 km to 70 km based on user needs.

Cartosat-1 (IRS-P5) Cartosat-1 was launched by ISRO in the year 2005, and its applications are mainly be toward cartography and high-resolution mapping. Cartosat-1 carries two panchromatic (PAN) cameras that take black and white stereoscopic pictures of the Earth in the visible region of the electromagnetic spectrum. The swath of high-resolution PAN cameras is 30 km with spatial resolution of 2.5 m. The cameras are mounted on the satellite in such a way that near simultaneous imaging of the same area from two different angles is possible. This facilitates the generation of accurate three-dimensional maps. The cameras are steerable across the direction of the satellite's movement to facilitate the imaging of an area more frequently. The images taken by Cartosat-1 cameras are compressed, encrypted, formatted, and transmitted to the ground stations. The images are reconstructed from the data received at the ground stations. Cartosat-1 also carries a solid-state recorder with a capacity of 120 gigabits to store the images taken by its cameras. The stored images can be transmitted when the satellite comes within the visibility zone of a ground station.

Cartosat-2 It was launched by ISRO in the year 2007, and it carries a state-of-the-art panchromatic (PAN) camera that take black and white pictures of the Earth in the visible region of the electromagnetic spectrum. The swath covered by these high-resolution PAN cameras is 9.6 km, and their spatial resolution is less than 1 m. The satellite can be steered up to 45 degrees along as well as across the track. Cartosat-2 is an advanced remote sensing satellite capable of providing scene-specific spot imagery. The data from the satellite are useful for detailed mapping and other cartographic applications at cadastral level, urban and rural infrastructure development, and management.

QuickBird QuickBird provides highly accurate and even higher-resolution imagery with panchromatic imagery at 60–70 cm resolution and multispectral imagery at 2.4 and 2.8 m resolutions. QuickBird's global collections of images greatly facilitate applications like infrastructure development and utilities management.

2.9.1.3 Superspectral Imaging Sensors

MODIS is a key instrument on aboard of the Terra (EOS AM) and Aqua (EOS PM) satellites. Both Terra MODIS and Aqua MODIS together are able to view the entire Earth's surface every 1–2 days. MODIS acquires the images in 36 spectral bands with spatial resolutions ranging from 250 to 1 km. Bands 1 and 2 provide 250 m resolution images in the red and near-infrared regions, bands 3–7 provide 500 m resolution multispectral images in the visible and infrared regions and bands 8–36 provide hyperspectral coverage in the visible, reflected infrared, and thermal infrared regions, with a 1 km resolution.

2.9.1.4 Hyperspectral Remote Sensing Sensors

In addition to traditional multispectral imagery, some new and experimental systems such as AVIRIS and MODIS are capable of capturing hyperspectral data. These systems cover a similar wavelength range to multispectral systems but in much narrower bands. This dramatically increases the number of bands available for image classification. Moreover, hyperspectral signature libraries can be created in lab conditions and contain hundreds of signatures for different types of land covers, including many minerals and other Earth materials. Thus, it should be possible to match signatures to surface materials with great precision. However, environmental conditions and natural variations in materials make this difficult. In addition, classification procedures have not been developed for hyperspectral data to the degree they have been for multispectral imagery. As a consequence, multispectral imagery still represents the major tool of remote sensing today.

The signature is generated by the information contained in the numerous spectral bands acquired by the sensor. In general these technologies collect imagery with spatial resolutions ranging from 1 to 10 m. Hyperspectral imagery is typically collected and represented as a data cube with spatial information collected in the X-Y plane and spectral information represented in the Z direction. Hyperspectral cubes are generated from airborne sensors like the NASA's Airborne Visible Infrared Imaging Spectrometer (AVIRIS) or from satellites like NASA's EO-1 Hyperion. The Hyperion sensor has been functional since November 2000, but no commercial spaceborne hyperspectral sensors are in orbit till date. However, for many development and validation studies, handheld sensors are being used.

Hyperspectral remote sensing uses the practice of spectroscopy to examine images of the Earth's surface. Although hyperspectral remote sensing sometimes applies the techniques of classical spectroscopy to the study of atmospheric gases and pollutants, for example, more commonly it applies these techniques to the making of precise, accurate, detailed spectral measurements of the Earth's surface. Such data have accuracy and detail sufficient to begin to match observed spectra to those stored in databases known as spectral libraries. Instruments for hyperspectral remote sensing differ from those of conventional spectroscopy in that they gather spectra not only for point targets but also for areas—not for stars or laboratory samples but for regions of the Earth's surface. Instruments for hyperspectral remote sensing differ from other remote sensing instruments in terms of their extraordinarily fine spectral, spatial, and radiometric resolutions and their careful calibration. Some hyperspectral instruments collect data in 200 or more channels at 10–12 bits. Because of their calibration and ability to collect data having fine detail, such instruments greatly extend the reach of remote sensing not only by extending the range of applications but also by defining new concepts and analytical techniques. Although the techniques of classical spectroscopy can be used in hyperspectral remote sensing to examine, for example, atmospheric gases, hyperspectral remote sensing typically examines very detailed spectra for images of the Earth's surface, applies corrections for atmospheric effects, and matches them to spectra of known features.

2.9.2 Thermal Remote Sensing Sensors

The sensors, which operate in thermal range of electromagnetic spectrum, record the energy emitted from the Earth features in the wavelength range of 3000–5000 nm and 8000–14,000 nm. In NOAA's Polar Orbiting Environmental Satellite series, the AVHRR sensor is a broadband, 4- (AVHRR/1), 5- (AVHRR/2), or 6- (AVHRR/3) channel scanning radiometer in the visible, near-infrared, and thermal infrared portions of the electromagnetic spectrum. AVHRR image data have two spatial resolutions: ~1.1 km for local area coverage (LAC) and 5 km for global area coverage (GAC). It acquires data along a 2400-km-wide swath each day. AVHRR collects five bands: red, near-infrared, and three thermal infrared. For studying very large areas, a resampled version with resolution of about 4 km is also available and is termed global area coverage (GAC). AVHRR may be "high" spatial resolution for meteorological applications, but the images portray only broad patterns and little detail for terrestrial studies. However, they do have a high temporal resolution, showing wide areas on a daily basis and are therefore a popular choice for monitoring large areas. Thermal remote sensing is very useful for fire detection and thermal pollution. e.g., the last five bands of ASTER and band 6 of Landsat ETM+ operates in thermal range. Thermal infrared sensor (TIRS) of Landsat-8 OLI collect the data in thermal bands 10 and 11 at 100m spatial resolution. These datasets are useful in providing more accurate surface temparatures.

2.9.3 Microwave Remote Sensing Sensors

A microwave remote sensor records the backscattered microwaves in the wavelength range of 1 mm to 1 m of electromagnetic spectrum. Most of the microwave sensors are active sensors, having their own sources of energy, e.g., RADARSAT. These sensors have edge over other type of sensors, as these are independent of weather and solar radiations. The microwaves are the electromagnetic waves with frequencies between 109 and 1012 Hz. Radar is an active microwave remote sensing system. The system illuminates the terrain with electromagnetic energy, detects the scattered energy returning from the terrain (called radar return), and then records it as an image. Intensity of radar return, for both aircraft- and satellite-based systems, depends upon radar system properties and terrain properties. The radar equation expresses the relationship between the radar parameters, the target characteristics, and the received signal.

In the active microwave remote sensing, information about the object's physical structure and electrical property is retrieved by analyzing the backscattering field. The microwave signatures of the object are governed by the sensor parameters (frequency, polarization, incidence angle) and the physical (surface roughness, feature orientation) and electrical (dielectric constant) property of the target. The influence of frequency on radar scattering is governed by terrain properties. A given surface will appear very rough at higher frequency, compared to lower frequency. Thus, generally, backscattering coefficient increases with increasing frequency. In addition, the signal penetration depth increases with wavelength in microwave region. The use of multifrequency data allows distinction between the roughness types. The backscattering also depends on the polarization of the incident wave. A vegetation canopy consisting of short vertical linear scatter over a rough surface can be considered as short vertical dipoles. In such case, vertically polarized incident wave interacts strongly with canopy. The multiple scattering and volume scattering occur from a complex surface.

RISAT (Radar Imaging Satellite) is a microwave remote sensing mission with synthetic aperture radar (SAR) operating in C band and having a 6×2 m planar active array antenna based on trans-receiver module architecture. SAR is an all-weather imaging sensor capable of taking images in cloudy and snow-covered regions and also both during day and night.

2.10 Applications of Satellite Remote Sensing Data

With the availability of wide range of spatio-temporal, radiometric and spectral resolutions, applications of satellite remote sensing data in land resources mapping, monitoring and management are being increased in the recent past. The technological advancement in data processing capabilities of computers and sophisticated image processing techniques further augmented the applications of satellite remote sensing data. The remote sensing datasets are being widely used in land resources based applications like soil survey and mapping, land use/land cover mapping, wasteland mapping, hydrology, agriculture, crop damage and stress assessment, watershed management, land use planning, urban development, ecology and environmental monitoring. The availability of temporal satellite remote sensing data like MODIS, Landsat, Sentinel-2 in public domain has transformed the remote sensing based applications like terrain analysis, soil resource inventory, land use/land cover mapping, land degradation assessment and environmental studies. The high resolution satellite sensors data like Sentinel-2, IRS-P6, Cartosat-1A, 1B and Google Earth have immense potential in land resource mapping particularly in delineation of distinct landforms, optimizing soil profile location and reducing the time of surveyors in the field. Conjunctive use of remote sensing data and collateral information like lithology, physiography has enabled mapping soils at different scales, ranging from 1:250,000 to 1:10,000 scale. High resolution stereo data like Cartosat-1A and IB are found to be useful in generating information on land resources at 1:10,000 scales for micro-level land use planning. In applications, like crop acreage estimation, crop condition assessment, crop yield forecasting, cropping system analysis and precision farming, beside satellite sensors data, various other ground based agro-physical/environmental datasets need to be integrated with remote sensing data in GIS. High resolution satellite sensors data has a great role to play in precision farming, in the way of mapping the spatial variability. Microwave remote sensing holds potential applications in the fields of agriculture, soils, forestry, earth sciences, snow, hydrology and disaster management. Airborne and field based hyperspectral sensors data provide ample opportunities in the fields of soil properties studies, mineral exploration, change detection, validation and development of spectral library.

2.11 Conclusions

Remote sensing data facilitates the study of various Earths' surface features in their spatial relation in synoptic view to delineate the required features and phenomena. With the increased amount of spectral data, scientists have the ability to derive specific information about the Earth's surface from an image pixel using diagnostic or characteristic absorption features in its spectral signature. The remote sensing satellites provide repetitive coverage of the Earth, and this temporal information is very useful for studying landscape dynamics, phenological variations of vegetation, and change detection analysis. Remote sensing process made it possible to gather information about the area when it is not possible to do ground survey like in mountainous and inaccessible areas. Since information about a large area can be gathered quickly, the remote sensing-based techniques save time of fieldwork and resources. Remote sensing technologies such as satellites and aerial sensors have immense potential in gathering of information about planet Earth's physical, chemical, and biological systems supplemented with ground-based observations. These technologies enable direct observation of the land surface at repetitive intervals, and therefore it allows mapping of the extent, monitoring of the changes, and management of the resources. Since it is a cost-effective technology, user can minimize the field observations, and also a large number of users can share and use the same data.

References

Campbell JB (1996) Introduction to remote sensing. Taylor & Francis, London

Campbell JB, Wynne RH (2011) Introdcution to remote sensing, 5th edn. Taylor & Francis, London, pp 31–56

Colwell RN (ed) (1983) Manual of remote sensing, 2nd edn. Vol I: theory, instruments and techniques. American Society of Photogrammetry and Remote Sensing ASPRS, Falls Church

Curran PJ (1985) Principles of remote sensing. Longman Group Limited, London

Elachi C (1987) Introduction to the physics and techniques of remote sensing. Wiley Series in Remote Sensing, New York. http://www.ccrs.nrcan.gc.ca/ccrs/learn/tutorials/fundam/chapter1/chapter1_1_e.html

Jenson JR (2007) Remote sensing of the environment: an earth resource perspective. Person Prentice Hall

Joseph G (1996) Imaging sensors. Remote Sens Rev 13:257–342

Lillesand TM, Kiefer R (1993) Remote sensing and image interpretation, 3rd edn. Wiley, New York

Lillesand TM, Kiefer RW (1994) Remote sensing and image interpretation. Wiley

Reddy GPO (2012) Principles of digital image processing, In: Reddy GPO, Sarkar D (eds) Remote sensing and GIS in digital terrain analysis and soil-landscape modeling, NBSS&LUP Publ. No. 152, pp 86–97

Sabins Jr, Floyd F (1978) Remote sensing: principles and interpretation. W. H, Freeman and Company

Sabins FF (1997) Remote sensing and principles and image interpretation. WH Freeman, New York

Chapter 3
Geographic Information System: Principles and Applications

G. P. Obi Reddy

Abstract Geographic information system (GIS) is a computer-assisted system for capture, storage, retrieval, analysis, and display of spatial data and nonspatial attribute data. The data can be derived from different sources such as survey data, geographical/topographical/aerial maps, satellite data, or archived data. Data can be in the form of locational data (such as latitudes/longitudes) or tabular (attribute) data. The introduction of modern technologies has led to an increased use of computers and information technology in all aspects of spatial data handling. The handling of spatial data involves data acquisition, storage and maintenance, processing, analysis, and output. GIS is a computerized database management system for capturing, storing, validating, maintaining, analyzing, displaying, and managing spatially referenced data with a primary function to integrate data from a variety of sources. GIS constitutes hardware, software, data, people, and methods. Advanced GIS can address questions related to location, condition, trends, patterns, and modeling. Manual digitization of existing paper maps, existing digital data, aerial photographs, remote-sensing satellite imagery, and global positioning system (GPS) are promising sources for data input in GIS. Before generation of any spatial data, one has to understand the available data types, data analysis procedures, and their capabilities in GIS to get realistic outputs. GIS is being widely used as spatial analysis tool for effective and efficient means of data generation, management, analysis, and display. Recent advances in computer technology, remote sensing, GPS, and communication technology augmented the development of enhanced functions of GIS in the field of data generation, management, analysis, and display. Data derived from remote sensing, GPS, and field surveys could be effectively used in GIS in mapping, monitoring, and management of land resources and dissemination of information at appropriate destinations in various location-based applications.

Keywords Geographic information system · Nonspatial data · Raster · Spatial data · Vector

G. P. O. Reddy (✉)
ICAR-National Bureau of Soil Survey & Land Use Planning, Nagpur, India

G. P. O. Reddy, S. K. Singh (eds.), *Geospatial Technologies in Land Resources Mapping, Monitoring and Management*, Geotechnologies and the Environment 21,
https://doi.org/10.1007/978-3-319-78711-4_3

3.1 Introduction

Geographic information system (GIS) is a computer system capable of capturing, storing, analyzing, and displaying geographically referenced information (Burrough 1986). GIS is a tool to efficiently capture, store, update, manipulate, analyze, and display all forms of geographically referenced data. Many other definitions of a GIS have been evolved in different areas and disciplines (Clarke 1995, 2001; Burrough 1986, 1998; Duecker 1979; Goodchild 1992; Star and Estes 1990). According to Duecker (1979), GIS is a special case of information systems where the database consists of observations on spatially distributed features, activities, or events, which are definable in space as points, lines, or areas. GIS manipulates data about the points, lines, and areas to retrieve data for ad hoc queries and analyses. A true GIS is designed to accept, organize, statistically analyze, and display diverse types of spatial information that are geographically referenced to a common coordinate system of a particular projection and scale. GIS is a potential tool for handling voluminous remotely sensed data and has capability to support spatial statistical analysis (Singh et al. 2003); storage, management, and modeling of input data and presentation of model results in the form of thematic maps.

GIS is a database management system, which effectively store, retrieve, manipulate, analyze, and display spatial information of both cartographic and thematic origin (Reddy 2012). GIS can handle large volumes of spatial data derived from a variety of sources such as field surveys, aerial surveys, and space remote sensing, in addition to the already existing maps and reports. Remote-sensing satellite imagery and GPS are promising sources for data input in GIS. Before generation of any spatial data, one has to understand the available data types, data analysis procedures, and their capabilities in GIS to get realistic outputs. This involves bringing together diverse information from a variety of sources on a common platform. Major components of GIS are data input, data encoding, data management, data analysis, data manipulation, and data presentation or output. The geocoded spatial data defines an object that has an orientation and relationship with other objects in two- or three-dimensional space, also known as topological data, and stores in topological database. On the other hand, attribute data stored in a relational database describes the objects in detail. GIS links these two databases by manipulating a one-to-one relationship between records of object of location in the topological database and records of the object attribute in relational databases by using end-user defined common identification index or code. Advances in computer technology, remote sensing, GPS, and communication technology are the driving forces for the development of enhanced functions of GIS in data generation, management, analysis, and display.

3.2 History of GIS

Research work on spatial sciences and GIS began in the late 1950s. The data analysis of geographic locations was being done by computers in government organizations and universities in the USA during the 1950s and 1960s. Canada was the pioneer in the development of GIS as a result of innovations dating back to the early 1960s. The first true operational GIS was developed by Dr. Roger Tomlinson, Department of Forestry and Rural Development, Canada, who was also known as the "father of GIS." It was called as Canada Geographic Information System (CGIS) and was used to store, analyze, and manipulate land-related data. In 1964, a laboratory of Computer Graphics and Spatial Analysis was established at the Harvard Graduate School of Design by Howard T. Fisher. This organization developed a number of important theoretical concepts of spatial data handling. By the early 1980s, M&S Computing (later Intergraph) and Environmental Systems Research Institute (ESRI) emerged as commercial vendors of GIS software. Of late, a number of organizations and universities have been doing research in the field of GIS and developing user-friendly software. Now there are a number of free, open-source GIS packages available, which run in a wide range of operating systems and perform specific tasks. Evolution of GIS has transformed and revolutionized and open the new vistas in many disciplines in which scientists, surveyors, planners, engineers, and managers are extensively using this wonderful geographical tool for their database generation, management, analysis, and display.

3.3 Definition of GIS

There are many definitions of GIS, which essentially explain the capabilities of GIS. Different people defined GIS according to capability and purpose for which it is applied. Some of the well-accepted working definitions of GIS are given below.

- GIS as a "Set of tools for collecting, storing, retrieving at will, transforming and displaying spatial data from the real world for a particular set of purposes" (Burrough 1987)
- GIS as a "computer - assisted system for the capture, storage, retrieval, analysis and display of spatial data, within a particular organization" (Stillwell and Clarke 1987)
- GIS as a computer-based system that provides four sets of capabilities to handle georeferenced data, viz., data input, data management (data storage and retrieval), manipulation analysis, and data output (Aronoff 1989)
- GIS as a "computer based system for the efficient input, storage, manipulation, analysis, representation and retrieval of all forms of spatially indexed and related descriptive data" (Jackson 1992)

All definitions of GIS recognize that spatial data are unique because they are linked to maps. GIS consists of a database, map information, and a computer-based link between them. It stores data about the world as a collection of thematic layers, a pictorial representation of which is given in to be linked together in spatial domain using geographical reference system. Users can visualize, explore, and analyze data by location, revealing hidden patterns, relationships, and trends that are not readily apparent in spreadsheets or statistical packages. As an extremely powerful and versatile tool, GIS had proven as invaluable in solving many real-world problems, which include mapping, monitoring, and management of land resources.

3.4 Philosophy and Components of GIS

The proliferation of GIS is explained by its unique ability to assimilate data from widely divergent sources, to analyze trends over time, and to spatially evaluate impacts caused by development. There are five types of questions that a sophisticated GIS can answer, which are related to location, condition, trends, patterns, and modeling. GIS constitutes of five components, which include hardware, software, data, people, and method. *Hardware* consists of the computer system on which the GIS software will run. The computer forms the backbone of the GIS hardware. Scanner and digitizer are the main input, and printers and plotters are the most common output devices for a GIS hardware setup. *GIS software* provides the functions and tools needed to store, analyze, and display geographic information. Many application-specific GIS software are available depending upon the user needs. *Data* in geographic perspective and related tabular data can be collected in-house or purchased from a commercial data provider. The digital map forms the basic data input for GIS. Tabular data related to the map objects can also be attached to the digital data. A GIS will integrate spatial data with other data resources and can even use a DBMS, used by most organizations to maintain their data and to manage spatial data. *People* who have expertise in handling GIS can design, maintain, and analyze data in GIS to perform various tasks. The people who use GIS can be broadly classified into two classes: the CAD/GIS operator, whose work is to vectorize the map objects, and the use of this vectorized data to perform query, analysis, or any other work which is the responsibility of a GIS engineer/user. *Methods* are essential for successful GIS operation according to a well-designed plan and business rules, which are the models and operating practices unique to each organization. There are various techniques used for map creation and further usage for any project. The map creation can either be automated raster to vector creator or can be manually vectorized using the scanned images.

3.5 Geographic Coordinate Systems

Coordinate system must be defined to specify locations on the Earth surface. The familiar XY coordinate pair of a typical graph is an example of a coordinate system. Universal Transverse Mercator (UTM), State Plane Coordinate System (SPCS), and Longitude and Latitude are commonly used coordinate systems in GIS. A geographic coordinate system (GCS) uses a three-dimensional spherical surface to define locations on the Earth. A GCS is often incorrectly called a datum, but a datum is only one part of a GCS. A GCS includes an angular unit of measure, a prime meridian, and a datum. A point is referenced by its longitude and latitude values. Longitude and latitude are angles measured from the Earth's center to a point on the Earth's surface. The angles often are measured in degrees.

Maps are flat, but the surfaces they represent are curved. Transforming three-dimensional space onto a two-dimensional map is called "projection." Projection is the translation of spherical coordinates onto a planar surface, while a datum is the ellipsoid, or "figure of the Earth" that approximates the actual shape of the Earth, and is used in the transformation equation. Each of the wide variety of projections (and datums) has specific characteristics that make them useful for specific mapping purposes. A datum is the geometric, 3-D "figure of the Earth" which is used as the basis for projecting onto a planar surface. The World Geodetic System of 1984 (WGS84) is the common datum used by the GPS configuration and is essentially identical to the NAD83.

For data at the regional and state level, the most common projections are UTM and State Plane (which comprises several projections, varying by state). Therefore, for different types of maps, different projections have been evolved in accordance with the scale and purpose of the map. This process inevitably distorts at least one of these properties: shape, area, distance, direction, and often more. There is no ideal map projection, but representation for a given purpose can be achieved. The selection of projection is made on the basis of the location and the extension of the feature of the globe, the shape of the boundary to be projected, the deformations or distortions of a map to be minimized, and the mathematical model to be applied to preserve some identity of graphical features. Based on these characteristics, the utility of the projection is ascertained.

Georeferencing assigns coordinates from a known reference system, such as latitude/longitude, UTM, or State Plane, to the page coordinates of a raster (image) or a planar map. Georeferencing raster data allows it to be viewed, queried, and analyzed with other geographic data. Converting the coordinates of a map/image from one system to another typically involves shifting, rotating, scaling, skewing, or projecting them. Also known as rectification, the conversion process requires resampling of values.

3.6 Geographic Data Representation in GIS

Any geographic data can be represented by points, lines, areas, and continuous surfaces in GIS. *Points* are the simplest type of spatial data. They are zero-dimensional objects with only a position in space but no length. *Lines* (also termed segments or arcs) are one-dimensional spatial objects. Besides having a position in space, they also have a length. *Areas* (also termed polygons) are two-dimensional spatial objects with not only a position in space and a length but also a width (in other words they have an area). *Continuous surfaces* are three-dimensional spatial objects with not only a position in space, a length, and a width but also a depth or height (in other words they have a volume).

3.7 Data Types

In GIS, any geographic data can be depicted in three basic forms, i.e., map data, attribute data, and image data. *Map data* contains the location and shape of geographic features. Maps use three basic shapes to present real-world features: points, lines, and areas (called polygons). *Attribute data* (tabular) is the descriptive data that GIS links to map features. Attribute data is collected and compiled for specific areas like states, census tracts, cities, and so on and often comes packaged with map data. Attributes are descriptive data providing information associated with the geometrical data. Attributes are usually managed in external or internal GIS database management systems (DBMS). The databases use the corresponding coordinates or identification numbers to link the attributes to the geometrical data. *Image data* ranges from satellite images and aerial photographs to scanned maps (maps that have been converted from printed to digital format).

3.8 Data Views in GIS

GIS is most often associated with maps. A map, however, is only one way user can work with geographic data in a GIS and only one type of product generated by a GIS. GIS can provide a great deal on more problem-solving capabilities than using a simple mapping tool. GIS can be viewed in three ways, i.e., database, map, and model view. *Database view* in GIS is a unique kind of database of the world—a geographic database (geodatabase). Fundamentally, GIS is based on a structured database that describes the world in geographic terms. *Map view* in GIS is a set of intelligent maps and other views that show features and feature relationships on the Earth's surface. *Model view* in GIS is a set of information transformation tools that derive new geographic datasets from existing datasets. These geo-processing functions take information from existing datasets, apply analytic functions, and write results into new derived datasets.

3.9 Principal Functions of GIS

Data capture, database update and management, geographic analysis, and presenting results are four principal functions of GIS. Data capture is basically assimilation of data from different data sources like manual digitization, scanning of analog maps, paper maps, and existing digital datasets. Database update and management follows after data are collected, where GIS provides facilities, which can store and maintain data. Data integration and conversion are only a part of the input phase of GIS. What is required next is the ability to analyze geographic data, interpret, and analyze the collected information quantitatively and qualitatively. One of the exciting aspects of GIS technology is that once data is processed, it can be visualized and presented in different ways.

3.10 Spatial Data Models in GIS

Traditionally spatial data has been stored and presented in the form of a map. The spatial data models are approaches for storing the spatial location of geographic features in a database. Data models help to convert real-world geographical variation into discrete objects in GIS. It represents the linkage between the real-world domain of geographic data and computer representation of these features. The three basic data models that GIS uses are vector, raster, and triangulated irregular network (TIN).

3.10.1 Vector Data Model

Vector data models represent geographic phenomena with points, lines, and polygons. Points are pairs of x,y coordinates, lines are sets of coordinate pairs that define a shape, and polygons are sets of coordinate pairs defining boundaries that enclose areas. Coordinates are usually pairs (x,y) or triplets (x,y,z, where z represents a value such as elevation). The coordinate values depend on the geographic coordinate system in which the data is stored. Discrete features, such as customer locations and data summarized by area, are usually represented using the vector model. Some of the important vector data formats are shapefile, geodatabase, DLG (Digital Line Graph), Digital Exchange Format (DXF), SVG (Scalable Vector Graphics), ArcInfo

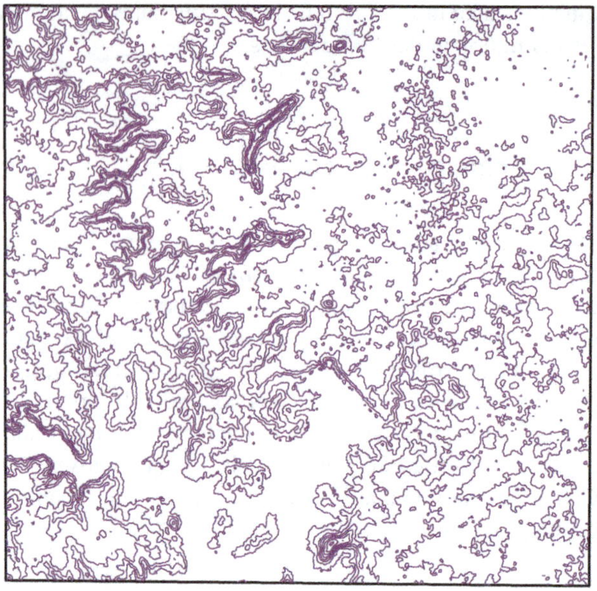

Fig. 3.1 Vector model of contours at 10 m interval in the part of Central India

Coverage, ArcInfo Interchange (e00), SDTS (Spatial Data Transfer Standard), TIGER (Topologically Integrated Geographic Encoding and Referencing), etc. Vector model of contours at 10 m interval in the part of Central India is shown in Fig. 3.1.

3.10.2 Raster Data Model

A raster model otherwise known as a raster dataset (image) is, in its simplest form, a matrix (grid) of cells. The grid cells are organized and accessed by *rows* and *columns*. Each cell has a width and height and is a portion of the entire area represented by the raster. The dimension of the cells can be as large or as small as necessary to represent the area and the features within the area, such as a square kilometer, square meter, or even square centimeter. The cell size determines how coarse or fine the patterns or features in your extent will appear. The smaller the cell size, the more detail the area will have. However, the greater the number of cells, the longer it will take to process, and it will require more storage space. Raster files are most often used for digital representations of aerial photographs, satellite images, scanned paper maps, and other applications with very detailed images. Some of the important raster data formats are Standard Raster Format, Tagged Image File Format (TIFF), GeoTIFF, Graphics Interchange Format (GIF), Joint Photographic Experts

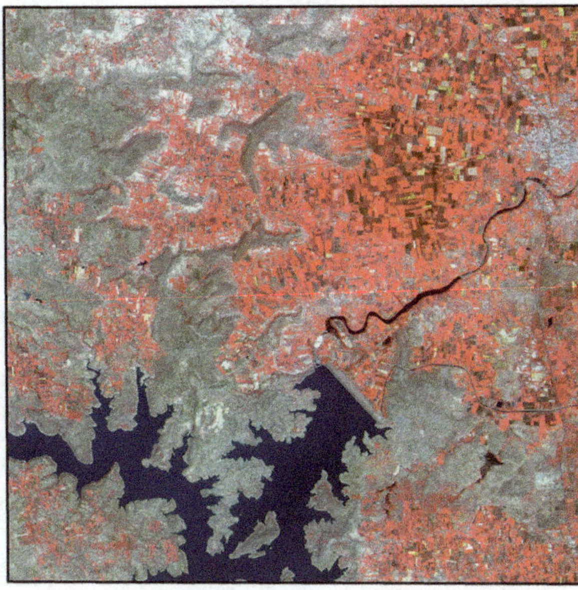

Fig. 3.2 Raster model of IRS-P6-LISS-IV data of part of Central India

Table 3.1 Advantages and disadvantages of vectors and raster data models

Vector model	Raster model
Advantages	*Advantages*
Compact data structure	Simple data structure
Efficient for network analysis	Good for statistical and spatial analysis
Efficient projection transformation	Easy and efficient overlaying
Accurate map output	Compatible with RS imagery
	Spatial variability is efficiently represented
	Simple for computer programming
	Same grid cells for several attributes
Disadvantages	*Disadvantages*
Complex data structure	Inefficient use of computer storage
Difficult overlay operations	Errors in perimeter and shape
High spatial variability	Difficult network analysis
Inefficiently represented	Inefficient projection transformations
Not compatible with remote sensing imagery	Loss of information when using large cells
	Datasets are very large

Group (JPEG), PostScript, band interleaved by line (BIL), band interleaved by pixel (BIP), ArcInfo GRID, DEM (digital elevation model), DRG (digital raster graphic), etc. Raster model of IRS-P6-LISS-IV data of part of Central India is shown in GIS (Fig. 3.2).

Fig. 3.3 TIN model generated from SRTM 30 m DEM of part of Central India

Both the vector and raster models for storing geographic data have unique advantages and disadvantages (Table 3.1). Modern GIS packages are able to handle both the models.

3.10.3 Triangulated Irregular Network (TIN)

In TIN model, the world is represented as a network of linked triangles drawn between irregularly spaced points with x, y, and z values. TINs are an efficient way to store and analyze surfaces. Heterogeneous surfaces that vary sharply in some areas and less in others can be modeled more accurately, in a given volume of data, with a triangulated surface than with a raster. That is because many points can be placed where the surface is highly variable and fewer points can be placed where the surface is less variable. In using only the points necessary, TINs also provide a more efficient method to store data. TIN model generated from SRTM 30 m DEM of part of Central India is shown in GIS (Fig. 3.3).

3.11 Data Conversion in GIS

Data conversion is the process of converting data from one format into another, whether it is from one data model to another or from one data format to another. The important data conversion formats in GIS are raster to vector (vectorization) and

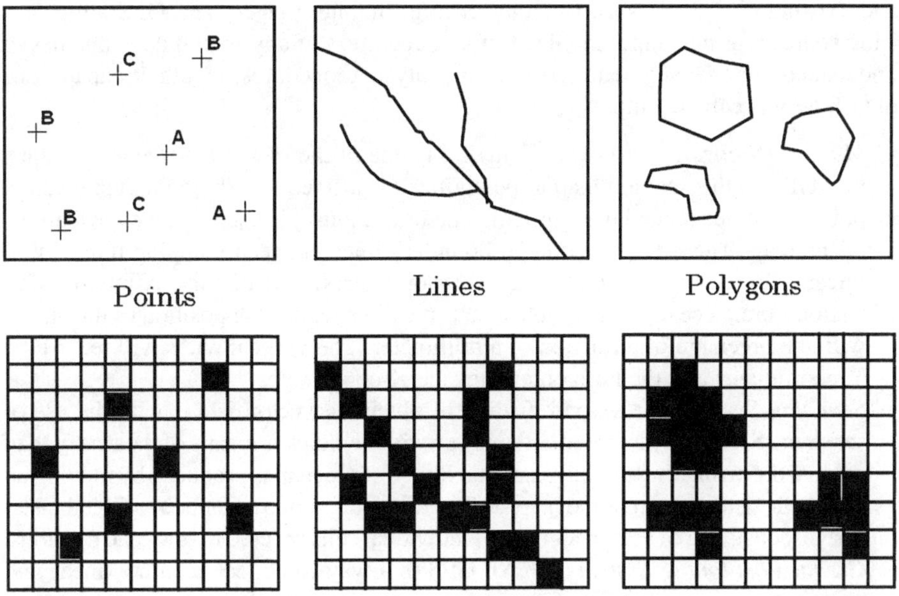

Fig. 3.4 Data conversion from vector to raster data in GIS

vector to raster (rasterization). Vectorization refers to conversion from raster to vector data, which is often called raster vector conversion. Vectorization is not very easy as compared with rasterization, because vector format needs topological structure. This algorithm is useful to convert raster image to vector data with the coordinates, but it is not sufficient because the algorithm will not build topological structure. Conversion between raster and vector data is very useful in practical applications of GIS. Rasterization refers to conversion from vector to raster data (Fig. 3.4). Raster format is more convenient to produce color-coded polygon maps such as color-coded land use map. Rasterization is also useful to integrate GIS with remote sensing because remote sensing images are in raster format.

3.12 Database Generation in GIS

GIS is being widely used as spatial analysis tool for effective and efficient means of data generation and management, analysis, and display. The true value of GIS can only be realized if the proper tools to collect spatial data and integrate them with attribute data are available. Data used in GIS often come from many types and are stored in different ways. A GIS provides tools and a method for the integration of different data into a format to be compared and analyzed. Data sources are mainly obtained from manual digitization and scanning of aerial photographs, paper maps, and existing digital datasets. Remote-sensing satellite imagery and GPS are promising sources for data input in GIS. Before generation of any spatial data, one has to understand the available data types, data analysis procedures, and their capabilities in GIS to get realistic outputs.

- *Manual Digitization*: Manual digitizing is one of the methods for entering maps into GIS. In this method map to be digitized is affixed to a digitizing table, and a pointing device (called the digitizing cursor or mouse) is used to trace the features of the map. These features can be boundary lines between mapping units, other linear features (rivers, roads, etc.), or point features (sampling points, rainfall stations, etc.) The digitizing table electronically encodes the position of the cursor with the precision of a fraction of a millimeter. The vertical wires will record the Y-coordinates and the horizontal ones the X-coordinates.
- *Scanning System*: The second method of obtaining vector data is with the use of scanners. Scanning (or scan digitizing) provides a quicker means of data entry than manual digitizing. In scanning, a digital image of the map is produced by moving an electronic detector across the map surface. The output of a scanner is a digital raster image, consisting of a large number of individual cells ordered in rows and columns.
- *Global Positioning Systems (GPS)*: GPS is a way to collect accurate linear and point location data for GIS. Depending on the unit, the locational accuracy can reach to the millimeter. Combined with attribute data entered at the time of collection, GPS is a rapid and accurate method of data collection.
- *Digital Image Processing*: Geodatasets can be derived from digital imagery processing for GIS. Most commonly satellite imagery is used in the process either through supervised or unsupervised classification algorithms to generate digital databases, and that can be used in GIS for further analysis in conjunction with other relevant datasets.

3.12.1 Data Input

The potential of GIS can only be realized if the proper tools used to collect, develop, and integrate spatial data with attribute data. GIS provides tools and methods for integration of different datasets into a format to be compared and analyzed. Digitization of analog maps, remote-sensing satellite imageries, and GPS are the promising data input sources for GIS. Manual digitization using digitizing table was the most common method for entering maps into GIS till the recent past. Now this method is almost absolute, and it was replaced with computer-based onscreen digitization. In this procedure, analog maps have to be first scanned using the scanner, or existing digital image can be taken as input to digitize the features like lines, points, and polygons in digital form. If it scanned digital image, it has to be first registered/georeferenced in GIS before digitizing the features. Figure 3.5 shows the steps involved in typical database generation in GIS.

Before data generation in GIS, maps and images should be geometrically rectified. The process begins by registering the raw maps and images to known

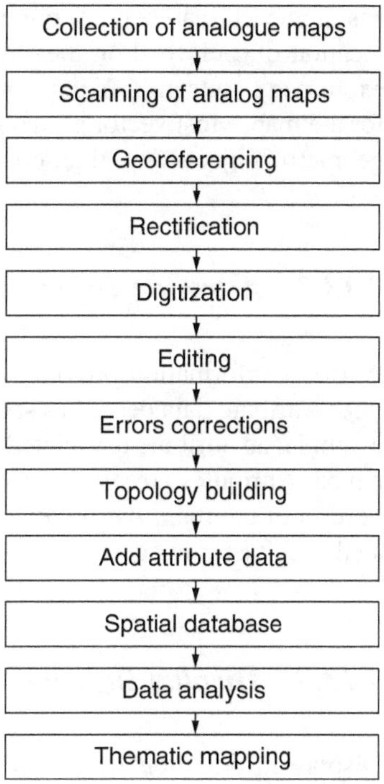

Fig. 3.5 Steps in creating GIS-based spatial database

coordinates. Appropriate projection system must be used to project the maps because each projection is particularly appropriate for certain uses. In GIS, uncertainties and errors are intrinsic to spatial data and need to be addressed properly before performing any spatial data integration, analysis, and thematic mapping. Data accuracy is often grouped according to thematic accuracy, positional accuracy, and temporal accuracy occurring at various stages in spatial data generation and handling.

3.13 Types of Information in a Digital Map

Any digital map is capable of storing much more information than a paper map of the same area, but it's generally not clear at first glance just what sort of information the map includes. Three general types of information can be included in digital maps.

3.13.1 Geographic Information

Geographic information provides the position and shapes of specific geographic features. The geographic information in a digital map provides the position and shape of each map feature. For example, a road map's geographic information is the location of each road on the map. In a vector map, a feature's position is normally expressed as sets of X, Y pairs or X, Y, Z triples, using the coordinate system defined for the map. Most vector geographic information systems support three fundamental geometric objects, i.e., line, point, and polygon.

3.13.2 Attribute Information

Attribute information provides additional non-graphic information about each feature. Attribute data describes specific map features but is not inherently graphic. For example, an attribute associated with a road might be its name or the date it was last paved. Attributes are often stored in database files kept separately from the graphic portion of the map. Attributes pertain only to vector maps; they are seldom associated with raster images.

3.13.3 Display Information

Display information describes how the features will appear on the screen. The display information in a digital map dataset describes how the map is to be displayed or plotted. Common display information includes feature colors, line widths, and line types (solid, dashed, dotted, single, or double); how the names of roads and other features are shown on the map; and whether or not lakes, parks, or other area features are color coded.

3.14 Metadata

Metadata can be defined as "data about data." Metadata is additional information (besides the spatial and tabular data) that is required to make the data useful. Metadata represents a set of characteristics about the data that are normally not contained within the data itself. Metadata could include an inventory of existing data, definitions of the names and data items, a keyword list of names and definitions, an index of the inventory and the keyword list for access, a record of the steps performed on the data, a documentation of the data structures and data models used, and a recording of the steps used on the data for analysis. Spatial metadata is important because it not only describes what the data is, but also reduce the size of spatial datasets. Metadata not only helps find data, but once data has been found, it also tells how to interpret and use data. Publishing metadata facilitates data sharing. Sharing data among organizations stimulates cooperation and integrated approach in spatial database management, dissemination, utility, and related policy issues.

3.15 Presenting Results/Outputs

One of the most exciting aspects of GIS technology is the variety of different ways in which the information can be presented once it has been processed by GIS. Traditional methods of tabulating and graphing data can be supplemented by maps and three-dimensional images. Visual communication is one of the most fascinating aspects of GIS technology and is available in a diverse range of output options. The final maps of GIS are of high cartographic quality and are brought out using a wide range of devices. Plotters are output devices for making copies of geographic data on paper or film. A printer is an output device that prints an electronically stored document on print media such as paper or transparencies.

3.16 Cartographic Principles and Map Composition in GIS

Traditionally, maps have been created to serve two main functions. The first function has been to store information. Creating a map has been a way to record information for future reference. The second function has been to provide a picture to relay spatial information to a user. There are many types of maps, each with general and possibly specific requirements. However, in GIS, user is having the flexibility in map composition using various map composition tools. These guidelines have been organized into seven areas that one can use as a checklist for map composition in GIS.

- *Purpose*: Typically, a map does not have more than one purpose. Trying to communicate too much in one map—having more than one purpose for the map—tends to blur the message and confuse the map reader. Using two or more maps, each focused on a single message, is always a better strategy.
- *Audience*: Are you designing a map for a few readers or for a large audience of hundreds or millions of people? It's better to target your map to the person least prepared to understand your map's message.
- *Size and scale*: The physical size of a map relative to the geographic extent shown on the map will dictate the scale of the map and determine how you will represent the actual size and number of features shown on the map.
- *Media*: Media also plays an important role, because a map printed on newspaper could not show fine details clearly, whereas one printed on high-quality paper can show finer details. In addition, the details on a digital map could vary depending on the viewing program.
- *Focus*: It refers to where the designer wants the map reader to first focus. Typically, cool colors (blues, greens, and light gray) are used for background information and warm colors (red, yellow, black) are used to capture the reader's attention.
- *Integrity*: If the data was produced by another organization, it is customary to give that organization credit on the map.

- *Balance*: How does map look on the page or screen? Are the parts of the map properly aligned? The body of the map should be the dominant element. Try to avoid large open spaces. Be flexible in where you place elements (i.e., not all titles need to go at the top).
- *Completeness*: A map generally should contain some basic elements, such as a title, legend, scale bar, and north arrow; however, there are exceptions. For example, if a graticule exists, it is not necessary to place a north arrow. Basically, the relevant information needs to be placed for the readers to fully understand the map.

3.17 Map Elements

The first feature that should be placed into the map layout is the frame line. This line is essentially a bordering box that surrounds all the map elements described hereafter. All of these map elements should be balanced within the frame line. To balance a map, ensure that neither large blank spaces nor jumbled masses of information are present within the map. Neat lines are border boxes that are placed around individual map elements. By definition, neat lines must occur within the frame line. Both frame lines and neat lines are typically thin, black-lined boxes, but they can be altered to match the specific aesthetics of an individual map. Taken together, these map elements should work together to achieve the goal of a clear, ordered, balanced, and unified map product. Since modern GIS software allow users to add and remove these graphic elements with little effort, care must be taken to avoid the inclination to employ these components with as little forethought as it takes to create them. The described following map elements provide guidance to compose and balance the mapped area in GIS:

- *Map body*: The mapped area is the primary geographic component of the overall map. One can display more than one image in primary mapped area within the document. The mapped area contains all of the features and symbols used to represent the spatial phenomena being displayed. The mapped area is typically bordered with a neat line. Insets can be thought of as secondary mapped areas, each encased within their own neat line. Insets often display the primary mapped area in relation to a larger area.
- *Title*: All maps should have a title, and it is often placed on a map layout as text. The title is one of the first map elements to catch the viewer's eye, so care should be taken to most effectively represent the intent of the map with this leading text. The title should clearly and concisely explain the purpose of the map and should specifically target the intended viewing audience.
- *Legend*: It lists the symbology used within the map and what it represents. The legend provides a self-explanatory definition for all symbols used within the mapped area. Although placement of the legend is variable, it should be placed within the white space of the map and not in such a way that it masks any other

map elements. The symbols representing mapped features should be to the left of the explanatory text. Placing a neat line around the legend helps to bring attention to the element and is recommended but not necessary. If a large legend is unavoidable, it is acceptable to place it outside of the map's frame line.

- *Scale*: It provides readers with the information they need to determine distance. A map scale is a ratio, where one unit on the map represents some multiple of that value in the real world. The three primary representations of scale are the representational fraction, verbal scale, and bar scale. The scale indicator should not be prominently displayed within the map as this element is of secondary importance.
- *Projection*: A mathematical formula that transforms feature locations from the Earth's curved surface to a map's flat surface. Projections can cause distortions in distance, area, shape, and direction; all projections have some distortion. Therefore, the projection type is often placed on the map to help readers determine the accuracy of the measurement information they get from the map.
- *Direction*: Always direction shows with north arrow. A map may show true north and magnetic north. This element is inserted in the map layout view. To assist in clarifying orientation, a graticule can also be included in the mapped area. Most maps are made such that the top of the page points to the north. Orientation is most often indicated with a north arrow, of which there are many stylistic options available in current GIS software packages.
- *Data source*: It is the bibliographic information for the data used to develop the map.

3.18 GIS Applications in Land Resource Management

GIS has become a core part of many disciplines like geology, climatology, geography, statistics, archaeology, oceanography, etc. Most of the GIS applications involve handling spatial data in a well-organized database for data retrieval, updating, and transformation applications. Remote sensing data of course is one of the main sources of data for many GIS applications. Data processing methodologies in any spatial data handling systems are dependent upon the data structures used in GIS. Using different data models, the vector-based data processing methods employed in GIS are based on polygon-oriented algorithms (Green et al. 1985), while the algorithms used in remote sensing are pixel oriented. GIS applications have immense potential in various aspects of land resource management like digital terrain analysis, geomorphology mapping, soil resource inventory, land use/land cover analysis, watershed hydrology, soil erosion studies, prioritization of watersheds, soil suitability evaluation, spatial data interpolation, landscape ecological planning, urban planning, decision support systems, etc.

3.19 Conclusions

GIS is used by multi-disciplines a tool for spatial data handling in a geographic environment. Basic elements of GIS consist of hardware, software, data, and people. GIS is considered one of the important tools for decision-making in problem-solving environment dealing with geo-information. Georeferenced data include a *spatial* (geometrical or graphical) component describing the location or spatial distribution of geographic phenomenon and an *attribute* component used to describe its properties. The disciplines like geography, geology, pedology, agronomy, soil and water conservation, cartography, remote sensing, photogrammetry, surveying, geodesy, statistics operation research, computer science, mathematics, and civil engineering associated in land resource management could be effectively used in GIS. Data derived from latest remote sensing, GPS, and information technologies could be effectively used GIS environment for mapping, monitoring, and management of land resources and dissemination of information at appropriate destinations in various applications.

References

Aronoff S (1989) Geographic information systems: a management perspective. WDC Publications, Ottawa

Burrough PA (1986) Principles of geographical information systems for land resource assessment. Clarendon Press, Oxford

Burrough PA (1987) Principles of geographical information systems for land resources assessment. Clarendon Press, Oxford 193 p

Burrough PA (1998) Dynamic modeling and geocomputation. In: Longley PA, Brookes SM, McDonnell, MacMillan B (eds) Geocomputation: a primer. Wiley, Chichester, pp 165–191

Clarke KC (1995) Analytical and computer cartography, 2nd edn. Prentice Hall, Upper Saddle River

Clarke (2001) Getting started with geographic information systems. Prentice Hall, Upper Saddle River

Dueker KJ (1979) Land resource information systems: a review of fifteen years experience. Geo-Processing 1:105–128

Goodchild MF (1992) Geographical information science. Int J Geogr Inf Syst 6(1):31–45

Green NP, Finch S, Wiggins J (1985) The 'state-of-art' in geographical information systems. Area 17(4):295–301

Jackson MJ (1992) Integrated geographical information systems. Int J Remote Sens 13(6–7):1343–1351

Reddy GPO (2012) Principles and concepts of GIS. In: Reddy GPO, Sarkar D (eds) Remote sensing and GIS in digital terrain analysis and soil-landscape modeling, pp 69–77

Singh R, Sahoo P, Rai A (2003) Use of remote sensing and GIS technology in agricultural surveys. From http://www.gisdevelopment.net/application/agriculture/cropping/mio368.htm

Star J, Estes J (1990) Geographical information systems: an introduction. Prentice Hall, Englewoods Cliffs

Stillwell J, Clarke G (eds) (1987) Applied GIS and spatial analysis. Wiley, West Sussex

Chapter 4
Global Positioning System: Principles and Applications

G. P. Obi Reddy

Abstract In recent times, global positioning system (GPS) has totally replaced the traditional surveying and mapping tasks such as triangulation, traverse, controls, etc., GPS is being extensively used in the areas of surveying, engineering, monitoring positions, and navigation. GPS provides the position, velocity, and timing information that enable many applications in human daily life. GPS-based surveying systems allow surveys to increase accuracy in mapping over conventional surveying techniques. GPS system consists of three segments, i.e., the space segment, comprising the satellites and the transmitted signals; control segment, the ground facilities carrying out the task of satellite tracking, orbit computations, telemetry, and supervision necessary for the daily control of the space segment; and user segment, the entire spectrum of application equipment and computational techniques that are available to the users. GPS play a significant role in data collection and surveying, mapping, and monitoring of land resource surveys on periodical basis.

Keywords Control segment · Differential global positioning system · Global positioning system · Space segment · User segment

4.1 Introduction

Global positioning system (GPS) has revolutionized research in the areas of surveying, engineering, monitoring positions, and navigation (Noviline et al. 1993). Today, GPS has totally replaced the traditional surveying and mapping tasks such as triangulation, traverse, controls, etc., For mapping, where surveying accuracy better than a meter is required, specialized differential global positioning system (DGPS) techniques have been developed. These techniques are being employed increasingly for detailed topographic-, soil-, geological-, engineering-, and environmental-related surveys at different scales. GPS-based surveying systems allow surveys to increase

G. P. O. Reddy (✉)
ICAR-National Bureau of Soil Survey & Land Use Planning, Nagpur, India

G. P. O. Reddy, S. K. Singh (eds.), *Geospatial Technologies in Land Resources Mapping, Monitoring and Management*, Geotechnologies and the Environment 21,
https://doi.org/10.1007/978-3-319-78711-4_4

accuracy in mapping over conventional surveying techniques. The information generated by conventional methods, remote sensing and GPS techniques is being widely used to create database in geographic information system (GIS) environment for performing integrated analysis to generate environment-friendly and sustainable action plans. There are three major segments of GPS, they are i) space segment, which consists of 24 satellites at an altitude of approximately 20,200 km above the Earth surface; (ii) ground control segment, in which five ground stations are distributed around the Earth for monitoring and controlling the satellite system continuously; and (iii) user segment, which includes all those who use GPS tracking equipment to receive GPS signals to satisfy positioning requirements (Kaplan and Hegarty 2006). In any land resource surveys, GPS play a significant role in field data collection and surveying and monitoring the location-based databases on periodical basis.

GPS provide current position, anywhere and at any time on the globe, with a reasonable degree of accuracy. GPS demonstrated a significant benefit to the civilian community who are applying GPS to a rapidly expanding number of applications. The different methods of observations with GPS include absolute positioning, relative positioning in translocation mode, relative positioning using differential GPS technique, and kinematic GPS surveying technique. GPS provides relatively high-positioning accuracies, from tens of meters down to the millimeter level. The capability of GPS in determining the velocity and time to an accuracy is commensurate with position. The signals are available to users anywhere on the globe: in the air, on the ground, or at sea. It is a positioning system with no user charges that simply requires the use of relatively low-cost hardware. It is an all-weather system, available 24 h a day. The position information is in three dimensional, that is, vertical as well as horizontal information is provided.

4.2 History of GPS

In the early part of 1960s, the US Department of Defense (DOD), the National Aeronautics and Space Administration (NASA), and the Department of Transportation (DOT) were interested in developing satellite systems for three-dimensional position determination. The optimum system was viewed to have the attributes of global coverage, continuous/all-weather operation, ability to serve high-dynamic platforms, and high accuracy. Development work on GPS was initiated in 1973 by US Department of Defense with an aim to develop an all-weather, 24 h, global positioning system to support the positioning requirements. The system was therefore designed to replace the large variety of navigational systems already in use, and great emphasis was placed on the system's reliability and survivability. In short, number of stringent conditions had to be met, suitable for all classes of platform, aircraft (jet to helicopter), ship, land (vehicle-mounted to handheld), and space (missiles and satellites), able to handle a wide variety of dynamics, real-time positioning, velocity, and time determination capability to an appropriate accuracy;

the positioning results were to be available on a single global geodetic datum and highest accuracy to be restricted to a certain class of user, resistant to jamming (intentional and unintentional) and redundancy provisions, to ensure the survivability of the system, passive positioning system that does not require the transmission of signals from the user to the satellite(s), able to provide the service to an unlimited number of users, low cost, and low power; therefore as much complexity as possible should be built into the satellite segment and total replacement of terrestrial-based systems. This led to a design of GPS based on the following essential concepts: a one-way ranging system, in which the satellites transmit signals but are unaware of who is using the signal (no receiving function). As a result the user cannot easily be detected or charged for using the system (civilian context). A system makes range-like measurements with the aid of pseudorandom binary codes modulated on carrier signals. Satellite signals are unaffected by cloud and rain. A multiple satellite system ensures there is always a sufficient number of satellites visible simultaneously anywhere on the globe and at any time.

4.3 Space-System Reliability

The US space program had by 1973 demonstrated the reliability of space hardware. In particular, the Transit system had offered important lessons. The Transit satellites were originally designed to last 2–3 years in orbit, yet some of the satellites have operated well beyond their design life. In fact Transit continued to perform reliably for over 25 years. The American Navstar (Navigation Satellite Time and Ranging) GPS system contains 24 satellites. There are four satellites on each of six orbital planes. These planes are inclined at 55° with respect to the equator and are equally spaced in right ascension. The satellite orbits are close to circular (to within 2%) with an orbital period chosen so that a satellite completes exactly two orbits while the Earth rotates 360° (one sidereal day). This ensures that the satellite trajectory on the Earth exactly repeats itself twice daily.

4.4 Atomic Clock Technology

With the development of atomic clocks, a new era of precise timekeeping had commenced. However, before the GPS program was launched, these precise clocks had never been tested in space. The development of reliable, stable, compact, space-qualified atomic frequency oscillators (rubidium and then cesium) was therefore a significant technological breakthrough. The advanced clocks now being used on the GPS satellites routinely achieve long-term frequency stability in the range of a few parts in 1014 per day (about 1 sec in 3,000,000 years!). This long-term stability is one of the keys to GPS, as it allows for the autonomous, synchronized generation and transmission of accurate timing signals by each of the GPS satellites without continuous monitoring from the ground.

4.5 Quartz Crystal Oscillator Technology

In order to keep the cost of user equipment down, quartz crystal oscillators are being used (similar to those used in modern digital watches), rather than using atomic clocks as in the GPS satellites. Besides their low cost, quartz oscillators have excellent short-term stability.

4.6 Satellite Tracking and Orbit Determination

Successful operation of GPS, as well as the Transit system, depends on the precise knowledge and prediction of a satellite's position with respect to an earth-fixed reference system. Tracking data collected by ground monitor stations is analyzed to determine the satellite orbit over the period of tracking (typically 1 week). This reference ephemeris is extrapolated into the future, and the data is then uploaded to the satellites. Prediction accuracies of the satellite coordinates, for 1 day, at the few meter level have been demonstrated.

4.7 Spread-Spectrum Technology

The ability to track and obtain any selected GPS satellite signal (a receiver will be required to track number of satellites at the same time), in the presence of considerable ambient noise, is a critical technology. This is now possible using spread-spectrum and pseudorandom-noise coding techniques.

4.8 Large-Scale Integrated Circuit Technology

To realize the desired low cost, low power, and small size necessary for much of the user equipment, the GPS program relies heavily on the successful application of very large-scale integration (VLSI) circuits and powerful computing capabilities built onto them.

4.9 Different Segments of GPS

The GPS system consists of three segments, i.e., the *space segment*, comprising the satellites and the transmitted signals; the *control segment*, the ground facilities carrying out the task of satellite tracking, orbit computations, telemetry, and supervision necessary for the daily control of the space segment; and the *user segment*, the entire spectrum of application equipment and computational techniques, which are available to the users.

4.9.1 The Space Segment

The space segment consists of the constellation of spacecraft and the signals broadcast by them, which allow users to determine position, velocity, and time. The basic functions of the satellites are to receive and store data transmitted by the control segment stations, maintain accurate time by means of several onboard atomic clocks, transmit information and signals to users on two L-band frequencies, provide a stable platform and orbit for the L-band transmitters. Figure 4.1 shows the constellation of GPS satellites in space.

Each GPS satellite transmits a unique navigational signal centered on two L-band frequencies of the electromagnetic spectrum, permitting the ionospheric propagation effect on the signals to be eliminated. At these frequencies the signals are highly directional and so are easily reflected or blocked by solid objects. Clouds are easily penetrated, but the signals may be blocked by foliage (the extent of blockage is dependent on the type and density of the leaves and branches and whether they are wet or dry). The satellite signal consists of two L-band carrier waves, the ranging codes modulated on the carrier waves, and the so-called navigation message.

Fig. 4.1 Constellation of GPS satellites

Modulated onto the carrier waves are the pseudorandom noise (PRN) ranging codes and navigation message for the user. The primary function of the ranging codes is to permit the *signal transit time* (from satellite to receiver) to be determined. The transit time when multiplied by the velocity of light then gives a measure of the receiver-satellite "range" (in reality the measurement process is considerably more complex). The navigation message contains the satellite orbit information, satellite clock parameters, and pertinent general system information necessary for real-time navigation to be performed. All signal components are derived from the output of a highly stable atomic clock. Each GPS satellite is equipped with several cesium and rubidium atomic clocks.

4.9.2 The Control Segment

The control segment consists of facilities necessary for satellite health monitoring, telemetry, tracking, command and control, satellite orbit and clock data computations, and data uplinking. There are five ground facility stations: Hawaii, Colorado Springs, Ascension Island, Diego Garcia, and Kwajalein. All are owned and operated by the US Department of Defense and perform the following functions.

All five stations are *Monitor Stations*, equipped with GPS receivers to track the satellites. The resultant tracking data is sent to the Master Control Station. Colorado Springs is the *Master Control Station* (MCS), where the tracking data are processed in order to compute the satellite ephemerides and satellite clock corrections. It is also the station that initiates all operations of the space segment, such as spacecraft maneuvering, signal encryption, satellite clock-keeping, etc. Three of the stations (Ascension Is., Diego Garcia, and Kwajalein) are *Upload Stations* allowing for the uplink of data to the satellites. The data includes the orbit and clock correction information transmitted within the navigation message, as well as command telemetry from the MCS.

Each of the upload stations can view all the satellites once a day. All satellites are therefore in contact with an upload station three times a day, and new navigation messages as well as command telemetry can be transmitted to the GPS satellites every 8 h if necessary. The computation of (a) the satellite orbits or "ephemerides" and (b) the determination of the satellite clock errors is the most important function of the control segment. The first is necessary because the GPS satellites function as "orbiting control stations" and their coordinates must be known to a relatively high accuracy, while the latter permit a significant measurement bias to be reduced. The GPS satellites travel at high velocity (of the order of 4 km/sec) but within a more or less regular orbit pattern. After a satellite has separated from its launch rocket and it begins orbiting the Earth, its orbit is defined by its initial position and velocity and the various force fields acting on the satellite. In the case of the gravitational field for a spherically symmetric body (a reasonable approximation of the Earth at the level of about one part in 103), this produces an elliptical orbit, which is fixed in space – the *Keplerian ellipse*. Due to the effects of the other non-spherical gravitational components of the earth's gravity field, and nongravitational forces, which perturb the orbit, the actual trajectory of the satellite departs from the ideal Keplerian ellipse.

The most significant forces that influence satellite motion are the spherical and non-spherical gravitational attraction of the Earth; the gravitational attractions of the sun, moon, and planets; the atmospheric drag effects; the solar radiation pressure (both direct and albedo effects); and the variable part of the Earth's gravitational field arising from the solid earth and ocean tides.

The GPS broadcast ephemeris, as represented in the navigation message, is actually a combination of all of the above orbit representations. The orbital ephemerides are expressed in the reference system most appropriate for positioning, which is an *earth-fixed* reference system such as WGS84. Hence, the control segment has the function of propagating the satellite datum, which users connect to via the transmitted satellite ephemerides. The behavior of each GPS satellite clock is monitored against GPS time, as maintained by an ensemble of atomic clocks at the GPS Master Control Station.

4.9.3 The User Segment

This is the part of the GPS system with which we are most concerned – the space and control segments being largely transparent to the operations of the navigation function. Each day new applications are being identified, each with its unique requirements with regard to accuracy of the results, reliability, operational constraints, user hardware, data processing algorithms, latency of the GPS results, etc. To make sense of the bewildering range of GPS applications, it may be useful to classify them according to the following:

1. Land, sea, and air navigation and tracking, including en route as well as precise navigation, collision avoidance, cargo monitoring, vehicle tracking, search and rescue operations, etc. While the accuracy requirement may be modest and the user hardware is generally comparatively low cost, the reliability, integrity, and speed with which the results are needed are generally high.
2. Surveying and mapping, on land, at sea, and from the air. Includes geophysical and resource surveys, GIS data capture surveys, etc. The applications are of relatively high accuracy, for positioning in both the static and moving receiver mode, and generally require specialized hardware and data processing software.
3. Military applications. Although these are largely mirrored by civilian applications, the military GPS systems are generally developed to "military specifications," and a greater emphasis is placed on system reliability.
4. Recreational uses, on land, at sea, and in the air. The primary requirement is for low-cost instruments which are very easy to use.
5. Other specialized uses, such as time transfer, attitude determination, spacecraft operations, atmospheric studies, etc. Obviously such applications require

specially developed, high-cost systems, often with additional demanding requirements such as real-time operation, etc.

4.10 GPS Instrumentation

The following components of a generic GPS receiver can be identified:

- *Antenna* and *Preamplifier*: Antennas used for GPS receivers have broad beam characteristics; thus they do not have to be pointed to the signal source like satellite TV receiving dishes. The antennas are compact, and a variety of designs are possible. There is a trend to integrating the antenna assembly with the receiver electronics.
- *Radio-Frequency Section* and *Computer Processor*: The RF section contains the signal processing electronics. Different receiver types use somewhat different techniques to process the signal. There is a powerful processor onboard not only to carry out computations such as extracting the ephemerides and determining the elevation/azimuth of the satellites, etc. but also to control the tracking and measurement function within modern digital circuits and in some cases to carry out digital signal processing.
- *Control Unit Interface*: The control unit enables the operator to interact with the microprocessor. Its size and type vary greatly for different receivers, ranging from a handheld unit to soft keys surrounding an LCD screen fixed to the receiver "box."
- *Recording Device*: In the case of GPS receivers intended for specialized uses such as the surveying, the measured data must be stored in some way for later data processing. In the case of ITS applications such as the logging of vehicle movement, only the GPS-derived coordinates and velocity may be recorded.
- *Power Supply*: Transportable GPS receivers these days need low-voltage DC power. The trend toward more energy-efficient instrumentation is a strong one, and most GPS receivers operate from a number of power sources, including internal NiCad or Lithium batteries, external batteries such as wet-cell car batteries, or from mains power.

4.11 Absolute and Relative Positioning

The GPS system was designed to provide users with two levels of performance. The highest accuracy was reserved for military users, who had access to the coded signals on both L-band frequencies. This is referred to as the Precise Positioning Service (PPS) and is intended to give absolute accuracies of the order of 10–20 m. A lower level of accuracy was to be provided to the general (civilian) user by the Standard Positioning Service (SPS). The accuracy was intended to be an order of magnitude worse. However, after extensive testing in the early to mid-1980s, it was found that the SPS proved to be far more accurate than expected, achieving accuracies in the few decameter range rather than the expected 100 m level of accuracy. GPS positioning accuracy can be expressed in a number of ways, firstly in an absolute sense, with respect to a coordinate system such as WGS84. This coordinate system is realized first through the coordinates of the monitor stations (of the control segment) and subsequently transferred to users via the (changing) coordinates of the GPS satellites. As the satellite coordinates are essential for the computation of user position, any error in these values will directly affect the quality of the position determination. This sets a lower bound for the magnitude of GPS absolute positioning error. The Precise Positioning Service can deliver positioning accuracies of the order of 10–20 m, mainly because of the lower level of measurement noise, and the data is unaffected by unmodelled ionospheric delay error.

4.12 Differential Global Positioning System (DGPS)

GPS is by far the most accurate global navigation system ever devised. But even its incredible accuracy can be boosted using a technique called "differential GPS." With differential GPS, GPS can achieve measurement accuracy better than a meter. The accuracy of the results is based on the idea that if we put a GPS receiver on the ground in a known location, we can use it to figure out exactly what errors the satellite data contains. It acts like a static reference point. It can then transmit an error correction message using a radio transmitter to any other GPS receivers that are in the local area, and they can use that error message to correct their position solutions. The concept works because the satellites are so high up that any errors measured by one receiver will be almost exactly the same for any other receiver in the same locale. Because of the simplicity of the GPS signal, this single correction factor, in effect, takes care of all the possible errors in the system, whether they are from the receiver clocks, the satellite clocks, the satellite's position, or the ionospheric and atmospheric delays. Differential GPS service can be provided by a number of sources such as satellite-based augmentation services (SBAS) like WAAS (Wide Area Augmentation Service) offered in North American or a regional SBAS like *GPS*-aided Geo augmented navigation (GAGAN) offered by the Indian Government over the Indian subcontinent.

DGPS is a method to improve the positioning or timing performance of GPS using one or more reference stations at known locations, each equipped with at least one GPS receiver (Sickle 2008). The reference station(s) provides information to the end user via a data link that may include corrections to the end user's pseudorange measurements, corrections to GPS satellite-provided clock and ephemeris data, or data to replace the broadcast clock and ephemeris information, raw reference station measurements (e.g., pseudorange and carrier phase), integrity data (e.g., "use" or "don't use" indications for each visible satellite or statistical indicators of the accuracy of provided corrections), and auxiliary data including the location, health, and meteorological data of the reference station(s).

DGPS techniques may be categorized in different ways: as *absolute* or *relative* differential positioning; as *local area*, *regional area*, or *wide area*; and as *code based* or *carrier based*. Absolute differential positioning is the determination of the user's position with respect to an earth-centered earth-fixed (ECEF) coordinate system. This is the most common goal of DGPS systems. For absolute differential positioning, each reference station's position must be accurately known with respect to the same ECEF coordinate system in which the user position is desired. Relative differential positioning is the determination of the user's position with respect to a coordinate system attached to the reference station(s), whose absolute ECEF position(s) may not be perfectly known.

DGPS systems may also be categorized in terms of the geographic area that is to be served. The simplest DGPS systems are designed to function only over a very small geographic area (i.e., with the user separated by less than 10–100 km from a single reference station). To effectively cover larger geographic regions, multiple reference stations and different algorithms are typically employed.

One final categorization of DGPS systems is between so-called code-based or carrier-based techniques. Code-based DGPS systems rely primarily on GPS code (i.e., pseudorange) measurements, whereas carrier-based DGPS systems ultimately rely primarily on carrier-phase measurements (Olynik et al. 2002).

4.13 GPS RTK Surveys

Real-time kinematic (RTK) surveying is an advanced form of relative GPS carrier-phase surveying in which the base station transmits its raw measurement data to rover(s), which then computes a vector baseline from the base station to the rover. This computation is done nearly instantaneously, with minimal delays between the time of the base station measurements and the time these are used for baseline processing at the rover (ideally a few seconds). The precision of real-time kinematic (RTK) baselines can be almost as good as the precision of static carrier-phase baselines. If the base station coordinates are accurately known, this will usually result in accurate rover positions. The combination of fast and precise positioning, one-man operation, and wide work areas has resulted in RTK becoming an impressively powerful tool for some survey applications. Like other survey techniques, RTK does not solve every survey problem. RTK is only suitable for environments with reasonably good GPS tracking conditions (limited obstructions, multipath, and RF noise) and with continuously reliable communication from the base station to the rover. The limitation described above has initiated the development of a network-based RTK GPS, in which the data from an entire network of RTK GPS base stations is considered in real time to allow fast and accurate rover initializations over a larger area. These techniques have been developing since the early 2000s, and the results look promising.

4.14 Factors Affecting GPS Accuracy

The accuracy with which a user receiver can determine its position or velocity, or synchronize to GPS system time, depends on a complicated interaction of various factors. In general, GPS accuracy performance depends on the quality of the pseudorange and carrier-phase measurements as well as the broadcast navigation data.

- *Satellite Clock Error*: The GPS satellites contain atomic clocks that control all onboard timing operations, including broadcast signal generation. Although these clocks are highly stable, the clock correction fields in the navigation data message are sized such that the deviation between SV time and GPS time may be as large as 1 ms (ARINC 2004).
- *Ephemeris Error*: Estimates of ephemerides for all satellites are computed and uplinked to the satellites with other navigation data message parameters for rebroadcast to the user. Ephemeris errors are generally smallest in the radial (from the satellite toward the center of the Earth) direction. The components of ephemeris errors in the along-track (the instantaneous direction of travel of the satellite) and cross-track (perpendicular to the along-track and radial) directions are much larger.
- *Atmospheric Effects*: The propagation speed of a wave in a medium can be expressed in terms of the index of refraction for the medium. The index of refraction is defined as the ratio of the wave's propagation speed in free space to that in the medium.
- *Ionospheric Effects*: The ionosphere is a dispersive medium located primarily in the region of the atmosphere between about 70 km and 1000 km above the Earth's surface. Within this region, ultraviolet rays from the sun ionize a portion of gas molecules and release free electrons. These free electrons influence electromagnetic wave propagation, including the GPS satellite signal broadcasts.
- *Multipath and Shadowing Effects*: One of the most significant errors incurred in the receiver measurement process is multipath. Multipath errors vary significantly in magnitude depending on the environment within which the receiver is located, satellite elevation angle, receiver signal processing, antenna gain pattern, and signal characteristics.
- *Selective Availability*: Selective availability is an induced error. It is by far the worse error than any of these "natural" sources of error. The idea behind it is to make sure that the accuracy of GPS should not misuse. Selective availability involves "dithering" the satellite clocks and/or falsifying the ephemeris data broadcast in the satellite telemetry message. Selective availability was removed from the GPS system in 1999 so that accurate position information could be made available to the users at all times. The significant errors in GPS and their respective sources are satellite clock errors, orbital errors, ionospheric and atmospheric errors, multipath errors, receiver errors and induced errors, or selective availability.

4.15 GPS Applications in Land Resources Management

As far as mapping and monitoring of land resources are concerned, GPS applications have immense potential and handy to the surveys to increase the positional accuracy over the conventional surveying techniques. GPS allows the surveyors to accurately navigate to specific locations in the field to collect samples or monitor the phenomena. In recent times, GPS-based applications are being widely used for detailed topographic, geological, soil, land, water, engineering and environmental related surveys at different scales. Data generated through GPS is being used as a field check for remote sensing data and also plays a supplementary and complementary role for analysis of remote sensing data. GPS-based applications have immense potential in precision farming particularly in field mapping, soil sampling, tractor guidance, variable rate applications, yield mapping etc. In survey and mapping, where accuracy better than a meter is required, specialized DGPS techniques could be used.

4.16 Conclusions

The information generated by conventional methods, remote sensing and GPS techniques is being used to create database in GIS environment for performing integrated analysis to generate environment friendly and sustainable action plans. In any land resource surveys, GPS play significant role in field data collection and surveying and monitoring the location-based databases on periodical basis. Differential-mode positioning relies upon an established control point. The reference station is placed on the control point, a triangulated position, and the control point coordinate. This allows for a correction factor to be calculated and applied to other roving GPS units used in the same area and in the same time series.

References

ARINC (2004) Research Corporation, NAVSTAR GPS space segment/Navigation user interfaces, Interface Specification, IS-GPS-200D (Public Release Version), ARINC Research Corporation, Fountain Valley, CA

Kaplan ED, Hegarty CJ (2006) Understanding GPS: principles and applications, 2nd edn. Artech House, Boston, p 703

Noviline JRM, Sundaram A, Natarajan T (1993) Wasteland development using GIS techniques. Int J Remote Sens 14(17):3249–3257

Olynik M, Petovello MG, Cannon ME, Lachapelle G (2002) Temporal variability of GPS error sources and their effect on relative positioning accuracy. In: Proceedings of the ION NTM-2002, January 28–30, San Diego, CA, pp 877–888

Sickle JV (2008) GPS for land surveyors, 3rd edn. CRC Press/Taylor & Francis, Boca Raton

Chapter 5
Geo-ICDTs: Principles and Applications in Agriculture

Saurabh Suradhaniwar, Soumyashree Kar, Rohit Nandan, Rahul Raj, and Adinarayana Jagarlapudi

Abstract Geographical information, communication and dissemination technologies (Geo-ICDTs) is an innovative initiative that integrates state-of-the-art technologies for geospatial information collection and rapid dissemination. It ensembles core emerging technologies that lay out the platform for spatial decision-making, geo-computation and location-based services (LBS). In the past few decades, rapid developments in geolocation-based platforms and services have made significant contributions towards emerging markets and applications like spatial data infrastructure (SDI), digital earth observations (EO), precision agriculture, location-based commerce (l-commerce), mobile commerce (m-commerce), e-commerce, e-governance, etc. These technologies have also indispensably effected the institutionalization of e-agriculture in the agricultural sector (the primary driver of economy across nations), which thrives with improved productivity and sustainability (adaptive to climate change). However, Geo-ICDTs face adamant challenge in the form of developing, implementing, integrating and steering adaptability among end-users. Understanding stochastic behaviour of these parameters requires capturing real-time/near real-time data from several sources, such as sensor networks, remote sensing, crowdsourcing, experimental setups and lab-based studies. These requirements necessitate development of a "system of things" infrastructure that can capture location-specific data from several sources and can communicate with each other, thus evolving as an integrated system. This system of things is often referred as the "Internet of things (IoT)", which works under the framework of Geo-ICDT. This chapter closely discusses the MMA (monitoring, management and adaptation) framework, its components and their implementation in precision agriculture.

Keywords Geographic information system (GIS) · Information, communication and dissemination technologies · Global positioning system (GPS) · Internet of things (IoT) · Precision agriculture

---

S. Suradhaniwar · S. Kar · R. Nandan · R. Raj · A. Jagarlapudi (✉)
Centre of Studies in Resources Engineering, Indian Institute of Technology Bombay, Powai, Mumbai 400076, India
e-mail: adi@iitb.ac.in

G. P. O. Reddy, S. K. Singh (eds.), *Geospatial Technologies in Land Resources Mapping, Monitoring and Management*, Geotechnologies and the Environment 21,
https://doi.org/10.1007/978-3-319-78711-4_5

5.1 Introduction

ICDT roots from the concepts of networked devices (which we principally referred to as the "Internet") and applicability of application-oriented web services. With evolution of the Internet, state-of-the-art communication technologies took a humongous leap towards development of wireless connectivity of systems that became the face of modern-day Internet of things. In conjunction to this, availability of large-scale geospatial data from various sources, such as earth-observing satellites (EO), global positioning systems (GPS) and geotagged resources enriched the geospatial domain knowledge by augmenting geographical attribute to conventional natural resource monitoring systems. These advancements contributed towards development of Geo-ICDT systems that could integrate communications and geospatial knowledge for development of geo-Internet of things (Geo-IoT).

As discussed above, Geo-ICDTs inherently rely on spatial-positioning (geospatial) technologies to integrate geospatial information with ICDT. Thus, intuitively it's the amalgamation of two primary technologies: ICDT and spatial-positioning technology. Several spatial-positioning technologies such as GPS, differential GPS (DGPS), assisted GPS (AGPS), wide-area augmentation system (WAAS), network-based, radio-based positioning system (NBPS and RBPS), etc. are available for embedding geographical information with ICDT systems. Technologies like real-time locating system (RTLS) and radio-frequency identification (RFID) inherit the principles of RBPS for providing location information. Apart from inheriting geographical information, robustness in design of ICDT systems also plays a vital role in operationalizing Geo-ICDT systems for user-specific applications. ICDTs (information collection and dissemination systems) belong to a generic class of instrumentation and technologies (hardware and software), with sole purpose of collecting, analysing and disseminating timely information for rapid/ informed decision-making. These technologies cover a wide spectrum of state-of-the-art systems like wireless sensor networks, remote sensing-based data assimilation, scientific instrumentation, experimental setups, in situ/in vitro data collection, etc. (Lindblom et al. 2017). It is evident that most real-world applications and decision-making depend on accuracy of data collected and fine-tuned processing methodologies in order to generate meaningful information. Thus, generally based on application, a subclass or combination of these ICDTs and geospatial technologies can be selected for monitoring and management of targeted recourses.

With geographical (or geospatial) information and ICDT systems in place, next big challenge in design of Geo-ICDT system is seamless integration of these systems. This integration became feasible with emergence of spatial databases that could store both, spatial and nonspatial information, which can be queried similar to a conventional database. These spatial databases help in extracting spatial information based on spatial queries while giving due importance to nonspatial attributes. Such integrated system requires sophisticated geographic information system (GIS) software interfaces that provide several in-built tools for spatial analyses. In addition, open-source GIS packages provide a wide and open platform for GIS developers to develop, test and publish new set of tools for efficient spatial and nonspatial analysis. Another challenge in Geo-ICDT systems is inherent heterogeneity of data acquisition systems. Conventional data integration for technologically diverse systems required trained personnel to understand this heterogeneity and assimilate the information based on domain specialization. Evolution of machine-to-machine (M2M) communication protocols and interoperable systems aided this integration of inherently autonomous and diverse ICDT systems. One prominent effort towards ensuring interoperability of systems was undertaken by Open Geospatial Consortium (OGC). OGC ensured interoperability between systems by standardizing the access to these ICDT systems. Standards like sensor web enablement (SWE) provide comprehensive guidelines for designing and mode of operation of ICDT systems using simple web service-based interface (Ueyama et al. 2017).

5.2 Components of Geo-ICDT Architecture

The eminence of Geo-ICDTs majorly depends on three important components: monitoring, managing and adaptation. Real-world applications necessitate regular monitoring for data generation and collection. Complexity and comprehensiveness of these datasets largely depends on the quality and quantity of information generated from the ICDT systems. For example, data complexity of application pertaining to flood hazards may be entirely different from those for precision agriculture. Complexity of data generation also depends on the selection and capabilities of ICDT platforms involved. Generally, science-based extensive experimentation and setups generate large volumes of complex data that are further conceptualized through generic (statistical) and sophisticated data models for better understanding of underlying physical phenomena and interactions (Janssen et al. 2017). In contrast to this, long-term monitoring usually generates large data sets with moderate complexity that eventually aids the development of more complex models. As a rule of thumb, complexity of data and data-generating systems (ICDTs) increases with increase in complexity of intended applications. Finally, these datasets should also have significant geographical (spatial and temporal) importance such that the technologies can be scaled for global applications. Further in this chapter, we discuss state-of-the-art ICDTs and technologies that constitute elements of monitoring in Geo-ICDTs.

Apart from collecting data, Geo-ICDTs are also responsible for managing the resources. In real-world problem domain, almost all Geo-ICDTs supplement the process of decision-making by modeling and validating physical phenomenon using existential data. Due to this, Geo-ICDTs applications should align themselves with objectives and desired outcomes pertaining to the phenomenon or resource of interest. For example, monitoring of plant disease severity and monitoring of soil moisture patterns require different premeditated methodologies such that the data generated through ICDT systems is valuable for the application. Moreover, there can always be a significant overlap of data requirements for closely related application

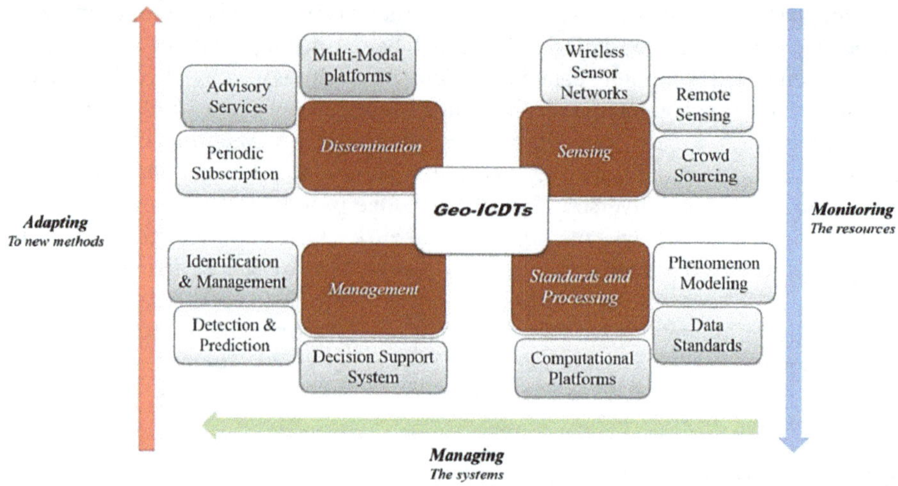

Fig. 5.1 Architecture of Geo-ICDT

domains (e.g. plant disease and plant phenology). Thus, intelligent monitoring is a crucial asset for Geo-ICDT systems and implementations. Finally, Geo-ICDTs are responsible for seamless dissemination of knowledge generated through monitoring and management.

Geo-ICDTs are complex to visualize in totality due to interdisciplinary nature of technologies, problem domains and evolutionary solutions. Figure 5.1 presents a concept-based architectural overview of Geo-ICDT and its components. Elements of Geo-ICDT are amalgamation of different domains and technologies that are often seen as independent entities. However, application of Geo-ICDTs requires a more holistic view and an integrating glue to bind together the different pieces of architecture. To formalize it Geo-ICDT components, i.e. monitoring, management and adaptation, are further classified into subclasses: sensing, standards, management and dissemination (Fig. 5.1).

5.2.1 Sensing

Sensing infrastructure serves as the backbone of any Geo-ICDT system. *Wireless sensor networks* help in collecting interrupted observations for resource of interest (resource that is monitored). They are also a cost-effective way of generating data related to observed phenomenon. Incorporating *remote/satellite-based sensing* is important for Geo-ICDTs as sensor networks have limited coverage and cannot scale to large areas due to increased cost (cost of sensors, systems and maintenance). Remote sensing methods help in improving scale (spatial extent) and frequency (temporal) of observations through satellite-based products. However, based on area of interest, WSNs can provide data at a finer temporal resolution as compared to remote sensing-based observations. *Crowdsourcing platforms* target end-user or community-based information assimilation through multimodal systems (Minet et al. 2017). Participatory sensing is also important as it directly involves the stakeholders and end-users in the process of data generation. For example, monitoring of events like flood or drought, both wireless sensor networks and satellite-based sensing help the decision-makers to make informed decisions. Local information can also be collected through crowdsourcing platforms like mobile-based, geotagged pictures of waterlogged areas, broken or damaged road networks, etc.

5.2.2 Standard and Processing

Sensing component of Geo-ICDT architecture does not understand (or rather is not designed to) the underlying interrelation between observations and hence cannot generate meaningful information by itself. Phenomenon modelling (data-based models, process-based models, etc.) helps in deriving meaningful information from observations. *Data standards* ensure the quality of data that is being collected through sensing platforms. These standardized datasets are important for seamless working of data mining and exploratory models (as used above) to generate unbiased results while capturing the variability in observed phenomenon. Apart from quality-based standards, Geo-ICDTs also need to define standard for collection, storing and retrieving data (Sun et al. 2017). This standardization facilitates data exchange between interoperable systems by predefined data representation constraints. Such geospatial standards are proposed by organizations such as OGC, to globalize the exchange of geospatial data, ensuring worldwide interoperability. Further, with enhancement in sensing platforms and inherent complexity, amount of data generated by these systems becomes overwhelming. Thus, it is important to identify suitable *computational platforms* that can handle this data stream and efficiently process this big data. Finally, it seems evident that sensing and standards/processing are crucial for *monitoring the resources.*

5.2.3 Management

Managing natural resources is a non-trivial task. Efficient management practices evolve from a combination of long-term observations, data-science-based processing and phenomenon modelling. Defining management component for any Geo-ICDT system is tricky due to innumerable application areas and viable management practices. In the similar context, Geo-ICDT *decision support systems* (DSS) are built upon data collected from sensing systems, processing and standardization. Lindblom et al. (2017) have very diligently noted the relevance of DSS in promoting sustainable agriculture. These DSS offer a unique interface for informed decision-making, while the decision choices are heavily reinforced with supporting data and informatics. Apart from this, *detection and prediction* of resources/processes/systems under different environmental or operational conditions is crucial for efficient management practices as it provides completeness to data-based predictions by providing future trends. *Identification and management* forms the final block for management that makes provisions for actual ground validation and management of predictions/problems identified by Geo-ICDTs. Consequently, standards/processing and management aid in *managing the systems.*

5.2.4 Dissemination

One important service of Geo-ICDTs is to disseminate the information generated by ICDTs systems. *Periodic subscription* provides periodic updates to end-users for ahead-of-time precautionary measures. *Advisory services* provide expert advisory service for end-users when the information from ICDTs is more scientific in nature and requires domain knowledge. *Multimodal platforms* like mobile, the Internet, app-based services, etc. provide a means of disseminating the information and advisory from ICDTs to end-users (Sawant et al. 2016). Management and dissemination combined Geo-ICDTs help end-users to "Adapt to new methods (more scientifically)".

5.3 Geo-ICDT in Agriculture: The MMA Approach

This section formally introduces the application of Geo-ICDT systems in agricultural domain. As discussed in previous sections, developments in Geo-ICDT system have significantly contributed to several fields such as hydrology, geology, remote sensing, climate change, urban studies, atmospheric sciences, to name a few. One such domain where Geo-ICDTs have made a massive contribution is agriculture. In the last few decades, conventional agriculture is rapidly being replaced by more advanced and technology-driven precision agriculture (PA) practices. In precision agriculture, farming practices are governed by location-specific observations, rapid modeling capabilities and variable rate application (VRA) with what-when-where methodology. These advancements and availability of rich information allows PA systems to capture the intra-field variability. Figure 5.2 shows a conceptual outline of systems, methods, services and advisory that constitutes a Geo-ICDT system in PA. It shows an overview of components and services for a multi-institutional and interdisciplinary project called GrIDSense (groundwater, irrigation and disease sensing), which employs ICT in water and pest/disease management for yield improvement in horticulture (citrus), funded from the Department of Electronics and IT, Government of India.

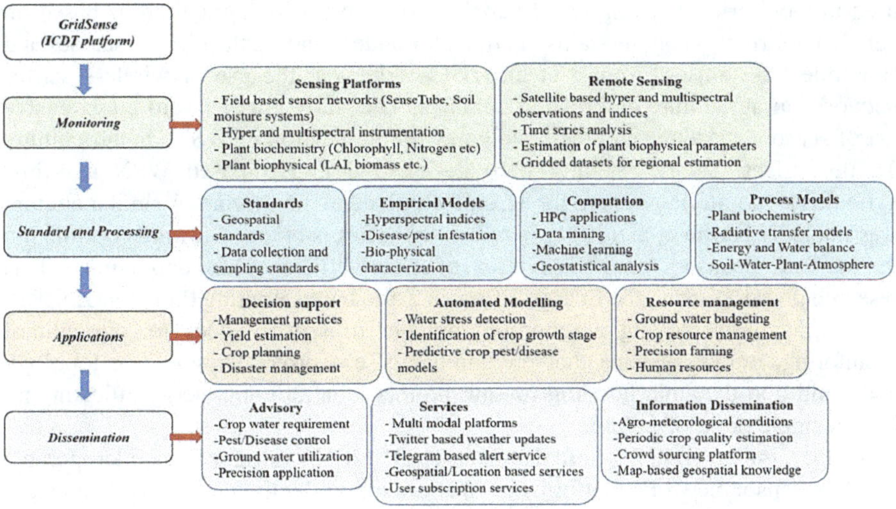

Fig. 5.2 GrIDSense as Geo-ICDT platform

5.3.1 *Sensing*

Sensing, standards, and processing together functionalize this component and serve as the backbone of any Geo-ICDT system. This section lays down some notable and prominent literature that formalizes the importance of monitoring in the agricultural domain.

5.3.1.1 Wireless Sensor Networks (WSNs)

Wireless sensor networks are undoubtedly one of the most prominent examples of ICDT systems. With several flavours such as underground autonomous, underwater, terrestrial sensor networks, WSNs provide a robust and economical solution for data generation and collection in any application domain. WSNs generally provide point-based observations for a resource of interest (or application/phenomenon), which are usually not scalable. However, they are a cost-effective way of generating data in a distributed way (Jain et al. 2018).

For more than a decade, precision agriculture (PA) has become one of the most significant domains for implementation of Geo-ICT. The focus has however been catapulted towards improvisations in the sustainability and productivity of different crops with the development of cost-effective, energy-efficient and fault-tolerant wireless sensor networks (WSNs). In agriculture, WSN applications traverse from crop/pest disease monitoring (Riquelme et al. 2009; Tripathy et al. 2011) to crop water management (Sudharsan et al. 2012; Ojha et al. 2015) to weather-based advisory services (IMD 2016; ISRO 2015; mKRISHI 2016). WSNs are extensively used in monitoring environmental variables since precision agriculture is based on detailed information on the status of crops, for understanding the plant processes and subsequent responses. Abbasi et al. (2014) reviewed the use of wireless sensor networks in agriculture and related domains. The study puts forward an extensive list of sensors, systems, and methodologies for application of WSNs in agriculture. To this effect, Sahitya et al. (2016) designed a ZigBee-based WSN real-time agricultural monitoring system for agrometeorological monitoring. A similar attempt was made by Karim et al. (2011) in which the effort is oriented towards optimizing the number of sensors required (in terms of energy-efficiency) for deployment. This research proposes using the Energy-Efficient Zone-based Routing Protocol (EEZRP) that works by deploying a minimum number of sensor nodes for agricultural monitoring, unlike systems that predominantly use protocols, e.g. tree topology and multi-hop dynamic flooding routing protocol that are not energy-efficient and have a single point of failure.

In the similar context, Suradhaniwar et al. (2016) designed an interoperable wireless sensor network platform for spatio-temporal estimation of soil moisture and temperature using open-source hardware and software platforms. This highlights the fact that though there could be a plethora of WSN implementation worldwide, a WSN's suitability and feasibility is determined based on how energy-efficiently do the nodes communicate and disseminate data. A prominent attempt is made in this regard by Sawant et al. (2017) to address the standardization of data discovery, access and sharing aspects of a WSN, which otherwise impedes the integration of data across various distributed sensors. The study adopts a framework based on Open Geospatial Consortium (OGC) standards for sensor web enablement (SWE), which includes sensor plug and play, remote monitoring, tools for crop water requirement estimation, pest, disease monitoring and nutrient management functionalities (Fig. 5.3).

5.3.1.2 Remote-/Satellite-Based Sensing

Sensor networks usually have a limited coverage due to increased cost (cost of sensors, systems, and maintenance) associated with dense sensor deployment. In the context of agriculture, both WSN and remote sensing methods have their own advantages and disadvantages. WSN can provide point-based observations at very high temporal resolution, while they lack coverage. Due to this, WSNs are usually employed for cross-validating or as ground validation site for remote sensing products. Remote sensing serves with global data (both raw/processed) for applications such as retrieval of biophysical parameters for vegetation (Kar et al. 2017), change detection, soil moisture-analysis-based studies, phenology detection of both forests and agricultural crops, environmental monitoring, etc. One of the most important aspects of remote sensing-based agriculture monitoring is numerous vegetation indices that are derived from remote sensing data (Thenkabail et al. 2013; Huete et al. 2002; Baret and Guyot 1991; Candiago et al. 2015; Balzarolo et al. 2016; Gitelson 2016). To aid this, a satellite-based drone monitoring toolkit has

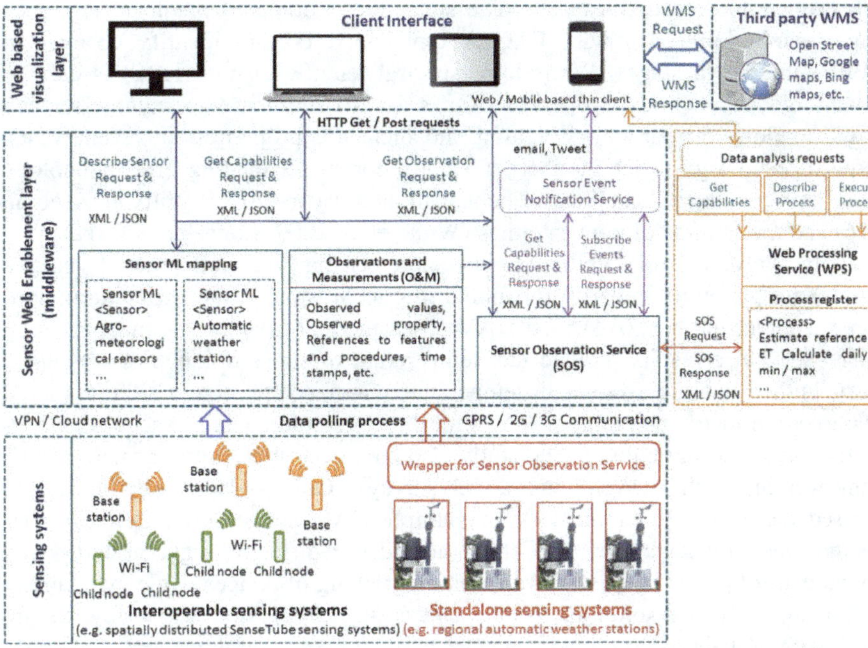

Fig. 5.3 SenseTube (Sawant et al. 2015): architecture for interoperable agrometeorological observation and analysis platform. (Adapted from Sawant et al. 2017)

been developed by Wardlow et al. (2016) for remote sensing of water resources, disasters, and urban studies.

The US Drought Monitoring (USDM) system incorporates several tools to get several drought indices values and other in situ measurements. Such studies are primarily done using NOAA-AVHRR, Landsat satellites, Moderate Resolution Imaging Spectroradiometer (MODIS), etc. which provide reliable global coverage. There is an immense amount of information about water cycle variables including soil moisture and groundwater retrieved using gravity field observations from NASA's Gravity Recovery and Climate Exchange (GRACE). Khanal et al. (2017) have diligently elaborated the applications and implications of thermal remote sensing in precision agriculture while trying to standardize the workflow to monitor soil and crop using thermal remote sensing. There are also several data-fusion-based studies reported which exploit multiple satellite sensor data for a common purpose. Since understanding plant evapotranspiration (at varying temporal scales) is of paramount importance, Semmens et al. (2016) have multi-scale, multi-sensor fusion methodology to estimate vineyard evapotranspiration. The study has reportedly combined data from GOES, MODIS, and Landsat-8 to model surface energy fluxes with respect to the ground flux measurements to identify the correspondence of spatial distribution of ET with yield estimates. Sawant et al. (2016) used time-series analysis to estimate crop water requirement (CWR) and crop growth stage for citrus crop using Landsat archives data. The study then compared predicted reference evapotranspiration (ET_0) with FAO recommended ET_0 to identify changes in CWR with growth stages. While hyperspectral remote sensing (HRS) is used to derive target biochemical properties (Kar et al. 2016), microwave remote sensing (MRS) exploits the target's structural and dielectric properties. In recent years, there's been a significant number of studies for understanding crop phenology using high-resolution satellites (Thenkabail and Lyon 2016; Willkomm et al. 2016) and medium-resolution satellites (White et al. 2009; Lopez et al. 2012).

As an alternative for satellite-based remote sensing, unmanned aerial vehicles (UAVs) are gaining popularity for monitoring local-/small-scale agricultural landscapes. Bellvert et al. (2014) used UAV-based thermal imaging for mapping crop water stress. The results showed that temperature of stressed canopies was up to 7.5 °C higher than the average air temperature. Grenzdörffer et al. (2008) explored the photogrammetric potential of UAVs for direct georeferencing of acquired imagery for forest and agriculture. The study also compared the utility of micro-UAVs (light weight) with conventional aerial surveys. Gómez-Candón et al. (2014) assessed the accuracy for mosaicking multiple UAV images used for early site-specific weed management. The study considered different flight altitudes and ground control points (GCPs) for efficient mosaicking of images while maintaining ortho-image spatial resolution. Honkavaara et al. (2012) used a new low-weight UAV with a Fabry-Perot interferometer-based hyperspectral camera and high-resolution small-format consumer camera to generate stereoscopic 2-D images for agriculture management. The study also discussed processing of high-density point clouds and hyperspectral reflectance image mosaics (reflectance signatures), which are further used as inputs in agricultural application.

5.3.1.3 Crowdsourcing Platforms

Participatory sensing (PS) or crowdsourcing is one of the fastest growing ways of collecting geotagged data through end-users. Combining with the ever-expanding market penetration of mobile phones and availability of on-board sensors (such as GPS, geotagging, accelerometer, gyroscope, etc.) are driving the sensing community towards a new paradigm called *participatory sensing* in which local citizens voluntarily collect and share information from their local environment simply using personal mobile devices.

In the context of precision agriculture, participatory sensing has shown unprecedented results. Using PS, farm owners can easily report any disease/pest infestations, nutrient deficiencies, etc. to agricultural experts using simple mobile interface. With geotagging, experts can exactly locate a tree/farm for further visits if required. With similar model, Mohite et al. (2015) presented a unique rural participatory sensing with a rewarding mechanism for crop monitoring (RuPS). The proposed application allows farm owners to map their respective fields with additional information such as crop type, soil condition, nutrient application schedules, intercropping patterns, etc. It also allows farmers to capture and tag pictures of any disease or pest infestation. Participatory sensing does have a fallback associated with it. In lack of any rewarding mechanism for contributors, quality and quantity of data fall rapidly. To counter this, Jaimes et al. (2012) proposed a recurrent reverse auction scheme that selects a representative subset of users based on a given location and a fixed budget. Van de et al. (2015) utilized the potential of PS for testing the climatic adaptability of different varieties of sorghum and cowpea in sub-Saharan Africa. They selected a large group of farmers to test and evaluate up to 20 varieties of cowpea with morphological evaluation data and dissimilar climatic conditions. PS from farmers helped the researchers to rapidly and economically collect data from different locations.

5.3.2 Standard and Processing

This subsection puts forwards importance of standards and processing in the agricultural domain. It also briefly describes each component of sensing and processing element of MMA architecture.

5.3.2.1 Phenomenon Modelling

Sensing infrastructure of Geo-ICDT systems is designed with the sole purpose of collecting data. These systems do not understand the underlying interrelation between observations and targeted physical process and hence cannot generate meaningful information from data. Due to this, phenomenon modelling (data-based models, process-based models, etc.) becomes an important part of Geo-ICDT architecture. It helps to derive meaningful information from data while using physical processes.

However, phenomenon modelling in agriculture is difficult to comprehend. Crops or plants are physically complex structures that involve a thorough understanding of crop biology, physiology, bio-chemistry, phenology, genotypic information etc. Apart from this, crops or in general, agriculture has deep-rooted nexus with other global and local systems such as local hydrology, soil dynamics, climate variability, pest/disease infestations, etc. Accurate modelling of plant dynamics also depends on management practices and recommendations. In the context of agriculture, Jones et al. (2017) put forward an excellent collection of agricultural systems, data models, and knowledge products for state-of-the-art agricultural science and systems. In a similar study, Makela et al. (2000) put forward a collection of process-based models for forest ecosystems and current challenges for practical implementation. In one study related to climate dynamics in agriculture, Storkey et al. (2014) studied a process-based approach for predicting the effect of climate change in the distribution of invasive allergenic plant in Europe. Rosenzweig et al. (2014) put forward an analysis of agricultural risks with climate change using a global gridded crop model (GGCMs) intercomparison. The study compared seven GGCMs, five global climate models and four representative concentration pathways for accessing the results of climate change on agriculture. Ramirez-Villegas et al. (2013) presented a similar analysis using the EcoCrop model and a case study of grain sorghum for assessing the impact of climate change. To increase crop production under climate variability, Campbell et al. (2014) proposed a climate-smart agriculture framework to include the aspects of climate resilient agriculture.

One of the most prominent works in the field of modelling leaf optical behaviour was done by Jacquemoud and Baret (1990). The study proposed a plat model for leaf structure and studied changes in optical properties of these layers across the electromagnetic spectrum. This work laid a path for several studies (Feret et al. 2008; Zarco-Tejada et al. 2001; Ceccato et al. 2001; Dai et al. 2004; Guan et al. 2016) in estimating plant biophysical characterization. A similar study was done by Verhoef (1984) that introduced another important optical (radiative transfer) model to study light scattering by leaf layers and applied it as a canopy reflectance model.

Evapotranspiration (ET) is undoubtedly one of the key aspects of modelling and unravelling plant behaviour. Hargreaves and Samani (1982) first proposed a method for estimating potential evapotranspiration in plants, popularly called "Hargreaves method" and was named after the author. Allen et al. (1998) published a landmark study on crop evapotranspiration using energy balance equation and guidelines for computing crop water requirement. After several years, this remains a de facto benchmark for comparing crop water requirement and ET estimation using remote sensing techniques.

In the context of phenomenon modelling in agriculture, it is important to understand the importance of crop simulation models (CSM). CSMs can be broadly classified into predictive (or descriptive) and explanatory (Palosuo et al. 2011; Cervigni et al. 2013; Di Paola et al. 2016) domains. In case of predictive models, the aim is to provide crop yield predictions under environmental conditions, thereby enabling it with vital information extraction functionality for the decision-makers (in terms of agricultural yield, production scenarios, and management options). In the second case, models are built and constantly developed to help researchers to understand one or more specific plant environment interactions or a specific eco-physiological process as well, e.g. photosynthesis or leaf area expansion. Such types of models are generally mechanistic, i.e. are focused on one or more specific system's component(s) usually detailed with mathematical equations. Popular models such as DSSAT (Jones et al. 2003), APSIM (Keating et al. 2003), hybrid maize (Yang et al. 2004), AquaCrop (Raes et al. 2009), etc. are site-specific models and are driven by specific environmental variables. Drewry et al. (2010a) addressed the role of structural and ecophysiological properties of soybean and maize crop. The study presented a comprehensive model that coupled different mass, energy and momentum equations using mass and energy conservation principle for vertical distribution of canopy. The model incorporated both C3 and C4 photosynthetic pathways and resolves the vertical radiation, thermal and environmental regimes within the canopy. In extension to this study, Drewry et al. (2010b) proposed use of free air carbon enrichment (FACE) to address the roles of ecophysiological, biochemical and structural plant acclimation with respect to canopy-scale exchange of

CO_2, water vapour and energy through the application of a multilayer canopy-root-soil model (MLCan) capable of resolving changes induced by elevated CO_2 through the canopy and soil systems.

5.3.2.2 Data Standards

Reliability of data and models largely depends on the quality of data that has been collected. Due to this, it is important to define a set of standards that regularize the process of data collection and analyses. In this context, agriculture is no different. With numerous parameters that can favourably or adversely affect the intended agricultural output, standardization is important. Standards help in maintaining the quality of data that is being collected through sensing platforms. As discussed before, geospatial standards aid exploring and exchange of geospatial data. This also helps in achieving interoperability between heterogeneous systems. Botts and Robin (2007) proposed one of the prominent works for enabling interoperability in ICDT systems by defining a comprehensive set of rules for connecting the ICDT systems and web services using sensor web enablement (SWE) framework. The study also introduced the notion of SensorML that standardized the way sensors and systems are defined in an interoperable environment. A study by Nash et al. (2009) provided a review on applications of geospatial web services in precision agriculture. The work puts forward use of OGC standards and use-cases that demonstrate how standards can improve interoperability between systems. Sawant et al. (2016) proposed an interoperable agrometeorological observations and analysis platform for precision agriculture with crop water requirement as case study. The study proposed a novel sensing platform with standardization using OGC sensor web enablement (SWE). It also proposed several additional functionalities of ICDTs such as plug-and-play interface, remote monitoring, crop water estimation, pest/disease monitoring, etc.

Precision agriculture being a coalition of several distributed platforms, it is important to device a standard for defining this heterogeneity. In connection to this, Maurya and Jain (2016) proposed a fuzzy-based sensor network protocol for precision agriculture for efficient management of sensor nodes. Murakami et al. (2007) proposed an infrastructure for development of distributed services for precision agriculture using XML and service-oriented architecture. In similar context, Schuster et al. (2011) proposed an XML-based auxiliary language with semantic interoperability for machine-to-machine communications in agricultural systems. With innovations in geo-coded linked data, semantic interoperability in agricultural systems led way for methods to encode domain knowledge into ontology models (Chen et al. 2003; Fileto et al. 2003). Going a step ahead, Bonacin et al. (2016) developed ontology models for analysis of agricultural and climate changes in water resources using interoperable models. In a similar effort, Hu et al. (2010) proposed ontology model AgOnt, which incorporated agricultural terminologies for Internet-of-things (IoT) methodology. Lauser et al. (2006) illustrated conversion of traditional thesaurus in agriculture (AGROVOC) to a new system, the agricultural ontology service concept server (AOS/CS). This conversion helped in developing a multilingual repository of concepts in agricultural domain providing ontological relationships and a rich, semantic terminology. Apart from embedding semantic knowledge for agricultural domain, it is also important to share the information generated in standardized form. Han et al. (2012) developed a web service-based application for exploring and dissemination of US conterminous geospatial cropland data products for decision support. Montanarella et al. (2010) proposed a method for sharing soil information at global scale by standardizing data format and enable interoperability between different Geo-ICDT systems that shared location-specific soil information. While contributing to sharing and generation of soil-specific information, Du et al. (2016) developed online soil moisture retrieval and sharing system using geospatial web-enabled services. The work laid forward use of Beidou navigation satellite system-reflectometry (BDS-R) that can be extended to global navigation satellite system (GNSS). From the above examples, it's evident that standards and implementations ensure seamless working of Geo-ICDT system across applications and across regional boundaries while maintaining the data accuracy.

5.3.2.3 Computational Platforms

With the evolution of sensing platforms amount of data generated in precision agriculture has become overwhelming. In modern-day agriculture, remotely sensed data (satellite remote sensing) constitutes the largest fraction of humongous data repositories that are used for processing, analysis, and modelling. Even simple farm processes such as GPS-based farm operations, variable rate applicators, sprinklers and irrigation systems, precision harvesting, crop-specific pest control, etc. require tremendous data processing and analysis to enable automation in agriculture. Apart from this, modelling of complex processes such as evapotranspiration, soil water balance, crop genotype and phenology (high-throughput phenotyping), soil-water-atmosphere-plant (SWAP) interactions, hyperspectral and multispectral analyses, plant biophysical characterization, image processing for plant health, etc. require tremendous computing capabilities. In addition to this, storage and querying of the huge data stream generated from the different system are also challenging. Computing infrastructure in precision agriculture can be realized using cloud computing, processing as a service, local or remote computation resources, on-board computing with advanced agricultural machinery, etc.

In precision agriculture capturing the variability at regional and local scales remains a computationally significant task. Studies involving remote sensing data extraction at different spatial and temporal scales become infeasible without high-performance computing (HPC). Zhou et al. (2016) showed the importance of HPC-based cloud computing by retrieving soil moisture from inversion of remote sensing data along the Yangtze River, China. The issues with large-scale RS-data-based processing, predictions and modelling spike up with large collection of data-dependent small tasks with ordering constraint. Ma et al. (2015) discussed challenges and opportunities in processing of remote sensing-based big-data analyses. The study also underlines the challenges in accessing, downloading, storing and processing large-scale global high-resolution data for different applications. Host et al. (2008) used HPC for simulating aboveground interactions of individual trees in forest patches. The study used functional-structural plant model that integrates detailed physiological and phenological processes to simulate the growth of individual plants. Kalluri et al. (2010) used HPC for characterizing land surface anisotropy using global-scale advanced very high-resolution radiometer (AVHRR) data. The study used three different models, simple linear, semiempirical and temporal, to estimate bidirectional distribution function (BRDF) for different land cover types. To facilitate the RS-based computations, Van Den et al. (2012) proposed a platform for time-series analysis of RS satellite imagery using HPC. The study also employed model-based change detection using MODIS time series. Leroy et al. (2012) designed TriAnnot, a high-performance-based computing platform for automated annotation of plant genomes. The modular architecture of TriAnnot allows annotation and masking of transposable elements, structural and functional annotation of protein-coding genes with an evidence-based quality indexing. Zhao et al. (2013) employed hybrid approach by combining grid and cloud computing to analyse wheat production, soil carbon and nitrogen dynamics in Australia's cropping regions at high resolution.

5.3.3 *Management*

5.3.3.1 Decision Support Systems

As seen in previous sections, sensing generates data, processing generates predictions and uses data for associated models. In continuation to this, decision support systems (DSS) help in making informed decisions (based on data and processing in previous steps). DSS assists the decision-maker in making informed decisions (for potential problems) without exposing him/her to the underlying complexity of ICDT systems. To aid the process of developing DSS, it is important to understand the issues, methods and tools for environmental decision support systems (EDSS) and practices that can benefit the farming community (Kropff et al. 2001; Fountas et al. 2006; Matthies et al. 2007; Adinarayana et al. 2015). The decision support system for agrotechnology transfer (DSSAT) (Jame and Cutforth 1996; Jones et al. 2003; Hoogenboom et al. 2015) allows estimation of production, resource use and risk associated with different crop production practices. It also allows users to combine the technical knowledge contained in crop growth models with economic considerations and environmental impact. DSSAT integrates crop simulation models, databases for soil, weather and crops and an interactive interface for users to generate different simulation scenarios for decision-making. Carberry et al. (2002) designed a participatory platform and DSS called FARMSCAPE (farmers', advisers', researchers', monitoring, simulation, communication and performance evaluation) for farming community in Australia. The application was intended to benefit from tools such as soil characterization and sampling, climate and weather forecast models and crop simulation modelling. In the similar context, Cain et al. (2003) designed a participatory DSS for agriculture using Bayesian approach for Deduru Oya river basin in Sri Lanka. Rao et al. (2007) outlined the design and development of a web-GIS DSS to support the Conservation Reserve Program (CRP) by US Department of Agriculture (USDA). The system integrates automated feature information retrieval system (AFIRS) and SWAT tools (soil and water assessment tool). Mirschel et al. (2016) designed LandCaRe-DSS that combines model-based interactive spatial information and decision support system. It also provides multi-model simulations at the regional level by using complex long-term impacts of climate change. Apart from this, it also suggests sustainable agricultural adaptation strategies such as crop rotation, soil tillage, fertilization, irrigation, price and cost changes, etc. at the local or farm level. Adinarayana et al. (2008) developed a prototype distributed collaboration tool, called GramyaVikas to assist the rural extension community in decision-making (such as crop statistics, priority watershed rehabilitation, land use/land cover for terrestrial forestry, etc.) in more interactive, integrated and coordinated manner. The tool integrated spatial and nonspatial information based in Geo-ICDTs and provided an interactive map-based interface for information access. Navarro et al. (2016) designed a smart DSS for irrigation management based on crop, climatic variables, soil and human expertise. Mahaman et al. (2003) developed diagnostic advisory rule-based expert system for integrated pest management (DIARES-IPM) in *solanaceous* crops. The tool supported automated identification of insects, diseases, nutritional deficiencies and beneficial insects for *solanaceous* crop and suggested appropriate treatments. Herwitz et al. (2004) designed an agricultural surveillance and decision support system using imagery from unmanned aerial vehicles (UAVs).

5.3.3.2 Detection and Prediction

Management of resources depends on accurately detecting, precisely predicting and segregating the anomalies from data- and process-driven models. It provides completeness to data-based predictions as done in *sensing* and *processing* and thus provides a holistic view of the problem (not as data crunching). With quality data processing, detection of changes in observed phenomenon and prediction of future trends become achievable. In this context, Tripathy et al. (2014) designed a multimodel data mining and knowledge discovery platform for understanding leaf spot dynamics in groundnut plant. Sudharsan et al. (2013) identified and compared two crop simulation models for different varieties of rice crop under semiarid conditions. The study used and fine-tuned the model parameters for rice production under semiarid regions of India. Badnakhe et al. (2015) used leaf optical models and spectroscopic analysis for identification of disease stress in the citrus plant. In general, detection and prediction of agroclimatic changes with respect to change in plant biophysical/chemical changes, climate resiliency, pest/disease management, etc. are not independent of models and systems discussed in previous sections. Huang and Asner (2009) employed remote sensing for monitoring spread of invasive plant species using spatial, temporal and spectral properties. Almost all models (empirical, process driven, exploratory, predictive, simulation) consider detection and prediction of important model parameters (through multi-model sensing, remote sensing, historical data, gridded and regional data sets, etc.) and perform a sensitivity analysis of variables to account for variability.

5.3.3.3 Identification and Management

DSS systems help in making informed decisions based on data (Lindblom et al. 2017). Detection and prediction help in observing and predicting the important variables. Finally, it is important to identify (e.g. identifying affected areas in flood or drought) and manage (e.g. providing citizens with alternative route maps during flooding) the systems and resources. Again for most models and systems, outcomes of those are evidently related to management practices for spatially non-varying agriculture. However, some models (mostly process-driven) provide a location-independent analysis for an associated agricultural domain.

5.3.4 Dissemination

5.3.4.1 Periodic Subscription

Geo-ICDT and its components, i.e. sensing, standards and processing and management, generate meaningful information using exemplified mining of data. However, it is important to disseminate this information to inculcate adaptability among end-users. As discussed in previous sections, most ICDT systems and models provide an easy-to-access interface for regular data dissemination. These systems use various platforms and services such as cloud-based data portals (Minet et al. 2017), web data repositories, on-demand data processing portals, multi-level scientific data product repositories, periodic and historical data storage and analysis, time-series estimations, regional and field-based variability analysis, impact assessments of local and global climate variability, etc. End-users can subscribe to these services for effective management decision-making. (e.g. subscribing to weather services for agricultural managing).

5.3.4.2 Advisory Services

In agricultural domain, apart from regular/periodic subscription services that link the subscribers to information, end-users may also require expert advisory services. These services are generally problem-specific and require domain experts/advisory for suggestions on associated management practices. Several countries across the globe have a special advisory committee and government organizations that issue crop-specific advisory services. However, these recommendations are usually broad in nature and rarely fill in the requirements of precision agriculture. Due to this, one important responsibility of Geo-ICDT systems is to generate precise and location-specific advisory for stakeholders. Chipeta (2006) discussed the importance of demand-driven advisory services for agriculture. The study presents an extension framework for common framework on agricultural extension of the Neuchatel initiative. Birner et al. (2009) have put forward a conceptual framework and methods for pluralistic agricultural advisory services, which also proposes the use of "best-fit" approaches in contrary to the "ideological" take on agricultural advisory.

5.3.4.3 Multimodal Platforms

One important aim of ICDT dissemination systems is to make subscription and advisory services easily accessible. In modern day, multi-model platforms such as mobiles and tablets help end-users easily access and connect to the Internet. This plays a crucial role in timely and periodic access to information generated from ICDT systems. With availability of social media platforms such as WhatsApp, Facebook, Twitter, etc., important information can easily reach a large group of end-users, stakeholders and policymakers. Apart from this, ICDT systems are generally accompanied with easy-to-use graphical user interface (GUI) that can be used to access the required information. In a similar effort, Sawant et al. (2016) used OGC event notification services for generating notifications and alerts for citrus cultivators. They also developed a map-based web interface using web map services for disseminating location-based information. Talking of social media platforms, Cho and Park (2011) analysed the Twitter activity of Ministry for Food, Agriculture, Forestry and Fisheries (MFAFF), a government body of Korean government. They employed social network analyst and content analysts to understand the use of Twitter as information distribution platform. For involvement of social media in agriculture, Rhoades and Aue (2010) coined the term "social agriculture" and discussed the adaptability of social media by agriculture educators and broadcasters. However, the study suggested that most organizations involved in agricultural information dissemination are website based and rarely use micro-blogging platforms. Zhang et al. (2016) studied the growth and adaptation of agricultural ICDTs in China. The study also discusses the use of different information dissemination models, viz. web base, voice-based services, text (SMS) based, interactive video learning and conferencing, etc. in agricultural domain. Finally, it can be said that though management practices and decision-making are crucial, timely dissemination of associated information is also equally important for adaptation of these management decisions/policies. Thus, ICDTs present new methods and systems for solving conventional problems in resource management. To motivate the community to adopt new (and more scientific methods) methods, regular information dissemination must be adopted.

5.4 Conclusions

The chapter briefly discusses the importance of Geo-ICDTs in various application domains and emphasizes the integration of geospatial technologies with ICDT systems in lieu of generating unique and novel solutions to real-world problems. It also provides an overview of principles and concepts of Geo-ICDTs with the backdrop of technological developments that laid the path for development of modern-day geo-Internet of things (Geo-IoT). It also introduces a comprehensive and detailed MMA methodology for monitoring, managing and steering adaptation for new state-of-the-art technologies with special importance to geographical attribution to resources. Later, the chapter presents a detailed discussion on the utility of Geo-ICDT and MMA architecture in the agricultural domain. This work also elaborates a compilation of notable studies and research articles that reinforce the usability of Geo-ICDTs (and MMA model) in precision agriculture. However, it is evident and reasonable that this work comprehends only a handful of studies and merely scratches the surface of possibilities and innovations that can be incorporated into the architecture.

References

Abbasi AZ, Islam N, Shaikh ZA (2014) A review of wireless sensors and networks' applications in agriculture. Comp Stand Interf 36(2):263–270

Adinarayana J, Azmi S, Tewari G, Sudharsan D (2008) GramyaVikas: a distributed collaboration model for rural development planning. Comput Electron Agric 62(2):128–140

Adinarayana J, Sudharsan D, Ninomiya S, Hirafuji M and Kiura T (2015) GeoSense: Decision Support System for Precision Faming in India. Wulfania Multi-disciplinary Journal, 2(2):22–47

Allen RG, Pereira LS, Raes D, Smith M (1998) Crop evapotranspiration-Guidelines for computing crop water requirements. FAO Irrigation and drainage paper 56. FAO, Rome, 300(9), D05109

Badnakhe MR, Durbha S, Adinarayana J (2015) Disease stress detection on citrus using a leaf optical model and field spectroscopy. In: SPIE remote sensing. International Society for Optics and Photonics, pp 963709–963709

Balzarolo M, Vicca S, Nguy-Robertson AL, Bonal D, Elbers JA, Fu YH, Grünwald T, Horemans JA, Papale D, Peñuelas J, Suyker A (2016) Matching the phenology of net ecosystem exchange and vegetation indices estimated with MODIS and FLUXNET in-situ observations. Remote Sens Environ 174:290–300

Baret F, Guyot G (1991) Potentials and limits of vegetation indices for LAI and APAR assessment. Remote Sens Environ 35(2–3):161–173

Bellvert J, Zarco-Tejada PJ, Girona J, Fereres E (2014) Mapping crop water stress index in a 'Pinot-noir' vineyard: comparing ground measurements with thermal remote sensing imagery from an unmanned aerial vehicle. Precis Agric 15(4):361–376

Birner R, Davis K, Pender J, Nkonya E, Anandajayasekeram P, Ekboir J, Cohen M (2009) From best practice to best fit: a framework for designing and analyzing pluralistic agricultural advisory services worldwide. J Agric Educ Ext 15(4):341–355

Bonacin R, Nabuco OF, Junior IP (2016) Ontology models of the impacts of agriculture and climate changes on water resources: scenarios on interoperability and information recovery. Futur Gener Comput Syst 54:423–434

Botts M, Robin A (2007) OpenGIS sensor model language (SensorML) implementation specification. OpenGIS Implement Spec OGC 7(000)

Cain JD, Jinapala K, Makin IW, Somaratna PG, Ariyaratna BR, Perera LR (2003) Participatory decision support for agricultural management. A case study from Sri Lanka. Agric Syst 76(2): 457–482

Campbell BM, Thornton P, Zougmoré R, Van Asten P, Lipper L (2014) Sustainable intensification: what is its role in climate smart agriculture? Curr Opin Environ Sustain 8:39–43

Candiago S, Remondino F, De Giglio M, Dubbini M, Gattelli M (2015) Evaluating multispectral images and vegetation indices for precision farming applications from UAV images. Remote Sens 7(4):4026–4047

Carberry PS, Hochman Z, McCown RL, Dalgliesh NP, Foale MA, Poulton PL, Hargreaves JNG, Hargreaves DMG, Cawthray S, Hillcoat N, Robertson MJ (2002) The FARMSCAPE approach to decision support: farmers', advisers', researchers' monitoring, simulation, communication and performance evaluation. Agric Syst 74(1):141–177

Ceccato P, Flasse S, Tarantola S, Jacquemoud S, Grégoire JM (2001) Detecting vegetation leaf water content using reflectance in the optical domain. Remote Sens Environ 77(1):22–33

Cervigni R, Valentini R, Santini M (eds) (2013) Toward climate-resilient development in Nigeria. World Bank Publications

Chen G, Lu RQ, Jin Z (2003) Constructing virtual domain ontologies based on domain knowledge reuse. J Softw 14(3):350–355

Chipeta S (2006) Demand driven agricultural advisory services, Green booklet series Neuchatel initiative

Cho SE, Park HW (2011) Government organizations' innovative use of the internet: the case of the twitter activity of South Korea's Ministry for Food, Agriculture, Forestry and Fisheries. Scientometrics 90(1):9–23

Dai Y, Dickinson RE, Wang YP (2004) A two-big-leaf model for canopy temperature, photosynthesis, and stomatal conductance. J Clim 17(12):2281–2299

Di Paola A, Valentini R, Santini M (2016) An overview of available crop growth and yield models for studies and assessments in agricultusre. J Sci Food Agric 96(3):709–714

Drewry DT, Kumar P, Long S, Bernacchi C, Liang XZ, Sivapalan M (2010a) Ecohydrological responses of dense canopies to environmental variability: 2. Role of acclimation under elevated CO2. J Geophys Res Biogeosci 115(G4)

Drewry DT, Kumar P, Long S, Bernacchi C, Liang XZ, Sivapalan M (2010b) Ecohydrological responses of dense canopies to environmental variability: 1. Interplay between vertical structure and photosynthetic pathway. J Geophys Res Biogeosci 115(G4)

Du W, Chen N, Yan S (2016) Online soil moisture retrieval and sharing using geospatial web-enabled BDS-R service. Comput Electron Agric 121:354–367

Feret JB, François C, Asner GP, Gitelson AA, Martin RE, Bidel LP, Ustin SL, Le Maire G, Jacquemoud S (2008) PROSPECT-4 and 5: advances in the leaf optical properties model separating photosynthetic pigments. Remote Sens Environ 112(6):3030–3043

Fileto R, Liu L, Pu C, Assad ED, Medeiros CB (2003) POESIA: an ontological workflow approach for composing web services in agriculture. VLDB J Int J Very Large Data Bases 12(4):352–367

Fountas S, Wulfsohn D, Blackmore BS, Jacobsen HL, Pedersen SM (2006) A model of decision-making and information flows for information-intensive agriculture. Agric Syst 87(2):192–210

Gitelson AA (2016) 15 remote sensing estimation of crop biophysical characteristics at various scales. Hyperspectral remote sensing of vegetation, p 329

Gómez-Candón D, De Castro AI, López-Granados F (2014) Assessing the accuracy of mosaics from unmanned aerial vehicle (UAV) imagery for precision agriculture purposes in wheat. Precis Agric 15(1):44–56

Grenzdörffer GJ, Engel A, Teichert B (2008) The photogrammetric potential of low-cost UAVs in forestry and agriculture. Int Arch Photogramm Remote Sens Spat Inf Sci 31(B3):1207–1214

Guan K, Berry JA, Zhang Y, Joiner J, Guanter L, Badgley G, Lobell DB (2016) Improving the monitoring of crop productivity using spaceborne solar-induced fluorescence. Glob Chang Biol 22(2):716–726

Han W, Yang Z, Di L, Mueller R (2012) CropScape: a web service based application for exploring and disseminating US conterminous geospatial cropland data products for decision support. Comput Electron Agric 84:111–123

Hargreaves GH, Samani ZA (1982) Estimating potential evapotranspiration. J Irrig Drain Div 108(3):225–230

Herwitz SR, Johnson LF, Dunagan SE, Higgins RG, Sullivan DV, Zheng J, Lobitz BM, Leung JG, Gallmeyer BA, Aoyagi M, Slye RE (2004) Imaging from an unmanned aerial vehicle: agricultural surveillance and decision support. Comput Electron Agric 44(1):49–61

Honkavaara E, Kaivosoja J, Mäkynen J, Pellikka I, Pesonen L, Saari H, Salo H, Hakala T, Marklelin L, Rosnell T (2012) Hyperspectral reflectance signatures and point clouds for precision agriculture by light weight UAV imaging system. ISPRS Ann Photogramm Remote Sens Spat Inform Sci I-7:353–358

Hoogenboom G, Jones JW, Wilkens PW, Porter CH, Boote KJ, Hunt LA, Singh U, Lizaso JL, White JW, Uryasev O, Ogoshi R (2015) Decision support system for agrotechnology transfer (DSSAT). Version 4.6. 1.0 (www. DSSAT. net). DSSAT Foundation, Prosser, Washington

Host GE, Stech HW, Lenz KE, Roskoski K, Mather R (2008) Forest patch modeling: using high performance computing to simulate aboveground interactions among individual trees. Funct Plant Biol 35(10):976–987

Hu S, Wang H, She C, Wang J (2010, October) AgOnt: ontology for agriculture internet of things. In: International conference on computer and computing technologies in agriculture. Springer, Berlin/Heidelberg, pp 131–137

Huang CY, Asner GP (2009) Applications of remote sensing to alien invasive plant studies. Sensors 9(6):4869–4889

Huete A, Didan K, Miura T, Rodriguez EP, Gao X, Ferreira LG (2002) Overview of the radiometric and biophysical performance of the MODIS vegetation indices. Remote Sens Environ 83(1): 195–213

IMD (2016) Automatic Weather Station (AWS) Lab. India MeteorologicalDepartment, Pune. http://www.imdaws.com/ViewAwsData.aspx. Accessed on 12 Jan 2016

ISRO (2015) Bhuvan Indian geo-platform of Indian Space Research Organization (ISRO), National Remote Sensing Centre (NRSC). http://bhuvan.nrsc.gov.in/bhuvan. Accessed on 26 Aug 2017

Jacquemoud S, Baret F (1990) PROSPECT: a model of leaf optical properties spectra. Remote Sens Environ 34(2):75–91

Jaimes LG, Vergara-Laurens I, Labrador MA (2012, March) A location-based incentive mechanism for participatory sensing systems with budget constraints. In: Proceedings of IEEE international conference on 'Pervasive Computing and Communications (PerCom), pp 103–108

Jain N, Madan S, Malik SK (2018) Review of multiparameter techniques for precision agriculture using wireless sensor network. J Network Commun Emerg Technol (JNCET) 8(2). www.jncet.org

Jame YW, Cutforth HW (1996) Crop growth models for decision support systems. Can J Plant Sci 76(1):9–19

Janssen SJ, Porter CH, Moore AD, Athanasiadis IN, Foster I, Jones JW, Antle JM (2017) Towards a new generation of agricultural system data, models and knowledge products: information and communication technology. Agric Syst 155:200–212

Jones JW, Hoogenboom G, Porter CH, Boote KJ, Batchelor WD, Hunt LA, Wilkens PW, Singh U, Gijsman AJ, Ritchie JT (2003) The DSSAT cropping system model. Eur J Agron 18(3): 235–265

Jones JW, Antle JM, Basso B, Boote KJ, Conant RT, Foster I, Keating BA (2017) Toward a new generation of agricultural system data, models, and knowledge products: state of agricultural systems science. Agric Syst 155:269–288

Kalluri SNV, Zhang Z, JaJa J, Liang S, Townshend JRG (2010) Characterizing land surface anisotropy from AVHRR data at a global scale using high performance computing. Int J Remote Sens 22(11):2171–2191

Kar S, Rathore VS, Sharma R, Swain SK (2016) Classification of river water pollution using hyperion data. J Hydrol 537:221–233

Kar S, Mandal D, Bhattacharya A, Adinarayana J (2017, July) Temporal analysis of Touzi parameters for wheat crop characterization using L-band AgriSAR 2006 data. In: Geoscience and Remote Sensing symposium (IGARSS), 2017 I.E. International, pp 3909–3912. IEEE

Karim L, Nasser N, El Salti T (2011, March) Efficient zone-based routing protocol of sensor network in agriculture monitoring systems. In: International conference on communications and information technology (ICCIT), 2011, pp 167–170. IEEE

Keating BA, Carberry PS, Hammer GL, Probert ME, Robertson MJ, Holzworth D, Huth NI, Hargreaves JN, Meinke H, Hochman Z, McLean G (2003) An overview of APSIM, a model designed for farming systems simulation. Eur J Agron 18(3):267–288

Khanal S, Fulton J, Shearer S (2017) An overview of current and potential applications of thermal remote sensing in precision agriculture. Comput Electron Agric 139:22–32

Kropff MJ, Bouma J, Jones JW (2001) Systems approaches for the design of sustainable agro-ecosystems. Agric Syst 70(2):369–393

Lauser B, Sini M, Liang A, Keizer J, Katz S (2006) From AGROVOC to the agricultural ontology service/concept server. An OWL model for creating ontologies in the agricultural domain. In: Dublin core conference proceedings. Dublin Core DCMI

Leroy P, Guilhot N, Sakai H, Bernard A, Choulet F, Theil S, Reboux S, Amano N, Flutre T, Pelegrin C, Ohyanagi H (2012) TriAnnot: a versatile and high performance pipeline for the automated annotation of plant genomes. Front Plant Sci 3

Lindblom J, Lundström C, Ljung M, Jonsson A (2017) Promoting sustainable intensification in precision agriculture: review of decision support systems development and strategies. Precis Agric 18(3):309–331

Lopez-Sanchez JM, Cloude SR, Ballester-Berman JD (2012) Rice phenology monitoring by means of SAR polarimetry at X-band. IEEE Trans Geosci Remote Sens 50(7):2695–2709

Ma Y, Wu H, Wang L, Huang B, Ranjan R, Zomaya A, Jie W (2015) Remote sensing big data computing: challenges and opportunities. Futur Gener Comput Syst 51:47–60

Mahaman BD, Passam HC, Sideridis AB, Yialouris CP (2003) DIARES-IPM: a diagnostic advisory rule-based expert system for integrated pest management in Solanaceous crop systems. Agric Syst 76(3):1119–1135

Mäkelä A, Landsberg J, Ek AR, Burk TE, Ter-Mikaelian M, Ågren GI, Oliver CD, Puttonen P (2000) Process-based models for forest ecosystem management: current state of the art and challenges for practical implementation. Tree Physiol 20(5–6):289–298

Matthies M, Giupponi C, Ostendorf B (2007) Environmental decision support systems: current issues, methods and tools. Environ Model Softw 22(2):123–127

Maurya S, Jain VK (2016) Fuzzy based energy efficient sensor network protocol for precision agriculture. Comput Electron Agric 130:20–37

Minet J, Curnel Y, Gobin A, Goffart JP, Mélard F, Tychon B, Wellens J, Defourny P (2017) Crowdsourcing for agricultural applications: a review of uses and opportunities for a farmsourcing approach. Comput Electron Agric 142:126–138

Mirschel W, Wenkel KO, Berg M, Wieland R, Nendel C, Köstner B, Topazh AG, Terleev VV, Badenko VL (2016) A spatial model-based decision support system for evaluating agricultural landscapes under the aspect of climate change. In: Novel methods for monitoring and managing land and water resources in Siberia. Springer, Cham, pp 519–540

mKRISHI (2016) Mobile based agro advisory system. TCS Innovation Labs Mumbai, Tata Consultancy Services http://tcsinnovations.com. Accessed on 25 Jan 2016

Mohite J, Karale Y, Gupta P, Kulkarni S, Jagyasi B, Zape A (2015, March) RuPS: rural participatory sensing with rewarding mechanisms for crop monitoring. In: Pervasive computing and communication workshops (PerCom Workshops), 2015, IEEE International Conference. IEEE, pp 378–383

Montanarella L, Wilson P, Cox S, McBratney A, Ahamed S, MacMillan R, Jacquier D, Fortner J (2010, May) Developing SoilML as a global standard for the collation and transfer of soil data and information. In: Geophysical research abstracts, vol 12

Murakami E, Saraiva AM, Ribeiro LC, Cugnasca CE, Hirakawa AR, Correa PL (2007) An infrastructure for the development of distributed service-oriented information systems for precision agriculture. Comput Electron Agric 58(1):37–48

Nash E, Korduan P, Bill R (2009) Applications of open geospatial web services in precision agriculture: a review. Precis Agric 10(6):546–560

Navarro-Hellín H, Martínez-del-Rincon J, Domingo-Miguel R, Soto-Valles F, Torres-Sánchez R (2016) A decision support system for managing irrigation in agriculture. Comput Electron Agric 124:121–131

Ojha T, Misra S, Raghuwanshi NS (2015) Wireless sensor networks for agriculture: the state-of-the-art in practice and future challenges. Comput Electron Agric 118:66–84

Palosuo T, Kersebaum KC, Angulo C, Hlavinka P, Moriondo M, Olesen JE, Patil RH, Ruget F, Rumbaur C, Takáč J, Trnka M (2011) Simulation of winter wheat yield and its variability in different climates of Europe: a comparison of eight crop growth models. Eur J Agron 35(3):103–114

Raes D, Steduto P, Hsiao TC, Fereres E (2009) AquaCrop the FAO crop model to simulate yield response to water: II. Main algorithms and software description. Agron J 101(3):438–447

Ramirez-Villegas J, Jarvis A, Läderach P (2013) Empirical approaches for assessing impacts of climate change on agriculture: the EcoCrop model and a case study with grain sorghum. Agric For Meteorol 170:67–78

Rao M, Fan G, Thomas J, Cherian G, Chudiwale V, Awawdeh M (2007) A web-based GIS decision support system for managing and planning USDA's conservation reserve program (CRP). Environ Model Softw 22(9):1270–1280

Rhoades E, Aue K (2010, February) Social agriculture: adoption of social media by agricultural editors and broadcasters. In: Proceedings of the annual meeting of the southern association of agricultural scientists, agricultural communication section, Orlando, FL

Riquelme JL, Soto F, Suardíaz J, Sánchez P, Iborra A, Vera JA (2009) Wireless sensor networks for precision horticulture in southern Spain. Comput Electron Agric 68(1):25–35

Rosenzweig C, Elliott J, Deryng D, Ruane AC, Müller C, Arneth A, Boote KJ, Folberth C, Glotter M, Khabarov N, Neumann K (2014) Assessing agricultural risks of climate change in the 21st century in a global gridded crop model intercomparison. Proc Natl Acad Sci 111(9):3268–3273

Sahitya G, Balaji N, Naidu CD (2016, July) Wireless sensor network for smart agriculture. In: Proceedings of IEEE-2016 2nd international conference on applied and theoretical computing and communication technology (iCATccT), pp 488–493

Sawant SA, Durbha SS, Adinarayana J (2015) SenseTube: a syntactically standardized web enabled wireless sensor network system for precision agriculture applications, Indian patent file number: 2236/MUM/2015 (status: patent pending)

Sawant SA, Chakraborty M, Suradhaniwar S, Adinarayana J, Durbha SS (2016) Time series analysis of remote sensing observations for citrus crop growth stage and evapotranspiration estimation. Int Arch Photogramm Remote Sens Spat Inf Sci 41:147–152

Sawant S, Durbha SS, Adinarayana J (2017) Interoperable agro-meteorological observation and analysis platform for precision agriculture: a case study in citrus crop water requirement estimation. Comput Electron Agric 138:175–187

Schuster EW, Lee HG, Ehsani R, Allen SJ, Rogers JS (2011) Machine-to-machine communication for agricultural systems: an XML-based auxiliary language to enhance semantic interoperability. Comput Electron Agric 78(2):150–161

Semmens KA, Anderson MC, Kustas WP, Gao F, Alfieri JG, McKee L, Prueger JH, Hain CR, Cammalleri C, Yang Y, Xia T (2016) Monitoring daily evapotranspiration over two California vineyards using Landsat 8 in a multi-sensor data fusion approach. Remote Sens Environ 185:155–170

Storkey J, Stratonovitch P, Chapman DS, Vidotto F, Semenov MA (2014) A process-based approach to predicting the effect of climate change on the distribution of an invasive allergenic plant in Europe. PLoS One 9(2):e88156

Sudharsan D, Adinarayana J, Tripathy AK, Ninomiya S, Hirafuji M, Kiura T, Desai UB, Merchant SN, Reddy DR, Sreenivas G (2012) *GeoSense*: a multimode information and communication system. ISRN Sens Network 2012:1

Sudharsan D, Adinarayana J, Reddy DR, Sreenivas G, Ninomiya S, Hirafuji M, Kiura T, Tanaka K, Desai UB, Merchant SN (2013) Evaluation of weather-based rice yield models in India. Int J Biometeorol 57(1):107–123

Sun Z, Di L, Zhang C, Fang H, Yu E, Lin L, . . . Guo L (2017, August) Building robust geospatial web services for agricultural information extraction and sharing. In Agro-Geoinformatics, 2017, 6th international conference on IEEE, pp 1–4

Suradhaniwar S, Sawant SA, Badnakhe M, Durbha SS, Adinarayana J (2016) An interoperable wireless sensor network platform for spatio-temporal soil moisture and soil temperature estimation. In: Agro-Geoinformatics (Agro-Geoinformatics), 2016 fifth international conference on IEEE, pp 1–6

Thenkabail PS, Lyon JG (eds) (2016) Hyperspectral remote sensing of vegetation. CRC Press, Boca Raton

Thenkabail PS, Mariotto I, Gumma MK, Middleton EM, Landis DR, Huemmrich KF (2013) Selection of hyperspectral narrowbands (HNBs) and composition of hyperspectral twoband vegetation indices (HVIs) for biophysical characterization and discrimination of crop types using field reflectance and Hyperion/EO-1 data. IEEE J Select Topic Appl Earth Observ Remote Sens 6(2):427–439

Tripathy AK, Adinarayana J, Sudharsan D, Merchant SN, Desai UB, Vijayalakshmi K, Reddy DR, Sreenivas G, Ninomiya S, Hirafuji M, Kiura T (2011, December) Data mining and wireless sensor network for agriculture pest/disease predictions. In: Information and communication technologies (WICT), 2011 World Congress on. IEEE, pp 1229–1234

Tripathy AK, Adinarayana J, Vijayalakshmi K, Merchant SN, Desai UB, Ninomiya S, Hirafuji M, Kiura T (2014) Knowledge discovery and leaf spot dynamics of groundnut crop through wireless sensor network and data mining techniques. Comput Electron Agric 107:104–114

Ueyama J, Faiçal BS, Mano LY, Bayer G, Pessin G, Gomes PH (2017) Enhancing reliability in wireless sensor networks for adaptive river monitoring systems: reflections on their long-term deployment in Brazil. Comput Environ Urban Syst 65:41–52

Van de Gevel JMJ, Kiambi D, Fadda C (2015) Using a crowdsourcing approach to test sorghum and cowpea varieties for climate adaptability. Procedia Environ Sci 29:243–244

Van Den Bergh F, Wessels KJ, Miteff S, Van Zyl TL, Gazendam AD, Bachoo AK (2012) HiTempo: a platform for time-series analysis of remote-sensing satellite data in a high-performance computing environment. Int J Remote Sens 33(15):4720–4740

Verhoef W (1984) Light scattering by leaf layers with application to canopy reflectance modeling: the SAIL model. Remote Sens Environ 16(2):125–141

Wardlow B, Anderson T, Tadesse C, Hain W, Rodell M, Thenkabail PS (2016) Remote sensing of drought: emergence of a satellite-based monitoring toolkit for the United States. In: Remote sensing of water resources, disasters, and urban studies. CRC Press, pp 367–398

White MA, Beurs D, Kirsten M, Didan K, Inouye DW, Richardson AD, Jensen OP, O'keefe JOHN, Zhang G, Nemani RR, Leeuwen V (2009) Intercomparison, interpretation, and assessment of spring phenology in North America estimated from remote sensing for 1982–2006. Glob Chang Biol 15(10):2335–2359

Willkomm M, Bolten A, Bareth G (2016) Non-destructive monitoring of rice by hyperspectral in-field spectrometry and uav-based remote sensing: case study of field-grown rice in North Rhine-Westphalia, Germany. Int Arch Photogramm Remote Sens Spat Inform Sci 41

Yang HS, Dobermann A, Lindquist JL, Walters DT, Arkebauer TJ, Cassman KG (2004) Hybrid-maize—a maize simulation model that combines two crop modeling approaches. Field Crop Res 87(2):131–154

Zarco-Tejada PJ, Miller JR, Noland TL, Mohammed GH, Sampson PH (2001) Scaling-up and model inversion methods with narrowband optical indices for chlorophyll content estimation in closed forest canopies with hyperspectral data. IEEE Trans Geosci Remote Sens 39(7): 1491–1507

Zhang Y, Wang L, Duan Y (2016) Agricultural information dissemination using ICTs: a review and analysis of information dissemination models in China. Inform Proces Agric 3(1):17–29

Zhao G, Bryan BA, King D, Luo Z, Wang E, Bende-Michl U, Song X, Yu Q (2013) Large-scale, high-resolution agricultural systems modeling using a hybrid approach combining grid computing and parallel processing. Environ Model Softw 41:231–238

Zhou L, Chen N, Chen Z, Xing C (2016) ROSCC: an efficient remote sensing observation-sharing method based on cloud computing for soil moisture mapping in precision agriculture. IEEE J Select Topics Appl Earth Observ Remote Sens 9(12):5588–5598

Chapter 6
Digital Image Processing: Principles and Applications

G. P. Obi Reddy

Abstract Digital image processing is an important part in digital analysis of remote sensing data. It allows one to enhance image features of interest while attenuating irrelevant features of a given application and then extract useful information about the scene from the enhanced image. It comprises the four basic steps, which include image correction/restoration, image enhancement, image transformation, and image classification. Image restoration is basically aimed to compensate the data errors, noise, and geometric distortions introduced during the scanning, recording, and playback operations. Image enhancement helps to alter the visual impact that the image has on the interpreter, that improve the information content and information extraction ability by utilizing the decision-making capability of the computer in order to recognize and classify the pixels on the basis of their digital signatures. The information extracted by comparing two or more images of an area that were acquired at different times. Unsupervised classification distinguish the patterns in the reflectance data and groups them into a pre-defined number of classes without any prior knowledge of the image. Whereas, in supervised classification, the user trains the computer, guiding it what type of spectral characteristics to look for and what type of land cover they represent, and guides the image processing software how to classify certain features. Change-detection analysis provides information about the changes in different seasons or dates. In post-classification analysis, image smoothing and accuracy assessment are important to generate the output from digital image processing.

Keywords Digital image processing · Image enhancement · Image restoration · Image classification

G. P. O. Reddy (✉)
ICAR-National Bureau of Soil Survey & Land Use Planning, Nagpur, India

G. P. O. Reddy, S. K. Singh (eds.), *Geospatial Technologies in Land Resources Mapping, Monitoring and Management*, Geotechnologies and the Environment 21,
https://doi.org/10.1007/978-3-319-78711-4_6

6.1 Introduction

Digital image processing (DIP) involves the modification of digital data for improving the image qualities with the aid of computer. Digital imaging is a process aimed to recognize objects of interest in an image by adopting advanced computing techniques with the aim to improve image quality parameters (Fu 1994; Kulkarni 1994, 2001; Ritter and Wilson 2001). The digital image processing helps in maximizing clarity, sharpness, and details of features of interest toward information extraction and further analysis. The image classification, therefore, forms an important tool for examination of the digital images. Many types of remote sensing images are recorded in digital form and then processed by computers to produce images for interpreters to study. Remote sensing data is represented digitally in the form of matrixes of pixels or cells, where each cell corresponds to a number. The number represents the intensity of reflected or emitted energy on the Earth's surface in the resolution area that the cell represents. These matrices of cells form raster images where the digital number is represented by a brightness level (in a grayscale image) or a color (if a color palette is used).

Digital image classification comprises the four basic steps, namely, image correction/restoration, image enhancement, image transformation, and image classification. *Image correction/restoration* is essential when image data recorded by sensors on a satellite or aircraft contain errors related to geometry and brightness values of the pixels. These errors are corrected using suitable mathematical models, which are either definite or statistical models. *Image enhancement* is the modification of image, by changing the pixel brightness values, to improve its visual impact. Image enhancement techniques are performed by deriving the new brightness value for a pixel either from its existing value or from the brightness values of a set of surrounding pixels. *Image transformation* normally performs on multispectral character of image data to spectrally transform it to a new set of image components or bands with a purpose to get some information more evident or to preserve the essential information content of the image (for a given application), with a reduced number of transformed dimensions. The pixel values of the new components are related to the original set of spectral bands via a linear operation. *Image classification* is the important objective of image classification procedures to automatically categorize all pixels in an image into different land cover classes or themes. A pixel is characterized by its spectral signature, which is determined by the relative reflectance in different wavelength bands. Multispectral classification is an information extraction process that analyzes these spectral signatures and assigns the pixels to classes based on similar signatures.

6.2 Digital Image Processing Techniques

Image processing methods may be grouped into three functional categories: image restoration, image enhancement, and information extraction; these are defined below together with lists of typical processing techniques.

6.2.1 Image Restoration

It is processing of recovering the original image by removing the noise from the image. It compensates for data errors, noise, and geometric distortions introduced during the scanning, recording, and playback operations. Image restoration processes are designed to recognize and compensate for errors, noise, and geometric distortion introduced into the data during the scanning, transmission, and recording processes. The objective is to make the image resemble the original scene. Image restoration is relatively simple because the pixels from each band are processed separately. Image restoration processing refers to operations that are required prior to the main data analysis. The two main types of image restoration processes are *radiometric correction* and *geometric correction.*

6.2.1.1 Radiometric Correction

Radiometric restoration refers to the removal or diminishment of distortions in the degree of electromagnetic energy registered by each detector. A variety of agents can cause distortion in the values recorded for image cells. Radiometric restoration is the removal of distortions in the amount of electromagnetic energy received by the satellite so that the reading reflects the true intensity reflected or emitted by the surface. This kind of correction is needed because of atmospheric attenuation of energy before it reaches the sensor and because of sensor irregularities such as striping, scan line dropping, and random noise. Some of the most common distortions for which correction procedures exist include restoring periodic line dropouts, restoring periodic line striping, filtering of random noise, and correcting for atmospheric scattering.

6.2.1.2 Restoring Periodic Line Dropouts

At times, data from one of the sensors of the satellite are missing because of a recording problem. On the particular scan line is a string of zeros that plots as a black line on the image. These are called periodic line dropouts. The first step in the restoration process is to calculate the average digital number (DN) value per scan line for the entire scene. The average DN value for each scan line is then compared with this scene average. Any scan line deviating from the average by more than a designated threshold value is identified as defective. The next step is to replace the defective lines. For each pixel in a defective line, an average DN is calculated using DNs for the corresponding pixel on the preceding and succeeding scan lines. The average DN is substituted for the defective pixel; the resulting image is a major improvement. This restoration program is equally effective for random line dropouts that do not follow a systematic pattern.

6.2.1.3 Restoring Periodic Line Striping

For each spectral band, the detectors were carefully calibrated and matched before the satellite was launched. With time, however, the response of some detectors may drift to higher or lower levels; as a result every scan line recorded by that detector is brighter or darker than the other lines. The general term for this defect is periodic line striping. Valid data are present in the defective lines but must be corrected to match the overall scene. One restoration method is to plot histograms for the DNs recorded by each detector and compare these with a histogram for the entire scene. For each detector the mean and standard deviation are adjusted to match values for the entire scene. Another restoration method plots a histogram of DNs for each of the detectors. Deviations in mean and median values for the histograms are used to determine corrections for detector differences.

6.2.1.4 Filtering of Random Noise

The periodic line dropouts and striping are forms of nonrandom noise that can be restored by simple means. Random noise, on the other hand, requires more sophisticated restoration method. Random noise occurs as individual pixels with DNs that are much higher or lower than the surrounding pixels. In the image these pixels produce bright and dark spots that mar the image. These spots also interfere with information extraction procedures such as classification. Random-noise pixels may be removed by digital filters. Low-pass filter, moving window operations, and high-pass filters are generally using this purpose.

6.2.1.5 Correcting for Atmospheric Scattering

Atmospheric scattering produces haze, which results in low image contrast. The contrast ratio of an image is improved by correcting for this effect (Campbell, 2002). There are two techniques to determine the correction factor for different bands. The first restoration technique employs an area within the image that has shadows caused by irregular topography. The second restoration technique also requires that the image has some dense shadows or other areas of zero reflectance.

6.2.1.6 Geometric Distortions

In any mapping process, it is essential to ensure that any form of remotely sensed imagery be accurately registered to the proposed map base. During the scanning process, a number of systematic and nonsystematic geometric distortions are introduced into image data. These distortions are corrected during production of the matter images. The geometric corrections must be applied before plotting images. Geometric correction is the removal of distortions in the shape of the image due to sensor-earth geometry variations. For example, the skew correction is needed to account for the fact that the Earth is moving, while the image is being captured. The scanner distortion correction is needed to account for the fact that the instantaneous field of view covers is more territory at the end of scan lines than in the middle. Certain types of these errors are systematic (such as scan skew, mirror-scan velocity variance, panoramic distortion, platform velocity, and perspective geometry), and they can easily be corrected using sensor characteristics and ephemeris data. Registration of image can usually be achieved through the use of a systematic rubber-sheet transformation process that gently warps an image (through the use of polynomial equations) based on the known positions of a set of widely dispersed control points.

On the other hand, unsystematic errors (such as roll, pitch, and yaw of the platform and/or altitude variance) must be corrected by more difficult methods using ground control points (GCPs). Errors due to altitude variance become more dramatic in topographically diverse landscapes. GPCs are references points whose correct coordinates are known and can be matched to points in the distorted image to correct image distortions. The difference in the location between the two points is used to compute the transformation matrix needed to rectify the image. This process is called image to ground geocorrection (if the GPC's are from maps or ground GPS recognizance) or image to image geocorrection (if the GPC's are from a reference digital image). The geocorrection algorithm follows three basic steps. Firstly, the user identifies the X,Y coordinates of several GCPs. Ideally, these points should be spread evenly throughout the image, especially concentrating on the edges. Secondly, the software solves equations that describe the relationship between the two coordinate systems so it can produce an equation for the conversion of the X, Y coordinates from the old reference system to the new one. Thirdly, rubber-sheet transformation is used to convert the image to the new reference system. The process involves both spatial interpolation and brightness interpolation. This process is called rubber-sheet transformation because it can be visualized by the incorrectly oriented image being printed on to a sheet of rubber and then this sheet being laid over the correctly oriented one and stretched and pulled until it lines up (Fig. 6.1).

Georectification, the conversion of the data to real-world coordinates, is also a geometric correction that is essential if one wants to import data layers from other sources to be used with the satellite imagery.

Nonsystematic Distortions Nonsystematic distortions are not constant because they result from variations in the spacecraft attitude, velocity, and altitude and therefore are not predictable. Variations in spacecraft velocity cause distortion in

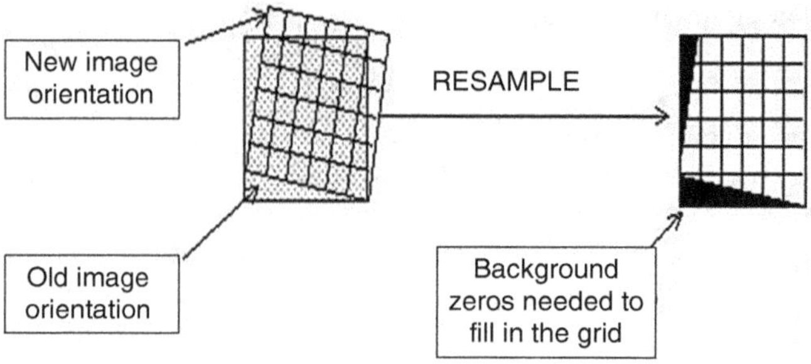

Fig. 6.1 Resampling to a new orientation

the along-track direction only and are known functions of velocity that can be obtained from tracking data. Variations in attitude (roll, pitch, and yaw) and altitude of the spacecraft cause nonsystematic distortions that must be determined for each image in order to be corrected. The correction process employs geographic features on the image, called GCPs, whose positions are known. Differences between actual GCP locations and their positions in the image are used to determine the geometric transformations required to restore the image. The various resampling methods are described by Rifman (1973), Goetz and others (1975), and Bernstein and Ferneyhough (1975).

Systematic Distortions Geometric distortions whose effects are constant and can be predicted in advance are called systematic distortions. Scan skew, cross-track distortion, and variations in scanner mirror velocity belong to this category.

6.2.2 Image Enhancement

Image enhancement is concerned with the modification of images to make them more suited to the capabilities of human vision. It alters the visual impact that the image has on the interpreter in a fashion that improves the information content. Enhancement is the modification of an image to alter its impact on the viewer. Generally enhancement distorts the original digital values; therefore enhancement is not done until the restoration processes are completed. In image enhancement, contrast enhancement; intensity, hue, and saturation transformations; density slicing; edge enhancement; and image transforming are the important techniques.

6.2.2.1 Contrast Enhancement

Image enhancement is a procedure used to improve the appearance of the imagery to make visual interpretation and analysis easier. The most widely used of these is contrast enhancement. Contrast is the range of brightness values present in an image.

These values can theoretically range from 0 to 255, but in many remotely sensed images, the initial range is much less than this. If an image starting with a range from only 40–90 is stretched so that it ranges from 0 to 255, the difference between features becomes accentuated. There are two types of contrast enhancement: linear contrast enhancement and nonlinear contrast enhancement. To produce an image with the optimum contrast ratio, it is important to utilize the entire brightness range of the display medium, which is generally film. The limited range of brightness values accounts for the low contrast ratio of the original image. Three of the most useful methods of contrast enhancement are described in the following sections.

Contrast stretching is the most common image processing routines for improving scene quality. Contrast stretching, which involves altering the distribution and range of digital numbers (DN), is usually the first and commonly a vital step applied to image enhancement. Both casual viewers and experts normally conclude from direct observation that modifying the range of light and dark tones (gray levels) in a photo or a computer display is often the single most informative and revealing operation performed on the scene. Contrast stretching by computer processing of digital numbers (DNs) is a common operation, and the reassignment of DN values is based on the particular stretch algorithm chosen.

The fundamental concepts that underlie how and why contrast stretching is carried out are shown in Fig. 6.2 (Lillesand and Kiefer 1999).

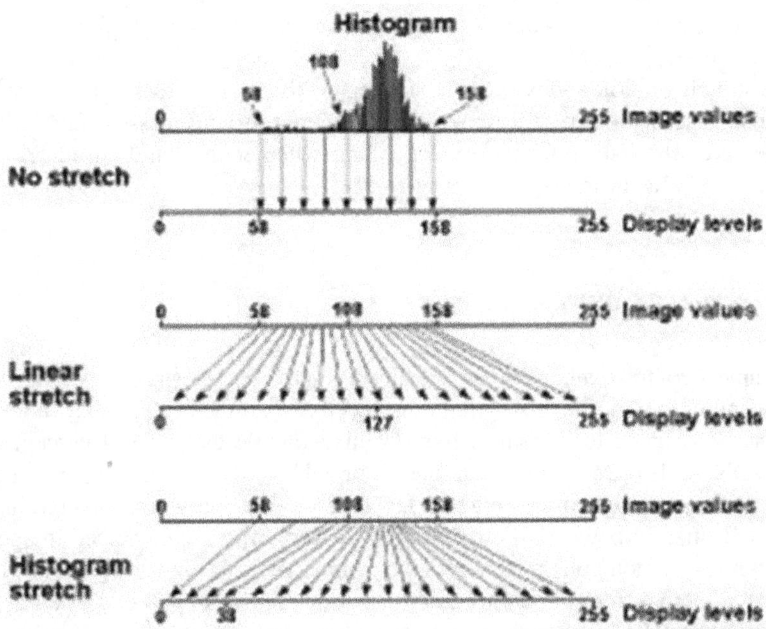

Fig. 6.2 Show the principle of contrast enhancement. Assume an output device capable of displaying 256 gray levels. The histogram shows digital values in the limited range of 58–158. (Source: Lillesand and Kiefer 1999)

In the top plot (a), the DN values range from 60 to 158 (out of the limit available of 0–255). But below 108 there are few pixels, so the effective range is 108–158. When displayed without any expansion (stretch), as shown in plot b, the range of gray levels is mostly confined to 40 DN values, and the resulting image is of low contrast—rather flat. In plot c, a linear stretch involves moving the 60 value to 0 and the 158 DN to 255; all intermediate values are moved (stretched) proportionately. This is the standard linear stretch. But no accounting of the pixel frequency distribution, shown in the histogram, is made in this stretch, so that much of the gray level variation is applied to the scarce to absent pixels with low and high DNs, with the resulting image often not having the best contrast rendition. In d, pixel frequency is considered in assigning stretch values. The 108–158 DN range is given a broad stretch to 38–255, while the values from DN 107 to 60 are spread differently—this is the histogram-equalization stretch. In the bottom example, e (labeled "special stretch"), some specific range, such as the infrequent values between 60 and 92, is independently stretched to bring out contrast gray levels in those image areas that were not specially enhanced in the other stretch types.

Linear expansion of DN's into the full scale (0–255) is a common option. Other stretching functions are available for special purposes. These are mostly nonlinear functions that affect the precise distribution of densities (on film) or gray levels (in a monitor image) in different ways, so that some experimentation may be required to optimize results. Commonly used special stretches include (1) piecewise linear, (2) linear with saturation, (3) logarithmic, (4) exponential, (5) ramp cumulative distribution function, (6) probability distribution function, and (7) sinusoidal linear with saturation.

The stretch version is obviously easier for picking out differences in gray levels that seem to some degree representative of different ground classes (Fig. 6.3). The stretches are variously informative depending whether the bands used are in the visible, the SWIR, or the thermal IR interval.

6.2.3 Linear Contrast Stretch

The simplest contrast enhancement is called a linear contrast stretch. A DN value in the low end of the original histogram is assigned to extreme black, and a value at the high end is assigned to extreme white. In this example the lower 4 percent of the pixels ($DN < 49$) are assigned to black, or $DN = 0$, and the upper 4 percent ($DN > 106$) are assigned to white, or $DN = 255$. The remaining pixel values are distributed linearly between these extremes, as shown in the enhanced image. The linear contrast stretch greatly improves the contrast of most of the original brightness values, but there is a loss of contrast at the extreme high and low end of DN values.

Linear contrast enhancement is also known as contrast stretching. It is best used on remotely sensed images with a Gaussian or near-Gaussian histogram. The simplest form of contrast stretching is the minimum-maximum linear contrast stretch. In this method, the minimum brightness value in the image is replaced

Fig. 6.3 The left image is an unstretched image and the right is a simple stretch

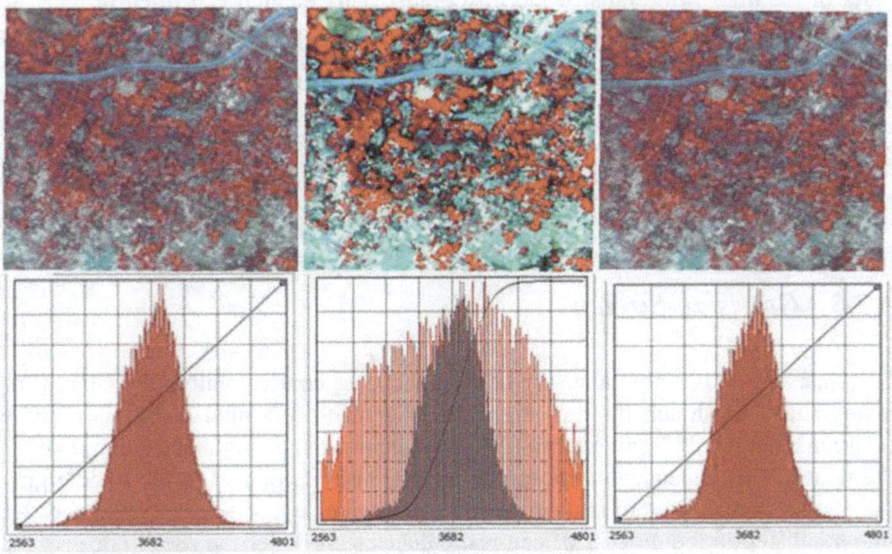

Fig. 6.4 Original image, histogram equalization, and min-max stretch

with 0, the maximum brightness value is replaced with 255, and all the intermediate values are scaled proportionately (Fig. 6.4).

On the other hand, a percentage linear contrast stretch is used; different portions of the histogram can be stretched as needed, instead of the entire thing. This can be used to increase the contrast of an image only in specific portions in the electromagnetic spectrum and is useful if the analyst is interested in seeing more detail in specific feature. Below the first one is original image, the second is a histogram equalization stretch, and the third is a min-max stretch.

6.2.4 Nonlinear Contrast Stretch

In uniform distribution stretch (or histogram equalization), the original histogram has been redistributed to produce a uniform population density of pixels along the horizontal DN axis. This stretch applies the greatest contrast enhancement to the most populated range of brightness values in the original image. One of the most common forms of nonlinear image contrast enhancement is histogram equalization. This method redistributes all the pixel values of an image so that there are approximately an equal number of pixels to each of the user-specified output gray scale classes. This technique increases contrast in the most populated range of brightness values (the peaks of the histogram) and reduces the contrast in the very light or dark parts of the image (the tails of the histogram). This method can often provide an image with the most contrast of any enhancement technique. However, it has one significant drawback. Each value in the input image can end up having several values in the output image, so that objects in the original scene lose their correct relative brightness value. For instance, pixels that are very dark or bright will be changed into a few gray scales. So, histogram equalization is not good if there are clouds in the data or terrain shadows.

6.2.5 Gaussian Stretch

Gaussian stretch is a nonlinear stretch that enhances contrast within the tails of the histogram. This stretch fits the original histogram to a normal distribution curve between the 0 and 255 limits, which improves contrast in the light and dark ranges of the image. This enhancement occurs at the expense of contrast in the middle gray range; the fracture pattern and some of the folds are suppressed in this image. An important step in the process of contrast enhancement is for the user to inspect the original histogram and determine the elements of the scene that are of greatest interest. The user then chooses the optimum stretch for his needs.

6.2.6 Intensity, Hue, and Saturation Transformations

The additive systems of primary colors are red, green, and blue (RGB). An alternate approach to color is the intensity, hue, and saturation system (IHS), which are useful

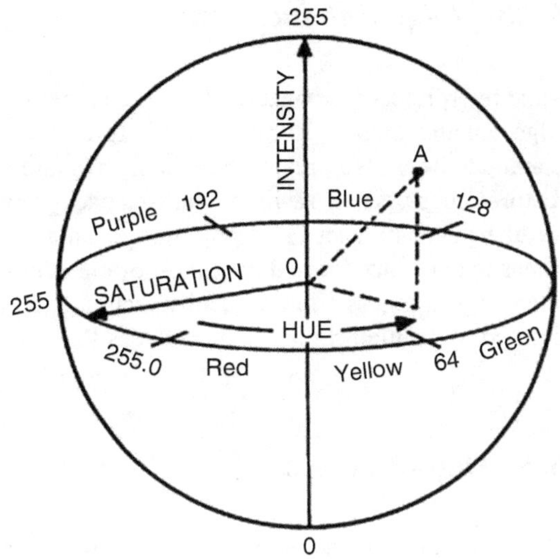

Fig. 6.5 Intensity, hue, and saturation coordination system

because it presents colors more nearly as the human observer perceives them. The IHS system is based on the color sphere in which the vertical axis represents intensity, the radius is saturation, and the circumference is hue. The intensity (I) axis represents brightness variations and ranges from black (0) to white (255); no color is associated with this axis. Hue (H) represents the dominant wavelength of color. Hue values commence with 0 at the midpoint of red tones and increase counterclockwise around the circumference of the sphere to conclude with 255 adjacent to 0. Saturation (S) represents the purity of color and ranges from 0 at the center of the color sphere to 255 at the circumference (Fig. 6.5). A saturation of 0 represents a completely impure color, in which all wavelengths are equally represented and which the eye will perceive a shade of gray that ranges from white to black depending on intensity. Intermediate values of saturation represent pastel shades, whereas, high values represent purer and more intense colors. In Fig. 6.5 the color at point A has the intensity (I) of 195, hue (H) of 75, and saturation (S) of 135.

6.2.7 Density Slicing

Density slicing converts the continuous gray tone of an image into a series of density intervals, or slices, each corresponding to a specified digital range. Digital slices may be displayed as separate colors or as areas bounded by contour lines. This technique emphasizes subtle gray scale differences that may be imperceptible to the viewer.

6.2.8 Edge Enhancement

Most interpreters are concerned with recognizing linear features in images. Through edge enhancement, natural linear features like faults, joints, and lineaments and manmade linear features such as highways and canals can be mapped. Some linear features occur as narrow lines against a background of contrasting brightness; others are the linear contact between adjacent areas of different brightness. In all cases, linear features are formed by edges. Some edges are marked by pronounced differences in brightness and are readily recognized. More typically, however, edges are marked by subtle brightness differences that may be difficult to recognize.

6.3 Digital Filters

By using filtering technique, various types of noise from the image can be removed. Digital filters have been developed specifically to enhance edges in images and fall into two categories: directional and nondirectional (Keys 1981).

6.3.1 Nondirectional Filters

Laplacian filters are nondirectional filters because they enhance linear features having almost any orientation in an image. The exception applies to linear features oriented parallel with the direction of filter movement; these features are not enhanced.

6.3.2 Directional Filters

Directional filters are used to enhance specific linear trends in an image. Haralick (1984) describes the concept of directional filters. A typical filter consists of two kernels, each of which is an array of three-by- three pixels. The filter is demonstrated by applying it to the array of nine pixels over the original data set.

6.3.3 Low-Pass Filters

Image enhancements that de-emphasize or block the high spatial frequency detail are low-frequency or low-pass filters. The simplest low-frequency filter evaluates a particular input pixel brightness value, BVin, and the pixels surrounding the input pixel and outputs a new brightness value, BVout, that is the mean of this convolution. The size of the neighborhood convolution mask or kernel (n) is usually 3×3, 5×5, 7×7, or 9×9. The simple smoothing operation will, however, blur the image, especially at the edges of objects. Blurring becomes more severe as the size of the kernel increases.

6.3.4 High-Pass Filters

High-pass filtering is applied to imagery to remove the slowly varying components and enhance the high-frequency local variations. Brightness values tend to be highly correlated in a nine-element window. Thus, the high-frequency filtered image will have a relatively narrow intensity histogram. This suggests that the output from most high-frequency filtered images must be contrast stretched prior to visual analysis.

6.4 Image Transformation

Image transformation refers to operations that are applied to multiple spectral bands within the image. Basically, it is another term for spectral math. Layers representing different bands can be added, subtracted, multiplied, or divided to produce a new image that will better highlight certain features in the scene. For example, image subtraction can be used to identify changes between data collected on different dates. Image division or spectral ratioing is one of the most common transforms applied, because it helps to highlight variations in the spectral responses of surface covers. For example, vegetation indices are spectral ratios that are used to map the presence of vegetation as well as measuring the amount or condition of vegetation within each pixel. The most common vegetation index is the normalized difference vegetation index (NDVI), which can be calculated according to the formula NDVI = Xnir-Xred/Xnir+Xred. This ratio can range from -1 to +1.

6.4.1 Principal Component Images

Principal components analysis (PCA) has traditionally been used in remote sensing as a means of data compaction. The principal components transformation, originally known as the Karhunen-Loeve transformation (Loeve 1955), is used to compress multispectral data sets by calculating a new coordinate system. For a typical multispectral image band set, it is common to find that the first two or three components are able to explain virtually all of the original variability in reflectance values. Later components thus tend to be dominated by noise effects. By rejecting these later components, the volume of data is reduced with no appreciable loss of information.

PCA is a linear transformation technique related to factor analysis. Given a set of image bands, PCA produces a new set of images, known as components, which are uncorrelated with one another and are ordered in terms of the amount of variance they explain from the original band set. The mathematical operation makes a linear combination of pixel values in the original coordinate system that results in pixel values in the new coordinate system. Any three principal component images can be combined to create a color image by assigning the data that make up each image to a separate primary color. The principal components transformation has several advantages: (1) most of the variance in a multispectral data set is compressed into one or two PC images; (2) noise may be relegated to the less-correlated PC images; and (3) spectral differences between materials may be more apparent in PC images than in individual bands.

PCA is a mathematic technique that uses statistical methods to decorrelate the data and reduce redundancy. PCA is used as method of data compression from number of bands into same number of principle components (PC) with transformation of image information in decreasing order of magnitude such that the first two to three PC contains nearly 95% of input details. Below the plotted TM Bands 1 and 2 are shown in Fig. 6.6.

The slightly scattered data form a narrow plot that is almost a straight line. The two bands are said to be *correlated*, that is, as Band 1 varies, so does Band 2, and either could be used in place of the other. But, there are 7 bands in the TM data set, and some are sufficiently different from others as to behave as though uncorrelated. All available bands participate in the calculations to produce a best FCC, wherein all three PCs are least correlated to each other.

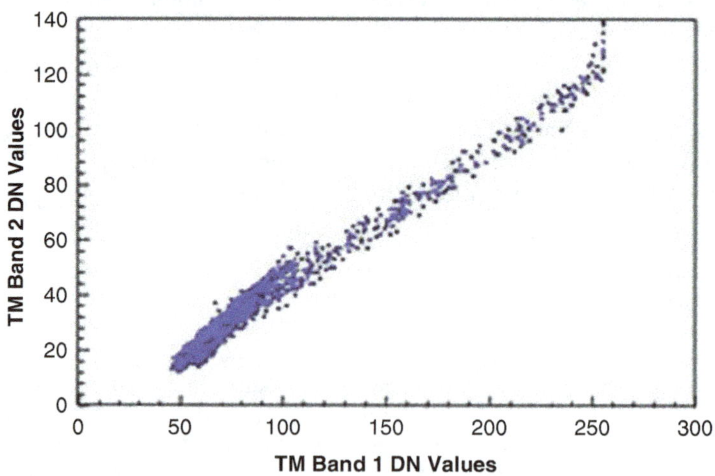

Fig. 6.6 PCA plotted for TM Bands 1 and 2

6.4.2 *Image Ratioing*

Ratio images can be generate by dividing the DN in one band by the corresponding DN in another band for each pixel, stretching the resulting value and plotting the new values as an image. In a ratio image, the black and white extremes of the gray scale represent pixels with the greatest difference in reflectivity between the two spectral bands. The darkest signatures are areas where the denominator of the ratio is greater than the numerator. Conversely the numerator is greater than the denominator for the brightest signatures. Where denominator and numerator are the same, there is no difference between the two bands. In addition to ratios of individual bands, a number of other ratios may be computed. An individual band may be divided by the average for all the bands, resulting in normalized ratios. Another combination is to divide the difference between two bands by their sum, for example, (band 4 - band 5)/(band 4 + band 5). The Landsat 8 OLI image of 30 May 2015 and derived NDVI of part of Indo-Gangetic Plains of India is shown in Fig. 6.7.

6.5 Image Fusion Techniques

Image fusion is the process of combining relevant information from two or more images into a single image. It helps in sharpening the images, enhancing certain features that are not visible in either of the images, replacing the defective data, and complementing the data sets for better decision-making. Most of these satellite

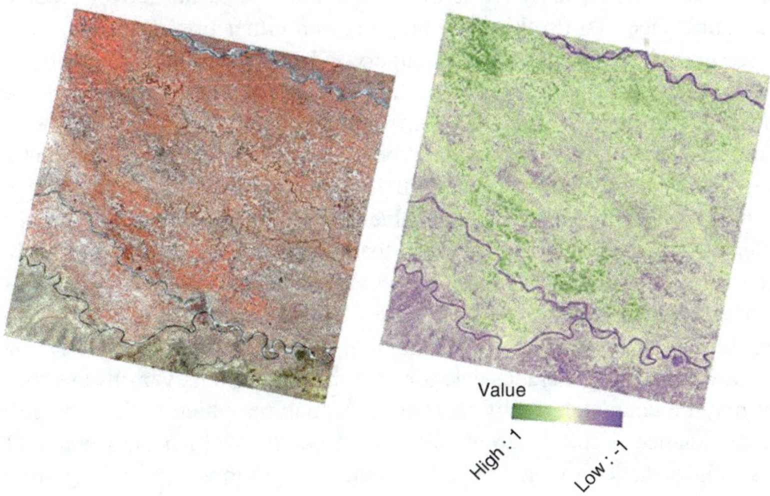

Fig. 6.7 Landsat 8 OLI data (30 May 2015) and derived NDVI of part of Indo-Gangetic Plains of India

sensors operate in two modes: *multispectral* mode and the *panchromatic* mode. The *panchromatic* mode corresponds to the observation over a broad spectral band (similar to a typical black and white photograph), and the *multispectral* (color) mode corresponds to the observation in a number of relatively narrower bands. For example, in the IRS-1D LISS-III operates in the multispectral mode. It records energy in the green (0.52–0.59 μm), red (0.62–0.68 μm), near infrared (0.77–0.86 μm), and mid-infrared (1.55–1.70 μm). In the same satellite PAN operates in the panchromatic mode. This multispectral mode data has a better *spectral resolution* than the panchromatic mode, whereas panchromatic data has better *spatial resolution* than the multispectral data. In image fusion process, user can keep maximum spectral information from the original multispectral image while increasing the spatial resolution from the panchromatic image (Ranchin et al. 2003).

6.6 Information Extraction from Satellite Image

Information extraction from satellite image involves producing principal component images, ratio images, multispectral classification, and change-detection images. Image restoration and enhancement processes utilize computers to provide corrected and improved images for study by human interpreters. The computer makes no decisions in these procedures. However, processes that identify and extract information do utilize the computer's decision-making capability to identify and extract specific pieces of information. A human operator must instruct the computer and must evaluate the significance of the extracted information.

Multispectral image classification is a technique used to generalize remote sensing data, where cells are assigned to one of a number of surface cover groups based on their reflectance. To do this the computer can either use an unsupervised or a supervised classification algorithm. Unsupervised classification distinguishes pattern in the reflectance data and groups them into a pre-defined number of classes without any prior knowledge of the image. Supervised classification is a technique where the user trains the computer, telling it what type of spectral characteristics to look for and what type of land cover they represent. Usually this is done by the user simply specifying a mean and range of digital values for each class, though sometimes more sophisticated statistical methods are used that take into account the prior probability that each class might exist in an area.

In order to make use of the multitude of digital data available from satellite imagery, it must be processed in a manner that is suitable for the end user. For many projects this processing includes categorizing the land into its various use functions. A pixel may be characterized by its spectral signature, which is determined by the relative reflectance in the different wavelength bands. Multispectral image classification is an information extraction process that analyzes these spectral signatures and then assigns pixels to categories based on similar signatures. The two major approaches to multispectral classification are unsupervised classification and supervised classification.

6.6.1 Unsupervised Classification

The goal of unsupervised classification is to automatically segregate pixels of a satellite image into groups of similar spectral character. Classification is done using one of several statistical routines generally called "clustering" where classes of pixels are created based on their shared spectral signatures. Clusters are split and/or merged until further clustering doesn't improve the explanation of the variation in the scene. The success of clustering is measure by the "between cluster" versus "within cluster" variability, maximizing the former and minimizing the latter. The computer separates the pixels into classes with no direction from the analyst.

Unsupervised classification is the identification of natural groups, or structures, within multispectral data by the algorithms programmed into the software. There is no extensive prior knowledge of the region that is required for unsupervised classification unlike supervised classification that requires detailed knowledge of the area. The opportunity for human error is minimized with unsupervised classification because the operator may specify only the number of categories desired and sometimes constraints governing the distinctness and uniformity of groups. Many of the detailed decisions required for supervised classification are not required for unsupervised classification creating less opportunity for the operator to make errors. Unsupervised classification allows unique classes to be recognized as distinct units. Supervised classification may allow these unique classes to go unrecognized and could inadvertently be incorporated into other classes creating error throughout the entire classification.

In unsupervised classification, the software does most of the processing on its own generally resulting in more use categories than the user is interested in. This is the point where the user has to make decisions on which categories can be grouped together into a single land use category. In either case additional image processing may be used to help determine which method is better for a given situation. It must be kept in mind that maps are simple attempts to represent what actually exists in the world and are never completely accurate.

In unsupervised classification any individual pixel is compared to each discrete cluster to see which one it is closest to. A map of all pixels in the image, classified as to which cluster each pixel is most likely to belong, is produced. This then must be interpreted by the user as to what the color patterns may mean in terms of classes; this requires some knowledge of the scene's feature/class/material content from general experience or personal familiarity with the area imaged. Unsupervised classification techniques do not require the user to specify any information about the features contained in the images. The user simply identifies which bands should be used to create the classifications and how many classes to categorize the features. The land use/land cover of part of Indo-Gangetic Palins of India derived from Landsat 8 OLI data using unsupervised classification is shown in Fig. 6.8. In this classification, decisions need to be made concerning which land cover types each

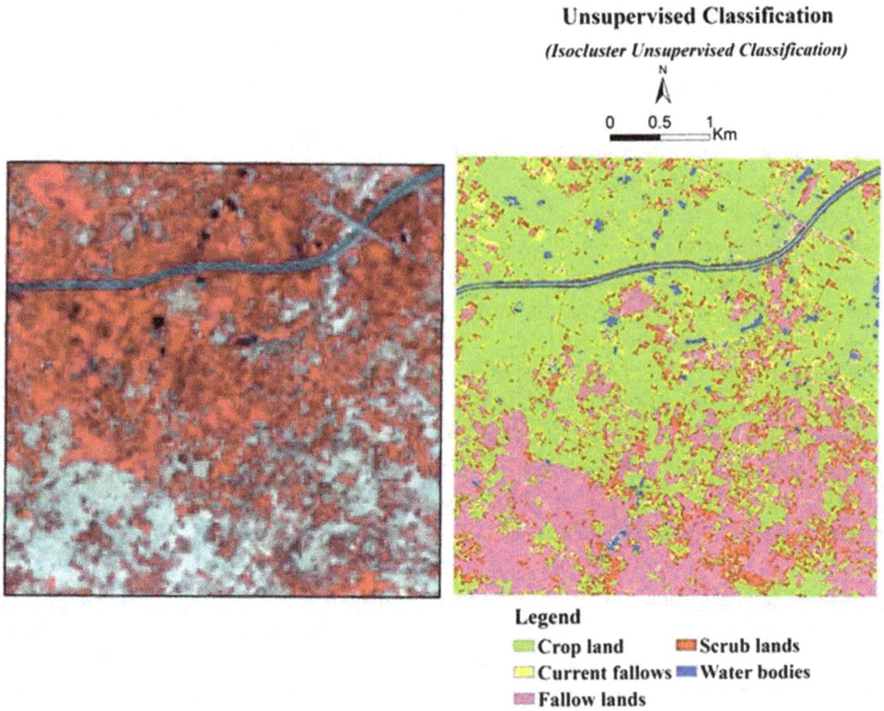

Fig. 6.8 Landsat 8 OLI image and land use/land cover of part of Indo-Gangetic Plains of India derived using unsupervised classification

category falls within. To make these decisions, other materials and knowledge of the area are useful. Ground truthing what is seen in the digital image with what was actually present at the time in the image was recorded makes this task more efficient and more accurate. If this knowledge is not available, scientific reasoning could be used to group the various land use/land cover categories.

6.6.2 *Supervised Classification*

In supervised land classification, the individual processing the imagery guides the image processing software to help it decide how to classify certain features. The analyst defines on the image a small area, called a training site, which is representative of each terrain category or class. Spectral values for each pixel in a training site are used to define the decision space for that class. After the clusters for each training site are defined, the computer then classifies all the remaining pixels in the scene. The first step in the supervised classification is to select *training sites* for each of the terrain categories. Some categories are represented by more than one training site in order to cover the full range of reflectance characteristics. Supervised classification is the process of using samples of known identity to classify pixels of unknown identity.

Supervised classification is tied to specific areas of known identity, provided by selecting training areas. Supervised classification is not faced with the problem of matching spectral categories on the final map with the informational categories of interest. The operator may be able to detect serious errors by examining training data to determine whether they have been correctly classified. In supervised training, it is important to have a set of desired classes in mind and then create the appropriate signatures from the data.

Supervised classification method involves "training" the computer to recognize the spectral characteristics of the features that the user would like to identify on the map (Jensen 1996). Once the user identified the training areas, the user can run the software to put the pixels into one of the feature classes or leave in "unclassified." In supervised classification, spectral signatures are developed from specified locations in the image. These specified locations are given the generic name "training sites" and are defined by the user. Generally a vector layer is digitized over the raster scene. The vector layer consists of various polygons overlaying different land use types. The image below shows the raster image seen earlier with the addition of several training sites outlined on top of it. The training sites help to develop spectral signatures for the outlined areas. The land use categories of interest could be water, crop, grassland, forest, etc.

Supervised classification is much more accurate for mapping classes but depends heavily on the cognition and skills of the image specialist. The strategy is simple: the specialist must recognize conventional classes or meaningful classes in a scene from prior knowledge, such as personal experience with what's present in the scene or, more generally, the region it's located in, by experience with thematic maps or by on-site visits. This familiarity allows the individual(s) making the classification to choose and set up discrete classes and, then, assign them category names. As a rule, the classifying person also locates specific training sites on the image—either a print or a monitor display—to identify the classes. The resulting *training sites* are areas representing each known land cover category that appear fairly homogeneous on the image. In the computer display, one must locate these sites and circumscribe them with polygonal boundaries drawn using the computer mouse. For each class thus outlined, mean values and variances of the DNs for each band used to classify them are calculated from all the pixels enclosed in each site. More than one polygon is usually drawn for any class. The classification program then acts to cluster the data representing each class. When the DNs for a class are plotted as a function of the band sequence, the result is a *spectral signature* or *spectral response* curve for that class. The multiple spectral signatures so obtained are for all of the materials within the site that interact with the incoming radiation. Classification now proceeds by statistical processing in which every pixel is compared with the various signatures and assigned to the class whose signature comes closest. A few pixels in a scene do

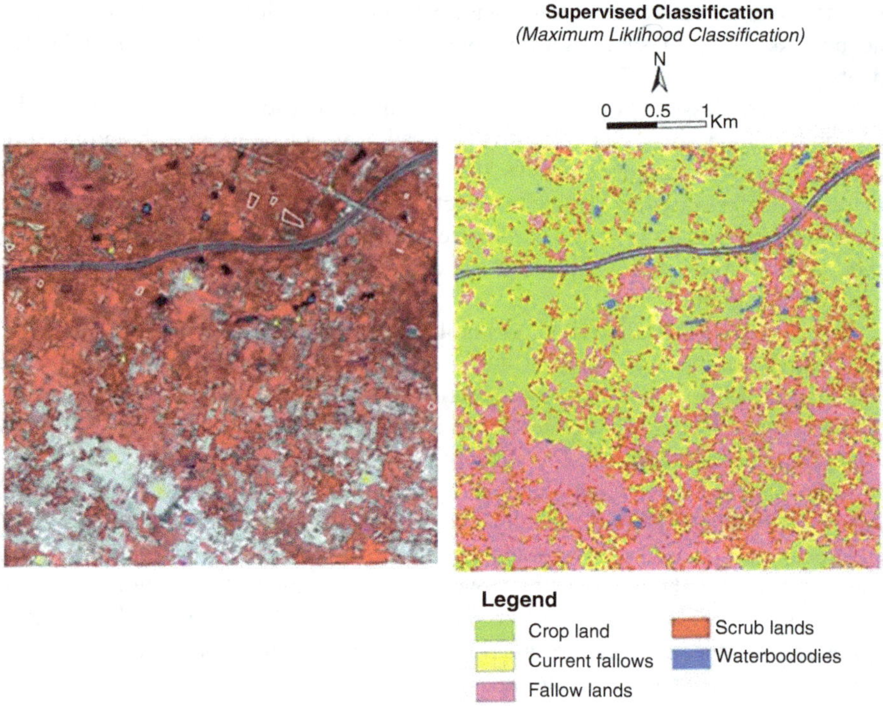

Fig. 6.9 Training sets shown on false color (Bands 4,3,2) composite of Landsat 8 OLI data and land use/land cover of part of Indo-Gangetic Plains of India derived from supervised classification

not match and remain unclassified, because these may belong to a class not recognized or defined. Training sites are shown as color polygons traced on the false color composite (FCC) (Bands 4,3,2) of Landsat 8 OLI data, and land use/land cover classes of part of Indo-Gangetic Plains of India derived from the image using supervised classification are shown in Fig. 6.9.

In a supervised classification, the interpreter knows before what land use/land cover classes are present and where they are present within the scene. These are located on the image, areas containing examples of the class are circumscribed (making them training sites), and the statistical analysis is performed on the multiband data for each such class. Instead of clusters then, one has class groupings with appropriate discriminate functions that distinguish each (it is possible that more than one class will have similar spectral values but that is unlikely when more than three bands are used because different classes/materials seldom have similar responses over a wide range of wavelengths). All pixels in the image lying outside training sites are then compared with the class discriminates derived from the training sites, with each being assigned to the class it is closest to—this makes a map of established classes (with a few pixels usually remaining unknown), which can be reasonably accurate (but some classes present may not have been set up; or some pixels are misclassified).

6.7 Image Classification Algorithms

6.7.1 Nonparametric Classification Techniques

A nonparametric classifier uses a set of nonparametric signatures to assign pixels to a class based on their location, either inside or outside the area in the feature space image. A nonparametric signature is based on area of interest (AOI) that defines in the feature space image for the image file being classified (Sabins 1987).

6.7.2 Parallelepiped

This is called as "box decision rule classifier." In this procedure two image bands are used to determine the training area of the pixels in each band based on maximum and minimum pixel values (Richards 1995). Although parallelepiped is the most accurate of the classification techniques, it is not the most widely used because it has several disadvantages. The most important disadvantage is that it can leave many unclassified pixels. Another disadvantage of this classification method is that it can have overlap between training pixels.

In the parallelepiped decision rule, the data file values of the candidate pixel are compared to upper and lower limits. These limits can be either the minimum and maximum data file values of each band in the signature, the mean of each band, plus and minus a number of standard deviations, or any limits that user specify, based on user knowledge of the data and signatures. The feature space decision rule determines whether or not a candidate pixel lies within the nonparametric signature in the feature space image. When a pixel's data file values are in the feature space signature, then the pixel is assigned to that signature's class as two-dimensional feature space classification. The polygons in the image are AOIs used to define the feature space signatures.

6.7.3 Parametric Classification Techniques

Parametric methods of supervised classification take a statistical approach. A parametric signature is based on statistical parameters of the pixels that are in the training sample or cluster. In addition to the standard attributes for signatures, a parametric signature includes the number of bands in the input image, the minimum and maximum data file value in each band for each sample or cluster, the mean data file value in each band for each sample or cluster, the covariance matrix for each sample or cluster, and the number of pixels in the sample or cluster.

6.7.3.1 Maximum Likelihood Classification (MLC)

This classification method uses the training data as a means of estimating means and variances of the classes, which are then used to estimate probabilities. Maximum likelihood classification considers not only the mean or average values in assigning classification but also the variability of brightness values in each class (Richards 1993). It is the most powerful of the classification methods as long as accurate training data is provided. Therefore, this method requires excellent training data. An advantage of this method is that it provides an estimate of overlap areas based on statistics. This method is different from parallelepiped that uses only maximum and minimum pixel values. The maximum likelihood decision rule is based on the probability that a pixel belongs to a particular class. The basic equation assumes that these probabilities are equal for all classes and that the input bands have normal distributions. The maximum likelihood algorithm assumes that the histograms of the bands of data have normal distributions. If this is not the case, you may have better results with the parallelepiped or minimum distance decision rule or by performing a first-pass parallelepiped classification.

6.7.3.2 Mahalanobis Distance

Mahalanobis distance classification is similar to minimum distance classification, except that the covariance matrix is used in the equation. Variance and covariance are figured in so that clusters that are highly varied lead to similarly varied classes and vice versa. For example, when classifying urban areas—typically a class whose pixels vary widely—correctly, classified pixels may be farther from the mean than those of a class for water, which is usually not a highly varied class. The Mahalanobis distance decision rule uses the covariance matrix in the equation. Variance and covariance are figured in so that clusters that are highly varied will lead to similarly varied classes and vice versa. The Mahalanobis distance algorithm assumes that the histograms of the bands have normal distributions. If this is not the case, user may gain better results with the parallelepiped or minimum distance decision rule or by performing a first-pass parallelepiped classification.

6.7.3.3 Minimum Distance

The minimum distance decision rule calculates the spectral distance between the measurement vector for the candidate pixel and the mean vector for each signature. This classification method derives distance between any pair of pixels after defining training data. The minimum distance classifier can be used as a supplement to the parallelepiped classification method, which can leave unanswered pixels. The classification technique takes pixels of known identity and then includes pixels closest to it as training pixels. Like the other methods of classification, this method uses two bands to evaluate the training data.

6.8 Separability and Error Matrices

Transformed divergence (TD) has upper and lower bounds. If the calculated divergence is equal to the appropriate upper bound, then the signatures can be said to be totally separable in the bands being studied. A calculated divergence of zero means that the signatures are inseparable. A separability listing is a report of the computed divergence for every class pair and one band combination. The listing contains every divergence value for the bands studied for every possible pair of signatures. The separability listing also contains the average divergence and the minimum divergence for the band set. These numbers can be compared to other separability listings to determine which set of bands is the most useful for classification.

6.9 Change-Detection Analysis

Change-detection analysis provides information about the changes in different seasons or dates. The information is extracted by comparing two or more images of an area that were acquired at different times. The first step is to register the images using corresponding ground control points. Following registration, the digital numbers of one image are subtracted from those of an image acquired earlier or later. The resulting values for each pixel will be positive, negative, or zero; the latter indicates no change. The next step is to plot these values as an image in which a neutral gray tone represents zero. Black and white tones represent the maximum negative and positive differences, respectively. Contrast stretching is employed to emphasize the differences. Change-detection processing is also useful for producing difference images for other remote sensing data, such as between nighttime and daytime thermal IR images. Temporal changes of NDVI derived from two different data images of Landsat 8 OLI data of the part of Indo-Gangetic Plains of India are shown in Fig. 6.10.

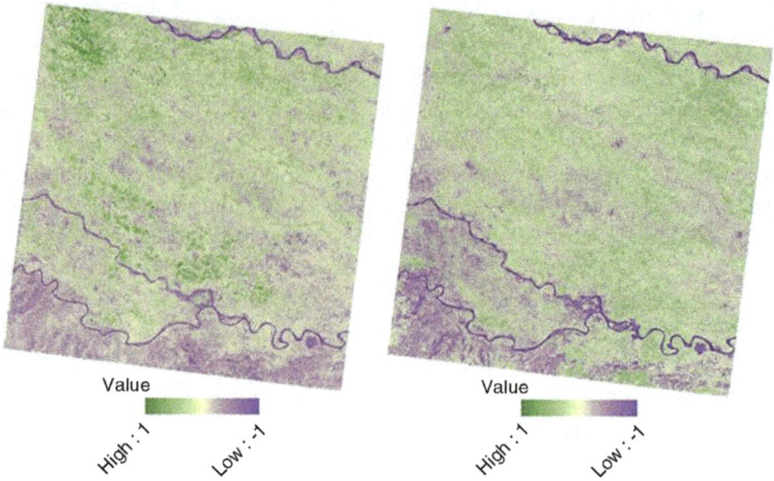

Fig. 6.10 Temporal changes of NDVI derived from Landsat 8 OLI data of 3 May 2015 and 5 Oct 2015 of part of Indo-Gangetic Plains of India

6.10 Post-classification Analysis

6.10.1 Smoothing

Most classifications have a problem with "salt and pepper," i.e., single or small groups of mis-classified pixels, as they are "point" operations that operate on each pixel independent of its neighbors:

- Majority filtering: replaces central pixel with the majority class in a specified neighborhood (3 x 3 window) (con: alters edges)
- Eliminate: clumps "like" pixels and replaces clumps under size threshold with majority class in local neighborhood (pro: doesn't alter edges)

6.10.2 Accuracy Assessment

Various techniques to assess the accuracy of the classified output by comparing the true identity of land cover derived from reference data (observed) vs. the classified (predicted) for a random sample of pixels (Lillesand and Kiefer 2000; Foody 2002)

6.10.3 Errors of Omission vs. Commission

Error of omission: pixels in class 1 erroneously assigned to class 2; from the class 1 perspective, these pixels should have been classified as class 1 but were omitted.

Error of commission: pixels in class 2 erroneously assigned to class 1; from the class 1 perspective, these pixels should not have been classified as class but were included.

6.10.4 Accuracy Assessment Measures

Overall accuracy: divide total correct (sum of the major diagonal) by the total number of sampled pixels; can be misleading and should judge individual categories. *Producer's accuracy:* measure of omission error; total number of correct in a category divided by the total in that category as derived from the reference data. *User's accuracy:* measure of commission error; total number of correct in a category divided by the totals that were classified in that category.

6.11 Conclusions

Digital image processing allows one to enhance image features of interest while attenuating detail irrelevant to a given application and then extract useful information about the scene from the enhanced image. There are three major functional categories of image processing, which are image restoration to compensate for data errors, noise, and geometric distortions introduced during the scanning, recording, and playback operations. Image enhancement helps to alter the visual impact that the image has on the interpreter, that improve the information content and information extraction ability by utilizing the decision-making capability of the computer in order to recognize and classify the pixels on the basis of their digital signatures. In all of these operations, the user should be aware of the trade-offs involved. However, it should also be noted that technical advances in software and hardware are steadily increasing the volume and complexity of the digital image processing that can be performed, often at a reduced unit cost.

References

Bernstein R, Ferneyhough DG (1975) Digital image processing. Photogramm Eng 41:1465–1476

Campbell JB (2002) Introduction to remote sensing, 3rd edn. Taylor and Francis, London

Foody GM (2002) Status of land cover classification accuracy assessment. Remote Sens Environ 80:185–201

Fu LM (1994) Neural networks in computer intelligence. McGraw-Hill, New York

Goetz AFH and others (1975) Application of ERTS images and image processing to regional geologic problems and geologic mapping in northern Arizona: Jet Propulsion Laboratory Technical Report, Pasadena, pp 32–1597

Haralick RM (1984) Digital step edges from zero crossing of second directional filters: IEEE transactions on pattern analysis and machine intelligence, V PAMI-6, pp 58–68

Jensen JR (1996) Introductory digital image processing: a remote sensing perspective. Prentice Hall, Glenview

Keys R (1981) Cubic convolution interpolation for digital image processing. IEEE Acoust Speech Signal Proc 29:1153–1160

Kulkarni AD (1994) Artificial neural networks for image understanding. ITP, New York

Kulkarni AD (2001) Computer vision and fuzzy-neural systems. Prentice Hall, Upper Saddle River

Lillesand TM, Kiefer RW (1999) *Remote sensing and image interpretation*, 4th edn. Wiley, New York

Lillesand TM, Kiefer RW (2000) Remote sensing and digital image interpretation. Wiley, New York 724 p

Loeve M (1955) Probability theory: D. van Nostrand Company, Princeton

Ranchin T, Aiazzi B, Alparone L, Baronti S, Wald L (2003) Image fusion—the ARSIS concept and some successful implementation schemes. ISPRS J Photogramm Remote Sens 58(1–2):4–18

Richards JA (1993) Remote sensing digital image analysis: an introduction, 2nd edn. Springer, Berlin/New York

Richards JA (1995) Remote sensing digital image analysis: an introduction. Springer, Berlin/New York, pp 265–290

Rifman SS (1973) Digital rectification of ERTS multispectral imagery: symposium on significant results obtained from ERTS-1. NASA SP-327:1131–1142

Ritter R, Wilson JN (2001) Computer vision algorithms in image algebra. CRC, Boca Raton

Sabins FF Jr (1987) Remote sensing: principles and interpretation. W.H. Freeman, New York

Chapter 7
Spatial Data Management, Analysis, and Modeling in GIS: Principles and Applications

G. P. Obi Reddy

Abstract Recent advances in spatial data development and management in the framework of geographic information system (GIS) have created a new era in the field of land resource management. Spatial database generally refers to any set of data describing the semantic and spatial properties of real-world phenomena. Manual digitization of paper maps, existing digital datasets, satellite remote-sensing imageries, global positioning system (GPS), field surveys, internet, etc. are the promising data input sources for spatial data development in GIS. Geographic information, attribute information, and display information are the three important types of information embedded in digital maps. Relational database management system (RDBMS) is the most effective and efficient data storage and management model in spatial database generation and management in GIS. Geographic analysis in GIS facilitates the study of various processes by developing and applying suitable models. The objective of geographic analysis is to transform data into useful information to satisfy the requirements or objectives of decision-makers. GIS provides tools and methods for the integration of different data into a format to be compared and analyzed. Analysis models comprise simple user-defined views to complex stochastic models, like suitability analysis, network analysis, optimization, allocation, etc. Geometric modeling has immense potential in generating buffers, calculating areas and perimeters, and calculating distances between features. Spatial analysis in GIS helps to identify trends on the data, create new relationships from the data, and view complex relationships between datasets to make better decisions. Vector and raster overlay operations are two different types of overlays in GIS depending upon data structures. GIS provides different ways in which the information can be presented once it is analyzed and processed by GIS.

Keywords Spatial database · Overlay analysis · RDBMS · Spatial data analysis · Spatial data modeling

---

G. P. O. Reddy (✉)
ICAR-National Bureau of Soil Survey & Land Use Planning, Nagpur, India

G. P. O. Reddy, S. K. Singh (eds.), *Geospatial Technologies in Land Resources Mapping, Monitoring and Management*, Geotechnologies and the Environment 21,
https://doi.org/10.1007/978-3-319-78711-4_7

7.1 Introduction

Spatial information is always related to geographic space. The handling of spatial data usually involves the process of data acquisition, storage, analysis, and output. Creation of spatial database in geographic information system (GIS) has become a very effective tool to aid and facilitate management decision-making (Burrough 1987). A collection of files that are used for complex information organization is called a database. In recent times, GIS is being widely used as spatial analysis tool for effective and efficient means of data generation and management, analysis, and display (Reddy 2012; Reddy et al. 2016, 2017). The true value of GIS can only be realized if the proper tools to collect spatial data and integrate them with attribute data are available. Data used in GIS often come from many types and are stored in different ways. Manual digitization of paper maps, existing digital data, remote-sensing satellite imagery, GPS, internet, etc. are promising data input sources for development of digital databases in GIS. Before generation of any spatial data, one has to understand the available data types, data analysis procedures, and their capabilities in GIS to get realistic outputs. Geospatial database management system has advantage over the conventional database systems as it is optimized to store and query data that is related to objects' location, including geometrical features (points, lines, and polygons).

GIS provides tools and methods for the integration of different databases into a format to be compared and analyzed. Spatial analysis is a vital part of GIS and can be used for many applications like site suitability, natural resource monitoring, environmental disaster management, etc. Spatial modeling infers the use of spatial characteristics and methods in manipulating data. The use of GIS spatial modeling tools in several resource activities has helped to quantify processes and define models for deriving analysis products. Spatial data analysis and modeling in GIS can make calculations that are too tedious to do manually, and the output of analysis and model provides digital data that is useful for many kinds of applications. They can be used to reveal patterns and trends in data that may not be otherwise apparent, and such information helps the decision-makers and planners to make decisions. Vector, raster-based analysis functions and arithmetic, logical, and conditional operations can be used based on the recovered derivations. The raster data model has become the primary spatial data source for analytical modeling with GIS. The raster data model is well suited to the quantitative analysis of numerous data layers. To facilitate these raster modeling techniques, most GIS softwares employ a separate module specifically for cell processing. The advance and efficient technique like "spatial data mining" is available to find out the knowledge from huge geospatial dataset for extracting unknown spatial relationship, trends, or patterns, not stored explicitly in spatial database (Shekhar et al. 2003). In spatial classification, methods such as decision trees, artificial neural network, remote sensing, and spatial autoregression could be used to find out the group of spatial objects together (Goyal et al. 2017).

7.2 Database in GIS

Database in GIS should be viewed as a representation of model of the real world developed for any speicific application. GIS technology utilizes two basic types of data, i.e., spatial data (coordinate and projection information for spatial features) and attribute data (characteristics about spatial data).

7.2.1 Spatial Data

Spatial data describes the absolute and relative location of geographic features. Spatial data can be point, line, polygon, or pixel. Spatial data includes location, shape, size, location, and orientation. Spatial data is usually stored as coordinates and topology. Spatial database is often updated, accessed, manipulated or analysed by using appropriate tools of GIS.

7.2.2 Attribute Data

Attribute data is often referred to as tabular data, and it describes characteristics of the spatial features. These characteristics can be quantitative and/or qualitative in nature. Attribute data is stored in the form of tables, and it is a database component that contains a series of rows and columns, where each row, or record, represents a geographic feature—such as a parcel, power pole, highway, or lake—and each column, or field, describes a particular attribute of the feature—such as length, depth, area, etc. Tables are stored in a database—for example, INFO, Access, dBASE, FoxPro, Oracle, or SQL Server. From a table, one can identify features with particular attributes and select them on the map. Over time, attribute data can also update to reflect changes to geographic features, for example, creating a new subdivision in the district boundary, and subsequently it also reflects in the attribution table.

7.3 Spatial Data Relationships in GIS

7.3.1 Geo-Relational Database Model

The geo-relational approach involves abstracting geographic information into a series of independent layers or coverages, each representing a selected set of closely associated geographic features (e.g., roads, land use, river, settlement, etc.). Each layer has the theme of a geographic feature, and the database is organized in the thematic layers. With this approach users can combine simple feature sets representing complex relationships in the real world.

7.3.2 Topological Data Structure

Topology is the spatial relationship between connection and adjacent coverage features. Topological relationships are built from simple elements into complex elements: line, points, and areas. Topology is defined as the mathematical procedure for explicitly defining spatial relationships between the data (connectivity or adjacency of points or lines in a GIS). The topological data structure logically determines exactly how and where points and lines connect on a map by means of nodes. The computer stores such information in various tables of the database structure. In digital maps or GIS, topological data structures provide additional intelligence for manipulating, analyzing, and using the information stored in a database. The order of connectivity defines the shape of an arc or polygon. Storing information in a logical and ordered relationship, missing information is readily apparent, data are stored efficiently, and large datasets can be processed quickly.

7.4 Attribute Database Models in GIS

Data within a GIS environment are stored with attribute databases. A database is a collection of information about things and their relationships to each other. A variety of different data models exist for the storage and management of attribute data. The most common attribute database models are tabular, hierarchical, network, and relational and object oriented.

7.4.1 Tabular Model

The simple tabular model stores attribute data as sequential data files with fixed formats, for the location of attribute values in a predefined record structure. It lacks any method of checking data integrity, as well as being inefficient with respect to data storage, e.g., limited indexing capability for attributes or records, etc.

7.4.2 *Hierarchical Model*

The hierarchical database organizes data in a tree structure. Data is structured downward in a hierarchy of tables. The data is stored as records, which are connected to one another through links. A record is a collection of fields, with each field containing only one value. Several records or files are hierarchically related with each other, and a set of links connect all record types in a tree structure. The advantages of hierarchical model are high speed of access to large datasets and eases of updating. However, the disadvantage is that linkages are only possible vertically but not horizontally or diagonally. Further, it is oriented for datasets that are very stable, where primary relationships among the data change infrequently or never at all. The quadtree, that is used to access a small part of a large raster image or map area, is a type of hierarchical model. The records at different levels in hierarchical model are shown in Fig. 7.1.

7.4.3 *Network Model*

The data in the network model are represented by collection of records, and relationships among data are represented by links, which can be viewed as pointers. Any record in the network model can be linked to any other. This model allows for

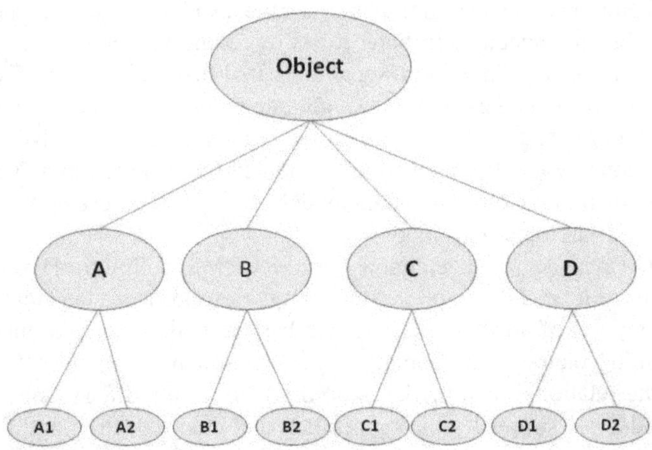

Fig. 7.1 Hierarchical model shows the records at different levels

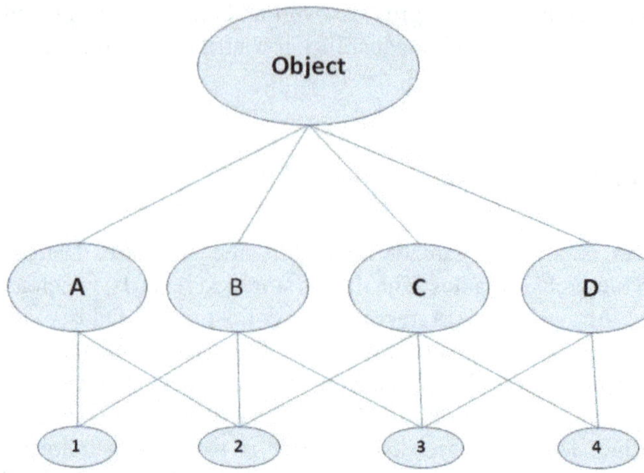

Fig. 7.2 Records and relationships in network model

children to have more than one parent. The vector mode is more suited to network analysis than the raster model. Records and relationships in network model are shown in Fig. 7.2.

7.4.4 Relational Database

The relational data model is conceived as a series of tables, with no hierarchy nor any predefined relations. Each table is identified by a unique table name and is organized by rows and columns. Each column within a table also has a unique name. Columns store the values for a specific attribute. Rows represent one record in the table. In a GIS each row is usually linked to a separate spatial feature. Accordingly, each row would comprise of several columns, each column containing a specific value for that geographic feature. The relation between the various tables should be made by the user. This is done by identifying a common field in two tables, which is assigned as the flexibility than in the other two data models. However, accessing the database is slower than with the other two models.

Relational Database Management System (RDBMS) follows relational database model and make it possible to pose complex queries, produce statistical summaries and tabular reports of attribute data. It also has the ability to make map analyses, often combining elements from many layers (Eastman 1992). Due to its greater flexibility, the relational data model is used widely in all GIS systems. Relational database model has simplicity in organization and data modeling and flexibility in joining tables in an ad hoc manner. This model has efficiency of storage data and minimized redundant data.

7.4.5 *Object-Oriented Model*

The object-oriented database model manages data through objects. An object is a collection of data elements and operations that together are considered a single entity. The object-oriented database is a relatively new model. This approach has the attraction that querying is very natural, as features can be bundled together with attributes at the database administrator's discretion. This model uses functions to model spatial and nonspatial relationships of geographic objects and the attributes. The model generates objects, classes, and superclasses through classification, generalization, association, and aggregation. This approach holds many operational benefits with respect to geographic data processing in GIS.

7.5 Spatial Database Query

The selective display and retrieval of information from a database are among the fundamental requirements of GIS. The ability to selectively retrieve information from GIS is an important facility. Database query simply asks to see already stored information. Basically there are two types of query in GIS: they are query by attribute and query by geometry. The procedure followed in query by attribute and query by geometry in GIS is shown in Fig. 7.3. The attribute database, in general, is stored in a table with a unique code linked to the geometric data. This database can be searched with specific characteristics. However, more complex queries can be made with the help of SQL. GIS can carry out a number of geometric queries. There are five forms of primitive geometric query: viz., query by point, query by rectangle, query by circle, query by line, and query by polygon. More complex queries can be developed, which uses both geometric and attributes search criteria together.

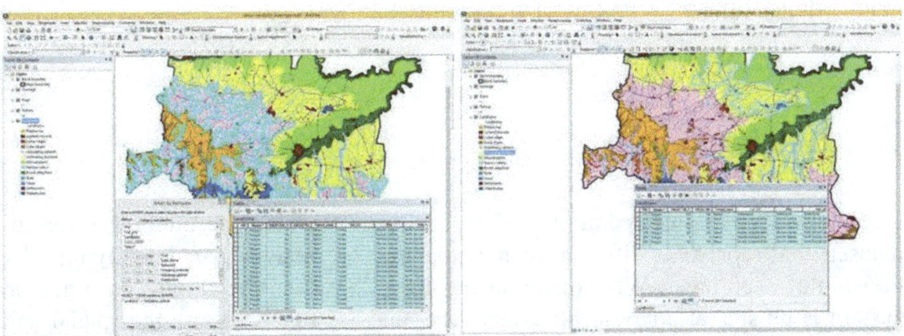

Fig. 7.3 Query by attribute and query by geometry in GIS

7.6 Analysis of Geographic Data

The heart of GIS is the analytical capabilities of the system. What distinguish the GIS system from other information systems is its spatial analysis functions. Geographic analysis facilitates the study of real-world processes by developing and applying models. The analysis functions use the spatial and nonspatial attributes in the database to answer questions about the real world. Geographic analysis facilitates the study of real-world processes by developing and applying models. Such models illuminate the underlying trends in geographic data and thus make new information available. In data analysis the most common operations carried out by GIS are database query, map algebra, and distance- and context-related analysis. The objective of geographic analysis is to transform data into useful information to satisfy the requirements or objectives of decision-makers. Spatial analysis helps to identify trends in the data, create new relationships from the data, and view complex relationships between datasets to make better decisions. In data analysis, the most common operations carried out by GIS are database query, map algebra, and distance- and context-related analysis. Buffer zone creation and reclassification are some of the important techniques in geographic analysis.

Buffer Zone Creation Distance operator in GIS is an example for analysis of geographic data by using buffer zone creation. It shows the proximity or nearness from any point, line, or polygon. Using these operations, the characteristics of an area surrounding in a specified location can be generated. This kind of analysis is called proximity analysis and is used whenever analysis is required to identify surrounding geographic features. The buffer operation generates polygon feature types irrespective of geographic features and delineates spatial proximity.

Reclassification Reclassification is a method of changing the attribute values without altering the geometry of the map. In fact it is a database simplification process that aims at reducing the number of categories of attribute data layer. Accordingly, features adjacent to one another that have a common value will be treated and appear as one class. Reclassification is an attribute generalization technique. After reclassification, the common boundaries between polygons with identical attribute values are dissolved and rebuilt the topology.

7.7 Overlay Analysis

Overlay analysis is an operation in GIS for superimposing the multiple layers of datasets that represents different themes together for analyzing or identifying relationship of each layer. What distinguishes the GIS system from other information system is its spatial analysis functions. The analysis functions use the spatial and nonspatial attributes in the database to answer questions about the real world. Spatial analysis in GIS includes all of the transformations, manipulations, and methods, which can be applied to geographic data to add value to them, to support decisions, and to reveal patterns and anomalies that are not visible in raw data. In the overlay analysis, new spatial datasets are created in GIS by merging data from two or more input data layers. In overlay analysis, topological overlay, spatial overlay, and criterion-based overlay techniques are important. Based on data structures, vector overlay and raster overlay are the two important overlay techniques.

7.7.1 Topological Overlay

Topological overlay is an analysis procedure for determining the spatial coincidence of geographic features. Append, union, identity, intersect, update, clip, split, and erase are some of the important topological overlay techniques.

- *Append*: Appending is used to merge together multiple datasets that represent the same thematic data but are contiguous.
- *Union*: A topological overlay of two polygon coverages, which preserves features that fall within the spatial extent of either input datasets, i.e., all features from both coverages are retained.
- *Identity*: This operation overlays polygons and keeps all input layer features and only those features from the analysis layer that overlap the input layer. The resultant layer has the same spatial features as that of the input layer. In the case of polygon overlays, the number of polygon in the output layer will always be larger in number than the input layer.
- *Intersect*: The topological integration of two spatial datasets that preserves features that fall within the spatial extent common to both input datasets. The resultant layer will keep those portions of the first input layer features, which fall within the second input layer polygons.
- *Update*: It replaces overlapping parts of the input layer with features from the update layer.
- *Clip*: It creates a new map that includes only those features of the input layer that fall within the area extent of the clip map. The input layer may be points, lines, or polygons, but the clip layer must be polygon layer. This operator is used to extract a smaller dataset from a larger dataset.
- *Split*: It divides the input coverage into two or more coverages. A series of clip operation can be performed, and each resultant layer contains only those portions of the input layer that are overlapped by the polygon satisfying the specified criteria.

- *Erase*: This feature can be used to erase polygons, lines, or points from the input features in GIS. A line-erase feature can be used to erase lines or points from the input features; a point-erase feature can be used to erase points from the input features.

7.7.2 Spatial Overlay

In spatial overlay analysis, two or more themes can be combined to form a new spatial feature (both geometric and attribute features are combined). Spatial overlay is accomplished by joining and viewing together separate datasets that share all or part of the same area. The result of this combination is a new dataset that identifies the spatial relationships. Three types of overlay can be performed: polygon-polygon, line-polygon, and point-polygon.

- *Polygon-on-polygon overlay*: This process merges overlapping polygons from two input layers in GIS to create new polygons in an output layer. The result of a polygon-on-polygon overlay is an output layer containing new polygons with merged attributes (i.e., those attributes from each of the two overlapping polygons). In this overlay, it is necessary to keep in mind that area should be common to input features.
- *Line-in-polygon overlay*: Polygon features of one input layer can be overlaid on lines (arcs) of another input layer in GIS. A line can be made up of many segments, line-in-polygon analyses, and therefore identifies which polygon (if any) contains each line or line segment. The result of a line-in-polygon overlay is a new layer containing lines with additional attributes (i.e., those attributes of the polygon within which the line falls).
- *Point-on-polygon overlay*: Point features of one input layer can be overlaid on polygon features of another input layer in GIS. Point-in-polygon analyses identify the polygon within which each point falls. The result of a point-in-polygon overlay is a set of points with additional attributes (i.e., those attributes of the polygon which the point lies within). This kind of overlay operation can be used to calculate number of points located in each of the polygon.

During the process of overlay, the attribute data associated with each feature type id merged. The resulting table will contain both the attribute data. The process of overlay will depend upon the modeling approach the user needs. One might need to carry out a series of overlay procedures to arrive at the conclusion, which depends upon the criterion (Fig. 7.4).

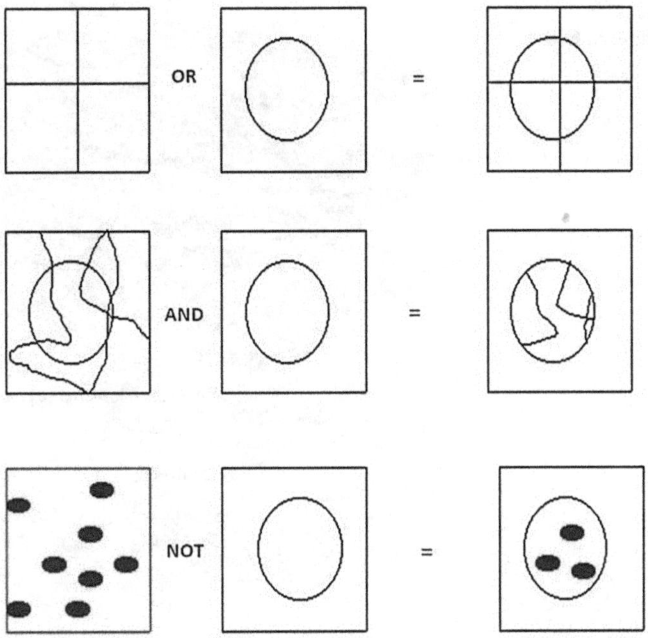

Fig. 7.4 Polygon-on-polygon overlay: difference between a topologic overlay and a graphic over plot

7.7.3 *Overlay Based on Data Structure*

Based on data structures, vector overlay and raster overlay are the two overlay techniques in GIS.

7.7.3.1 Vector Overlay

The vector overlay, however, is far more difficult and complex and involves more processing. In simple vector overlay, the layers were overlaid without assigning any weightage either for layer or classes. Weighted overlay is a technique for applying a common scale of values to diverse and dissimilar input to create an integrated analysis. Geographic problems often require the analysis of many different factors. The weighted overlay process reclassifies values in the input rasters onto a common evaluation scale of suitability or preference, risk, or some similarly unifying scale. The input rasters are weighted by importance and added to produce an output raster. The weighted overlay process accepts only discrete rasters (integer values) as input.

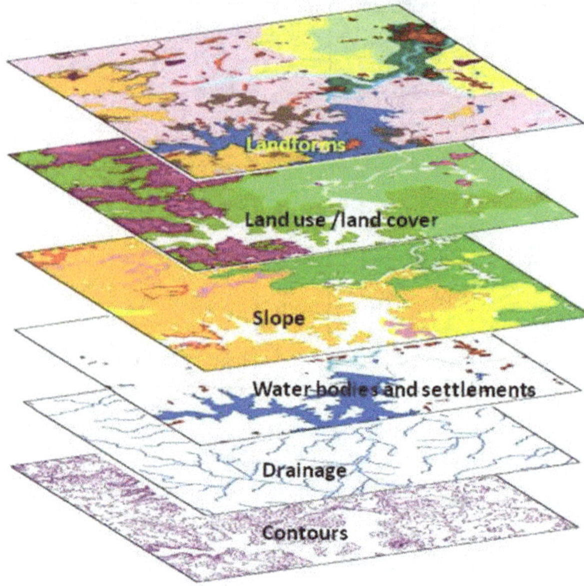

Fig. 7.5 Typical vector overlay procedure in GIS

Continuous rasters should and must be reclassified to discrete rasters before they can be used. The typical vector overlay procedure performed in delineation of landforms by integrating contours, drainage, waterbodies and settlements, slope, and land use/land cover is shown in Fig. 7.5.

During vector overlay, map features and the associated attributes are integrated to produce new composite maps. Logical rules can be applied to how the maps are combined. Vector overlay can be performed on different types of map features: viz., polygon-on-polygon overlay, line-in-polygon overlay, and point-on-polygon overlay.

There are some difficulties in geographic analysis, which includes lack of required datasets. Maintenance of spatial relationships in the input datasets is also important to get the accurate areas and shapes in outputs. Inherent uncertainties in the datasets due to scale are also another difficulty in integrated analysis. When using the spatial datasets from different sources, it is difficult to make data sources compatible easily. Selection of suitable model for the analysis of specific objective is also important to get the desirable outputs. The data file produced as a result of polygon overlay may be considerably larger than the original because lines have been split into smaller segments and new nodes and polygons have been created.

7.7.3.2 Raster Overlay

Raster overlays are relatively simple compared to their vector counterparts and require much less computational power (Burrough 1983). In raster overlay, the pixel or grid cell values in each map are combined using arithmetic and Boolean operators to produce a new value in the composite map. The maps can be treated as arithmetical variables and perform complex algebraic functions, and it is called as map algebra. Despite their simplicity, it is important to ensure that all overlain rasters are coregistered (i.e., spatially aligned), cover identical areas, and maintain equal resolution (i.e., cell size). If these assumptions are not met, the analysis will either fail, or the resulting output layer will be flawed (Chrisman 2002). The mathematical raster overlay is the most common overlay method. As an example, the rasters of digital elevation model (DEM), satellite data, plan curvature, profile curvature, and total curvature overlaid using raster overly model in GIS to delineate the distinct landforms in part of Katol tehsil of Nagpur district, Maharashtra, Central India is shown in Fig. 7.6.

Weighted overlay process reclassifies the values in the input rasters onto a common evaluation scale of suitability or preference, risk, or some similarly unifying scale. The input rasters are weighted by importance and added to produce an output raster. The weighted overlay process accepts only discrete rasters as input. Continuous rasters should and must be reclassified to discrete rasters before they can be used.

In raster overlay, if two grids are aligned and have the same grid cell size, then it is relatively easy to perform overlay operations. A new layer of values is produced from each pair of coincident cells. The values of these cells can be added, subtracted, divided, or multiplied, the maximum value can be extracted, mean value can be calculated, a logical expression can be computed, and so on.

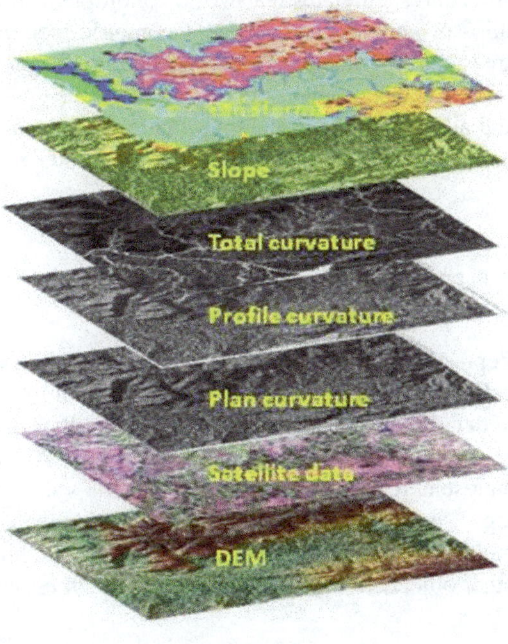

Fig. 7.6 Typical raster overlay procedure in GIS

7.7.4 Overlay Analysis with Logical Operators

The concept of map logic can be applied during overlay. Logical operators are based on point-by-point or cell-by-cell analysis. The most important of this group is the overlay analysis. In the raster-based analysis, either the logical or arithmetic operators are used. The logical operators are Boolean functions. There are basically four types of Boolean operators: viz., OR, AND, NOT, and XOR. Basic arithmetic operators in raster overlay operations are addition, subtraction, division, and multiplication.

7.8 Spatial Modeling

Spatial modeling is an analytical process conducted in conjunction with a GIS in order to describe basic processes and properties for a given set of spatial features. The objective of spatial modeling is to be able to study and simulate spatial objects or phenomena that occur in the real world and facilitate problem solving and planning. Due to the inherent complexity of the world and the interactions in it, models are created as a simplified, manageable view of reality. Spatial models help to understand, describe, or predict how things work in the real world. There are two main types of models: representation model, which represents the objects in the landscape, and process model, which simulates processes in the landscape. Set of analytical procedures simulates real-world conditions within a GIS using their spatial relationships of geographic features. Geometric modeling (generating buffers, calculating areas and perimeters, and calculating distances between features), coincidence modeling (topological overlay), and adjacency modeling (pathfinding, redistricting, and allocation) are three important categories of spatial modeling functions that can be applied to geographic features within a GIS.

7.8.1 Representation Models

Representation models try to describe the objects in a landscape. The way representation models are created in a geographic information system (GIS) is through a set of data layers. These data layers will be either raster or feature data. Raster layers are represented by a rectangular mesh or grid, and each location in each layer is represented by a grid cell, which has a value. Cells from various layers stack on top of each other, describing many attributes of each location. The representation model attempts to capture the spatial relationships within an object and between the other objects in the landscape. Along with establishing the spatial relationships, the

GIS representation model is also able to model the attributes of the objects. Representation models are sometimes referred to as data models and are considered descriptive models.

7.8.2 *Process Models*

Process models attempt to describe the interaction of the objects that are modeled in the representation model. The relationships are modeled using spatial analysis tools. Since there are many different types of interactions between objects, process modeling is sometimes referred to as cartographic modeling. Process models can be used to describe processes, but they are often used to predict what will happen if some action occurs. Some process models are simple, while others are more complex. A process model should be as simple as possible to capture the necessary reality to solve your problem. Even more complexity can be added by adding logic, combining multiple process models.

There are many types of process models to solve a wide variety of problems. A set of conceptual steps can be used to build a model. Some include surface modeling, distance modeling, hydrologic modeling, suitability modeling, etc. Suitability modeling in GIS helps to find out the optimum locations. Distance modeling helps to find out the distance like what is the minimum distance between the two areas. Buffering is the best example for distance modeling. In buffering process, it creates buffer polygons to a specified distance around the input features. An optional dissolve can be performed to remove overlapping buffers. Using these operations, the characteristics of an area surrounding in a specified location are evaluated. This kind of analysis is called proximity analysis and is used whenever analysis is required to identify surrounding geographic features. The buffer operation will generate polygon feature types irrespective of geographic features and delineates spatial proximity. Hydrologic modeling helps to find out the directions like water flow in hydrological analysis. Surface modeling helps to find out the different level of information like what is the pollution level for various locations.

7.9 Conclusions

In development of spatial databases in GIS, often data, comes from different formats and sources like manual digitization and scanning of aerial photographs, remote-sensing satellite imageries, paper maps, and existing digital datasets. GIS provides tools and methods for integration of different data into a format to be compared and analyzed. In data analysis, query, map algebra, and distance- and context-related analysis are the most common operations carried out in GIS. GIS-based spatial analysis includes the transformations, manipulations, and methods to analyze the database and the anomalies that are not visible in raw data. In GIS simulation modeling, a set of analytical procedures helps to simulate real-world conditions within a GIS using their spatial relationships of geographic features. Spatial modeling is a vital part of GIS and can be used for many applications like site suitability, natural resource monitoring, environmental disaster management, etc.

References

Burrough PA (1983) Geographical information systems for natural resource assessment. Oxford University Press, New York

Burrough PA (1987) Principles of geographical information systems for land resources assessment (corrected printing). Clarendon Press, Oxford 193 p

Chrisman N (2002) Exploring geographic information systems, 2nd edn. Wiley, New York

Eastman JR (1992) IDRISI User's guide, version 4.0 rev.1. Clark University Graduate School of Geography, Worcester

Goyal H, Sharma C, Joshi N (2017) An integrated approach of GIS and spatial data Mining in big Data. Int J Comp Appl 169(11):1–6

Reddy GPO (2012) Conceptual design of soil information system – a geoportal. In: Reddy GPO, Sarkar D (eds) Remote sensing and GIS in digital terrain analysis and soil-landscape modeling, NBSS&LUP Publ. No. 152, pp 217–226

Reddy GPO, Sarkar D, Mandal C, Srivastava R, Bhattacharyya T, Naidu LGK, Sidhu GS, Baruah U, Singh SK, Singh RS, Nair KM, Sen TK, Chandran P, Sahoo AK, Srinivas S, Kumar N, Chavan S (2016) Digital soil resource database and information system. In: Roy PS, Dwivedi RS (eds) Geospatial technology for integrated natural resources management. Yes Dee Publishing Pvt Ltd, Chennai, pp 370–406

Reddy GPO, Singh SK, Kumar N, Mondal C, Srivastava R, Maji AK, Sarkar D (2017) Geospatial technologies in development of soil information system and prototype geoportal. In: Reddy GPO, Patil NG, Chaturvedi A (eds) Sustainable management of land resources – an Indian perspective. Apple Academic Press, Oakvile/Waretown, pp 411–442

Shekhar S, Zhang P, Huang Y, Vatsavai RR (2003) Trends in spatial data mining

Chapter 8
Geostatistics: Principles and Applications in Spatial Mapping of Soil Properties

Nirmal Kumar and N. K. Sinha

Abstract Soil properties show a high degree of spatial variability due to the combined effect of physical, chemical, or biological processes that operate within the soil profile with different intensities at different scales. Knowledge of spatial variability of soil properties in the form of spatial continuous surfaces is important for identifying suitable zones for agricultural land management. To accurately describe the spatial variability, a very intensive survey is required. These intensive surveys are labor- and time-consuming. Soil maps can also be viewed as a source of spatial information on soil properties. However, these maps are prepared using pedological and morphological studies of pedons and are themselves based on point observations. The soil physical and chemical properties may be found to vary within a soil map unit. These maps also lack the required information of all the soil properties. Another method to get a spatial continuous raster of soil properties is to correlate them with satellite data, particularly hyperspectral data. This method requires high computations and sophisticated methodologies. In such conditions, spatial interpolation methods provide tools to analyze the spatial variabilities of soil properties and provide a spatial continuous layer by estimating the values of unsampled sites using data from existing point observations within the same region. This chapter reviews some of the popular spatial interpolation techniques used in soil science.

Keywords Geostatistics · Spatial interpolation · Spatial variability · Soil properties

N. Kumar (✉)
ICAR- National Bureau of Soil Survey & Land Use Planning, Nagpur, India

N. K. Sinha
ICAR- Indian Institute of Soil Science, Bhopal, India

G. P. O. Reddy, S. K. Singh (eds.), *Geospatial Technologies in Land Resources Mapping, Monitoring and Management*, Geotechnologies and the Environment 21,
https://doi.org/10.1007/978-3-319-78711-4_8

8.1 Introduction

A key feature of soil is the variation in soil properties with depth and across the landscapes as it is formed as a result of the influences of climate, plants, and time acting on geologic parent material in different landscape positions. Soil properties also vary significantly within a landscape or within a field across shorter distances. These variations in soil properties are often caused by small changes in topography and agronomic management practices. These spatial variabilities necessitate the use of a spatial continuous data (or spatial continuous surfaces) on soil properties in planning effective location-specific managements. They are, however, usually, not always readily available and often difficult and expensive to acquire, as the data collected on field surveys are typically from point sources. Thus, the values of an attribute at unsampled points need to be estimated based on the existing point data (Collins and Bolstad 1996; Hartkamp et al. 1999). In such instances, spatial interpolation methods provide tools to fulfill such task by estimating the values of an environment variable at unsampled sites using data from point observations within the same region. Predicting the values of a variable at points outside the region covered by existing observations is called extrapolation (Burrough and McDonnell 1998). The rationale behind spatial interpolation is the observation that points close together in space are more likely to have similar values than points far apart (Tobler's law of geography). Estimations of nearly all spatial interpolation methods can be represented as weighted averages of sampled data (Li and Heap 2008). The weights may be defined based on the distance of the point in consideration (of unknown value) from the points of known values. The methods considering only the distances are called non-geostatistical methods, such as inverse distance weighting (IDW). However, in case of soils, the properties also vary with the directions along with the distance. In these cases, geostatistical interpolation methods are used which consider both the distances and the directions from the point of unknown value from the points of known values. The spatial interpolation methods, including geostatistics, have been developed for and applied to various disciplines. Selection of an appropriate spatial interpolation method depends on the data, variable, and underlying assumptions of the spatial interpolation methods (Zhou et al. 2007).

Many factors including sample size, sampling design, and data properties affect the estimations of the methods (Hengl et al. 2004; Li and Heap 2008). The performance of the spatial interpolation methods is better when the sample density is greater (Englund et al. 1992; Isaaks and Srivastava 1989; Stahl et al. 2006). The kriging predictor does not strictly require that the data follow a normal distribution. But if the data follow a normal distribution, it gives the best result (Negrieros et al. 2010). Irregularly spaced sampling design and stratification would improve the accuracy of estimation (Collins and Bolstad 1996). However, there are no consistent findings about how these factors affect the performance of the spatial interpolators (Hengl et al. 2017; Zhou et al. 2007; Li and Heap 2008). This chapter aims to provide guidelines and suggestions useful to environmental scientists, especially in soil sciences, on the spatial interpolation of soil physical and chemical data by comparing the features of commonly applied spatial interpolators. This covers several aspects of spatial interpolation, which includes brief descriptions of the commonly used spatial interpolation methods in soil property mapping, error estimation of the applied model, and application of interpolation methods in mapping of soil properties.

8.2 Spatial Interpolation Methods

Numerous methods have been developed for spatial interpolation in various disciplines. The easiest way to predict the value of an attribute at any unsampled location would be assigning the value of the nearest sampled point to it. The other easier and more acceptable method would be taking the average value of more than one sampled locations. However, as the nearer things are more similar, assigning weights to the values of sampled points based on the distances of the sampled point from the point of interest would give a better result. For interpolating soil properties – which vary not only with distances but also with direction – a method which assigns weights to the values of sampled points based on distances as well as the direction would be preferable. The latter method is called geostatistical method of interpolation, while the former methods are non-geostatistical. A combination of both geostatistical and non-geostatistical methods is also available. The most popular methods in these categories are discussed in the following sections.

8.3 Non-geostatistical Interpolators

8.3.1 Nearest Neighbors

The nearest neighbor (NN) method predicts the value of an attribute at an unsampled point based on the value of the nearest sample by drawing perpendicular bisectors between sampled points (n), forming such as Thiessen (or Dirichlet/Voronoi) polygons (V_i, $i = 1,2,\ldots, n$). This produces one polygon per sample, and the sample is located in the center of the polygon, such that in each polygon all points are nearer to its enclosed sample point than to any other sample points (Isaaks and Srivastava 1989; Webster and Oliver 2001). The estimations of the attribute at unsampled points within polygon V_i are the measured value at the nearest single sampled data point x_i, that is, $z\hat{}(x_0) = z\ (x_i)$. The weights are:

$$\lambda_i = \begin{cases} 1 & \text{if} \quad x_i\ \varepsilon\ V_i, \\ 0 & \text{otherwise.} \end{cases} \tag{8.1}$$

All points (or locations) within each polygon are assigned the same value (Ripley 1981; Webster and Oliver 2001). A number of algorithms exist to generate the polygons, including pycnophylactic interpolation (Burrough and McDonnell 1998).

8.3.2 Inverse Distance Weighting

The inverse distance weighting (IDW) method estimates the values of an attribute at unsampled points using a linear combination of values of sampled points weighted by an inverse function of the distance from the point of interest to the sampled points (Hartkamp et al. 1999). The assumption is that sampled points closer to the unsampled point are more similar to it than those further away in their values. The weights can be expressed as:

$$\lambda_i = \frac{\frac{1}{d_i^p}}{\sum_{i=1}^{n} {}^{1}/_{d_i{}^p}} \tag{8.2}$$

where d_i is the distance between x_0 and x_i, p is a power parameter, and n represents the number of sampled points used for the estimation. The main factor affecting the accuracy of IDW is the value of the power parameter (Isaaks and Srivastava 1989). Weights diminish as the distance increases, especially when the value of the power parameter increases, so nearby samples have a heavier weight and have more influence on the estimation, and the resultant spatial interpolation is local (Isaaks and Srivastava 1989). The choice of power parameter and neighborhood size is arbitrary (Webster and Oliver 2001). The most popular choice of p is 2, and the resulting method is often called inverse square distance or inverse distance squared (IDS). The power parameter can also be chosen on the basis of error measurement (e.g., minimum mean absolute error, resulting in the optimal IDW) (Collins and Bolstad 1996). The smoothness of the estimated surface increases as the power parameter increases, and it was found that the estimated results become less satisfactory when p is 1 and 2 compared with p is 4 (Ripley 1981). IDW is referred to as "moving average" when p is 0, "linear interpolation" when p is 1, and "weighted moving average" when p is not equal to 1 (Burrough and McDonnell 1998).

The most common problem with IDW is that it may produce bull's-eyes on sampled positions which can be seen in Fig. 8.1. Otherwise, it's an easy method and helps to understand the influence of distance on the interpolated parameter. The IDW works well when the sample points are regularly spaced (Isaaks and Srivastava 1989). In IDW, the interpolated values are never outside the range of sampled values.

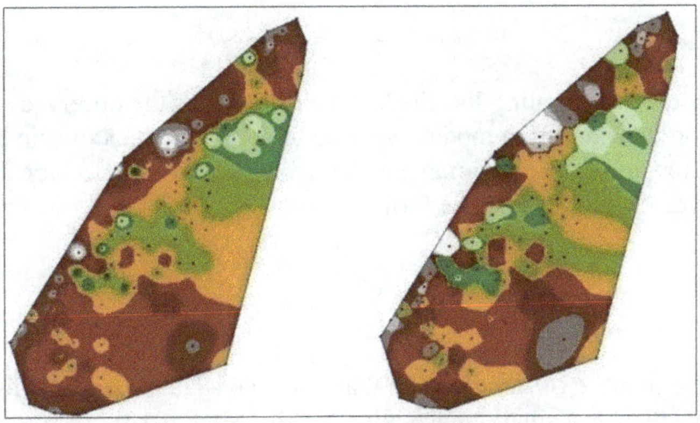

Fig. 8.1 Comparison of IDW interpolation with different power parameters, i.e., 2 and 10

8.4 Geostatistics

Geostatistics originated from the mining and petroleum industries, starting with the work by Danie Krige in the 1950s and was further developed by Georges Matheron in the 1960s. In both industries, geostatistics are successfully applied to solve cases where decisions concerning expensive operations are based on interpretations of sparse data located in space. Geostatistics has since been extended to many other fields related to the earth sciences, e.g., hydrogeology, hydrology, meteorology, oceanography, geochemistry, geography, soil sciences, forestry, and landscape ecology.

Geostatistics is a branch of applied statistics that quantifies the spatial dependence and spatial structure of a measured property and, in turn, uses that spatial structure to predict values of the property at unsampled locations. These two steps typically involve spatial modeling (variography) and spatial interpolation (kriging).

8.4.1 Semivariance and Variogram (Spatial Modeling)

The first step is to analyze the spatial correlation of the dataset. It is assumed that spatial autocorrelation exists in the dataset, i.e., the values of the particular variable are correlated with itself and the distance vector (separation distance) decides the amount of correlation between any two pairs of data. The correlation is expressed in terms of semivariance. Each pair of sampling points has a semivariance $\gamma(x_i, x_j)$ and is given by the equation:

$$\lambda(x_i, x_j) = \frac{1}{2}\left[z(x_i) - z(x_j)\right]^2 \tag{8.3}$$

where x denotes the sampling locations and z (x) denotes the observed value.

To develop the variogram model, we need to derive the relationship between the distance vector and the semivariance. The average semivariance for a separation h (distance vector) is given by the formula:

$$\widehat{\lambda}(\mathrm{h}) = \frac{1}{2n(h)} \sum_{i=1}^{n(h)} \left[z(x_i) - z(x_{i+h})\right]^2 \tag{8.4}$$

where n is the number of sample pairs that have the same separation distance h. In practice, h is taken as a small range called bin (Rossiter 2012).

Then we model this relationship in the form of a variogram and use it for spatial prediction. The semivariance is plotted as a function of distance, and a suitable curve/model is fitted to express this relationship (Fig. 8.2). Four commonly used variogram models are spherical, exponential, linear, and Gaussian (Li and Heap 2008). The variogram is also called as semivariogram.

There are three major parameters that describe the relationship between distance and the semivariance. They are range, sill, and nugget (Fig. 8.2). Range is the separation distance at which the observed values become independent of each other, i.e., the distance at which there is no spatial dependence. Beyond the range value, there will be no change in semivariance with an increase in separation distance

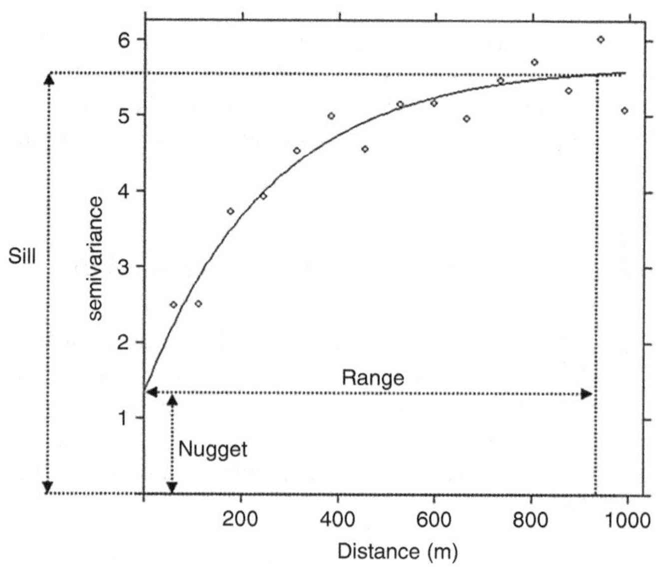

Fig. 8.2 Experimental variogram fitted to exponential model

h. The sill is the maximum semivariance and it happens in the range. The nugget is the semivariance at which the variogram curve intercepts the y axis, and it reflects the sampling errors and the spatial variance at a shorter distance than the minimum sample spacing (Li and Heap 2008). The range value can help in identifying the search window size to be used in the spatial interpolation (Burrough and McDonnell 1998). There is one more parameter called partial sill which is the difference between the sill and the nugget variance.

8.4.2 Kriging Estimator

Once the experimental variogram is ready, we can predict the value at an unknown point using Eq. 8.1. Kriging is a generic name for a family of generalized least -squares regression algorithms, used in recognition of the pioneering work of Danie Krige. All kriging estimators are variants of the basic equation, which is a slight modification of Eq. 8.1.

8.4.2.1 Simple Kriging

The estimation of simple kriging (SK) is based on a slightly modified Eq. (8.1), leading to Eq. (8.6) as:

$$\widehat{Z}(x_0) = \sum_{i-1}^{n} \lambda_1 Z(x_i) + \left[1 - \sum_{i=1}^{n} \lambda_i\right]\mu \tag{8.5}$$

where μ is a known stationary mean. The parameter μ is assumed constant over the whole domain and calculated as the average of the data (Wackernagel 2003). SK is used to estimate residuals from this reference value μ given a priori and is therefore sometimes referred to as "kriging with known mean" (Wackernagel 2003).

8.4.2.2 Ordinary Kriging

The ordinary kriging (OK) is similar to SK, and the only difference is that OK does not assume stationarity. Armstrong (1989) suggests that OK accounts for the local fluctuations of the mean by limiting the area of stationarity of the mean to the local neighborhood, which means that the mean may vary in the area and does not remain constant. OK estimates the local constant mean, then performs SK on the corresponding residuals, and only requires the stationary mean of the local search window (Goovaerts 1999). In OK, there is a condition that the sum of the weights must be equal to 1, which is not the case in SK (Goovaerts 1997).

Thus, $\left[1 - \sum_{i=1}^{n} \lambda_i\right] = 0$, and the kriging estimator is as Eq. 8.1. Armstrong (1989) further notes that OK better estimates resources, where datasets have large areas with low values and large areas with high values. Local means appear more meaningful in a situation where the global mean is not constant.

8.4.2.3 Block Kriging

The block kriging (BK) is a generic name for estimation of average values of the primary variable over a segment, a surface, or a volume of any size or shape (Goovaerts 1999). It is an extension of OK and estimates a block value instead of a point value by replacing the point-to-point covariance with the point-to-block covariance (Wackernagel 2003). Essentially, BK is block OK and OK is "point" OK.

8.4.2.4 Cokriging

Cokriging is an interpolation technique that uses information about the spatial patterns of two different but spatially correlated properties to interpolate only one of the properties (Vauclin et al. 1983). Typically, cokriging is used to map the property that is more difficult or expensive to measure (z_2) based on its spatial dependence on a property that is easier or less expensive to measure (z_1). An example of this is the interpolation by cokriging of sparsely sampled soil test P levels using intensively sampled soil organic C content values (Bhatti et al. 1991; Mulla 1997). Another example might be the interpolation by cokriging of sparsely sampled soil moisture content levels (Yates and Warrick 1987).

8.4.2.5 Indicator Kriging

Punctual kriging, block kriging, and cokriging are all useful for estimating the value of a property at an unsampled location. The arithmetic average of all interpolated values is an estimate of the mean. Oftentimes, rather than estimating the mean, it is necessary to estimate the proportion of values at a site which is above or below some critical value. For example, the extent of cleanup and remediation required at a site contaminated with trace metals depends upon the proportion of the site where the soil concentrations of trace metals exceed statutory limits. The kriging methods discussed up to this point are unsuitable for making such estimates. Indicator kriging is an approach for estimating the proportion of values that fall within specified class intervals or the proportion of values that are below a threshold level (Journel 1986; Isaaks and Srivastava 1989). The basic approaches in indicator kriging are (1) to transform measured values into indicator variables, (2) to estimate the semivariogram for the indicator transformed values, (3) to use simple kriging to interpolate the indicator transformed values across the study site, and (4) to compute the average of the kriged indicator values. This average is the proportion of values that is less than the threshold.

8.5 Factors Affecting Interpolation

The interpolation procedure includes acquiring the suitable data, selecting suitable interpolation method, and estimating the error in prediction. Before going for interpolation, there are two primary concerns. First is identifying a suitable number of sampling location and their spatial pattern, i.e., the sampling technique. Second is evaluation of the data for its statistical distribution pattern such as detection of outliers, the skewness and kurtosis of the data, checking for normality of the data, the transformation needed, if any, etc.

8.5.1 Sampling Strategy

The collection of soil samples is as important as its laboratory analysis and spatial interpolation. The geographic coordinates of the sampling sites are also very important since it is a requirement for the spatial interpolation of the collected soil information. The geographic coordinates of the sampling locations could be collected with the help of global positioning system (GPS). There are mainly three different methods for the collection of soil samples. A regular grid with square, triangular, or hexagonal elements is most often used method of sampling (Webster and Burgess 1984). For a regular grid sampling program, the most efficient placement of sample locations is in the center of each grid cell (Webster and Burgess 1984). Sample spacings for these grid cells should be less than of the range for the semivariogram (Flatman and Yfantis 1984). This kind of sampling can capture the field variability in a better way, when there is no prior information about the area to be studied. When the study area is uniform in its characteristics, samples may be collected randomly. When there are marked variations in the area, zones may be defined based on soils and land characteristics like slope, soil texture, depth, etc., and any of the above sampling strategies may be applied within the zones (Dinkins and Jones 2008; Robinson and Metternicht 2006). This sampling technique is called as zone sampling.

Figure 8.3 is drawn using a subset data from the open-source data (meuse.all) available with the R statistical programing package that shows different sampling techniques.

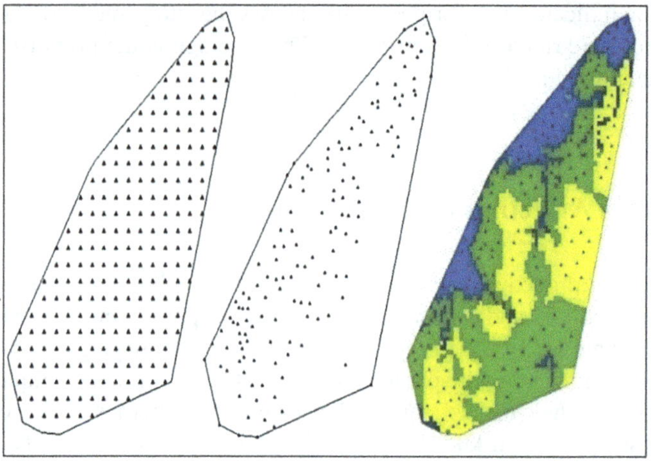

Fig. 8.3 Different kinds of sampling procedures: (**a**) grid sampling, (**b**) random sampling, and (**c**) zone sampling

8.5.2 *Sample Size*

The estimating precision of an interpolation method depends heavily on sample number and density (Liu et al. 2006). Webster and Oliver (2001) and Burrough and McDonnell (1998) suggested a minimum of 50 samples to create a variogram, and low sample sizes result in noisy variograms which contribute little to explain the spatial structure of the dataset. Rossiter (2012) suggested at least 100 to 150 sampling points with a good distribution of separations to create a meaningful variogram. Oliver and Webster (2014) considered sample size the most important factor that determines the reliability, or accuracy, of the empirical variogram. They suggested a minimum number of 100 samples for an accurate estimate of the variogram.

8.5.3 *Data Distribution and Presence of Outliers*

Generally, a random spatial process follows a normal/Gaussian distribution, and the statistical models assume that the data follow a normal distribution. The kriging predictor does not strictly require that the data follow a normal distribution. But if the data follow a normal distribution, it gives the best result (Negrieros et al. 2010). So the first step is to analyze if the data follow a normal distribution. The easiest method is to plot the histogram and see it follows a normal distribution curve. One can also use Q-Q plot or box plot for the purpose. Normality tests such as Shapiro-Wilk test (Shapiro and Wilk 1965) and Anderson-Darling test (Anderson and Darling 1954) are also available for testing the normality of the data. If it is not following a normal distribution, one can use any data transformation techniques, for example, log transformation, to force the data to have a normal distribution.

The other most common problem when we attempt for interpolation is the presence of outliers in the data. An outlier is a data value that is either very large or very small compared to the rest of the data. These may be valid values or a result of malfunctioning of monitoring equipment or human error. The presence of outliers makes it very difficult to build a valid kriging model. The outliers in the data may be detected by identifying values – four standard deviations in either side of the mean (Cahn et al. 1994; Wani et al. 2013). Other common approach to identify the outliers is to use standard score (z score). A sample is identified as outlier if its standard score is ±3.

8.5.4 *Identification of Suitable Interpolation Method*

It is important to refer that there is no single interpolation method that can be applied to all situations. Some are more exact and useful than others but take longer to calculate. They all have advantages and disadvantages. In practice, selection of a particular interpolation method should depend upon the sample data, the type of surfaces to be generated, and the tolerance of estimation errors.

8.6 Error Estimation

Error assessment is the final phase of soil nutrient mapping. It gives a picture of the quality of the nutrient map you have produced. Just like any other modeling process, spatial interpolation techniques have also resulted in errors while making the predictions. We can make a check on the map using an independent validation dataset. There are several error assessment methods to evaluate the performance of spatial interpolation methods. The most common methods used are mean error (ME), mean absolute error (MAE), mean squared error (MSE), and root mean squared error (RMSE), and the equations to find these measures are given below (Li and Heap 2008).

$$\mathrm{ME} = \frac{1}{n}\sum_{i=1}^{n}(Pi - Oi) \tag{8.6}$$

$$\mathrm{MAE} = \frac{1}{n}\sum_{i=1}^{n}\mid Pi - Oi \mid \tag{8.7}$$

$$\mathrm{MSE} = \frac{1}{n}\sum_{i=1}^{n}(Pi - Oi)^2 \tag{8.8}$$

$$\text{RMSE} = \sqrt{\frac{1}{n}\sum_{i=1}^{n}(Pi - Oi)^2} \tag{8.9}$$

where n is the number of samples, P is the predicted value, and O is the observed value.

Leave one out cross validation (LOOCV) is another commonly used method to evaluate the performance of the spatial interpolation techniques. Here, you leave out one point from the dataset and predict the value at this particular point using rest of the data. Then compute the difference between predicted and the actual value at that point, i.e., the residual. Do this iteratively for all the points, and summarize the residuals using any method of prediction accuracy (Rossiter 2012).

8.7 Geostatistics in Spatial Mapping of Soil Properties

8.7.1 Spatial Mapping of Soil Physical Properties

Physical properties play a key role in fertility and are a basic component of soil quality (Amézquita et al. 2004), and their evaluation has been proposed to define specific management units (Cucunubá-Melo et al. 2011). Spatial variability of soil physical properties within or among agricultural fields is inherent in nature due to both geologic and pedologic factors of soil formation. However, management practices such as tillage, irrigation, and fertilizer application may also induce variability within the field and may further interact with each other across different spatial and temporal scales and are further modified locally by erosion and deposition processes (Iqbal et al. 2005; Tuffour et al. 2016).

Among the various soil physical properties, saturated hydraulic properties and related measures are reported to have the highest statistical variability (Biggar and Nielsen 1976). Vieira et al. (1981) used variogram, kriging, and cokriging techniques to determine the magnitude of spatial variation. Vauclin et al. (1983) used classical and geostatistical techniques to study spatial viability of sand, silt, and clay contents and available water content (AWC). Reza et al. (2016) studied the degree of spatial variability of sand, silt, and clay contents and water content at field capacity (FC), permanent wilting point (PWP), and available water content (AWC) of alluvial floodplain soils of India. Results showed that soil physical properties exhibits large variability with greatest variation in sand content (68%). Exponential and spherical models fit well for the soil physical properties. The nugget/sill ratio indicates except clay all other soil physical properties were moderate spatially dependent (37–70%). Sağlam et al. (2011) prepared semivariograms and corresponding kriging maps of soil texture, soil organic matter (SOM), bulk density (BD), saturated hydraulic conductivity (SHc), and available water content (AWC) for alluvial and adjacent colluvial soils formed under ustic moisture regime of Turkey. Haruna and Nkongolo (2013) studied spatial variability of soil physical properties in a clay-loam soil cropped to corn and soybean in Missouri, USA. The soil properties considered were bulk density (BD), volumetric water content (VWC) and gravimetric water content (GWC), volumetric air content (VAC), total pore space (TPS), air-filled pore space (AFPS) and water-filled pore space (WFPS), the relative gas diffusion coefficient (DIFF), and the pore tortuosity factor (TORT). The semivariogram analysis showed the presence of a strong (≤25%) to weak (>75%) spatial dependence of soil properties. Camacho-Tamayo et al. (2013) developed the soil physical properties based on management zones in Oxisol of Colombia. The soil properties considered for the development of management zones were saturated hydraulic conductivity, bulk density, and soil particle density.

8.7.2 Spatial Mapping of Soil Chemical Properties

Soil nutrients are essential for plant growth. The status of soil nutrients is changing continuously with the crop growth, state of development, yields and arrangement of crops, fertilizers application and anthropic factor, etc. Due to the combined effect of these factors, soil nutrients and their distribution patterns differentiate in different areas (Xu et al. 2013). The conventional method of uniform application is not scientifically suitable and efficient to apply fertilizer in places with different soil nutrients. Hence, it is imperative to monitor changes in soil nutrients for site-specific management that leads to less waste, improvement on the margin of growers, and less impact on the environment. As discussed in earlier sections, geostatistics provides a scientific basis to predict the unsampled area by various means. Worldwide, researcher applied the theory of geostatistics to generate soil nutrient map and their management zones.

Hengl et al. (2017) produced spatial prediction for 15 nutrients (organic carbon, total nitrogen, total phosphorus, extractable phosphorus, potassium, calcium, magnesium, sulfur, sodium, iron, manganese, zinc, copper, aluminum, and boron) in sub-Saharan Africa at 250 m spatial resolution. In this study, model fitting and prediction were undertaken using an ensemble of two machine learning algorithms (Hengl et al. 2017), ranger (random forest) (Wright and Ziegler 2017) and xgboost (gradient boosting tree) (Chen and Guestrin 2016), as implemented in the R environment for statistical computing. The produced maps of soil macro and micronutrients of the region could be further used for delineating areas of nutrient deficiency/sufficiency relative to nutrient requirements and as an input to crop modeling. Bishop and McBratney (2001) compared several prediction methods, the statistical (generalized additive model, regression tree, multiple linear regression), geostatistical (ordinary kriging), and hybrid methods(regression kriging, kriging with external drift), for mapping of the soil cation exchange capacity (CEC), using auxiliary variables (terrain attributes, aerial photos, Landsat TM sensor 5, data of crop yields, and soil electrical conductivity) in different combinations. The root mean square error (RMSE) was used as an index to assess the quality of prediction. The results showed that CEC was best predicted by kriging with external drift, multiple linear regressions, and the generalized additive model. The choice of a prediction model for soil properties depends on several factors such as the availability of soil data and environmental covariates, size and environmental characteristics of the area mapped, the computer run time, ease of model implementation and result interpretation, as well as the desired mapping accuracy (McBratney et al. 2000; Junior et al. 2014).

In intensive cultivated Trans-Gangetic Plains of India, Shukla et al. (2016) mapped soil organic carbon, pH, and plant available micronutrients, viz., extractable zinc (Zn), copper (Cu), manganese (Mn), and iron (Fe), through geostatistical methods. They observed deficiencies of available Fe, Zn, Mn, and Cu deficiencies (including acute deficiencies) in 28, 15, 14, and 13% of soil samples, respectively. Further, soil pH showed significant and negative correlations with the concentrations of extractable Zn, Cu, Mn, and Fe, whereas the correlation was significant and positive with soil organic carbon (SOC) concentration in this study. Shukla et al. (2017) delineated four micronutrient-based management zones following the principle of geostatistics and combining with principal component-based fuzzy clustering technique and recommended several management practices to improve crop productivity in these delineated areas. Shit et al. (2016) used classical ordinary kriging interpolation for spatial visualization of pH, electrical conductivity (EC), phosphorus (P), potassium (K), and organic carbon (OC) in Medinipur Sadar block of Paschim Medinipur district in West Bengal, India. However, spatial interpolation solely based on ordinary kriging method produces large uncertainty in the prediction of soil spatial distribution because of the impacts of land use, topographic characteristics, and other factors (Stacey et al. 2006). In recent years, some methods have been proposed to solve this problem, such as the regression-kriging method, the geographically weighted regression (GWR) method, and the geographically weighted regression-kriging (GWRK) method (Harris et al. 2010; Kumar et al. 2012). However, among these methods, only the regression-kriging method can incorporate topographic factors, vegetation coverage, and other elements and thereby improve the accuracy of spatial prediction (Hengl et al. 2004; Peng et al., 2013). Regression kriging is the combination of multivariate regression and kriging. It takes into consideration the spatial autocorrelation of the variable of interest, the correlation between the variable of interest and auxiliary variables (e.g., remotely sensed images are often relatively easy to obtain as auxiliary variables), and the unbiased spatial estimation with minimized variance (Meng et al. 2013). Peng et al. (2013) tested regression-kriging method to analyze spatial soil organic carbon and total nitrogen distributions at the small watershed. The results indicated that geostatistical characteristics of soil organic carbon and total nitrogen concentrations in the watershed were closely related to both land use type and spatial topographic structure and that regression-kriging is suitable for investigating the spatial distributions of soil organic carbon and total nitrogen in the complex topography of the watershed.

8.8 Conclusions

Information on spatial variation of physical and chemical properties of soil is a necessary input in precision agriculture and other soil management decisions. This information helps to match the input with the crop demand and thus help to conserve the resources. But, traditionally we have been generating point information on the soil properties, and its value at non-sampled area needs to be estimated by interpolation methods. A number of methods are available and being used, and the appropriate selection depends on a number of factors that should be considered, given that there is no one best spatial interpolation method.

References

Amézquita E, Thomas RJ, Rao IM, Molina DL, Hoyos P (2004) Use of deep-rooted tropical pastures to build-up an arable layer through improved soil properties of an Oxisol in the Eastern Plains (Llanos Orientales) of Colombia. Agric, Ecosyst Environ 103:269–277

Anderson TW, Darling DA (1954) A test of goodness-of-fit. J Am Stat Assoc 49:765–769

Armstrong M (1989) Basic linear Geostatistics. Springer, Berlin/Heidelberg/New York

Bhatti AU, Mulla DJ, Frazier BE (1991) Estimation of soil properties and wheat yields on complex eroded hills using geostatistics and thematic mapper images. Remote Sens Environ 37:181–191

Biggar JW, Nielsen DR (1976) Spatial variability of the leaching characteristics of a field soil. Water Resour Res 12:78–84

Bishop TFA, McBratney AB (2001) A comparison of prediction methods for the creation of field-extent soil property maps. Geoderma 103:49–60

Burrough PA, McDonnell RA (1998) Principles of geographical information systems. Oxford University Press, Oxford 333 pp

Cahn MD, Hummel JW, Brouer BH (1994) Spatial analysis of fertility for site-specific crop management. Soil Sci Soc Am J 58:1240–1248

Chen T, Guestrin C (2016) XGBoost: a scalable tree boosting system. In: Proceedings of the 22nd acmsigkdd international conference on knowledge discovery and data mining. ACM, 2016

Collins FC, Bolstad PV (1996) A comparison of spatial interpolation techniques in temperature estimation, proceedings, third international conference/workshop on integrating GIS and environmental modeling, Santa Fe, NM. National Center for Geographic Information and Analysis, Santa Barbara, Santa Barbara

Cucunubá-Melo JL, Álvarez-Herrera JG, Camacho-Tamayo JH (2011) Identification of agronomic management units based on physical attributes of soil. J Soil Sci Plant Nutr 11:87–99

Camacho-Tamayo JH, Sanabria YR, Santana LM (2013) Management units based on the physical properties of an Oxisol. J Soil Sci Plant Nutr 13(4):767–785

Dinkins CP, Jones C (2008) Soil sampling strategies. A self-learning resource from MSU Extension. MT200803AG New 4/08

Englund E, Weber D, Leviant N (1992) The effects of sampling design parameters on block selection. Math Geol 24:329–343

Flatman GT, Yfantis AA (1984) Geostatistical strategy for soil sampling: the survey and the census. Environ Monit Assess 4:335–349

Goovaerts P (1997) Geostatistics for natural resource evaluation. Oxford University Press, New York

Goovaerts P (1999) Geostatistics in soil science: state of the art and perspectives. Geoderma 89:1–45

Harris P, Fotheringham AS, Crespo R, Charlton M (2010) The use of geographically weighted regression for spatial prediction: an evaluation of models using simulated data sets. Math Geosci 42:657–680

Hartkamp AD, De Beurs K, Stein A, White JW (1999) Interpolation techniques for climate variables. NRG-GIS series 99–01. Mexico City: CIMMYT

Haruna SI, Nkongolo NV (2013) Variability of soil physical properties in a clay-loam soil and its implication on soil management practices. ISRN Soil Sci 2013:1–8

Hengl T, Heuvelink G, Stein A (2004) A generic framework for spatial prediction of soil variables based on regression-kriging. Geoderma 120(1):75–93

Hengl T, Leenaars JGB, Shepherd KD, Walsh MG, Heuvelink GBM, Mamo T, Tilahun H, Berkhout E, Cooper M, Fegraus E, Wheeler I, Kwabena NA (2017) Soil nutrient maps of Sub-Saharan Africa: assessment of soil nutrient content at 250 m spatial resolution using machine learning. Nutr Cycl Agroecosyst 109:77–102

Iqbal J, Thomasson JA, Jenkins JN, Owens PR, Whisler FD (2005) Spatial variability analysis of soil physical properties of alluvial soils. Soil Sci Soc Am J 69:1338–1350

Isaaks EH, Srivastava RM (1989) Applied geostatistics. Oxford University Press, New York 561 pp

Journel AG (1986) Constrained interpolation and qualitative information — the soft kriging approach. Math Geol 18:269–286

Junior WDC, Chagas CDS, Lagacherie P, Filho BC, Bhering SB (2014) Evaluation of statistical and geostatistical models of digital soil properties mapping in tropical mountain regions. Braz J Soil Sci 38:706–717

Kumar S, Lal R, Liu DS (2012) A geographically weighted regression kriging approach for mapping soil organic carbon stock. Geoderma 189:627–634

Li J, Heap AD (2008) A review of spatial interpolation methods for environmental scientists. Geosci Aust Rec 2008(/23):1–137

Liu X, Clarke K, Herold M (2006) Population density and image texture: a comparison study. Photogramm Eng Remote Sens 72:187–196

McBratney AB, Odeh IOA, Bishop TFA, Dunbar MS, Shatar TM (2000) An overview of pedometric techniques for use in soil survey. Geoderma 97:293–327

Meng Q, Liu Z, Borders BE (2013) Assessment of regression kriging for spatial interpolation – comparisons of seven GIS interpolation methods. Cartogr Geogr Inf Sci 40:28–39

Mulla DJ (1997) Geostatistics, remote sensing and precision farming. In: Lake JV, Bock GR, Goode JA (eds) Precision agriculture: spatial and temporal variability of environmental quality. Wiley, Chichester, pp 100–114

Negrieros J, Painho M, Aguilar F, Aguilar M (2010) Geographical information systems principles of ordinary kriging interpolator. J Appl Sci 10(11):852–867

Oliver MA, Webster R (2014) A tutorial guide to geostatistics: computing and modelling variograms and kriging. Catena 113:56–69

Peng G, Bing W, Guangpo G, Guangcan Z (2013) Spatial distribution of soil organic carbon and total nitrogen based on GIS and geostatistics in a small watershed in a hilly area of northern China. PLoS One 8(12):e83592

Reza SK, Nayak DC, Chattopadhyay T, Mukhopadhyay S, Singh SK, Srinivasan R (2016) Spatial distribution of soil physical properties of alluvial soils: a geostatistical approach. Arch Agron Soil Sci 62:972–981

Ripley BD (1981) Spatial statistics. Wiley, New York 252 pp

Robinson TP, Metternicht G (2006) Testing the performance of spatial interpolation techniques for mapping soil properties. Comput Electron Agric 50:97–108

Rossiter DG (2012) A minimal introduction to geostatistics with R/gstat. University of Twente, Faculty of Geo-Information Science & Earth Observation (ITC), p 69

Sağlam M, Öztürk HS, Erşahin S, Özkan Aİ (2011) Spatial variation of soil physical properties in adjacent alluvial and colluvial soils under ustic moisture regime. Hydrol Earth Syst Sci Discuss 8:4261–4280

Shapiro SS, Wilk MB (1965) An analysis of variance test for normality (complete samples). Biometrika 52:591–611

Shit PK, Bhunia GS, Maiti R (2016) Spatial analysis of soil properties using GIS based geostatistics models. Model Earth Syst Environ 2:107. https://doi.org/10.1007/s40808-016-0160-4

Shukla AK, Behera SK, Lenka NK, Tiwari PK, Chandra P, Malik RS, Sinha NK, Singh VK, Patra AK, Chaudhary SK (2016) Spatial variability of soil micronutrients in the intensively cultivated Trans-Gangetic Plains of India. Soil Tillage Res 163:282–289

Shukla AK, Sinha NK, Tiwari PK, Prakash C, Behera SK, Lenka NK, Singh VK, Dwivedi BS, Majumdar K, Kumar A, Srivastava PC (2017) Spatial distribution and management zones for sulphur and micronutrients in Shiwalik Himalayan Region of India. Land Degrad Dev 28(3):959–969

Stacey KF, Lark RM, Whitmore AP, Milne AE (2006) Using a process model and regression kriging to improve predictions of nitrous oxide emissions from soil. Geoderma 135:107,117

Stahl K, Moore RD, Floyer JA, Asplin MG, McKendry IG (2006) Comparison of approaches for spatial interpolation of daily air temperature in a large region with complex topography and highly variable station density. Agric For Meteorol 139:224–236

Tuffour HO, Abubakari A, Bashagaluke JB, Djagbletey E (2016) Mapping spatial variability of soil physical properties for site-specific management. Int Res J Eng Technol 03:149–163

Vauclin M, Viera SR, Vachaud G, Nielsen DR (1983) The use of co-kriging with limited field soil observations. Soil Sci Soc Am J 47:175–184

Vieira SR, Nielsen DR, Biggar JW (1981) Spatial variability of field-measured infiltration rate. Soil Sci Soc Am J 45:1040–1048

Wackernagel H (2003) Multivariate Geostatistics: an introduction with applications. Springer, Berlin 387 pp

Wani M, Bhat M, Kirmani NA, Wani ZM (2013) Mapping of soil micronutrients in Kashmir agricultural landscape using ordinary kriging and indicator approach. J Indian Soc Remote Sens 41(2):319–329

Webster R, Burgess TM (1984) Sampling and bulking strategies for estimating soil properties in small regions. J Soil Sci 35:127–140

Webster R, Oliver M (2001) Geostatistics for environmental scientists. Wiley, Chichester 271 pp

Wright MN, Ziegler A (2017) Ranger: a fast implementation of random forests for high dimensional data in C++ and R. J Stat Softw 77(1):1–17

Xu Y, Dong D, Duan G, Yu X, Yu Z, Huang W (2013) Geostatistical analysis of soil nutrients based on GIS and geostatistics in the typical plain and hilly-ground area of Zhongxiang, Hubei Province. Open J Soil Sci 03:218–224

Yates SR, Warrick AW (1987) Estimating soil water content using cokriging. Soil Sci Soc Am J 51:23–30

Zhou F, Guo HC, Ho YS, Wu CZ (2007) Scientometric analysis of geostatistics using multivariate methods. Scientometrics 73:265–279

Part II
Geospatial Technologies in Land Resources Mapping

Chapter 9
Land Resource Inventory for Agricultural Land Use Planning Using Geospatial Techniques

S. K. Singh and S. Chatterji

Abstract ICAR-National Bureau of Soil Survey and Land Use Planning (NBSS&LUP), Nagpur, India, has taken major initiative to conduct land resource inventory (LRI) on 1:10000 scale at country level to provide vital inputs for site-specific soil and land resource management and agricultural land use planning. In LRI, different kinds of remote sensing data, such as digital terrain model (DTM), IRS-P6 LISS-IV data of 5.8 metre resolution, etc., were used. The present chapter discusses the conceptual approach of LRI and expected outputs. It involves systematic surveys of soils (cultivable land) on 1:10,000 scale and collection of other collateral data needed for scientific land use planning. ICAR-NBSS&LUP executed the programme in consortia mode by involving State Governments/State Departments of Agriculture, State Agricultural Universities, ISRO-National Remote Sensing Centre, State Remote Sensing Applications Centres and Soil and Land Use Survey of India. The said programme uses the latest time-efficient and cost-effective geospatial technologies and thereby ensures accuracy of the methodology. The nationwide survey helps to categorize the agricultural and non-agricultural areas in terms of their strengths, limitations and opportunities for appropriate use and threats from misuse/abuse. This also helps in developing perspective land use plans and monitoring their impact at macro (district/state) and micro level (village level). Special agricultural zones (present and potential) were delineated for giving focused support and services for major agricultural and horticultural production systems across the country. The importance of such land use planning could be further enhanced, if it is communicated/disseminated through well-designed geo-portal.

Keywords Geospatial techniques · Land resource inventory · Land resource management · Land use planning

S. K. Singh (✉) · S. Chatterji
ICAR-National Bureau of Soil Survey & Land Use Planning, Nagpur, India

G. P. O. Reddy, S. K. Singh (eds.), *Geospatial Technologies in Land Resources Mapping, Monitoring and Management*, Geotechnologies and the Environment 21,
https://doi.org/10.1007/978-3-319-78711-4_9

9.1 Introduction

The population of India is estimated to increase to 1.6 billion by 2050 from the current level of 1.2 billion, and the food grain production would need to be raised to 581 million tonnes from the present level of 241 million tonnes (Ghosh 2013). This additional food requirement will have to be produced from the projected cultivable land of 142 Mha (Kathpalia and Kapur 2011). However, the fact that per capita land holding will go down to an abysmally low 0.087 ha poses a grave concern (Sharma 2006). During the journey of about six decades of green revolution, about 121 million hectare land in the country suffered from various kinds of degradation (Maji et al. 2010). Recent data indicated that 64 out of 142 Mha of net sown area are affected with one or other kinds of degradation (NBSS&LUP 2015). In spite of exponential increase in fertilizer consumption, productivity of food grains was linear from 2009 onwards. The main reasons of stagnant agriculture productivity is inappropriate land use planning consisting of growing crops and applying technologies out of their capability/suitability domain. In this situation, increasing food production to 581 million tonnes in the next three to four decades from the present level of 241 million tonnes would be an arduous task (Ghosh 2013). Stagnant net sown area from 1980 onwards suggests a little possibility of horizontal expansion of agriculture. It is projected that crop land may be reduced by 20% in 2050 due to land degradation, urban expansion and conversion of crop land to nonagricultural land uses. The problem is expected to be compounded further due to the scarcity of water. At present, water availability for the country as a whole is around 2000 billion cubic metres (BCM), which will go down to 1500 BCM by 2050, thus downscaling India from a status of water sufficiency to water scarcity (Singh and Singh 2013).

Over the last few decades, there has been growing interest in watershed management, integrated agriculture system, precision farming, sustainable land use planning, ecosystem restoration, arresting degradation and ensuring food security (Lin et al. 2006). For executing such programmes, site-specific land resource information and situation-specific recommendations are prerequisite. Land resource inventory (LRI) on 1:10000 scale provides site-specific information needed for farm-/village-level planning. Land resource inventory programme has been taken in the country in a phased manner. In the first phase, 60 blocks representing 1 block in each agroecological subregion of India have been taken with a target to conduct survey in 3.3 Mha area in the duration of 4 years. During the year 2014–2015, mapping was taken up in 34 blocks of the country. LRI involves systematic surveys of soils (agricultural land) on 1:10000 scale and collection of other collateral data needed for scientific land use planning through geographic information system (GIS). The detailed database generated at farm level and its subsequent abstraction to village, mandal, taluk, district, state and country form the basis needed for prioritizing, initiating and executing any land-based developmental programmes. Application of remote sensing and GIS is an integral part of land resource inventory programmes.

In LRI, high-resolution remote sensing data of IRS-P6 LISS-IV and digital elevation model (DEM)/digital terrain model (DTM) have been widely used. A DEM can provide data that can assist the surveyor in mapping and deriving quantitative attributes of landform (Ardak et al. 2010; Sankar et al. 2010). DEM (NBSS&LUP 2015) is capable of providing very precise and quantified information on the degree, length and curvature of slope, which are much needed information for the execution of soil conservation programmes, irrigation planning and precision agriculture. Data on the degree, length and curvature of slope together with contours and drainage play an important role to quantify the water harvesting potentials (Lin et al. 2006). DEM also provides 3D perspective viewing of an area for precise delineation of landform and generation of contours at smaller interval. Thus, detailed characterization of soils together with added site-specific information has the potential to become the basis of any land-based planning programme. Integrated analysis and mining of data in GIS further enhance the capability of data to fulfil the requirement of research and development in the years to come.

Land use planning consisting of right land use, right technology on well-defined soil and site characteristics will be the cornerstone for arresting land degradation, mitigating climate change and ensuring food security (Singh et al. 2013a, b; Sharma et al. 2015). The value of the information would further widen if it is done by using the information generated through LRI and disseminated in the farthest and remotest village of the country through web geo-portal, where stakeholders could interact with the system and extract the desired information (Singh et al. 2013a, b). The present chapter discusses the need and relevance of LRI of the country on 1:10000 scale in generation of resource base information required for situation-specific agricultural land use planning and management of land resources. The conceptual model of LRI including the preparation of base map using remote sensing and GIS and generation of soil map (on 1:10000 scale) was also discussed. This chapter also highlights the utility of geo-portal in synthesizing database and mining the information from the database for their use towards the ultimate goal of sustainable agricultural land use planning and productivity enhancement.

9.2 Land Resource Management for Agricultural Land Use Planning: Challenges

9.2.1 Changing Bioclimatic Regimes

Vulnerability to climate change is a serious issue in land resource management for agricultural land use planning. It is expected that by 2050, there will be a drop in total precipitation by 20% and an increase in mean annual temperature by 1–3 °C (IPCC 2007). In view of the changing climate, it becomes important to revisit the

Table 9.1 Changing bioclimatic regime, India

Bioclimatic type	Area (1992) (Mha)	Area (1992) % TGA	Area (2015) (Mha)	Area (2015) % TGA	% change [increase (+) or decrease (−)]
Hyperarid	22.9	6.9	12.5	3.8	−45.41
Typic arid	22.7	8.3	24.6	7.5	+8.4
Semiarid (dry)	51.2	15.5	52.5	16.0	+2.50
Semiarid (moist)	72.2	21.9	93.5	28.5	+29.5
Subhumid (dry)	54.1	16.9	79.4	24.5	+46.7
Subhumid (moist)	39.8	12.0	14.8	4.5	−62.8
Subhumid (dry and moist transition)	21.0	6.8	–	–	–
Humid	16.6	5.0	28.9	8.8	+74.0
Per humid	20.5	6.2	21.9	6.7	–
Humid/per humid transition	1.8	0.5	–	–	–

bioclimatic map of India published in 1992 by ICAR-NBSS&LUP (Sehgal et al. 1992). Comparing it with the bioclimatic map of 2014, it is observed that changes in semiarid (dry) areas are insignificant, but semiarid (moist) areas have increased approximately by 30% from earlier estimated 72.2 million hectares (Table 9.1). This is at the expense of dry and moist subhumid areas. The dry subhumid areas have increased by approximately 47% from the earlier estimate of 54.1 Mha at the expense of moist and transitional subhumid areas of Maharashtra, Madhya Pradesh, Chhattisgarh (north west), Andhra Pradesh (central) and north fringes of Odisha, Jharkhand and Bihar (south). The humid area has increased by 74% from earlier estimated value of 16.6 Mha, which falls under coastal parts of West Bengal and Odisha, deltaic regions of Andhra Pradesh, Bengal basin, Tarai plains of West Bengal, Brahmaputra valleys and Western coastal plains (NBSS&LUP 2015). The study reveals the probable impact of climate change on land resource management for agricultural land use planning.

9.2.2 Underutilized Prime Agricultural Lands

Another issue affecting the agricultural land use planning is underutilized prime agricultural lands, which is defined as the land highly suitable for agriculture in a set of agroclimatic conditions. These are characterized as the land where moderately deep to deep (soil depth >75 cm), neutral to slightly alkaline (pH 6.5 to 8.5) soils occur on less than 8% slope. Total net sown area in the State of Maharashtra is 21.82 Mha. Out of which, 6.2 and 15.6 Mha are classified as prime and non-prime land, respectively. The study suggests that full potentiality of 70% of prime land is yet to be realized due to lack of assured irrigation, whereas, around 40% of the total non-prime land is managed out of their capability class due to the availability of irrigation. Both of these lead to extensive land degradation in the event of drought/excessive rainfall conditions (NBSS&LUP 2015).

9.2.3 Extensive Land Degradation in Crop Land Areas

In India, about 121 Mha area is estimated under different categories of degraded lands (Maji et al. 2010). The issue of estimation of extent and type of degradation prevailing only in the cultivated land in the country was not addressed for long. GIS-based approach has been followed to assess the degraded lands in different agricultural lands in the country. The study reveals that 64.8 Mha cultivated area is affected by different kinds of degradation. About 76% and 9.2% of the degraded area are affected by water and wind erosion, respectively, whereas salinity and sodicity affect 2.3 and 3.5% area of the cultivated land. Salinity and sodicity together with erosion are observed to affect another 0.04 and 1.54% area, respectively. Acidity and acidity plus water erosion affect 3.11 and 3.22% area, respectively (NBSS&LUP 2015).

9.2.4 Extensive Desertification in Arid Lands

Desertification mapping status in northern arid region of India indicated that the dominant processes of land degradation are wind, water, vegetal and salinization, which cover about 44.2, 11.2, 6.25 and 1.07% of total area of arid Rajasthan, respectively. The largest area under wind erosion is in Bikaner (25%, 25,332 km^2), followed by Jaisalmer (21%, 21,020 km^2), Churu (10%, 9790 km^2), Hanumangarh and Ganganagar district. More than 53% area of wind erosion was mapped in agricultural lands. Waterlogging was a localized problem, largest in Hanumangarh (143 km^2). Salinity was prevalent in 0.41% of the study area. Mining has degraded the land so far in 0.50% area (514 km^2), mostly in Jaisalmer (387 km^2) and Bikaner (123 km^2) districts (Kar et al. 2007; CAZRI 2008).

9.2.5 Extensive Multi-nutrient Deficiency

Soil nutrient mapping programme has been undertaken in different parts of the country. Extensive deficiencies of nitrogen, phosphorus, potassium, zinc and boron are reported. It is further observed that deficiencies of nutrient occur in groups of various combinations. The low balance of phosphorus, potassium and zinc occur extensively in Birbhum district of West Bengal (Fig. 9.1). On the integration of three layers in GIS, it was observed that the deficiency of these three nutrients occurs in seven combinations. Blanket application of fertilizers may not help in the present

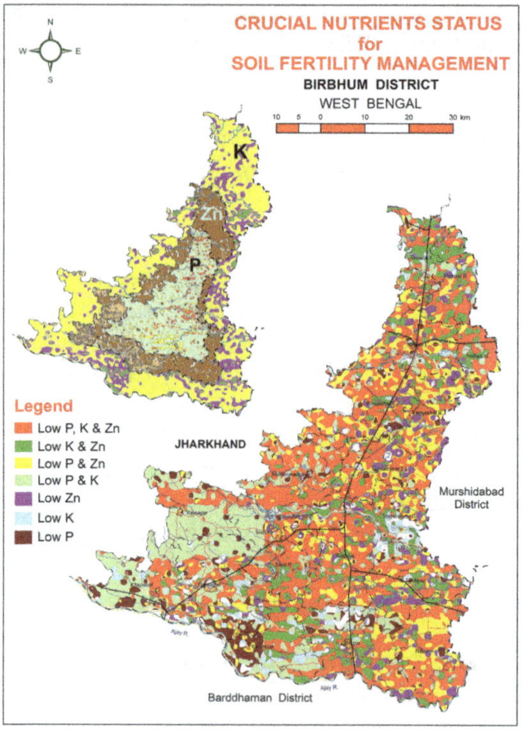

Fig. 9.1 Multi-nutrient deficiency in Birbhum district, West Bengal

situation of agriculture. Very precise and quantified resource base information is needed to enhance the productivity the present level (NBSS&LUP 2013).

9.3 Underutilized Land Resource Data

Studies conducted in different parts of the country indicate that the properties of soils play a major role for planning sustainable precision agriculture. It includes genesis, transformation, geochemistry of clay minerals, hydrological properties and morphological characteristics of soils (Singh et al. 1994, 2001a, b, 2002, 2003a, b, 2004); soil temperature and moisture regimes, bulk density, water holding capacity, moisture retention at field capacity and soil organic carbon, phosphorus and potassium stock (Singh et al. 2013a, b); soil organic carbon (SOC) potentials under different land use systems (Singh et al. 2005, 2007; Pandey et al. 2011); and oscillatory behaviour of SOC in different land uses (Singh et al. 2008), soil moisture and temperature regime (Singh et al. 2013a, b) and suitable crops and cropping sequences under set of soil and climatic conditions (Singh et al. 2007, 2013a, b). It shows that each and every crop and cropping sequence requires different types of biophysical conditions, which vary widely. Extensive degradation together with changing bioclimatic conditions further widens the variability. Uniform and blanket management recommendations certainly do not help in the present situation. There is a need of site- and situation-specific recommendation to break the prevailing yield plateauing of most of crops and cropping sequences. Therefore, large-scale mapping on 1:10000 scales of the biophysical conditions is conceptualized.

9.4 Conceptual Model for Land Resource Inventory on 1:10000 Scale

Soil survey is largely dependent on soil-landform relationship (Singh et al. 1994; Shyampura et al. 1994; Sharma et al. 1999; Gangopadhyay et al. 2012), and the map represents the static soil properties, which are acquired after a series of climatic episode (Lin et al. 2006). However, LRI is basically meant for developing sustainable land use plan, which is dynamic and dependent on present climatic conditions and the prevailing soil forming processes. Therefore, landscape ecological unit (LEU) is preferred over the landform as the basis of mapping. LEU is the assemblage of landform, slope and land use. Landform is the testimony of climatic events that occurred in the past, whereas, slope and land use represent the influence of present climatic conditions on the soil formation. The conceptual model of LRI programme is presented in Fig. 9.2. The assumption is if landform, slope and land use are identical, there is high possibility of getting similar kind of soils.

9.4.1 Remote Sensing Data in LRI

In LRI, different kinds of remote sensing data such as digital elevation model (DEM) of 10m resolution and IRS-P6 LISS-IV data of 5.8m resolution and other available datasets in public domain were used. In the undulating terrain, digital terrain model (DTM), which is the integral part of LRI, is prepared using Cartosat-1 data of 1 m resolution. Sahu et al. (2014) also used Cartosat-1 DTM and IRS P6 LISS-IV data of 5.8 m resolution for characterization of landforms, land use and land use/land cover (LULC) mapping in the basaltic terrain of Nagpur. In the flat terrain like Indo-Gangetic plains, Thar Desert of Rajasthan and coastal region of West Bengal, IRS-P6 LISS-IV data of 5.8 metre resolution is used. Recently the methodology was fine-tuned for the part of lower Gangetic plains using IRS LISS-IV data of 5.8 metre resolution (NBSS&LUP 2012). Stepwise methodology is explained in the subsequent paragraphs for the undulating terrain of southern Deccan plateau with the example of Indervelly Mandal, Adilabad district of Telangana State.

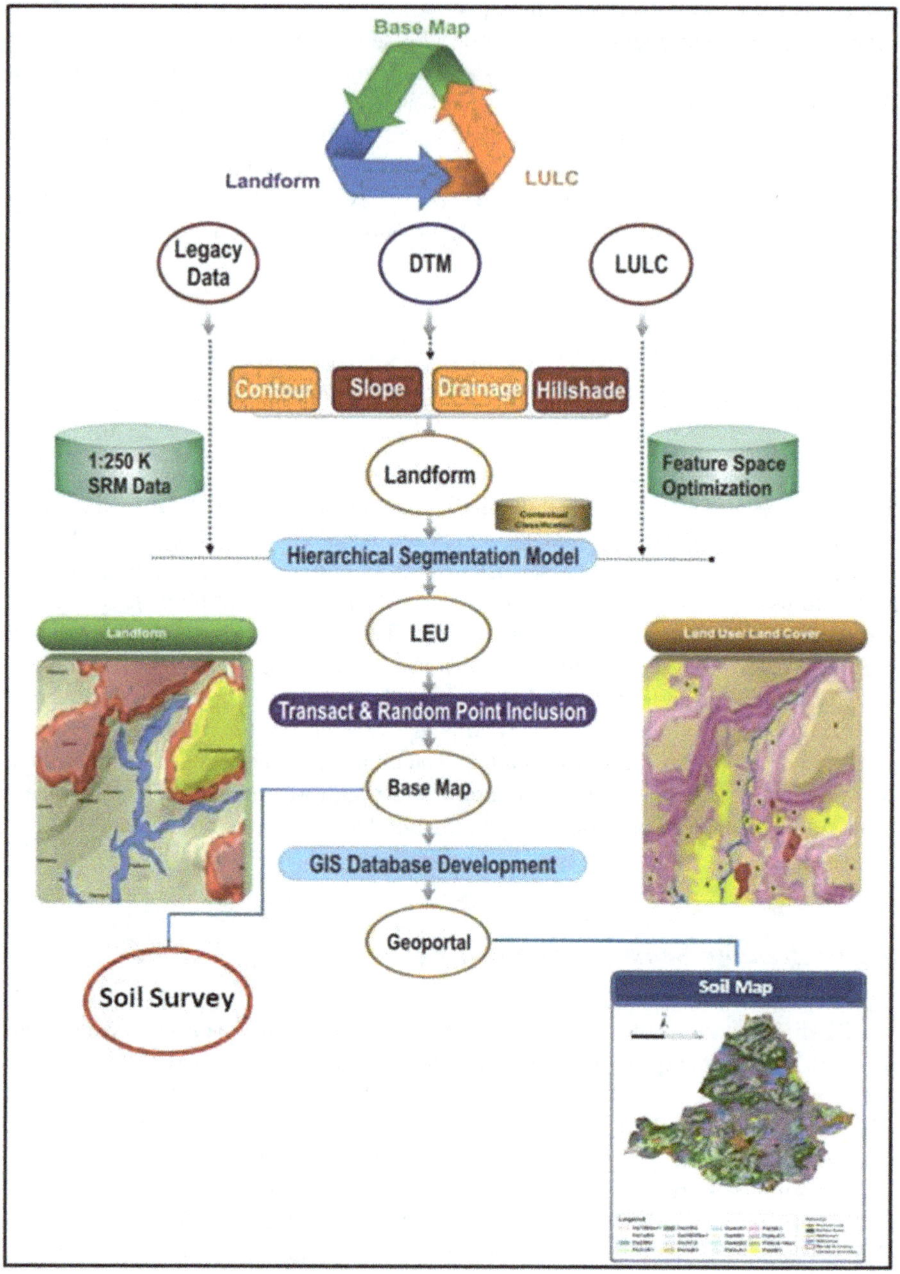

Fig. 9.2 Conceptual model of LRI programme

9.4.2 Digital Terrain Model (DTM)

Cartosat-1 stereo pair data of 1m resolution are processed to generate DTM of 10m spatial resolution. Steps including projection setup, sensor data reading, collection of GCPs and tie points, block adjustment, model computation (satellite math model) and epipolar image generation are followed using rigorous math model (Toutin's Model) before developing digital surface model (DSM) in Geomatica (version 14.0). Further, balancing algorithm and filters are applied to obtain the seamless mosaic and convert bare earth model DSM to DTM, respectively. Editing was done to smooth out the irregularities and create a quality output.

9.4.3 Slope

The object-based modelling using the spectral and spatial/contextual properties of pixels and segmentation process with interactive learning algorithm are used in the process. For extracting slope, the raster slope layer, the output of DTM is taken as input layer. The slope layer is classified in nine classes following the USDA-NRCS slope class threshold criteria. The criteria were fitted as fuzzy instead of hard rule using the less than and greater than "S-curve" membership function. Contextual filters are applied to generate smooth slope class zones.

9.4.4 Contour and Drainage

DTM is subjected to a series of hydro-enforcement process including reconditioning, sink and pit removal, flat and level waterbodies, flat and level bank to bank and gradient smoothening by DAT/EM and Arc Hydro tool, etc. for enriching the quality of the hydrological output such as slope, contour and drainage.

9.4.5 Landforms

Finally the terrain attributes like contours, drainage, slope and hill shade are treated as input layer for landform delineation. The landform classification process is hastened taking into consideration the slope class, hill shade, contour and auto-drainage pattern along with legacy mapping units of 1:2,50,000 scale.

9.4.6 Land Use/Land Cover (LULC)

LULC map has been developed using *rabi* season data of Cartosat-1-merged LISS-IV (2.5m) as well as high-resolution (0.5m) public domain imagery at the backend.

The delineation of subclasses, viz. single- and double-cropped areas, within the agriculture zone was carried out using novel LULC subclass classification algorithm. The merged data was segmented into spectrally homogeneous region using multi-resolution segmentation algorithm. The optimum scale parameter for segmentation of the layer is achieved through estimation of scale parameter (ESP) analysis tool. The point of interest lies where the local variance and rate of change are minimum in the graphical output. Feature space optimization as a data mining tool is applied to extract the double-cropped area based on certain number of layer variables and the vegetation index combination as obtained through the maximum separation distance.

9.5 Delineation and Characterization of Landscape Ecological Units

The integration of three secondary layers, i.e. landform, slope and land use/land cover, was carried out through the hierarchical object-based segmentation algorithm taking into consideration the area, morphology of the landform units and its relation with the neighbour objects to develop landscape ecological unit (LEU) map. The segmentation was accomplished in three levels. Level I segmentation is done based on the landform layer; in level II segmentation runs within each of the first-level segment based on fuzzy threshold-based slope class. Second-level intermediate outputs give rise to landform-slope unit. In level III the landform-slope segments of second level are further subdivided into landform-slope-land use unit, i.e. LEU, by incorporating the land use factor. The logical condition used to incorporate the land use factor is that the minimum overlap with the thematic polygon, i.e. level II segment, is more than or equal to 60%. The criteria ensure the continuity of LEU zone vis-à-vis soil boundary by ignoring negligible change in land use. Figure 9.3 explains the steps involved in the delineation of LEU.

A schema has been developed for delineating land ecological units. The interpretative unit consists of letters and numeral. Letters are designed for landforms;

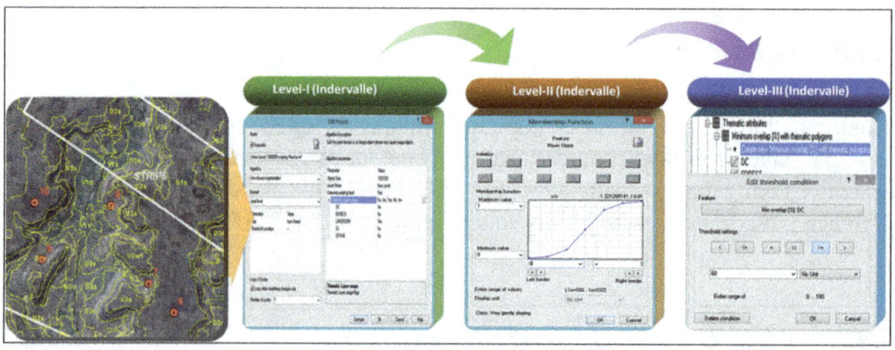

Fig. 9.3 Hierarchical object-based segmentation algorithm process for generating LEU maps

numeral represents slope class, followed by the letters for land use and land cover. In some places for segmenting the uniform pattern of landscape, image characteristics are also used for delineating LEUs particularly in Indo-Gangetic plains. The interpretative units are preceded with text and symbols to have linkage with physiographic, sub-physiographic and broad landform, defined during soil resource mapping of the country on 1:250,000 scale (Table 9.2).

Table 9.2 LEU interpretation (South Deccan Plateau: Gajwel Mandal, Medak district, Telangana state)

Physiography	Sub-physiography	Broad landform	Map unit	Legend description
Deccan Plateau (D)	South Deccan Plateau (Ds)	Granitic landscape – isolated hillocks	I	Isolated hillock
		Pediment	D2d	Very gently sloping pediment (double crop)
			D2s	Very gently sloping pediment (single crop)
			D2fa	Very gently sloping pediment (fallow)
			D3d	Gently sloping pediment (double crop)
			D3s	Gently sloping pediment (single crop)
			D3fa	Gently sloping pediment (fallow)
			D4fa	Moderately sloping pediment (fallow)
		Upland	U1d	Nearly level upland (double crop)
			U1s	Nearly level upland (single crop)
			U2d	Very gently sloping upland (double crop)
			U2s	Very gently sloping upland (single crop)
			U2fa	Very gently sloping upland (fallow)
			U3s	Gently sloping upland (single crop)
			U3fa	Upland, gently sloping upland (fallow)
			U4b	Moderately sloping upland (barren)
		Plains	Ap1d	Nearly level plain (double crop)
			Ap1s	Nearly level plain (single crop)
			Ap1fa	Nearly level plain (fallow)
		Basaltic landscape – valley	V1d	Nearly level valley (double crop)
			V1s	Nearly level valley (single crop)
			V1fa	Nearly level valley (fallow)
			V2d	Very gently sloping valley (double crop)
			V2fa	Very gently sloping plain (fallow), wasteland

9.6 Establishing Soil-Landscape Ecological Unit Relationship

Field studies are conducted to establish soil-LEU relationship by studying profiles, minipits and auger observation in well-defined strips. Numbers of strips are used to mark for covering entire units of LEU. The relationship between soil and LEU is described for Gajwel Mandal of Medak district in the State of Telangana (Table 9.3). For describing the relationship, the properties, which have significant bearing on management, are chosen. During the traverse of the area, majority of delineated LEU boundaries are checked and confirmed.

9.7 Soil Map

After extensive field work and soil correlation, soil series were established; phases of each soil series were defined. Phases of series include soil depth, surface texture, slope, erosion, gravelliness, salinity/sodicity and any other feature influencing management. The soil map consisting of series and phase for Gajwel Mandal, Medak district, Telangana State, is shown in Fig. 9.4. For collecting very precise and quantified information in the field and their subsequent transmission to the data centre for processing, ICAR-NBSS&LUP Nagpur has developed an android-based mobile application.

9.8 Database in NBSS Geo-portal

ICAR-NBSS&LUP initiated another programme on developing and deploying geo-portal for synthesizing present dataset of LRI, the legacy dataset of bureau and other organizations. It is a web-based platform deployed in a simple architecture with database server and application server to manage land resource database. Geo-portal contains the information in four modules, viz. polygon, point, raster and other nonspatial format (Singh and Singh 2013) for soils, climate, water resources, area-specific agro-techniques and socio-economic conditions of the farmers. Phases of soil series are redesignated as land management unit (LMU), which is linked with the cadastral map for giving farmer-wise and plotwise information on soils and land (Fig. 9.5). Depending on the authorization, the user is able to visualize soil information, upload maps, create new maps and merge them with other maps. The developed web-based NBSS geo-portal helps to acquire, process, store, distribute and improve the utilization and dissemination of geospatial data through Web Map Services (WMS).

Table 9.3 Soil-LEU relationship of Gajwel Mandal, Medak district, Telangana State

Broad landform	Pre-field interpretation	Post-field interpretation	Soil landform relationship
Pediment	D2s	D3s	Very shallow, well-drained, brown, loamy sand soils on gently sloping pediments, severe erosion
	D2s	D2s	Shallow, well-drained, dark yellowish brown, sandy clay soils on very gently sloping pediments with sandy clay loam surface and moderate erosion
	D2s	D1s	Moderately deep to deep, well-drained, strong brown, sandy clay soils on very gently sloping pediments with sandy loam to sandy clay loam surface and moderate erosion
	D2d	D2d	Moderately deep to deep, well-drained, strong brown, sandy clay soils on very gently sloping pediments with sandy loam to loam surface and moderate erosion
Upland	U2s	U2s	Moderately deep to deep, well-drained, strong brown, sandy clay soils on very gently sloping uplands with sandy loam surface and moderate erosion
	U2s	U2s	Deep to very deep, well-drained, brown, sandy clay on very gently sloping uplands with loamy sand surface and moderate erosion
	U2d	U2d	Deep to very deep, well-drained, brown, sandy clay soils on very gently sloping uplands with sandy clay surface and moderate erosion
	U1s	U1s	Deep to very deep, well-drained, brown, sandy clay soils on nearly level to level uplands with loamy sand to sandy loam surface and slight erosion
	U1d	U1d	Deep to very deep, well-drained, brown, sandy clay soils on nearly level to level uplands with loamy sand to sandy loam surface and slight erosion
Plains	AP1d	AP1d	Deep to very deep, moderately well-drained, dark grey, clayey soils on nearly level to level alluvial plains with sandy clay to clay surface and slight erosion
	AP1s	AP1s	Deep to very deep, moderately well-drained, dark grey, clayey soils on nearly level to level alluvial plains with sandy clay to clay surface and slight erosion
Valley	V1d	V1d	Deep to very deep, moderately well-drained, dark grey, clayey soils on nearly level to level valley with sandy clay surface and slight erosion
	VP1d	VP1d	Deep to very deep, moderately well-drained, dark grey, clayey soils on nearly level to level valley plains with clay loam to clay surface and slight erosion
	VP1s	VP1s	Moderately deep to deep, moderately well-drained, very dark grey, clayey soils on nearly level to level valley plain with sandy clay loam surface and slight erosion
	V1w1	V1w1	Deep to very deep, moderately well to imperfectly drained, dark grey, clayey soils on nearly level to level valley with clay loam surface and slight erosion

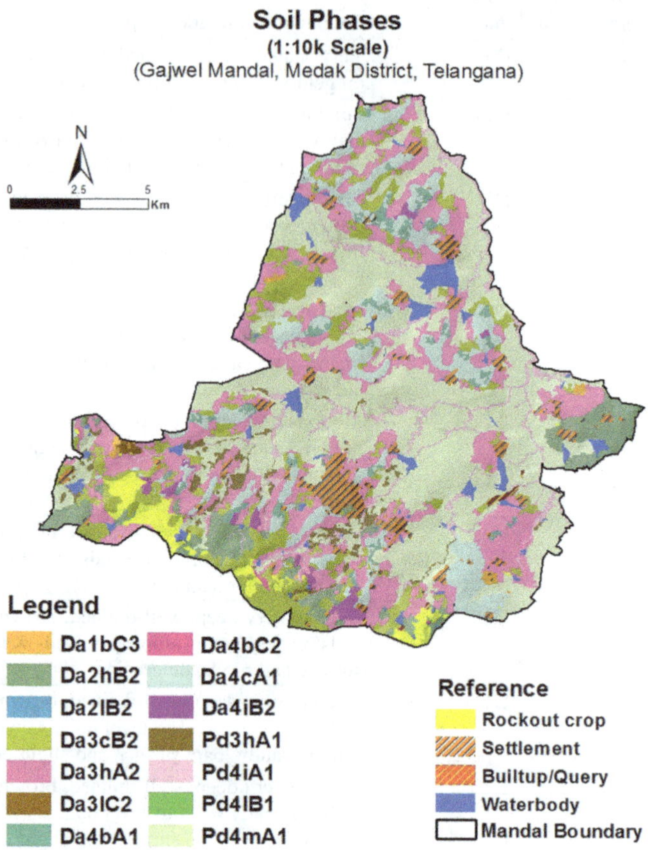

Fig. 9.4 Soil map of Gajwel Mandal of Medak district, Telangana State

9.9 LRI Data for Development of Soil Health Cards

Developing soil health cards includes designing sampling scheme, soil sample collection, analysis of soil health parameters in the laboratory, development of raster surfaces for different parameters using GIS techniques and attaching the surface layer with the cadastral information. Finally farmer-/plotwise information on soil health parameters is developed (NBSS&LUP 2015). In the present programme of LRI, samples have been collected at 325 × 325 metre grid interval for this purpose. Following such scheme, soil health cards were distributed for the farmers of Gajwel Mandal, Medak district in the State of Telangana.

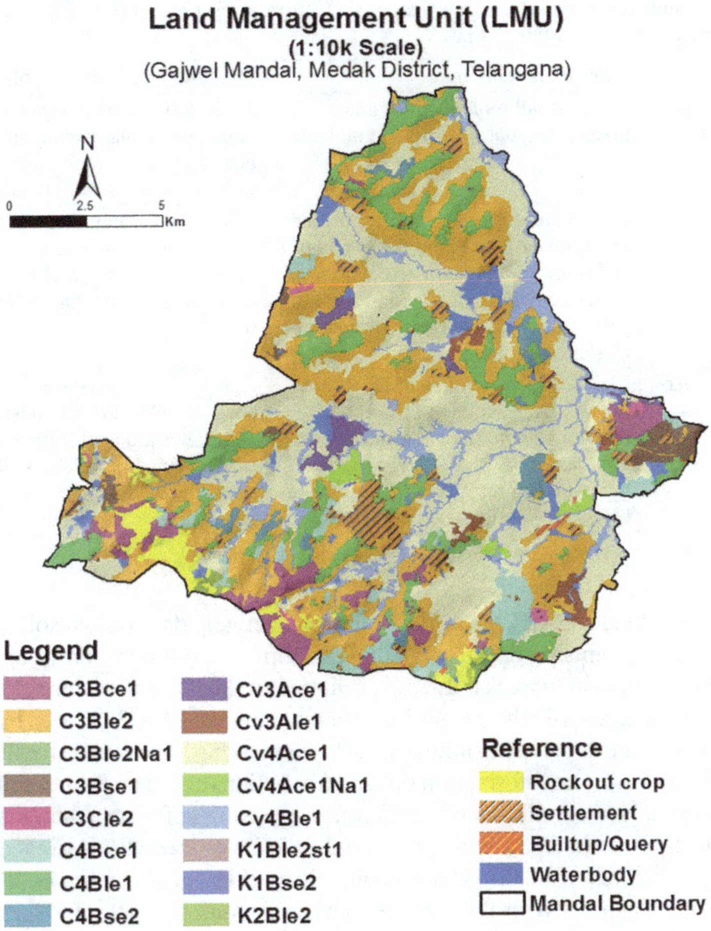

Fig. 9.5 Land management unit (LMU) map of Gajwel Mandal, Medak district, Telangana

9.10 LRI Database for Agricultural Land Use Planning

An interface is designed in the geo-portal, which integrates land resource information and agro-techniques and thus helps in identifying the best land use options for well-defined land management units. The land use options based on cost-benefit ratio for three well-defined land management units of Gajwel Mandal have been worked out. In LMU1 characterized with gravelly soils, sorghum is one of the most profitable options, whereas, in LMU2 grouped with deep black soils, cotton is the

Table 9.4 Land use options based on the cost-benefit ratio in three LMUs of Gajwel Mandal, Medak district, Telangana State, India

LMU	Short-term planning	Medium- and long-term planning
LMU1 (shallow gravelly soils)	Jowar, small millets, horse gram, green gram, black gram, cotton, maize	Agri-horti system comprising of custard apple, amla, jamun and ber; boundary plantation with *Gliricidia*; water harvesting farm ponds
LMU2 (deep black soils)	Cotton, maize, paddy, sunflower, red gram, jowar, maize/sunflower (short duration)-gram; cotton + red gram, millets, cowpea, green gram, black gram and vegetables like coriander, cluster bean, pumpkin, tomato, chilli	Agri-horti system (banana, mango, guava with drip); agri-horti-pasture system (*C. ciliaris*, stylo, *Brachiaria*); water harvesting farm ponds
LMU3 (moderately shallow brown soils)	Maize, cotton, sunflower, red gram, jowar, maize/sunflower (short duration)-gram, millets, paddy as aerobic rice cultivation, cowpea, green gram, black gram and vegetables like coriander, cluster bean, pumpkin	Agri-horti-pasture system (*C. ciliaris*, stylo, *Brachiaria* on bunds; mango, guava, custard apple, amla, jamun, ber and suitable MPTs; SWC measures and water harvesting farm ponds need to be created

most preferred land use. In LMU 3 defined for moderately dark brown soils, maize is one of the most promising options. On the medium- and long-term basis, agriculture is intensified with woody trees in LMU1, fruit trees in LMU2 and the mixture of both wood and fruit trees which are advised for planting in LMU3. Water harvesting mechanism is to be developed simultaneously for catering water needs for the cluster of farm (Table 9.4). The information on land use options could be further linked with the socio-economic conditions of the farmers by using software Automated Land Evaluation System (ALES) with geo-portal. ALES is specially designed to link biophysical features with the social economic profiles of the farmers. Table 9.5 illustrates the links between the socio-economic conditions of the farmers and the land use options.

9.11 Agricultural Land Use Planning at 1:10000 Scale: A Case Study

The Bali Island consisting of five villages (*Bijoynagar, Bijay Nagar, Bali, Amlamethi and Mathurakhud*) on 4430 ha area is located in Gosaba block, 24 Parganas (south) district (Sundarbans), West Bengal. Soils of Bali village were surveyed using IRS-P6 LISS-IV data and cadastral map. Occurrence of acid sulphate layers at different depths; acidity, salinity and poor drainage are the major soil constraints. Soils were classified into soil series Bali 1 and Bali 2 (Fig. 9.6). Acid

Table 9.5 Link between biophysical and socio-economic information

Suggested land use options based on B:C ratio					
LMUs	Marginal	Small	Semi-medium	Medium	Large
LMU1	Horse gram (2.13), sorghum (0.9), coconut (0.65)	Coconut (2.18), horse gram (1.72), sorghum (1.34), horse gram (2.47), sorghum (1.17), cotton (0.45)	Horse gram (3.66), mulberry (3.51), banana (1.34), sorghum (1.26), coconut (1.07), tomato (1.07)	Banana (4.6), beans (3.03), tomato (1.79), watermelon (1.65), horse gram (1.42), sorghum (1.03), cotton (0.9), coconut (0.31)	
LMU2	Mulberry (2.71), tomato (2.0), horse gram (1.98), chilli (1.63), sorghum (0.63), coconut (0.6)		Chilli (4.17), tomato (3.35), horse gram (0.86), cotton (0.44), sorghum (0.33)	Coconut (1.46), sorghum (0.85)	Coconut (4.92), tomato (2.13), watermelon (1.75)
LMU3	Cotton (0.83)	Horse gram (1.09), castor (0.65), sorghum (0.63)	Sorghum (1.04)		
LMU4	Coconut (3.09), tomato (1.8)	Sorghum (1.54), chilli (1.5), horse gram (1.45), tomato (0.83)	Horse gram (2.15), sorghum (0.95), coconut (0.32),	Horse gram (1.81), cotton (1.27), sorghum (0.75)	Coconut (3.09), tomato (1.8)
LMU5	Coconut (2.25)		coconut (0.93)		
LMU6		Mulberry (3.35), tomato (2.74), chilli (2.0), sorghum (1.73), horse gram (1.71), coconut (1.47)	Sorghum (1.36), coconut (0.99)		

sulphate layer occurred between the depths of 40 and 60 cm in Bali 1, whereas soils of Bali 2 contained the same between the depths of 80 and 100 cm. Based on the nature of constraints, three land management units were delineated (Fig. 9.7).

9.11.1 Package of Programmes

Farm pond technology was recommended for LMU3, shallow furrow and medium ridge for LMU2 and deep furrow with medium ridge in LMU1. After land shaping, vegetables and horticultural crops were recommended on the ridge, whereas, paddy

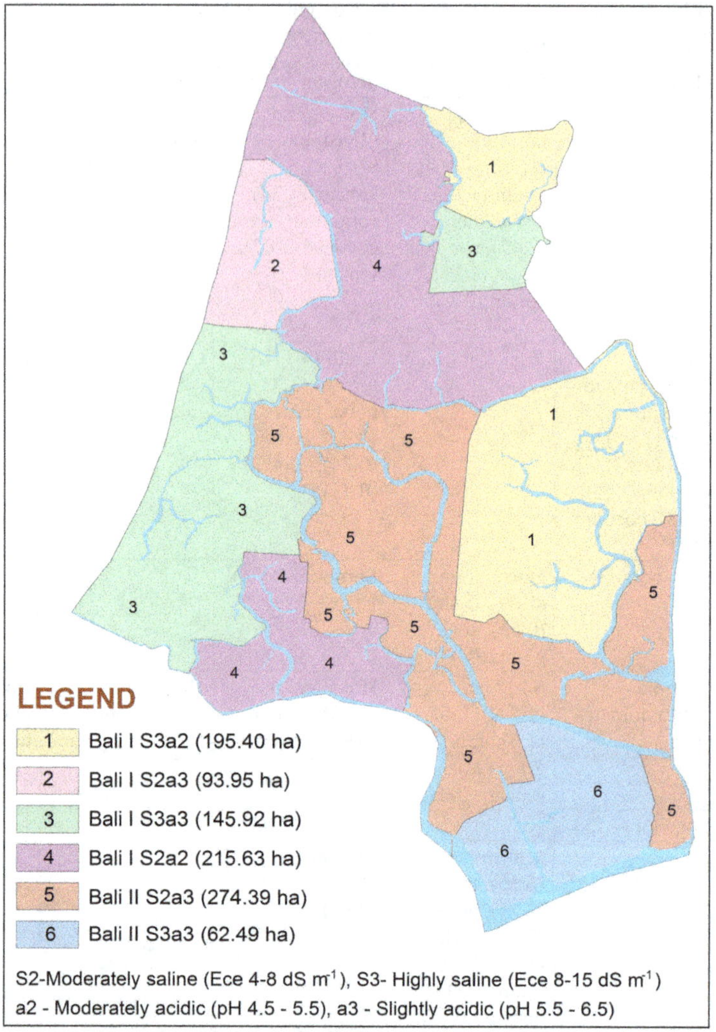

Fig. 9.6 Soil map of Bali village

cum fish was recommended in shallow and deep furrows. Farm ponds were exclusively used for fish farming.

9.11.2 Impact

The farm pond technology on 0.20, 0.27 and 0.39 ha under LMU3 generated employment to the tune of 285, 405 and 600 man-days, respectively, and enhanced the income from Rs. 2300 to Rs. 34,400, Rs. 3100 to Rs. 47,800 and Rs. 4500 to Rs. 68,900, respectively.

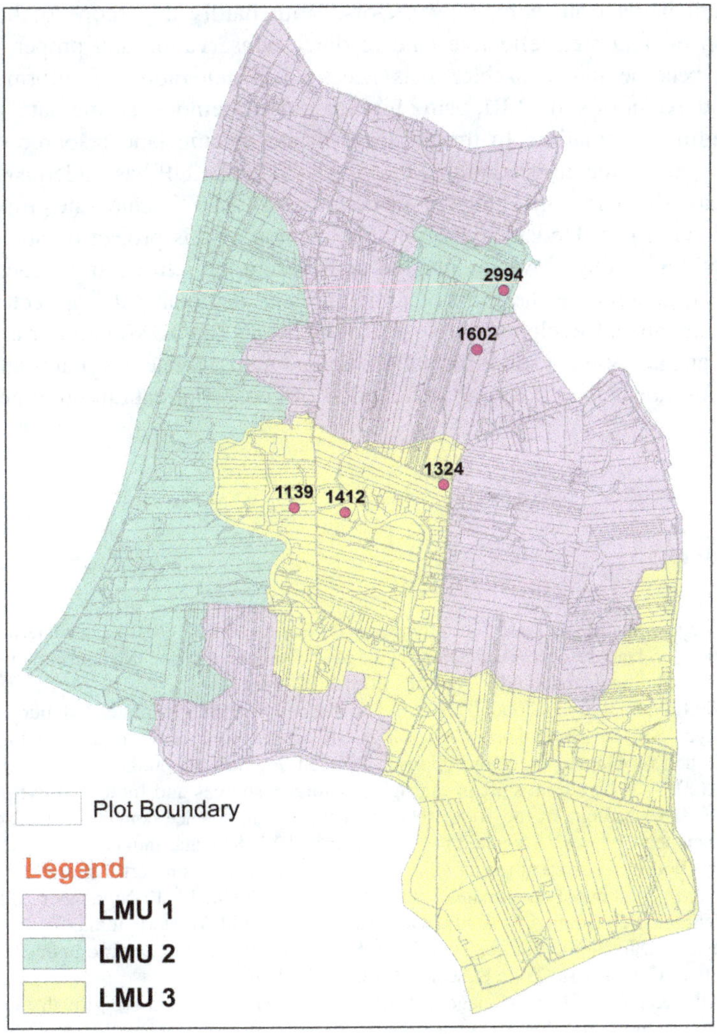

Fig. 9.7 Land management units in Bali village

9.12 Conclusions

Degradation of natural resources and public concern over environmental issues are shifting research priorities and funding towards broader issues, such as land, water, forests and biodiversity, pesticide safety and residue minimization, livestock waste management, water quality preservation and watershed protection. Success in meeting these challenges requires sharply increased skills and higher investments in research on natural resource management, social sciences and environmental issues. Another issue affecting the productivity is the rampant encroachment of prime agricultural lands from competing sectors. With hardly any scope of horizontal expansion of land area, effective land resource conservation and proper land use planning become indispensable. This necessitates generation of information on nature and extent of soils. LRI, using high-resolution remote sensing data, provides such needful information. In the absence of site-specific land resource data and situation-specific recommendations, the ICAR-NBSS&LUP has undertaken a project on Land Resources Inventory of the country on 1:10000 scale. The project plans to fill this vital gap. Understandably, the execution of this project on such a large scale would need huge investments, but considering the returns, it is generating or expected to generate in the form of judicious land use plans; the project assumes much significance. Developed datasets through NBSS BHOOMI geo-portal helps to visualize and analyse the soil information, upload maps, create new maps and merge them with other maps through dissemination of geospatial data through geo-portal services.

References

Ardak AS, Nagaraju MSS, Prasad J, Srivastava R, Barthwal AK (2010) Characterization and evaluation of land resources in Khapri village of Nagpur district, Maharashtra using high resolution satellite data and GIS. Agropedology 20:7–18

CAZRI (2008) Annual report 2006–07, Central Arid Zone Research Institute, Jodhpur

Gangopadhyay SK, Sarkar D, Sahoo AK, Singh SK (2012) Soils of the rainfed region of West Bengal, their problems, potentials appraisal. J Indian Soc Soil Sci 60:83–91

Ghosh PK (2013) A paradigm shift in managing natural resources and food security: strategy to action. In: Proceedings of the 100th session of the Indian science congress organised by the Indian Science Congress Association, 3–7 January, 2013, Kolkata, India

IPCC (2007) Fourth assessment report: climate change – a synthesis report. An assessment of the intergovernmental panel on climate change at Valencia, Spain, 12–17 November 2007

Kar A, Mohrana PC, Singh SK (2007) Desertification in arid Western India. In: Vittal KPR, Srivastava RL, Joshi NL, Kar A, Tiwari VP, Katju S (eds) Current scenario in dry land Indian perspective. Central Arid Zone Research Institute, Jodhpur

Kathpalia GN, Kapur R (2011) Management of land and water for inclusive growth : India 2050, Alternative futures: NRM South Asia

Lin H, Bouma J, Pachepsky Y, Western A, Thompson J, Genuchten R, Van V, H.J., & Lilly, A. (2006) Hydropedology: synergistic integration of pedology and hydrology. Water Resour Res. https://doi.org/10.1029/2005WR004085

Maji A, Reddy GPO, Sarkar D (2010) Degraded and waste land of India. Prasad R, Pathak PS (Eds.) Indian Council of Agricultural Research, New Delhi

NBSS&LUP (2012) Annual report 2011–12. National Bureau of Soil Survey and Land Use Planning, Nagpur

NBSS&LUP (2013) Annual report 2012–13. National Bureau of Soil Survey and Land Use Planning, Nagpur

NBSS&LUP (2015) Annual report 2014–15. National Bureau of Soil Survey and Land Use Planning, Nagpur

Pandey CB, Singh L, Singh SK (2011) Buttresses induced habitat heterogeneity increases nitrogen availability in tropical rainforests. For Ecol Manag 262:1679–1685

Sahu N, Reddy GPO, Kumar N, Nagaraju MSS, Srivastava R, Singh SK (2014) Characterization of landforms and land use/land Cover in basaltic terrain using IRS-P6 LISS-IV and Cartosat-1 DEM data: a case study. Agropedology 24:166–178

Sankar M, Sivasamy R, Dadhwal KS (2010) Land resource data and its evaluation for village level land use planning. Agropedology 20:1–6

Sehgal J, Mandal DK, Mandal C, Vadivelu S (1992) Agro ecological regions of India, NBSS Publ. No.24, 2nd edn. NBSS&LUP, Nagpur

Sharma N (2006) Population and economic growth as drivers of future land use in India. In: Browser F, BA MC (eds) Agriculture and climate beyond 2015, Volume 46 of the series environment and policy. Springer, Dordrecht

Sharma RK, Swami BN, Shyampura RL, Giri JD, Singh SK (1999) Characterization of some soils of Haldi Ghati region of Rajasthan in relation to land physiography. J Indian Soc Soil Sci 47:329–333

Sharma RP, Singh RS, Singh SK (2015) A review on agricultural land use planning: a case study of Shelwara district. Adv Appl Sci Res 6(8):125–133

Shyampura RL, Giri JD, Das K, Singh SK (1994) Soil physiographic relationship on transact in southern Rajasthan. J Indian Soc Soil Sci 42(4):622–625

Singh AK, Singh SK (2013) Strategies for arresting land degradation in South Asian countries. In: Sarkar D, Azad AK, Singh SK, Akter N (eds) Strategies for arresting land degradation in South Asian countries. SAARC Agricultural Centre, Dhaka

Singh SK, Das K, Shyampura RL (1994) Soil-physiographic relationship in basaltic terrain. J Indian Soc Soil Sci 42:663–664

Singh RS, Singh SK, Jain BL, Shyampura RL (2001a) Water retention characteristics of some Vertisols of Rajasthan. J Indian Soc Soil Sci 49:245–249

Singh SK, Baser BL, Shyampura RL (2001b) Variability in hydrological properties of Vertisols derived from two parent Materials. J Indian Soc Soil Sci 49:239–244

Singh SK, Baser BL, Shyampura RL (2002) Chemical composition and charge behavior of smectites in Vertisols of Rajasthan. J Indian Soc Soil Sci 50:106–111

Singh SK, Baser BL, Shyampura RL, Narain P (2003a) Phosphorus fractions and their relationship to weathering indices in Vertisols of Rajasthan. J Indian Soc Soil Sci 51:247–251

Singh SK, Baser BL, Shyampura RL, Pratap N (2003b) Genesis of lime nodules in vertisols of Rajasthan. J Indian Soc Soil Sci 51:273–278

Singh SK, Baser BL, Shyampura RL, Narain P (2004) Variations in morphometric properties of Vertisols in Rajasthan. J Indian Soc Soil Sci 52:114–119

Singh SK, Singh RS, Shyampura RL, Narain P (2005) Organic and inorganic carbon stock in soils of Rajasthan. J Indian Soc Soil Sci 53:281–287

Singh SK, Kumar M, Singh AK, Sharma BK, Tarafdar JC (2007) Carbon stock and organic carbon dynamics in soils of Rajasthan, India. J Arid Environ 68:408–421

Singh RK, Singh SK, Tarafdar JC (2008) Influence of crippoing sequence and nutrient management on soil organic carbon and nutrient status of Typic Rhodustalf. J Indian Soc Soil Sci 56:174–181

Singh SK, Kumar M, Pandey CB, Ghosh A, Mukhopadhyay S, Sarkar D (2013a) Difference in soil properties between irrigation and cropping sequence in Thar desert of Rajasthan. Arid Land Res Manag 27:17–31

Singh SK, Patil NG, Tiwari P, Chatterji S (2013b) Issue and strategies of natural resource management and land use planning in semi arid regions of India. Ann Arid Zone 52:1–13

Chapter 10
Open-Source Satellite Data and GIS for Land Resource Mapping

Nirmal Kumar, S. K. Singh, V. N. Mishra, G. P. Obi Reddy, and R. K. Bajpai

Abstract The capabilities of satellite data and geographic information system (GIS) to assess and monitor land resources are escalating with the technical advancements. However, the costs of these data and the GIS softwares are barrier to their use by broader community, particularly in developing countries. To maximize the scientific and societal benefits of the spatial data, open data and softwares are being promoted. Several satellite data from different sources with a wide array of spatial, spectral, temporal, and radiometric resolutions are now being made available free to all categories of users. To analyze the satellite data of widely differing characteristics and other spatial and nonspatial data from different sources, rapid advancements have been observed in free/open-source software (FOSS) development in the field of GIS. This chapter discusses the availability of open spatial data (satellite and vector) and open-source GIS softwares (OSGS). The viability of adoption of OSGS as an alternative to proprietary softwares has also been discussed.

Keywords Open-source GIS · Open data · Open standards · Remote sensing

10.1 Introduction

Remote sensing and GIS are undeniably important methodological tools for studies related to inventory, surveying, and monitoring of natural resources (Burrough 1986; Singh 1989; Trotter 1991; Brown 2004; Miller et al. 2013; Kumar et al. 2015). The satellite data, with a synoptic view of the Earth's surface, submeter to kilometer spatial resolutions, and hourly to monthly temporal resolutions, has been used as a significant information source for these studies (Kumar 2013; Dowman and Reuter 2016). Data continuity, affordability, and access are three important factors that

N. Kumar (✉) · S. K. Singh · G. P. O. Reddy
ICAR-National Bureau of Soil Survey & Land Use Planning, Nagpur, India

V. N. Mishra · R. K. Bajpai
Department of Soil Science & Agricultural Chemistry, IGKV, Raipur, India

G. P. O. Reddy, S. K. Singh (eds.), *Geospatial Technologies in Land Resources Mapping, Monitoring and Management*, Geotechnologies and the Environment 21,
https://doi.org/10.1007/978-3-319-78711-4_10

significantly affect the utility of satellite data (Turner et al. 2015). Data continuity relates to the maintenance of long-term satellite data products. A long-term data allows tracking the causes and spatial pattern of changes in the Earth's surface features. Earth observation programs, with standard imaging specifications and internationally agreed complementary satellite launches, can ensure data continuity with similar quality images. Data affordability arises from the cost of the imagery, and data access addresses the ability of researchers to discover, retrieve, manipulate, and extract value from satellite imagery as well as integrate with other datasets. GIS has become an integral tool for these processes. It allows integration and analysis of spatial data, collected from various sources (Burrough and McDonnell 1998; Worboys and Duckham 2004; Liu and Philippa 2009; Kumar 2013). The use of GIS has produced remarkable changes in the way and rate at which spatial data are generated, updated, analyzed, and disseminated, making production and analysis of geographic information very efficient.

However, the highly priced proprietary softwares and unavailability of public satellite data have significant impact on their use, particularly in developing countries. A large number of satellites are currently orbiting the Earth. The vast majority of these are relaying data to the Earth, and many are equipped with high-power cameras. But until recently they were available only to those who could pay for them. Nowadays, much satellite Earth observation data are open and available for download free of charge, especially at medium to low spatial resolution (Abdi 2013; Harris and Baumann 2015; Bakillah and Liang 2016; Dowman and Reuter 2016; Rocchini et al. 2017). New data policies promoting free and open access to satellite imagery are expanding the use of satellite imagery (Miller et al. 2013; Turner et al. 2015; Harris and Baumann 2015; Dowman and Reuter 2016). A detailed study of open data policies of different organizations, which relate in one way or the other to satellite Earth observation data, is presented by Harris and Baumann (2015). From late 2008 when the entire archive of Landsat imagery, including all new acquisitions, became available online at no cost to September 2012, 9 million scenes were downloaded, with more than 3 million scenes distributed in 2012 alone. Prior to the availability of no cost imagery, the most scenes sold in 1 year totaled 25,000 (Miller et al. 2013). The current level of use will likely increase among these users, particularly as it becomes better known that the imagery is available at no cost and as new use cases are identified.

With rapid developments in the capabilities of GIS tools, the use of satellite data is increasing. However, proprietary GIS softwares are highly priced with additional charges for specific modules. Many of such licensed softwares remain unutilized in various institutes and research centers because they require either extensive training or easy availability on multiple locations. These may not be compatible with all operating systems. All these problems can be solved by the use of free/open-source software tools. Open-source (OS) software is software where the source code is made available under a license that allows the modification and redistribution of the software at will. OSGIS is free of cost and can be installed on multiple operating systems, such as Windows, MacOS, and Linux. Several OSGIS software packages (GRASS, QGIS, SAGA, etc.) are available and freely accessible to the users—

covering all sectors of geospatial data handling (desktop GIS, remote sensing software, GIS libraries, spatial database management system, GIS servers, and web map server)—some of them having very good capabilities. The usage of these types of software is increasingly growing because of reasons such as security, open standards, vendor independence. Increasing download rates of open-source GIS softwares, their use cases, and financial support by government organizations to projects on open-source GIS explain their rise in popularity (Steiniger and Bocher 2009). This chapter discusses the availability of free satellite data and free/open-source GIS softwares.

10.2 Open Satellite Data

Free global satellite image data of spatial resolutions of 1 m to 1 km and spectral resolution and from PAN to hyperspectral are available from an increasing number of sources (Table 10.1). Here are the most important archives and other useful data:

10.2.1 Coarse Resolution Satellite Data

MODIS (Moderate Resolution Imaging Spectroradiometer) is an extensive program using sensors on two satellites that each provide complete daily coverage of the Earth. The data have a variety of resolutions: spectral, spatial, and temporal. It has 36 bands ranging from visible to thermal region. The spatial resolution varies from 250 m (bands 1–2) to 500 m (bands 3–7) and 1000 m (bands 8–36). Because the MODIS sensor is carried on both the Terra and Aqua satellites, it is generally possible to obtain images in the morning (Terra) and the afternoon (Aqua) for any particular location. Nighttime data are also available in the thermal range of the spectrum. These data are available freely for download from the http://earthexplorer.usgs.gov/, http://glovis.usgs.gov and http://reverb.echo.nasa.gov/reverb. While the user can order a daily scene for any specific date and location, many consolidated products have been developed with MODIS data. These include 8-day or 16-day composite images, a variety of indices, and a range of other global products with different time scales. Products are separated into four science discipline groups, namely, land, atmosphere, ocean, and cryosphere products. Processed MODIS data such as the Surface Reflectance products are normally segmented into tiles with an area of $10^\circ \times 10^\circ$ using a sinusoidal projection.

The Advanced Very High Resolution Radiometer (AVHRR) provides four- to six-band multispectral data from the NOAA polar-orbiting satellite series. There is fairly continuous global coverage since June 1979, with morning and afternoon acquisitions available. The resolution is 1.1 km at nadir. The number of available bands will depend on the satellite. It also provides Normalized Difference Vegetation Index (NDVI) Composites produced from multiple AVHRR daily observations

Table 10.1 A list of freely available satellite data

Sensor	Satellite	Resolution			Scene size	Availability
		Spatial	Spectral	Temporal		
Multispectral data						
Landsat MSS	Landsat 1–3 (bands 4–7)	60* m	4 bands, VNIR	18 days		July 1972 to October 1992 and June 2012 to January 2013
	Landsat 4–5 (bands 1–4)			16 days		
Landsat TM	Landsat 4–5	30 m (bands 1–5, 7)	7 bands, VNIR, SWIR, TIR	16 days	170 km × 183 km	July 1982 to May 2012
		120** m (band 6)				
Landsat ETM$^+$	Landsat 7	30 m (bands 1–5, 7)	8 bands, VNIR, SWIR, TIR, PAN	16 days	170 km × 183 km	July 1999 to May 2003 (SLC on)
		60** m (band 6)				May 30, 2003 till date (SLC off)
		15 m (band 8)				
Landsat OLI/TIRS	Landsat 8	30 m (bands 1–7, 9)	11 bands, VNIR, SWIR, TIR, PAN	16 days	170 km × 183 km	March 2013 till date
		100 m (bands 10, 11)				
		15 m (band 8)				
ALI	EO 1	30 m (bands 1–7)	10 bands, VNIR, SWIR, PAN	16 days	37 km × 42 km	May 2001 to March 2017
		10m (band 8)				
MODIS	Terra and Aqua	250 m (bands 1, 2)	36 bands	1–2 days	2330 km × 10 km	February 2000 till date
		500 m (bands 3–7)	VNIR, SWIR, TIR			
		1 km (bands 8–36)				

(continued)

Table 10.1 (continued)

Sensor	Satellite	Resolution			Scene size	Availability
		Spatial	Spectral	Temporal		
AVHRR	NOAA	1.1 km	6 bands VNIR, SWIR, TIR		2400 km × 6400 km	1981 till date
LISS-III	Resourcesat 1	24 m	4 bands VNIR, SWIR	24 days	140 km	Selected dates
AWiFS	Resourcesat 1	56 m	4 bands VNIR, SWIR	5 days	740 km	
MIS	Sentinel 2A	10 m (bands 2–4, 8) 20 m (bands 5–7, 8a, 11, 12) 60 m (bands 1, 9, 10)	13 bands VNIR, SWIR	10 days	100 km × 100 km	From October 2015
Hyperspectral data						
Hyperion	EO 1	30 m	220 bands VNIR, SWIR		7.7 km × 42 km	May 2001 to March 2017
HySI	IMS1	500 m	64 bands VNIR	24 days	128 km	Selected dates
Digital elevation models						
SRTM	Shuttle Radar Topography Mission	30 m	C-band SAR	–	1° × 1°	–
ASTER	ASTER	30 m	Optical	–	1° × 1°	–
Cartosat	Cartosat	30 m	Optical	–	1° × 1°	–
ALOS	ALOS	30 m	L-band SAR	–	1° × 1°	–
High-resolution data						
OrbView 3	GeoEye	1 m	PAN	3 day		Selected number of data

* Spatial Resolution: 68 m × 83 m (commonly resampled to 60 m)
** Resampled to 30 m

that have been composited together to create a nearly cloud-free image showing maximum greenness. An NDVI ratio is produced from bands 1 and 2 of the AVHRR composite to produce a derived NDVI band composite.

10.2.2 Medium-Resolution Satellite Data

Some of the most popular and valued remote sensing images of the Earth are obtained from the Landsat satellites, which have been orbiting the Earth for over 45 years. Landsat satellites provide multispectral, moderate-resolution land imagery of the whole globe free of cost from 1972 till today with no user restrictions. In 2008, the owners of the data—NASA and the US Geological Survey—agreed to provide free access to Landsat's archives. Single scene Tri-Decadal Global Landsat Orthorectified MSS, TM, ETM+, and ETM+ Pan-sharpened data may be browsed, searched, and downloaded through EarthExplorer or the USGS Global Visualization Viewer (GloVis). Mosaic data products, which are also available for Tri-Decadal Global Landsat Orthorectified TM and ETM+ Pan-sharpened data, may be searched and downloaded through EarthExplorer. Even the data from newly launched Landsat 8 are also downloadable freely.

Linear Imaging Self Scanning (LISS-III) and Advanced Wide Field Sensor's (AWiFS) data from Resourcesat 1 (IRS-P6) for certain dates have been made available from 2011 to be downloaded freely from National Remote Sensing Centre (NRSC) (http://bhuvan-noeda.nrsc.gov.in/download/download/download.php), limited to India. Data for other parts of the world can be searched and downloaded from USGS EarthExplorer. However, the coverage is not global. The LISS-III data have four bands with 23.5 m spatial resolution, whereas, AWiFS operates in four spectral bands identical to LISS-III, providing a spatial resolution of 56 m and covering a swath of 740 Km.

The European Commission provides free, full, and open access to a wealth of important environmental data gathered by Copernicus, Europe's Earth observation system. Copernicus is an Earth observation program operated by a partnership of the European Commission (EC), the European Space Agency (ESA), and the European Environment Agency (EEA). The Sentinel-2 mission consists of two satellites developed to support vegetation, land cover, and environmental monitoring. The Sentinel-2A satellite was launched by ESA on June 23, 2015, and operates in a sun-synchronous orbit with a 10-day repeat cycle. A second identical satellite (Sentinel-2B) was launched on March 7, 2017, and is expected to be operational for data acquisitions in 3–4 months. Together they will cover all Earth's land surfaces, large islands, and inland and coastal waters every 5 days. The data from the Sentinel-2 Multispectral Instrument (MSI) is available to free download from October, 2015. It acquires images in 13 spectral bands ranging from visible and near-infrared (VNIR) to shortwave infrared (SWIR) wavelengths along a 290-km orbital swath. Out of the 13 bands, 4 bands (blue, green, red, and NIR) are having spatial resolution of 10 m. Four narrow bands in red and NIR region for vegetation characterization and two larger SWIR bands for applications such as snow/ice/cloud detection or vegetation moisture stress assessment are having resolution of 20 m. The resolution of the other three bands mainly for cloud screening and atmospheric corrections is 60 m.

10.2.3 High-Resolution Satellite Data

Policy documents and legal instruments concerning open satellite data are predominantly in favor of full, free of charge, and open access to data (Harris and Baumann 2015). However, high-resolution satellite data have not been made open for reasons like financial viability, proprietary interests, defense and security, national legislations, etc. A limited number of scenes (only 179,981 worldwide) of OrbView PAN data are made available by USGS with no restrictions.

10.2.4 Hyperspectral Satellite Data

Hyperspectral remote sensing, also known as imaging spectroscopy, is a relatively new technology that is currently being investigated by researchers and scientists with regard to the detection and identification of minerals, terrestrial vegetation, and man-made materials. Multispectral data are acquired in a relatively small number (3–10) of spectral bands with broad spectral bandwidths (~10 nm), whereas hyperspectral data bands are numerous (10s–100s) and narrow (~10 nm). Many surface materials have unique absorption and reflectance characteristics that are only 10–20 nm wide; the detailed reflectance spectrum acquired by hyperspectral remote sensing makes it possible to identify and distinguish material and conditions on the ground in ways that are impossible even with very high-resolution multispectral imagery. As of now only two satellite-based hyperspectral data are available, namely, Hyperion (30 m resolution) onboard EO-1 satellite of NASA and HySi (500 m resolution) onboard IMS-1 of ISRO. The Hyperion data, which have been once purchased, are available for download freely from http://earthexplorer.usgs.gov/ and http://glovis.usgs.gov. The HySI data are available to download freely from http://bhuvan-noeda.nrsc.gov.in/download/download/download.php for India.

10.3 Digital Elevation Models (DEM)

The natural resource management researchers are increasingly aware of the importance of DEMs in their applications. DEMs are generated either from stereo pairs or from *radar* interferometer. Optical sensors like ASTER (the Advanced Spaceborne Thermal Emission and Reflection Radiometer) and Cartosat having in-track stereo vision are capable of generating DEMs. ASTER Global Digital Elevation Model (GDEM) at 30 m spatial resolution was developed jointly by the US National Aeronautics and Space Administration (NASA) and Japan's Ministry of Economy, Trade, and Industry (METI) and is available freely downloadable from a number of sources including http://earthexplorer.usgs.gov/, http://glovis.usgs.gov and http://reverb.echo.nasa.gov/reverb. The Cartosat DEM at 30 m resolution is available for download from http://bhuvan-noeda.nrsc.gov.in/download/download/download.php for India only with, some strategic area masked out.

The NASA Shuttle Radar Topographic Mission (SRTM) has provided DEMs for over 80% of the globe. The void filled data is distributed free of charge by USGS and is available for download. Earlier, these data were available at 1 arc second (approx. 30 m resolution) and 3 arc second (approx. 90 m resolution), respectively, for ISA and other countries. Recently, the 1 arc second data product is made available for all countries. However, DEM tiles above 50° north and below 50° south latitude are sampled at a resolution of 2 arc second by 1 arc second. The vertical error of the DEM is reported to be less than 16 m. The data currently being distributed by NASA/USGS (finished product) contains "no-data" holes where water or heavy shadow prevented the quantification of elevation. These are generally small holes, which nevertheless render the data less useful, especially in fields of hydrological modeling. The SRTM digital elevation data provided on http://srtm.csi.cgiar.org/ has been processed to fill data voids and to facilitate its ease of use by a wide group of potential users. This data is provided in an effort to promote the use of geospatial science and applications for sustainable development and resource conservation in the developing world.

On May 2016 the Japan Aerospace Exploration Agency (JAXA) released the digital surface model (DSM) dataset for global terrestrial region with a horizontal resolution of approx. 30 m (1 arc sec) free of charge. The dataset has been compiled using the Advanced Land Observing Satellite "DAICHI" (ALOS)—PALSAR's L-band. The dataset is published based on the DSM dataset (5-m mesh version) of the "World 3D Topographic Data," which is the most precise global-scale elevation data at this time, and its elevation precision is also at a world-leading level as a 30-m mesh version (http://www.eorc.jaxa.jp/ALOS/en/aw3d30/). The users are free to download and use unless they display that the original data is provided by JAXA by showing the copyright (© JAXA).

10.4 Open-Access Vector Data

Natural Earth is a public domain map dataset available at 1:10 m, 1:50 m, and 1:110 million scales. Natural Earth was built through a collaboration of many volunteers and is supported by NACIS (North American Cartographic Information Society) and is free for use in any type of project in any manner, including modifying the content and design, electronic dissemination, and offset printing. The data can be downloaded from http://www.naturalearthdata.com/downloads/. Another open-source vector database is OpenStreetMap®, licensed under the Open Data Commons Open Database License (ODbL) by the OpenStreetMap Foundation (OSMF) (Bakillah and Liang 2016). OpenStreetMap (OSM) is a collaborative project to create a free editable map of the world. Users are free to copy, distribute, transmit, and adapt the data, as long as credit is given to OpenStreetMap and its contributors. If the users alter or build upon this data, the result may be distributed only under the same license.

10.5 Open-Source GIS Software Packages

The criteria used for selecting suitable GIS software include cost, standards, functionality, and usability (Wang and Wang 2001; Ven et al. 2008; Chen et al. 2010; Fitzgerald 2011; Brovelli et al. 2017). Cost is the most important factor in selecting software packages (Keil and Tiwana 2006), particularly in developing countries. It includes the expenditure-associated software license, training, maintenance, software upgradation, add-on modules, and support services costs. Open-source softwares are available free including the source code and are protected under "free software licenses" to ensure the legal rights of the users and the freedoms of the software. The major free software license templates are General Public License (GPL), Lesser General Public License (LGPL), Eclipse Public License (EPL), Mozilla Public License (MPL), and Berkeley Software Distribution (BSD), a detail of which can be found in Steiniger and Bocher (2009). The availability of source code enables development and free availability of customized add-on applications and rapid identification, reporting, and fixation of bugs. For continued improvements, maintenance, and support, the users need not to be dependent on a single vendor or group as in case of proprietary softwares. Further, for each project, global community of users and developers is available for asking questions and advice. Support includes detailed documentation, forums, wikis, newsgroups, email lists, and live chat. None of this costs anything except time.

Open-source developers prefer to adhere to open standards like those of the Open Geospatial Consortium (OGC). This improves interoperability within and between open-source and proprietary software. The OGC is an international not for profit organization committed to making quality open standards for the global geospatial community. The members include representatives from government, commercial organizations, NGOs, and academic and research organizations. OGC are at the core of major infrastructures such as the European Union's INSPIRE spatial data infrastructure (SDI) (Bakillah and Liang 2016).

A GIS project involves spatial data collection/creation, conversion, data storage, manipulation, analysis, and visualization. Based on these functionalities, four major types of GIS softwares are found: desktop GIS, GIS libraries, GIS database management systems, and web GIS (Steiniger and Hunter 2013). Over the last two decades, an increasing number of free and open-source softwares have been developed in all categories (Table 10.2). Some desktop GIS such as GRASS also provides

Table 10.2 A list of popular GIS softwares in different categories

Software	Current version	Operating system	Tribe	License	Home page
Desktop GIS					
MapWindow GIS	MapWindow5.1.1.0/ Jun 14, 2016	Windows	C#	MPL	http://www.mapwindow.org
Quantum GIS	2.18.10 (Las Palmas)/ June 23, 2017	Linux, Mac OSX, Windows, Android	C++, Python, Qt	GNU GPL	http://www.qgis.org
uDig	2.0.0.RC1	Windows, Linux, Mac OSX	Java	EPL + BSD	http://udig.refractions.net
GRASS	GRASS GIS 7.2.1/ May 3, 2017	Cross platform	C, C++, Python, Tcl	GNU GPL	http://grass.itc.it/
SAGA	SAGA-5/ June 30, 2017	Cross platform	C++	GNU LGPL	http://www.saga-gis.uni-goettingen.de/
JUMP	OpenJUMP 1.11/ April 13, 2017	Cross platform	Java	GNU GPL	www.openjump.org
ILWIS	ILWIS 3.08.05/ September 3, 2015	Windows	C++	GNU GPL	http://52north.org/index.php?
gvSIG	gvSIG 2.3.1/ October 26, 2016	Windows, Linux, Mac OS*	Java	GNU GPL	http://www.gvsig.com/en
GIS libraries					
GDAL/ORG	GDAL/OGR 2.2.1/ June 23, 2017	Cross platform	C, C++	X/MIT	http://www.gdal.org/
Proj4	Proj.4 4.9.3/ September 2, 2016	Cross platform	C	MIT	http://proj4.org/
Geo Tools	GeoTools 17.1/ May 19, 2017	Cross platform	Java	GNU LGPL	http://www.geotools.org/
JTS	1.14 / September 23, 2015	Cross platform	Java	GNU LGPL	github.com/locationtech/jts
TerraLib	TerraLib: 4.2.2/ February 18, 2013	Linux, Windows	C++	GNU LGPL	http://www.terralib.org/
GIS database					
Post GIS	PostGIS 2.3.3/ July 1, 2017	Linux, Windows, Mac OS X, POSIX-compliant systems	C++	GNU GPL	http://postgis.net/
SpatiaLite	4.3.0a / September 7, 2015	Linux, Windows, Mac	C++	MPL, GPL,	https://www.gaia-gis.it/

(continued)

Table 10.2 (continued)

Software	Current version	Operating system	Tribe	License	Home page
		OS X, POSIX-compliant systems		LGPL tri-license	fossil/ libspatialite/ home
Web GIS					
MapServer	7.0.6 / June 21, 2017	Cross platform	C/ C++	X/MIT	www. mapserver. org
GeoServer	2.11.1/ May 19, 2017	Linux, Windows, Mac OS X, POSIX-compliant systems	Java	GNU GPL	http:// geoserver. org
Mapnik	3.0.12/ September 08, 2016	Cross platform	C++	GNU LGPL	http:// mapnik.org/
MapGuide Open Source	3.0/ October 11, 2016	Windows; Linux	C++	GNU LGPL	http:// mapguide. osgeo.org/
deegree	3.3.20/ September 15, 2016	Cross platform	Java	GNU LGPL	http://www. deegree.org/
Mobile GIS					
gvSIG Mobile	gvSIG Mobile 0.3/ April 23, 2010	Windows Mobile 5.0 or higher, Window XP	Java	GNU GPL	http://www. gvsig.com/ en/products/ gvsig-mobile/
QGIS Experimental	QGIS Experimental/ April 23, 2017	Android	C++	GNU GPL	http://www. qgis.org

*MAC OS X 64 bits: Only portable version

remote sensing functionality such as raster processing, image correction, enhancement, transformation, classification, geographic object extraction (Blaschke et al. 2008), and vectorization (i.e., raster to vector conversion). Mobile version of some of the softwares is also available such as QGIS and gvSIG (Table 10.2). A detailed categorization based on functionality of open-source GIS softwares may be found in Salinas and Lajara (2009) and Steiniger and Hunter (2013). This review is restricted to desktop GIS. Some of the open-source GIS softwares such as GRASS, QGIS, SAGA, and ILWIS are at par with proprietary softwares like Environmental Systems Research Institute (ESRI) ArcGIS desktop in general-purpose GIS functionalities (Steiniger and Hay 2009; Steiniger and Bocher 2009; Chen et al. 2010; Akbari and Rajabi 2013). These general functionalities include data access to raster, vector, table, and database, vector editing and manipulations, vector and raster analysis, and presentations in the form of maps, charts, plots, and tables (Eldrandaly 2007; Steiniger and Hay 2009; Steiniger and Bocher 2009). For advanced GIS analysis such as terrain analysis and geostatistics, special extensions has to be purchased in case of proprietary softwares (ESRI's 3D analyst, spatial analyst, and geostatistical analyst). Terrain analysis can be very well accomplished with tools in GRASS (r. geomorphon), QGIS (terrain models), ILWIS (no extension), etc. For geostatistics, GRASS and SAGA offer coupling with the R statistics package (Bivand and Neteler 2000; Bivand 2002, 2007; Bivand et al. 2008; Brenning 2008), while ILWIS provides a large set of statistic functions, integrated.

There are more than 350 OSGIS projects listed on http://www.opensourcegis.org/ updated in December 2013. Most of these applications were believed to be used by relatively technically sophisticated users (Lerner and Tirole 2002; Nichols and Twidale 2003). However, many of these have now reached a mature stage and are being used by novice to expert users (Steiniger and Hay 2009). The acceptance of FOSS for GIS depends on its usability. ISO/IEC 9126-1991 defines three sub-criteria of usability: understandability, learnability, and operability. The usability parameters for OSGIS softwares may be installation easiness, suitability for all platform, graphical user interface (GUI) suitability, ability to import/export different data format, interacting with existing data (legacy integration), user support, and easy and understandable steps to do analysis. Most of the matured OSGIS softwares (GRASS, SAGA, QGIS, etc.) fulfill these criteria and show excellent usability. Descriptions and capabilities of some of the most popular and matured desktop OSGIS softwares are given in the following sections.

10.5.1 Integrated Land and Water Information System (ILWIS)

ILWIS, initially developed and distributed by Netherlands based ITC Enschede (International Institute for Geo-Information Science and Earth Observation) for its researchers and students, is now distributed under the terms of the GPL since 2007 (http://www.itc.nl/ilwis/downloads/ilwis33.asp). It combines raster (image analysis), vector, and thematic data operations in one comprehensive software program on the desktop. ILWIS delivers a wide range of possibilities, including import/ export, user-friendly digitizing, editing, analysis and display of data as well as quality map composition. It also offers advanced functions like image processing, hydrological modeling, digital terrain analysis, geostatistical and statistical analysis, database, and similar operations. It even allows the user to analyze stereo photographs from a monitor with a stereoscope mounted on the screen. The program runs only on Windows. However, the ILWIS versions ILWIS 3.3 with the patch331.exe and ILWIS 3.4 can work in Linux via WINE (http://www.winehq.org/), a free and open-source compatibility layer that allow programs developed for Windows to run on Linux.

10.5.2 Geographic Resources Analysis Support System (GRASS)

GRASS is easily the oldest of the open-source GIS software products. Like ILWIS, GRASS also is an open-source raster and vector program, and it is available for a variety of operating systems like Windows, Linux, Mac, etc. GRASS is used for geospatial data management and analysis, image processing, graphics/maps production, spatial modeling, and visualization (Neteler and Mitasova, 2008). Originally written as a raster analysis system, GRASS has had vector analysis capabilities added to it as well. GRASS can import a wide range of formats, using both the GDAL and OGR libraries for data import. GRASS also has the ability to directly read attribute and spatial data from PostGIS/ PostgreSQL. It can be coupled with R (R core team 2016) for geostatistical analysis (Dasgupta et al. 2012; Steiniger and Hay 2009). GRASS GIS is an official project of the Open Source Geospatial Foundation (OSGeo). There are more than 400 modules available in the latest stable GRASS software release. GRASS is currently used in academic and commercial settings around the world, as well as by many governmental agencies and environmental consulting companies. Neteler and Mitášová (2008) and Neteler et al. (2012) provide detailed information about the software capabilities, design and technical developments, and its uses. In over three decades of development, GRASS has become a high-quality cutting-edge GIS with an almost unparalleled depth of functionalities directly within the main software package (Neteler et al. (2012)).

10.5.3 Quantum GIS (QGIS)

QGIS is an official project of the OSGeo. It is not only a desktop GIS, but provides a spatial file browser, a server application, and web applications. It runs on Linux, Unix, Mac OSX, Windows, and Android and supports numerous vector, raster, and database formats and functionalities. QGIS provides a continuously growing number of capabilities provided by core functions and plugins. QGIS supports PostGIS and Shapefiles as vector data sources. QGIS uses OGR as a data import bridge, so support of all OGR formats is also available. QGIS supports DEM, ArcGrid, ERDAS, SDTS, and GeoTIFF raster formats. GRASS and SAGA are strong in raster analysis and are comparable with ArcInfo, but are lesser user-friendly, or fewer vector graphics drawing and editing functions. QGIS is the preferred visualization environment for GRASS users providing a modern user interface and map element symbology editor. GRASS integration provides access to GRASS GIS databases and functionalities. QGIS provides tools in spatial analysis, geoprocessing, geometry, and data management with special focus on hydrological and geomorphological analysis. The integration of a WMS and WFS server into QGIS allows a seamless crossover from desktop to the web. Extended functionalities can be created through plug-ins. The growing number of plug-ins is being created by developer communities using libraries provided or using C++ or Python. A well-elaborated training manual may be found in http://docs.qgis.org/2.14/en/docs/ training_manual/index.html.

10.5.4 System for Automated Geoscientific Analyses (SAGA)

SAGA GIS is specially designed for easy and effective implementation of spatial algorithms. It offers very comprehensive and growing algorithms for raster applications. It also provides object-oriented image classification of satellite data (Kothe and Bock 2006). SAGA GIS can be used together with other GIS software like Kosmo or QGIS to get better vector data- and map-producing capabilities. SAGA GIS modules can be executed from within the statistical data analysis software R (R core team 2016) in order to integrate statistical and GIS analyses. Several modules also focus on DEMs and terrain analysis, like analytical hill shading, visibility analysis, local geomorphometry and geomorphographic classifications, terrain parameters related to hydrology, channel network and watershed basin extraction, and the creation of profiles and cross-sectional diagrams. Except for the SAGA Application Programming Interface (API) (licensed under LGPL), most SAGA source codes have been licensed under the GPL.

10.6 Conclusions

The open movement, which promotes the use of open-source licenses for data and software, is gaining popularity now. The cost of data and GIS softwares is an important factor affecting the adoption of geospatial technologies. The open data and OSGS remove this barrier. However, the most important consideration for adoption of OSGS is the availability of source codes which provide flexibility to the users to customize it according to their need. The OSGS provides more flexibility to the users but requires more efforts to use, whereas, the opposite is true for proprietary software in general. However, many of the matured OSGS provides equally good usability. Compliance to open standards and a large community of developers and users to support further adds to their usability.

References

Abdi AM (2013) Integrating open access geospatial data to map the habitat suitability of the declining corn bunting (*Miliaria calandra*). ISPRS Int J Geogr Inf 2:935–954

Akbari M, Rajabi MA (2013) Evaluation of desktop free/open source GIS software based on functional and non-functional capabilities. Tehnički vjesnik 20:755–764

Bakillah M, Liang S (2016) Open geospatial data, software and standards. Open Geospat Data Softw Stand 1:1–2

Bivand R (2002) Development of the GRASS/R interface: GIS and statistical data analysis. In: Benciolini B, Ciolli M, Zatelli P (eds) Proceedings of open source free software GIS – GRASS users conference. University of Trento, Trento

Bivand R (2007) Using the R–Grass interface. OSGeo J 1:36–38

Bivand R, Neteler M (2000) Open source geocomputation: using the R data analysis language integrated with GRASS GIS and PostgreSQL data base systems. In: Proceedings of the 5th conference on GeoComputation, 23–25 August 2000. University of Greenwich, London

Bivand R, Pebesma E, Gómez-Rubio V (2008) Applied spatial data analysis with R, UseR! Series. Springer, New York

Blaschke T, Land S, Hay GJ (eds) (2008) Object-based image analysis—spatial concepts for knowledge-driven remote sensing applications. Springer, Berlin

Brenning A (2008) Statistical geocomputing combining R and SAGA: the example of landslide susceptibility analysis with generalized additive models. In: Boehner J, Blaschke T, Montanarella L (eds) SAGA – seconds out (= Hamburger Beitraege zur Physischen Geographie und Landschaftsoekologie), 19, pp 23–32

Brovelli MA, Minghini M, Moreno-Sanchez R, Oliveira R (2017) Free and open source software for geospatial applications (FOSS4G) to support Future Earth. Int J Digital Earth 10:386–404

Brown G (2004) Mapping spatial attributes in survey research for natural resource management: methods and applications. Soc Nat Resour 18:17–39

Burrough PA (1986) Principles of geographical information systems for land resources assessment. Oxford University Press, New York, p 194

Burrough PA, McDonnell RA (1998) Principles of geographical information systems. Oxford University Press, Oxford 333 pp

Chen D, Shams S, Carmona-Moreno C, Leone A (2010) Assessment of open source GIS software for water resources management in developing countries. J Hydro-Environ Res 4:253–264

Dasgupta A, Kumar U, Mukhopadhyay C, Ramachandra TV (2012) GRASS with R: an introductory tutorial to open source GIS & statistical computing software for geospatial analysis. In: Proceedings of first national conference on "open source geospatial resources to spearhead development and growth" 25–27th October 2012, IIIT Hyderabad, India

Dowman I, Reuter I (2016) Global geospatial data from Earth observation: status and issues. Int J Digital Earth 10:328–341

Eldrandaly K (2007) GIS software selection: a multi-criteria decision making approach. Appl GIS 3:1–17

Fitzgerald B (2011) Open source software adoption: anatomy of success and failure. In: Koch S (ed) Multi-disciplinary advancement in open source software and processes, pp 1–23

Harris R, Baumann I (2015) Open data policies and satellite Earth observation. Space Policy 32:44–53

http://bhuvan-noeda.nrsc.gov.in/download/download/download.php. Accessed on 15 Sept 2017

http://docs.qgis.org/2.14/en/docs/training_manual/index.html. Accessed on 16 Oct 2017

http://earthexplorer.usgs.gov/. Accessed on 20 Nov 2017

http://glovis.usgs.gov. Accessed on 15 Nov 2017

http://reverb.echo.nasa.gov/reverb. Accessed on 10 Nov 2017

http://srtm.csi.cgiar.org/. Accessed on 10 Oct 2017

http://www.eorc.jaxa.jp/ALOS/en/aw3d30/. Accessed on 12 Oct 2017

http://www.itc.nl/ilwis/downloads/ilwis33.asp. Accessed on 5 Oct 2017

http://www.naturalearthdata.com/downloads/. Accessed on 14 Oct 2017
http://www.opensourcegis.org/. Accessed on 10 Aug 2017
http://www.winehq.org/. Accessed on 5 Oct 2017
Keil M, Tiwana A (2006) Relative importance of evaluation criteria for enterprise systems: a conjoint study. Inf Syst J 16:237–262
Köthe R, Bock M (2006) Development and use in practice of SAGA modules for high quality analysis of geodata. Göttinger Geographische Abhandlungen 115:85–96
Kumar N (2013) Remote sensing and GIS applications in land use/land cover analysis. In: Reddy GPO, Sarkar D (eds) Remote sensing and GIS in digital terrain analysis and soil-landscape modeling, NBSS&LUP pub. No. 152, pp 254–263
Kumar N, Sinha NK, Mohanty M (2015) Integrating remote sensing data in crop process models. In: Mohanty M, Sinha NK, Hati KM, Chaudhary RS, Patra AK (eds) Crop growth simulation modelling and climate change. Scientific Publishers, Jodhpur, pp 212–221
Lerner J, Tirole J (2002) Some simple economics of open source. J Ind Econ 50:197–234
Liu JG, Philippa JM (2009) Essential image processing and GIS for remote sensing. Wiley
Miller HM, Richardson L, Koontz SR, Loomis J, Koontz L (2013) Users, uses, and value of Landsat satellite imagery—results from the 2012 survey of users: U.S. Geological Survey Open-File Report 2013–1269, 51 p
Neteler M, Mitasova H (2008) Open source GIS: a GRASS GIS approach, 3rd edn. Springer, Berlin
Neteler M, Bowman MH, Landa M, Metz M (2012) GRASS GIS: a multi-purpose open source GIS. Environ Model Softw 31:124–130
Nichols DM, Twidale MB (2003) The usability of open source software. First Monday 8:1–6
R Core Team (2016) R: A language and environment for statistical computing. R Foundation for Statistical Computing, Vienna URL https://www.R-project.org/
Rocchini D, Petras V, Petrasova A, Horning N, Furtkevicova L, Neteler M, Leutner B, Wegmann M (2017) Open-access and open-source for remote sensing training in ecology. Ecol Inf 40:57–61
Salinas JGS, Lajara MM (2009) Current panorama of the FOSS4G ecosystem. Upgrade 10:43–51
Singh A (1989) Digital change detection techniques using remotely-sensed data. Int J Remote Sens 10:989–1003
Steiniger S, Bocher E (2009) An overview on current free and open source desktop GIS developments. Int J Geogr Inf Sci 23:1345–1370
Steiniger S, Hay GJ (2009) Free and open source geographic information tools for landscape ecology. Eco Inform 4:183–195
Steiniger S, Hunter AJS (2013) The 2012 free and open source GIS software map – a guide to facilitate research, development, and adoption. Comput Environ Urban Syst 39:136–150
Trotter CM (1991) Remotely-sensed data as an information source for geographical information systems in natural resource management a review. Int J Geogr Inf Syst 5:225–239
Turner W, Rondinini C, Pettorelli N, Mora C, Leidner AK, Szantoi Z, Buchanan G, Dech S, Dwyer J, Herold M, Koh LP, Leimgruber P, Taubenboeck H, Wegmann M, Wikelskim M, Woodcock C (2015) Free and open-access satellite data are key to biodiversity conservation. Biol Conserv 182:173–176
Ven K, Verelst J, Mannaert H (2008) Should you adopt open source software. IEEE Softw 25:54–59
Wang H, Wang C (2001) Open source software adoption: a status report. IEEE Softw 18:90–95
Worboys M, Duckham M (2004) GIS: a computing perspective. CRC Press, Boca Raton

Chapter 11
Remote Sensing and GIS in Digital Terrain Modeling

G. P. Obi Reddy

Abstract Topography is an important land surface characteristic that affects most aspects of the water balance in a catchment, including the generation of surface and subsurface runoff, the flow paths followed by water as it moves down and through hillslopes, and the rate of water movement. Topographic attributes derived from digital elevation models (DEMs) and automated terrain analyses are increasingly used in terrain analysis and geomorphological research. DEM is convenient for representing the continuously varying topographic surface of the Earth, and it is a common data source for terrain analysis and other spatial applications. The utility of the DEM is evidenced by widespread availability of satellite-based DEMs at different resolutions and by the ever-increasing list of uses from DEM. Common terrain attributes, which could be computed from a DEM include slope gradient, slope aspect, slope curvature, upslope length, specific catchment area, compound topographic index (CTI) etc. One of the most limiting factors of the use of the DEM is its accuracy and spatial resolution. DEM of different resolutions could be used to derive DEM-based attributes, which could be used to investigate and evaluate resources like soil, water, vegetation, etc., in given landscape. In digital terrain modeling, predictive relationships developed at one scale might not be much useful for prediction of variables at different scales. That may limit the use of terrain variables developed for large scale in small-scale studies.

Keywords Digital terrain analysis · Digital elevation model · Geographic information system · Remote sensing · Terrain attributes

---

G. P. O. Reddy (✉)
ICAR-National Bureau of Soil Survey & Land Use Planning, Nagpur, India

G. P. O. Reddy, S. K. Singh (eds.), *Geospatial Technologies in Land Resources Mapping, Monitoring and Management*, Geotechnologies and the Environment 21,
https://doi.org/10.1007/978-3-319-78711-4_11

11.1 Introduction

Digital elevation model (DEM) is the most basic and interesting geographical data type. DEM is defined as any digital representation of the continuous variation of relief over space (Burrough 1986), and it is a common data source for many terrain analyses and other spatial applications. Digital elevation data are typically compiled and stored in one of three data structures: (1) point elevation data on a regular grid, (2) point elevation data in triangulated irregular networks, and (3) digitized contour line data. DEMs are increasingly used for visual and digital analysis of topography, landscapes, and landforms, as well as modeling of surface processes (Dikau et al. 1995; Giles 1998; Tucker et al. 2001; Reddy et al. 2017). Selection of suitable source of DEM and its resolution play a key role in deriving various terrain parameters. A major disadvantage of the use of low-resolution DEM is the loss of important small-scale features that can seriously affect the modeling results. If the input DEM is at a higher grid resolution, during the transformation or resampling of the original DEM data to a lower model resolution, important topographic details are lost mainly as a result of averaging. If the input DEM is already at a low resolution, it does not represent the actual on-ground topographic features, which might significantly affect the accuracy and reliability of the results from the modeling exercise. Applications of geographic information system (GIS) provide opportunities for modeling, analyzing, and displaying geographical phenomena connected with topography and relief. Much valuable information in the form of attributes related to the surface of a real terrain can be gained by interpretation of terrain models.

Information about terrain and associated landforms are extremely important for landscape evaluation, soil resource mapping, erosion studies, land use systems inventories, land suitability studies, land use planning, hazard prediction, and regional planning. The classic ways to incorporate relief units into a landscape assessment are to delineate them during field survey or using stereo images. This approach is relatively time-consuming, and the results depend on subjective decisions of the interpreter and are, therefore, neither transparent nor reproducible. A digital terrain model (DTM) is a continuous surface that, besides the values of height as a grid, also consists of other elements that describe the topographic surface, such as slope or skeleton (Podobnikar 2005). Different techniques for the generation of DTMs have been developed since their inception in the 1950s (Miller and Laflamme 1958; Doyle 1978). Until the end of the 1990s, high-quality terrain data were acquired mainly photogrammetrically using aerial photographs and manual stereo measurements.

The widespread distribution and extended graphical capabilities of GIS as well as the availability of high-resolution remote sensing data and DEMs have led to the revolutionary changes in digital terrain analysis (Paron and Claessens 2011; Smith et al. 2011; Reddy et al. 2012). DEMs are increasingly used for visual and digital analysis of topography, landscapes, and landforms, as well as modeling of surface processes (Reddy et al. 2013). DEM offers the most common method for extracting vital topographic information and even enables the modeling of flow across topography, a controlling factor in distributed models of landform processes. To accomplish this, the DEM must represent the terrain as accurately as possible, since the accuracy of the DEM determines the reliability of the geomorphometric analysis. With the availability of advance satellites like Cartosat-1 and Cartosat-2 and LiDAR (Light Detection and Ranging), the automatic generation of a DEM from remotely sensed data with sub-pixel accuracy is possible. The derivation of topographic attributes relies on digital elevation datasets that may be acquired from satellite imagery, digitizing the contour lines on topographic maps or conducting ground surveys (Wilson and Gallant 2000). Common terrain attributes that are readily computed from a DEM include slope, gradient, slope aspect, slope curvature, upslope length, specific catchment area, and compound topographic index (Moore et al. 1993a).

11.2 Digital Elevation Models: Sources and Resolutions

DEM data have been used to derive hydrological features, which serve as inputs to various models. The important DEM data sources are ground surveys, airborne photogrammetric data capture, existing cartographic surveys (topo maps), airborne laser scanning, and stereoscopic or radar-based satellite imagery (Reddy 2012a). Most of the currently available digital elevation datasets are the product of photogrammetric data capture (Moore et al. 1991). These sources rely on the stereoscopic interpretation of aerial photographs or satellite imagery using manual or automatic stereoplotters (Carter 1988). Additional elevation datasets can be acquired by digitizing the contour lines on topographic maps and conducting ground surveys. The advent and widespread use of global positioning system (GPS) provides many new and affordable opportunities for the collection of large numbers of elevation datasets. Stereo-correlated DEMs can be created from two complementary images, aerial photographs, or satellite images (Schenk 1989). Raw data in the form of stereo photographs or field survey are not readily available to potential end users of a DEM. Many users rely on published topographic maps or available DEMs at different spatial resolutions produced by different agencies. The DEM creation methods, sources, nominal resolutions, and relative vertical resolutions are shown in Table 11.1.

DEMs can be generated by different sources and methods such as space- and airborne optical stereo data, interferometric SAR (InSAR) data, space- and airborne radar and laser alimentary (e.g., LiDAR), and topographic maps. The accuracy of the DEM derived from remote sensing depends on several factors, such as spatial resolution of satellite data, positional and vertical accuracy of ground control points (GCPs), and contrast and cloud cover in optical satellite images. GCPs can be computed for DEM generation mainly from three methods: highly precise DGPS measurements during fieldwork, existing rectified topographic maps, and known satellite positions and rotation angles, where differential global positioning system (DGPS) measurements and topographic maps are not available (Kääb et al. 2003).

Table 11.1 DEM sources and their resolutions

DEM creation method	DEM type	Nominal resolution (m)	Relative vertical resolution (m)
Spaceborne photogrammetric	Terra (ASTER)	30	2
	SPOT 5		10
	IKONOS	30	15
	HRSC	2	20
	HRISE	50	01
Spaceborne IfSAR	SRTM-c-band	90	10
	SRTM-x-band	30	6
	ERS tandem	25	20
	Terra SAR	12	20
	Terra ASAR	15	10
Airborne	IF SAR	1–5	21
	LiDAR	<2	<0.25

Source: Smith and Pain (2009)

11.3 Digital Elevation Models: Data Representation

Data representation can be classified using the following three types of models:

1. Raster model is seen in two variants. The first one considers a cell as a facet enclosed by four points of a raster net, each of which can have different grid. The resulting model is therefore formed by warped quadrangles. The second one considers a cell as an object representing a rectangular facet integrally, and the assigned value represents an attribute for the whole surface of a pixel – this variant is most used in raster-oriented GIS (Kraus 2000).
2. Polyhedron model, where elementary facets are triangles, which adjoin to each other, thus creating a polyhedron adjoining to a terrain. The polyhedron vertices are points on a terrain area; area interpolation is usually done linearly triangle-by-triangle. This approach, denoted as a triangulation or triangulated irregular network (TIN), is currently the most widespread with vector-oriented GIS.
3. Plate model has many features common to the polyhedron model. The terrain is again divided into smaller surfaces, which do not have to be only flat; they can be curved in a certain way. Areas are described by polynomic functions, which concur so fluently to guarantee a continuity of derivations to a certain advance system, e.g., Bezier curve (Pfeifer & Pottmann 1996).

11.4 Digital Elevation Models: Data Quality

Before generating the DTM, one has to know the expected quality and what quality is required with regard to the respected standards. These two factors enable regular production and usability of the DTM. A priori assessments are based mostly on analyses of the datasets and methods for the DTM production, while the a posteriori methods are based on the final DTM. The higher the resolution, the more difficult is the evaluation of input data quality and the assessment of the resulting DTM. Several studies indicate that the effort is proportional to the square of the inverse value of horizontal resolution. High-resolution DTMs are thus more prone to errors. Visual methods can be very important for the evaluation of spatial data and can balance some weaknesses of statistical methods. They are still underused for at least three reasons. Visual approaches being qualitative are generally more neglected than statistical ones, which are considered to be more objective. The other reasons for the lower acceptance of visual methods lie in the insufficient graphical capabilities of computers until recently and in the longer tradition of using statistical methods. Finally, visualization of spatial data has traditionally been part of cartography.

One of the DTM quality assessment goals is to fulfill the requirements of spatial data standards. The International Organization for Standardization (ISO) distinguishes five elements of data quality, completeness, logical consistency, and three types of accuracy (positional, temporal, and thematic). The *accuracy* is defined as a difference between the value of a variable, as it appears in a dataset, and the value of the variable in the data model. The difference between absolute and relative accuracy can be distinguished in terms of nature of the data. The position (horizontal or vertical) of the objects could be assigned to absolute accuracy and the irregularity of the shapes of objects to the relative accuracy, that is, morphologically relative to a general position. The term *precision* is considered as a component of accuracy, related to the scale, resolution, and also to the generalization of datasets (Podobnikar 2008). The quality DEMs are essential in digital terrain modeling and applications in hydrological and geomorphological studies.

11.5 Digital Elevation Models: Data Resampling

The three most commonly used resampling techniques could be applied on DEMs. They are nearest-neighbor interpolation, bilinear interpolation, and cubic convolution (Mitchell & Netravali 1988). The nearest-neighbor method simply assigns the value of the single closest observation to each cell. Once the location of the cell's center on the output grid is located on the input DEM, the nearest-neighbor assignment will determine the location of the closest cell center on the input grid, identify the value that is associated with the cell, and assign that value to the cell that the output cell center is associated with. The bilinear interpolation takes a weighted average of the nearest four input cells around the transformed point to determine the output cell. This method results in a smoother surface than the nearest-neighbor method. In the cubic convolution technique, the output cell value is computed by fitting a smooth surface to the nearest 16 input cells. It tends to smooth data even more than the bilinear interpolation.

11.6 Digital Elevation Models: Data Structures

DEMs are usually organized into one of three data structures—(1) regular grids, (2) triangulated irregular networks, and (3) contours—depending on the source and/or preferred method of analysis. Square-grid DEMs have emerged as the most widely used data structure during the past decade because of their simplicity and ease of computer implementation (Moore et al. 1991). These advantages offset at least three disadvantages. First, the size of the grid mesh will often affect the storage requirements, computational efficiency, and the quality of the results (Collins and Moon 1981). Second, square grids cannot handle abrupt changes in elevation easily, and they will often skip important details of the land surface in flat areas (Carter 1988). Third, the computed upslope flow paths will tend to zigzag across the landscape and increase the difficulty of calculating specific catchment areas accurately (Zevenbergen and Thorne 1987). Several of these obstacles have been overcome in recent years. For example, there is no generic reason why regular DEMs cannot represent shape well in flat areas, so long as the terrain attributes are calculated by a method that respects surface drainage.

The advent of several new compression techniques has reduced the storage requirements and improved computational efficiency of DEMs (Kidner and Smith 1992). DEMs with grid sizes of 500, 100, 30, 10, and even 1 m are increasingly available for different parts of the globe (USGS 1993). Triangulated irregular networks (TINs) are based on triangular elements (facets) with vertices at the sample points and have widespread use (Moore et al. 1991). These facets consist of planes joining the three adjacent points in the network and are usually constructed using Delaunay triangulation (Weibel and Heller 1991). TINs can easily incorporate discontinuities and may constitute efficient data structures because the density of the triangles can be varied to match the roughness of the terrain (Moore et al. 1991). The third structure incorporates the stream tube concept (Onstad and Brakensiek 1968) and divides landscapes into small, irregularly shaped polygons (elements) based on contour lines and their orthogonals (O'Loughlin 1986). This structure is used most frequently in hydrological applications because it can reduce complex three-dimensional flow equations into a series of coupled one-dimensional equations in areas of complex terrain (Moore and Foster 1990).

11.7 GIS and Remote Sensing in Digital Terrain Analysis

DEMs can be generated from stereopairs derived from remote sensing data or aerial photographs or can be generated from digital terrain elevation data. DEM-derived products can be readily combined with remote sensing data. DEM is also very useful for discriminating land use and land cover classes during the digital processing of remote sensing data. The interfacing of GIS, DEM, and remote sensing provides a new and exciting capability to analyze the dynamics of land-use change. Nellis et al. (1990) used GIS to classify and assess changes in vegetation in a landscape. Availability of DEM data from high-resolution satellites like Cartosat-1 and Cartosat-2 and LiDAR opened new avenues in digital terrain analysis (Reddy 2012b), large-scale landscape mapping, soil resource mapping, forestry applications, hydro-geomorphological mapping (Reddy 2012c), dynamic landscape simulation modeling, etc.

11.8 Digital Terrain Models: Terrain Modeling Techniques

Digital terrain models have become an integral part of digital processing of spatial geographical information. In GIS applications, they can be used for modeling, analyzing, and representation of phenomena connected to topography and relief. Terrain modeling techniques can be classified using the following criteria:

- Data structure – basic and most frequent classification according to the description of the surface using raster models with different pixel (or facet) type, or the triangular grid (TIN).
- Mathematical model – classification according to the methods of description of the surface between facets; it is, in fact, a method of interpolation of surface values.
- Method of model generation – basic classification is based on the assumption of automatic generation of values or possibility of semiautomatic processing with the operator's help.

11.9 Digital Terrain Models: Spatial Surface Interpolation

Source (terrain) data represent quite unorganized data elements (random values). For the creation of DTM, it is necessary to introduce topological relations among these elements as well as interpolate these spatial data into a continuous surface with defined behavior. DTMs are usually based on two main data sources, i.e., data from photogrammetric measurements and data from digital elevation sets. Recently there has also been available DTM from laser scanning (or radar altimetry or radar interferometry). In the first case, however, the quantitative parameters are influenced by the way the data is obtained – the accuracy of determining terrain level from a photogrammetric analysis is limited by open visibility of the ground – and in the second case, the model is negatively influenced by the spatial distribution of contour lines and point elevations (Klimanek 2006).

Interpolation in the process of a terrain modeling makes for calculation of values in places where they were not measured. Most frequently it is the calculation of a grid for a given point or pixel, the calculation of location at interpolation of contour lines, or the change of resolution, in some case the change of data structure. The problem covers generally applicable statistic procedures and methods, which are specifically modified for the need of the relief terrain modeling. Most frequently applied methods include inverse distance weighting (IDW) method, triangulation with linear interpolation, spline methods, radial function methods, and kriging and conditional stochastic simulation.

11.10 Digital Terrain Attributes

Topographic attributes derived from digital elevation models and automated terrain analyses based on them are increasingly used in geomorphological research (Reddy et al. 2017). DEM is convenient for representing the continuously varying topographic surface of the Earth, and it is a common data source for terrain analysis and other spatial applications. The utility of the DEM is evidenced by the widespread availability of digital topographic data and by the ever-increasing list of uses for and products from DEM. Common terrain attributes, which could be computed from a DEM include slope gradient, slope aspect, slope curvature, upslope length, specific catchment area, and the compound topographic index (CTI), a hydrologically based index that is related to zones of surface saturation (Reddy 2012d).

To date, numerous methods have been suggested for an enumeration of DTM, commonly differentiated in primary and secondary terrain attributes. Primary local terrain attributes such as slope, aspect, and various curvature measures yield a local geomorphometric terrain description, whereas the so-called primary complex terrain attributes describe the regional spatial interrelation of a grid cell within the broader neighborhood of the entire DTM domain. Secondary terrain attributes such as the terrain wetness index (TWI), the sediment transport index (STI), or the stream power index (SPI) are the most prominent examples for combining different primary terrain attributes (Moore et al. 1993a).

11.11 Digital Elevation Models and Terrain Parameters

DEM is used to derive quantitative measures of terrain parameters, also called terrain parameterization. This is a process of quantitative description of terrain by terrain parameters, which can be derived using various algorithms that quantify morphological, hydrological, ecological, and other aspects of a terrain. Extracted terrain parameters can then be used, for example, to improve mapping and modeling of soils, vegetation, land use, and geomorphological and geological features. There are relatively simple and easy to derive terrain parameters (the slope gradient, aspect, curvature), and there are some more complex ones, which are derived with the combined use of the primary terrain parameters. The primary features are direct descriptors of the terrain features, while secondary features describe more complex characteristics of the landform, which are linked to certain terrain-regulated processes, like stream power index (SPI) or the compound topographic index (CTI). These features can be used to estimate potential soil loss or sedimentation and also for calculating "terrain-adjusted" climatic variables, like temperature, solar irradiation, long-wave surface radiation, and reflected radiation, which are important factors in the energy balance of the surface and thus in the soil formation (Wilson and Galant 2000).

Usually primary attributes can be distinguished, which are computed directly from the DEM and secondary or compound attributes that involve combinations of primary attributes and constitute physically based or empirically derived indices that can characterize the spatial variability of specific processes occurring in the landscape (Moore et al. 1991, 1993a). To date, numerous methods have been suggested for an enumeration of DTM, commonly differentiated in primary and secondary terrain attributes. Primary topographic attributes, which can be computed from digital terrain analysis, are shown in Table 11.2.

Table 11.2 Primary topographic attributes, which can be computed from DEM

Attribute	Definition	Significance
Altitude	Elevation	Climate, vegetation, potential energy
Upslope height	Mean height of upslope area	Potential energy
Aspect	Slope azimuth	Solar insolation, evapotranspiration, flora and fauna distribution and abundance
Upslope slope	Mean slope of upslope area	Runoff velocity
Dispersal slope	Mean slope of dispersal area	Rate of soil drainage
Catchment slope	Average slope over the catchment	Time of concentration
Upslope area	Catchment area above a short length of contour	Runoff volume, steady-state runoff rate
Dispersal area	Area downslope from a short length of contour	Soil drainage rate
Catchment area	Area draining to catchment outlet	Runoff volume
Specific catchment area	Upslope area per unit width of contour	Runoff volume, steady-state runoff rate, soil characteristics, soil-water content, geomorphology
Flow path length	Maximum distance of water flow to a point in the catchment	Erosion rates, sediment yield, time of concentration
Upslope length	Mean length of flow paths to a point in the catchment	Flow acceleration, erosion rates
Dispersal length	Distance from a point in the catchment to the outlet	Impedance of soil drainage
Catchment length	Distance from highest point to outlet	Overland flow attenuation
Profile curvature	Slope profile curvature	Flow acceleration, erosion/deposition rate, geomorphology
Plan curvature	Contour curvature	Converging/diverging flow, soil-water content, soil characteristics
Tangential curvature	Plan curvature multiplied by slope	Provides alternative measure of local flow convergence and divergence
Elevation percentile	Proportion of cells in a user-defined circle lower than the center cell	Relative landscape position, flora and fauna distribution and abundance

Source: Moore et al. (1991)

11.12 GIS for Extraction of Primary Terrain Variables from DEM

Much valuable information in the form of attributes related to the surface of a real terrain can be gained by interpretation (analysis) of terrain models. These analyses can proceed in two levels, though they occur in combination more frequently. Either it is a visual (graphic) interpretation or purely quantitative analysis, outputs can be used in other components of GIS, or they represent an input into other models (e.g., hydrological or erosive). Evans (1972) divides the analysis of geomorphometric parameters into general geomorphometrics (slope, aspect, curvature) and specific geomorphometrics (e.g., topographic shape classification, hydrologic modeling, viewshed analysis). GIS-based extraction of different terrain variables from DEM is briefly discussed below.

11.12.1 Slope and Aspect

Slope is defined as gradient (maximum) of height. Gradient in a defined point is the angle measured from the horizontal line to the tangent plane in this point. Slope can be calculated as slope in the direction of axis x, slope in the direction of axis y, and maximal slope. The slope of relief is a significant natural pattern of the landscape, because it distributes free available energy. Through a suitable selection of slope categories and their cartographic representation, it is possible to display edges of a terrain as well as terrain with predominant slope, e.g., dendrite shapes of valleys dug in leveled surfaces, fault slopes, cuestas, alluvial planes, etc. Aspect is usually defined as azimuth, which is a horizontal angle expressed in degrees included between geographic north and slope. Aspect distributes incident sunlight. Depending on the orientation of the Earth's surface in respect of the cardinal points (and slope), energy and moisture are redistributed, which generates a wide range of processes and affects their course and duration. Normally suitable colors used to reflect the respective cartographic and environmental standards of aspect of area. Warm colors are used to mark richer southern areas; cold colors are used for more humid and energetically poorer northern areas. Areas with a slope up to 3° are marked with different colors (white and gray) as they are considered a plane, excluded from DTM analysis for the purposes of defining the aspect. The impact of spatial resolution of DEM on derived slope classes generated from Cartosat-1 10 m, SRTM 30 m, and SRTM 90 m DEMs of part of basaltic terrain of Central India is depicted in Fig. 11.1.

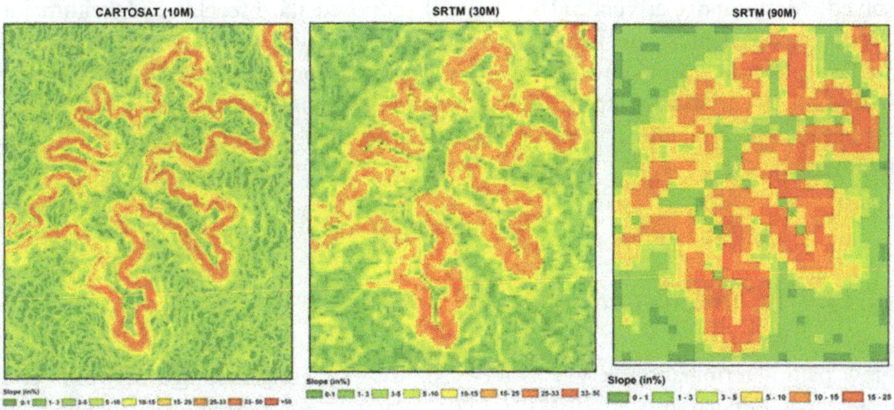

Fig. 11.1 Slope generated from Cartosat-1 10 m, SRTM 30 m, and SRTM 90 m DEMs of part of basaltic terrain of Central India

11.12.2 Hillshade

Computing of slope and aspect is closely connected with the procedures for creating analytical shading of relief. That is highlighting relief by displaying illuminated areas and shaded areas by the selected light source when defining its azimuth (angle from which the terrain is illuminated ranging from 0 to 360°) and its height above horizon (elevation ranging from 0 to 90°). There are many computation methods for the determination of surface reflectance – the most common one is Lambert's reflectance model, which supposes ideal reflectance of rays from the surface, which then appears as equally lit from all angles of view. More complicated models, such as the Lommel-Seeliger Law, are based on the analysis of variance of rays in dependence on the surface and distance. In cartography, it is common to use illumination from the northwest, which gives the impression of reading a map with the light coming from the front and from the left that means with the azimuth of 315° and the elevation of 30°.

11.12.3 Flow Direction

Currently, the most commonly used procedures for calculating the flow directions from raster DEMs are based on Jenson and Domingue's (1988) algorithm. This procedure uses a depression-filling technique to treat flat areas and depressions. However, two problems were encountered when this method was applied to realistic landscapes. The first problem is that in some cases more than one outflow point from a depression or flat area exists after the area is filled (Jenson and Domingue 1988). The second problem is that looping depressions located on a flat surface cannot be solved. Martz and Garbrecht (1998, 1999) proposed the breaching algorithm for treating closed depressions in a DEM. Flow length is a measure of the distance along the flow path (determined by the flow direction grid) from a given cell to its drainage basin outlet. Variations with flow length may lead to significant change to the time lag between precipitation and flow peak discharge, thus resulting in a much different hydrograph.

11.12.4 Flow Accumulation

Flow path determination serves as the basis for extracting drainage networks and watershed delineation from DEMs in hydrologic modeling. Slope is a fundamental topographic property used in all flow path algorithms. Slope, together with aspect, determines the flow direction from which the upslope contributing area, representing the accumulated area draining into a given grid cell can be calculated. Flow direction and flow accumulation derived from Cartosat-1 10 m DEM of part of basaltic terrain of Central India are shown in Fig. 11.2.

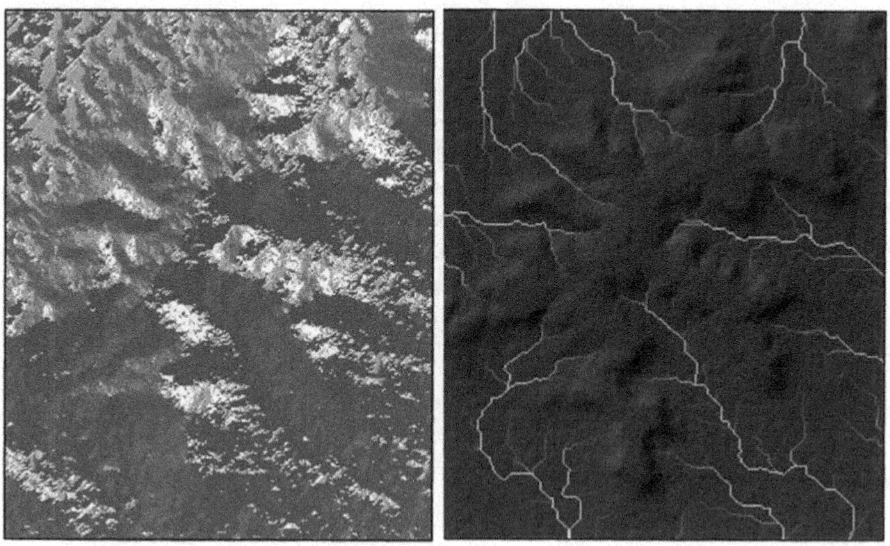

Fig. 11.2 Flow direction and flow accumulation derived from Cartosat-1 10 m DEM of part of basaltic terrain of Central India

11.12.5 Drainage Network Extraction

Accurate delineation of drainage networks is a prerequisite for many land resource mapping and management issues (Paik 2008; Liu and Zhang 2011). Drainage networks are traditionally obtained by manually digitizing stream channels from maps or airborne photographs. A drainage network can be extracted from a DEM with an arbitrary drainage density or resolution (Tarboton et al. 1991). The "constant-threshold" method could be used to calculate the number of upstream elements, i.e., the number of cells that contribute surface flow to any particular cell. Cells with catchment numbers greater than a given threshold are considered on the flow path. The smaller the chosen threshold, the more complicated are the channels obtained. The characteristics of the extracted network depend on the definition of channel sources on the digital land surface topography. Once the channel sources are defined, the essential topology and morphometric characteristics of the drainage network are implicitly defined because of their close dependence on channel source definition. Thus, the proper identification of channel sources is critical for extraction of a representative drainage network from DEMs. The extracted stream network in hydrologic analyses is important because the network indirectly determines the hillslope travel distance and network link lengths. The characteristics of the extracted network depend extensively on the definition of channel initiations on the digital landscape. Once the channel initiations are defined, the essential topology and morphometric characteristics of the corresponding downstream drainage network are implicitly predefined because of their close dependence on channel initiation definition. Thus, the identification of channel sources is critical for extraction of a representative drainage network from DEMs (ASCE 1999). Due to the increasing availability of grid DEMs, numerous research studies have been carried out to automate the extraction of drainage networks to analyze the quantitative drainage characteristics of the terrain using GIS techniques (Bertolo 2000; Reddy et al. 2002, 2004a, b, 2017; Vijith and Satheesh 2006; Valeriano et al. 2006; Kattimani and Prasad 2016; Sameena et al. 2009; Pareta and Pareta 2011; Singh et al. 2014; Sahu et al. 2016). Hillshade and extracted drainage from Cartosat-1 10 m DEM of part of basaltic terrain of Central india are shown in Fig. 11.3.

11.12.6 Terrain Curvature

Apart from the slope and exposition, another general geomorphological analysis is the curvature (degree of curvature). Curvature comprises two components (Evans 1980): vertical curvature (profile curvature) kv (rate of change at slope) and horizontal curvature (plan curvature or tangential curvature) kh (rate of change at contour lines). In the results, the negative values mean convex shapes and positive values mean concave ones. Due to the fact that this characteristic is represented by two values, in practice these two are often merged into one complex variable, which

Fig. 11.3 Hillshade and extracted drainage from Cartosat-1 10 m DEM of part of basaltic terrain of Central India

is the average curvature or accumulation curvature or total curvature. Profile curvature and total curvature extracted from Cartosat-1 10 m DEM of part of basaltic terrain of Central India are shown in Fig. 11.4.

11.12.7 Terrain Hypsometry

Hypsometry is being used as an indicator of the geomorphic form of the catchments and landforms. Computationally, it refers to finding the distribution of elevations as a function of area occupied by each contour interval within a terrain unit. Langbein and Basil (1947) first introduced the idea of hypsometry and later extended by Strahler (1952). Using dimensionless parameters, such as proportionate area and proportionate altitude, hypsometric curves can be plotted, described, and compared irrespective of the true absolute scale. Curves show distinctive differences both in sinuosity of form and in proportionate area below the curve, termed as the hypsometric integral (HI). The geometry of hypsometric curve and the magnitude of the hypsometric integral together describe the stage of evolution of a landscape (Strahler 1957). Hence, the shape of the curve, the hypsometric integral, and the regression parameters of the best-fit statistical curves can be conveniently used as descriptive parameters for the purpose of comparison and classification and to determine the stage of a landscape in the evolutionary process. Based on hypsometric integral, a

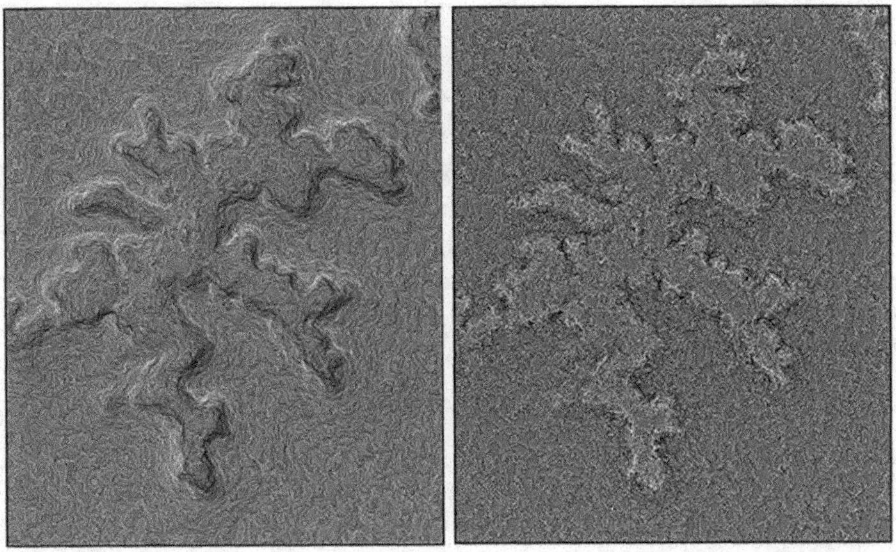

Fig. 11.4 Profile curvature and total curvature extracted from Cartosat-1 10 m DEM of part of basaltic terrain of Central India

landscape can be genetically classified in terms of the stage of evolution as youthful (HI > 0.5), mature or in equilibrium (HI = 0.5), or old (HI < 0.5) (Strahler 1957).

11.13 GIS for Extraction of Secondary Terrain Variables from DEM

Secondary terrain attributes such as the topographic position index (TPI), terrain wetness index (TWI), the sediment transport index (STI), or stream power index (SPI) of Moore et al. (1993b) as the probably most prominent examples are subsequently inferred by combining different primary terrain attributes.

11.13.1 Topographic Position Index (TPI)

Topographic position index (TPI) calculates grids from elevation grids and provides a simple and repeatable method to classify the landscape into slope position and landform category (Jenness 2006). TPI is the basis of the classification system and is simply the difference between a cell elevation value and the average elevation of the neighborhood around that cell. Positive values mean the cell is higher than its surroundings, while negative values mean it is lower. TPI is naturally very scale-dependent.

11.13.2 Topographic Wetness Index (TWI)

The topographic wetness index (TWI) is frequently used to assess the soil moisture conditions quantitatively, and it is the most commonly used indicator for static soil moisture content as well. The index was generally extracted from DEM, and the formula is expressed as

$$w = \ln\ (As/\tan\beta) \tag{11.1}$$

where As means the specific catchment area (catchment area divided by the cell width in slope direction) and β means the local slope in the steepest downslope direction of the terrain in degrees. Then tanβ is the local slope angle of the specific grid, which is used to replace approximately the local hydraulic gradient under steady-state conditions. However, TWI only has the physical significance based on the runoff flowing by gravity and doesn't consider other factors. For example, as for some cells with the same catchment area and slope, the soil moisture content varies in different aspects and terrain positions. This result was directly caused by the differences in temperature, heat, and the soil physical and chemical properties which were induced by the solar radiation variance.

11.13.3 Stream Power Index (SPI)

Stream power is the time rate of energy expenditure and has been used extensively in studies of erosion, sediment transport, and geomorphology as a measure of the erosive power of flowing water. SPI is a measure of the erosive power of water and can be used to identify suitable locations for soil conservation measures so as to reduce the effect of concentrated surface runoff (Moore et al. 1993b; Burrough and McDonnell 1998):

$$\text{SPI is expressed as SPI} = \text{AS} \times \tan(\beta) \tag{11.2}$$

where AS is the specific catchment area and β is the local slope gradient measured in degrees (Moore et al. 1988; Park et al. 2013). High SPI values represent areas on the landscape where high slopes and flow accumulations exist and thus areas where flows can concentrate with erosive potential. Moore et al. (1991) concluded that threshold values of these indices are likely to vary from place to place because of differences in soil properties.

11.13.4 Sediment Transport Index (STI)

The sediment transport index characterizes the process of erosion and deposition (Moore et al. 1993b). It calculates a spatially distributed sediment transport capacity, and it explicitly accounts for flow convergence and divergence (Moore and Wilson 1992). STI is computed using the equation described by Moore et al. (1993b):

$$\mathrm{STI} = \left(\frac{\mathrm{A_f}}{22.13}\right)^{0.6}\left(\frac{\sin\beta}{0.0896}\right)^{1.3} \tag{11.3}$$

where $\mathrm{A_f}$ is the specific catchment area draining through the point and β is the representative local slope angle. STI reflects the erosive power of the overland flow and was derived by considering the transport capacity limiting sediment flux and catchment evolution erosion theories (Pradhan and Kim 2014).

11.14 Digital Elevation Models: Data Accuracy

The accuracy of DEM and DEM-derived products depends on (a) the source of the elevation data, including the techniques for measuring elevation either on the ground or remotely, the locations of samples, and the density of samples; (b) the methods used to create the DEM from this elevation data; (c) the data model, or structure of the elevation data grid, contour, and triangular irregular network; (d) the horizontal resolution and vertical precision at which the elevation data is represented; (e) the topographic complexity of the landscape being represented; and (f) the algorithms used to calculate different terrain attributes.

The scale of terrain features represented by a DEM is dependent upon the DEM resolution – the horizontal spacing of points in the elevation grid. The accuracy of DEM and DEM-derived products may be critical when the DEM data are used for environmental modeling and prediction of the spatial distribution of hydrological, geomorphological, and biological properties. These scale issues are particularly important in hydrology and hydrologic modeling. Kalma and Sivapalan (1995) found that as DEM resolution decreased, there was a trend for decreasing total flow lengths and decreasing drainage density total length of channels per area of watershed. Considering the multiple uses of DEM data, especially for use in predictive models, it is important to consider the accuracy of the topographic input data that are used. The source of a DEM, horizontal resolution, and vertical precision all affect this DEM accuracy and, consequently, the accuracy of predictions based on DEM-derived data. This accuracy becomes increasingly important as we extend the use of DEM data for spatial prediction of soil attributes, especially because at the field or watershed scale, these pedological relationships are mainly driven by differences in topography. Different DEMs for the same landscape may produce different landscape predictions because of model coefficients, model variables, and model structure.

11.15 Digital Elevation Models: Impact of Scale

The issue of scale impacts on all aspects of landscape mapping and its use for catchment characterization. First, there are a number of terms that are used for catchments or parts of catchments of different sizes, e.g., catchment, sub-catchment, and regional catchment. Second, landscapes and their constituent parts come in various sizes; the internal scale of a landscape and its processes are critical to understanding and characterizing catchments. And thirdly, the data from which maps are compiled range in scale and resolution. Consideration of scale is thus unavoidable when describing catchments. Thus, at its simplest, small things cannot be shown on maps of big areas; this is the simple problem of choosing a map scale appropriate for the size of things the map is to show. At a slightly more complex level, landscape scale can control useful map scale. Also, controls on landscape character operate at different scales, and therefore mapping criteria will change with scale.

11.16 Digital Elevation Models: Resolutions and Challenges

Raster data resolution ranges from kilometers (NOAA) through 10s of meters (SRTM) to submeter (LiDAR). On-ground solutions can be guided only by data with a resolution that is at least as fine as the scale at which landscape processes operate. Data collected at a coarser scale can be used only as a guide to where more detailed work should be undertaken. Which data resolution is chosen will depend on the scale of the landscape to be mapped, the scale of the controlling processes, and the proposed application of the final product. For detailed mapping image data should be at a resolution that equals or is better than the scale of landform and regolith processes. Thus, the appropriate scale for a particular landscape can only be determined by geomorphic analysis of landform shape and processes; in most cases this will mean ground survey.

Several analytical challenges arise when comparing terrain attributes across this range of spatial resolutions. First, the exact locations of grid points that are to be compared may not coincide at different spatial resolutions. In this situation, spatially aggregated comparisons of data resolutions are inappropriate, especially in rugged mountainous landscapes where terrain characteristics often display enormous variation over short horizontal distances. Second, the population of grid points is small at a coarse resolution, implying unstable statistics. Third, spatial autocorrelation between neighboring sample points may be stronger at fine resolutions because of close sample distances (Wilson et al. 1998). The first challenge was overcome by adopting a resampling procedure equivalent to retaining every nth grid point after consistently dropping all the other points to produce DEMs of coarser resolutions. As a result, all grid points on a coarser resolution DEM can be matched to original grid points on finer-resolution DEMs. The second and third challenges were overcome because this resampling procedure limited the output resolution to be an integer multiple of the input resolution. As long as terrain attributes based only on the grid points of the coarsest resolution are compared, the sample size will be constant at all compared resolutions, and the influence of variable spatial autocorrelation is minimized because the distance between neighboring sample points is uniform for each resolution in the comparison.

11.17 Conclusions

The study indicates that selection of suitable source of DEM and its resolution play an important role in deriving the various terrain parameters. The low-resolution DEMs are useful to derive the parameters at regional scale. A major disadvantage of the use of low-resolution input data is the loss of important small-scale features that can seriously affect the modeling results. If the input DEM is already at a low resolution, it does not represent the actual on-ground topographic features, which might significantly affect the accuracy and reliability of the results from the modeling exercise. GIS provide opportunities for modeling, analyzing, and displaying geographical phenomena connected with topography and relief. The grid resolution of DEMs can profoundly influence the spatial patterns of attributes derived from them and also influence models built from these attributes. In general, higher-resolution DEMs produced much better results than the lower-resolution DEMs. Hence, appropriate DEM resolution is important in order to extract the heterogeneity of landscape. The terrain features, like slope or aspect, which are recognized as leading forces of the soil formation within a relatively small area, show significant relationship with soil attributes but often represent low predictive value when used individually. However, when these terrain variables are combined in model, the predictive value can raise relatively high. There is a good and commonly accepted toolkit of digital terrain variables, but the need to develop new variables and approaches to improve the capability of soil-landscape modeling and decrease the unexplained portion of the soil-landscape relationship is still evident.

References

ASCE (1999) GIS modules and distributed models of the watershed. ASCE, Virginia, USA, 24

Bertolo F (2000) Catchment delineation and characterization: a review. Technical report EUR 19563 EN, Joint Research Centre European Commission, Ispra, Italy, 30pp

Burrough PA (1986) Principles of geographical information systems for land resource assessment. Clarendon Press, Oxford

Burrough PA, McDonnell RA (1998) Principles of geographical information systems. Oxford University Press, New York

Carter JR (1988) Digital representations of topographic surfaces. Photogramm Eng Remote Sens 54(11):1577–1580

Collins SH, Moon FGC (1981) Algorithms for dense digital terrain models. Photogramm Eng Remote Sens 47(1):71–76

Dikau R, Brabb EE, Mark RK, Pike RJ (1995) Morphometric landform analysis of New Mexico, Zeitschrift für Geomorphologie, N.F., Suppl-Bd, 101:109–126

Doyle FJ (1978) Digital terrain models: an overview. Photogramm Eng Remote Sens 44:1481–1487

Evans IS (1972) General geomorphometry, derivatives of altitude, and descriptive statistics. In: Chorley RJ (ed) Spatial analysis in geomorphology. London, Methuen, pp 17–90

Evans IS (1980) An integrated system for terrain analysis for slope mapping. Z Geomorphol 36:274–295

Giles PT (1998) Geomorphological signatures: classification of aggregated slope unit objects from digital elevation and remote sensing data. Earth Surf Process Landf 23:581–594

Jenness J (2006) Topographic Position Index (tpi_jen.avx) Extension for ArcView 3.x, v. 1.3a. Jenness Enterprises

Jenson SK, Domingue's JO (1988) Extarcting topographic structures from digital elevation data geographic information system analysis. Photogramm Eng Remote Sens 54(11):1593–1600

Kääb A, Huggel C, Paul F, Wessels R, Raup B, Kieffer H, Kargel J (2003) Glacier monitoring from ASTER imagery: accuracy and applications. EARSEL eProc 2:43–53

Kalma JD, Sivapalan M (1995) Scale issues in hydrological modeling. Wiley, New York

Kattimani JM, Prasad TJR (2016) Cartosat digital elevation model (DEM) to drainage extraction techniques of Vrushabhavati basin of Karnataka, India using remote sensing and GIS techniques. Int J Adv Res 4(7):2008–2013

Kidner DB, Smith DH (1992) Compression of digital elevation models by Huffman coding. Comput Geosci 18(8):1013–1034

Klimanek M (2006) Optimization of digital terrain model for its application in forestry. J For Sci 52(5):233–241

Kraus K (2000) Photogrammetrie, Band 3, Topographische Informationssysteme. Dümmler Verlag, in Druck

Langbein WB, Basil W (1947) Topographic characteristics of drainage basins. USGS Water-Suppl Pap 968-C:125–158

Liu X, Zhang Z (2011) Drainage network extraction using LiDAR-derived DEM in volcanic plains. Area 43(1):42–52

Martz LW, Garbrecht J (1998) The treatment of flat areas and closed depressions in automated drainage analysis of raster digital elevation models. Hydrol Process 12:843–855

Martz LW, Garbrecht J (1999) An outlet breaching algorithm for the treatment of closed depressions in a raster DEM. Comput Geosci 25:835–844

Miller CL, Laflamme RA (1958) The digital terrain model: theory and application. Photogramm Eng Remote Sens 24:433–422

Mitchell D, Netravali A (1988) Reconstruction filters in computer graphics. ACM Comput Graph 22:221–228

Moore ID, Foster GR (1990) Hydraulics and overland flow. In: Anderson MG, Burt TP (eds) Process studies in hillslope hydrology. Wiley, New York, pp 215–252

Moore ID, Wilson JP (1992) Length-slope factors for the revised universal soil loss equation: simplified method of estimation. J Soil Water Conserv 47:423–428

Moore ID, O'loughlin EM, Burch GJ (1988) A contourbased topographic model for hydrological and ecological applications. Earth Surf Process Landf 13(4):305–320

Moore ID, Grayson RB, Ladson AR (1991) Digital terrain modeling: a review of hydrological, geomorphological, and biological applications. Hydrol Process 5:3–30

Moore ID, Lewis A, Gallant JC (1993a) Terrain attributes: estimation method and scale effects. In: Jakeman AK, Beck MB, McAleer MJ (eds) Modeling change in environmental systems. Wiley, New York, pp 30–38

Moore ID, Gessler PE, Nielsen GA, Peterson GA (1993b) Soil attribute prediction using terrain analysis. Soil Sci Soc Am J 57:443–452

Nellis MD, Lulla KM, Jensen J (1990) Interfacing geographic information system and remote sensing for rural land-use analysis. Photogramm Eng Remote Sens 56(3):329–331

O'Loughlin EM (1986) Prediction of surface saturation zones in natural catchments by topographic analysis. Water Resour Res 22(5):794–804

Onstad CA, Brakensiek BL (1968) Watershed simulation by stream path analogy. Water Resour Res 4(5):965–971

Paik K (2008) Global search algorithm for nondispersive flow path. J Geogr Res 113(F04001):9

Pareta K, Pareta U (2011) Quantitative morphometric analysis of a watershed of Yamuna basin, India using ASTER (DEM) data and GIS. Int J Geomatics Geosci 2:248–269

Park S, Choi C, Kim B, Kim J (2013) Landslide susceptibility mapping using frequency ratio, analytic hierarchy process, logistic regression, and artificial neural network methods at the Inje area, Korea. Environ Earth Sci 68:1443–1464

Paron P, Claessens L (2011) Makers and users of geomorphological maps. In: Smith MJ, Paron P, Griffiths J (eds) Geomorphological mapping: methods and applications. Elsevier, London

Pfeifer N, Pottmann H (1996) Surface models on the basis of a triangular mesh – surface reconstruction. Int Arch Photogramm Remote Sens XXXI, IWG III/IV, 638–643, Vienna, Austria

Podobnikar T (2005) Production of integrated digital terrain model from multiple datasets of different quality. Int J Geogr Inf Sci 19(1):69–89

Podobnikar T (2008) Simulation and representation of the positional errors of boundary and interior regions in maps. Moore A, Drecki I (eds) Geospatial vision: new dimensions in cartography, pp 141–169, Lecture notes in geoinformation and cartography

Pradhan AMS, Kim YT (2014) Relative effect method of landslide susceptibility zonation in weathered granite soil: a case study in Deokjeok-ri Creek, South Korea. Notational Hazards 72(2):1189–1217

Reddy GPO (2012a) Digital elevation models- sources and resolutions. In: Reddy GPO, Sarkar D (eds) Remote sensing and GIS in digital terrain analysis and soil-landscape modeling, NBSS&LUP Publ. No. 152, pp 121–125

Reddy GPO (2012b) Principles and applications of digital terrain analysis. In: Reddy GPO, Sarkar D (eds) Remote sensing and GIS in digital terrain analysis and soil-landscape modeling, NBSS&LUP Publ. No. 152, pp 129–135

Reddy GPO (2012c) Advances of GIS and remote sensing in hydro-geomorphology. In: Reddy GPO, Sarkar D (eds) Remote sensing and GIS in digital terrain analysis and soil-landscape modeling, NBSS&LUP Publ. No. 152, pp 184–193

Reddy GPO (2012d) Extraction of terrain variables from DEM. In: Reddy GPO, Sarkar D (eds) Remote sensing and GIS in digital terrain analysis and soil-landscape modeling, NBSS&LUP Publ. No. 152, pp 140–146

Reddy GPO, Maji AK, Gajbhiye KS (2002) GIS for Morhophometric analysis of river basins. GIS India 11(9):9–14

Reddy GPO, Maji AK, Gajbhiye KS (2004a) Drainage morphometry and its influence on landform characteristics in basaltic terrain – a remote sensing and GIS approach. Int J Appl Earth Obs Geoinf 6:1–16

Reddy GPO, Maji AK, Chary GR, Srinivas CV, Tiwary P, Gajbhiye KS (2004b) GIS and remote sensing applications in prioritization of river sub basins using morphometric and USLE parameters— a case study. Asian J Geoinf 4(4):35–49

Reddy GPO, Maji AK, Das SN, Srivastava R (2012) Development of GIS based seamless mosaic of SRTM elevation data of India to analyze and characterize the selected geomorphic parameters. Project report. NBSS&LUP, Nagpur 54p

Reddy GPO, Nagaraju MSS, Ramteke IK, Sarkar D (2013) Terrain characterization for soil resource mapping in part of semi-tract of Central India using high resolution satellite data and GIS. J Indian Soc Remote Sens 41:331–343

Reddy GPO, Kumar N, Sahu N, Singh SK (2017) Comparative evaluation of automatic drainage extraction using ASTER and Cartosat-1 DEM in basaltic terrain of Central India. Egyptian J Remote Sens Space Sci 21:95. https://doi.org/10.1016/j.ejrs.2017.04.001

Sahu N, Reddy GPO, Kumar N, Nagaraju MSS, Srivastava R, Singh SK (2016) Morphometric analysis in basaltic terrain of Central India using GIS techniques: a case study. Appl Water Sci 7(5):2493–2499

Sameena M, Krishnamurthy J, Jayaraman V, Ranganna G (2009) Evaluation of drainage networks developed in hard rock terrain. Geocarto Int 24(5):397–420

Schenk AF (1989) Determination of DEM using iteratively rectified images. Photogrammetry Technical Report No 3. Department of Geodetic Science and Surveying, The Ohio State University. Columbus, Ohio

Singh P, Gupta A, Singh M (2014) Hydrological inferences from watershed analysis for water resource management using remote sensing and GIS techniques. Egyptian J Remote Sens Space Sci 17:111–121

Smith MJ, Pain CF (2009) Applications of remote sensing in geomorphology. Prog Phys Geogr 33(4):568–582

Smith MJ, Griffiths J, Paron P (eds) (2011) Geomorphological mapping: methods and applications. Developments in earth surface processes. Elsevier, London

Strahler AN (1952) Hypsometric (area-altitude) analysis of erosional topography. Geol Soc Am Bull 63:1117–1142

Strahler AN (1957) Quantitative analysis of watershed geomorphology. Trans Am Geophys Union 33:913–920

Tarboton DG, Bras RF, Rodriguez-Iturbe I (1991) On the extraction of channel networks from digital elevation data. Hydrol Process 5:81–100

Tucker GE, Catani F, Rinaldo A, Bras RL (2001) Statistical analysis of drainage density from digital terrain data. Geomorphology 36:187–202

USGS (1993) Digital elevation models, Data user guide 5. US Geological Survey, Reston, Virginia

Valeriano MM, Kuplich TM, Storino M, Amaral BD, Mendes JN Jr, Lima DJ (2006) Modeling small watersheds in Brazilian Amazonia with shuttle radar topographic mission-90m data. Comput Geosci 32:1169–1181

Vijith H, Satheesh R (2006) GIS based morphometric analysis of two major upland sub-watersheds of Meenachil river in Kerala. J Indian Soc Remote Sens 34(2):181–185

Weibel R, Heller M (1991) Digital terrain modelling. In: Maguire DJ, Goodchild MF, Rhind DW (eds) Geographical information systems: principles and applications. Longman, London, pp 269–297

Wilson JP, Gallant JC (2000) Terrain analysis: principles and applications. J.W. Wiley, p 469

Wilson JP, Spangrud DJ, Nielsen GA, Jacobsen JS, Tyler DA (1998) Global positioning system sampling intensity and pattern effects on computed topographic attributes. Soil Sci Soc Am J 62:1410–1417

Zevenbergen LW, Thorne CR (1987) Quantitative analysis of land surface topography. Earth Surf Process Landf 12:47–56

Chapter 12
Remote Sensing and GIS for Geomorphological Mapping

G. P. Obi Reddy

Abstract Geomorphology is the study of the landforms, their processes, form and sediments at the surface of the earth. Geomorphological maps are very effective tools in management of land resources and help in various types of resources inventory, mapping and management. Running water, wind, glaciers, karst and sea waves are powerful erosional and depositional agents, which are acting over a long period of time and produce systematic changes leading to sequential development of landforms. The traditional mapping approaches emphasized qualitative interpretation, as frequently dictated by the inherent limitations associated with fieldwork, paucity of digital data and lack of prior field/geographic experience and domain knowledge. Recent advances in remote sensing, geographic information system (GIS) and availability of satellite-based digital elevation models and developments in numerical modelling capabilities enhance the ability to understand the surface processes more clearly in the field of geomorphology. Geomorphological mapping and analysis of various processes using advance tools like remote sensing and GIS act as preliminary tools for land resources inventory, mapping and management, geomorphological and geological risk management, as well as providing baseline data for other applied sectors of environmental research such as landscape ecology, soil science, hydrology and forestry, etc.

Keywords Remote sensing · Geomorphological mapping · Fluvial landforms · Aeolian landforms · Karst landforms · Glacial landforms · Coastal landforms

12.1 Introduction

Geomorphology is the scientific study of the geometric features of the earth's surface. Although the term is commonly restricted to those landforms that have developed at or above sea level, geomorphology includes all aspects of the interface

---

G. P. O. Reddy (✉)
ICAR-National Bureau of Soil Survey & Land Use Planning, Nagpur, India

G. P. O. Reddy, S. K. Singh (eds.), *Geospatial Technologies in Land Resources Mapping, Monitoring and Management*, Geotechnologies and the Environment 21,
https://doi.org/10.1007/978-3-319-78711-4_12

between the solid earth, the hydrosphere and the atmosphere. Geomorphological mapping plays an important role in understanding the earth surface processes, geochronology, natural resources, natural hazards and landscape evolution (Blaszczynski 1997; Bishop and Shroder 2004; Smith et al. 2011; Reddy 2012). Geomorphological processes comprise the physical and chemical interactions between the earth's surface and the natural forces acting upon it to produce landforms. Several related landforms together make up landscapes, and each landform has its own physical shape, size and materials and is a result of the action of certain geomorphic processes and agent(s). Actions of most of the geomorphic processes and agents are slow, and hence, the results take a long time to take shape. The efficiencies of these processes are determined by such variables as geology, climate, vegetation, base level and human influences. Due to changes in climatic conditions and vertical or horizontal movements of landmasses, either the intensity of processes or the processes themselves might change leading to new modifications in the landforms. Landform evolution implies stages of transformation of either a part of the earth's surface from one landform into another or transformation of individual landforms after they are once formed. A geomorphological map can act as a preliminary tool for land management and provides baseline data to other applied sectors of environmental research such as landscape geology, geography, soil science, hydrology, forestry, ecology, environmental sciences, etc. (Cooke and Doornkamp 1990a; Dramis et al. 2011).

Geomorphological mapping plays an important role in understanding earth surface processes, geochronology, natural resources, natural hazards and landscape evolution (Blaszczynski 1997; Reddy et al. 2004; Bishop and Shroder 2004). It involves the partitioning of the terrain into conceptual spatial units/entities based upon criteria that include morphology (form), genetics (process), composition and structure, chronology, environmental system associations (land cover, soils, ecology), as well as spatial topological relationships of surface features (landforms). Geomorphological mapping has been based upon integration of multidisciplinary information from the field, remotely sensed data and cartographic map products. Regional-scale geomorphology and physiographic analysis and mapping (Baker 1986) were based upon the interpretation of photography and smaller-scale maps to classify terrain types/features at the regional (physiographic) scale. Detailed geomorphological mapping is being carried out based on surveying and other *in-situ* measurements, although detailed large-scale geomorphological maps did not exist for many areas. These traditional mapping approaches emphasized the qualitative interpretation, as frequently dictated by the inherent limitations associated with fieldwork, paucity of digital space–time data and human a priori field/geographic experience and domain knowledge.

Traditionally, terrain and landform mapping was undertaken by utilizing topographical map, aerial photographs and field surveys (Dent and Young 1981; Pain 1985). Aerial photographs are used as remote sensing data in terrain mapping programmes in British Columbia (Slaymaker 2001) and landslide mapping in Hong Kong (Dai and Lee 2002) and fieldwork for large- and medium-scale mapping (Mantovani et al. 1996). The satellite images are unique resources for geomorphological mapping. Different spatial information such as land cover, hydrology and digital elevation model (DEM) integrated in a GIS allows interpretation and analysis of geomorphological features more precisely than satellite data alone. Application of various image enhancement techniques could significantly improve the visual interpretability of an image by increasing the apparent distinction between the features on the land (Sabins 2000). Digital image processing techniques such as generation of false colour composites (FCC), band ratioing and principal component analysis (PCA) help to evaluate geological, structural and geomorphologic characteristics of the terrain. Many researchers have also used Landsat datasets for identification of terrain features, landform, geology and geomorphology (Verstappen 1977; Kayan and Klemas 1978; Pain 1985; Bocco et al. 2001). Recent advances in geospatial technologies, which include remote sensing, GIS, global positioning system (GPS), digital elevation models as well as developments in numerical modelling of surface processes, have revolutionized the field of geomorphology mapping and analysis of the processes (Shroder and Bishop 2003; Bishop and Shroder 2004).

12.2 Geomorphological Processes

A geomorphological process includes the interactions between physical components of the atmosphere, the earth (lithosphere) and water (hydrosphere) with the biological communities of plants and animals and especially humans that modify landforms. The landforms are produced by a consequence of the interaction of relative uplift (endogenetic) on the earth's surface and down wearing (exogenetic) by weathering and erosional processes. The struggle between these opposing processes is reflected at any particular point in time by the landforms that appear on the earth. The processes, which influence the characteristics of landforms, are activity of running water (fluvial) and moving ice (glaciers), work of underground water (Karst), the transport and deposition of dust and sand by wind (aeolian), beach and shoreline processes that involve the mechanics and effects of waves and current (littoral) and the study of mass movement of material through downhill creep of soil/ rock and by landslide and mud flows. Detailed study of these processes form a base for classification of landforms, which is a major aspect of geomorphological mapping.

12.3 Evolution of Landforms

Changes on the surface of the earth owe mostly to erosion by various geomorphic agents. Of course, the process of deposition too, by covering the land surfaces and filling the basins, valleys or depressions, brings changes in the surface of the land. Deposition follows erosion, and the depositional surfaces too are ultimately subjected to erosion. Running water, groundwater, glaciers, wind and sea waves are powerful erosional and depositional agents shaping and changing the surface of the earth aided by weathering and mass wasting processes. These geomorphic agents acting over long periods of time produce systematic changes leading to sequential development of landforms. Each geomorphic agent produces its own assemblage of landforms, and each geomorphic process and agent leaves their distinct imprints on the landforms they produce. The study of evolution of landforms will reveal the process and agent, which has made or has been making those landforms. As the geomorphic agents are capable of erosion and deposition, two sets – erosional or destructional and depositional or constructional – of landforms are produced by them. The distinct landforms develop by the action of each of the geomorphic agents depending upon especially the type and structure, i.e. folds, faults, joints, fractures, hardness and softness, permeability and impermeability, etc. come under structure of rocks. There are some other independent controls like (i) stability of sea level, (ii) tectonic stability of landmasses, and (iii) climate, which influence the evolution of landforms. Any disturbance in any of these three controlling factors can upset the systematic and sequential stages in the development and evolution of landforms.

12.4 Geomorphological Mapping

Recognizing the importance of geomorphological mapping, the 18th Congress of the International Geographical Union (IGU) held at Rio de Janeiro in 1956 and 19th Congress of IGU held at Stockholm in 1958 formed a subcommission on geomorphological mapping to (1) introduce and develop a standardized methodology of geomorphological mapping, (2) adopt a uniform system to ensure compatibility and (3) demonstrate the applications of geomorphological mapping in local and regional economic planning in order to facilitate a rational utilization of the earth's surface (Gilewska 1966; Gilewska and Klimek 1968). In 1968, IGU developed Unified Key mapping system for international detailed geomorphological mapping (Demek et al. 1972; Unified Key 1968) and subsequently Unified Key developed for geomorphological mapping at medium scale (Demek and Embleton 1978). During the same time, the International Institute for Aerial Survey and Earth Sciences (ITC) also published an international mapping system for various landscapes (Verstappen and Van Zuidam 1968) and presents information about morphometry/morphography, processes/genesis, age and lithology. The legend and manual for geomorphological mapping was published in 1971 and 1972, respectively (Bashenina et al. 1971; Demek et al. 1972). Demek et al. (1972) suggested that a large-scale geomorphological map could have a scale of 1:10,000–1:50,000 and occasionally up to 1:100,000. UNESCO (Anonymous 1976) suggested a scale of 1:5000–1:10,000 to qualify a map as 'large-scale'. Demek and Embleton (1978) reported that geomorphological mapping should describe the surface of earth's crust and its contact with the hydrosphere, the atmosphere, the pedosphere and the biosphere. Cooke and Doornkamp (1990b) reported that large-scale maps form the strongest scientific source of geomorphological information and are the best explanatory presentation of landforms and landscape development. Klimaszewski (1990) suggests that a detailed geomorphological map should mainly present the landforms, while information on lithology and hydrography should be supplements.

In geomorphological survey, mapping and classification, the morphogenetic and morphodynamic criteria are extensively used (Verstappen and van Zuidam 1968). In fact, the interaction between form and process is the core of geomorphology (Evans 1998), and form characteristics are key components of geomorphological systems (Ahnert 1998). Lane et al. (1998) in his extensive review underlined the importance of the form in the relief assessment for a variety of purposes. Verstappen (1983) noted that four different types of data could be collected for mapping landforms that include morphographic, morphogenetic, morphometric and morphochronologic information. Many authors noted that land cover or land systems (Cooke and Doornkamp 1990b), hydrology (Gustavsson et al. 2006), surface materials, lithology and structure (Gustavsson 2006), morphometry and morphography (Gustavsson et al. 2006; Fookes et al. 2007), sediment-transfer cascades (Chorley and Kennedy 1971; Shroder and Bishop 2004), surface-process regimes (Cooke and Doornkamp 1990b) and chronology (mapping the age of landforms) should be part of geomorphological mapping activity.

In detailed geomorphological mapping, it may be appropriate to use contours as a base for mapping surface form. Surface features can then be superimposed, providing full detail of the shape of the terrain. Slope length and direction are usually important in geomorphological mapping to distinguish the features like cliffs, upslopes, downslopes, concave or convex units. Surface materials would typically be incorporated in the geomorphological map. They may be mapped according to type or grade, or may incorporate relevant features such as the extent of jointing or fracturing to understand the slope stability or groundwater resources. Other features such as soil depth, porosity, permeability or strength may also be shown depending on the specific purpose of the study. The surface processes are generally mapped indirectly by means of interpreting landform and surface materials. The type of landform and symbol used will therefore differ markedly, according to the types of process operating in the region of interest. The age of a landform is often difficult to ascertain. It is generally more useful in the interpretation of long-term landscape evolution than in applied settings. However, in some contexts, the age may be a useful proxy for different material properties. In some cases, land system mapping is used at regional to national scales for a variety of purposes, including terrain evaluation, resource evaluation, route-corridor alignment, hazard assessment or avoidance and conservation planning. The approach depends on the evaluation of patterns of topography, soils and vegetation that recur within a particular landscape (Cooke and Doornkamp 1990a). The land system is divided into units and then elements or facets.

12.5 Geomorphological Mapping at Different Scales

The issue of spatial and temporal scale in geomorphological mapping is well known (Quattrochi and Goodchild 1997; Tate and Atkinson 2001; Sheppard and McMaster 2004). Obviously, practical issues of scale are associated with data representation, collection, analysis, modelling and presentation. Spatial scales are varied and include concepts of *geographic scale,* which represents aerial coverage or the size of objects, *cartographic scale,* which represents aerial distribution and detail of information presented, *measurement scale,* which represents the smallest area over and which data can be collected or represented to maintain distinguishable parts of an object, *operational scale,* which represents the scale at which processes, feedback mechanisms and systems operate and *computational scale*, representing the scale at which data are analysed. Dikau (1989) and Brändli (1996) proposed hierarchical level, which essentially describes a vertical structure of levels, and that a subsystem at any level is spatially constrained by a higher level. It can be used to describe the complexity of scale associated with a landscape, and the theory has been proposed as a basis for modelling and geomorphological mapping. Fookes et al. (2007) underlined that the detail that can be shown and the form of representation of specific features in geomorphological mapping will depend on the scale of map to be drawn. The map scale will in turn be determined by the aim and type of study carried out and may vary from a regional assessment at 1:1,00,000 scale or coarser resolution to detail mapping or detailed plan at 1:2000 scale.

12.6 Remote Sensing in Geomorphological Mapping

Recent developments in remote sensing and GIS have vastly improved the accuracy in mapping of several geomorphological variables. With the advent of digital image processing, GIS and GPS, it is now possible to map the geomorphological features in greater details than before and to interpret them for newer information. Further, use of digital remote sensing, GIS and digital elevation models (DEM) in geomorphological mapping is cost-effective and time efficient (Batten 2001; Bocco et al. 2001). Reddy and Maji (2003) reported that analysis of IRS-ID LISS-III data in conjunction with distinct lithological units, drainage pattern and contour information improves the capability in delineation and characterization of geomorphological units. Remote sensing data has been used in the analysis of terrain mapping and geomorphic features by the analysis of colour composites (Pain 1985; Novak and Soulkellis 2000; Bocco et al. 2001). The false colour composites, i.e. blue, green and red bands of Landsat (with bands 7, 5 and 4), display good contrast for analyses of landform (Pain 1985). Verstappan (1977) used Landsat ERTS-1 band 5 and 7 data and concluded that some landform types were better distinguished than others, mainly because of variations on the correlation between geomorphic units and vegetation. Band 7 of Landsat ERTS-1 was the most valuable for identifying geologic formations, tectonic fault lines and geomorphology slope contrast (Kayan and Klemas 1978). Pain (1985) used Landsat MSS for landform mapping in Australia and existing land use classification as a reference. Bocco et al. (2001) concluded that application of Landsat data helps to quickly identify and map terrain features at reconnaissance scale (1:250,000) and semi-detailed (1:50,000) levels.

Recent advances in remote sensing and GIS, as well as developments in numerical modelling of surface processes, have revolutionized the field of geomorphology (Shroder and Bishop 2003; Bishop and Shroder 2004). The synoptic view provided by satellite remote sensing offers technologically the appropriate method to map distinct landforms. Geospatial technologies can be used to address some of the conceptual issues such as heterogeneous surface composition with fuzzy classification membership (Warner and Shank 1997), indeterminant boundaries and features (Burrough 1989; Usery 1996; Burrough et al. 2000; Smith et al. 2000; Deng and Wilson 2008), hierarchical organization and spatial analysis using object-oriented technology (Ralston 1994; Brändli 1996; Schmidt and Dikau 1999), scale dependence of properties and patterns using geostatistics (Tate and Wood 2001) and objective mapping using different analytical approaches such as descriptive statistics, inferential statistics, artificial intelligence and analytical reasoning technologies. Geomorphological maps can act as a preliminary tool for land management and geomorphological and geological risk management, as well as providing baseline data for other applied sectors of environmental research such as landscape ecology, forestry or soil science (Cooke and Doornkamp 1990a; Dramis et al. 2011; Paron and Claessens 2011). Many authors used remote sensing data in geomorphological mapping in fluvial, arid, glacial, karst and coastal environments.

12.7 Geomorphological Mapping in India

In India, under national mission on geomorphological and lineament mapping on 1:50,000 scale, ISRO-National Remote Sensing Agency (presently known as National Remote Sensing Centre) and Geological Survey of India (GSI) have developed a new genesis-based three-level classification system (Anonymous 2009). In this system, the first level addressed the genetic aspect (origin) of landforms, the second level grouped the landforms on the basis of their broad current processes, and the third level provided the actual mapped landforms. In India many authors used both aerial photographs and satellite remote sensing data in geological, geomorphological and structural mapping (Krishnamurthy and Srinivas 1995; Reddy et al. 2004). Detailed information on various geomorphological properties can be extracted accurately from high-resolution satellite data and large-scale aerial photographs. Further detailed information can be generated on geomorphological units and their characteristics in GIS.

12.8 Geomorphological Classification and Legend Systems

Geomorphic classification is the categorization and description of the nature, origin and development of landforms. The framework of classification system is that a geomorphic unit can be classified based collectively on its origin and development (process), on its general structure and shape (landform), on measurements of its dimensions and characteristics (morphometry) and on the presence and status of process overprinting (geomorphic generation). In the past, a number of classification systems have been devised, in which a general objective has been to describe and group landforms according to the processes that shaped or influenced them. In addition, some classification systems incorporate the stage of development of landforms as an aspect of their evolutionary development through geologic time and take into consideration the lithology of rocks, the attitude of strata, the presence of faults and joints and factors of broad influence, including regional uplift and climatic change. The geomorphic classification system also needs to be hierarchical in order to address related landforms at various scales; to illustrate the relationships to both higher-order and more finite geomorphic processes, landforms and landform components; and to link at all scales (USDA 1993). The taxonomic schema for describing landform patterns includes how they were formed, the environment in which they were formed and their composition.

In the early twentieth century, a large number of symbols and legends have been developed for geomorphological mapping. The concepts and cartographic conventions created basically related to the terrain configuration of the surveyed region and the scientific focus and aim of the map. Obviously, more complex terrains have the greater diversity of symbols and colours (Verstappen 2011). In general geomorphological maps and their legend systems can be differentiated into maps showing a single aspect of geomorphology, most commonly morphology, and analytical maps that encompass the full information of a landscape including processes, morphogenesis and even lithology (Knight et al. 2011; Verstappen 2011). An overview of different legend systems is provided by Otto et al. (2011), and it highlights how different countries have developed their own systems, either formally or informally, to suit their own needs. Most of the interest in geomorphological mapping has centred on the development of various mapping systems for use in environmental management (Cooke and Doornkamp 1974); the most detailed systems have been developed in Europe, where different countries utilize different procedures. Despite attempts at international standardization (Demek 1972; Demek and Embleton 1978), the major problem remains the correlation and harmonization of various mapping schemes.

12.9 Geomorphological Mapping in Different Environments

Geomorphological mapping involves the partitioning of the terrain into conceptual spatial units/entities based upon criteria that include morphology (form), genetics (process), composition and structure, chronology, environmental system associations (land cover, soils, ecology), as well as spatial topological relationships of surface features (landforms). Landform elements are commonly described by five main attributes: slope, topographic position, dimension, 'mode of geomorphological activity' and 'geomorphological agents' (Klingseisen et al. 2007). The genesis of landforms due to various geomorphic agents like fluvial, arid, glacial, karst and coastal is briefly discussed below.

12.9.1 Geomorphological Mapping in Fluvial Environment

The surface run-off starts commonly in the form of a thin sheet of water in motion – rain wash/sheet wash. This accumulates and begins its downward journey; it becomes a stream. Further down a number of streams unit to form a river. Along its seaward course, a river produces a few tributaries and developed a river system gradually. The water flowing in a river erodes the land, brings about its chemical decay, denudes the country rocks, transports the rock debris and finally deposits the transported material under favourable conditions. The fluvial cycle operates in three well-defined stages. At the stage of 'youth', the flowing water predominantly erodes and tributaries gradually develop. At the stage of 'maturity', the long profiles of rivers are attained equilibrium, drainage system is well evolved, and the power of erosion is also much reduced and is balanced with its transporting capacity. At the stage of 'old', the river valley becomes very wide due to lateral cutting, and the process of erosion becomes extremely slow while deposition phase dominates. The country is reduced to a '*peneplain*', which is a region of low relief exhibiting very gentle undulations. Isolated hillocks made up of harder, durable and resistant rocks occur here and there on the peneplains, which are known as monadrocks, and they are the remnants of the pre-existing country.

12.9.1.1 Geomorphological Processes in Fluvial Environment

The important fluvial processes of running water are hydraulic action, abrasion, attrition, solution, transportation and deposition. *Hydraulic action* is basically breaking down of the rock masses essentially due to continuous impact of water moving with appreciable velocity along the channel. *Abrasion* is the process of wear and tear of rock material. *Attrition* is the process of mechanical breakdown of the transported rock fragments due to impact among themselves. Abrasion and attrition always work together and produce fine rock particles, which ranges in size from boulder–pebbles to silt–clay and are readily transported downstream along with flowing water. *Transportation* can be defined as the size and amount of load, and the velocity of stream determines their transporting power. Transportation power is directly proportional to the sixth power of stream velocity. However, *deposition* occurs primarily during the periods of low flow and form various depositional landforms.

12.9.1.2 Fluvio-denudational Landforms in Fluvial Environment

Stream Capture Drainage progressively or abruptly diverted from one basin to another as a stream is beheaded by headward erosion. It is a progressive movement of drainage network towards headward side of terrain through the activity of erosion by drainage streams. In this process parallel streams formed at different elevation with different base levels (drainage captures by lower stream). Some streams have structural advantage, i.e. degrading more rapidly in softer rock than the hard rock.

Plateau A plateau is an extensive, elevated region with a fairly flat top surface. Plateaus are generally dominated by a structure of horizontal rock layers. Many striking plateaus exist in the deserts and semiarid regions of the world. Colorado Plateau of the United States is a tectonically uplifted desert plateau, where streams and their tributaries respond to uplift by cutting narrow and steep-sided canyons.

Canyon In the upper part of the course, the process of erosion becomes very conspicuous in downcutting the valley floor. Therefore, the river passes through a narrow but deep valley, which may develop in softer rocks known as 'gorge' or 'canyon'. Numerous waterfalls and rapids also develop on the upper parts. The canyon walls consist of horizontal layers of alternating resistant and erodible rocks; differential weathering and erosion exert a strong influence on the canyon walls. Canyons in these areas tend to have stair-stepped walls, with near-vertical cliffs marking the resistant layers, and weaker rocks form different slopes.

Mesa Weathering and erosion will eventually reduce the extent of a caprock until only flat-topped, steep-sided mesas remain. A mesa has a smaller surface area than a plateau and is roughly as broad across as it is tall. Mesas are relatively common landscape features in Deccan Plateau region of India.

Butte Due to erosional processes of the caprock from all sides, a mesa may be reduced to a butte, which is a similar, flat-topped erosional remnant but with a smaller surface area than a mesa. Mesas and buttes in a landscape are generally evidence that uplift occurred in the past and that weathering and erosion have been extensive since then. Variations in the form of the slope extending down the sides of buttes, mesas and plateaus are related to the height of the cliff at the top, which is controlled by the thickness of the caprock in comparison to the size of landform features.

12.9.1.3 Fluvio-depositional Landforms in Fluvial Environment

Natural Levees During floods streams commonly overflow their channels, and when this water retreats, ridge-like deposits of silt and clay are formed along the fringes of the flooded channels known as 'natural levees'. Flood plain is the portion of a river valley, which is readily submerged underwater during floods. Slough are depressions upon the flood plains of meandering rivers due to the tendency of the overflowing water to follow a shorter course. The process of repeated filling up and cutting of channels upon the valley floors is known as 'braiding'.

Alluvial Fans In the earlier stages of development of a river valley, its transverse profile is 'V' shaped. With maturity of the river, its width increases in comparison with its depth, and the transverse profile is modified to a more or less flat base with gently sloping *abutments*. Accumulation of boulders and pebbles at a point where running water enters into the plains are known as 'alluvial fans', where rock fragments are arranged in a radiating fan-like pattern.

Oxbow Lakes Streams and rivers never follow a straight course. They move in curved, more or less zigzag paths, and the curvature along the course is known as meanders. Meanders are present since running water has the tendency to follow the direction of maximum slope of the landforms. During the later stages, however, the meanders become more conspicuous due to deposition of sediments along the inner curve of the lore-existing meander and proportionate excavation along the outer curve. In this manner the river ultimately takes an extremely round about course, and during floods the running water may cut straight through a meander and may follow a shorter horseshoe or oxbow lakes. Sometimes the major portion of the course of the river is abandoned due to shifting, which is termed as paleo-channel.

Deltas Deltas are low-lying land mass formed at the mouth of a river due to deposition of sediments and are commonly swampy in nature. In the earlier stage of formation of delta, the fine rock waste accumulate at the basement on which delta grows, and they are known as 'bottomset beds'. During later stages large quantity of silt, sand and clay are deposited in the form of layers and are known as forest beds. Thin layers of silt and clay are deposited over the forest beds, which are termed as '*topset beds*'. Thus, the delta structure is made up of three types of beds.

12.10 Landform Mapping in Fluvial Environment: A Case Study

Terrain characterization and landform mapping of a region are prerequisite to conduct soil resource inventory, develop landscape–soil relationship, land degradation mapping and other environmental applications (Reddy et al. 1999, 2002). Landform mapping under fluvial environment of Basltic Terrain of Central India has been carried out. In the study, IRS-P6 LISS-IV (5.8 m) satellite data and SRTM DEM (30 m) digital terrain database have been used in detail landform mapping. Slope is an important consideration in landform analysis and mapping. The slope analysis using SRTM DEM (30 m) of the study area shows that level to nearly level (0–1%) slopes occupy 28.4% of the study area. Major parts of the area are covered under very gently sloping (1–3%), and it accounts for 33.6% of the study area. Gently sloping (3–5%) area is associated with upland areas and accounts for 19.3%, whereas moderate slopes (5–10%) are encountered in 9.3% of the study area. Steep

Table 12.1 Slope classes of the study area

S.No	Class	Area (ha)	% Area
1	Level to nearly level (0–1%)	4313.4	28.4
2	Very gently sloping (1–3%)	5098.4	33.6
3	Gently sloping (3–5%)	2931.9	19.3
4	Moderately sloping (5–10%)	1406.8	9.3
5	Moderately steep sloping (10–15%)	783.9	5.2
6	Steeply sloping (15–25%)	447.7	3.0
7	Very steeply sloping (25–33%)	189.6	1.2
	Total	**15171.8**	**100.0**

(15–25%) and very steep (25–33%) slopes were noticed in 3.0 and 1.2% of the study area, respectively (Table 12.1). Landform analysis was carried out based on the visual interpretation of high-resolution IRS-P6 LISS-IV data with the help of image elements such as shape, tone or colour, pattern, shadow, association and texture in conjunction with DEM. The analysis shows nine distinct landform units namely plateau top, isolated mounds, linear ridges, scarp slopes, undulating uplands, undulating lowlands, alluvial plains, narrow valleys and main valley floor in the study area (Fig. 12.1 and Table 12.2). The land use/land cover of the study area shows that double- and single-cropped area occupy about 35.0 and 20.0 of the area, respectively. Nearly 29.0% of the study area is under deciduous forest cover.

Plateau Top This landform unit occupies the highest position in the study area, and elevation varies ranging from 580 to 662 m above mean sea level (MSL) and covered with an area of 1453.2 ha (9.6%). The general slope is varying between 0–1 and 1–3 per cent. However, on the fringes of this unit, they exceed 5–8%. Rain splash, rillwash and slopewash processes dominate rather than concentrated surface run-off. Sheet wash processes are also observed on the very gently sloping areas. Hence, typical summit areas are devoid of fingertip tributaries. Regolith cover is very thin with numerous core stones of varying dimensions spread throughout this unit. Weathering limited condition is responsible for such a situation.

Isolated Mounds This unit occurs mainly in middle and lower portion of the area. Their elevation is around 560–660 m above MSL and covered with an area of 93.1 ha (0.6%). These mounds are generally restricted, much dissected, denuded and appears bevelled on its fringes. By parallel retreat of slopes due to back wearing and circum-denudation, these mounds might have been denuded and detached from the extensive basaltic plateau. The summits of these mounds are marked by the relative absence of concentrated run-on/run-off. It is observed that rains plash is the dominant geomorphic process followed by sheet wash. Numerous core stones are spread throughout this unit, and they vary in their size and dimension, which indicates the severity of erosion.

Linear Ridges Linear ridges are sinuous, narrow, highly disintegrated by mechanical weathering processes and are remnants of land reduction process formed due to

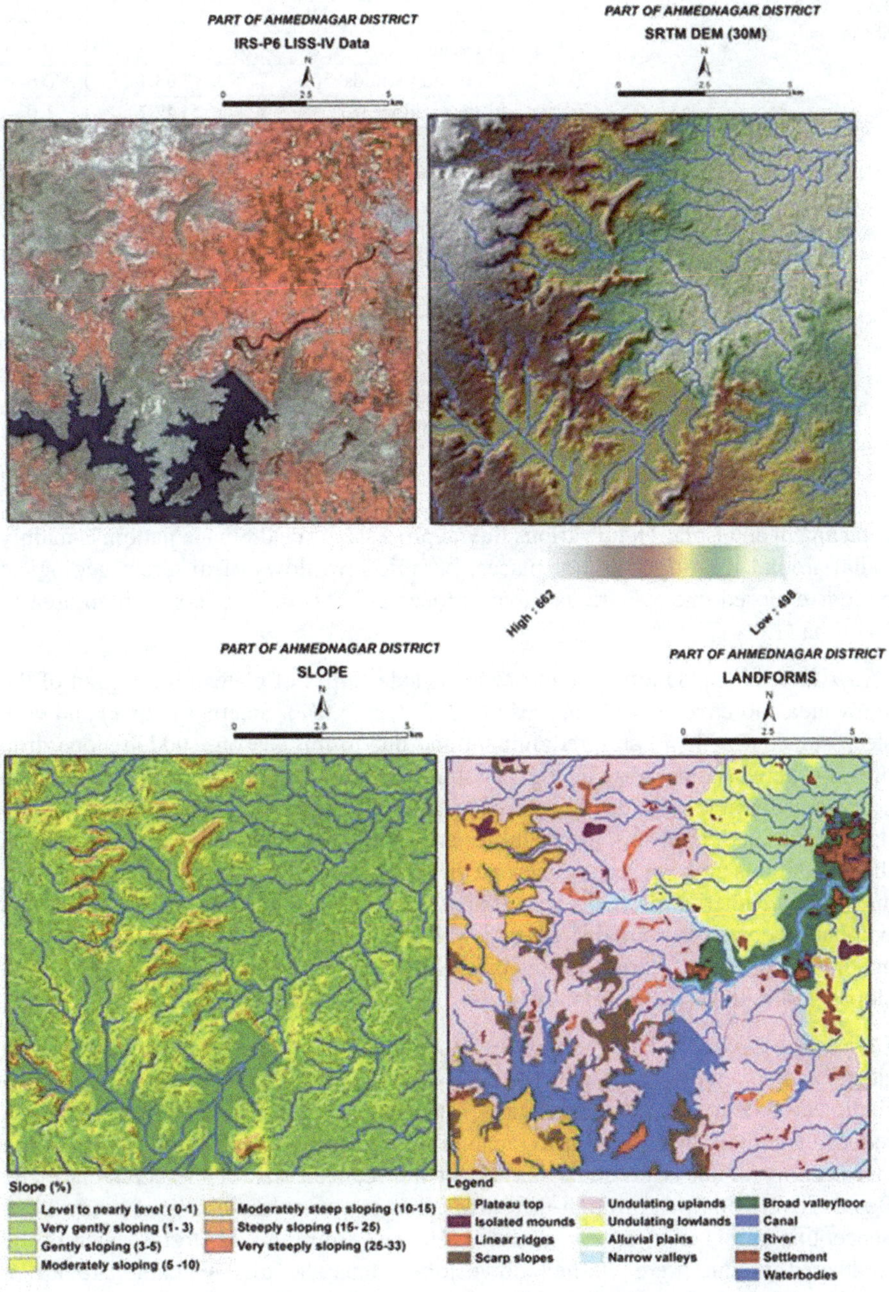

Fig. 12.1 IRS P6 LISS-IV data, SRTM-30 m data, slope and landform maps of part of Ahmednagar district of Maharashtra

Table 12.2 Landform units of the study area

S.No	Class	Area (ha)	% TGA
1	Plateau top (Pt)	1453.2	9.6
2	Isolated mounds (Im)	93.1	0.6
3	Linear ridges (Lr)	147.7	1.0
4	Scarp slopes (Ss)	1212.0	8.0
5	Undulating uplands (Uu)	6787.1	44.7
6	Undulating lowlands (Ul)	1768.4	11.7
7	Alluvial plain	906.5	6.0
8	Narrow valleys (Nv)	299.0	2.0
9	Broad valleyfloor (Bvf)	598.4	3.9
10	River	102.7	0.7
11	Canal	10.4	0.1
12	Settlements	485.6	3.2
13	Water bodies	1310.7	8.6
	Total	**15174.8**	**100.0**

detachment and isolation by various fluvial processes. The drainage pattern is mainly radial around the ridges and at places, basaltic lava flows of different geological periods exposed due to various fluvial processes. This unit covers with an area of 147.7 ha (1.0%).

Scarp Slopes This landform unit occurs as side slopes of plateau-top region of the study area and covered with an area of 1212.0 ha (8.0%). Steep (15–25%) and very steep slopes (25–33%) are predominent and due to this reason, sudden slope drop prevails between the plateau top and undulating uplands. Rigorous slope wash processes dominate due to the presence of the extreme steep slope. Moderate to deep gullies and, at places, proto-gullies are common, which led to extreme erosion throughout this unit. Moreover, increase in gradient also favours the removal of fines and other weathering products very easily; hence, the entire escarpments are strewn with larger corestones and disintegrated piles of basalt pebbles. Continuous incision and removal of weathering products resulted in the formation of hard and nose slopes within the escarpments.

Undulating Uplands This unit occurs immediately below the plateau top and scarp slopes and covered with an area of 6787.4 ha (44.7%). The slope angles are much shallower having concave curvatures. The surface spread of pebbles are much reduced and their dimensions also much diminished. Increased chemical weathering and decomposition is the main reason for the reduced size of the corestone here. Besides overland flow, through-flow also emerges out in this subunit as return flow; hence, this unit is always over saturated. However, during rain storms much of the weathered products are washed downslopes; thus the fills available are all of transitory in nature. Colluviation from the above-occurring units and entrainment of sediments are common features in this unit.

Undulating Lowlands This unit was noticed between 500 and 520 m above MSL and covered with area of this unit is 1768.4 ha (11.7%). This is the zone of obstruction of drainage networks wherein the fourth-order deeper colluvic-alluvial fills. The slope angles are much gentler; hence, this unit is having very gentle undulations and increased area. The weathering front is observed to be much deeper and is more than 150 cm. A few first-order drainages are originating in this subunit, and their lengths are much more when compared with those occurring on the undulating uplands. Another remarkable feature observed in this subunit is the total absence of corestones on the surface as well as on the subsurface horizons. It may be because of the fact that, due to increased chemical weathering, these fragments totally disintegrated, decomposed and assimilated into the colluvic-alluvial fills.

Alluvial Plains Alluvial plains are relatively flat landform and created by the deposition of highlands eroded due to weathering and water flow in study area. The sediment from the hills is transported to the lower plain over a long period of time. It identified on the imageries dark, reddish, moderate to fine texture due to agricultural activities. Alluvial deposits of the area constitute gravel-, sand-, silt- or clay-sized unconsolidated material. This unit covers with an area of 906.5 ha (6.0%).

Narrow Valleys This unit occurs on the lowest portion of the study area with an area of 299.0 ha (2.0%). Because of its narrowness and linear nature, the elevation in this unit is varying upstream ranges from 500 to 580 m above MSL. The well-entrenched nature of this narrow drainage floor is indicative of total integration of drainage network. The valley walls are placed more than 2–3 m deep and show alluvial fills. Vertical accretionary deposits are also available on the hairpin bends indicating lateral shifting of bed material load in the geological past.

Broad Valley Floor This unit is adjacent to a stream or river that stretches from the banks of its channel to the base of the enclosing valley walls and experiences flooding during periods of high discharge. It includes the floodway, which consists of the stream channel and adjacent areas that actively carry flood flows downstream and the flood fringe, which are areas inundated by the flood but which do not experience a strong current. This unit covers with an area of 598.4 ha (3.9%).

12.11 Geomorphological Mapping in Arid (Aeolian) Environment

In regions of comparatively dry climate, the efficiency of the blowing wind becomes most conspicuous in bringing about dust and sandstorms, which badly remove a portion of the mantle. The velocity and direction of wind directly influence earth's surface processes, as there is little vegetation to slow its velocity at ground level and to bind the soil, and the loose, dry sediments are easier to erode. The greater temperature ranges experienced in deserts also generates stronger and more consistent winds. Unlike stream and glacial erosion, wind can only erode fine sediment, but it is not confined to channels, and the erosion affects vast areas. As far as India is concerned, the arid region is spread in 38.7 Mha area out of which 31.7 Mha is under hot arid zone and 7 Mha is under cold arid zone. In India, during 2012–2015, Central Arid Zone Research Institute (CAZRI) prepared geomorphological map of arid western Rajasthan using a classification system developed at 1:50,000 scale in collaboration with ISRO (Indian Space Research Organization) and GSI (Geological Survey of India). The mapping involved digital interpretation of the false colour composite (FCC) of images of 2005–2006 from IRS1D LISS-III, followed by field verification of the mapped units and analysis of sediments. Kar (1993) recognized 9 major dune types and 23 subtypes in Thar Desert and then grouped under the old dune system and the new dune system (Pandey et al. 1964; Vats et al. 1976; Singh 1982). Several phases of dune-building activities have also been identified from the morphological characteristics of dune sediments within the desert and its eastern margin. The efficiency of aeolian processes increases with decreasing rainfall from east to west, as well as with the increasing wind speed in that direction (Kar 1993).

12.11.1 Geomorphological Processes in Arid (Aeolian) Environment

In arid environment, deflation, abrasion, attrition, saltation and suspension processes are prominent. *Deflation* (L. deflare = to blow away) can be defined as the process of removal of all loose, light particles from the earth's surface. The process of removal of loose soil or rock particles during storms along the course of the wind has been defined as 'deflation'. Continual deflation of fine material can result in an accumulation of coarser material, which then protects the land from further deflation and a 'gibber plain' (desert pavement) result. Alternatively, 'deflation basins' or 'blowouts' may occur until the water table is reached, where the moist ground and resulting vegetation prevents further erosion. *Abrasion* process is wear and tear of the exposed country rocks (mainly by sand blasting). It is concentrated close to the ground, where most of the material is transported. Close to the ground, objects are smooth and polished, and large rock formations are undercut. Larger rocks have their windward faces flattened and polished and are termed blowing wind which generally causes erosion by three distinct processes deflation. *Attrition* can be defined as that the wind-borne particles traveling is suspension condition collide with one another and causes a further grinding of the particles. *Saltation* is a process of transport of sand grains by wind in a series of leaps along a low, curved trajectory. In case of sand particle transportation, it takes place due to the forward movement of the grains in a series of jumps, and the process is described as '*saltation*'. *Suspension* is a process that being finer in grain size the dust particles are conveniently lifted up in air and continue to move in suspension due to the turbulence present in the air currents.

12.11.1.1 Denudational Landforms in Arid Environment

Due to wind abrasion, the exposed irregular surface of a rock mass is gradually converted into a plane surface, which may be smooth and polished.

Pedestal Rocks Blasts of wind cause more of abrasion near the surface than what is possible in higher horizons. Vertical columns of rocks are more readily worn out towards their lower portions, and as a result 'pedestal rocks' are formed.

Desert Pavement The process of deflation leads to the development of 'desert pavement' – which are made up of a layer of residual pebbles and cobbles stream upon the surface while the intervening finer particles have been removed.

Yardangs The wind causes the development of 'U' shaped troughs within relatively softer rocks. The ridges left between such successive troughs are known as '*yardangs*'. They are generally small in size and extending from tens of kilometres long and up to 100 m high to small structures a few metres in length.

Ventifacts Ventifacts are pebbles of rock, which have developed some plane surfaces due to wind abrasion. The rock particulars, which travel along with the blowing wind, are commonly more concentrated near the surface than higher up in the atmosphere. These are sometimes mistaken for man-made artefacts. Soft rock strata are more easily eroded, leading to 'honeycombed rocks'.

Reg A surface covering of coarse gravel, pebbles, and/or boulders from which all sand and dust have been removed by wind and water.

Desert Varnish A dark, shiny coating that forms on rock surfaces that are exposed to desert air for a long time, often associated with regs.

Hamada A barren surface of consolidated material, usually consisting of exposed bedrock but sometimes of cemented sedimentary layers.

Inselbergs Isolated hills, known as inselbergs, form in areas of relatively homogeneous, resistant rock. Horizontal strata may form dissected plateaux, mesas and buttes.

Bornhardt A rounded inselberg composed of very resistant rock that stands above the surrounding terrain because of differential erosion and weathering.

Pediments Gently inclined bedrock platform that extends outward from a mountain front, usually in an arid region a residual surface formed by erosion not deposition. Sloping away from the uplands are 'pediments' – thinly and incompletely covered with alluvial material and containing the courses of ephemeral streams. Pediments are bedrock surfaces, rather than alluvial fans, but can superficially resemble them, having a concave surface typical of fluvial landforms. They are thought to be the product of sheet erosion during flash floods, which erode laterally as well as vertically, but the exact way they are formed has not been firmly established.

12.11.1.2 Depositional Landforms in Arid Environment

Dust and sand is deposited if the winds lose their velocity, obstructions create wind shadows, or the particles moving by saltation strike a softer surface and so lose energy. Sediments deposited by wind are called 'aeolian' (wind) deposits. Typical deposits include:

Loess Loess are deposits of windblown silt (rather than sand). Loess deposits produce productive farmlands but make a poor foundation material. It has a light, open structure that is relatively strong when dry, but when wetted it tends to settle, crack and become more consolidated.

Dunes Dunes assume a characteristic cross-section with a gently sloping windward side and a steeper downwind side. Generally particles rolled and moved by saltation up the shallower upwind face and then rolled down the sheltered 'slip face'. Migrating dunes can be a major problem; they migrate over roads, forests, etc. Dunes on beach roads can be a constant problem, with the usual remedy being to try to stabilize the dunes with vegetation. Desert roads are harder to protect due to the harsh climate, which supports little vegetation, especially on dry and shifting dunes.

Barchans The wind-borne particles are dropped down whenever the velocity is reduced to an appreciable extent. The deposits thus formed are either in the form of irregular mounds or have some definite shape. The deposits of sand are commonly described as dunes; they may vary in considerable dimensions. Dunes formed within wind shadow zones are described as shadow dunes. The dunes are found to have a gentle slope along the windward side and a comparatively steep gradient along the leeward side. Dunes, which are more or less crescent-shaped, are commonly described as 'barchans'.

12.11.2 Geomorphological Mapping in Glacial Environment

Geomorphological mapping in glacial environments is the study and mapping of landforms produced by glacial and fluvioglacial processes in the areas of present glaciers as well as in the areas covered by glaciers in the past. Glacier mapping is to represent the spatial morphology of glacier terrain and geomorphological features. The Himalayan Mountains are characterized by the presence of peaks, which are crowned permanently with snow. These peaks necessarily lie above the snow line, which represents in any region the lowest limit of perpetual snow. The altitude of the snow line varies more or less inversely as the latitude of the place concerned and is to some extent modified by the local climatic conditions and topographic pattern. DEMs are valuable tools for mapping, modelling, analysing and visualizing of various glacio-geomorphic phenomena in glacial environment (Etzelmüller and Sollid 1997; Etzelmüller et al. 2001). DEMs play an important role in the preparation of ortho-images in high mountain terrain (Finsterwalder 1984), in estimating glacier hypsometry (Brocklehurst and Whipple 2004) and in derivative of various terrain parameters such as slope, plan and profile curvature and aspects, which can be further used for geomorphological mapping in glacial environment.

12.11.3 Geomorphological Processes in Glacial Environment

Above the snow line continuous precipitation and accumulation of snow gradually gives rise to a 'snow field'. Periodic additions of snow in course of time increase the thickness of any snow field. At this stage the prevailing moisture and the weight of the overlying snow brings about a conversion of loose snowflakes into small granules of ice. Such granular mass of ice are known as 'neve' or 'firn'. With further addition of snow to the field, neve or firn is gradually transformed into a compact block of ice. In any snow field therefore loose flakes sive by neve or firn followed by compact ice. Snow fields slowly gain in thickness and do not ordinarily cause the ice to move on level grounds. But along the hill slopes, the increasing weight of the growing mass of ice at length is sufficient to make the ice flow or creep downwards. Such bodies of slowly moving ice are called '*glaciers*'. Sometimes a portion of the creeping mass of ice may break off and slide down under gravity forming an '*avalanche*'.

12.11.3.1 Denudational Landforms in Glacial Environment

Typical denudational landforms formed due to ice in glacial environment include:

Quarrying and Frost Wedging Despite their slow movement and limited extent in space, the geomorphological work done by glaciers by no means is less important. Glaciers conveniently erode the highlands, transport the products of erosion and deposit them suitably, thus grading the landmasses lying within their access. Glacial erosion takes place due to 'abrasion', 'quarrying' and 'frost wedging'.

Frost Wedging The glaciers may under favourable conditions cause uprooting of blocks of rocks forming the valley floors, and such excavations are known as 'quarrying' and are most common in regions where the country rocks are very much fractured. The process of breaking down of the country rocks due to alternate melting and freezing of ice within openings in rocks has been described as 'frost wedging', which causes shattering of rocks, thereby rendering them extremely weak and very much susceptible to erosion. The process of abrasion, quarrying and frost wedging operate hand in hand and mould the valley suitably offering a characteristic pattern and profile of its own. Unlike the 'V'-shaped river valleys, the transverse profile of all glaciated valleys is necessarily 'U' shaped.

Cirque Cirques are common occurrence of landform in the glaciated mountain regions and the most recurrent worldwide feature in the glacial landscape. Frost wedging followed by quarrying of the shattered rocks renders the head of the glaciated valleys blunt and steep; they are described as 'cirques', and the process, which leads to their formation, has been defined as 'nivation'. With the growth of adjacent radiating cirques along the adjacent slopes of the mountains, the peaks become narrower and on the long run give rise to a sharp and irregular ridge known as 'arete' or 'combridge'. A pyramid-like peaks with steep slopes are produced due to enlargement of three or more neighbouring cirques in glacial environment.

Roches Mountannes They have smooth rock surfaces on all the sides, which have been produced by glacial abrasion. The boulders and pebbles held firmly within the mobile mass of solid ice rub themselves against the floor of the valley and thus cause its gradual abrasion. During abrasion the boulders and pebbles worn out and in course of time are found to have been polished, striated or grooved. Due to firm grip of ice on them, the boulders and pebbles cannot have any rolling motion and therefore develop polished/grooved faces only. Small mounds on the sides of the valley floors with fine striation on their surface due to glacial abrasion are described as 'roches mountannes'.

Hanging Valley The amount of erosion caused by a glacier is a function of its size, shape, rate of flow and the quantity of rock fragments, which it can utilize in abrading the valley and the nature of the bedrock. No two glaciers are therefore likely to excavate their valleys to the same extent. Thus when a smaller glacier joins a larger one as its tributary, they do not have their valleys meeting at the same level. The valley of the larger glacier is much deeper than that of the smaller tributary. The tributary valley appears to hang above the floor of the main valley occupied by the larger glacier. They are called '*hanging valleys'*.

Fiords At high latitudes, the glaciers often excavate their valleys up to the sea level. These glacial valleys occurring along the coasts and occupied subsequently by the sea itself are known as 'fiords'. Within fiords glaciers come in contact with marine water, and blocks of ice are found to break off from the mass of the glacier. This process of wastage of glacier ice is known as 'calving', and the dislodged blocks of ice float on the sea in the form of icebergs.

Arêtes and Horns The mechanical disintegration of rocks in glacial environment results in the development of certain peculiar forms like arête, which is a knife-edged, sharp and narrow crest of a glaciated mountain. The slender shape of an arête commonly occurs due to removal of most of the material from in between two cirques, due to their advanced growth. Their sharp edges are attributed to wedging by frost action. A horn is a pyramidal multifaceted projection formed in the glacial mountainous region.

12.11.3.2 Depositional Landforms in Glacial Environment

Glacial Drift The materials undergoing glacial transport and all deposits of glacial origin constitute what is known as 'glacial drift'.

Till Unsorted heaps of glacial deposits made up of an assemblage of rock fragments and particles of widely varying dimensions are known as *till.*

Deglaciated Valleys The deglaciated valleys are formed due to the retreat of a glacier from its maximum extent. Cross profile of well-developed deglaciated valley appears in U shape, and side walls are considerably steep, approaching vertical in place, and the valley floor is broad and flat. Widening and deepening deglaciated valleys depend upon the ability of the glacier to erode valley walls and cut into the rock of its bed and lithology of its bedrocks. The valley floors are filled with reworked outwash sediments and marked by a series of rock basins.

Lateral Moraines Lateral moraines are formed as the linear ridges produced from the dumping of till with allied outwash deposits in the trough between the glaciers and valley sides in the glacial environment. They are one of the commonest depositional features in glacial environment. During their downhill journey, valley glaciers always deposit some rock debris in the form of ridges along their lateral margins and known as *lateral moraines.*

Medial Moraines When two valley glaciers meet and together form a large one, they commonly give rise to a medial moraine, which results due to coalescene of two lateral moraines.

Terminal Moraines A major part of the glacial drift, however, accumulates at the terminus of the glacier in the form of ridges known as terminal or end moraines. Terminal moraines are the ridge of till that marks the maximum limit of glacier advance. A well-developed terminal moraine indicates the past climate features that the ice remained inactive for a considerable period of time. The series of terminal moraines indicates the recession of the glacier and are called as recessional moraines. These are basically crescent-shaped glacial landforms with the convex side extending down the valley and appear as arcuate ridge.

12.11.4 Glacio-fluvial Landforms in Glacial Environment

Drumlins In regions where the deposit is made up mainly of clayey materials, it occurs in the form of a series of smooth mounds known as drumlins, which appears like plain and more or less elliptical with their long axes lying parallel to the direction of flow of the pre-existing glacier.

Outwash Plains The terminal moraines are readily worked upon by the melt water and gives rise to stratified glacio-fluvial deposits. Stratified deposits formed by melt water and occurring further down the valley beyond the end moraine are known as outwash plain.

Kettles Outwash plains usually contain a number of circular or elongated depressions known as 'kettles', which are formed due to enclosure of blocks of ice within glacial drifts, which subsequently melt in course of time.

Kame Terraces Glaciated valleys often contain well-defined terraces of glacio-fluvial origin known as kame terraces. Kame is an isolated hill mound of stratified drift deposited in an opening within or between ice blocks or in a moulin. If the opening was between ice and a valley wall, the resulting form is called a kame terrace.

Eskars Stratified glacio-fluvial deposits occurring in the form of meandering ridges are called esker. Eskar is a sinuous or meandering ridge of stratified sand and gravel commonly with very well-rounded pebbles and cobbles. Eskars form in open channels on glaciers and in tunnels within or beneath them. They often appear as sinuous ridges with lengths up to several kilometres and heights of tens of metres.

Verves Glacio-fluvial deposits in the form of very thin laminated layers occurring on the nearly lakes are called verves, which are indicators of the advance and retreat of glaciers.

Debris/Talus Cones The deglaciated areas generally marked with several debris/talus cones. These talus cones are formed by several glacio-fluvial processes, which include avalanches, mass wasting, hill slope evolution, etc. The loose debris formed due to glacial deposition provides the soft target for the movement of the large boulders during these processes.

Glacio-fluvial Deltas Glacio-fluvial deltas are landforms formed due to transported sand and gravel material by glacial meltwater and deposited in standing water. Subsequent to deglaciation, the landforms have been separated from the original body of water either by land uplift or lake-level lowering.

12.11.5 Geomorphological Processes in Karst Environment

Karst is 'terrain with distinctive hydrology and landforms arising from a combination of high rock solubility and well-developed secondary porosity' (Ford and Williams 1989). Karst as a type of landscape found on carbonate rocks (limestone, dolomite, marble) or evaporites (gypsum, anhydrite, rock salt) is characterized by a suite of landforms comprising springs, dolines, caves, collapsed sinkholes and carbonate depositional landforms (Gunn 2004; Ford and Williams 2007). Solution becomes a process of both weathering and erosion, wherein the weathered products are removed from the surface and carried away in solution. Temperature, precipitation, biologic activity and amount and seasonality of run-off are all climatic factors that affect the intensity of karst processes. Application of remote sensing and GIS improves the generation of geomorphological maps and helps to monitor and mitigate karst geohazards (De Carvalho Jr et al. 2014; Pardo-Igúzquiza et al. 2013). The fusing of different satellite datasets at different resolutions and analysis in GIS is proven to provide better results in karst morphological studies and landform mapping in karst environment (Siart et al. 2009).

Solution Pits On exposed rock (bare karst, pavement karst), horizontal surfaces develop shallow solution pits (rain pits, makatea). On surfaces inclined a few degrees from the horizon, sheet run-off produces flat solution facets (solution bevels). Solution runnels (Rinnenkarren) are branching channels in networks first formed under soil cover but, having formed, are perpetuated by run-off on bare rock surfaces.

Dolines The fundamental geomorphic component of karst topography is the doline, or limestone sink. Whether due to joint control (the most common cause), differential solubility or random events such as cave collapse, localized areas of karst terrain are lowered more rapidly than the surrounding area and form closed depressions. They range in size from shallow soil depressions a few metres in diameter and a metre deep to major landforms several kilometres in diameter and hundreds of metres in depth.

Uvalas A genetic progression can be visualized in which dolines progressively abstract surface run-off into groundwater circulation and leave networks of dry valleys as relict surface forms. The term uvala was given to a compound doline, or a chain of intersecting dolines. A karst window is similar, in that an unroofed segment of an underground stream channel becomes a surface valley in the window and passes underground again downstream.

Poljes The karst topography is on carbonate rocks that are folded and faulted, often in proximity to insoluble rocks. Structural control becomes a major factor of landscape evolution. Large blind valleys or karst valleys, enclosed by either soluble or insoluble rocks, usually elongated along tectonic axes, are called poljes.

Stalactites Stalactites develop downwards and grow from dripping walls and ceilings. The basic form is a straw stalactite.

Stalagmites It grows from the floor, their exact form (columnar or conical) depending upon drip rates, water hardness and the cave atmosphere.

12.11.6 Geomorphological Mapping in Coastal Environment

Coastal geomorphology by definition is the study of the morphological development and evolution of the coasts as it acts under the influence of waves, winds, currents and sea level changes. Bhaskara Rao and Vaidyanadhan (1975) studied the coastal features between Pudimadaka and Visakhapatnam of east coast of India. Waves are one of the most significant forces in shaping the coastline. There are two main types of wave – constructive waves and destructive waves.

Constructive waves are low-energy waves that tend to arrive at the coast at a rate of less than 8 waves per minute. Constructive waves are small in height. They have a strong swash and a weak backwash. This means that constructive waves tend to deposit material and build up a beach.

Destructive waves have much higher energy and tend to arrive at the coast at a rate of more than 8 per minute. They are much larger in height than constructive waves, often having been caused by strong winds and a large fetch. Destructive waves have a weak swash but a strong backwash so they erode the beach by pulling sand and shingle down the beach as water returns to the sea. This means that less beach is left to absorb wave energy.

12.11.6.1 Geomorphological Processes of Coastal Environment

Hydraulic action occurs when waves striking a cliff face compress air in cracks on the cliff face. This exerts pressure on the surrounding rock and can progressively splinter and remove pieces. Over time, the cracks can grow, sometimes forming a cave. The splinters fall to the seabed where they are subjected to further wave action. *Attrition* occurs when waves cause loose pieces of rock debris (scree) to collide with each other, grinding and chipping each other, progressively becoming smaller, smoother and rounder. Scree also collides with the base of the cliff face, chipping small pieces of rock from the cliff, or has a corrasion (abrasion) effect, similar to sandpapering. *Corrasion* (abrasion) occurs when waves break on cliff faces and slowly erode it. As the sea pounds cliff faces, it also uses the scree from other wave actions to batter and break off pieces of rock from higher up the cliff face which can be used for this same wave action and attrition. *Corrosion* is nothing but solution/chemical weathering occurs when the sea's pH (anything below pH 7.0) corrodes rocks on a cliff face. Limestone cliff faces, which have a high pH, are particularly affected in this way. Wave action also increases the rate of reaction by removing the reacted material.

12.11.6.2 Denudational Landforms in Coastal Environment

Erosional coasts typically exhibit high relief and rugged topography. The different types of coastal erosional landforms are discussed below.

Sea Cliffs The most widespread landforms of erosional coasts are sea cliffs. These very steep to vertical bedrock cliffs range from only a few metres high to hundreds of metres above sea level. Cliffs that extend to the shoreline commonly have a notch cut into them where waves have battered the bedrock surface.

Wave-Cut Platforms At the base of most cliffs along a rocky coast, one finds a flat surface at about the mid-tide elevation. This is a benchlike feature called a wave-cut platform, or wave-cut bench. It is a gently sloping surface produced by wave erosion and extends outward into the sea from the base of a sea cliff. These platforms occur at some height above MSL and formed a result of constant wave action on the rocky cliffs of the head land.

Sea Stacks Erosion along rocky coasts occurs at various rates and is dependent both on the rock type and on the wave energy at a particular site. A sea stack is a small isolated usually steep-sided rocky mass or island near cliffy shore often detached from the head land by wave erosion assisted by sub-aerial weathering and are remnants of retreated sea cliff (Bloom 2003). These are erosional remnants on the horizontal wave-cut surface called sea stacks, and they provide a spectacular type of coastal landform.

Sea Caves It is a cavity or opening in the base of a sea cliff excavated by wave action along the weak zones in an easily weather-able rock (Bloom 2003). This is formed as a result of scooping of rock material by constant attack of waves on rocks.

Natural Bridge Another spectacular type of erosional landform of coastal environment is the natural bridge or sea arch, which forms as the result of different rates of erosion typically due to the varied resistance of bedrock (Bloom 2003). These archways may have an arcuate or rectangular shape, with the opening extending below water level. The height of an arch can be up to tens of metres above sea level. Most natural arches form as a narrow ridge, walled by cliffs. They become narrower from erosion, with a softer rock stratum under the cliff-forming stratum that is gradually eroding out until the rock shelters, thus formed and meet underneath the ridge.

12.11.6.3 Depositional Landforms in Coastal Environment

Depositional coasts may experience erosion at certain times and places due to such factors as storms, depletion of sediment supply, and rising sea level. The latter is a continuing problem as the mean annual temperature of the earth rises and the ice caps melt. Nevertheless, the overall, long-range tendency along these coasts is that of sediment deposition. Depositional coasts can be described in terms of three primary large-scale types: (1) deltas, (2) barrier island/estuarine systems and (3) strand plain coasts. The latter two have numerous features in common. The important depositional geomorphological features in coastal environment are discussed below.

Deltas An accumulation of sediment at the mouth of a river extending beyond the trend of the adjacent coast is called a delta. Deltas vary greatly in both size and shape, but they all require that more sediment is deposited at the river mouth than can be carried away by coastal processes. A delta also requires a shallow site for accumulation – namely, a gently sloping continental shelf.

Barrier Island/Estuarine Systems Many depositional coasts display a complex of environments and landforms that typically occur together. Irregular coasts have numerous embayments, many of which are fed by streams. Such embayments are called estuaries, and they receive much sediment due to run-off from an adjacent coastal plain. These barrier islands are typically separated from the mainland and may have lagoons, which are long, narrow, coastal bodies of water situated between the barrier and the mainland.

Strand Plain Coasts Some wave-dominated coasts do not contain estuaries and have no barrier island system. These coasts, however, do have beaches and dunes and may even have coastal marshes. The term strand plain has been applied to coasts of this sort.

Coastal Sand Dunes There are several specific landforms representative of coastal environments that are common and associated with deltas, barrier island/estuarine systems and strand plain coasts. Especially prominent among these are beaches and dunes. They are the primary landforms on barrier islands, strand plain coasts, and many deltas, particularly the wave-dominated variety.

Beaches A consideration of the beach must also include the seaward adjacent nearshore environment because the two are intimately related. The nearshore environment extends from the outer limit of the longshore bars that are usually present to the low-tide line. In areas where longshore bars are absent, it can be regarded as coincident with the surf zone. The beach extends from the low-tide line to the distinct change in slope and/or material landward of the unvegetated and active zone of sediment accumulation. It may consist of sand, gravel or even mud, though sand is the most common beach material. The thickness of the individual layers of beaches varies from a few cm to a few mm.

Coastal Dunes Immediately landward of the beach are commonly found large, linear accumulations of sand known as dunes. They form as the wind carries sediment from the beach in a landward direction and deposits it wherever an obstruction hinders further transport. Sediment supply is the key limiting factor in dune development and is the primary reason why some coastal dunes are quite small whereas others have large dunes. As per the classification of Smith (1954), dunes are transverse, crescent-shaped and parabolic in nature.

12.12 Conclusions

Geomorphological mapping and analysis of various processes in fluvial, arid, glacial, karst and coastal environments plays an important role in understanding various earth surface processes, landscape evolution, geochronology, structural characteristics, natural resources inventory and mapping and natural hazards. It involves the partitioning of the terrain into conceptual spatial units/entities based upon criteria that include morphology, genesis, composition and structure, chronology, environmental system associations and their spatial topological relationships of landforms. The data representation, collection, analysis, modelling and presentation of results in geomorphological mapping depend on the spatial and temporal scale adopted. Integration of multidisciplinary information from the field, remotely sensed data, digital elevation models, GIS and GPS improve the accuracy in geomorphological mapping. Recent advances in remote sensing, GIS and modelling capabilities of surface processes revolutionized the field of geomorphological mapping. The synoptic view provided by satellite remote sensing and digital analysis techniques in conjunction with DEMs offers technologically the appropriate approach to map distinct geomorphic units in different environments.

References

Ahnert F (1998) Introduction to geomorphology. Arnold, London

Anonymous (1976) Engineering geological maps. A Guide to their interpretation UNESCO Press, Paris

Anonymous (2009) Manual for national geomorphological and lineament mapping on 1:50,000 scale. Geological Survey of India and Indian Space Research Organisation, Government of India. National Remote Sensing Centre, Hyderabad 116 p

Baker VR (1986) Introduction: regional landforms analysis. In: Short M, Blair RW Jr (eds) Geomorphology from space, a global overview of regional landforms. NASA, SP 486, Washington DC, pp 1–26

Bashenina NV, Blagovolin NS, Demek J, Dumitrashko NV, Ganeshin GS, Gellert JF, Leontyev OK, Mirnova AV, Scholz E (1971) Legend to the international geomorphological map of Europe 1: 2,500,000,. 5th version. Czechoslovak Academy of Sciences, Institute of Geography, Brno 30 p

Batten P (2001) A new approach for landscape mapping. Proceedings of the 6th International Conference on Geocomputation. University of Queensland, Brisbane, Australia, 24–26 Sept 2001

Bhaskara Rao VU, Vaidyanadhan R (1975) Photo geomorphic study of coastal features between Visakhapatnam and Pudimadaka in Andhra Pradesh. Photonirvachak 3(1):43–46

Bishop MP, Shroder JF Jr (eds) (2004) GIS science and mountain geomorphology: Overview, feedbacks, and research directions. Springer-Praxis, Chichester, pp 1–31

Blaszczynski JS (1997) Landform characterization with geographic information systems. Photogramm Eng Remote Sens 63(2):183–191

Bloom AL (2003) Geomorphology – A systematic analysis of late Cenozoic landform, 3rd edn. Prentice Hall, Upper Saddle River

Bocco G, Mendoza M, Velázquez A (2001) Remote sensing and GIS-base regional geomorphological mapping – A tool for land use planning in developing countries. Geomorphology 39:211–219

Brändli M (1996) Hierarchical models for the definition and extraction of terrain features. In: Burrough PA, Frank AU (eds) Geographic objects with indeterminate boundaries. Taylor & Francis, London, pp 257–270

Brocklehurst S, Whipple KX (2004) Hypsometry of glaciated landscapes. Earth Surf Process Landf 29(7):907–926. https://doi.org/10.1002/esp.1083

Burrough P (1989) Fuzzy mathematical methods for soil survey and land evaluation. J Soil Sci 40:477–492

Burrough P, van Gaans P, MacMillan R (2000) High-resolution landform classification using fuzzy k-means. Fuzzy Sets Syst 113(1):37–52

Chorley RJ, Kennedy BA (eds) (1971) Physical geography: a systems approach. Prentice Hall, London

Cooke RU, Doornkamp JC (1974) Geomorphology in environmental management. An introduction. Oxford University press, Oxford, pp 22–44

Cooke RU, Doornkamp JC (1990a) Geomorphology in environmental management. A new Introduction, 2nd edn. Clarendon Press, Oxford

Cooke RU, Doornkamp JC (1990b) Mapping geomorphology (Cooke, Doornkamp eds.). Clarendon Press, Oxford, pp 22–63

Dai FC, Lee CF (2002) Landslide characteristics and slope instability modelling using GIS, Lantau Island, Hong Kong. Geomorphology 42:213–228

De Carvalho OA Jr, Guimarães RF, Montgomery DR, Gillespie AR, Trancoso Gomes RA, de Souza Martins É, Silva NC (2014) Karst depression detection using ASTER, ALOS/PRISM and SRTM-derived digital elevation models in the Bambuí Group, Brazil. Remote Sens 6:330–351

Demek J (ed) (1972) Manual of detailed geomorphological mapping. Academia, Praha

Demek J, Embleton C (eds.) (1978) Guide to medium-scale geomorphological mapping. E. Schweizerbart'sche Verlagsbuchhandlung (Nägele u. Obermiller), Stuttgart

Demek J, Embleton C, Gellert JF, Verstappen HT (eds) (1972) Manual of detailed geomorphological mapping. Academia, Publishing House of the Czechoslovak Academy of Sciences, Prague 343 p

Deng Y, Wilson JP (2008) Multi-scale and multi-criteria mapping of mountain peaks as fuzzy entities. Int J Geogr Inf Sci 22:205–218

Dent D, Young A (1981) Soil survey and land evaluation. George Allen & Unwin, London

Dikau R (1989) The application of a digital relief model to landform analysis in geomorphology. In: Raper J (ed) Three dimensional applications in geographical information systems. Taylor and Francis, London, pp 51–77

Dramis F, Guida D, Cestari A (2011) Nature and aims of geomorphological mapping. In: Smith MJ, Paron P, Griffiths J (eds) Geomorphological mapping: methods and applications. Elsevier, London, pp 39–74

Etzelmüller B, Sollid JL (1997) Glacier geomorphometry – an approach for analyzing long-term glacier surface changes using grid-based digital elevation models. Ann Glaciol 24:135–141

Etzelmüller B, Ødegård RS, Berthling I, Sollid JL (2001) Terrain parameters and remote sensing in the analysis of permafrost distribution and periglacial processes: principles and examples from southern Norway. Permafr Periglac Process 12:79–92

Evans IS (1998) What do terrain statistics really mean? In: Lane S, Richards K, Chandler J (eds) Landform monitoring, modelling and analysis. Wiley, Chichester, pp 119–138

Finsterwalder R (1984) Utilization of orthophotographs in the mapping of high mountain regions. Mt Res Dev 4:315–318

Fookes PG, Lee EM, Griffiths JS (eds) (2007) Engineering geomorphology: theory and practice. CRC Press, New York

Ford D, Williams P (1989) Karst geomorphology and hydrology. Chapman and Hall, London

Ford DC, Williams PW (2007) Karst geomorphology and hydrogeology, 2nd edn. Wiley, Chichester

Gilewska S (1966) IGU commission on applied geomorphology. Zeitschrift für Geomorphologie NF 10:191–192

Gilewska S, Klimek M (1968) Project of the unified key to the detailed geomorphological map of the world. Foelia Geographica, Ser. Geographica-Physica, vol II. Polska Akademia Nauk, Kraków

Gunn J (2004) Encyclopedia of caves and karst science. London, Fitzroy Dearborn

Gustavsson M (ed) (2006) Development of a detailed geomorphological mapping system and GIS database in Sweden. Acta University, Uppsala

Gustavsson M, Kolstrup E, Seijmonsbergen AC (2006) A new symbol-and-GIS based detailed geomorphological mapping system: renewal of a scientific discipline for understanding landscape development. Geomorphology 77:90–111

Kar A (1993) Aeolian processes and bed forms in the Thar Desert. J Arid Environ 25:83–96

Kayan I, Klemas V (1978) Application of LANDSAT Imagery to studies of structural geology and geomorphology of the Mentese region of southwestern Turkey. Remote Sens Environ 7:51–60

Klimaszewski M (1990) Thirty years of geomorphological mapping. Geogr Pol 58:11–18

Klingseisen B, Metternicht G, Paulus G (2007) Geomorphometric landscape analysis using a semi-automated GIS-approach. Environ Model Softw:1–13

Knight J, Mitchell W, Rose J (2011) Geomorphological field mapping. In: Smith MJ, Paron P, Griffiths J (eds) Geomorphological mapping: methods and applications. Elsevier, London, pp 151–188

Krishnamurthy J, Srinivas G (1995) Role of geological and geomorphological factors in ground water exploration: a study using IRS LISS data. Int J Remote Sens 16(14):2595–2618

Lane SN, Richards KS, Chandler JH (eds) (1998) Landform monitoring, modelling and analysis. Wiley, Chichester

Mantovani F, Soeters R, Westen VCJ (1996) Remote sensing technique for landslide studies and hazard zonation in Europe. Geomorphology 15:213–225

Novak ID, Soulkellis N (2000) Identifying geomorphic features using LANDSAT-5/TM data processing techniques on Lesvos, Greece. Geomorphology 34:101–109

Otto JC, Gustavsson M, Geilhausen M (2011) Cartography: design, symbolisation and visualisation of geomorphological maps. In: Smith M, Paron P, Griffiths J (eds) Geomorphological mapping: methods and applications. London, Elsevier, pp 253–296

Pain CF (1985) Mapping of landforms from LANDSAT imagery: an example from Eastern New South Wales, Australia. Remote Sens Environ 17:55–65

Pandey S, Singh S, Ghose B (1964) Orientation, distribution and origin of sand dunes in the central Luni basin. In: Proceedings, symposium on problems of Indian Arid Zone. CAZRI, Jodhpur, pp 84–91

Pardo-Igúzquiza E, Durán JJ, Dowd PA (2013) Automatic detection and delineation of karst terrain depressions and its application in geomorphologic mapping and morphometric analysis. Acta Carsoligica 42:17–24

Paron P, Claessens L (2011) Makers and users of geomorphological maps. In: Smith MJ, Paron P, Griffiths J (eds) Geomorphological mapping: methods and applications. Elsevier, London, pp 75–106

Quattrochi DA, Goodchild MF (eds) (1997) Scale in remote sensing and GIS. CRC Press, Boca Raton

Ralston BA (1994) In: Fotheringham, Rogerson (eds) Object oriented spatial analysis. Taylor & Francis, London, pp 165–185

Reddy GPO (2012) Geomorphological processes and evolution of landforms. In: Reddy GPO, Sarkar D (eds) Remote sensing and GIS in digital terrain analysis and soil-landscape modelling. National Bureau of Soil Survey and Land Use Planning No. 152, Nagpur, pp 26–35

Reddy GPO, Maji AK (2003) Delineation and characterization of geomorphological features in a part of lower Maharahstra metamorphic plateau, using IRS-ID LISS-III data. J Indian Soc Remote Sens 31(4):241–250

Reddy GPO, Shekinah DE, Maurya UK, Thayalan S, Prasad J, Ray SK, Bhasker BP (1999) Landscape-soil relationship in part of Bazargaon plateau, Maharashtra. Geogr Rev India 61 (3):280–291

Reddy GPO, Maji AK, Srinivas CV, Velayutham M (2002) Geomorphological analysis for inventory of degraded lands in a river basin of basaltic terrain, using remote sensing data and Geographical Information Systems. J Indian Soc Remote Sens 30(1&2):15–31

Reddy GPO, Maji AK, Gajbhiye KS (2004) Drainage morphometry and its influence on landform characteristics in a basaltic terrain, Central India – a remote sensing and GIS approach. Int J Appl Earth Obs Geoinf 6:1–16

Sabins FF (2000) Remote sensing principles and interpretation, 3rd edn. W.H. Freeman and Company, New York, p 494

Schmidt J, Dikau R (1999) Extracting geomorphic attributes and objects from digital elevation models — semantics, methods, future needs. In: Dikau R, Saurer H (eds) GIS for earth surface systems: analysis and modelling of the natural environment. Gebrüder Borntraeger Berlin, Stuttgart, pp 153–173

Sheppard E, McMaster RB (eds) (2004) Scale and geographic inquiry. Blackwell Publishing, Malden

Shroder JF Jr, Bishop MP (2003) A perspective on computer modeling and fieldwork. Geomorphology 53:1–9

Shroder JF Jr, Bishop MP (2004) Mountain geomorphic systems. In: Bishop MP, Shroder JF Jr (eds) Geographic information science and mountain geomorphology. Springer, Berlin, pp 33–73

Siart C, Bubenzer O, Eitel B (2009) Combining digital elevation data (SRTM/ASTER), high resolution satellite imagery (Quickbird) and GIS for geomorphological mapping: a multi-component case study on Mediterranean karst in Central Crete. Geomorphology 112:106–121

Singh S (1982) Types and formation of sand dunes in the Rajasthan Desert. In: Sharma HS (ed) Perspectives in Geomorphology, vol. 4. Concept, Delhi, pp 165–183

Slaymaker O (2001) The role of remote sensing in geomorphology and terrain analysis in the Canadian cordillera. Int J Appl Earth Obs Geoinf 3:11–17

Smith HTU (1954) Coastal dunes, Coastal geography conference, February 1954 office of Naval Research, pp 51–56

Smith GR, Woodard JC, Heywood DI, Gibbard PL (2000) Interpreting Pleistocene glacial features from SPOT HRV data using fuzzy techniques. Comput Geosci 26:479–490

Smith MJ, Griffiths J, Paron P (eds) (2011) Geomorphological mapping: methods and applications. Elsevier, London

Tate JN, Atkinson PM (eds) (2001) Modelling scale in geographical information science. Wiley, Chichester

Tate JN, Wood J (2001) In: Tate A (ed) Fractals and scale dependencies in topography. Wiley, Chichester, pp 35–52

Unified Key (1968) Project of the unified key to the detailed geomorphological map of the world. Folia Geographica, Ser. Geographica-Physica, II. Polska Akademia Nauk. Kraków

USDA Forest Service (1993) National hierarchical framework of ecological units. ECOMAP, Washington, DC 12 pp

Usery EL (1996) In: Burrough, Frank (eds) A conceptual framework and fuzzy set implementation for geographic features. Taylor and Francis, London, pp 71–85

Vats PC, Singh S, Ghose B, Kaith DS (1976) Types, orientation and distribution of sand dunes in Bikaner District. Geograp Obs 12:69–75

Verstappen HT (1977) Remote sensing in geomorphology. Elseviers, Amsterdam

Verstappen HT (1983) Applied geomorphology (geomorphological surveys for environmental development). Elsevier, Amsterdam 442 p

Verstappen HT (2011) Old and new trends in geomorphological and landform mapping. In: Smith MJ, Paron P, Griffiths J (eds) Geomorphological mapping: methods and applications. London, Elsevier, pp 13–38

Verstappen HT, Van Zuidam RA (1968) ITC textbook of photo-interpretation, VII: 2 –ITC system of geomorphological survey. ITC, Delft

Warner T, Shank M (1997) An evaluation of the potential for fuzzy classification of multispectral data using artificial neural networks. Photogramm Eng Remote Sens 63(11):1285–1294

Chapter 13 Geospatial Technologies for Semiautomated Baseline Database Generation for Large-Scale Land Resource Inventory

S. Chattaraj, S. K. Singh, S. K. Ray, V. Ramamurthy, A. Daripa, and G. P. Obi Reddy

Abstract The goal of land resource inventory is to enable the lab-to-land transfer of agro-technology on a sustainable basis through identification of homogeneous soil management units. The identification of homogeneous landscape ecological unit (LEU) boundaries for soil mapping through conventional methods is time-consuming and laborious. Hence, it is necessary to develop a semiautomated geospatial framework for delivering reliable soil resource information to the users on time. In the present chapter, the approach for semiautomation in landform delineation using high-resolution IRS Cartosat-1 and LISS-IV data was discussed. Cartosat-1 stereopair data are processed to generate the digital terrain model (DTM) of 10 m spatial resolution. The digital terrain analysis was carried out to generate contour, drainage, slope, and hillshade for landform delineation in two distinct terrain conditions. *Object-based slope classification algorithm* is developed by following USDA-NRCS slope class thresholds to hasten the process of landform identification. The land use/land cover (LULC) map of the area is generated based on the *rabi* season data of Cartosat-1 merged LISS-IV (2.5 m) as well as high-resolution (0.5 m) public domain imagery at the backend so as to get the reliable land use boundary at cadastral level through feature optimization algorithm in eCognition software using near-infrared (NIR) and Normalized Difference Vegetation Index (NDVI) data. The integration of three secondary layers, i.e., landform, slope, and LULC, are achieved through the *hierarchical object-based segmentation algorithm* to develop landscape ecological unit (LEU) map. The logical automation algorithm developed at each stage assists in optimizing sampling intensity, which leads to a considerable saving of man power, labor, cost, and time.

S. Chattaraj (✉) · S. K. Singh · A. Daripa · G. P. O. Reddy
ICAR-National Bureau of Soil Survey & Land Use Planning, Nagpur, India

S. K. Ray
ICAR-National Bureau of Soil Survey & Land Use Planning, Regional Centre, Jorhat, India

V. Ramamurthy
ICAR-National Bureau of Soil Survey & Land Use Planning, Regional Centre, Bengaluru, India

G. P. O. Reddy, S. K. Singh (eds.), *Geospatial Technologies in Land Resources Mapping, Monitoring and Management*, Geotechnologies and the Environment 21,
https://doi.org/10.1007/978-3-319-78711-4_13

Keywords Baseline database · Digital terrain modeling · Object-based digital terrain classification

13.1 Introduction

The application of satellite remote sensing data products for small- and medium-scale soil mapping is widely accepted (Soil Survey Division Staff 1995), but its use in large-scale soil mapping is restricted till date not only for the limited availability of high-resolution data but also due to lack of understanding the too much details present in the high-resolution data. Large-scale soil mapping is mostly done following conventional methods that are time-consuming and expensive and have low repetitive value especially in difficult and inaccessible terrain. However, with the recent advances in satellite data processing and analysis, availability of high-resolution satellite data like IRS-R2 LISS-IV data (5.8 m) and Cartosat-1 can now be utilized well for large-scale soil mapping. Srivastava and Saxena (2004) discussed the technique of large-scale soil mapping (1:12,500 scale) in a basaltic terrain with a PLU approach and differentiated soil types using topographic information available in the Survey of India toposheet and LULC information from IRS-1C PAN merged data of two seasons (*kharif* and *rabi*). Similar exercise was also carried out by Nagaraju et al. 2014 using Cartosat-1 and IRS-R2 LISS-IV data. However, the traditional way of landform extraction by an interpreter through the topographic maps, aerial photograph, or satellite imagery followed by ground truthing is accepted and appropriate. In one hand, the traditional way is relatively time-consuming, and the results are subjected to interpreter's biasness, and also not reproducible. On the other hand, pixel-based digital landform and LULC mapping of high-resolution data results in noise at the larger scale owing to the presence of minute details in the data. That's why object-based image analysis (OBIA) comprising of image objects, i.e., groups of pixels that are similar to one another based on a measure of spectral properties (i.e., color), size, shape, and texture, as well as context from a neighborhood surrounding the pixels, has gained increasing attention in landform and LULC research from the last decade (Drăguţ and Blaschk 2006; Eisank et al. 2011; d'Oleire-Oltmanns et al. 2013; Chattaraj et al. 2017).

The goal of soil/land resource inventory is to identify and delineate homogeneous soil patterns formed within a complex, heterogeneous soil-forming environment to enable the lab-to-land transfer of agro-technology in a sustainable basis. Successful mapping of soil resources on large scale is highly dependent on precise information of landforms (the testimony of past climate as well as topographic factors), slope, and LULC (the indicators of present climate and management conditions). It is realized that the identification and delineation of homogeneous landscape ecological unit boundaries for soil mapping through conventional methods is time-consuming. Further, the experienced man power to carry out soil survey is also declining rapidly. Hence, it is necessary to develop a semiautomated geospatial framework so that reliable information on soil resource is delivered to the users in time. However, the key to successful knowledge-based modeling depends on how effectively the implicit knowledge understanding on the target objects is transformed into explicit decision rules (Cheng and Han 2016). The present chapter discusses the approach for semiautomation in slope, landforms, and LULC classification for generating the Hierarchical Landscape Ecological Unit (LEU) segmentation model using high-resolution Cartosat-1 and IRS-R2 LISS-IV data.

13.2 Methodology Framework

The overall methodology flow diagram is presented in Fig. 13.1.

The steps involved are:

First step is the generation of digital terrain model (DTM) particularly in the undulating terrain using Cartosat-1 data of 1 m resolution. The primary terrain attributes, namely, contour, drainage, hillshade, slope, and curvatures, are derived from DTM, which have been used as input layers for developing precise and quantified data on landforms (Fig. 13.1a).

Second step is the generation of LULC maps using IRS-R2 LISS-IV data of 5.8 m resolution. Derived LULC map superimposed on landform and slope map to develop Landscape Ecological Unit (LEU) map, the base map of soil/land resource inventory at larger scale. LEUs are defined by a set of symbol D2s, D4w1, U4w4, D2d, etc., consisting of letters and numerals. First letter in capital is the landform, second numeral is slope class, and third letter and numeral is LULC (Fig. 13.1b).

Third and final step is the extensive traversing and ground truth collection through mini-pits and profile investigations in well-defined strips representing assemblage of LEUs. Establishing phases of soil series and developing soil-landform relationship are the next part of third step. However, the third step is beyond the scope of the present paper. Hence, the development of object-based models for delineation of LEUs is the prime focus of the chapter (Fig. 13.1c).

13.2.1 Semiautomated Modeling

13.2.1.1 Digital Terrain Modeling

Cartosat-1 stereo pair data were processed to generate the digital terrain model (DTM) of 10 m spatial resolution using rigorous math model (Toutin's Model). In the model, OrthoEngine of Geomatica version 14.0 is used to generate DTM following the sequence of steps, namely, projection setup, sensor data reading, collection of GCPs and tie points, block adjustment, model computation (Satellite Math Model), epipolar image generation, and digital surface model (DSM)

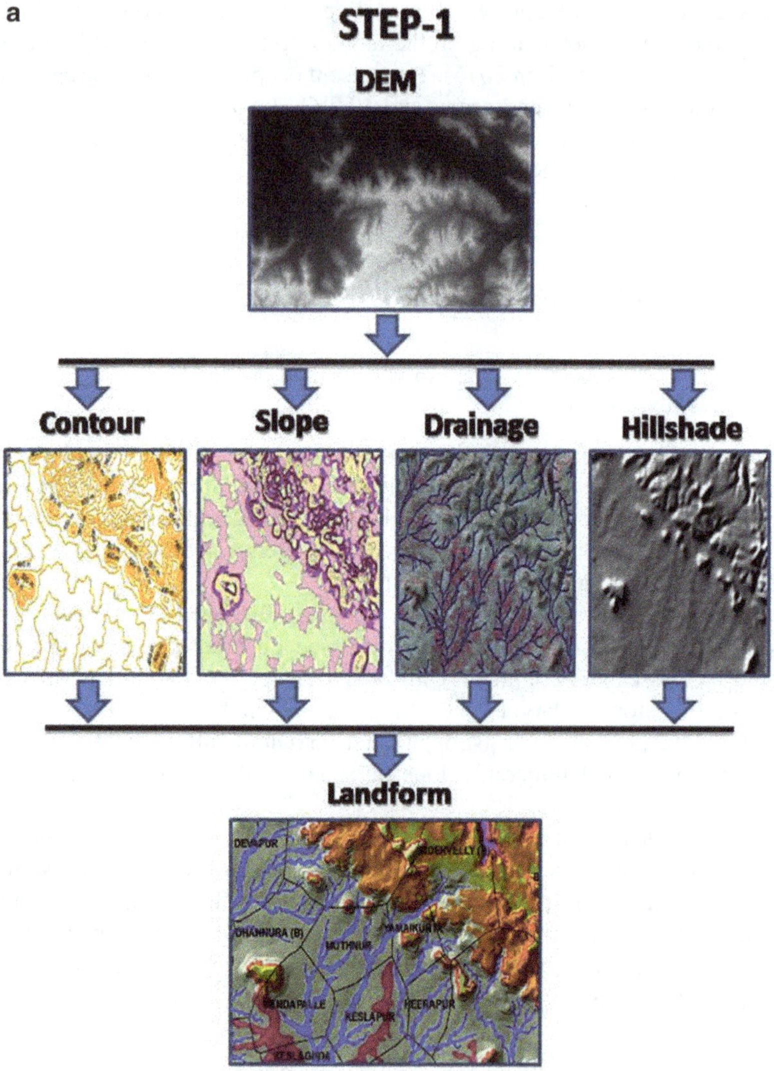

Fig. 13.1 (**a**)–(**c**) Steps in large-scale soil/land resource inventory

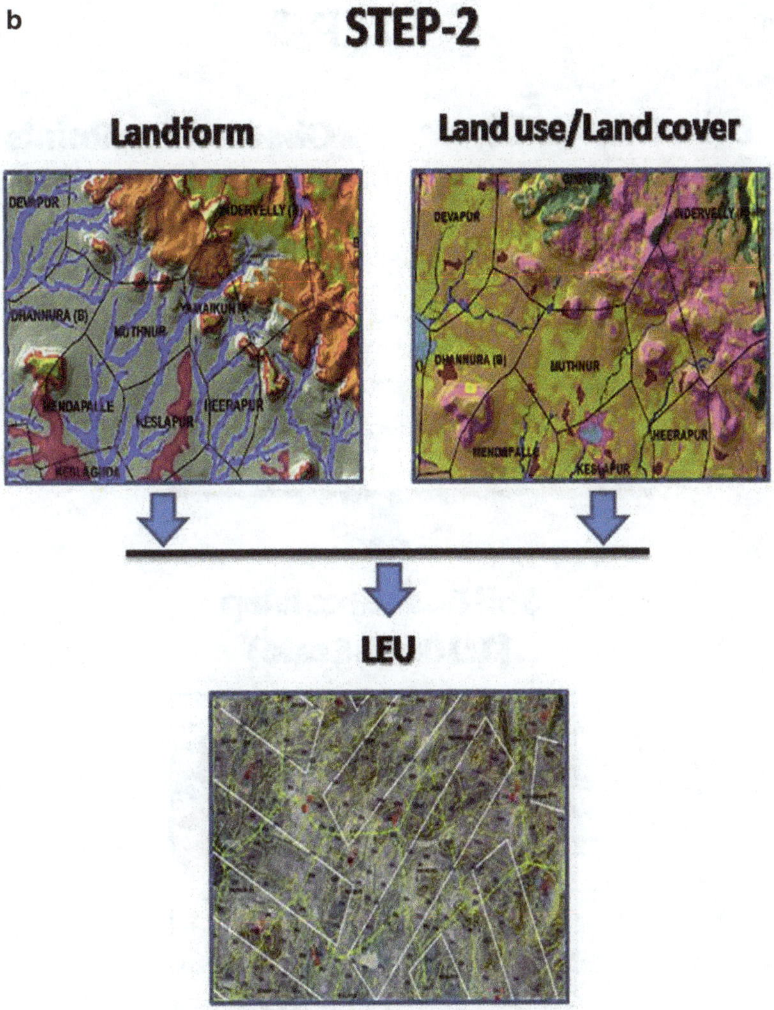

Fig. 13.1 (continued)

extraction. Balancing algorithm is applied to obtain the seamless mosaic DSM height. Filtering is done to convert bare earth model, DSM to DTM. Editing is done to smooth out the irregularities and create a quality output. RMSE statistics report is also generated to evaluate the accuracy of the DTM output (Fig. 13.2).

Further, DTM is subjected to a series of hydro-enforcement process including reconditioning, sinks and pit removal, flat and level water bodies, flat and level bank to bank, and gradient smoothening by DAT/EM and Arc Hydro tool, etc. This is

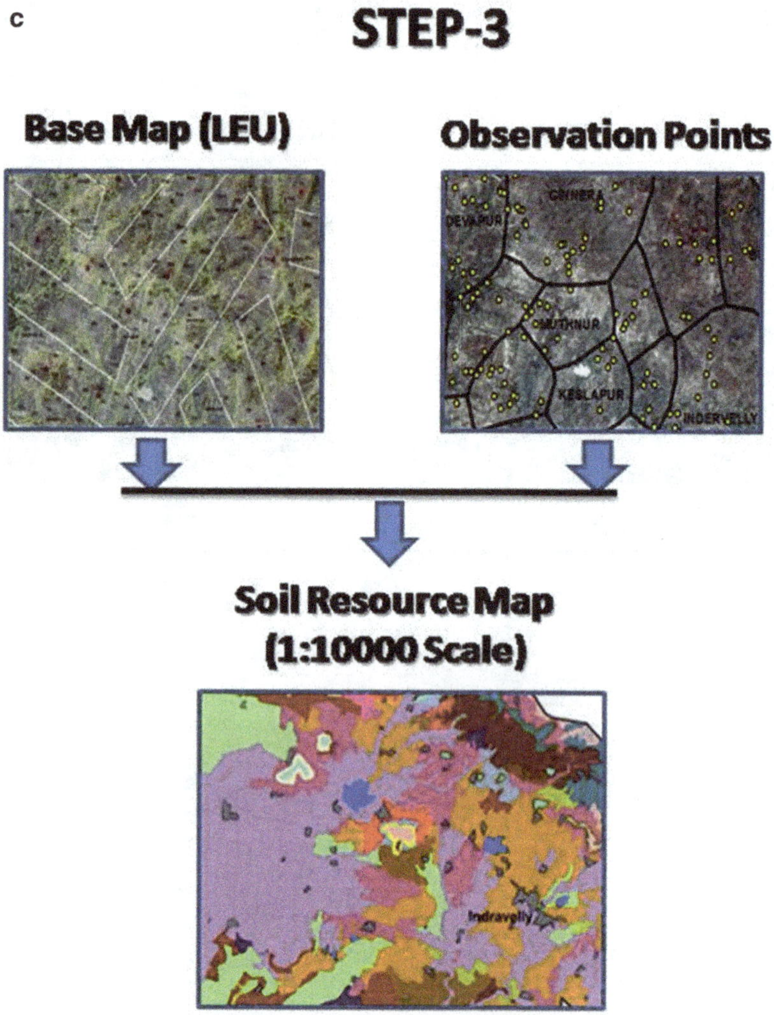

Fig. 13.1 (continued)

essentially needed to enrich the quality of the hydrological output such as slope, contour, and drainage (Romstad and Etzelmuller 2012). This altogether needed to improve the accuracy of landform mapping (Fig. 13.3).

13.2.1.2 Object-Based Digital Terrain Classification Models

The pixel-based classification procedure analyzes only the spectral properties, but the spatial or contextual information is lacking. Pixel-based methods applied to high-resolution images give a "salt and pepper" effect that contribute to the inaccuracy of

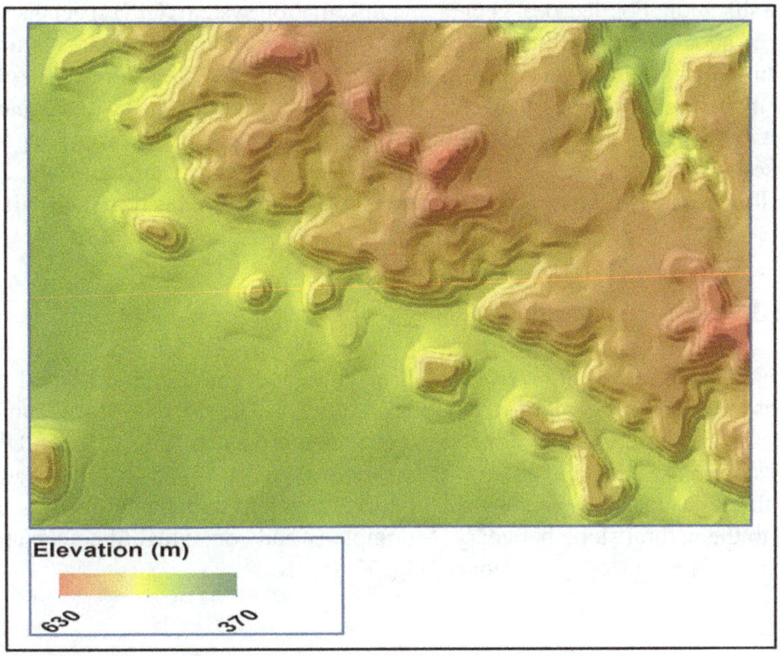

Fig. 13.2 DTM of 10 m resolution for a part of Indervelly block, Adilabad district, Telangana

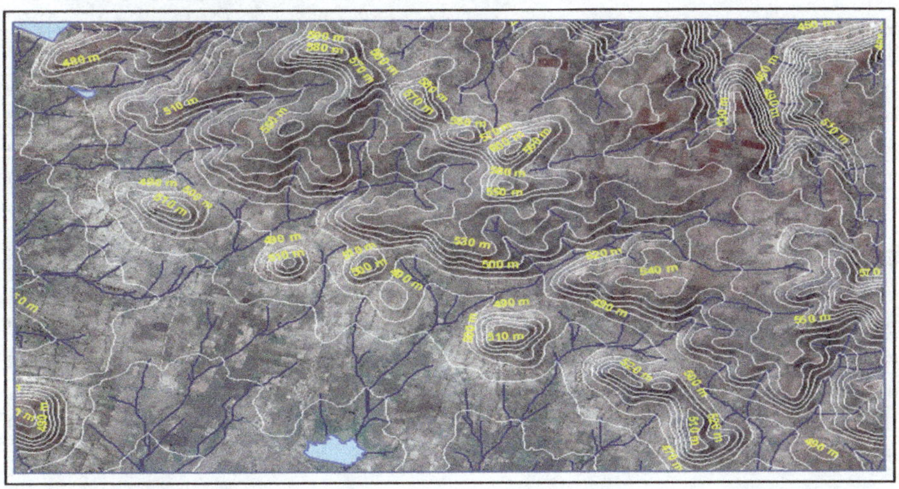

Fig. 13.3 Contour (10 m) and auto-drainage derived from DTM for a part of Indervelly block, Adilabad district, Telangana

the classification. For decades, geographic information system (GIS) specialists have theorized about the possibility of developing a fully or semiautomated classification procedure that would be an improvement over pixel-based procedures. The object-based modeling by taking into consideration the spectral and spatial/contextual properties of pixels and segmentation process with interactive learning algorithm promises to be more accurate than the pixel-based methods (Camargo et al. 2011). The following object-based semiautomated models are developed in the study.

13.2.1.3 Slope Classification Model

The raster slope layer output of DTM is taken as input in the object-based image analysis in the environment of eCognition® software. The slope layer was subjected to chessboard segmentation. Nine slope classes are created following the USDA-NRCS slope class threshold criteria. The criteria is fitted as fuzzy instead of hard rule using the less than and greater than "s-curve" membership function so as to get closer to the natural slope boundary. Morphology and contextual filters are applied to generate smooth slope class zones (Fig. 13.4).

13.2.1.4 Landform Classification Model

Case Study-I

The terrain attributes derived through digital terrain analysis of DTM layer, i.e., contour, drainage, slope, and curvature, are treated as input for landform delineation. The landform classification process is hastened taking into consideration the slope class zone, hillshade, contour, and auto-drainage pattern along with legacy physiography unit of 1:250k. The table (Table 13.1) below illustrates an example of logical rule set used for different landform units occurring in the Indervelly block of Telangana state (Fig. 13.5).

Case Study-II

The similar kind of exercised is carried out in the northeastern hilly region of Ri-Bhoi district, Meghalaya, where the objects resulting from segmentation are partitioned into subdomains based on thresholds given by the mean values of elevation and standard deviation of elevation, respectively, following the modeling approach given by Drăguţ and Eisank (2012). The layer variable thresholds are modified as per the local condition. The rule set window (Fig. 13.6) and resultant landform (Fig. 13.7) are presented below.

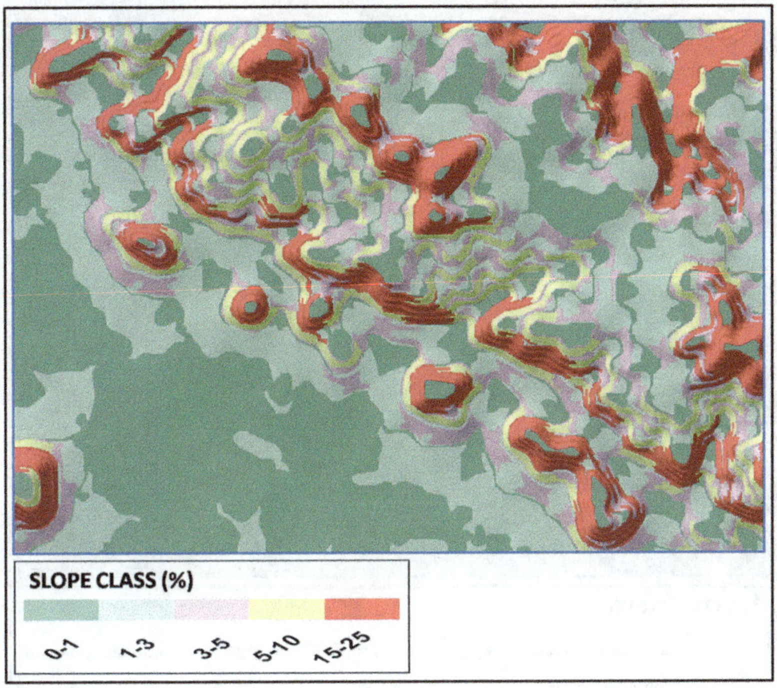

Fig. 13.4 Slope class zone derived from DTM for a part of Indervelly block, Adilabad district, Telangana

Table 13.1 Logical rule set used for different landforms

Landform	Logical ruleset condition
1. Undissected plateau	Slope range, 0–5%
	Relative boarder to escarpment, >80%
	Existence of drainage = false
	Relative topographic position = upper
2. Pediment	Side slope of plateau/upland
	Slope range, >1 to <15%
	Profile curvature = convex
	Presence of erosive features
3. Valley	Existence of drainage = true
	V-shaped contour with decreasing elevation gradient
	Profile curvature = concave
	Relative topographic position = lower

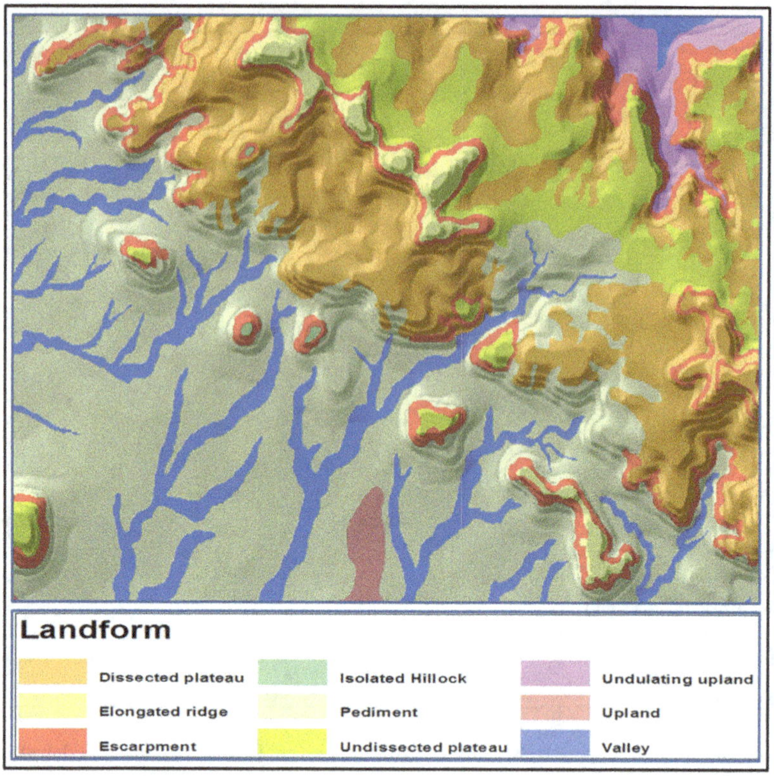

Fig. 13.5 Landform map on 1:10000 scale derived from DTM as a part of Indervelly block, Adilabad district, Telangana

Fig. 13.6 Rule set algorithm for delineating landform in Ri-Bhoi district, Meghalaya

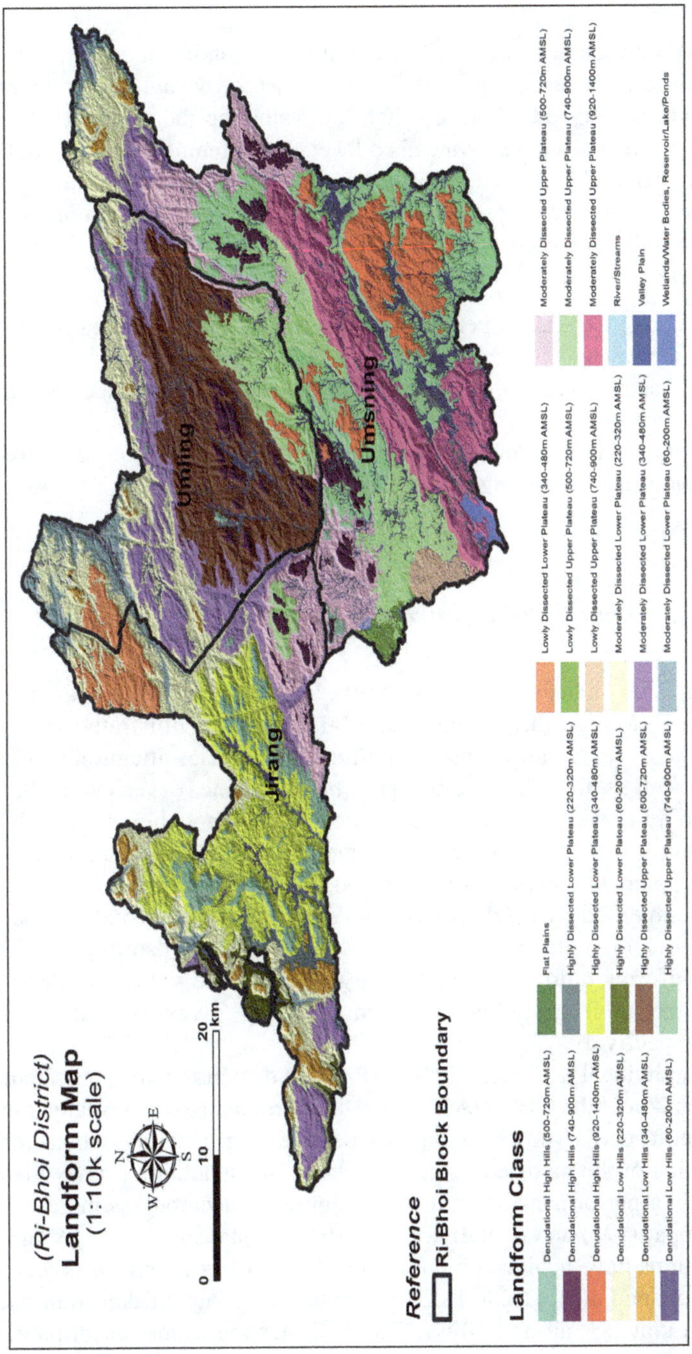

Fig. 13.7 Landform map derived from DTM for Ri-Bhoi District of Meghalaya state

13.3 Accuracy Assessment

To determine the accuracy of the OBIA-based digital landform output, independent reference landform is needed to be delineated based on visual interpretation in the study area (i.e., watershed). This will help in evaluating the accuracy of landform modeling on a larger scale, i.e., watershed level. The reference landform delineation is completed manually by using the background information of IRS-P6 LISS-IV imagery and DTM-based output of slope, contour and drainage pattern on a shaded relief layer. Finally, the accuracy is assessed based on the following three measures (d'Oleire-Oltmanns et al. 2013):

1. User's accuracy (UA), the percentage of correctly classified area from the total classified area
2. Producer's accuracy (PA), the percentage of correctly classified area from the total reference
3. Detection rate, the percentage of reference data that have been detected by the classification (also including partial detection)

13.3.1 An Example of Accuracy Assessment

Visual illustration of classified and reference landforms of Tandulwani watershed of Katol tehsil, Maharashtra (Chattaraj et al. 2017) are illustrated in Fig. 13.8. The top image of each landform section illustrates the classification results (solid color fill) as well as the reference polygon (black outlines) overlayed with shaded relief draped as a base layer. The three color-coded insets below in each section display examples of good matches (green box), as well as underestimations (yellow box) and overestimations (red box). Similar approach of illustration was also documented by d'Oleire-Oltmanns et al. (2013). Visual comparison of different landform segments at the two chosen scales illustrates the GEOBIA modeled landforms hold good even at larger scale. The values of classification accuracies and their graphical representation are given in Table 13.2 and Fig. 13.9, respectively.

For each individual landform, the UA, PA, and detection rate were calculated. A visual comparison of the GEOBIA modeled landform map to the visually interpreted reference landforms is shown in Fig. 13.10 revealing a highly satisfactory areal extent matching of the modeled output. Similarly, the quantitative assessment result of the modeling performance documents a high-level accuracy as indicated by the excellent kappa (0.91) and overall accuracy (92.8%) statistics. The UA and PA for all the landforms have achieved more than 90% accuracy except for lower alluvial plain and upper pediment (Table 13.2). It is noteworthy that the detection rate for all the landform units are around 100%. This indicates the sound performance of the

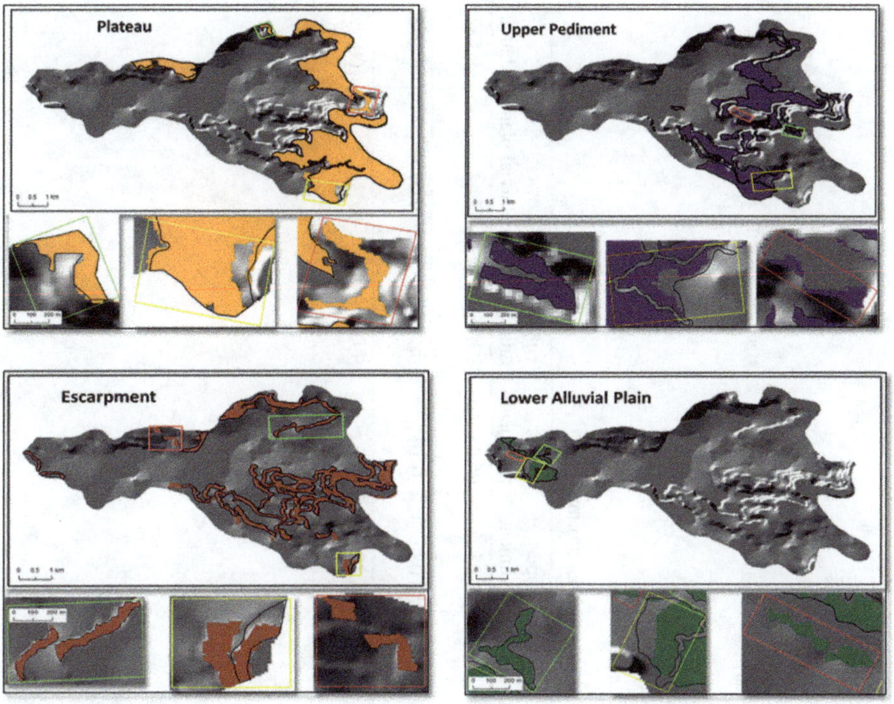

Fig. 13.8 Classification results for major landform units (solid color fill) and the corresponding reference landform units (black outlines) are illustrated. A shaded relief layer is displayed in the background. Three color-coded insets show examples of (green) good matches between classification and reference, (yellow) underestimations of reference, and (red) overestimations of reference

knowledge-based modeling to capture the existence of the landform units occurring in the watershed including partial detection.

13.4 Object-Based Land Use/Land Cover Classification Model

The LULC map was prepared based on the current *rabi* season data of Cartosat-1 merged LISS-IV (2.5 m) as well as high-resolution (0.5 m) public domain imagery at the backend so as to get the reliable land use boundary at cadastral level. The delineation of subclasses, viz., single- and double-cropped areas within the agriculture zone, was done using novel *LULC subclass classification algorithm* (Fig. 13.11). The merged data was segmented into spectrally homogeneous region using multiresolution segmentation algorithm. The optimum scale parameter for segmentation of the layer was achieved through estimation of scale parameter (ESP)

Table 13.2 Accuracy statistics of the GEOBIA modeled landforms in relation to reference landforms

Accuracy measures	Calculations	Isolated hillock	Plateau	Escarpment	Upper pediment	Lower pediment	Upper alluvial plain	Lower alluvial plain	Channel
User's accuracy	(Overlap area/classified area) *100	94.86	93.43	86.81	92.35	96.48	93.16	87.44	90.37
Producer's accuracy	(Overlap area/reference)*100	97.24	96.9	91.1	88.08	91.01	98.44	79.97	97.03
Detection rate	(Amount of classified refrence/ total reference)*100	100	100	100	96.47	98.2	99.4	97.3	100

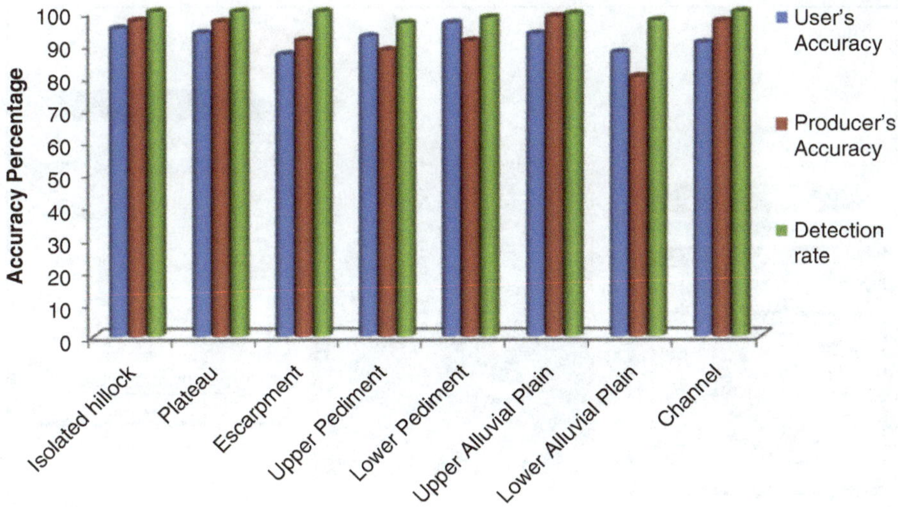

Fig. 13.9 Graphical representation of classification accuracy report across different landform units in the watershed

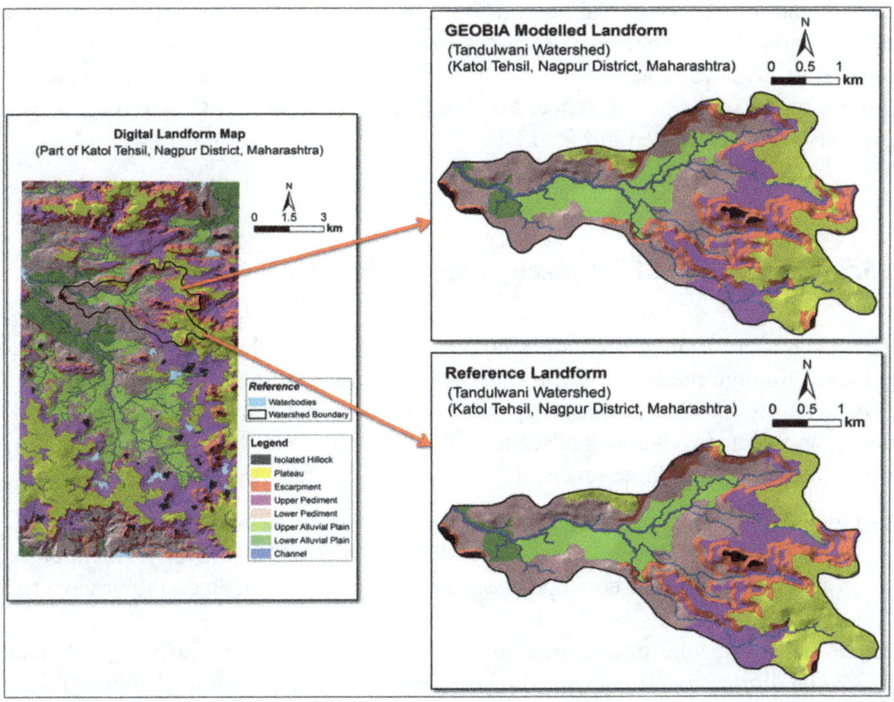

Fig. 13.10 Landform map of the study area as well as the modeled and reference landform map of the Tandulwani watershed as validation site

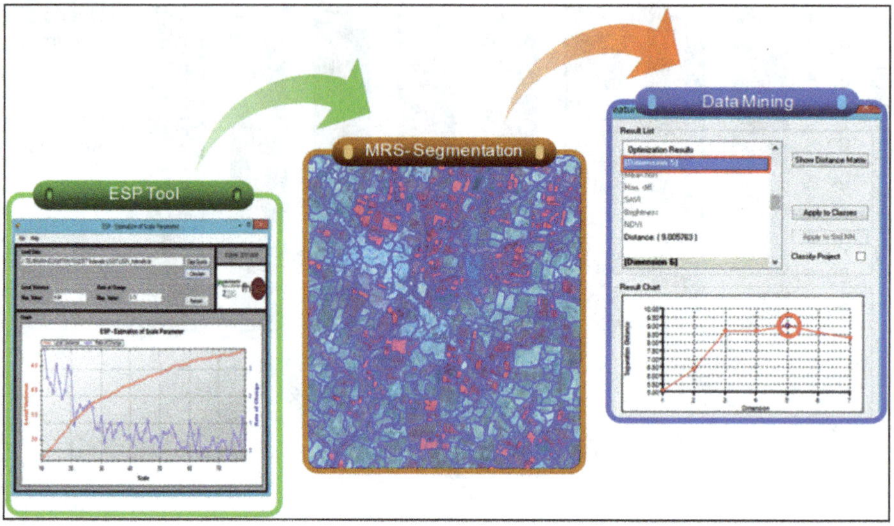

Fig. 13.11 LULC subclass classification model workflow

analysis tool. The point of interest lies where the local variance and rate of change are minimum in the graphical output. The data mining technique, i.e., feature space optimization, was applied to extract the double-cropped area based on certain number of layer variables and vegetation indices combination as obtained through the maximum separation distance. Following such scheme LULC map for a part of Indervelly block is given in Fig. 13.12.

13.5 Hierarchical Landscape Ecological Unit (LEU) Model

The integration of three secondary layers, i.e., landform, slope, and land use, was achieved through the *hierarchical object-based segmentation algorithm* taking into consideration the area, morphology of the landform units, and its relation with the neighbor objects to develop landscape ecological unit (LEU) map. The segmentation was accomplished in three levels:

1. Level-I: First level segmentation was done based on the landform layer.
2. Level-II: This segmentation was run within each of the first level segment based on fuzzy threshold-based slope class. Second level intermediate output gave rise to landform-slope unit.
3. Level-III: The landform-slope segments of second level were further subdivided into landform-slope-land use unit, i.e., LEU, by incorporating the land use factor. The logical condition used to incorporate the land use factor is that the minimum

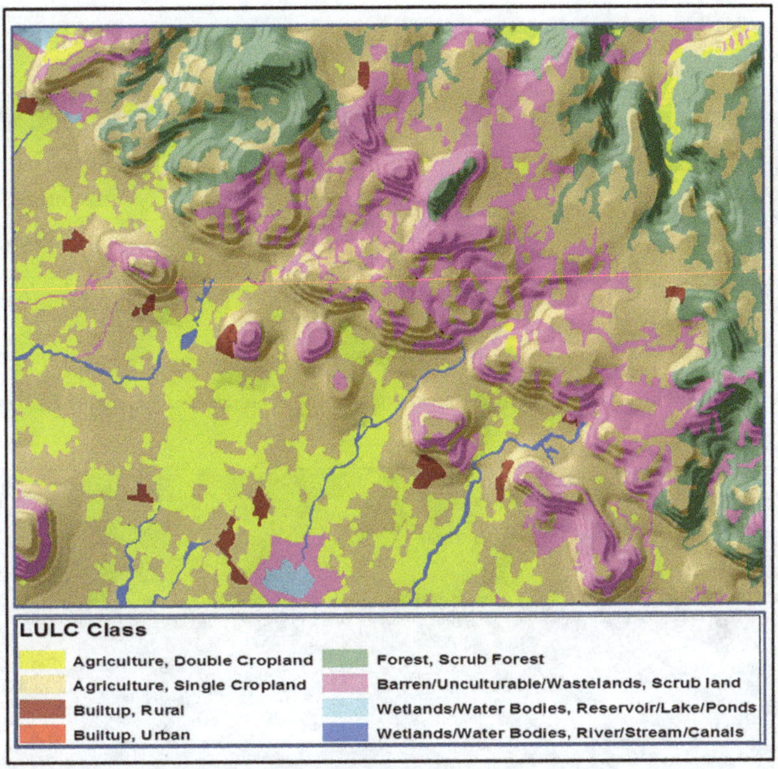

Fig. 13.12 Land use/land cover map on 1:10000 scale for a part of Indervelly block, Adilabad district, Telangana

overlap with the thematic polygon, i.e., level-II segment, will be more than or equal to 60%. The criteria ensure the continuity of LEU zone vis-à-vis soil boundary by ignoring negligible change in land use. Figure 13.13 explains the steps involved in the delineation of LEU.

13.6 Base Map in LRI Project

This LEU map has been used as base for developing soil-landform relationship for mapping soils on 1:10000 scales. Transacts were demarcated in GIS-based geo-database framework by assimilating the legacy data of 1:250 k scale and expert knowledge as shown in Fig. 13.14.

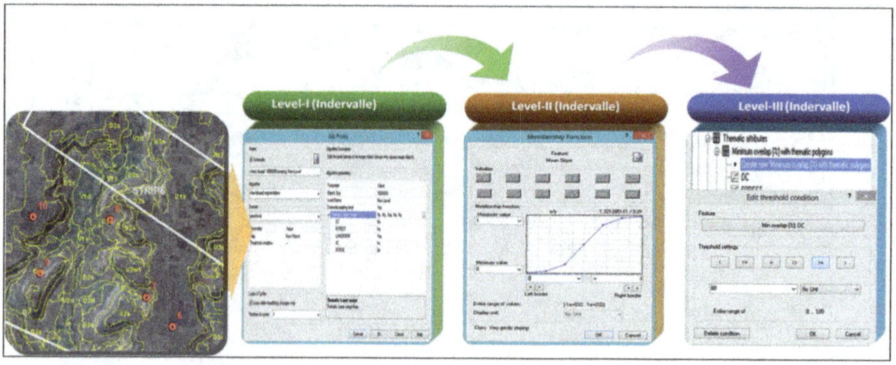

Fig. 13.13 Hierarchical object-based segmentation algorithm process for generating LEU maps

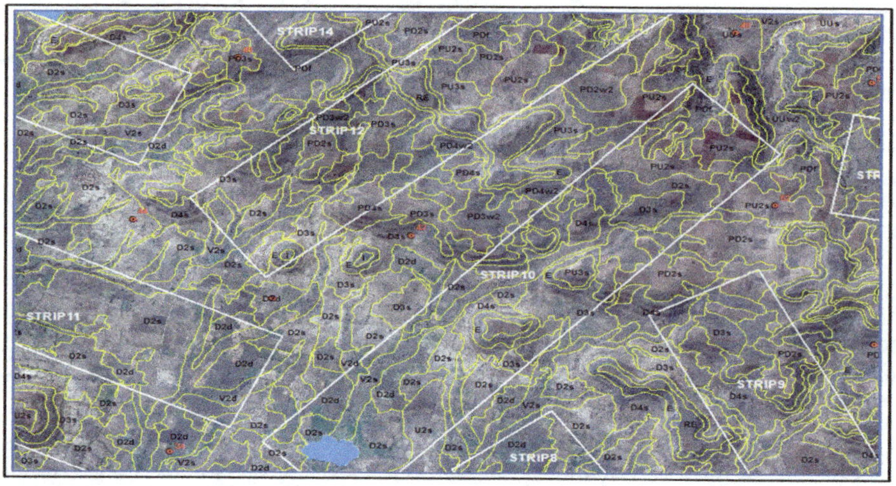

Fig. 13.14 Base map on 1:10000 scales for a part of Indervelly block, Adilabad district, Telangana

13.7 Conclusions

The logical automation algorithm developed at each stage results in considerable reduction in time for base map preparation. This will assist in optimizing sampling intensity, which leads to a considerable saving of man power, labor, cost, and most importantly the time. Finally, a hierarchical geo-database structure having unified schema is proposed for deploying in the *National Soil Geo-portal* to disseminate the information in a user-friendly way.

References

Camargo FC, Almeida T, Florenzano C, Heipke R, Feitosa, Costa G (2011) ASTER/Terra imagery and a multilevel semantic network for semi-automated classification of landforms in a subtropical area. Photogramm Eng Remote Sens 77:619–629

Chattaraj S, Srivastava R, Barthwal AK, Giri JD, Mohekar DS, Reddy GPO, Daripa A, Chatterji S, Singh SK (2017) Semi-automated object-based landform classification modelling in a part of the Deccan Plateau of central India. Int J Remote Sens 38(17):4855–4867

Cheng G, Han J (2016) A survey on object detection in optical remote sensing images. ISPRS J Photogramm Remote Sens 117:11–28

d'Oleire-Oltmanns S, Eisank C, Drăguţ L, Blaschke T (2013) An object-based workflow to extract landforms at multiple scales from two distinct data types. IEEE Geosci Remote Sens Lett 10(4):947–951

Drăguţ L, Blaschke T (2006) Automated classification of landform elements using object-based image analysis. Geomorphology 81(3–4):330–344

Drăguţ L, Eisank C (2012) Automated object-based classification of topography from SRTM data. Geomorphology 141-142:21–33

Eisank C, Drăguţ L, Blaschke T (2011) A generic procedure for semantics-oriented landform classification using object-based image analysis. In: Geomorphology 2011 conference paper. Elsevier, Redlands, pp 125–128

Nagaraju MSS, Kumar N, Srivastava R, Das SN (2014) Cadastral level soil mapping in basaltic terrain using Cartosat-1-derived products. Int J Remote Sens 35(10):3764–3781

Romstad B, Etzelmuller B (2012) Mean-curvature watersheds: a simple method for segmentation of a digital elevation model into terrain units. Geomorphology 139/140:293–302

Soil Survey Division Staff (1995) Soil survey manual, USDA handbook no. 18, revised edn. Scientific Publishers, Jodhpur

Srivastava R, Saxena RK (2004) Techniques of large scale soil mapping in basaltic terrain using satellite remote sensing data. Int J Remote Sens 25(4):679–688

Chapter 14
Pedogeomorphic Analysis in Land Resource Inventory: A Case Study from Central India

B. P. Bhaskar

Abstract In the study, the importance of pedogeomorphic analysis in land resource inventories at various scales, mutual contributions of pedology and geomorphology, classification of geoforms, application of remote sensing and geographic information system (GIS) in geomorphology mapping were discussed. This article describes how pedogeomorphic analysis was used in detailed soil surveys in basaltic uplands of Darwha plateau in Yavatmal district, Maharashtra, in developing soil-landscape models that control and explain the shrink-swell soil pattern within a land region having similar agricultural management requirements for cotton-based cropping systems. Soil-landscape relationships in Darwha plateau were studied in detailed along the transects comprised of lithic/typic Ustorthents on plateau tops to lithic/typic/vertic Haplustepts on pediments and leptic to Typic Haplusterts on pediplains. The key pedogenic processes recognized in shrink-swells were identified as development of vertic features and $CaCO_3$ accumulation in deeper layers. Both field and laboratory analyses confirmed that the soils within landscapes form a continuum but that recognition of the soil groupings was more practical and feasible in mapping soils in the region. A combination of IRS-P6 and Cartosat-I fused data, field examination of soils and their characterization was used to recognize landscape-related soil groupings to develop farm scale maps that can support multiple land use applications and decision-making.

Keywords Basaltic clay soils · Cotton-based cropping systems · Darwha plateau · Pedogeomorphic analysis · Soil series · Soil transects

---

B. P. Bhaskar (✉)
ICAR-National Bureau of Soil Survey & Land Use Planning, Nagpur, India

G. P. O. Reddy, S. K. Singh (eds.), *Geospatial Technologies in Land Resources Mapping, Monitoring and Management*, Geotechnologies and the Environment 21,
https://doi.org/10.1007/978-3-319-78711-4_14

14.1 Introduction

Pouquet (1966) was the first one to use the word geopedology, emphasizing the pedologic component, and implements geopedology as an approach to soil survey and to erosion/soil conservation studies. In contrast, Tricart (1972) was one of the first authors to use the word pedogeomorphology. It is difficult to find any theoretical statement on soil-geoform relationships, except the reference that is usually made to classic models such as the hillslope model of Ruhe (1975) and the soil equation of Jenny (1941, 1980). The importance of landscape has been tacitly accepted by soil scientists (Bridges 1982) and cited as 'catena' concept by Milne (1935). Relief influences pedogenesis in a passive way through its form or the characteristics of topography and in active way through the processes of relief formation, which directly controls soil formation (Michel 1973). Hammond (1954, 1964) developed a macro-landform classification procedure to identify landform types for the United States and combines three important parameters (i.e. slope, relief and profile type) to identify different landform or terrain types. Interrelationship between landform and soils in a defined toposequence is generally studied by establishing pedogeomorphological relationship (Dent and Young 1981; Birkeland 1990). The analysis of landform-soil relationship, in association with drainage and elevation properties, can be effective to understand spatial patterns of soil attributes in similar geological, climatic and terrain conditions (Bell et al. 1994; Gessler et al. 1995). Such relationships were used in determining uniform mapping units for soil survey and for soil management and soil productivity determinations (Daniels and Hammer 1992; McGregor and Thompson 1995).

14.2 Importance of Pedogeomorphic Analysis in Land Resource Inventory

Very few studies of pedogeomorphic studies on 1:10,000 scale were reported in literature as in case studies of Roodepoort area, Transvaal of South Africa, for understanding soil – depth relationships (Munnik et al. 1984, 1986). In the similar line, the distribution patterns of the footslope soils could be explained by means of a simple catenary model (Verster 1987) by the Shepstone form comprising greyish coarse sands abruptly overlying reddish, apedal, mesotrophic clays and loams.

The catenary relationships in basaltic landscapes of Central India were reported in central India (Bhattacharjee et al. 1977; Gawande and Biswas 1977 and Bhaskar et al. 2001). The majority of the Indian Vertisols occurs in lower physiographic areas (Pal and Deshpande 1987; Pal et al. 2009) but spatially associated with red ferruginous soils (Bhattacharyya et al. 1993; Raja et al. 2010) under semiarid climates (Pillai et al. 1996). Landscape configuration features follow a fairly definite soil pattern, which influences soil morphology and physicochemical properties on the basaltic terrain (Pofali et al. 1979). The landscape-soil relationship in part of the

Bazargaon plateau, Maharashtra, showed that the soil-physiography relationship plays a vital role in understanding genesis of the soil and assessment of their potential for various kinds of uses (Reddy et al. 1999). Innumerable studies have been reported on soil mapping with visual interpretation of remotely sensed data (Venkataratnam 1984; Prasad et al. 1990), and particularly, 1:50,000 scale have been found very useful for soil mapping at different levels (Ahuja et al. 1992; Reddy et al. 2003; Bhaskar 2015).

Therefore, need is being felt for fresh soil mapping at larger scale with fine level of details to use them for developmental planning in agriculture (Ravisankar and Thamappa 2004; Rao et al. 2004; Dobos and Hengle 2009). The developmental planning envisage preparation of soil resource inventory, understanding of resource constraints and suggesting feasible alternatives for agriculture development (Krishnamurthy and Adiga 2008). The agri-research farms at regional level are important centres of technology transfer and needs detailed soil reports (Bhaskar and Gajbhiye 1997). Once soil information is documented, these soils are used to conduct applied research and involve farming community through participatory approach for sustainable farming. In recent times (Shabbir et al. 2004; Bhaskar et al. 2010 and Prasad et al. 2012), pedologists emphasized on importance of soil survey at farm level to map diverse soil types within the farm and among different farm sites. The orthorectified satellite imageries may serve as base for ground truth coupled with transect walks (Gobin et al. 2000). Using IRS-Resourcesat-1 LISS-IV + Cartosat-1 merged data in Mohammad village of Nalgonda district Andhra Pradesh, soil resource information was generated and prepared action plan (Wadodkar and Ravishankar 2011). On similar lines, with IRS-Resourcesat-1 + Cartosat data, the soil map was generated at phase level for deriving action plan (Nagaraju et al. 2014).

14.3 Concepts and Principles in Pedogeomorphic Analysis

Soil geomorphology was first studied by the US National Cooperative Soil Survey (NCSS) programme in the 1930s and developed interest among geographers, geologists and soil scientists on understanding the relationships between soils and landforms (Effland and Effland 1992). Ruhe (1975) described hilly landscapes having landscape elements such as summit, shoulder and footslope with sequence of soils over these landscape elements. Similarly, Schoeneberger and Wysocki (2001) described flat broad areas as *talfs* ('flat' spelled backwards), and flat surfaces occur on stepped or 'tread-and-riser' landscapes. For pedogeomorphic analysis, it is imperative to understand the ten fundamental concepts as enumerated by Thornbury (2004) as listed below:

1. The physical processes and their intensities in different geologic time.
2. Geological structure is a dominant factor in the evolution of landforms.
3. The geomorphic processes operate at differential rates.

4. Geomorphic processes leave their distinctive imprint upon landforms.
5. Different erosional agents act upon the earth's surface.
6. Complexity of geomorphic evolution is more than simplicity.
7. Little of the earth's topography is older than tertiary and most of it no older than Pleistocene.
8. Proper interpretation of present-day landscapes.
9. Proper understanding of the varying importance of the different geomorphic processes.
10. Present-day landscapes attain its maximum usefulness by historical extension.

Mutual Contributions The review of literature shows three significant contributions found in development of soil-geomorphology relationship, which are

1. *The mechanistic Jenny's model* (Jenny 1941) explaining qualitatively soil development is condensed in the equation:

$$S = f(cl, o, r, p, t, \ldots)$$

The review work of Yaalon (1975) contemplates at the most monofactorial qualitative or semi-quantitative description of relationships between soil-forming factors and soil state. Therefore, climofunctions, organofunctions, topofunctions, lithofunctions and chronofunctions have been proposed, disregarding the role of interactions between soil-forming factors themselves, whose recognition could conversely be a useful work so that more detailed spatial patterns on soils are available.

2. *Concepts of soil-landscape model* is mainly focused on the following points that has to be verified during field investigations (Hudson 1992):

   - Soil-landscape units are natural terrains resulting from five soil factors conventionally cited in the functional equation for soil formation.
   - Soil-landscape units have a predictive spatial relationship to one another.
   - In a given soil survey area, there are relatively few soil-landscape units replicated again and again.
   - Generally, the more different two adjacent soil-landscape units are, the more abrupt and striking the discontinuity separating them.
   - A distinctive, relatively homogenous soil cover develops on each soil-landscape unit.
   - The boundaries between distinct soil-landscape units can be observed and mapped as discontinuities on the earth's surface. As a result, they can be delineated accurately by trained mappers.
   - Since the boundaries between distinctly different landscape units tend to be abrupt and prominent, the boundaries between their associated soils tend to be abrupt and prominent.
   - As a general rule, the more different two adjacent soils are, the easier it is to locate the boundary between them accurately and precisely.

- Within a given soil-landscape unit, soil variation at the human scale of perception, is mostly cyclic.

3. *The environmental correlation model*

   McKenzie and Ryan (1999) proposed a modified version of Jenny's functional factorial model in which soil is fitted in space domain by means of an environmental correlation model of the form:

$$S = f(Cl, T, PM, M, \ldots) \tag{14.1}$$

where *S* is the soil property or land quality; *Cl, T* and *PM* are explanatory variables from spatial layers of climate, terrain and parent material, respectively; while *M* represents other miscellaneous auxiliary layers from multispectral sensing and land management, etc.

McBratney et al. (2003) proposed a review of approaches in making digital soil maps using GIS co-variable layers and proposed a generic quantitative framework into which various methods of spatial inference can take place. They replaced the Jenny's *clorpt* model with the following one in which factors of soil formation are treated quantitatively and an objective high-resolution numerical handling of soil spatial variability is given:

S = f (s, c, o, r, p, a, n) where S is a soil class or attribute; s, refers to soil punctual data; c, climate; o, organisms; r, (relief) topography and land surface parameters; p, parent material; a, age factor; and n, spatial position on ground.

14.4 Criteria for Classifications of Landforms

A set of assumptions are formulated as basis for structuring a taxonomic systems for geoforms in an abridge version of Zink (2013) as described below:

- The *landform* is defined as terrain features formed by natural processes, which have a defined composition and a range of physical and visual characteristics that occur wherever the form is found and whatever is the geographic region (Way 1973).
- The objects that are classified are the geoforms, or geomorphic units, which are identified on the basis of their own characteristics, rather than by reference to the factors of formation.
- Classes of geoforms are arranged hierarchically to reflect their level of membership to the geomorphic landscape.
- The genesis of the geoforms is taken into consideration preferably at the lower levels of the taxonomic system. At higher levels, the use of more objective rather descriptive attributes is privileged, in parallel with the criteria of pattern recognition implemented in photo and image interpretation.

- The dimensional characteristics are subordinate attributes and are not diagnostic for the identification of the geoforms.
- The names of the geoforms are often derived from the common language, and some of them may be exposed to controversial interpretation. Priority is given here to those terms that have greater acceptation by their etymology or usage.
- The geographic distribution of the geoforms is not a taxonomic criterion. The chorology of the geoforms is reflected in their cartography and in the structure of the geomorphic map legend.
- Toponymic designations can be used as phases of the taxonomic units

14.5 Remote Sensing and GIS Applications in Landform Classification

The increasing availability of high-resolution satellite imagery leads to a growing number of landscape research applications using remote sensing and GIS (Florinsky 1998; Walsh et al. 1998; Ehlers et al. 2002). In India, a critical review was made and reported in remote sensing applications in geomorphic mapping by Rao (2002). Integrating satellite, aircraft and terrestrial remote sensing systems to achieve a scale-dependent set of observations can be achieved through operational systems and current technologies. Satellite remote sensing have undergone a technological revolution with a massive amount of spatial data with high spectral and temporal resolution via various sensors and satellite missions (Benediktsson et al. 2012). However, in order to exploit the full potential of the data, semiautomated and automated mapping techniques are necessary. The land surface can be divided into a hierarchy of landscapes, landforms and landform elements, where the landform element is the smallest unit, indivisible at the given resolution and bounded by topographic discontinuities (Pike 2000).

Recently automated and semiautomated mapping techniques of volcanic landforms have focused on delineating the bases of volcanic edifices (Euillades et al. 2013) and in basaltic terrain of Katol tehsil, Nagpur, using geospatial object-based image analysis (GEOBIA) technique with good statistical agreement (Chattaraj et al. 2017). An object-based image analysis (OBIA) approach is a two-step classification technique: (1) segmentation of an image creating segments/objects, each of which consists of a group of contiguous grid cells from a raster layer, followed by (2) rule-based classification of the individual objects (Benz et al. 2004). Pixel-based classification, on the other hand, is only based on the pixel value and ignores spatial autocorrelation (Van Den Eeckhaut et al. 2012). The DEM (digital elevation model) approach was found to be useful tool to define profile curvature that influences erosion and deposition of material (Pedersen and Grosse 2014).

14.6 Land Resource Inventory: A Case Study from Central India

14.6.1 Geographical Settings

The study site lies between 20°16′02″ to 20°17′20″ N and 77°40′51″ to 77°44′57″ covering 81253.37 ha (Fig. 14.1). The Adan-Arunavati basin, comprising the entire Darwha tehsil, the south-western parts of Yavatmal and parts of Kelapur tehsil, lies between two distinctive eastward running spurs or the Buldhana plateau at average elevations of 370 m–460 m and forms the only broad alluvial valleys of the uplands. The flat-topped water-divide separates the Adan and Arunavati river valleys, at heights of about 400 m. The land as a whole slopes gently to the south-east towards the Penganga, into which river the Adan-Arunavati finally empties. However, along the Penganga, the descent is through a steep slope to the narrow valley base. Darwha has the least area under uncultivable and culturable wastes wherein *Kharif* jowar (*Sorghum bicolor*) and cotton (*G. hirsutum*) account for more than four-fifths of the area. Pigeon pea (*Cajanus cajan*) and chick pea (*Cicer arietinum*) among pulses and groundnut (*Arachis hypogaea*) are the minor crops. Patches of fairly dense mixed forests dot the hills and water-divides, particularly in the west and along the Penganga slopes. The climate is semiarid hot with ustic soil moisture regime and isohyperthermic soil temperature regime. June, July, August and September are rainy months, and mean annual rainfall of the study area is 939.38 mm with rainy days of 51 days (Fig. 14.2). The contribution of rainfall from south-west monsoon

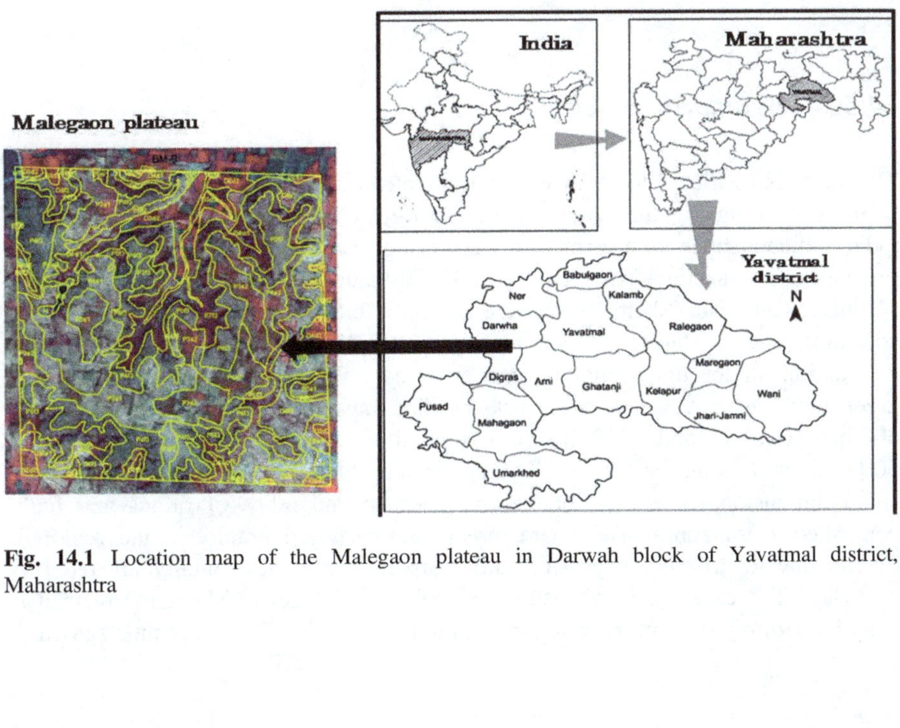

Fig. 14.1 Location map of the Malegaon plateau in Darwah block of Yavatmal district, Maharashtra

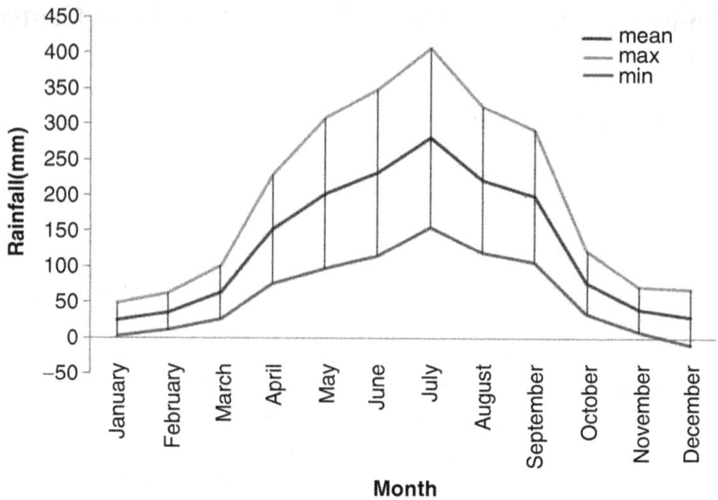

Fig. 14.2 Monthly rainfall distribution in Darwha block

(June to September) accounts to 87%, of which July contributes 29.8%–17.46% in September. The number of rainy days during south-west monsoon varies from 14.2 days in July to 8.2 days in September but less than 1 rainy day during December to February. The basaltic plateaus and pediment surfaces with varying degrees of erosional processes are often subjected to seasonal water stress during drought period.

14.6.2 Geomorphic Setup

The study is restricted to Darwha basaltic plateau comprising distinct geomorphic features of strong structural control. The flat-topped hills, ridges and plateaus as seen today indicate subaerial erosion and weathering of massive basaltic boulders comprising horizontal flows with red boles. These features show the evidence of post-Mesozoic tectonism (Sheth 1999) and erosional cycles of altrimetric variations in the adjoining parts of study area (Subramanyam 1974). These landscapes subjected to denudation all through Cenozoic and erosional cycles with tectonic factor have given three distinct topographic surfaces like trappean highlands (not exceeding 500 m–600 m), uplands (350 m–450 m) and alluvial plains (Adan river basin below 300 m towards south-east trend). The trappean highlands have flat-topped plateaus; small dimensions of mesas, buttes and cuestas; and relics of ridges and fault-controlled scarp zones. The characteristic trappean topography in the region is mainly due to differential erosion and scarp retreat of horizontally layered lava sequence. The escarpment plateau steep cliff assemblages at Malegaon plateau in Darwha strongly support the opinion of Alavi and Merh (1991) in pointing towards differential weathering rates of basaltic layers having different lithologies; by and large the faulting, vertical uplift and subsidences are the dominant players controlling topographic features. Similar kind of geomorphological features were reported in basaltic landscapes of Vidharbha regions by various workers (Reddy et al. 1999, 2003; Bhaskar et al. 2015).

14.6.3 Pedomorphic Analysis for Land Resource Inventory

The detailed soil survey on 1:10,000 scale was made with the interpretation of Indian remote sensing satellite (IRS-P6) LISS-IV merged with Cartosat-1 data to draw soil boundaries on the imagery to speed up the field work and laying out field traverses to verify ground features with image characteristics. It was reported that four important steps (FAO 1967 and Vink 1968) are involved in peogeomorphic analysis, which includes image interpretation that precedes conventional soil survey so as to prepare preconceived legend of soil maps, ground verification that was made for image interpretation units during field work where soil legend tend towards soil-landform map, preinterpretation reconnaissance procedure that was made to enhance local reference level in advance especially for the correlation between soils of the study area and elements or patterns as recognized in the imagery and interpretation with full check procedure that demands high level of accuracy and precision of personal reference level and methodology of soil survey expert making interpretations.

To achieve the objectives of detailed land resource inventory, soil window concepts (Fraser et al. 2014) were adopted for Darwha tehsil at 1:10,000 scale. These windows approximately covering 1600 (ha) were selected for thorough investigation of correlations between photo-morphic units recognized and soils identified on various landforms. The soil window concept began with review of literature and in consultation with farmers and farm consultants during reconnaissance survey of Yavatmal district (Bhaskar et al. 2014). Initial field-based work was undertaken in Darwha block to develop soil-landscape models that describe soil patterns at the farm scale and in response to broader-scale gradients in parent material and rainfall. Subsequently, soil profiles were studied with the aid of a soil auger and excavation pits along transects from the upper footslope to the valleybottom. The relative elevation and position of each sampling site were recorded using hand-held global positioning system (GPS). The soil morphology was described in detail, using the techniques and terminology described by Schoenberger et al. (2012) and classified as per Soil Survey Staff (2014). Horizon-wise soil samples for laboratory analyses were taken and air dried to pass through <2 mm sieve. For particle size distribution, the pipette and sieve method of Laker and Dupreez (1982) was used. The pH and electrical conductivity (EC) were measured on a 1:2.5 soil water suspension. Readily oxidizable organic carbon was determined according to the method of Walkley and Black (1934) as described by Jackson (1973). The ammonium-saturated clays were prepared (Chapman 1965) by allowing the starting sample to stand overnight in 1 N ammonium acetate solution, which had been adjusted to pH 7, after which the samples were leached with 1 N NH_4CL. Excess ammonium salts were removed by washing with isopropyl alcohol, after which the clays were dried and ready for determination of CEC. The concentrations of sodium, potassium, calcium, magnesium and sodium in IN ammonium acetate extratable bases and DTPA-extractable mícronutrient concentrations (10 g soil with 25 ml of 0.025MDTPA + 0.5 triethanolamine + 0.05M$CaCl_2$ solution) of Zn, Cu, Mn and Fe were determined using atomic absorption spectrophotometer (Lindsay and Norvell 1978).

14.6.4 Classification of Landform Mapping

For the purpose of this study (land form classification), the basaltic landscapes were viewed with respect to the three main slope components based on steepness and (length of the slope) and defined these units as plateau tops (crest), pediplains and pediments (middleslope and footslope). The northern region of Darwha is characterized based on elevation into three classes as plateaus, pediments and escarpments. These land units were further segmented into four slope classes as the crest, upper slope, middle slope and lower slope and then classified into three classes on the basis of land use such as single cropped, double cropped and forest area. The scheme of landform classification followed in the study area is shown in Fig. 14.3.

The structure of basaltic landform at Malegaon plateau was divided into plateaus, escarpments and pediments (Fig. 14.4). According to landform classification, the Malegaon site is broadly divided into three basic units, viz. plateaus (872.8 ha, 54.5% of total area), pediments (393.4 ha, 24.6% of area) and escarpments (302.3 ha, 18.9% of area) with further subdivisions made on the basis of slope (%) and land use. Within the plateaus, 25.45% of area falls under C slopes (5–10% slopes) with further

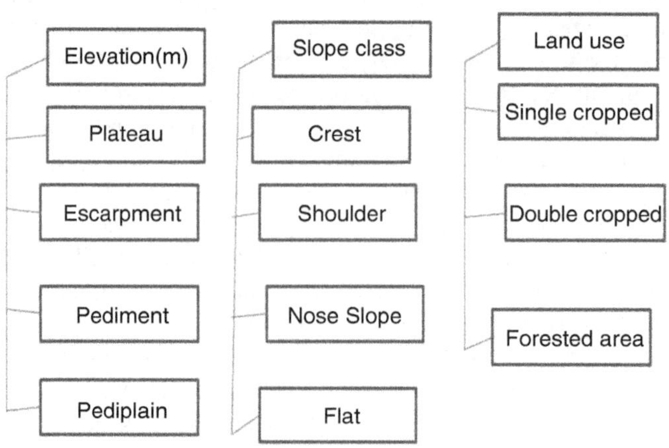

Fig. 14.3 Scheme of landform classification based on morphology

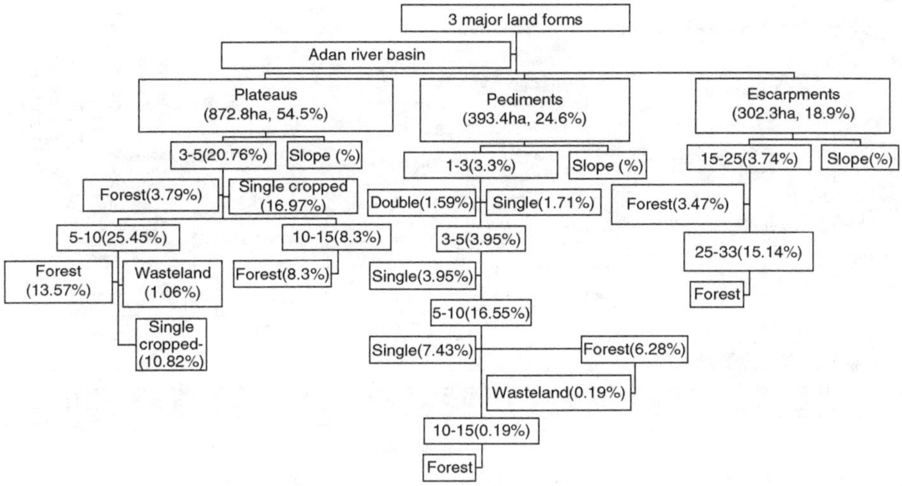

Fig. 14.4 Structure of basaltic landscape system in Malegaon plateau (BM-9)

division of 13.57% of area under forest, 10.82% of area under single crop and 1.06% of area under wasteland. Likewise, the pediments having C slopes (5–10% slopes) cover 16.55% of area with 7.43% of area under single crop and 6.28% under forest, whereas escarpments have 15.14% of area under F slopes (25–33%) with forest cover.

14.6.5 Soil-Landscape System

The land systems at Chopadi to Malegaon plateau at an elevation of 400–540 m above mean sea level in a line transect diagram of 3 km are depicted in Fig. 14.5. The transect is selected along the topographic gradient of each site and depicted the close association of soil types with land use systems existing within the agroecological zone of particular interest. The transect in the window area of Chopadi to Malegaon plateau is divided into pediments, plateaus and upper pediplains at first-order level. The dominant soils on pediments are Chopadi (Typic Calciustepts), Hatola (Lithic Ustorthents) and Chikhali (Leptic Calciusterts) whereas Malegaon series (Lithic Calciustepts), Waghari (Typic Haplustepts) and Arunavati series (Typic Calciustepts) on plateaus and Nagdhari (Fluventic Haplustepts) and Lakhi (Lithic Ustorthents) on upper pediplains.

The distribution of dominant soils according to land unit at Malegaon plateau is given in Table 14.1. Based on these data, as well as the percentage occurrence of each category, the expected frequencies for the dominant soils in relation to land units and, in turn, the results for the χ^2 test with standardized coefficient and contingency coefficient were calculated. In the Malegon plateau, the well-drained Chopadi (Leptic Caliustepts) and Lakhi (Lithic Ustorthents) occur mainly on the

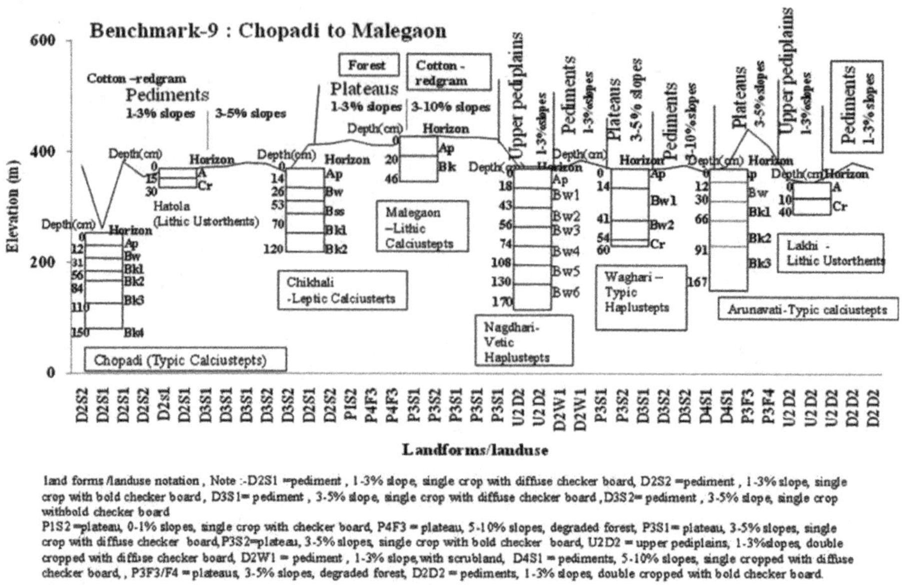

Fig. 14.5 Soil transect diagram at Malegaon plateau

crests and middleslopes, Arunavati on pediments/pediplains and Nagdhari on pediplains, and they have a fairly strong association between soil series and land unit relationship as well as a significant with contingency coefficient of 0.49 and standardized coefficient of 0.55. The similar kind of pedogeomorphic analysis was carried out extensively at a scale of 1:10,000. The statistical relationships were significant between certain soil forms and site characteristics, in various parts of Roodepoort area, Transvaal, South Africa (Munnik et al. 1984, 1986).

The seven dominant soil series at Malegaon plateau show that these series are grouped under four depth classes as per Sehgal (1992). The Lakhi series occurring at an elevation of above 400 m have mean soil depth of 24.6 ± 13 cm for Lakhi with coefficient of variation (CV) of 53% (Table 14.2). As elevation decreases from 400 m, the Malegaon series confined to 396 ± 23 m elevation have mean soil depth of 73 ± 2.5 cm (moderately shallow) with CV of 3.4% followed by Waghari series at 386 ± 16.4 m with mean soil depth of 60.36 ± 32.9 cm with CV of 54.6%. The deep Chopadi series at 351 ± 44 m with mean soil depth of 115 ± 37 cm and CV of 33% and Nagdhari series (having mean of 132.3 ± 11.1 cm with CV of 8.39%. The Malegaon series occupies 55.33% of total area in association with Chopadi (18.97%), Chikhali (12.04%) and Lakhi (6.39%). From the literature (Furley 1971; Veneman and Bodine 1982), it is evident that other systematic relationships also exist between soil properties and landscape elements. Some of these are very aptly explained by the toposequence and catena concepts in Jayakwadi command area, Maharashtra (Bhaskar et al. 2001). It is further evident that there is a strong polynominal relationship between soil depth and elevation with coefficient of determination ($R^2 = 0.45^{**}$, Fig. 14.6, total number of observations (N) =39).

Table 14.1 Observed and expected frequencies of the dominant soils on each land unit

	Land units								
	Plateaus		Pediments		Escarpments		Pediplains		
Soil series	Observed	Expected	Observed	Expected	Observed	Expected	Observed	Expected	Total
Chapodi	13	11.89	8	8.14	2	5.94	12	9.25	**35**
Chikhali	5	6.11	8	4.19	3	2.94	2	4.75	**18**
Malegaon	6	7.13	4	4.89	7	3.43	4	5.55	**21**
Nagdhari	2	5.43	4	3.72	–	2.62	10	4.23	**16**
Waghari	8	8.15	4	5.58	7	3.92	5	6.34	**24**
Arunavati	2	6.45	8	4.42	3	3.11	6	5.02	**19**
Lakhi	18	8.83	1	6.05	4	4.25	3	6.87	**26**
	54 (34)		37 (23%)		26 (17%)		42 (26%)		159

Table 14.2 Descriptive statistics of soil depth in relation to elevation and soil series

Soil series	Area (ha)	(%)	Elevation (m) mean ± SD	CV (%)	Soil depth (cm) mean ± SD	CV (%)
Chopadi	235.6	18.97	351 ± 45	12.6	115 ± 37	32
Chikhali	149.48	12.04	356 ± 23.6	6.2	89 ± 26.4	29.7
Malegaon	687.2	55.33	396 ± 24	5.9	73 ± 2.5	3.4
Nagdhari	56.9	4.58	345 ± 13.4	3.9	132.33 ± 11.11	8.39
Waghari	5.42	0.44	386 ± 16.4	4.3	60.36 ± 32.9	54.6
Arunavati	27.29	2.20	367 ± 25.8	7.2	94.29 ± 22.54	23.91
Lakhi	79.3	6.39	413 ± 22	5.3	25 ± 13	53

Fig. 14.6 Polynominal relation between soil depth and elevation

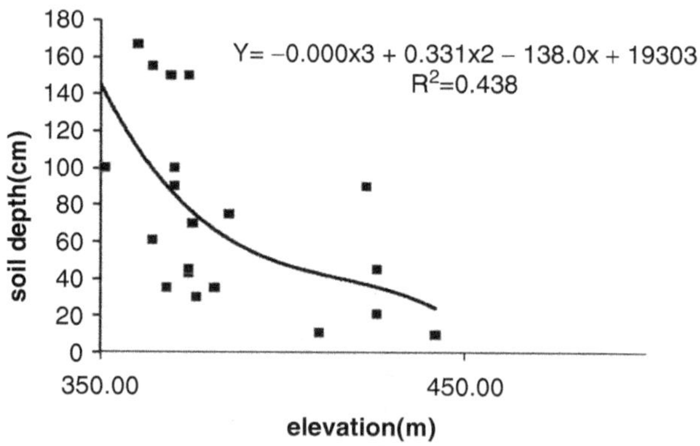

As there is a dearth of information on toposequence research in Central India, in practice, it is difficult to determine research in India at large scale, and the results of this study could only be used as an example to work out further in basaltic landscapes. Based on pedomorphic studies at 1:10,000 scale of Darwha plateau, it is concluded that the occurrence of morphogenetically similar kind of shrink-swell soils with dark colour topsoils (low chroma) has minimal horizon differentiation and carbonate accumulation (high value) in Ck horizons (Soil Survey Staff 2014). The basaltic landscapes under study have the shrink-swell soil associations under the land units as plateaus-pediments-escarpments-pediplains sequence with strong association between soils and land units. Obviously, to verify these findings, the toposequence research must be taken up in window areas of other agroecological regions.

14.6.6 Morphology and Chemical Characteristics

The selected morphological and chemical properties of seven soil series are presented in Table 14.3. The Chopadi series (P1) on plateaus and on pediments have very dark greyish brown in Ap and Bw horizons to dark yellowish brown

Table 14.3 Morphology and chemical characteristics of soils

Soil series/ taxonomy	Hori-zon	Depth (cm)	Matrix colour (moist)	Tex-ture	pH	EC (dSm1)	OC (%)	CEC (cmol/ kg)	DTPA extractable (mg/kg)			
									Cu	Zn	Fe	Mn
P1. Chopodi – Typic Calciustepts												
	Ap	0–12	10YR3/2	cl	8.22	0.27	0.57	26.23	3.09	0.38	6.51	5.55
	Bw	12–31	10YR3/2	cl	8.29	0.23	0.57	30.03	3.61	0.01	1.54	6.13
	Bk1	31–56	10YR3/3	scl	8.42	0.15	0.22	22.84	0.78	0.03	1.63	9.95
	Bk2	56–81	10YR3/4	scl	8.35	0.20	0.21	25.38	1.34	0.17	1.33	10.41
	Bk3	81–110	10YR4/3	scl	8.23-	0.34	0.19	24.67	0.98	0.12	2.34	5.34
	Bk4	110–150	10YR4/3	scl	8.15	0.45	0.15	23.45	0.78	0.21	2.56	4.67
P2. Chikhali – Leptic Calciusterts												
	Ap	0–12	10YR3/2	c	8.2	0.14	0.66	41.73	4.0	1.52	6.4	8.1
	Bw1	12–34	10YR3/2	c	8.1	0.11	0.53	44.30	3.5	0.81	6.5	9.1
	Bss1	34–62	10YR2/2	c	8.2	0.14	0.46	46.30	3.6	0.79	6.3	7.5
	Bss2	62–95	10YR6/4	cl	8.1	0.11	0.36	37.80	3.1	0.76	6.6	7.6
	Bk	95–120	10YR5/6	c	8.2	0.14	0.26	32.60	4.0	0.52	5.3	4.5
P3. Malegaon – Lithic Calciustepts												
	Ap	0–20	10YR3/3	c	8.32	0.04	0.48	23.4	2.57	0.45	6.53	2.75
	Bk	20–40	10YR5/4	c	8.42	0.07	0.23	25.6	1.65	0.35	5.32	3.84
P4. Nagdhari – Fluventic Haplustepts												
	A	0–18	10YR 6/4	sicl	8.2	0.24	1.7	46.9	0.7	0.59	1.2	3.3
	Bw1	18–43	10YR4/4	sicl	8.3	0.22	1.2	48.7	1.4	0.52	1.0	1.0
	Bw2	43–64	10YR4/4	sicl	8.3	0.21	1.1	44.3	1.3	0.39	1.4	0.3
	Bw3	64–85	10YR3/4	sicl	8.3	0.24	0.6	42.6	1.3	0.58	1.1	1.0
	Bw4	85–110	10YR3/4	sicl	8.4	0.21	0.5	47.8	1.5	0.34	1.0	0.9
P5. Waghari – Typic Haplustepts												
	Ap	0–10	10YR3/1	c	8.23	0.178	0.92	23.45	3.84	0.20	6.79	11.98

(continued)

Table 14.3 (continued)

Soil series/ taxonomy	Hori-zon	Depth (cm)	Matrix colour (moist)	Tex-ture	pH	EC (dSm1)	OC (%)	CEC (cmol/kg)	DTPA extractable (mg/kg)			
									Cu	Zn	Fe	Mn
	Bw1	10–27	10YR3/1	c	8.26	0.003	0.81	20.53	3.64	0.04	2.51	9.44
	Bw2	27–42	10YR3/1	c	8.3	0.003	0.8	23.14	3.83	0.05	2.45	9.57
P6. Arunavati – Typic Calciustepts												
	Ap	0–21	10YR3/3	c	8.26	0.23	0.63	29.19	3.15	0.40	6.91	3.75
	Bw	21–38	10YR3/3	c	8.36	0.04	0.19	29.61	2.67	0.01	3.59	4.45
	Bk1	38–59	10YR3/4	gcl	8.45	0.23	0.34	27.07	1.96	0.11	2.45	3.96
	Bk2	59–92	10YR4/4	gscl	8.40	0.02	0.37	25.8	1.71	0.09	4.16	5.84
	BC	92–130	10YR4/4	gls	8.66	0.02	0.66	27.07	0.63	0.14	0.37	2.80
P7. Lakhi – Lithic Ustorthents												
	Ap	0–20	10YR3/2	gcl	8.1	0.002	0.91	28.34	4.35	0.29	9.45	11.43

(10YR 3/4 to 4/4) in calcium carbonate-enriched Bk horizons. The Chikhali series (P2) have very dark grey Ap-BW horizons but changed to very dark brown (10YR2/2) to light yellowish brown slickensided horizons (10YR6/4) over yellowish brown Bk horizons. The Malegaon series (P3) has dark brown (10YR3/3) Ap horizon and yellowish brown (10YR5/4) Bw horizon with lithic contact with in 50 cm. In Nagadhari series (P4), the matrix colour changes from light yellowish brown (10YR6/4) in Ap horizons to dark yellowish brown (10YR3/4). The Waghari series (P5) has uniformly very dark grey matrix (10YR3/1), whereas, Arunavathi series (P6) has dark brown to dark yellowish brown (10YR4/4) Bk horizons and very dark greyish brown Ap horizons in Lakhi series on plateau tops (P7). The dark colour (low chroma) could be related to the strong impregnation of profile by organic matter during pedogenesis or to prolong waterlogging (Ahmad and Marmut 1996; Bhaskar et al. 2014). These soils have clay loam to sandy clay loam with depth in Chopadi (P1), uniformly clay with depth in P3 and P5, clay to clay loam in P2, uniformly silty clay loam with depth in P4, clay to gravelly clay loam in P6 and gravelly clay loam in P7. The depth functions of two dominant soil series: Arunavati (P6) and Nagdhari (P4) is presented in Fig. 14.7. The particle size distribution shows that silt is dominant fraction in Nagdhari series on pediplains (P4) exceeding more than 50% with depth whereas gradual decrease in clay (35.2–27.5%) and irregular trends of sand (10.2–18.9%). The Arunavati series on pediments (P6) shows irregular distribution of clay (more than 35% throughout depth), gradual increase of sand from 27.6 to 42.6% and uniform depth distribution of silt in B horizons. All the soils under study are moderately alkaline in reaction with an organic carbon of less than 1% and cation exchange capacity (CEC) of 20.53–48.7 cmol/kg. The CEC is rated as high to very high (Ilaco 1985) with strong positive relation with silt and organic carbon as expressed in regression equation with F value of 6.89.

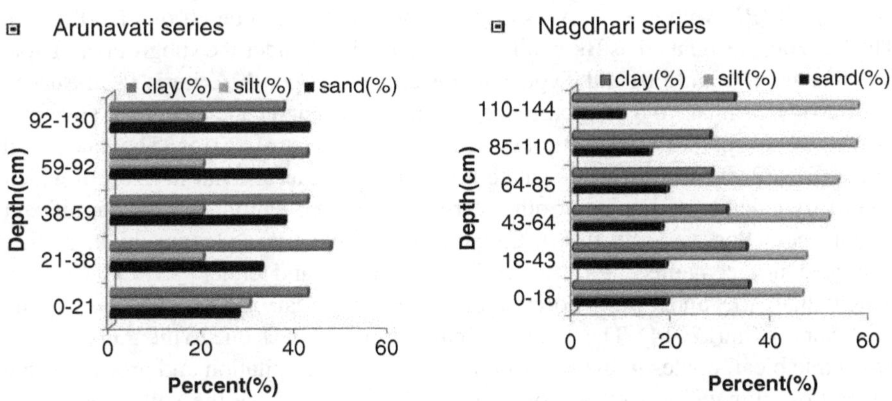

Fig. 14.7 Depth functions of particle size distribution

$$\mathrm{CEC(cmol/kg)} = 21.67 + 4.64(\text{organic carbon}, \%) + 0.32(\text{silt}, \%) R^2 = 0.37^{*}*, F$$

The DTPA-extractable micronutrient status shows gradual decreasing trends of Cu and Zn except in Nagdhari series, whereas Fe and Mn show irregular trends. Considering the critical limits for micronutrients, all soils are deficient in Zn (<0.6 mg/kg) and for Fe in subsoils (<4.5 mg/kg, Malewar 2005). The Cu and Mn status is sufficient throughout depth. The correlation studies show that there is strong significant (at 1% level) negative relationship of DTPA-extractable Fe with pH and clay with coefficient of determination ($R^2 = 0.56^{**}$) and F value of 15.08. This relationship is expressed in a regression equation as given under:

$$\text{DTPA extractable Fe (mg/kg)} = 104.47 - 12.73\,(\text{pH}) + 0.13\,(\text{Clay}, \%)$$

The DTPA-extractable Mn shows a significant negative relation with silt content at 1% level and expressed in regression equation as:

$$\text{DTPA extractable Mn (mg/kg)} = 9.26 - 0.14\,(\text{silt}\%,) R^2 = 0.37^{*}*, F = 14.76$$

14.6.7 Depth Function of Calcium Carbonate and CEC/Clay Ratio

The amount and per cent distribution of $CaCO_3$ content in three major soil series are depicted in Fig. 14.8. The depth trend of $CaCO_3$ is gradual increase and reached up to 41.06% in Arunavati series (P6) and 23.4% in Chikhali (P2) but irregular in case of Chopadi (P1) with 6.37% in Bk1 horizon to 14.76% at a depth of 110–150 cm. The horizons designated as Bk in all soils but classified under the subgroup as Typic Calciustepts (P1 and P6) and Typic Calciustert (P2) as per the criteria (Soil Survey Staff 2014). Similar kind of occurrence of Calciustepts and Calciusterts in Yavatmal district were reported in Basaltic landscapes (Bhaskar et al. 2015). The calcic soil series (P1, P2, P3 and P6) have calcic horizon defined an illuvial horizon in which secondary calcium carbonate or other carbonates have accumulated to a significant extent' (Soil Survey Staff 1996). Calcic horizons must be $\geq$15 cm thick, neither indurated nor cemented, $\geq$15% $CaCO_3$ by weight, and either $\geq$5% $CaCO_3$ by weight than the underlying horizon or $\geq$ 5% secondary carbonates by volume (Soil Survey Staff 2014). The genesis of calcic horizons was due to the translocation of soluble bicarbonates in the soil solution and their precipitation and processes that depend on climate, as well as soil relief and stability where the soils are formed (Bachman and Machette 1977). Secondary carbonate concretions and nodules are commonly seen in shrink-swell soils of Central India usually in landscape positions that favour the buildup of Ca and Mg ions in the soil solution (Hendricks 1991). The very rich alkaline and alkali carbonate contents in the profiles are due to the low

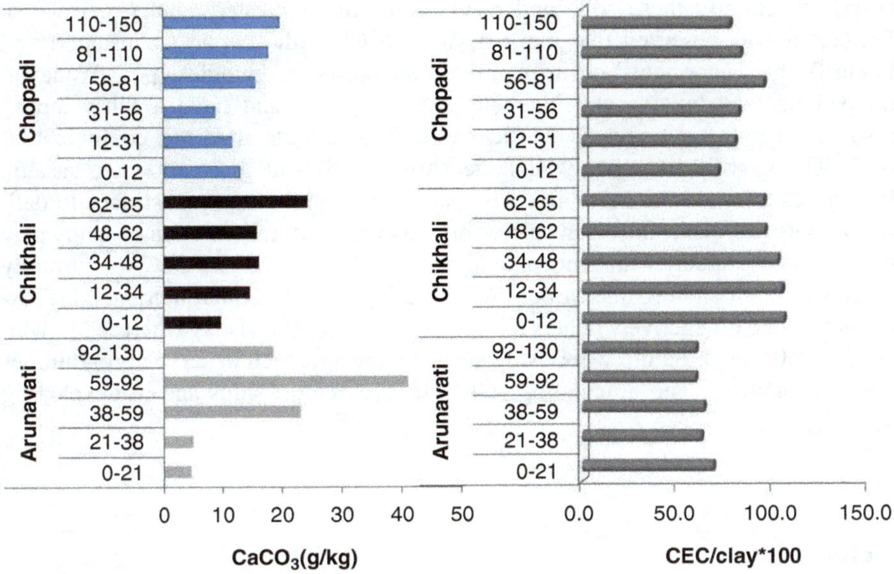

Fig. 14.8 Depth function of calcium carbonate (g/kg) and CEC/clay*100

topographic positions and strongly contrasted climate with mean rainfall of 939 mm and 51 rainy days and moderate lixiviation (process of separating a soluble substance from one that is insoluble, by washing with some solvent, as water, leaching) (Fig. 14.8). The use of calcic horizons as the parameter for calculating the depth to carbonate horizon is reasonable (Birkeland 1984). Royer (1999) reported considering 1168 soil profiels from US Natural Resources Conservation Service, the depth to carbonate horizon was correlated with mean annual precipitation of <760 mm ($p < 0.05$, Royer 1999). In the study area, the calcic horizons are formed at shallow depth due to high intensity of seasonal rains with short duration and prolonged dry spells that favour the accumulation of carbonate in the profiles (Bhaskar et al. 2017). The CEC to clay ratio (Fig. 14.8) shows increasing depth trends with values of 0.59–1.1 indicating higher proportion of smectite with possibility of feldspars (Shaw et al. 1998; Bhaskar et al. 2015). Nagdhari is classified at subgroup level as Fluventic Haplustepts and Lakhi as Lithic Ustorthents.

14.7 Conclusions

The pedogemorphic analysis and land resource inventory of Darwha tehsil, Malegaon plateau, Yavatmal district, Maharashtra, on 1:10,000 scale show that there was strong association of depth of occurrence of shrink-swell soils in relation to elevation ($R^2 = 0.44$**) and the thickness of variations in $CaCO_3$ accumulation as criteria to classify these soils under subgroups of Calciusterts and Calciustepts. Further, it was observed that the well-drained Chopadi (Leptic Caliustepts) and Lakhi (Lithic Ustorthents) occur mainly on the crests and middleslopes, Arunavati on pediments/pediplains and Nagdhari on pediplains and have a fairly strong association with contingency coefficient of 0.49 and standardized coefficient of 0.55. The slightly to strongly alkaline shrink-swell soil association on basaltic landscapes have similar morphogenetic pathways with dark grey at surface to dark brown/dark yellowish brown in calcic horizons with textural variation of sandy clay to silty clay loam/clay with depth and high to very high CEC. The CEC was strongly related with silt and organic carbon ($R^2 = 0.37^{**}$). Among micronutrients, DTPA-extractable Fe is negatively related ($R^2 = 0.56^{**}$) with pH and clay whereas Mn with silt ($R^2 = 0.37^{**}$). Future work using soil window approach in survey programmes helps in capturing and articulating soil-landscape relationships and in developing soil mapping.

References

Ahmad N, Marmut A (1996) Vertisols and technologies for their management. Elsevier, Amsterdam

Ahuja RL, Manchandra ML, Sangwan BS, Goyal VP, Agarwal RP (1992) Utilization of remote sensing data for soil resource mapping and its interpretation of land use planning of Bhiwani district Haryana. J Indian Soc Remote Sens 20(2&3):105–120

Alavi SA, Merh SS (1991) Morphotectonic analysis of South Gujarat landscape. Proc Indian Natl Sci Acad 57A(6):683–698

Bachman GO, Machette MN (1977) Calcic soils and calcretes in the southwestern United States. United States of America, U.S. Geological Survey (USGS), pp 77–797

Bell JC, Cunningham RL, Havens MW (1994) Soil drainage class probability mapping using a soil-landscape model. Soil Sci Soc Am J 58:464–470

Benediktsson JA, Chanussot J, Moon WM (2012) Very high-resolution remote sensing: challenges and opportunities. Proc IEEE 100(6):1907–1910

Benz UC, Hofmann P, Willhauck G, Lingenfelder I, Heynen M (2004) Muliti resolution, object oriented fuzzy analysis of remote sensing data for GIS ready information. ISPRS J Photogramm Remote Sens 58:239–258

Bhaskar BP (2015) Landscape planning for agridevelopment at regional scale: an example from cotton growing Yavatmal district, Maharashtra, India. J Agric Environ Int Dev 109(2):235–269

Bhaskar BP, Gajbhiye KS (1997) Soils of Jayakwadi irrigation project, minor-4, report no. 535, pp 73

Bhaskar BP, Anantwar SG, Challa O, Bharambhe PR, Velayutham M (2001) Catenary variations in properties of shrink-swell soils in minor 4 of Jayakwadi irrigation project in Parbhani District. J Geol Soc India 57(5):429–434

Bhaskar BP, Raja P, Gajbhiye KS, Maji AK, Singh SR, Anantwar SG, Nimkar AM (2010) Topographic appraisal for irrigation suitability in a part of Jayakwadi command area, Parbhani District, Maharashtra. J Indian Soc Soil Sci 58(4):363–370

Bhaskar BP, Sarkar D, Mandal C, Bobade SV, Gaikwad MS, Gaikwad SS et al (2014) Reconnaissance soil survey of Yavatmal district, Maharashtra, India. NBSS Publication. No.1059, NBSS&LUP, Nagpur, pp 208

Bhaskar BP, Satyavati PLA, Mandal C, Singh SK (2015) Analysis of Aridityin relation to climate and sustainability of cotton production in India. Climate change and sustainable technologies. Environ Sci Engg 12:84–114

Bhaskar BP, Chaturvedi A, Bhattacharyya T, Gaikwad SS (2017) Land resource inventory and evaluation for agricultural land use planning in semiarid ecosystem of Western India. In: Reddy GPO, Patil NG, Chaturvedi A (eds) Sustainbale management of land resources- an Indian perspective. Apple Academic Press, Pleasant, pp 47–76

Bhattacharjee JC, Landey RJ, Kalabande AR (1977) A new approach in the study of Vertisols morphology. J Soil Indian Soc Sci 25:221–232

Bhattacharyya T, Pal DK, Deshpande SB (1993) Genesis and transformation of minerals for in the formation of red (Alfisols) and black (Inceptisols and Vertisols) soils on Deccan Basalt in the Western Ghats, India. J Soil Sci 44:159–171

Birkeland PW (1984) Soils and geomorphology. Oxford University Press, New York

Birkeland PW (1990) Soil geomorphic research-a selective overview. In: Kneuenfer PLK, Mac Fadden LD (ed) "Soils and landscape evaluation", Geomorphology 3:207–224

Bridges EM (1982) Techniques of modern soil survey. In: Bridges EM, Davidson DA (eds) Principles and applications of soilgeography. Longman, London

Chapman HD (1965) Cation exchange capacity. In: Black CA (ed) Methods of soil analysis, part 2, Number 9 in the series agronomy. American Institute of Agronomy, Madison, pp 891–901

Chattaraj S, Srivastava R, Barthwal AK, Giri JD, Mohekar DS, Reddy GPO, Daripa A, Chatterji S, Singh SK (2017) Semi-automated object-based landform classification modelling in a part of the Deccan Plateau of central India. Int J Remote Sens 38(17):4855–4867

Daniels RB, Hammer RD (1992) Soil geomorphology. Wiley, New York

Dent D, Young A (1981) Soil survey and land use planning. George Allen and Unwin Ltd, London

Dobos E, Hengle T (2009) Soil mapping applications. Dev Soil Sci 33:461–479

Effland ABW, Effland WR (1992) Soilgeomorphology studies in the US soil surveyprogram. Agric Hist 66:189–212

Ehlers M, Janowsky R, Gaehler M (2002) New remote sensing concepts for environmental monitoring. Proceedings of SPIE 4545, 1–12 (Bellingham)

Euillades LD, Grosse P, Euillades PA (2013) NETVOLC: an algorithm for automatic delimitation of volcano edifice boundaries using DEMs. Comput Geosci 56:151–160

FAO (1967) The state of food and agriculture. FAO, Rome, pp 1–200

Florinsky IV (1998) Accuracy of land topographical variables derived from digital elevation models. Int J Geogr Inf Sci 12:47–61

Fraser S, Jones H, Hill R, Hainsworth S, Tikkisetty B (2014) Soil windows-unravelling soils in the landscape at the farm scale. In: Currie LD, Christensen CL (eds) Nutrient management for the farm, catchment and community. http://flrc.massey.ac.nz/publications.html. Occasional report 27. Fertilizer & Lime Research Centre, Massey University, Palmerston North, New Zealand

Furley PA (1971) Relationships between slope form and soilproperties developed over chalk parent materials. In: Brunsden D (ed) Slopes: forms and processes, Institute of British geographers special publication no.3. Institute of British Geographers, Kensington Gore, pp 141–163

Gawande SP, Biswas TD (1977) Characterisation and classification of black soils developed on basaltin relation to microrelief. J Indian Soc Soil Sci 25(3):233–238

Gessler EE, Morre ID, McKenzie NJ, Ryan EJ (1995) Soil-landscape modelling and spatial pridiction of soil attributes. Int J Geogr Inf Syst 9:421–432

Gobin A, Campling P, Deckers J, Feyen J (2000) Integrated toposequence analysis to combine local and scientific knowledge systems. Geoderma 97:103–123

Hammond EH (1954) Small scale continental landform maps. Ann Assoc Am Geogr 32–42(2):44

Hammond EH (1964) Analysis of properties in landform geography: an application to broad scale landform mapping. Ann Assoc Am Geogr 54:11–19

Hendricks DM (1991) Genesis and classification of arid region soils. In: Skujins J (ed) Semiarid lands and deserts: soil resource and reclamation. Marcel Dekker, New York, pp 33–80

Hudson BD (1992) The soil survey as paradigm-based science. Soil Sci Soc Am J 56:836–841

Ilaco BV (1985) Agricultural compendium: for rural development in the tropics and subtropics. Elsevier, Amsterdam, p 738
Jackson ML (1973) Soil chemical analysis. Prentice Hall of India Pvt. Ltd, New Delhi, p 498
Jenny H (1941) Factors of soil formation. McGraw-Hill, New York
Jenny H (1980) The soil resource. Origin and behaviour, Ecological studies 37. Springer, New York
Krishnamurthy YVN, Adiga S (2008) Remote sensing and GIS for land use planning. http://www.ncap.res.in/upload_files/workshop/ws5_chapter10.pdf
Laker MC, Dupreez CC (1982) An investigation into the accuracy of hydrometers for soil particle size determination. Agroplantae 14:17–22
Lindsay WL, Norvell WA (1978) Development of a DTPA soil test for zinc, iron, manganese, and copper. Soil Sci Soc Am J 42:421–428
Malewar GU (2005) Micronutrient stresses in soils and crops:serious sickness and clinical approaches for sustainable agriculture. J Indian Soc Soil Sci 53(4):484–499
McBratney AB, Mendonça Santos ML, Minasny B (2003) On digital soil mapping. Geoderma 117(1–2):3–52
McGregor DFM, Thompson DA (eds) (1995) Geomorphology and land management in a changing environment. Wiley and Sons, Chichester
McKenzie NJ, Ryan PJ (1999) Spatial prediction of soil properties using environmental correlation. Geoderma 89:67–94
Michel P (1973) Les basins des fleuves Senegal et Ques. Mem.de l'. O.R.S.T.O.M. 63. (Paris), pp 1–752
Milne G (1935) Some suggested units of classification and mapping for East African soils. Soil Res 4:183–198
Munnik MC, Verster E, Rooyen TC (1984) Pedogeomorphic aspects of the Roodepoort area, Transvaal: soil depth-slope relationships. South Afr J Plant Soil 1:61–66
Munnik MC, Verster E, van Rooyen TH (1986) Soil-site relationships in the Roodepoort area, Transvaal. South Afr J Plant Soil 3:35–40
Nagaraju MSS, Kumar N, Srivastava R, Das SN (2014) Cadastral level soil mapping in basaltic terrain using Cartosat-1 derived products. Int J Remote Sens 35(10):3764–3781
Pal DK, Deshpande SB (1987) Characteristics and genesis of minerals in some benchmark Vertisols of India. Pédologie 37:259–275
Pal DK, Dasog GS, Bhattacharyya T (2009) Pedogenetic processes in cracking clay soils (Vertisols) in tropical environments of India: a critique. Indian Soc Soil Sci 57:422–432
Pedersen GBM, Grosse P (2014) Morphometry of subaerial shield volcanoes and glaciovolcanoes from Reykjanes Peninsula, Iceland: effects of eruption environment. J Volcanol Geotherm Res 282:115–133
Pike RJ (2000) Geomorphometry — diversity in quantitative surface analysis. Prog Phys Geogr 24:1–20
Pillai M, Pal DK, Deshpande SB (1996) Distribution of clay minerals and their genesis in ferruginous and black soils occurring in close proximity on Deccan basalt plateau of Nagpur district, Maharashtra. J Indian Soc Soil Sci 44:500–507
Pofali RM, Bhattacharya JC, Landey RJ (1979) Landscape anlysis for soil mapping in basaltic terrain. Geograph Rev India 41(3):267–276
Pouquet J (1966) Initiation géopédologique. Les sols et la géographie. SEDES, Paris
Prasad CR, Thayalan S, Reddy RS, Reddy PSA (1990) Use of Landsat imagery for mapping soil and land resources for devolopment planning in parts of north Kamataka of India. Int J Remote Sens 11(10):1889–1900
Prasad J, Sarkar D, Gupta RK, Tiwary P, Bhaskar BP, Chandran P, Ray SK, Bhattacharyya T, Gaikwad SS, Nimkar AM, Anantwar SG (2012) Resource Inventory of Khamaria seeds farm in Jabalpur, Madhya Pradesh for Borlaug Institute of South Asia (BISA). NBSS&LUP Publications, Nagpur, p 90

Raja P, Bhaskar BP, Deepak Malpe, Anantwar SG, Tapaswi PM (2010) Geochemistry of salt affected soils in Purna valley, Maharashtra, India: their paleoclimatic implications. J App Goechem 12(3):491–501

Rao DP (2002) Remote sensing application in geomorphology. Trop Ecol 43(1):49–51

Rao BRM, Fyzee MA, Wadodkar MR (2004) Utilityof remote sensing data for mapping soils at various scales and levels. In: Venkatratnam L, Ravisankar T, Sudarshana R (eds) Soils and crops. NRSA Publications, Hyderabad

Ravisankar T, Thamappa SS (2004) Satellite data interpretation and analysis for soil mapping. In: Venkatratnam L, Ravisankar T, Sudarshana R (eds) Soils and crops. NRSA Publication, Hyderabad

Reddy GPO, Shekinah DE, Maurya UK, Thayalan S, Prasad J, Ray SK, Bhaskar BP (1999) Landscape-soil relationship in part of Bazargaon plateau, Maharashtra. Geograph Rev India 61(3):280–291

Reddy MGR, Reddy GPO, Maji AK, Nageshwara Rao K (2003) Landscape analysis for pedo-geomorphological characterization in part of basaltic terrain, Central India using remote sensing and GIS. J Indian Soc Remote Sens 31(4):271–282

Royer DL (1999) Depth to pedogenic carbonate horizon as a paleoprecipitation indicator? Geology 27(12):1123–1126

Ruhe RV (1975) Geomorphology. Geomorphic processes and surficial geology. Houghton Mifflin, Boston

Schoeneberger PJ, Wysocki DA (2001) A geomorphic description system (version 3.0). National Soil Survey Center. US Departmentof Agriculture, Lincoln

Schoeneberger PJ, Wysocki DA, Benham EC, Soil Survey Staff (2012) Field book for describing and sampling soils. Version 3.0. Natural Resources Conservation Service, National Soil Survey Center, Lincoln

Sehgal JL (1992) Soil series criteria and norms. Technical Bulletin. NBSS Publications 36. NBSS&LUP. ICAR, Nagpur

Shabbir AS, Samira ASO, Muayyad EJ, Shihab A, Hanna AR (2004) Soil survey for farm level planning in northern Kuwait. Kuwait J Sci Eng 31(1):43–57

Shaw RJ, Coughlan KJ, Bell LC (1998) Root zone sodicity. In: Sumner NE, Naidu R (eds) Sodic soils: distribution, properties, management, and environmental consequences. Oxford University Press, New York, pp 95–106

Sheth HC (1999) A historical approach to continental flood basalt volcanism: insights into pre-volcanic rifting, sedimentation, and early alkaline magmatism. Earth Planet Sci Lett 168:19–12

Soil Survey Staff (1996) Soil survey laboratory methods manual. Ver. 3.0. USDA/NRCS, soil survey investigations report no. 42. U.S. Government Printing Office, Washington, DC

Soil Survey Staff (2014) Keys to soil taxonomy, 12 th edn. USDA-NRCS, Washington, DC

Subramanyam V (1974) A quantitative analysis of two drainage basins around Sagar, Madhya Pradesh. Indian Nat Sci Acad 40(1):76–99

Thornbury WD (2004) Principles of geomorphology, 2nd edn. Wiley and Sons, London, p 631

Tricart J (1972) Landforms of the humid tropics, forests and savannas, 1st edn. Longman, London

Van Den Eeckhaut M, Kerle N, Poesen J, Hervás J (2012) Object-oriented identification of forested landslides with derivatives of single pulse LiDAR data. Geomorphology 173–174(0):30–42

Veneman PLM, Bodine SM (1982) Chemical and morphological soil characteristics in a New Engiand drainagetoposequence. Soil Sci Soc Am J 46:359–363

Venkataratnam L (1984) Monitoring and managing soil and land resources in India using remotely sensed data. Proceedings on 3rd Asian agriculture symposium, Chiangmai, Thailand

Verster E (1987) Soils derived from granite in two Mt. Garnet toposequences, North Queensland, Australia. South Afr J Plant Soil 4:35–42

Vink APA (1968) Role of physical geography in integrated surveys in developing countries. J Econ Soc Geogr 59(6):294–312

Wadodkar M, Ravishankar T (2011) Soil resource data base at village level for developmental planning. J Indian Soc Remote Sens 30(4):529–536
Walkley A, Black IA (1934) An examination of Degtjareff method for determining soil organic matter and a proposed modification of the chromic acid titration method. Soil Sci 37:29–37
Walsh S, Butler D, Malanson G (1998) An overview of scale, pattern, process relationships in geomorphology: a remote sensing and GIS perspective. Geomorphology 21:183–205
Way DS (1973) Terrain analysis. A guide to site selection using aerial photographic interpretation. Dowden, Hutchinson & Ross, Stroudsburg
Yaalon DH (1975) Conceptual models in pedogenesis: can soil-forming functions be solved? Geoderma 14:189–205
Zinck JA (2013) Geopedology-elements of geomorphology for soil and geohazard studies. ITC, Enschede

Chapter 15
Land Resource Inventory and Mapping: Tools and Techniques

S. K. Gangopadhyay, G. P. Obi Reddy, D. C. Nayak, and S. K. Singh

Abstract Nature, extent, and reliability of spatial data on land resources become the imperative need of the present day for scientific utilization of land resources and sustainable management. In land resource inventory and mapping, the applications of remotely sensed data and geographic information system (GIS) are found to be not only timesaving but also economic in generation of base maps and conducting of soil survey. Further, high-resolution remote sensing and applications of GIS have made the task of land resource inventory and mapping cost effective and time efficient. Depending upon the objective, method and intensity of land resource surveys and scale, the type of satellite data to be selected to generate the base maps in order to show the details of survey information. A detailed soil survey provides sufficient information about various kinds of soils, including problematic or degraded soils, and is immensely useful for resource appraisal and development of alternative strategies for land use and site-specific agricultural development. As a case study, a detailed land resource survey was conducted in Piprakothi Block under Purba Champaran district of Bihar, and it shows that soils are very deep, well to somewhat poorly drained, and highly calcareous. Soils suffer from waterlogging, frequent flooding, drainage congestion, calcareousness, salinity, and multi-nutrient deficiency, affecting the crop yield. Soil-site suitability evaluation of sugarcane in the soils of the block indicates that 31.7% of the area is moderately suitable for the cultivation of sugarcane due to moderate limitation of soil fertility and 53.0% of the area is marginally suitable due to relatively low water availability and severe soil fertility limitations.

Keywords Detailed soil survey · Exploratory soil survey · Geographic information system · Land resource inventory · Reconnaissance survey · Remote sensing

---

S. K. Gangopadhyay (✉) · D. C. Nayak
ICAR-National Bureau of Soil Survey & Land Use Planning, Regional Centre, Kolkata, India

G. P. O. Reddy · S. K. Singh
ICAR-National Bureau of Soil Survey & Land Use Planning, Nagpur, India

G. P. O. Reddy, S. K. Singh (eds.), *Geospatial Technologies in Land Resources Mapping, Monitoring and Management*, Geotechnologies and the Environment 21,
https://doi.org/10.1007/978-3-319-78711-4_15

15.1 Introduction

Soil is one of the most valuable, life-supporting natural resources and an important component of a geosphere-biosphere system, which directly provides food, fuel, fiber, and fodder for a variety of human and livestock needs and indirectly contributes to regulations of water and nutrient supply to plants (FAO and ITPS 2015). The ever-burgeoning human and animal population coupled with the ever-shrinking per capita availability of land and water resources has resulted into a serious threat to the land resources in terms of quantity and quality resulting in poor agricultural productivity and ecôlogical sustainability. Therefore, basic and reliable data on soil resources have become an imperative need of the present day for scientific utilization of land resources for the posterity (Reddy et al. 2017). The optimal management of land resources with minimal adverse environmental impact is essential not only for sustainable development but also for human survival. A knowledge of soil in respect to their extent, distribution, characteristics, use potential, and limitation is, therefore, extremely important for optimizing land use. Soil resource inventory provides an insight into the potentialities and limitations of the mapped area for its optimum utilization.

The study and assessment of soil properties and their response to management is important in agriculture and forestry for decision-making in agricultural land use planning and various land development projects. Soil survey information is a prerequisite to map the soils into relatively homogeneous units (Deshpande and Sarkar 2009; Sarkar 2003). Soil survey depicts ideal combination of changes of soil properties occurring in the landscape. The usefulness of soil survey depends on two things: the accuracy with which soil properties are mapped and the relevance to the purpose in hand. The information collected from a soil survey helps in the development of land-use plans and evaluates the effect of land use on the environment (Deshpande and Sarkar 2009).

Soil survey is essentially a systematic study, description, classification, and mapping of soils of an area in their natural environment. Soil survey comprises a group of interlinked operations involving fieldwork to study the important characteristics of soils and the associated external land features, such as landform, natural vegetations, slope, etc., and laboratory analysis to support and supplement the field observations, correlations, and classification of soils into delineated taxonomic units (Deshpande and Sarkar 2009). Mapping of soils is establishing and drawing of soil boundaries of different kinds of soils on a standard base map. Soil survey interpretations consist of making predictions about the potential of soils for alternative uses, like agricultural crops, grasses, fruit, and forestry or plantation crops, and ascertaining their management requirements for sustained production (Sehgal 1996). Transfer of agro-technology from research stations to farmer's fields based on precise soil information will be of more effective and sustainable. Various authors have reported the applications of remote sensing and GIS in soil resource inventory (Dobos et al. 2000; Dwivedi 2001; Srivastava and Saxena 2004; Reddy et al. 2013a, b; Gangopadhyay et al. 2014, 2015), land capability classification (Rosser et al.

1974; Panhalkar 2011), soil suitability assessment (Velmurugan and Carlos 2009; Boonyanuphap et al. 2004; Gangopadhyay et al. 2010), land productivity assessment (Patil et al. 2010), quantification of soil loss (Lal et al. 1998; Karale et al. 1988; Olson et al. 1998; Singh and Dwivedi 1989; Reddy et al. 2004, 2016), and assessment of biophysical resources (Reddy et al. 2013a, b).

15.2 Purpose and Objectives of Soil Survey

Modern soil surveys basically aimed at providing comprehensive information about soils and an inventory of the soil resources of the area. They consist of studying and recording important characteristics of soils in field and in laboratory, classifying them into well-defined units, and demarcating their extent on a map. The broad general objectives of soil survey are both fundamental and applied in nature. Fundamentally, soil surveys help in expanding our knowledge and understanding of different soils, as regards to their genesis, development, classification, and nomenclature. The applied part of soil surveys includes interpretation of soil data for use in agriculture, pasture development, forestry, engineering, urban development, recreation, and others. It gives information needed for development of land-use plans under irrigated and rainfed conditions. Further, it helps in correlating with characteristics of soils of known behavior and predicting their adaptability to various uses and also their behavior and productivity under defined sets of management practices. It thus forms the very basis for land-use planning as per the potential and limitations of soils. In addition, soil map helps in determining the degraded soils (Karale et al. 1988) such as saline, alkaline, waterlogged, water and wind eroded, and so-called wastelands and suggesting soil and water conservation measures to ameliorate these degraded soils. Furthermore, soil surveys are useful for land settlement, rehabilitation, tax appraisal, and locating and designing highways, airports, and other engineering structures and in public sanitation works.

15.3 Development of Soil Survey Program in India

Soil surveys were conducted earlier for developing irrigated agriculture for promoting sugarcane cultivation in the then Madras Presidency and also for understanding the causes of soil salinity in Bombay Deccan area. At the same time, surveys were undertaken in Uttar Pradesh, and a linkage between genetic soil types and specific management practices for the major crops was suggested (GOI 1984). Based on the report of the Royal Commission on Agriculture in India (1928), it was reported that the large-scale soil surveys in India may be taken up for specific purposes in specific areas. A wider range of investigations including soil characteristics like drainage, soil reaction, exchangeable bases and location, or concentration of clay and salts in the profile were considered. These studies mark the recognition of the importance of profile characteristics, physical and chemical properties of the soil, as against the purely chemical studies of plant nutrients carried out earlier. Almost with similar objective but in a more comprehensive manner, soil surveys were carried out during 1953–1955 under Technical Cooperation Mission (TCM) project on "Determination of Soil Fertility and Fertilizer Use." The Stewart Report in 1947 recommended soil surveys and fertilizer trials with a view to correlate fertilizer responses to different soil types. However, soil surveys carried out under this scheme were preliminary and broad types and yield only some general information.

Systematic studies on soils based on genetic factors have been initiated by Raychoudhury during the 1950s. Raychoudhury and Mathur (1954) divided India into 16 major soil regions with 108 minor basic soil regions by integrating the effects of climate, vegetation, and topography on soil formation. Further, Raychoudhury (1964) and Raychoudhury and Govindarajan (1971) have subdivided soils of India into 27 major groups. In the year 1954, the government of India invited Dr. F.F. Reickens, a soil specialist of the USA, to advise on soil survey work to be taken up in India. Following his recommendations, the All India Soil Survey Scheme was initiated in 1956 at Indian Agricultural Research Institute (IARI), New Delhi, with four regional centers, to carry out reconnaissance soil survey, correlate and classify soils, and prepare a small-scale soil map of India. In 1958, the scheme was integrated with the "Land Use Survey Scheme" of the central soil conservation board and redesignated as the All India Soil and Land Use Survey (AIS&LUS) presently renamed as Soil and Land Use Survey of India (SLUSI). The scope and activities of the enlarged scheme were expanded to cover detailed soil and land-use surveys in the catchments of major river valley projects (RVPS). In 1969, AIS&LUS was bifurcated, and the portion remaining with the Department of Agriculture was entrusted with responsibilities of soil surveys for soil conservation and other developmental activities. The research aspects of soil survey, classification, and correlation including countrywide reconnaissance surveys and training were taken over by the Indian Council of Agricultural Research (ICAR), which was known as All India Soil Survey Unit of ICAR. Based on the presidential notification in 1973, this unit was converted into a directorate and was given the status of national bureau and renamed as National Bureau of Soil Survey and Land Use Planning (NBSS&LUP) in the year 1976 with its headquarters at Nagpur. As per the recommendations made by the task force in 1972 on land and soil resources for soil correlation, uniform nomenclature, and proper soil mapping and also following the presidential notification on December 15, 1973, the duties with respect to research, training, correlation, mapping, and interpretation were entrusted with the NBSS&LUP.

15.4 Basic Understanding from Soil Surveys

Unlike in other types of surveys, soil survey is not aimed to identify soils per se or delineate areas having absolutely same soil, but to separate areas having almost similar soils requiring similar management needs (Sehgal 1992). This means that within a certain range of soil properties, the management needs would not vary and there is likely to be similar response by plants of same species. There are thus ranges in soil properties, which can be standardized for a particular area and scale of mapping. This concept of identifying soils in the field based on their range in characteristics has shifted the earlier focus of studying soils in the field from pedon to pedon. Thus, polypedon has become the unit of soil, but locating the boundary of the polypedons in the field is a challenge for soil scientists. There is a close relationship existing between the various landforms and types of soils present in an area. Since, the properties of a soil change gradually from one place to another and governed by various factors and processes of soil formation, the change observed at the surface if any one of these features (like changes in slope, erosion, wetness, land use, etc.) can be used to identify the boundary of the soil (polypedon) in the field.

The concept of polypedon as a unit of soil with their range in characteristics and linking their boundary with changes observed in the landforms has helped the soil surveyors in identifying the soil and their boundaries in the field easily. Further, this concept serves the purpose of the land user better, because this approach divides the landform into more homogenous units suitable for better management of the land resources based not only on soil but also by all the site characteristics.

15.5 Types of Soil Surveys

Depending upon the objective, method and intensity of surveys, scale and type of base map used, and resulting detail and precision of mapping, the following types of soil surveys are recognized.

15.5.1 Exploratory Soil Survey

Surveys at 1:1,000,000 and 1:2,50,000 scale can be employed for large areas as general inventory maps of soils as a natural resource. At these small scales, surveys rely heavily upon satellite or extensive air photo interpretation as most boundaries are drawn in the laboratory on the evidence of interpretative methods. Ground inspection of soils is limited to approximately one per km^2 and takes the form of checking areas defined upon air photographs to see if the boundaries are valid and that the soils belong to particular great soil groups or orders. In addition to the general inventory purpose, these maps may be used for large-scale comparative exercises and for identifying sites for future work and macro-level planning for various agro-based developmental programs.

15.5.2 Reconnaissance Survey

The reconnaissance survey is undertaken to prepare resource inventory of large areas. Their major purpose is to identify possible areas for further intensive soil survey works which might be done before locating a new irrigation scheme. Reconnaissance survey enables to access broad potentialities of soils and recognition of areas that are suitable for intensive agriculture and helps in recognizing areas which require amelioration due to their suffering from erosion, acidity, salinity, and sodicity, which affect soil fertility and capacity for crop production. In reconnaissance soil survey, only few or part of the soil boundaries is actually checked in the field, and the remaining are extrapolated from the available information.

In reconnaissance survey, toposheets and satellite images, either alone or in conjunction, are used as a base for mapping soils. The scale of mapping can vary from 1:50,000 to 1: 500,000, but normally 1:50,000 or 1:100,000 scale is preferred. First the satellite images (1:50,000 scale) are interpreted, either visually or digitally, in conjunction with the available collateral data from toposheets and geological and land-use/land cover maps to identify the various landforms occurring in the area. Then based on slope, elevation, and image characteristics, the landforms are delineated into physiographic/photomorphic units. The delineated boundaries are transferred to 1:50,000 scale Topobase and used as a base map for field survey and mapping. Many authors conducted reconnaissance survey at district level (1:50,000 scale) for optimum land-use planning (Gangopadhyay et al. 2013; Nayak et al. 2014; Verma et al. 2015). The type of soil survey and potential satellite sensor data for the use at different scales is shown in Table 15.1.

For the survey, a base map of 1:100,000 or (smaller) (1:2,50,000) is used. Aerial photo or interpretation of remotely sensed data is the basis for physiographic boundary location. The land type, land use, texture, and tone of images are used as a basis for boundary delineation. These are checked in the field by a few random checks to develop and confirm correlation between soils and land types. The soils are identified by traversing represented areas to determine soil patterns and composition of map units. The boundaries plotted by interpretation of remote sensing data are confirmed by ground truth checks. The fieldwork will involve studies of profiles at an interval of 2–3 km or shorter depending upon the soil heterogeneity. Auger

Table 15.1 Type of soil survey and potential satellite sensor data

Type of survey	Scale of survey	Mapping unit	Sensor data used
Reconnaissance	1:2,50,000	Subgroup association	Landsat MSS LISS-I, IRS LISS-I
Semi-detailed reconnaissance	1:50,000	Families/series association	Landsat TM-2,3, IRS-1A, 1D-LISS-III, IRS IC LISS-III, Spot
Detailed	1:25000	Phases of soil series	IRS IC, LISS-III and PAN data, IRS P6 LISS-IV

sampling for every 1/4 to 1 km is carried out to study the variation in profile characters.

After rapid traverse of all the physiographic/photomorphic units, representative areas are selected for transect study and soil profile examination. In the selected transect or sample area, soil profiles are studied at close intervals based on the difference observed in slope, erosion, moisture content, presence of gravels, land use, etc. The soil profiles from transect study are grouped into soil series and preliminary legend with association of soil series as mapping unit is determined by translating each image interpretation-cum-physiographic unit in terms of soils. The preliminary physiographic soil legend forms the base for further survey and mapping. Any further investigation done in the field is only for the confirmation of this preliminary legend.

To test the validity of the preliminary legend, minipits, auger borings, road cuts, well cuts or gully cuts, and soil profiles are examined in the entire survey area. In the course of the survey, if any variations are observed, then the preliminary legend and whatever necessary the delineation is modified accordingly. This way, the legend is progressively modified and refined as the survey progresses and soil map is finalized for the area with associations, complexes, and undifferentiated groups as soil mapping units. The draft physiography-cum-soil map thus prepared for the area can be subjected to test its validity. Soil samples from typifying pedons are collected for laboratory characterization as per the guidelines. The field map in different sheets can be scanned and digitized, and final soil map can be generated using the GIS software.

The reconnaissance soil survey is being carried out following a three-tier approach comprising image interpretation, field survey and soil analysis, and cartography and printing.

The different steps involved in this type of survey are:

- Collection of available soil survey information
- Image interpretation, field survey, and mapping
- Laboratory analysis of soil
- Cartography/GIS and printing
- Interpretation for land-use planning

15.5.3 Detailed Reconnaissance Survey

This kind of survey combines both the detailed and reconnaissance surveys and is undertaken for understanding distribution of basic soil classes of series/families and their phases. The selected areas that have development potentials are surveyed in detail and the others at reconnaissance level.

15.5.3.1 Image Interpretation

The soil survey is carried out on 1: 50,000 scale by using the Survey of India toposheet and satellite imagery (FCC) as base maps. The geocoded IRS-ID LISS-III false-color composite (1:50,000) was visually interpreted for landform analysis based on the color, tone, texture, and pattern of false-color composite (FCC) and field visit. The information thus generated is being transferred to the toposheet on the same 1:50,000 scale, and the Topobase was prepared for undertaking the soil survey following the procedure as outlined in the *Soil Survey Manual* (Soil Survey Staff 1995). Out of the area surveyed at reconnaissance level, the selected areas that have development potentials are surveyed in detail. For this purpose, a cadastral map on 1:12500/1:10,000/1:4000 scale is usually employed. Interpretation of IRS-ID LISS-III fused with PAN/IRS-P6 LISS-IV false-color composite and its superimposition on the cadastral map helps in the delineation of landforms, which are further confirmed by traversing and ground truth checking. The Topobase thus prepared is being used for detailed soil survey of the selected area.

15.5.3.2 Soil Sampling/Fieldwork

Field surveys include the study of soil profiles in all the landform units to represent the soil-physiography relationship. The morphological features of representative pedons were studied up to a depth of 150 cm or more/lithic contact (for shallow soils), and the soil samples were collected from different horizons for laboratory analysis. Soil-physiography relationship was established, and a random study of soil profiles at different landform units was also carried out to confirm the soil-landform relationship. During soil survey, field reviews were conducted to correlate and classify the soils up to the soil series level (Soil Survey Staff 1999). The surface soil samples (0–25 cm) were also collected at an interval of 100 meter for preparing the soil fertility maps of the study area.

For a detailed survey of the selected portion of the study area, soil profiles are exposed to a depth of 150 cm in each and every landform, and the morphological properties are being studied in all the horizons. The representative soil samples from each and every horizon were collected and analyzed in the laboratory as per the standard procedure. For soil fertility mapping, the surface soil samples (0–25 cm) were also collected at an interval of 10 m from the selected area and analyzed to determine various soil macro- and micronutrients.

15.5.3.3 Soil Analysis

The soil samples collected from different genetic horizons were air-dried and analyzed for the physical and chemical properties, viz., particle size class, pH, organic carbon, cation exchange capacity, exchangeable bases and exchangeable acidity (H^+ and Al^{3+}), available macro- and micronutrients, etc., following standard procedures (Black 1965). Of the two series, the dominant one occupies 60% or more of the area; the subdominant series covers 30% or more of the area of the mapping units. Soils have been mapped in relation to landform units. The soil map shows the soil association numbers (polygon nos.) such as 1, 2, 3.... Each polygon number specifies a map unit representing the association of two soil series (dominant and subdominant).

15.5.3.4 Generation of Thematic Maps and Analysis Through GIS

For preparing the various thematic maps, viz., district boundary, drainage and water bodies, road network, contours and heights, soil map, etc., the satellite data and SOI toposheets are used as inputs to Geo-Media Professional GIS (Ver. 3.0) system. The maps are subsequently transformed to vector layers. These covers are projected and transformed to real-world coordinates, edited, and assigned polygon IDs, and topopolygon is built for areas and polygons. Using the available profile-wise soil information (soil morphological and physicochemical properties), attribute/nonspatial data could be compiled at mapping unit level. This attribute data is then linked to the spatial data. These are subsequently made into individual covers of soils by adding items in the polygon attribute table (PAT) of soil coverage following map generalization procedures. These are subsequently made into individual covers using "dissolve operation" thus forming different derivative thematic maps. Thus, various thematic maps could be prepared in GIS based on dominant soils of the mapped unit.

15.5.4 Semi-detailed Survey

This survey comprises very detailed study of some selected strips cutting across many physiographic units for developing correlation between physiographic units and soils. Once the correlation is developed and is found to be valid by random checking, the rest of the area is checked at random, and soil boundaries based on physiographic units are delineated. The base maps commonly used for detailed soil surveys are cadastral maps of about 1:5000–1:8000 scale, topographical maps of 1:25000–1:2,50,000 scale, aerial photographs of 1:10,000–1:50,000 scale, or Indian remote sensing data on 1:25000–1:2,50,000 scale. Aerial photo interpretation may have its part to play in boundary location at these scales, but ground observation becomes increasingly important as the scale becomes larger. Thus maps compiled at 1:1,00,000 would have soil associations as their mapping units, but maps at scales larger than 1:50,000 would have soil series as the major mapping unit. A detailed soil survey provides sufficient information about various kinds of soils, including problematic or degraded soils – soil surveys published at scales between 1:1,00,000 and 1:50,000 scale are used for general resource appraisal purposes at regional level and to establish areas for detailed investigations. The semi-detailed surveys provide a basis for developing alternative strategies for land use, settlement, or agricultural development. The relationship between the scale of soil mapping and frequency of field observations is shown in Table 15.2. The scale, minimum area covered, and minimum number of observations to be studied at different scales are shown in Table 15.3.

15.5.5 Grid Soil Survey

Grid survey is generally employed where a survey of detailed or very detailed nature is required in difficult terrain. It is most suitable for scales of soil mapping greater than 1:10,000 and is used for all intensive soil surveys. If the area is a plain (like in a delta or alluvial plain or if the slope gradient is negligible), then locating of soil profiles in transects is not possible. In such situations, either freestyle survey or grid type of survey is recommended to locate the profiles. In a grid survey, soil profiles are located at a predetermined interval irrespective of variations encountered in the field. Though the number of profiles studied will be more in a grid type of survey, it is the easiest method to follow by the surveyor. However, one major disadvantage of the method is that it can be inflexible in areas where the soil pattern is simple and it can lead to wastage of time and resources. It can also result in inspection taking place too close to roads or the rivers. The grid survey is also adopted for mapping small areas, such as micro-watershed or an agricultural research station. With the use of global positioning system (GPS) and computers, grid survey at geo-referenced points (with latitude and longitude) has been found to be of great value in digitization of database and generating several thematic maps of practical value.

15.5.6 Free Soil Survey

At semi-detailed and reconnaissance scale of soil mapping, the most economical method of producing a soil map is by "free survey." As the surveyor walks across the field, he positions his soil observations according to his interpretation of the landscape and soil boundaries according to morphological breaks and changes of slope. The selection of observation points depends on the assumption that change in physiography (as observed by aerial photographs) and other surface features such as soil color, vegetation, and land use are indicative of differences in soil characteristics. The density of observations can be varied as the mapper concentrates on confirming the inferred boundaries and checking the uniformity of the soil within each boundary. On large scales, the inferred boundaries are often accepted as soil boundaries, and very little efforts are made to find soils within the physiographic

Table 15.2 Relationship of the scale of soil map and frequency of field observations

Kind of survey	Scale of base map	Area represented by 1 cm^2 on map (ha)	Distance between field observations	Frequency of observation	Mapping unit	Accuracy of soil boundaries
Reconnaissance	1:2,50,000	625	2.5 km	1 in 625 ha	Association of soil groups or soil families	Almost all boundaries are inferred
	1:1,00,000	100	1.0 km	1 in 100 ha		
Semi-detailed	1:50,000	25	500 m	1 in 25 ha	Soil series or association of soil series	Some boundaries are checked (sample areas); most inferred
Detailed low intensity	1:10,000	1	100 m	1 per ha	Phases of soil series	Almost all boundaries are checked through the course of traversing
High intensity	1:5000	0.25	50 m	4 per ha	Phases of soil series	All boundaries are checked through the course of traversing

Table 15.3 Scale, area of 1 cm^2 on the map, and minimum number of observations per km^2

Scale of map	Size of 1 cm^2 area	Minimum number of observations per km^2
1:2500	0.000625 km^2	8000
1:10,000	0.01 km^2 (1 ha)	400
1:25,000	0.0625 km^2	64
1:50,000	0.25 km^2	16
1:1,00,000	1 km^2	4
1:2,50,000	6.25 km^2	0.7
1:5,00,000	25 km^2	0.15
1:10,00,000	100 km^2	0.05

boundaries. On large scale, however, several new boundaries within physiographic boundaries are recognized depending upon the scale of mapping. Based on the differences in physiography, parent materials, drainage class, and profile development, surveyors group soils of an area into defined soil map units, the limits of which are rigidly defined as per criteria used. For general purpose, each soil map unit is built around a central concept covering a limited range in the values of several soil parameters. Sometimes, the soils are grouped at soil series level.

The great advantage of this method is that the surveyor is free to vary the intensity of his observation according to the intricacy of the soil pattern. This results in great accuracy when soil parameters are complex and does not waste time and energy when conditions are uniform. The use of free survey method necessitates a good base map or aerial photograph upon which the surveyor can work without any problem of location so that observations and boundaries are correctly placed. Experience has shown that free survey is best adopted for the production of maps between the scales of 1:1,00,000 and 1: 25,000.

15.5.7 *Detailed Soil Survey*

This type of survey is undertaken in priority areas such as pilot projects, agricultural stations/farms, micro-watershed, and areas for urban development. For this kind of soil survey, cadastral maps on 1:4000–1:8000 or aerial photos of 1:10,000 scale are used depending upon the intensity of survey and developmental need of the areas. The purpose of detailed soil survey is to delineate similar areas, which respond or expected to respond similarly to a given level of management. This is achieved by studying all the site (such as slope, erosion, drainage, salinity, occurrence of rock fragments) and soil characteristics (such as depth, texture, color, and soil reaction) in detail and delineating/grouping areas based on their similarity or differentiating characteristics into various management units.

15.5.7.1 Pre-field Activities

The scale and type of base map required for survey is very important because that will decide the amount of information that can be depicted on the map. The larger the scale used, the more will be the information available and vice versa. For detailed soil survey, cadastral maps showing field boundaries and survey numbers are the most suitable maps. High-resolution large-scale satellite images available from Indian remote sensing satellite (IRS)-1D or Resourcesat at 1:12,500 scale can be used as a base in conjunction with the cadastral maps for field survey. In addition to the base maps, 1:25,000 or 1: 50,000 scale toposheets and geological and landform maps available for the survey area can be used for traversing the area for preparing preliminary legend. At 1:25,000 scale, the smallest area, which can be conveniently shown true to scale, is 0.25 ha, and a soil boundary line on the map represents a zone of 5 m wide on the ground. When the scale is increased to 1:10,000 scale, the smallest area, which can be conveniently shown, is 0.625 ha, and at this scale a boundary line on the map represents a zone only 2 m wide on the ground.

Before starting fieldwork in the selected village/block, preliminary traverse of the surrounding areas is to be carried out with the help of the cadastral map, toposheets, or satellite images of the same scale. During the traverse, based on geology and landform position, the major landform divisions occurring in the block are to be delineated first. This is followed by the identification of various landforms occurring within each physiographic division based on shape, slope, and position in toposequence.

15.5.7.2 Field Activities

Soil boundaries are demarcated by actual traverse throughout the course. After careful delineation of the various landforms, updating the cadastral sheets with new permanent features, and familiarization of the area with survey numbers and field boundaries, intensive traversing of each landform (like ridges, uplands, lowlands/valleys, etc.) is undertaken to select representative areas for transect study. Transects are located across the slope and cover most of the variations or changes observed in a landform. In the selected transect, soil profiles are located at closely spaced intervals to take care of any change in the land features like break in slope, erosion, gravels, stones, etc. In the selected sites, soil profiles were to be opened up to 150 cm or hard substratum or rock, whichever comes earlier, and studied in detail for all their morphological properties. Detailed soil and site characteristics are to be recorded in the standard proforma for all the profile sites as per the guidelines given in the *Soil Survey Manual* (IARI 1971; Soil Survey Staff 1951, 1993). Apart from the transect study, if the distance between two transects are more, then few more soil profiles can be studied in between the transects, and in areas not covered by the transects, additional profiles at random can be studied to bring out all the possible variability of the survey area. Auguring can also be carried out to study soil properties whichever possible in the field.

The soil mapping can be done by intensive traversing covering all plots and survey numbers after grouping the soils into different soil series within each landform. During the traverse, starting from one transect area, boundary between two series can be established by checking for the differentiating characteristics of one series from the other through minipit, auger, road cut, well cut, or any other information available from the area. The exercise is continued and boundaries established for all the series identified in the area. During this exercise itself, within the delineated area of a soil series, based on variations in the surface texture, slope, erosion, presence of gravels, salinity, etc., phases of soil series can be separated, and their boundaries can be delineated on the cadastral map. The survey enables identification of soil units up to phases of soil series for planning development of individual parcels of land. Soils in detailed survey are examined in details and closer intervals so as to detect differences that can be significant in their use and management. A detailed soil survey furnishes detailed information required for proper assessment of soil properties, terrain features, erosional aspect, and related factors, which help to work out the details about use capability and management practices that would be needed for conservation and crop production.

15.5.7.3 Laboratory Analysis

Soil samples collected from different genetic horizons have to be analyzed for the physical and chemical properties, viz., particle size class, pH, organic carbon, cation exchange capacity, exchangeable bases and exchangeable acidity (H^+ and Al^{3+}), available macro- and micronutrients, etc., following standard procedures (Black 1965). Based on the morphological, physical, and chemical characteristics, soils can grouped into a series depending upon similarities in soil characteristics, and the extension of the soil and the name of the series is given in the name of the area, i.e., name of the village or adding numbers with the name of the village (Piprakothi-1, Piprakothi-2. . ., if the Piprakothi village has got two soil series) or the name of the river, etc. Again, the mapping units are prepared based on the different phases of soil series like slope, soil depth, surface texture, erosion, salinity/sodicity, gravelliness, etc. For example, the mapping unit Pip-sic-d5/A-e1 indicates that Pip represents the name of the soil series Piprakothi; sic, surface texture, silty clay; d5, very deep; A, nearly level; and e1, slight erosion. In detailed soil survey, the mapping units are phases of soil series.

15.5.7.4 Post-field Activities

The scale and type of base map required for survey is very important because that will decide the amount of information that can be depicted on the map. The larger the scale used, the more will be the information available and vice versa. For detailed soil survey, cadastral maps showing field boundaries and survey numbers are the most suitable maps. High-resolution large-scale satellite images available from IRS (IRS-1D fused with PAN or Resourcesat (IRS-P6 LISS-IV) at 1:10,000 scale can be used as a base in conjunction with the cadastral maps for field survey. In addition to the base maps, 1:25,000 or 1:50,000 scale toposheets, geological and landform maps available for the survey area can be used for traversing the area for preparing preliminary legend. At 1:25,000 scale, the smallest area, which can be conveniently shown true to scale, is 0.25 ha, and a soil boundary line on the map represents a zone of 5 m wide on the ground. When the scale is increased to 1:10,000 scale, the smallest area, which can be conveniently shown, is 0.625 ha, and at this scale, a boundary line on the map represents a zone only 2 m wide on the ground.

Before proceeding for soil survey work, the available secondary information pertaining to the area, viz., geology, geomorphology, groundwater depth, rainfall, and temperature, are collected. After procuring the remote sensing data, it is interpreted visually/digitally, and the information generated based on remote sensing data were superimposed on the cadastral map of the area, and thus the base map is produced for soil survey work. With the base map thus prepared, the surveyor has to go to the field and performs extensive traversing of the area to find out the different features observed on the ground, i.e., the changes in slope, direction of slope, surface stoniness, erosion status, and other morphological features like waterlogging, surface cracks, salinity/sodicity, wasteland areas, etc. During field traverse, the features observed in the field were compared with the base map prepared from the interpretation of satellite image, and if anything is found not matching with the map information, the ground truth was recorded, and finally the base map is modified, and thus the final base map is used for soil resource inventory.

Based on the soil-site characteristics, the soils were grouped into different soil series (soil series is the most homogeneous unit having similar horizons and soil properties and behaves uniformly for a given level of management). The important characteristics used to group the pedons into different soil series are soil depth, texture, color, amount and nature of coarse fragments present, calcareousness, presence of calcium carbonates, mottles, coats, stress features, nature of the substratum, horizon sequences, etc.; a significant difference in any one of these characteristics can be the basis for recognizing a different soil series. For the soil series identified in the study area, the characteristics that differentiate one series from another are listed in the form of a table. In any survey area, the differentiating characteristics between one soil to another will be only few in number, and this table helps the surveyor to identify and remember them easily.

15.6 Soil Mapping

Soil mapping is done by intensive traversing covering all plots and survey numbers after grouping the soils into different soil series within each landform. During the traverse, a boundary between two series is established by checking for the differentiating characteristics of one series from the other through a minipit, auger, road cut, well cut, or any other information available from the area.

15.7 Detailed Land Resource Survey: A Case Study

15.7.1 Study Area

The detailed land resource survey was conducted in Piprakothi Block under Purba Champaran district of Bihar which is located between 26°31′40″ to 26°36′20″ N latitude and 84°51′49″ to 85°02′07″ E and consists of 20 villages covering 6613 ha area representing nearly level to very gently sloping alluvial plain (AESR – 13.1) developed from the deposition of alluvium from the Nepal Himalayas and also from the Ganga river (Fig. 15.1). The mean maximum temperature of the area is 46 °C, and the mean minimum temperature is 05 °C, and the average rainfall is 1242 mm. Geology of the area is depositional landforms of recent origin and developed from the alluvium deposited by the Ganga river and its tributaries and also from Nepal Himalayas. The major landforms are young alluvial plain and active alluvial plain (meander plain, point bar, and oxbow lakes) with a flat topography sloping toward east. The major crops cultivated in this area are *kharif* paddy, wheat, sugarcane, maize, vegetables, and horticultural crops, viz., mango and litchi. The soils are very deep, well to somewhat poorly drained, and highly calcareous. The soils suffer from waterlogging, frequent flooding, drainage congestion, calcareousness, salinity, and multi-nutrient deficiency, affecting the crop yield.

15.7.2 Image Interpretation

IRS Resourcesat-II (IRS-P6) LISS-IV image of March 13, 2015 (Fig. 15.2), was interpreted based on the tone, texture, size, shape, and pattern for the delineation of landforms, which were confirmed by intensive traversing of the area. The image of the block was superimposed on the cadastral map of the block to prepare the base map for undertaking the soil survey following the procedure as outlined in the *Soil Survey Manual* (Soil Survey Staff 1995).

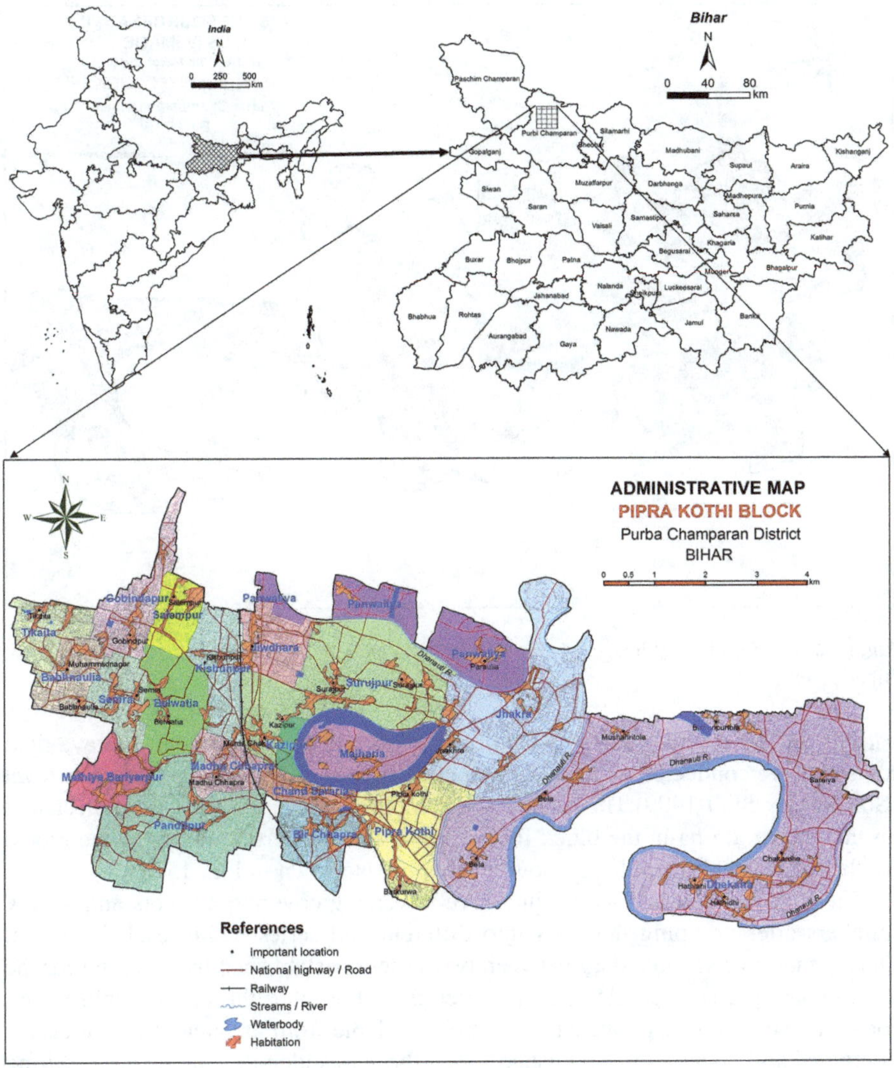

Fig. 15.1 Location map of Piprakothi Block, Purba Champaran district, Bihar

15.7.3 Soil Sampling/Fieldwork

Field surveys include the study of soil profiles in all the landform units. The morphological features of representative pedons were studied up to a depth of 150 cm or more/lithic contact (for shallow soils), and the soil samples were collected from different horizons for laboratory analysis. Soil physiographic relationship was established and random study of soil profiles at different landform units was also

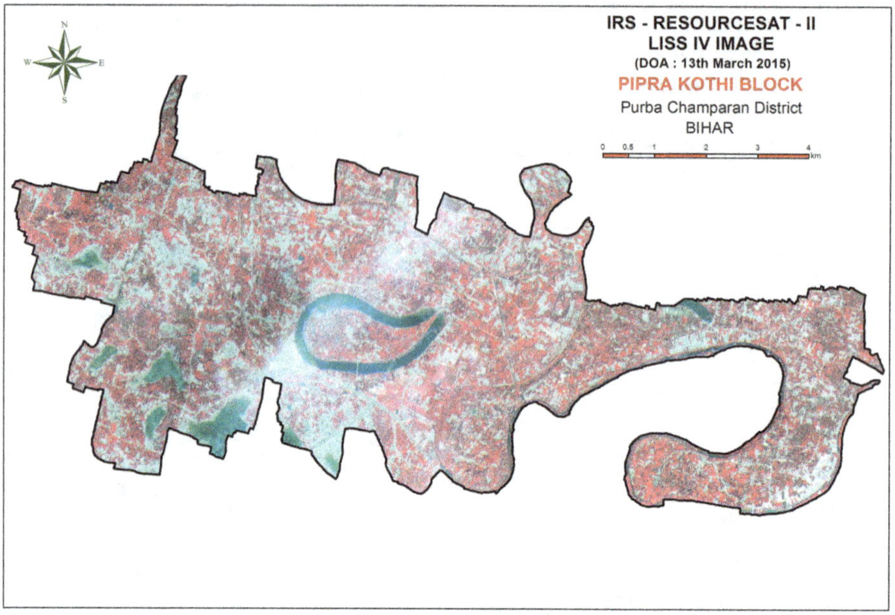

Fig. 15.2 IRS-P6 LISS-IV image (March 13, 2015) of Piprakothi Block, Purba Champaran district, Bihar

carried out to confirm the soil landform relationship. During soil survey, field reviews were conducted to correlate and classify the soils up to the soil series level (Soil Survey Staff 1999).The surface soil samples (0–25 cm) were also collected at an interval of 2.5 ha in the block for preparing the soil fertility maps of the block. Methodology adopted in land resource mapping is shown in Fig. 15.3.

The soil mapping is done by intensive traversing covering all plots and survey numbers after grouping the soils into different soil series within each landform. During the traverse, boundary between two series is established by checking for the differentiating characteristics of one series from the other through minipit, auger, road cut, well cut or any other information available from the area. The exercise is continued and boundaries established for all the series identified in the area. During this exercise itself, within the delineated area of a soil series, based on variations in the surface texture, slope, erosion, presence of gravels, salinity, etc., phases of soil series are separated and their boundaries delineated on the cadastral map.

15.7.4 Soil Analysis

The soil samples collected from different genetic horizons were air-dried and analyzed for the physical and chemical properties, viz., particle size class, pH, organic carbon, cation exchange capacity, exchangeable bases and exchangeable

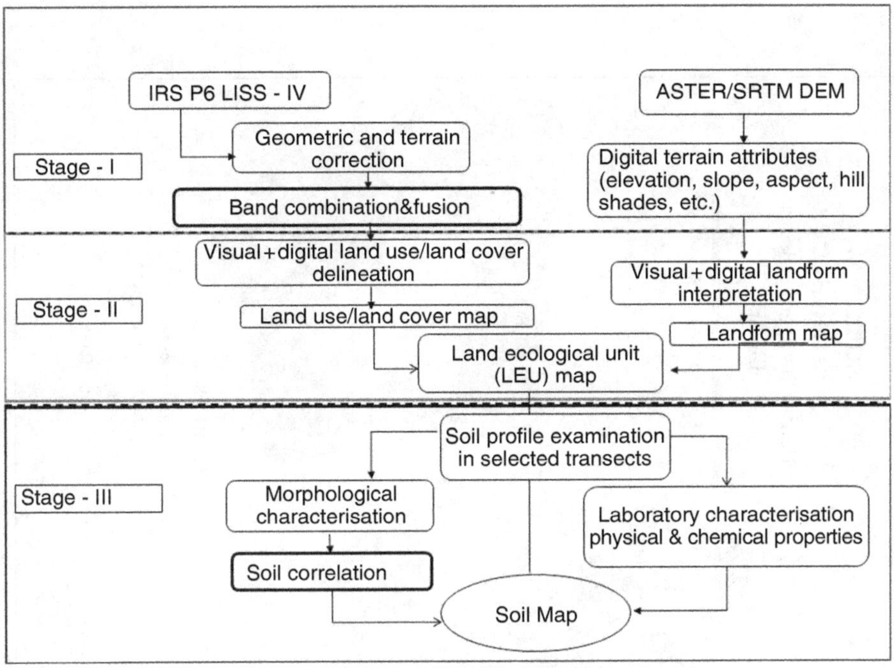

Fig. 15.3 Methodology adopted in land resource mapping

acidity (H^+ and Al^{3+}), available macro- and micronutrients, etc., following standard procedures (Black 1965). The soil map of Piprakothi Block was prepared on 1:10,000 scale with phases of soil series as mapping unit. A total of 7 soil series were identified and mapped into 13 mapping units (Fig. 15.4).

15.7.5 Soil Physical Characteristics

Soils of Piprakothi Block are very deep, moderately well to imperfectly drained, developed from alluvium of Ganga river as well as from the Nepal Himalayas, grayish brown to dark yellowish brown in color, and mostly silt loam to silty clay loam in texture. The particle size distribution of these soils indicates that these soils are dominant in silt and are mostly in the form of fine silt which restricts the infiltration and reduces the permeability of soils. The soils are very prone to flooding, and in *kharif* season, the low-lying areas are flooded extensively, and crops cannot be grown in these areas. The presence of redoximorphic features, viz., mottles in subsurface horizons, indicates fluctuation of groundwater in these soils. Paddy, wheat, maize, pulses, oilseeds, and sugarcane are mostly grown in this area.

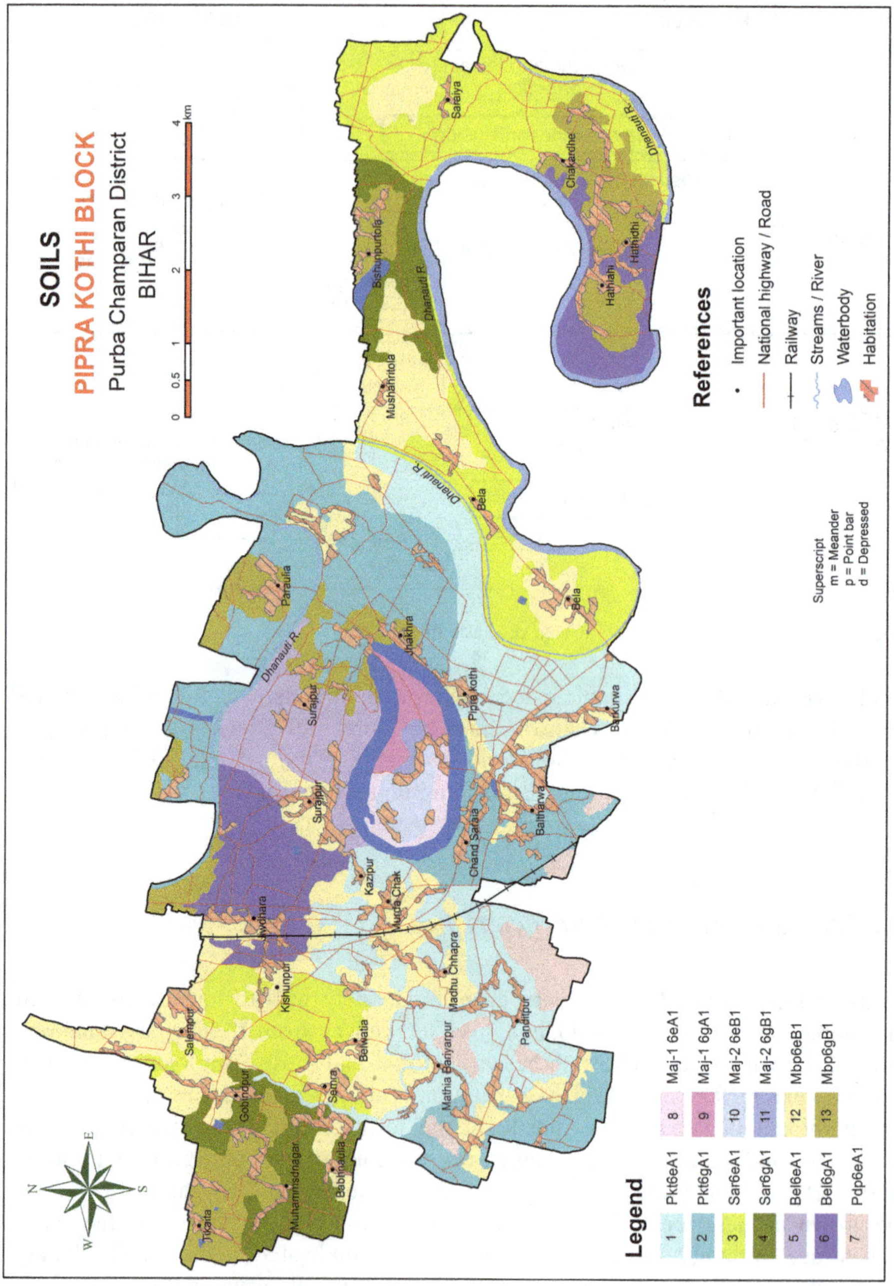

Fig. 15.4 Soil map of Piprakothi Block, Purba Champaran district, Bihar

15.7.6 Soil Chemical Characteristics

Soils are mostly alkaline, and the pH of the soils ranges from 7.0 to 9.9. However, the electrical conductance (EC) of the soils is low (0.1–0.97 dS m^{-1}), and also the EC_e (electrical conductance of the saturated paste of the soil) ranges from 1.4 to 5.8 dS m^{-1}. Soils are rich in silt (more than 65%), and the clay content varies from 13 to 37%. Soils are low to high in organic carbon (1.0–14.0 g kg^{-1}) and are low in CEC (2.0–9.9 cmol (p^+) kg^{-1}) with high base status (80–90%). Soils are highly calcareous, and the $CaCO_3$ content varies from 40 to 50%.

15.7.7 Land Capability Classification

Land capability classification is an interpretative grouping of soils mainly based on (i) the inherent soil characteristics, (ii) external land features, and (iii) environmental factors that limit the use of land (Dent and Young 1981). Information on the first two items is provided by a standard detailed soil survey. Scientific survey and classification of soils are the primary requirements for grouping soils according to their capability for use of varying intensity. Capability grouping is thus another aspect of soil survey work and is closely related to soil correlation. The taxonomic soil units established after field and laboratory studies and correlation are the ultimate management units, which afford specific information about the ability of the soil to respond to use, management, and plant growth.

The basic USDA land capability classification allocates land suited to cultivation to classes I to IV followed by land suited to grazing to classes V to VII and forestry to class VII, leaving class VIII for wildlife and recreation (Dent and Young 1981). In Piprakothi Block, soils are suffering from drainage congestion, calcareousness, and salinity and hence mostly grouped under the land capability subclasses IIws and IIIws indicating the limitations of drainage and soil.

15.7.8 Land Irrigability Classification

The interpretation of soil and land conditions for irrigation is concerned primarily with predicting the behavior of soils under the greatly altered water regime brought about by introduction of irrigation. For irrigation projects, interpretations are required to indicate the areas suitable for irrigation, crops that may be grown, yields that may be expected, water delivery requirements, land development needs, problem in drainage, and special reclamation practices. The soils are first grouped into soil irrigability classes according to their limitations for sustained use under irrigation, regardless of their location. Soil irrigability classes are useful to make groupings of soils according to their suitability for sustained use under irrigation. The classes are defined in terms of the degree of soil limitations. Soil irrigability classification is made on the basis of important soil characteristics, viz., soil texture, depth, available water-holding capacity, infiltration, and permeability. Soil irrigability classification of the soils of Piprakothi Block grouped them into two soil irrigability subclasses, viz., 2ds and 3ds indicating the major limitations of drainage and soil.

15.7.9 Soil-Site Suitability Evaluation

Each plant requires definite soil and climatic conditions for optimum growth. Since the availability of both water and plant nutrients is largely controlled by the physical and chemical properties and microenvironments of soils, the success and failure of any species in a particular area is, therefore, governed by the soil characteristics. The soil-site suitability studies for different plant species illustrate how soil depth, subsoil texture, salinity, and drainage conditions are related to the site quality. The objectives of various site evaluation studies are to predict and classify land for growth. Since such evaluation of land has gained high popularity in almost every land development program, the objective of this study is to develop criteria for soil-site suitability evaluation on the basis of the available soil and site characteristics, so that the soil maps can be interpreted in terms of suitability for agricultural crops and/or forest plantations.

The FAO panel for land evaluation suggested the classification of land for specific use (FAO 1976). The classification itself is presented in different categories or orders (S for suitable and N for unsuitable land). There are three classes (S1 to S3) under the orders S and two classes (N1 to N2) under the order N. The appraisal of the classes within an order is done according to the evaluation of land limitations or the main kinds of improvement measures required within classes. The limitations are climate (c), topography (t), wetness (w), salinity (n), soil fertility (f), and physical soil limitations (s). They are indicated by the symbol, for example, S2w, where w indicates wetness limitation (subclass). The limitation approach has successfully been used to provide a qualitative land evaluation based on general characteristics which are made available after a quality soil survey and a general study of other resources. Sys et al. (1993) proposed a scheme for evaluating the degree of limitation rating with the following general observations.

0 (No limitation)	Optimum crop growth
1 (Slight limitation)	Nearly optimum for crop growth
2 (Moderate limitation)	Moderate influence on crop growth decline
3 (Severe limitation)	Uneconomical for the suggested land use
4 (Very severe limitation)	Yield below profitable level not fit for suggested land use

The soil-site parameters considered for the purpose of evaluating land for general agricultural crops, horticultural crops, other plantation crops, and forestry are as follows. These are considered for defining suitable classes.

Soil-site characteristics	Related land, soil, and crop quality
Climate (c)	Available moisture
Rainfall	
Temperature	
Topography and landscape (t)	Resistance to erosion
Slope	Landscape position
Wetness (w)	Available moisture/soil aeration
Drainage	Landscape position
Flooding	Deficiency/toxicity of nutrients
Physical condition (s) of soil	Water availability/soil aeration/soil structure
Texture	Available space for root development
Depth	Landscape position
Groundwater table	Available moisture/soil aeration
Soil fertility (f)	
pH (soil reaction)	Availability of plant nutrients
Silt and clay content	Availability of moisture nutrients
Cation exchange capacity and base saturation	Availability of plant nutrients
Organic matter	Soil structure/availability of plant nutrients

Source: Sys et al. (1993)

The soils of Piprakothi Block, Purba Champaran district, Bihar, have been evaluated for soil-site suitability for sugarcane (Fig. 15.5). The analysis indicates that 31.7% of the area of the block is moderately suitable for the cultivation of sugarcane due to moderate limitation of soil fertility. About 53.0% of the area is marginally suitable due to relatively low water availability, which is a prime requirement for the crop, and severe soil fertility limitations. Only 2.0% of the area representing nearly depressed meander plain is not suitable for the crop as the area remains under deep water throughout the year.

15.8 Conclusions

Soil resource mapping through soil survey involves manpower to accomplish the job in a given time following the traditional methods of soil survey and mapping. During the soil resource mapping program, the introduction of remotely sensed data and their interpretation to generate the base maps for ground truth was found to be not only timesaving but also economic. With the advent of the modern tools of remote sensing and GIS, the task of completing of soil resource mapping has become very easy and fast. Besides this, the remotely sensed data with 5.8 m resolution has made resource mapping easy with less time. It is now high time that such remotely sensed data with even higher resolution and at a very large scale should be complemented

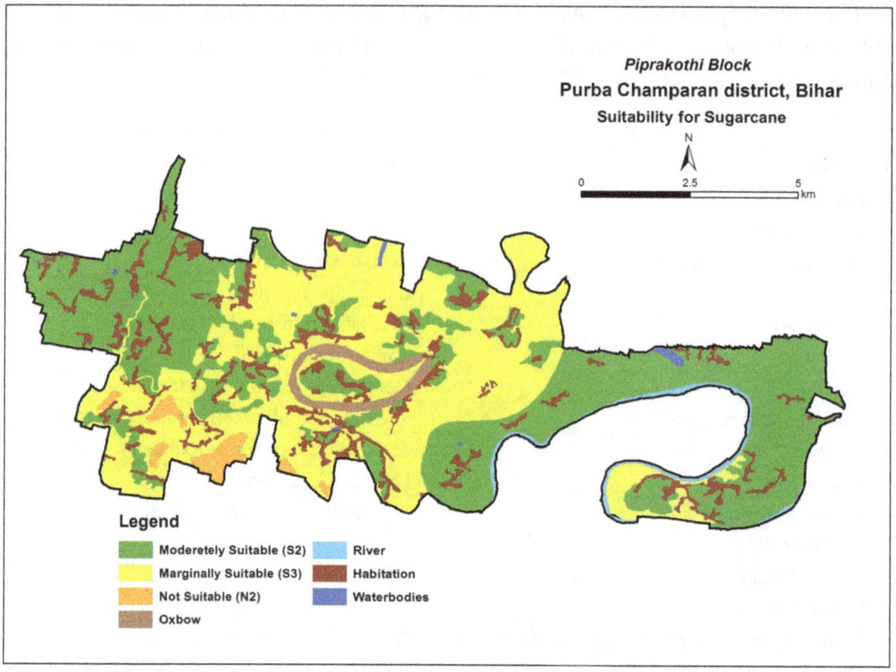

Fig. 15.5 Soil-site suitability for sugarcane in Piprakothi Block, Purba Champaran district, Bihar

with ground truth. With increasing GPS capability and its integration with GIS, the ability to collect accurate spatial data for soil survey field data collection becomes easy. Considerable production time is saved on the digital version of the soil map rather than on a hard copy map. Real-time mapping enables the creation or modification of digital product on site by a surveyor. With the integration of remote sensing data, GIS, and GPS, the surveyor is no longer constantly burdened to move place to place, and it saves time and labor and brings more accuracy in collecting the field data. Detailed land resource survey conducted in Piprakothi Block under Purba Champaran district of Bihar shows that soils are very deep, well to somewhat poorly drained, and highly calcareous. Soil-site suitability evaluation of sugarcane in the soils of block indicates that 31.7% of the area is moderately suitable, 53.0% of the area is marginally suitable, and 2.0% of the area is not suitable for sugarcane.

References

Black CA (1965) Methods of soil analysis, part-2. American Society of Agronomy, Inc, Madison, pp 849–1378

Boonyanuphap J, Wattanachaiyingcharoen D, Sakurai K (2004) GIS-based land suitability assessment for Musa (ABB group) plantation. J Appl Hortic 6(1):3–10

Dent D, Young A (1981) Soil survey and land evaluation. George Allen & Unwin, London

Deshpande SB, Sarkar D (2009). Review of soil survey in India. In: Bhattacharyya T, Sarkar D, Pal DK (eds) Soil survey manual. National Bureau of Soil Survey and Land Use Planning. India NBSS & LUP Publication no. 146, 400 pp

Dobos E, Micheli E, Baumgardner MF, Biehl L, Helt T (2000) Use of combined digital elevation model and satellite radiometric data for regional soil mapping. Geoderma 97:367–391

Dwivedi RS (2001) Soil resource mapping: a remote sensing perspective. Remote Sens Rev 20:89–122

FAO (1976) A framework for land evaluation, FAO soils bulletin 32. FAO, Rome

FAO, ITPS (2015) Status of the World's Soil Resources (SWSR) – main report. Food and Agriculture Organization of the United Nations and Intergovernmental Technical Panel on Soils, Rome

Gangopadhyay SK, Reddy GPO, Sarkar D, Srinivas CV, Khan QI (2010) Soil suitability evaluation using remotely sensed data and GIS- a case study from semi-arid tropics of India. Int J Geoinform 6(3):35–47

Gangopadhyay SK, Mukhopadhyay S, Singh SK, Sarkar D (2013) Soil resource inventory and land evaluation of Aurangabad district, Bihar (1: 50,000 Scale) for land use planning. ICAR- NBSS Publ. No. 1058, National Bureau of Soil Survey and Land Use Planning, Nagpur, India, 179p

Gangopadhyay SK, Reddy GPO, Sarkar D (2014) Erosion risk mapping in parts of semi-arid tropics of India using remote sensing and GIS. Indian J Soil Conserv 42(1):29–38

Gangopadhyay SK, Reddy GPO, Mukhopadhyay S, Singh SK (2015) Characterization of landforms and soils in complex toposequence of Subarnarekha catchment, Chhotanagpur plateau using remote sensing and GIS. Agropedology 25(1):95–109

GOI (1984) Report of the task force on land and soil resources. Planning Commission, New Delhi

IARI (1971) Soil survey manual. All India Soil and Land Use Survey Organization, IARI, New Delhi 121pp

Karale RL, Saini KM, Narula KK (1988) Mapping and monitoring ravines using remotely sensed data. J Soil Water Conserv 32(1):75

Lal R, Mokma D, Lowery B (1998) Relation between soil quality and erosion. In: Lal R (ed) Soil quality and soil erosion. CRC Press, Boca Raton, pp 237–257

Nayak DC, Chattopadhyay T, Mukhopadhyay S, Singh SK, Sarkar D (2014) Soil resource inventory and land evaluation of Rhotas district, Bihar (1: 50, 000 scale) for land use planning. ICAR- NBSS Publication No. 1078, National Bureau of Soil Survey and Land Use Planning, Nagpur, India, 130p

Olson KR, Mokma DL, Lal R, Schumacher TE, Lindstrom MJ (1998) Erosion impacts on crop yield for selected soils of the North Central United States. In: Lal R (ed) Soil quality and soil erosion. CRC Press, Boca Raton, pp 259–283

Panhalkar S (2011) Land capability classification for integrated watershed development by applying RS and GIS techniques. J Agric Biol Sci 6(4):46–55

Patil SS, Patil VC, Khalid AA (2010) Wheat acreage, productivity and production estimation through remote sensing and GIS techniques. Aust J Basic Appl Sci 4(8):3132–3138

Ray Choudhury SP (1964) Land resources of India, Vol 1. In: Indian soils-their classification, occurrence and properties, committee on natural resources, Planning Commission, Govt of India, New Delhi, pp 1–89

Raychoudhury SP, Govindarajan SV (1971) Soils of India. Indian Council of Agricultural Research Tech Bull (Agric) No 25, pp 1–45

Raychoudhury SP, Mathur LM (1954) A basic land resource map of India. Bulletin of National Institute of Sciences of India No 3

Reddy GPO, Maji AK, Chary GR, Srinivas CV, Tiwary P, Gajbhiye KS (2004) GIS and remote sensing applications in prioritization of river sub basins using morphometric and USLE parameters – a case study. Asian J Geoinfor 4(4):35–49

Reddy GPO, Nagaraju MSS, Ramteke IK, Sarkar D (2013a) Terrain characterization for soil resource mapping in part of semi-tract of Central India using high resolution satellite data and GIS. J Indian Soc Remote Sens 41(2):331–343

Reddy GPO, Sarkar D, Prasad J, Ramamurthy V (2013b) Geospatial modeling in assessment of biophysical resources for sustainable land resource management. Trop Ecol 54(2):227–238

Reddy GPO, Kurothe RS, Sena DR, Harindranath CS, Niranjana KV, Naidu LGK, Singh SK, Sarkar D, Mishra PK, Sharda VN (2016) Assessment of soil erosion in tropical ecosystem of Goa, India using universal soil loss equation, geostatistics and GIS. Indian J Soil Conserv 44(1):1–7

Reddy GPO, Patil NG, Chaturvedi A (eds) (2017) Sustainable management of land resources-an Indian perspective. Apple Academic Press, New York, p 769

Rosser J, Swartz GL, Dawson NM, Briggs HS (1974) A land capability classification for agricultural purposes. Division of Land Utilization. Techcnical Report, p 14

Royal Commission on Agriculture in India (1928) Royal Commission on agriculture in India report. Government Central Press, Bombay, p 755

Sarkar D (2003) Soil survey and mapping. In: Fundamentals and applications of pedology, 1st edn. Kalyani Publishers, New Delhi, 276pp

Sehgal JL (1992) Soil series criteria and norms. Technical bulletin, NBSS & LUP Publication No. 36, NBSS & LUP (ICAR), Nagpur

Sehgal J (1996) Pedology: concepts and applications, 1st edn. Kalyani Publishers, New Delhi, 485 pp

Singh AN, Dwivedi RS (1989) Delineation of salt affected soils through digital analysis of Landsat-MSS data. Int J Remote Sens 10:83–92

Soil Survey Staff (1951) Soil survey manual, agriculture handbook no. 18. U.S. Department of Agriculture, Washington, DC 503 pp

Soil Survey Staff (1993) Soil survey manual, agriculture handbook no. 18. US Department of Agriculture, Washington, DC 437 pp. 503 pp

Soil Survey Staff (1995) Soil survey manual, agriculture handbook no. 18. USDA, Jodhpur

Soil Survey Staff (1999) Soil survey staff, soil taxonomy – a basic system of soil classification for making and interpreting soil surveys, agricultural handbook, 2nd edn. USDA, Washington, DC, p 436

Srivastava R, Saxena RK (2004) Techniques of large scale soil mapping in basaltic terrain using satellite remote sensing data. Int J Remote Sens 25(4):679–688

Sys C, Van Ranst E, Debaveye J (1993) Land evaluation. Part III: crop requirements. International Training Center for Post Graduate Soil Scientists, Ghent University, Ghent 199 pp

Velmurugan A, Carlos GG (2009) Soil resource assessment and mapping using remote sensing and GIS. J Indian Soc Remote Sens 37(3):511–525

Verma TP, Singh RS, Giri JD, Naitam RK, Tailor BL, Singh R, Shyampura RL, Sarkar D, Singh SK (2015) Soil resource inventory and land evaluation of Chittaurgarh district, Bihar (1: 50,000 scale) for Land Use Planning. ICAR- NBSS Publ. No. 1091, National Bureau of Soil Survey and Land Use Planning, Nagpur, 223p

Chapter 16
Applications of Remote Sensing in Land Resource Inventory and Mapping

Rajeev Srivastava

Abstract Comprehensive information on soil resources in terms of type, extent, physical and chemical properties and limitations/capabilities is required for optimal management of land resources and monitoring changes in land qualities. The technological advancements in the remote sensing have revolutionized the land resource inventory and mapping process. The advantage of remote sensing data is that it provides synoptic view of the terrain, which enables to understand the relief, land use and drainage conditions for better delineation of landform-soil units. Further, digital elevation models (DEMs) have facilitated surface parameterization by attributes such as elevation, slope, aspect, flow accumulation, plan and profile curvature to obtain relief or surface topography units. Hyperspectral remote sensing and soil spectroscopy data can be analysed using statistical and chemometric techniques to derive information about wide variety of soil attributes, which can be used for digital soil mapping.

Keywords Digital elevation models · Digital soil mapping · Land resource inventory · Landform-soil mapping · Remote sensing

16.1 Introduction

Land resource inventory plays an important role in the management of land resources as it serves as a base for monitoring changes in land quality. Management of land resources on scientific principles is essential to maintain the present level of soil/land productivity and to prevent soil/land degradation. This calls for comprehensive knowledge on soil resources in terms of type, extent, physical and chemical properties and limitations/capabilities (Saxena et al. 2000). Therefore, in recent years increasing emphasis is laid on characterization of soils, accurate mapping of

R. Srivastava (✉)
ICAR-National Bureau of Soil Survey & Land Use Planning, Nagpur, India

G. P. O. Reddy, S. K. Singh (eds.), *Geospatial Technologies in Land Resources Mapping, Monitoring and Management*, Geotechnologies and the Environment 21,
https://doi.org/10.1007/978-3-319-78711-4_16

different kinds of soils and developing rational and scientific criteria for land evaluation and interpretation of soils for multifarious land uses.

With the advancement in space and information technology, remote sensing has emerged as an important and efficient tool for soil resource inventory and mapping. The advantage of remote sensing technique is that it provides a wide range of information available in the electromagnetic spectrum in a synoptic and more frequent manner to detect and map the land resources in spatial domain. Not only this, the remotely sensed data provide an opportunity to 'look back in time' for any comparison of the resources positioned between past and present. Further, observations made at different resolutions provide a better tool for mapping land resources at different scales as compared to traditional methods. Remote sensing technique has reduced the fieldwork to a considerable extent, and soil boundaries can be more precisely delineated than in conventional methods (Srivastava and Saxena 2004; Nagaraju et al. 2014; Giri et al. 2016). The remote sensing technology has been found to be more efficient and economical than conventional survey (Manchanda et al. 2002) and also offers possibilities for extending existing soil survey data sets to similar areas.

Remote sensing data can be used in various ways. It may be used to stratify the landscape into more or less homogeneous soil-landscape units for which soil composition can be assessed by ground-truth sampling, and it also facilitates mapping inaccessible areas by reducing the need for extensive time-consuming and costly field surveys. Remote sensing imagery can be used as a data source supporting digital soil mapping (Ben-Dor et al. 2008). The hyperspectral remote sensing data can also be analysed using empirical methods to derive information about soil properties. In the following sections, the application of remote sensing and soil spectroscopy in characterization of land/soil resources and its spatial mapping was briefly discussed.

16.2 Remote Sensing in Landform-Soil Mapping

Soil is described by the well-known state factor of soil formation (Jenny 1941) as a function of climate, organisms, relief, parent material and time, referred to as CLORPT. Soil surveyors mostly use the landform or physiography map (a combination of topography or relief, slope, parent materials and land use) as a base to draw soil boundaries. Landform maps are most commonly used to map soils because soil development often occurs in response to the underlying lithology and water movement in the landscape (Ballantine et al. 2005; McKenzie and Ryan 1999). The main objective of physiography/landform mapping is to delineate more or less homogeneous and mutually contrasting landform units (Hewitt 1993; Hudson 1992). Within these landform units, soils are studied to assign the most appropriate soil type or soil association. Srivastava and Saxena (2004) observed that in basaltic terrain, physiography and land use (PLU) unit prepared through visual interpretation of PAN merged LISS-III data correlated well with soils and helped in delineation of phases of soil series.

Physiography/landform maps are prepared through visual interpretation of remote sensing data in conjunction with Survey of India toposheets. Visual image interpretation is a task that relies on human analyst's knowledge and experience. Although the knowledge on general principles about remote sensing is fundamental for successful interpretation, the extensive knowledge and experience on the features of interest will in many ways enhance the ability of the interpreter to make correct feature identification and image interpretation.

In optical remote sensing, weather plays an important role in the availability of cloud-free satellite data. For landform-soil mapping through remote sensing, it is important that considerable part of the land surface is exposed to direct solar radiation, so that reflection from the surface soils could be recorded by the satellite sensor. For better interpretation of satellite image, it is also important that the image should have good contrast so that different land features could be delineated more precisely. In a study conducted jointly by NBSS&LUP and NRSA at Nagpur (representing part of Peninsular region), Gangtok, Sikkim (representing Eastern Himalayan mountain region), and Meerut, UP (representing the Indo-Gangetic plains), it has been observed that for landform-soil mapping, satellite data acquired during mid-March to April for Peninsular region, during February to March for Eastern Himalayan mountain region and between March and April for the Indo-Gangetic plains are most appropriate (Ravisankar and Srivastava 2009). While mapping salt-affected soils of the Indo-Gangetic plain, Verma et al. (1994) observed that satellite data acquired between the period first of March and first week of April is most appropriate as it provides better contrast between salt-affected soils and crop.

16.3 Digital Elevation Models

The digital elevation models (DEMs) – typically acquired by remote sensing – have facilitated in digital mapping of physiography/landform. Nowadays satellite-based DEMs like SRTM (90 m), SRTM (30 m) and ASTER (30 m) were easily accessible and even available for free (Farr 2000). Typically, the surface is parameterized by attributes such as elevation, slope, aspect, plan and profile curvature and flow accumulation (Moore et al. 1993) to obtain relief or surface topography units. These attributes quantify the role of topography in redistributing water in the landscape, which may affect the pedogenesis and thereby the soil characteristics (Wilson and Gallant 2000). A semiautomated geospatial object-based classification modelling approach to delineate landform units in the Deccan Plateau region of Central India was used by Chattaraj et al. (2017). Knowledge-based rules were framed to classify the primary terrain parameters of elevation, slope, profile curvature and drainage derived through Cartosat-1 DTM into different landform units.

The key constraint of this knowledge-based modelling is its limited adaptability to only localized conditions.

16.4 Vegetation Indices

The application of remote sensing to map soil properties in densely vegetated areas depends on indirect relations between vegetation and soil attributes. The temporal and spatial variations in vegetation indices have been found to be linked to prevailing climate, ecosystem, terrain and physical soil properties (Singh et al. 2004). The normalized difference vegetation index (NDVI) is one of the most common indicators of crop growth characteristics and, indirectly, of specific site qualities (Sommer et al. 2003). Dobos et al. (2000) found that the use of spectral indices such as NDVI in combination with a DEM often produced soil pattern delineations comparable to existing regional-scale soil and terrain data.

Soil properties that have been found related to NDVI are root zone soil moisture (Wang et al. 2007), soil texture and water holding capacity (Lozano-Garcia et al. 1991) and soil carbon and nitrogen content (Sumfleth and Duttmann 2008). NDVI poses serious problem in partly vegetated areas due to the influence of soil background reflectance on NDVI (Huete 1988; Tucker et al. 1985). To account such situations, several other vegetation indices, e.g. the soil-adjusted vegetation index (SAVI) (Huete 1988; Rondeaux et al. 1996), the transformed SAVI (TSAVI) (Rondeaux et al. 1996), the modified SAVI (MSAVI) and the global environment monitoring index (GEMI) (Qi et al. 1994; Rondeaux et al. 1996), have been developed.

16.5 Soil Reflectance Spectroscopy for the Assessment of Soil Properties

In soil survey, typically, large numbers of soil samples are collected and analysed in order to capture the spatial variability in soil properties. The laboratory analysis of a large number of samples is expensive and time-consuming and requires large amounts of labour and chemicals for performing this task (Rossel and McBratney, 1998). Research over the last few decades has demonstrated the ability of reflectance spectroscopy as a low-cost method for rapid prediction of soil physical, chemical and biological properties in the laboratory (Ben-Dor and Banin 1995; Shepherd and Walsh 2002). In visible near-infrared (VNIR) spectroscopy, soil reflectance characteristics are determined over the entire visible (400–700 nm) and near-infrared (700–2500 nm) region.

Soil constituents have unique absorption features in these regions mainly due to electronic transitions of atoms (visible region) and overtones and the combinations related to stretching and bending vibrations in molecular bonds in the infrared region (Dalal and Henry 1986; Clark 1999). This unique spectral feature of soil also referred as 'spectral signature' allows the assessment of different soil attributes. Due to the presence of many overlapping features, soil information cannot be directly derived from the soil spectra. Reflectance spectroscopy is based on the use of calibrations, coupled with chemometric techniques, which utilize absorbances at specific wavelengths to predict particular properties of a sample (Shepherd and Walsh 2002).

Studies carried out by several researchers amply demonstrated that under laboratory conditions several soil attributes, viz. soil organic carbon (Gomez et al. 2008; Shepherd and Walsh 2002; Srivastava et al. 2015), Fe_2O_3, SiO_2, Al_2O_3 (Genú and Demattê 2006; Nanni and Demattê 2006; Stoner and Baumgardner, 1981), salt and carbonates (Ben-Dor et al. 2002; Farifteh et al. 2006; Srivastava et al. 2017), sand, silt and clay (Chang and Laird 2002; Hahn and Gloaguen 2008; Salisbury and D'Aria 1992), can be determined by spectral analysis. However, airborne or spaceborne spectroscopy complicates the measurement due to atmospheric influences (Richter and Schläpfer 2002), lower spectral and spatial resolution, geometric distortions and spectral mixture of features (Kriebel 1978; Richter and Schläpfer 2002).

16.5.1 Prediction of Salt-Affected Soil Properties: A Case Study from Indo-Gangetic Plains

The following paragraphs discuss the study carried out by Srivastava et al. (2017) in the Indo-Gangetic plains of Haryana to predict the soil properties of salt-affected soils based on their reflectance characteristics.

For the study surface soil samples ($n = 806$) collected from three villages (Jagsi of Sonepat district, Kahni of Rohtak district and Waiser of Panipat district) under different types of salinity representative of the current spread of salt-affected soils over most districts in Haryana state were used. All the villages/sites were under rice-wheat cropping system and have light to moderate texture soil. Salient characteristics of soil properties are shown in Table 16.1.

Reflectance characteristics of soils were studied under laboratory condition using a FieldSpec Pro FR spectroradiometer (Analytical Spectral Devices Inc., Boulder, Colorado). The raw and first derivative reflectance spectra between 350 and 2500 nm wavelength for all the soil samples (Fig. 16.1) showed prominent absorption features around 1400, 1900 and 2200 nm that are known to be related to features of free water (~1400 and ~1900 nm) and clay lattice OH features (~1400 and ~2200 nm) which could be more responsive towards the presence of salt minerals in the salt-affected soils, the sensitivity of which is likely to depend upon the nature and amount of the hygroscopic minerals present in the soil samples. It was also observed that the intensity of absorption peaks near 1900 nm region wavelengths shift towards the higher wavelength with the increase in soil salinity. The statistical correlation of soil

Table 16.1 Descriptive statistical parameters of soil chemical properties

Soil property	*N*	Min	Max	Mean	SD	Percentile		
						25th	50th	75th
pH_s	806	6.9	9.1	8.13	0.4	7.9	8.2	8.4
EC_e ($mS\ m^{-1}$)	806	40	17,700	1297	2350	160	450	1180
$CaCO_3$ ($g\ kg^{-1}$)	806	1.0	130.0	16.0	13.5	8.0	12.5	20.0
SE Na^+ ($meq\ l^{-1}$)	806	1.9	2120.0	135.7	242.3	11.2	50.0	125.0
SE K^+ ($meq\ l^{-1}$)	806	0.03	159.0	2.6	11.8	0.2	0.4	0.9
SE $Ca^{2+} + Mg^{2+}$ ($meq\ l^{-1}$)	805	1.5	652.0	43.8	71.7	10.0	17.5	48.0
SE CO_3^{2-} ($meq\ l^{-1}$)	806	0	8.0	0.8	1.3	0.0	0.0	1.0
SE HCO_3^- ($meq\ l^{-1}$)	806	0	63.2	5.6	5.1	3.0	4.3	7.0
SE Cl^- ($meq\ l^{-1}$)	806	2.0	2110.0	122.7	265.2	8.0	25.8	92.3
Saturation extract SO_4^{2-} ($meq\ l^{-1}$)	796	0.3	524.0	45.3	65.4	6.7	26.1	48.3
SAR ($[meq\ l^{-1}]^{0.5}$)	805	0.72	574.6	23.9	31.9	5.3	16.3	29.1

Source: Srivastava et al. (2017)
SE Saturation extract

salinity parameters (ECe, saturated extract Na^+, $Ca^{2+} + Mg^{2+}$, Cl^- and SO_4^{2-}) with first derivative soil reflectance data showed high correlation between 1400 and 2400 nm which indicated its high sensitivity to change in salinity levels.

The calibration of soil properties, viz. pHs, cations (Na^+ and $Ca^{2+} + Mg^{2+}$), anions (Cl^- and SO_4^{2-}, CO_3^{2-} and HCO_3^-) and SAR (sodium adsorption ratio), against the first derivative reflectance wavebands between 1390 and 2450 nm region using partial least square regression technique showed that very good calibrations were obtained for ECe ($r^2 = 0.94$, RMSE = 5.86), $Ca^{2+} + Mg^{2+}$ ($r^2 = 0.89$, RMSE = 1.23), Na^+ ($r^2 = 0.91$, RMSE = 2.29), Cl^- ($r^2 = 0.93$, RMSE = 2.19), SO_4^{2-} ($r^2 = 0.76$, RMSE = 1.69) and SAR ($r^2 = 0.76$, RMSE = 1.04). The poor calibration of PLSR models for pHs, CO_3^{2-} and HCO_3^- could be due to narrow variations in soil samples as compared to the other soil parameters. The application of developed calibration models for different soil attributes also resulted in high r^2 and low RMSE for all the attributes except soil pHs, CO_3^{2-} and HCO_3^- in the independent validation data sets. The calibration models obtained for soil pHs ($r^2 = 0.57$), CO_3^{2-} ($r^2 = 0.28$) and HCO_3^- ($r^2 = 0.35$) did not show good prediction due to low variability in the data sets. The high RPD values (> 2.0) obtained for ECe, SAR, Na^+, $Ca^{2+} + Mg^{2+}$ and Cl^- indicate that these properties can be reliably predicted from the spectral models, whereas the spectral model of SO_4^{2-} (RPD = 1.76) needs further improvement with larger data sets. The regression coefficients of ECe and SAR models showed distinctly higher negative coefficient value around 1930 nm (Fig. 16.2), which indicate that this region is most important for prediction of soil salinity.

The study concludes that ECe, being the most commonly used indicator for measuring the degree of soil salinity, can be effectively retrieved or predicted through laboratory-based hyperspectral model. In addition to ECe, other salinity-

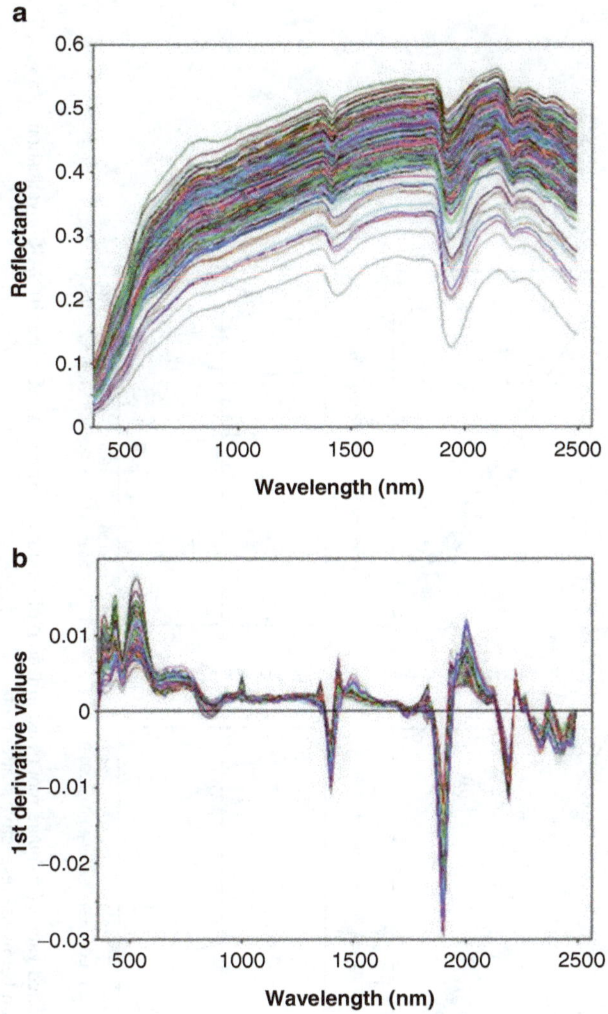

Fig. 16.1 Reflectance spectra (**a**) and first derivative equivalents (**b**) for all salt-affected soils under study. (Source: Srivastava et al. 2017)

related parameters, namely, saturated extract cations (Na^+ and Ca^{2+} + Mg^{2+}), selective anions (Cl^- and SO_4^{2-}) and SAR, might also be estimated through the hyperspectral models with reasonably good accuracy in the study area.

16.6 Soil Spectral Libraries

Spectral libraries are collections of reflectance spectra measured from materials of known composition, usually in the field or laboratory. Many investigators create spectral libraries for materials in their field sites as part of every project, to facilitate

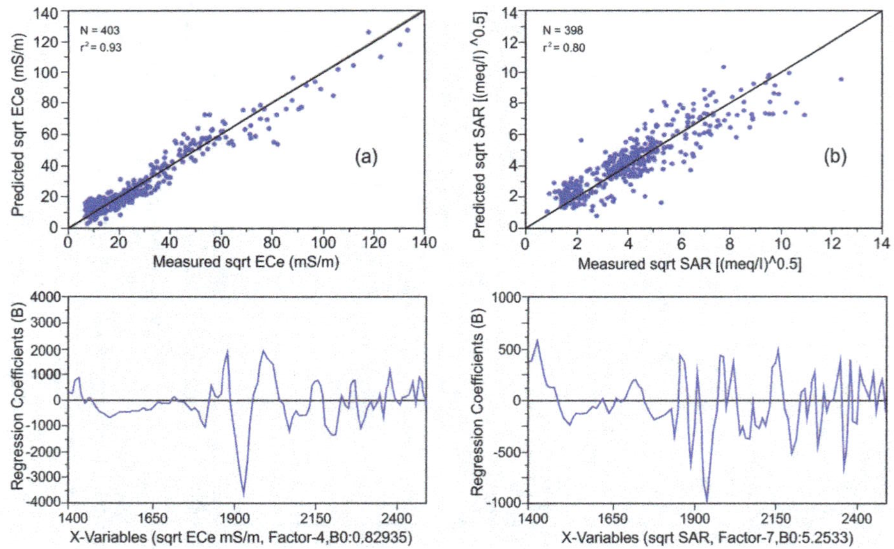

Fig. 16.2 Scatter plot of measured and predicted values (in square root transformation) of saturation extract (**a**) ECe and (**b**) SAR in validation data sets. The values of the regression coefficients are shown below the respective scatter plots

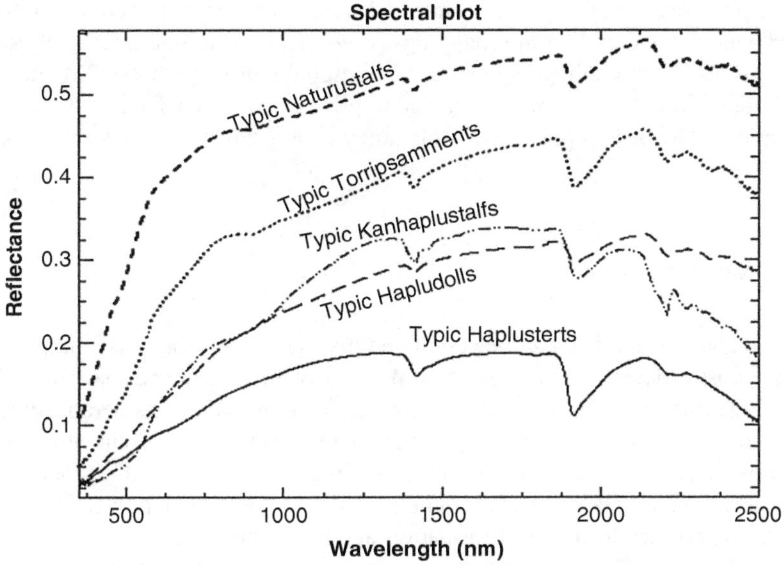

Fig. 16.3 Spectral reflectance characteristics of some selected soils of India

analysis of multispectral or hyperspectral imagery from those sites. Several high-quality spectral libraries are also publicly available. Spectra from libraries can guide spectral classifications or define targets to use in spectral image analysis.

Under the National Agricultural Technology Project (NATP), ICAR-NBSS&LUP (2005) prepared soil reflectance library of soils of India in the electro-magnetic spectral region of 350 nm–2500 nm using ASD spectroradiometer. The spectral library contains reflectance spectra measured under laboratory conditions of 128 soil samples collected from different physiographic and agroecological regions of the country. The spectral plot of five soils having different soil characteristics is shown in Fig. 16.3.

16.7 Digital Soil Mapping

Digital soil mapping (DSM) has been defined as 'the creation and population of spatial soil information by the use of field and laboratory observational methods coupled with spatial and non-spatial soil inference systems' (Lagacherie et al. 2007; McBratney et al. 2003). It allows for the prediction of soil properties or classes using soil information and environmental covariates of soil (Carré et al. 2007).

Typically, remote sensing provides information about the surface layer only. Bartholomeus et al. (2007) observed that accurate estimation of soil attributes is hampered if the pixels have over 20% vegetation cover. Another problem is that the spectral signatures of urban areas, roads and water surfaces do not contain information relevant for soil and terrain mapping. However, by combining remotely sensed imagery, DEM's and soil sample data using digital soil mapping (DSM) methods, spatial prediction of soil properties can be improved. An extensive review on digital soil mapping has been reported by McBratney et al. (2003).

16.8 Conclusions

From the above studies, it can be concluded that remote sensing has emerged as an effective cost-efficient technology for soil resource inventory and mapping of land resources at different scales. It has reduced the fieldwork to a considerable extent as compared to traditional methods. Further, it helps in precise delineation of soil boundaries and offers possibilities for extending existing soil survey data sets to similar areas. In vegetated areas, remote sensing-derived vegetation indices can be used as soil proxies to derive information about soil properties. The advantage of hyperspectral remote sensing and soil spectroscopy is that the data can be analysed using empirical methods to derive information about soil properties. A wide variety of soil attributes can be derived with use of statistical and chemometric analysis of spectroscopic data which can be used for digital soil mapping. However, most of the methods employed for retrieving soil attributes have been developed using local or regional correlation approaches, and may not scale for operational use over vast areas. Therefore, research on the improved integration of soil spectroscopy and remote sensing using scaling-based approaches is needed to make optimal use of all available data sources.

References

Ballantine JAC, Okin GS, Prentiss DE, Roberts DA (2005) Mapping North African landforms using continental scale unmixing of MODIS imagery. Remote Sens Environ 97(4):470–483

Bartholomeus H, Epema G, Schaepman ME (2007) Determining iron content in Mediterranean soils in partly vegetated areas, using spectral reflectance and imaging spectroscopy. Int J Appl Earth Obs Geoinf 9(2):194–203

Ben-dor E, Banin A (1995) Near-infrared analysis as a rapid method to simultaneously evaluate several soil properties. Soil Sci Soc Am J 59:364–372

Ben-Dor E, Patkin K, Banin A, Karnieli A (2002) Mapping of several soil properties using DAIS-7915 hyperspectral scanner data — a case study over clayey soil in Israel. Int J Remote Sens 23(6):1043–1062

Ben-Dor E, Taylor RG, Hill J, Demattê JAM, Whiting ML, Chabrillat S, Sommer S (2008) Imaging spectrometry for soil applications. In: Sparks DL (ed) Advances in agronomy, vol 97. Academic Press, Elsevier, pp 321–392

Carré F, McBratney AB, Mayr T, Montanarella L (2007) Digital soil assessments: beyond DSM. Geoderma 142(1–2):69–79

Chang CW, Laird DA (2002) Near-infrared reflectance spectroscopic analysis of soil C and N. Soil Sci 167(2):110–116

Chattaraj S, Srivastava R, Barthwal AK, Giri JD, Mohekar DS, Reddy GPO, Daripa A, Chatterji S, Singh SK (2017) Semi-automated object-based landform classification modelling in a part of the Deccan Plateau of Central India. Int J Remote Sens 38(17):4855–4867

Clark RN (1999) Spectroscopy of rocks & minerals, & principles of spectroscopy. In: Rencz AN (ed) Manual of remote sensing, volume 3, remote sensing for the earth sciences. Wiley, New York, pp 3–58

Dalal RC, Henry RJ (1986) Simultaneous determination of moisture, organic carbon, & total nitrogen by near infrared reflectance spectrophotometry. Soil Sci Soc Am J 50:120–123

Dobos E, Micheli E, Baumgardner MF, Biehl L, Helt T (2000) Use of combined digital elevation model and satellite radiometric data for regional soil mapping. Geoderma 97(3–4):367–391

Farifteh J, Farshad A, George RJ (2006) Assessing salt-affected soils using remote sensing, solute modelling, and geophysics. Geoderma 130(3–4):191–206

Farr TG (2000) The shuttle radar topography mission. In: IEEE aerospace conference proceedings, p 63

Genú AM, Demattê JAM (2006) Determination of soil attribute contents by means of reflected electromagnetic energy. Int J Remote Sens 27(21):4807–4818

Giri JD, Nagaraju MSS, Srivastava R, Singh DS, Nasre RA, Barthwal AK, Mohekar DS (2016) Accuracy assessment of large-scale soil map prepared by remote sensing approach. Int J Agric Stat Sci 12(1):229–237

Gomez C, Viscarra Rossel RA, McBratney AB (2008) Soil organic carbon prediction by hyperspectral remote sensing and field vis-NIR spectroscopy: an Australian case study. Geoderma 146(3–4):403–411

Hahn C, Gloaguen R (2008) Estimation of soil types by non linear analysis of remote sensing data. Nonlinear Process Geophys 15(1):115–126

Hewitt AE (1993) Predictive modelling in soil survey. Soils Fertilizers 56(3):10

Hudson BD (1992) The soil survey as paradigm-based science. Soil Sci Soc Am J 56(3):836–841

Huete AR (1988) A soil-adjusted vegetation index (SAVI). Remote Sens Environ 25(3):295–309

ICAR-NBSS&LUP (2005) Reflectance libraries for development of soil sensor for periodic assessment of state of soil resources. NATP Project Report (NBSS No. 835), National Bureau of Soil Survey & Land Use Planning, Nagpur

Jenny H (1941) Factors of soil formation, a system of quantitative pedology. McGraw-Hill, New-York

Kriebel KT (1978) Average variability of the radiation reflected by vegetated surfaces due to differing irradiations. Remote Sens Environ 7(1):81–83

Lagacherie P, McBratney AB, Voltz M (2007) Digital soil mapping: an introductory perspective. Developments in soil science, 31. Elsevier, Amsterdam

Lozano-Garcia DF, Fernandez RN, Johannsen CJ (1991) Assessment of regional biomass–soil relationships using vegetation indexes. IEEE Trans Geosci Remote Sens 29(2):331–339

Manchanda ML, Kudrat M, Tiwari AK (2002) Soil survey and mapping using remote sensing. Trop Ecol 43(1):61–74

McBratney AB, Santos ML, Minasny B (2003) On digital soil mapping. Geoderma 117(1–2):3–52

McKenzie NJ, Ryan PJ (1999) Spatial prediction of soil properties using environmental correlation. Geoderma 89(1–2):67–94

Moore ID, Gessler PE, Nielsen GA, Peterson GA (1993) Soil attribute prediction using terrain analysis. Soil Sci Soc Am J 57(2):443–452

Nagaraju MSS, Kumar N, Srivastava R, Das SN (2014) Cadastral-level soil mapping in basaltic terrain using Cartosat-1-derived products. Int J Remote Sens 35:3764–3781

Nanni MR, Demattê JAM (2006) Spectral reflectance methodology in comparison to traditional soil analysis. Soil Sci Soc Am J 70(2):393–407

Qi J, Chehbouni A, Huete AR, Kerr YH, Sorooshian S (1994) A modified soil adjusted vegetation index. Remote Sens Environ 48(2):119–126

Ravisankar T, Srivastava R (2009) Satellite imagery- their interpretation and applications in soil survey and mapping. In: Bhattacharyya, T. Sarkar, D. & Pal, D.K. (Eds), Soil Survey Manual, NBSS Pub 146, ICAR-National Bureau of Soil Survey & Land Use Planning, Nagpur, India, pp 59–71

Richter R, Schläpfer D (2002) Geo-atmospheric processing of airborne imaging spectrometry data. Part 2: atmospheric/topographic correction. Int J Remote Sens 23(13):2631–2649

Rondeaux G, Steven M, Baret F (1996) Optimization of soil-adjusted vegetation indices. Remote Sens Environ 55(2):95–107

Rossel RAV, McBratney AB (1998) Laboratory evaluation of a proximal sensing technique for simultaneous measurement of soil clay and water content. Geoderma 85:19–39

Salisbury JW, D'Aria DM (1992) Infrared (8–14 μm) remote sensing of soil particle size. Remote Sens Environ 42(2):157–165

Saxena RK, Verma KS, Chary GR, Srivastava R, Barthwal AK (2000) IRS-1C data application in watershed characterization and management. Int J Remote Sens 21(17):3197–3208

Shepherd KD, Walsh MG (2002) Development of reflectance spectral libraries for characterization of soil properties. Soil Sci Soc Am J 66:988–998

Singh D, Herlin I, Berroir JP, Silva EF, Simoes MM (2004) An approach to correlate NDVI with soil colour for erosion process using NOAA/AVHRR data. Adv Space Res 33(3):328–332

Sommer M, Wehrhan M, Zipprich M, Weller U, Castell W z, Ehrich S, Tandler B, Selige T (2003) Hierarchical data fusion for mapping soil units at field scale. Geoderma 112(3–4):179–196

Srivastava R, Saxena RK (2004) Technique of large-scale soil mapping in basaltic terrain using satellite remote sensing. Int J Remote Sens 25(4):679–688

Srivastava R, Sarkar D, Mukhopadhayay SS, Sood A, Singh M, Nasre RA, Dhale SA (2015) Development of hyperspectral model for rapid monitoring of soil organic carbon under precision farming in the indo-gangetic plains of Punjab, India. J Indian Soc Remote Sens 43(4):751–759

Srivastava R, Sethi M, Yadav RK, Bundela DS, Singh M, Chattaraj S, Singh SK, Nasre RA, Bishnoi SR, Dhale S, Mohekar DS, Barthwal AK (2017) Visible-near infrared reflectance spectroscopy for rapid characterization of salt-affected soil in the indo-Gangetic Plains of Haryana, India. J Indian Soc Remote Sens 45(2):307–315

Stoner ER, Baumgardner MF (1981) Characteristic variations in reflectance of surface soils. Soil Sci Soc Am J 45(6):5

Sumfleth K, Duttmann R (2008) Prediction of soil property distribution in paddy soil landscapes using terrain data and satellite information as indicators. Ecol Indic 8(5):485–501

Tucker CJ, Vanpraet CL, Sharman MJ, van Ittersum G (1985) Satellite remote sensing of total herbaceous biomass production in the Senegalese Sahel: 1980–1984. Remote Sens Environ 17(3):233–249

Verma KS, Saxena RK, Barthwal AK, Deshmukh SN (1994) Remote sensing technique for mapping salt affected soils. Int J Remote Sens 9:1901–1914

Wang X, Xie H, Guan H, Zhou X (2007) Different responses of MODIS-derived NDVI to root-zone soil moisture in semi-arid and humid regions. J Hydrol 340(1–2):12–24

Wilson JP, Gallant JC (2000) Digital terrain analysis. In: Wilson JP, Gallant JC (eds) Terrain analysis: principles and applications. Wiley, New York, pp 1–27

Chapter 17
Land Resource Inventory for Village-Level Agricultural Land Use Planning

Nisha Sahu, G. P. Obi Reddy, Rajeev Srivastava, Narayan Lal, and S. K. Singh

Abstract Timely and reliable information on soils with respect to their nature, extent and spatial distribution is vital for optimal utilization of natural resources on a sustained basis. The technological advances in the field of remote sensing, global positioning system (GPS) and geographic information system (GIS) have augmented the efficiency of land resource inventory and mapping. The recent advances in remote sensing have immense potential to explore the full range of spectral, spatial and temporal resolutions of high-resolution satellites in soil resource mapping and characterization. In the present study, based on landforms, slope, land use/land cover and ground truth, 37 physiography-land use units (PLU) were identified and described. Soil profiles were studied at representative sites on different PLUs for characterization of various sites and physical and chemical properties. PLU-soil relationship was developed by correlating soil-site characteristics and physical and chemical properties of soils. Six soil series were identified in major landforms, and soil map depicting phases of soil series was developed. The study revealed that the combined use of Cartosat-1 Digital Elevation Model (DEM) (10 m) and high-resolution IRS-P6 LISS-IV data will be of immense help in identifying distinct soil patterns for large-scale soil resource inventory for village-level agricultural planning.

Keywords Agricultural land use planning · Geographic information system · Global positioning system · High-resolution remote sensing · Land resource inventory · Soil survey

---

N. Sahu (✉) · G. P. O. Reddy · R. Srivastava · S. K. Singh
ICAR-National Bureau of Soil Survey & Land Use Planning, Nagpur, India

N. Lal
ICAR-National Research Centre on Litchi, Muzaffarpur, India

G. P. O. Reddy, S. K. Singh (eds.), *Geospatial Technologies in Land Resources Mapping, Monitoring and Management*, Geotechnologies and the Environment 21, https://doi.org/10.1007/978-3-319-78711-4_17

17.1 Introduction

Land is the platform for our living, and we make use of it for crop production, livestock keeping, forestry, housing, etc. Land resources, particularly soil and water, are limited in extent; their efficient and sustainable utilization is imperative, particularly when the population pressure is increasing alarmingly. A comprehensive land-based approach plays vital role to identify and prioritize target areas, which have high potential for success, selecting and disseminating the most appropriate sustainable land management (SLM) practices, supported by proper policies and financial mechanisms. The needs at farm level should be at the centre of sustainable land development processes (Mediterra 2016; Ziadat et al. 2015). Over-exploitation of natural resources is causing widespread damage to soil eco-environment. The multiuse of land involves various trade-offs that favour one use at the expense of others leading to land degradation (FAO and ITPS 2015). About 187.7 Mha representing 57.1% of the total geographical area (TGA) of the country is subjected by various forms of soil degradation by water, wind, chemical and physical agencies (Sehgal and Abrol 1994) and a threat to long-term productivity (Bhattacharyya et al. 2007; Pratap Narain 2008). On account of various forms of degradation, it has been estimated that the loss of 5.33 million tonnes of top soil occurs annually, which is equivalent to 5.37 to 8.4 million tonnes of plant nutrients (Dhruvanarayana and Babu 1983). Beside this, the human population explosions in the pursuit of meeting the fuel and fodder demand indiscriminately destroy vegetation cover. In order to maintain or enhance the present level of productivity in a limited area, it is at most important to manage land resources on scientific principles. For optimum utilization of available land resources on a sustainable basis, timely and reliable information on soils regarding their nature, extent and spatial distribution along with their potentials and limitations is important (Reddy et al. 2017, 2017a).

In this context, characterization and mapping of different types of soils and their interpretation attain greater importance. Accurate baseline information and methods to evaluate the quantity and quality of resources are prerequisites for mapping and characterization of soils (Laake 2000). It also provides adequate information in terms of landform, terrain, vegetation as well as characteristics of soils, which can be utilized for land resource management and development (Manchanda et al. 2002). The requirements for both accurate and timely information on resources had exploded considerably over the last decade for integrated resource management with village/watershed as a unit of planning by integrating information on geomorphology, land use, slope and soils (Srinivasa et al. 2008). The traditional methods of mapping and gathering information are expensive and time consuming due to large number of observations. However, advances in computer and information technology have introduced new group of tools, methods, instruments and systems. Rapid developments in new technologies such as remote sensing, GPS and GIS provide new approaches to meet the demand of resource planning and are successfully used in studying the various aspects of soils in spatial and temporal domain (Shrestha 2006; Yeung and Lo 2002). Liengsakul et al. (1993) estimated about 60–80% time is

Table 17.1 High-resolution (<20 m) satellite sensors and their characteristics

Satellite	Operation period	Spatial resolution (metres)	Swath (kms)	Actual revisit (in days)	Scene coverage (Km^2)
IRS IC	1995	5.8	70	24	4900
SPOT-2	1990	10/20	60	26	3600
SPOT-4	1998	10/20	60	26	3600
Terra	1999	15(VNIR)	60	16	3600
IKONOS-2	1999	1 (PAN)	11.3	3	121
		4 (NIR)			
Quickbird-2	1999	0.6(PAN)	16.5	3	400
		2.4(NIR)			
SPOT-5	2002	2.5/5/10/20	60	26	3600
OrbView-3	2003	1.0(PAN)	8	<3	64
		4.0 (NIR)			
Cartosat-1	2005	2.5	30	5	756
KOMPSAT-2	2006	1.0(PAN)	15	3	225
		4.0 (NIR)			
Cartosat-2	2007	0.8	9.6	4	92
GeoEye-1	2008	0.41(PAN)	15.2	<3	225
		1.65(NIR)			
WorldView-2	2009	0.46(PAN)	16.4	4	230
		1.84 (NIR)			
Resourcesat-2	2011	5.8	23.9	24	560

Source: Sahu et al. (2015)

saved using satellite imagery for soil mapping compared to manual methods. The application of satellite remote sensing data products for large- and medium-scale soil mapping are widely accepted and suggested utilization of high-resolution satellite data like IRS-P6 LISS-III (Velmurugan and Carlos 2009) and IRS-P6 LISS-IV (Walia et al. 2010; Reddy et al. 2013) for soil mapping. Various high-resolution satellites and their spatial resolutions are shown in Table 17.1.

17.2 Need for Land Resource Mapping at Large Scale

With the general acceptance of village as principal unit of planning of all developmental activities based on suitable utilization of locally available natural resources, the characterization and inventorization of land resources at village level assumes greater importance (Patil et al. 2010; Manchanda et al. 2002). The modern remote sensing technologies using sensors in the visible, infrared, thermal and microwave regions of the electromagnetic spectrum are of immense use in evaluation, monitoring and management of land, water and crop resources (Das et al. 2009). The potential utility of remote sensing data has been well recognized in mapping and assessing land attributes such as physiography, soils, land use/land cover and relief, which are prerequisite for planning soil conservation and watershed management programmes (Reddy and Maji 2003; Potdar et al. 2003; Surya et al. 2008; Velmurugan and Carlos 2009). At present, geospatial technologies play a vital role in generating consistent and timely spatial information for decision-making and planning from micro to macro level. Remote sensing has become an indispensable scientific tool for mapping and monitoring of natural resources (Kasturirangan et al. 1996) and frequently used in the characterization of the land for planning. The information generated with respect to geomorphology, soil and land use/land cover through remote sensing can be interpreted for various themes, viz. land capability, irrigability and crop suitability (Walke et al. 2012), etc. for better management and conservation of these resources on watershed/village basis (Bodhankar et al. 2002; Sharma et al. 2004). Large-scale soil database appears to be a major tool for land users and planners, although it is broadly acknowledged that GIS has the capability to examine both spatial and aspatial data as well as temporal aspects in a commercial and effective manner. The integrated remote sensing and GIS help in planning at large scale. Several studies have been carried out on potential use of remote sensing data for characterization and management of land resources at parcel level (Sarkar et al. 2006; Srinivasa et al. 2008; Elvis et al. 2009). In order to enhance the farm income with minimum investment, there is need to protect the land resources and manage them on sustainable basis. Remotely sensed data provides valuable and up-to-date spatial information on natural resources and physical terrain parameters. Using satellite data in natural resource management has proved to be an indispensable and decision-making tool, which can ensure optimum use of the resources and help in devising systems for judicious resource use and management practices (Hiese et al. 2011). The information thus generated can be analysed in a GIS that aids planners in decision-making as well as in scientific management of land resources. In the context of land use planning at micro level, geospatial techniques and various models have been developed for their effective use in sustainable development of natural resources by integration of various GIS layers, which further demonstrates that geospatial techniques help in generation of a reliable spatial and nonspatial databases (Kushwaha et al. 2010).

The soil maps are required on different scales to meet the requirements of planning at various levels. The larger the scale used, more will be the information available and vice versa. The coarse resolution data (spatial resolution 70 m or more) from IRS LISS-I, AWiFs and LANDSAT-MSS sensors were useful to prepare soil maps on 1:250,000 scale or smaller (Sehgal 1990). To map soils on 1:50,000 scale medium resolution data from LANDSAT TM, IRS LISS-II and SPOT-MLA are employed. Satellite data from IRS P6 LISS-IV, Cartosat-1 and Cartosat-2 and IKONOS are now being employed for detailed characterization of soils on 1:10,000 scale or larger (Dwivedi 2001). The smaller an area represented by pixel, the higher the resolution of the image (Dhinwa et al. 2010).

The high-spatial resolution remotely sensed data coupled with topographical data provides real-time and accurate information related to distinct geological formation and landforms. Dobos and Montanarella (2007) stated that the use of digital data sources, such as DEM and high-resolution satellite data, can speed up the compilation of digital soil databases and improve the overall quality, consistency and reliability of the database. However, current knowledge on soils and existing soil maps impart coarse scale information that is regional in nature than local and specific. Therefore, need is being felt for soil resource mapping at larger scale to generate fine level of detail (Ravisankar and Thamappa 2004 and Rao et al. 2004; Dobos and Hengle 2009) to use them for developmental planning in agriculture. Further, with the advent and availability of high-resolution satellite data from various Indian satellites like Resourcesat-1 and Cartosat-1/2, new vistas have been opened up for micro-level planning (Giri et al. 2016).

Village is measured to be a feasible micro-administrative unit for micro-level planning. India has a massive vacant potential of natural resources and familiar human resources in remote and rural areas; therefore, growth of spatial database infrastructure at village level will help in recording these resources into micro-level database. Village-level land use planning is the process of evaluating and proposing alternative uses of natural resources and to design, implement and revise land use plans in order to improve the living conditions of villagers (Ramamurthy et al. 2017). The optimal use of these natural resources depends mainly on the potential of people to utilize and manage them, their priorities, the socio-economic conditions and the carrying capacity of the natural resources. There are no specific tools at village level to assess land degradation dynamics, soil health or hydrological information, thereby limiting the application of adaptive management based on early detection of negative impacts.

The lack of tools has limited the usefulness of planning and decision-making to correct or mitigate the impact of current practices and to define land use policies by taking into account of local conditions and needs at village level, where land degradation is experienced the most (Manchanda et al. 2002). It is believed that this process only becomes effective when it is carried out in a participatory way, which means that the principal users of land, the villagers, are fully involved. To ensure full participation, it is important to consider the different socio-economic groups in a village (including gender), which have different interests and expectations. Land resources are under enormous pressure due to the fast-growing population in combination with stagnant land productivity and slow development of production sectors, which make less use of land resources. This has led to the expansion of settlements, agricultural areas, livestock grazing, tree cutting for fuel wood, water demand, etc. The increasing pressure on land for different uses is shown in Fig. 17.1. Village-level information system (VLIS) is a computer-based information system, which helps the decision-makers and planners to produce different eco-socio-economic views. Primary objectives are:

- To educate, train and promote informatics culture/experience to local officials
- To evaluate the need for appropriate common base map to be used by all rural development partners

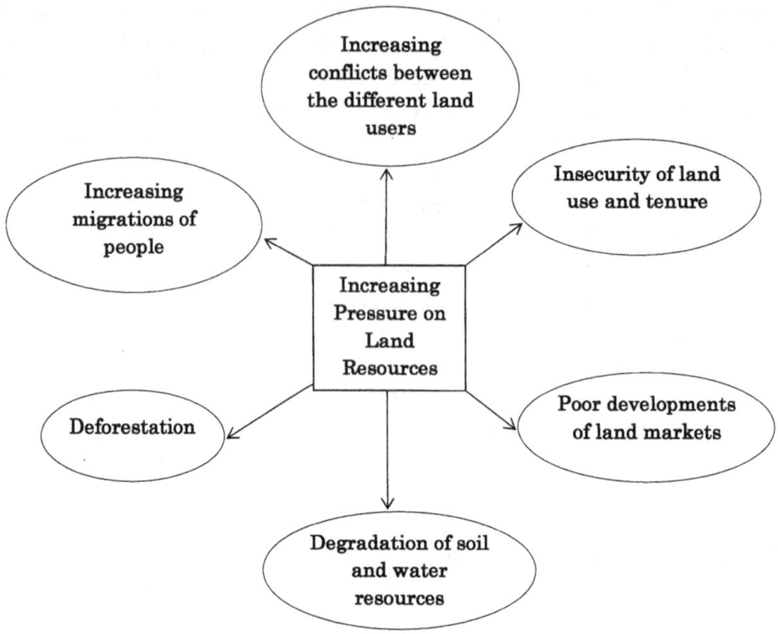

Fig. 17.1 Impact of increasing pressure on land for different uses

- To design own management packages by the rural extension communities according to local needs and conditions

VLIS tool is useful for grass-root level planning and future development. Also, the technology should be appropriate to the needs and resources available in the local area, be relatively simple yet should be able to perform the task, be cost-effective and be application- and user-oriented. Land resource mapping involves making knowledgeable decisions about land use and the environment. The resource conservation and development programmes are being taken up increasingly on parcel-level basis for better management.

17.3 Village-Level Land Resource Inventory: A Case Study

17.3.1 Study Area

Village-level land resource inventory was carried out in Miniwada Panchayat, Katol tehsil, 45 km to the west of Nagpur city of Maharashtra (Fig. 17.2). The panchayat includes three villages, namely, Miniwada, Mhasala and Malkapur, which lies between 21^0 08′ and 21^0 12′ north latitudes and 79^0 08′ to 79^0 15′ east longitudes

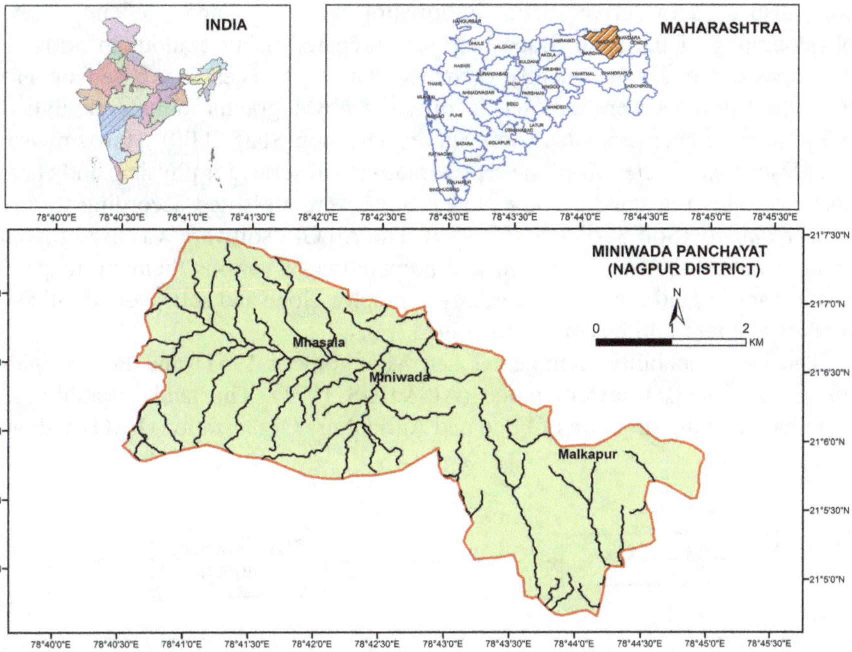

Fig. 17.2 Location map of Miniwada Panchayat, Katol tehsil, Nagpur district (Source: Sahu et al. 2016)

and covers an area of 1630 ha. The elevation ranges from 407 to 472 m above mean sea level (MSL).

The climate is mainly hot subtropical type with mean annual temperature of 28 °C and mean annual rainfall of 980 mm. The area qualifies for hyperthermic soil temperature regime and ustic soil moisture regime. The geology of the study area is covered by basaltic lava flows, commonly known as traps. The main field crops are cotton (*Gossypium* spp.), soybean (*Glycine max*), pigeon pea (*Cajanus cajan*), gram (*Cicer arietinum*), wheat (*Triticum aestivum*), etc. The natural vegetation comprises of teak (*Tectona grandis*), babul (*Acacia* spp.), palash (*Butea frondosa*), neem (*Azadirachta indica*), mahua (*Madhuca longifolia*), etc.

17.3.2 Methodology

GIS help in characterization of land resources and to prepare maps of land capability, land irrigability, various crop suitability and action plan for taking appropriate soil conservation and management measures for optimum utilization of the resources at parcel level. Two season (*rabi* and *kharif*) high-resolution satellite data were visually interpreted in association with Survey of India toposheet, and other ancillary data were also used to derive spatial information related to land use/land cover and physiography of the area. Slope map was prepared using contour information of the toposheet and subsequently corrected wherever necessary, based on ground truth. Soil profiles were exposed in different physiographic units and studied for morphometric characteristics (Soil Survey Division Staff 2000). Horizonwise soil samples were collected from the representative soil series for physical and chemical analysis following standard procedures. Soils were classified according to *Keys to Soil Taxonomy* (Soil Survey Staff 1999). The ArcGIS software was used for spatial and attribute database generation and preparation of various thematic maps. Thematic maps of land use/land cover, physiography, slope and soils were digitized and processed under GIS environment (Fig. 17.3).

The land capability (Klingebiel and Montgomery 1961) and land irrigability grouping of soils were computed (AIS&LUS 1971). The land suitability levels were based on the structure of Food and Agriculture Organization (FAO) and ranked

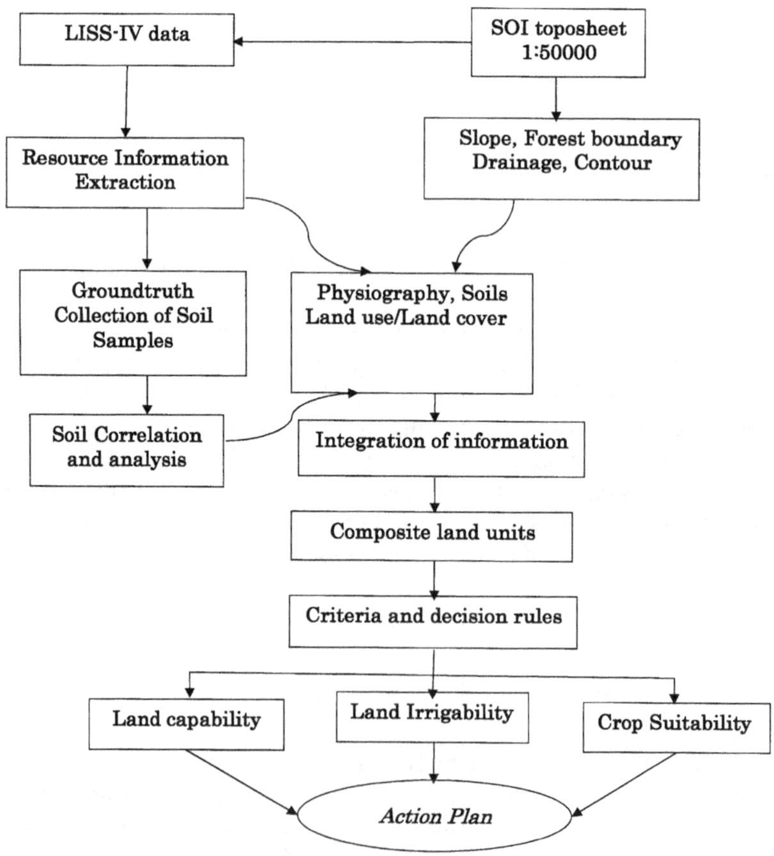

Fig. 17.3 Methodology for land resource mapping

as highly suitable (S1), moderately suitable (S2), marginally suitable (S3) and not suitable (N).

17.4 Results and Discussion

17.4.1 Terrain Analysis

Precise delineation of landforms is very important for parcel-level soil mapping (Martha et al. 2012; Nagaraju et al. 2014). Using hillshade information generated from the Cartosat-1 DEM and 3D perspective viewing of the area, various landforms were delineated based on visual interpretation. Slope information was derived from the high-resolution Cartosat-1 DEM and reclassified into different slope classes. Furthermore, the Cartosat-1-sharpened IRS-P6 LISS-IV data were used to segment the area into different land use/land cover classes.

17.4.2 Landforms Delineation

Using stereo vision, the area was characterized into plateau top (450–470 m), scarp slopes (450–460 m), plateau spurs (420–440 m), pediment (430–450 m), undulating plain (410–430 m), broad valley (410–430 m), narrow valleys (410–420 m) and floodplain (400–410 m). The major landforms were further subdivided based on elevation and topoposition. Pediments were subdivided into upper (440–450 m) and lower pediments (430–440 m), whereas the undulating plain was further subdivided into upper (420–430 m) and lower plains (410–420 m) (Fig. 17.4a).

17.4.3 Slope

After eliminating the speckle effects due to the high-resolution DEM, the raster slope map was reclassified into seven slope classes, viz. nearly level land (0–1%) covering 31.9%, very gently sloping land (1–3%) with 29.8% and gently sloping (3–5%) land with 11.9% area. Moderately sloping (5–10%), strongly sloping (10–15%), moderately steep to steep sloping (15–25%) and steep to very steep sloping (25–50%) lands occupy 8.2, 2.3, 7.3 and 8.6% of study area, respectively (Fig. 17.4b).

17.4.4 Land Use/Land Cover Analysis

Six land use/land cover classes were identified, namely, double crop, single crop, orchard, waste land with and without scrub and degraded forest based on visual

a

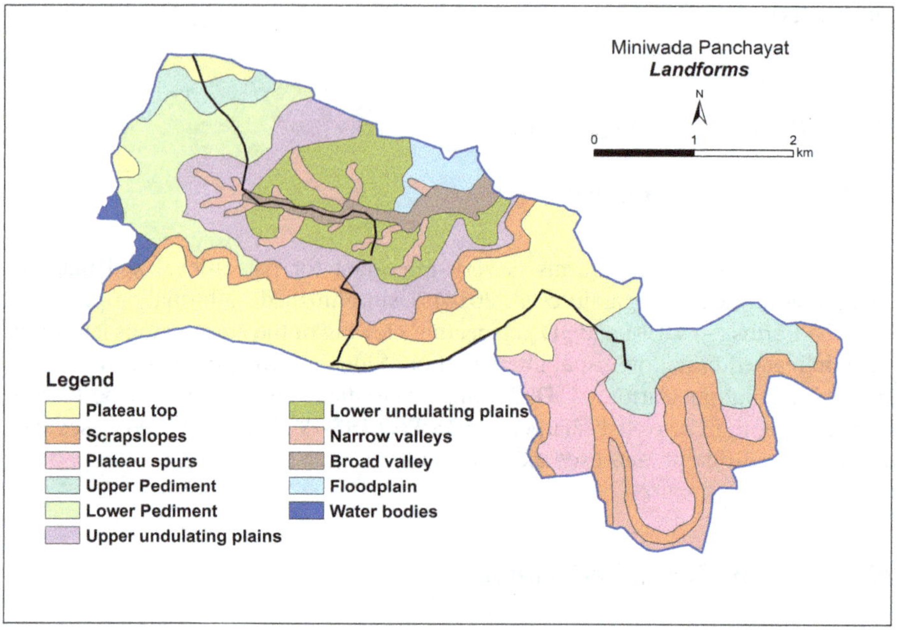

b

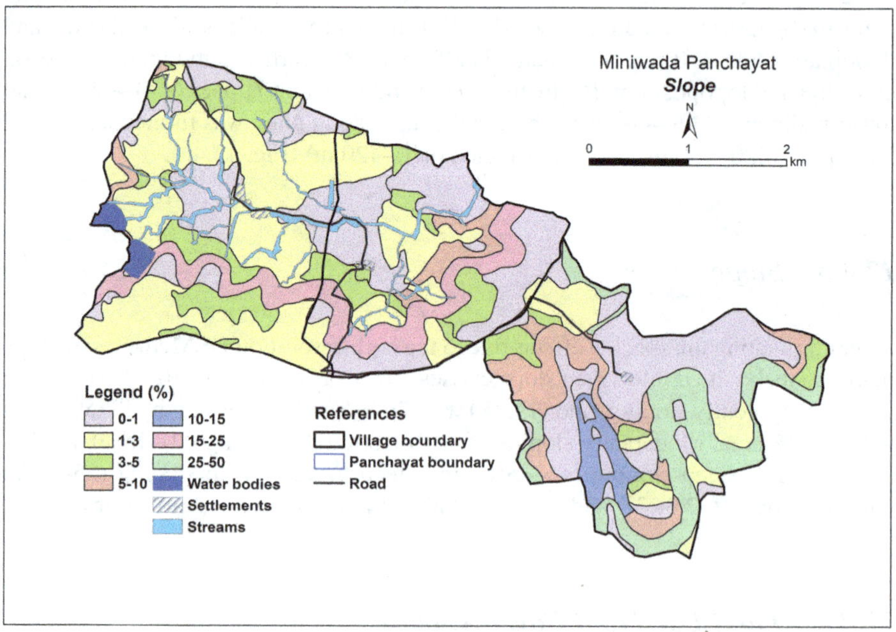

Fig. 17.4 Detailed mapping of (**a**) landforms, (**b**) slope, (**c**) land use/land cover (**d**) physiography-land use units, (**e**) profile locations and (**f**) phases of soil series of Miniwada Panchayat. (Source: Sahu et al. 2016)

c

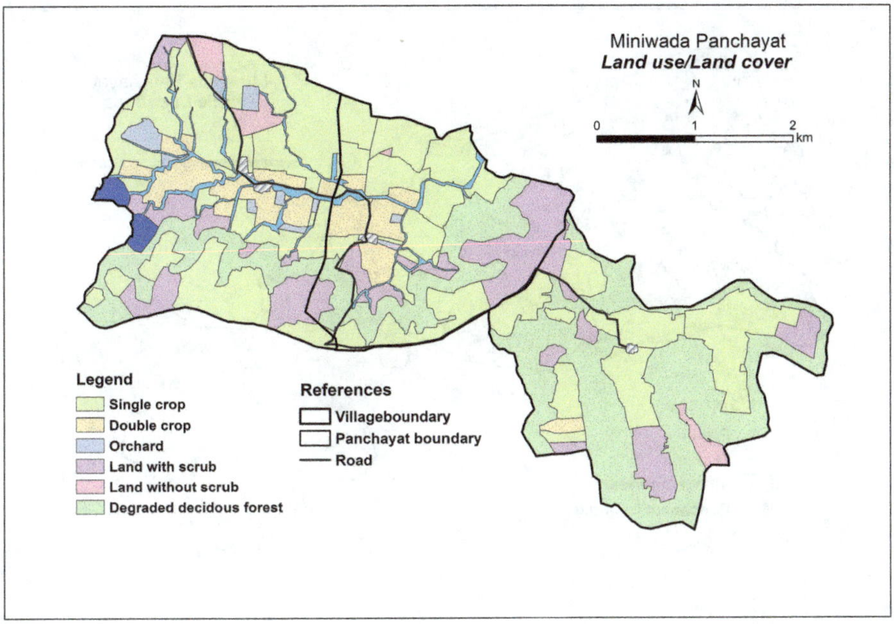

d

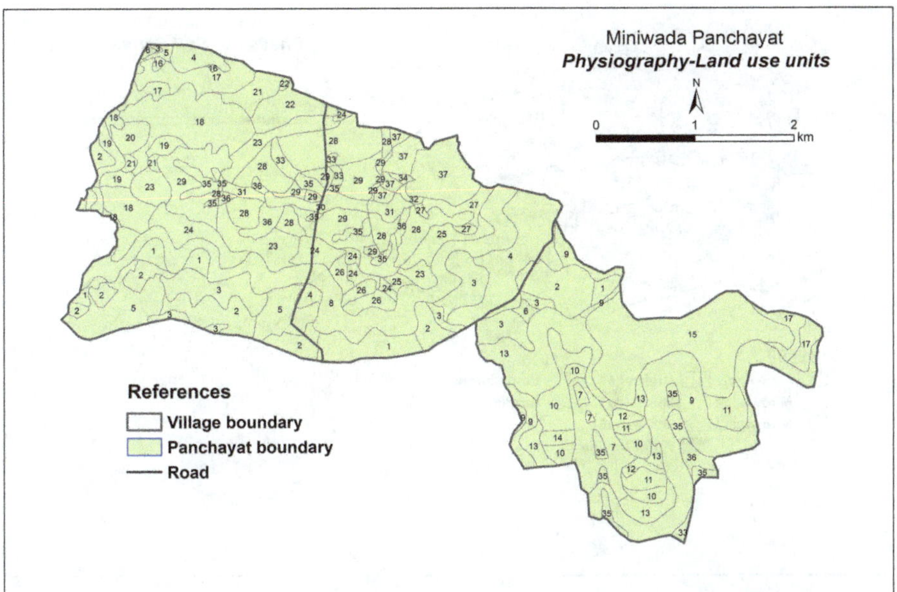

Fig. 17.4 (continued)

e

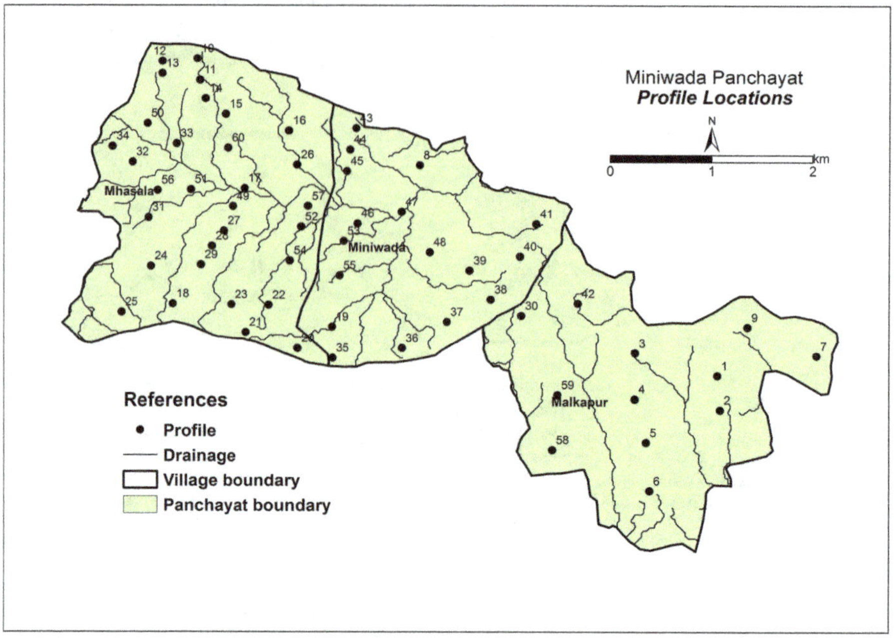

f

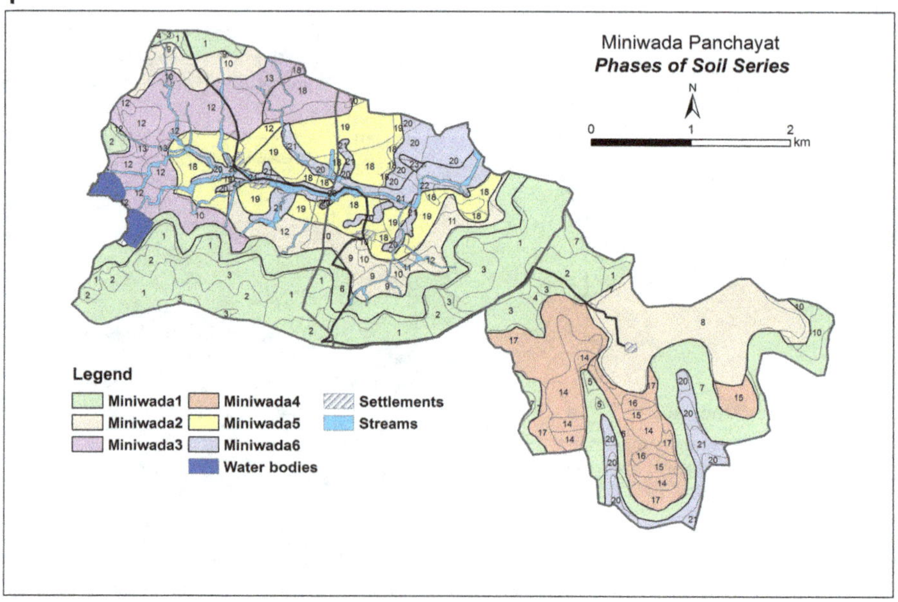

Fig. 17.4 (continued)

interpretation of two-season IRS-P6 LISS IV satellite data (Fig. 17.4c). Double-cropped area is mainly practiced in the landform which is dominant by depositional processes such as lower pediment, lower undulating plains and broad and narrow valleys with nearly level to gentle slopes. Area under wheat, gram and vegetables occupy 7.15% of TGA. Single crop occupies 41.55% of TGA and practiced in almost all landforms except scarp slopes due to strong to steep slopes. Soybean and cotton is spread in maximum area of cultivation. Orange orchards occupy 1.46% of TGA with nearly level to moderate slopes in lower pediment and upper undulating plains. Wasteland with and without scrub occupies an area of 13.26 and 2.16%, respectively. Degraded forest is prominent in scarp slopes covering an area of 30.25% with strong to steep slopes. The dominant vegetation comprises of teak (*Tectona grandis*), babul (*Acacia* spp.), palash (*Butea frondosa*), mahua (*Madhuca longifolia*), etc.

17.4.5 Delineation of PLU Units

The landform, slope and land use/land-cover maps were integrated in ArcGIS, and a PLU map was prepared. Based on integration, 37 PLU units were delineated in the study area (Fig. 17.4d), and the characteristics of each PLU unit were described (Table 17.2). On the plateau top, six PLU units (P11, P21, P31, P14, P24 and P45) were identified based on four slope classes (0–1, 1–3, 3–5 and 5–10%) and three land use/land cover classes (single crop, wasteland with scrub and without scrub). Three PLU units (S56, S66 and S76) were identified on the escarpment with three slope classes (10–15, 15–25 and 25–50%) and one land use/land cover class (degraded forest). Five PLU units (PS11, PS21, PS31, PS41 and PS12) were identified on the plateau spurs based on four slope classes (0–1, 1–3, 3–5 and 5–10%) and two land use/land cover classes (single crop and double crop). Three PLU units (UP11, UP21 and UP31) were identified on the upper pediment based on three slope classes (0–1, 1–3 and 3–5%) and one land use/land cover class (single crop). Four PLU units (LP21, LP22, LP23 and LP33) were identified on the lower pediment based on two slope classes (1–3 and 3–5%) and three land use/land cover classes (single crop, double crop and orchard). Five PLU units were identified on the upper undulating plain (UUP11, UUP21, UUP31, UUP41 and UUP24) based on four slope classes (0–1, 1–3, 3–5 and 5–10%) and two land use/land cover classes (single crop and wasteland with scrub). Lower undulating plains were further differentiated into three PLU units (LUP11, LUP21 and LUP12) based on the variation in slopes (0–1 and 1–3%) and land use (single and double crop). Three PLU units (B12, B22 and B32) were identified on the broad valley based on three slopes (0–1, 1–3 and 3–5%) and land use (double crop). Four PLU units (N21, N31, N12 and N22) were identified on the narrow valley based on three slopes (0–1, 1–3 and 3–5%) and land use (single and double crop). One PLU unit (F11) was identified on the floodplain based on slope (0–1%) and land use/land cover class (single crop). PLU-soil relationship was developed by correlating soil-site characteristics and physical and chemical

Table 17.2 Characteristics of PLU units of the study area

S. No.	PLU unit	Landform/slope/land use	Area (ha)	%
1.	P11	Plateau top with 0–1% slope, single crop	45.02	2.8
2.	P21	Plateau top with 1–3% slope, single crop	93.92	5.8
3.	P31	Plateau top with 3–5% slope, single crop	72.01	4.4
4.	P14	Plateau top with 0–1% slope, land with scrub	77.40	4.7
5.	P24	Plateau top with 1–3% slope, land with scrub	61.84	3.8
6.	P45	Plateau top with 5–10% slope, land without scrub	4.08	0.3
7.	S56	Scarp slopes with 10–15% slope, degraded forest	40.40	2.5
8.	S66	Scarp slopes with 15–25% slope, degraded forest	119.32	7.3
9.	S76	Scarp slopes with 25–50% slope, degraded forest	141.21	8.6
10.	PS11	Plateau spurs with 0–1% slope, single crop	45.27	2.7
11.	PS21	Plateau spurs with 1–3% slope, single crop	24.60	1.5
12.	PS31	Plateau spurs with 3–5% slope, single crop	7.79	0.5
13.	PS41	Plateau spurs with 5–10% slope, single crop	82.00	5.0
14.	PS12	Plateau spurs with 0–1% slope, double crop	4.91	0.3
15.	UP11	Upper pediment with 0–1% slope, single crop	123.52	7.6
16.	UP21	Upper pediment with 1–3% slope, single crop	1.93	0.1
17.	UP31	Upper pediment with 3–5% slope, single crop	52.93	3.2
18.	LP21	Lower pediment with 1–3% slope, single crop	80.25	4.9
19.	LP22	Lower pediment with 1–3% slope, double crop	17.36	1.1
20.	LP23	Lower pediment with 1–3% slope, orchard	9.85	0.6
21.	LP33	Lower pediment with 3–5% slope, orchard	17.02	1.0
22.	UUP11	Upper undulating plains with 0–1% slope, single crop	22.68	1.4
23.	UUP21	Upper undulating plains with 1–3% slope, single crop	59.98	3.7
24.	UUP31	Upper undulating plains with 3–5% slope, single crop	41.35	2.5
25.	UUP41	Upper undulating plains with 5–10% slope, single crop	17.83	1.1
26.	UUP24	Upper undulating plains with 1–3% slope, land with scrub	17.28	1.1
27.	LUP11	Lower undulating plains with 0–1% slope, single crop	12.76	0.8
28.	LUP21	Lower undulating plains with 1–3% slope, single crop	91.20	5.6
29.	LUP12	Lower undulating plains with 0–1% slope, double crop	86.91	5.3
30.	B12	Broad valley with 0–1% slope, double crop	12.90	0.8
31.	B22	Broad valley with 1–3% slope, double crop	11.24	0.7
32.	B32	Broad valley with 3–5% slope, double crop	2.08	0.1
33.	N21	Narrow valley with 1–3% slope, single crop	10.55	0.6
34.	N31	Narrow valley with 3–5% slope, single crop	1.55	0.1
35.	N12	Narrow valley with 0–1% slope, double crop	37.35	2.3
36.	N22	Narrow valley with 1–3% slope, double crop	19.93	1.2
37.	F11	Floodplain with 0–1% slope, single crop	50.72	3.1

properties of soils. Six soil series were identified mainly Miniwada-1(Mw-1), Miniwada-2(Mw-2), Miniwada-3(Mw-3), Miniwada-4(Mw-4), Miniwada-5 (Mw-5) and Miniwada-6(Mw-6) in major landforms covering an area of 38.68%, 16.11%, 11.22%, 10.78%, 12.13% and 10.41% of TGA, respectively, and soil map depicting 23 phases of soil series was developed (Fig. 17.4e, f).

17.5 Interpretation of Land Resources Data

According to land capability, lands suitable for cultivation are grouped into class I to class IV and from V to VIII for nonarable lands. The relative quality of the land decreases from I to IV (arable) and V to VIII. The land capability classes are further subdivided on the basis of the predominant hazard or limitation, viz. runoff (e), excess of water (w), root zone limitations (s) and climatic limitations (c), where soils have two kinds of limitations, both are indicated, the dominant one being used first. In land irrigability classification criteria, the lands are grouped into six irrigability classes according to degree of limitations for sustained use under irrigation. Subclasses of land irrigability class (s = soil, t = topography, d = drainage) are groups of land irrigability units that have the same kinds of dominant limitations for sustained use under irrigation.

17.6 Soil Suitability Evaluation

The suitability of soils was assessed as per FAO guidelines (Sys et al. 1991) with slight modification (NBSS & LUP 1994). According to the soil suitability criteria, the limitation levels of each characteristic (climate, site and soil) are defined and take into account the number and intensity of limitations. Five land suitability classes, viz. S1 (very suitable), S2 (moderately suitable), S3 (marginally suitable), NI (actually unsuitable and potentially suitable) and N2 (unsuitable) have been derived with covering an area of 22.99%, 42.50%, 9.06%, 14.99% and 10.01% of TGA, respectively. In the study, the soil-site suitability for cotton has been evaluated according to FAO framework, which provides a set of principles and concepts on which national and regional land evaluation systems can be constructed. Assessment of the land suitability needs integration of multidisciplinary database. A geospatial database on slope, soil depth, soil erosion, soil drainage, soil texture and other theme layers are overlay. Multi-criteria overlay analysis model was adopted to assign the class- and theme-wise weights for input parameters, and the weighted layers were integrated. In the composite layer, the higher the cumulative value, the least suitable the soil unit, whereas the lower the cumulative value, the higher the suitability of the unit.

17.7 Conclusions

Micro-level databases help to identify farm-specific problems and potentials using high-resolution satellite data in GIS environment. The variation in soil pattern was very well explained based on delineation of physiographic units considering landform, slope and land use. Seven major landforms plateau top, scarp slopes, plateau spurs, pediment, undulating plains, broad valley and narrow valley have been delineated using Cartosat-1 DEM, contour and hillshade. Detailed maps of slope and land use/land cover classes have been delineated. A total of 37 PLU units have been derived by integrating landform, slope and land use/land cover. Six soil series have been tentatively identified in major landforms and mapped into 37 mapping units with 24 phases of soil series. The study reveals that high-resolution IRS-P6 LISS-IV and DEM (10 m) derived from Cartosat-1 data will be of immense help in precise and faster mapping of soils as compared to the conventional method. However, it helps to guide agricultural planning and implementation of soil- and crop-related management technologies. It facilitates selection of potential crops, soil- and crop-specific fertilizers, reclamation of problem soil and use of suitable agricultural machineries and tools. Village-level soil resource database enables the local government in making policies on land use and agricultural production strategies and to formulate rural development programmes at farm level.

References

AIS & LUS (1971) Soil survey manual. IARI, New Delhi

Bhattacharyya T, Ram B, Sarkar D, Mandal C, Dhyani BL, Nagar AP (2007) Soil loss and crop productivity model in humid subtropical India. Curr Sci 93:1397–1403

Bodhankar RM, Srivastava R, Saxena RK, Prasad J (2002) Integrated approach of RS and GIS in characterization and evaluation of land resources for watershed management-A case study in (Dhyani et al.) Resource Conservation and Watershed Management: Technology Options and Future Strategies. IASWC, Dehradun, India, 435 p

Das DK, Bandopadhyay S, Chakraborty D, Srivastava R (2009) Application of modern techniques in characterization and management of soil and water resources. J Indian Soc Soil Sci 57:445–460

Dhinwa PS, Arya AS, Pathan SK, Vyas PR (2010) Generation of action plans at implementation level using IRS-P6, LISS-4 MX Data at 1:12,5000 Scale-A Case Study of Nalchha Microwatershed, Dhar District (M.P.): 1–4

Dhruvanarayana VV, Babu R (1983) Estimation of soil erosion in India. J Irri Drought Eng ASCE 109(4):419–434

Dobos E, Hengle T (2009) Soil mapping applications. Dev Soil Sci 33:461–479

Dobos E, Montanarella L (2007) The development of a quantitative procedure for soil scape delineation using digital elevation data for Europe. Chapter 9, digital soil mapping-an introductory perspective. Dev Soil Sci 31:107–117

Dwivedi RS (2001) Soil resource mapping: a remote sensing perspective. Remote Sens Rev 20:89–122

Elvis A, Prasad J, Nagaraju MSS, Srivastava R, Kauraw DL (2009) Use of remote sensing in characterisation and management of Dhamni micro-watershed of Chandrapur district of Maharashtra. J Indian Soc Remote Sens 37(1):129–137

FAO & ITPS (2015) Status of the World's Soil Resources (SWSR) – main report. Food and Agriculture Organization of the United Nations and Intergovernmental Technical Panel on Soils, Rome

Giri JD, Nagaraju MSS, Srivastava R, Singh DS, Nasre RA, Barthwal AK, Mohekar DS (2016) Accuracy assessment of large-scale soil map prepared by remote sensing approach. Int J Agric Stat Sci 12(1):229–237

Hiese N, Mich N, Yanthan M, Yeptho H, Moatoshi, Lotha R, Kruyie P, Suokhrie S, Murry B, Mor S (2011) Application of Geoinformatics in Land Resource Management at Micro-watershed level for sustainable development of Sanis, Wokha District, Nagaland. Geospatial World Forum, 18–21 January, Hyderabad, India

Kasturirangan K, Aravamudan R, Deekshatulu BL, Joseph G, Chandrasekhar MG (1996) Indian remote sensing satellite IRS IC. The beginning of new era. Curr Sci 70:495–500

Klingebiel AA, Montgomery PH (1961) Land capability classification. Agric. Handbook 210, SCS, USDA, Washington, DC

Kushwaha SPS, Mukhopadhyay S, Hari Prasad V, Kumar S (2010) Sustainable development planning in Pathri Rao sub-watershed using geospatial techniques. Curr Sci 98:1486

Laake PE (2000) Spatial tools for laymen, natural resource management by experts. Int Arch Photogramm Remote Sens 33:1593–1600

Liengsakul M, Mekpaiboonwatana S, Pramojanee P, Bronsveld K, Huizing H (1993) Use of GIS and remote sensing for soil mapping and for locating new sites for permanent cropland – A case study in the highlands of northern Thailand. Geoderma 60(1–4):293–307

Manchanda ML, Kudrat M, Tiwari AK (2002) Soil survey and mapping using remote sensing. Trop Ecol 43(1):61–74

Martha TR, Saha R, Vinod Kumar K (2012) Synergetic use of satellite image and DEM for identification of landforms in on Ridge-Valley Topography. Int J Geosci 3:480–489

Mediterra (2016) Zero waste in the Mediterranean. Natural Resources, Food and Knowledge/ International Centre for Advanced Mediterranean Agronomic Studies (CIHEAM) and Food and Agriculture Organization of the United Nations (FAO). Presses de Sciences Po, Paris

Nagaraju MSS, Kumar N, Srivastava R, Das SN (2014) Cadastral-level soil mapping in basaltic terrain using Cartosat-1 derived products. Int J Remote Sens 35(10):3764–3781

NBSS&LUP (1994) Soil-site suitability criteria for different crops. In: Proceedings of National meet on Soil-site suitability criteria for different crops. 31 p

Patil S, Nagaraju MSS, Srivastava R (2010) Characterization and evaluation of land resources of Basaltic terrain for watershed management using remote sensing and GIS. Indian J Soil Sci 38(1):16–23

Potdar SS, Srivastava R, Nagaraju MSS, Prasad J, Saxena RK (2003) Mapping of erosional soil loss in Nanda-Khairi watershed of Nagpur district of Maharashtra using remotely sensed data and GIS techniques. Agropedology 13:10–18

Pratap Narain (2008) Dryland management in arid ecosystem. J Indian Soc Soil Sci 56:337–347

Ramamurthy V, Singh SK, Chattaraj S, Ramesh Kumar SC, Reddy GPO (2017) Land resource inventory Towards Village level agricultural land use planning. In: Reddy GPO, Patil NG, Chaturvedi A (eds) Sustainable management of land resources – an Indian perspective. Apple Academic Press, Waretown, pp 643–668

Rao BRM, Fyzee MA, Wadodkar MR (2004) Utility of remote sensing data for mapping soils at various scales and levels. In: Venkatratnam L, Ravisankar T, Sudarshana R (eds) Soils and crops. NRSA Publication, Hyderabad

Ravisankar T, Thamappa SS (2004) Satellite data interpretation and analysis for soil mapping. In: Venkatratnam L, Ravisankar T, Sudarshana R (eds) Soils and crops. NRSA Publication, Hyderabad

Reddy GPO, Maji AK (2003) Delineation and characterization of geomorphological features in a part of lower Maharashtra metamorphic plateau, using IRS-ID LISS-III data. J Indian Soc Remote Sens 31(4):241–250

Reddy GPO, Nagaraju MSS, Ramteke IK, Sarkar D (2013) Terrain characterization for soil resource mapping using IRS-P6 data and GIS –a case study from basaltic terrain of Central India. J Indian Soc Remote Sens 41(2):331–343

Reddy GPO, Patil NG, Chaturvedi A (eds) (2017) Sustainable Management of Land Resources – an Indian perspective. Apple Academic Press, Waretown, p 769

Reddy GPO, Singh SK, Patil NG, Chaturvedi A (2017a) Land resource inventory, mapping and management – an indian perspective. In: Reddy GPO, Patil NG, Chaturvedi A (eds) Sustainable management of land resources – an Indian perspective. Apple Academic Press, Waretown ISBN-13: 978-1-77188-517-1, pp 1–30

Sahu N, Reddy GPO, Kumar N, Nagaraju MSS (2015) High resolution remote sensing, GPS and GIS in soil resource mapping and characterization: a review. Agric Rev 36:14–25

Sahu N, Singh SK, Reddy GPO, Kumar N, Nagaraju MSS, Srivastava R (2016) Large-scale soil resource mapping using IRS-P6 LISS-IV and Cartosat-1 DEM in basaltic terrain of Central India. J Indian Soc Remote Sens 44(5):811–819

Sarkar D, Gangopadhyay SK, Sahoo AK (2006) Soil resource appraisal towards land use planning using satellite remote sensing and GIS – a case study in Patloinala micro watershed, district Puruliya, West Bengal. J Indian Soc Remote Sens 35(1):21–30

Sehgal JL (1990) Soil series criteria and norms. Soil Series Bull. l, National Bureau of Soil Survey and Land Use Planning, Nagpur

Sehgal J, Abrol IP (1994) Soil degradation in India: status and impact. Oxford and IBH Publishing Co. Pvt Ltd, New Delhi

Sharma IP, Saxena RK, Nagaraju MSS, Srivastava R, Prasad J, Kumar S, Barthwal AK (2004) Utilization of remote sensing data for characterization of land resources of Khursapar village, Nagpur district, Maharashtra. Abstracts, 248 ISRS Annual Convention and National Symposium, held at Jaipur from November 3–5, 2004, pp 198–199

Shrestha HL (2006) Using GPS and GIS in participatory mapping of community forest in Nepal. Electron J Inf Syst Dev Countries 25(5):1–11

Soil Survey Division Staff (2000) Soil survey manual, (Indian Print) Handbook No 18. USDA, Washington, DC

Soil Survey Staff (1999) Keys to soil taxonomy, 8th edn. USDA, Washington, DC

Srinivasa SV, Givindaiah S, Gowda HH (2008) Prioritization of sub-watersheds for sustainable development and management of natural resources: an integrated approach using remote sensing, GIS and socio-economic data. Curr Sci 95:345–354

Surya JN, Sidhu GS, Lal T, Sharma JP (2008) Macro and micro level soil resource mapping for soil erosion assessment and conservation in Shiwalik region of Himalayas, India. Proceedings of 15th International Congress on "Soil and Water Conservation, Climate Change and Environmental Sensitivity" of the International Soil Conservation Organisation (ISCO) held at Hungry 1–8

Sys CE, Ranst V, Debayeve J (1991) Land evaluation, Part I and II. Re-edited volumes of Publication No. 7. General Administration of Cooperation and Development, Brussels

Velmurugan V, Carlos GG (2009) Soil resource assessment and mapping using remote sensing and GIS. J Indian Soc Remote Sens 37:511–525

Walia CS, Singh SP, Dhankar RP, Ram J, Kamble KH, Katiyar DK (2010) Watershed characterization and soil resource mapping for land use planning of Moolbari watershed, Shimla, H.P. in lesser Himalayas. Curr Sci 98(2):176–182

Walke N, Reddy GPO, Maji AK, Thayalan S (2012) GIS-based multi-criteria overlay analysis in soil- suitability evaluation for cotton: a case study in the black soil region of Central India. Comput Geosci 41:108–118

Yeung AKW, Lo CP (2002) Concepts and techniques of geographic information systems. Int J Geogr Inf Sci 17(8):819–820

Ziadat F, Berrahmouni N, Grewer U, Bunning S, Bockel L, Oweis T (2015) Reversing land degradation in the drylands: scaling out and monitoring proven sustainable land management options. In Griffiths J (ed.) Living land. United Nations Convention to Combat Desertification, pp 14–17. Available at www.unccd.int/Lists/SiteDocumentLibrary/Publications/Living_Land_ENG.pdf

Part III
Geospatial Technologies in Land Resources Monitoring

Chapter 18
Remote Sensing for Land Resource Monitoring and Management

Suresh Kumar

Abstract Remote sensing and geographic information system (GIS) have emerged as the most effective tools in generating up-to-date, reliable information on soil and land resources of large regions in a cost-effective and time-efficient manner. Advances in spatial, spectral, and radiometric resolutions of the remote sensing sensors are providing a wide opportunity to the user to characterize soil and land resources. Multispectral broadband remote sensing data have been used for mapping soil types/classes, whereas hyperspectral remote sensing data have been used in quantifying surface soil properties showing their potential application in soil health monitoring. Soil and terrain parameters derived from remote sensing or soil proxies can be used as secondary variables to improve the interpolation of existing soil data and in deriving digital soil map. Availability of temporal remote sensing data in various spatial and spectral resolutions has provided edge in characterizing, mapping, and monitoring of degraded lands. Several web resources have provided various thematic information of soil, land use/land cover, vegetation/forest cover, geomorphology, hydro-geomorphology, geology, terrain, groundwater level, drainage network, watershed boundary, climate, etc. at various scales. Land evaluation methods like land capability classification, parametric methods of land evaluation, FAO framework of land evaluation, agroclimatic suitability, soil quality index, etc. are commonly used to assess the potential of land employing biophysical information. Recent advances in remote sensing technology provide enormous opportunity to resource managers to monitor the state of land degradation as well as to assess the potential of land resources to ascertain their optimal use in insuring food security.

Keywords GIS · Land resource monitoring · Land evaluation · Land resources · Remote sensing

S. Kumar (✉)
ISRO-Indian Institute of Remote Sensing, Dehardun, India
e-mail: suresh_kumar@iirs.gov.in

G. P. O. Reddy, S. K. Singh (eds.), *Geospatial Technologies in Land Resources Mapping, Monitoring and Management*, Geotechnologies and the Environment 21,
https://doi.org/10.1007/978-3-319-78711-4_18

18.1 Introduction

Extensive and reliable information on soil and land resources are prerequisites for efficient and effective management of vital natural resources (Kumar 2017; Dwivedi 2001). Generation of spatial database on soil and land resources by conventional method is a time-consuming and highly expensive process. Satellite remote sensing by virtue of its synoptic, multispectral, and better temporal and spatial sampling capability has opened new vistas in the mapping technology. It provides real-time information about various natural resources over a large region in a short period of time. The soil and terrain information is required for land resource planning, monitoring of environmental impact, and policy-making for sustainable land use (Mulder et al. 2011).

Remote sensing and GIS have emerged as very powerful tools in soil and land resource management. GIS, global positioning system (GPS), and image processing software systems for processing RS data form the basic components of the geo-spatial information technology. The primary role of remote sensing in land resource management is to provide soil and terrain information as well as land under various land use/land cover types. It serves as the most effective tool in monitoring of land resources. GIS is used to store spatial information in a geocode, and it allows integration of spatial information of various thematic layers. GIS is an analytical tool capable to perform storing spatial operations, spatial queries, data linkages, data matching, and output generation. It utilizes remote-sensed data and also enhances the effectiveness of this data through correlation of remote sensing inputs with data already stored in GIS. The increasing availability of temporal remote sensing makes it highly useful to develop the land monitoring systems. Remote sensing satellites provide a repetitive and consistent view of earth facilitating the ability to monitor the earth system.

Land evaluation for agricultural planning is carried out based on land resource data integrated with a GIS technology. GIS coupled with satellite data provides decision-makers a unique view of the landscape that enables land managers to enhance efficient use of natural resources at national, state, and district levels (Bandyopadhyay et al. 2009). High-spatiotemporal remote sensing data emerged as the most potential tool in monitoring land resources in respect to their use (Thenkabail 2015), soil health, wetlands, and land degradation condition (Metternicht and Zinck 2003). Land degradation due to anthropogenic activities is a major challenge that needs to be monitored for their sustainable use and enhancing its potential. Remote sensing data and GIS technologies and their application in assessing and monitoring of land resource potential have been discussed in the present chapter.

18.2 Remote Sensing

Remote sensing technology is a powerful tool for generating large amount of data related to natural resources in a short time and can be a prominent source of information for GIS. Remote sensing is the science of obtaining and interpreting information from a distance, using sensors that are not in physical contact with the object being observed (Lillisand et al. 2015). Remote sensing employs passive and/or active sensors. Passive sensors are those which sense natural radiations, either reflected or emitted from the earth. The land surface images acquired by the onboard satellite sensors in different wavelengths (reflected, scattered, and/or emitted) are radiometrically and geometrically corrected before extraction of spectral information (Guo et al. 2016). Remote sensing data provides a synoptic view and a repetitive coverage with calibrated sensors to detect changes in land surface at different resolutions which is a best source of information compared to traditional methods. Satellite observations of land, oceans, and atmosphere, specifically during natural and human-induced hazards, have become crucial for protecting the global environment, reducing disaster losses, and achieving sustainable development (Bello and Aina 2014). It plays a significant role in providing geo-information in a spatial format and also in determining, enhancing, and monitoring the overall capacity of the earth.

Over the past 40 years, satellite remote sensing (SRS) has shown increased utility in providing state of soil, terrain, land use/land cover, vegetation, ecosystem, and other land resource information at various scales ranging from landscape to global spatial scales (Lulla et al. 2012). Several satellite missions, such as Landsat, ASTER, SPOT, IRS, IKONOS, and QuickBird series, are widely accessible and provide very economical spatial data to generate information of land resources ranging from regional to large scale (Toth and Jozkow 2016). Very high-spatial resolution (pixel size) satellite data such as IRS Cartosat-I and Cartosat-II, QuickBird, and IKONOS are offering high potential in deriving land information at large scale for detail planning of natural resources at grassroot level (Radhadevi et al. 2010) (Table 18.1).

18.3 Hyperspectral Remote Sensing

Multispectral broadband remote sensing data from sensors, such as SPOT HRV, Landsat MSS, Landsat TM, ASTER, and IRS satellite data, have been used for mapping soil as they do not provide sufficient information to characterize soil differences, because their 100–200 nm bandwidth is unable to resolve diagnostic spectral features of terrestrial materials (De Jong 1994). Hyperspectral remote sensing data provided large volumes of high spectral and spatial resolution data in a number of narrow contiguous spectral bands. The spectral characteristics of soils

Table 18.1 Salient characteristics of commonly used remote sensing satellites for land resource inventory

Remote sensing satellites	Spatial resolution (meter)	Spectral bands	Spectral range (μm)	Country
Multispectral				
Landsat TM	15–60	8	0.45–12.50	NASA, USA
SPOT-5	10 (PAN), 20 (MS)	5	0.43–1.75	SPOT image, France
ASTER	15–90	15	0.52–11.65	NASA, USA
MODIS	250–1000	36	0.40–14.40	NASA, USA
QuickBird	0.82 (PAN), 3.2 (MS)	1 (PAN), 4 (MS)	0.45–0.90	Digital lobe, USA
ALOS/PRISM	2.5	1 (PAN)		
IRS-P6				ISRO, India
AWiFS	56	4	0.52–1.70	
LISS-III	23.5	4	0.52–1.70	
IRS-Resourcesat				ISRO, India
LISS-IV	5.8	3	0.52–0.86	
Cartosat-1 and Cartosat-2	2.0	1 (PAN)	0.52-	ISRO, India
Hyperspectral				
Hyperion	30	242	0.40–2.5	NASA, USA
Microwave				
SRTM	30	C-band	4–8 cm	NASA, USA
		X-band	2.5–4 cm	
RADARSAT-2	3–100	C-band	4–8 cm	NASA, USA
ALOS-POLSAR scansar	100	L-band	15–30 cm	ESA
ALOS-POLSAR polarimetric	24–88	L-band	15–30 cm	ESA
SMOS (passive)	1000	L-band	1.4 GHz	ESA, Europe
RISAT-1 and RISAT-2	1–55	C-band	5.35 GHz	ISRO, India

are governed by the combined effect of physicochemical properties such as soil texture, moisture, structure, soil mineral composition, and organic matter (Baumgardner et al. 1985). Thus, it is difficult to obtain spectra representing a typical soil type, although it is possible to find pure spectra that develop due to certain soil processes such as soil salinity/alkalinity, erosion features, etc. Hyperspectral remote sensing data allowed better discrimination of land surface and helped to increase the accuracy in mapping of land surface features (Clark et al. 1990). Characteristic features for salt-affected soils are mostly observed in narrow wavelength bands (Weng et al. 2008). Several studies had shown that the hyperspectral data can be used to quantify characteristics of saline soils at various scales (Ben-Dor et al. 2002; Farifteh et al. 2008). Das et al. (2015) reviewed the potential of hyperspectral remote sensing data in soil studies and operational application in soil health monitoring in India. They felt the need of a dedicated spaceborne mission from polar orbits would provide further impetus to the application of high-resolution satellite remote sensing for soil assessment by providing quality hyperspectral data.

18.4 Microwave Remote Sensing

Microwave remote sensing comprises of both active and passive forms of remote sensing. It covers the spectral wavelength range from approximately 1 cm to 1 m. Because of their long wavelengths, compared to optical remote sensing, microwave radiation can penetrate through cloud cover, haze, dust, and all but the heaviest rainfall. Active microwave sensing radar transmits electromagnetic radiation to the earth surface and measures the amount of radiation scattered back at the same position as where the radiation was originally transmitted. Passive microwave sensing is similar in concept to thermal remote sensing. All objects emit microwave energy of some magnitude, but the amounts are generally very small. A passive microwave sensor detects the naturally emitted microwave energy within its field of view. This emitted energy is related to the temperature and moisture properties of the emitting object or surface. Applications of passive microwave remote sensing include meteorology, hydrology, and oceanography.

Remote sensing data from synthetic aperture radar (SAR) offers high potential in mapping and monitoring of land resources. They provide their own source of energy to record backscatter radiation through SAR system. Application of SAR data for mapping land has increased manyfold in the recent past. Spaceborne SAR systems European Remote Sensing (ERS-1 and ERS-2), Advanced Synthetic Aperture Radar (ASAR), Japanese Earth Resources Satellite (JERS-1), RADARSAT-1 and RADARSAT-2, Advanced Land Observation Satellite (ALOS-1), and RISAT-1 and RISAT-2 are commonly being used in land studies. SAR techniques are particularly useful as they provide to monitor soil parameters under any weather conditions (Ulaby et al. 1986; Kornelsen and Coulibaly 2013). The SAR data has applications for land use/land cover, vegetation, forest degradation, forest cover, crop type and monitoring, soil moisture and wetland mapping, etc. SAR data have been used in combination with optical data for improved crop classifications (Blaes et al. 2005) and mapping land management regimes (Price et al. 2002).

18.5 Land Resource Assessment

The components of the natural land unit can be termed land resources, including physical, biotic, environmental, infrastructural, social, and economic components, inasmuch as they are fixed to the land unit. Convention to Combat Desertification

(UNEP 1994) defined land as "a delineable area of the earth's terrestrial surface, encompassing all attributes of the biosphere immediately above or below this surface, including those of the near-surface climate, the soil and terrain forms, the surface hydrology (including shallow lakes, rivers, marshes, and swamps), the near-surface sedimentary layers and associated groundwater reserve, the plant and animal populations, the human settlement pattern and physical results of past and present human activity." Natural land units are characterized by land characteristics. Land characteristics are single attributes of the land that connote a certain behavior. A land quality is a complex or compound characteristic of the land which acts in a manner distinct, and largely independent, from the actions of other land qualities in its influence on the suitability of land for a specified kind of use (FAO 1976).

Information on land resources is vital for sustainable management of land resources. Land resource information generated using remote sensing data in the past are now increasingly being available on various web resources. The information of soil, land use/land cover, vegetation/forest cover, geomorphology, hydro-geomorphology, geology, terrain, groundwater level, drainage network, watershed boundary, climate, etc. are now available at various scales meeting the requirement of users from regional- to local-scale resource planning (Table 18.2).

18.6 Soil Resource Inventory

Satellite remote sensing data are being extensively used for soil resource inventory (Dwivedi 2001; Mulder et al. 2011). It is primarily used in segmenting landscape into more or less homogenous soil–landscape unit. Soils in each soil–landscape map unit are characterized by dominant soil type. The soil–landscape model (Catena concept) captures the relationships between the soils in the area and the different landscape units. The soil surveyor detects different soil formative environments through visual interpretation of geological, geomorphological, and topographical maps in conjunction with satellite images (Manchanda et al. 2002). The spatial extents of the soil formative environments are then used to delineate soil–landscape units known as physiographic units. The soil surveyor first establishes the soil–landscape model over the area through field investigation. Soil profiles are characterized and classified for each soil–landscape unit, and the field boundary of physiographic units is checked and compiled on the base map. The soil surveyor uses satellite data either by digital image analysis or through visual interpretation of satellite imagery (Hilwig and Karale 1973; Hilwig 1975). The standard false color composite (FCC) satellite imagery on 1:50,000 scale is widely being used for soil resource inventory in a semi-detailed and reconnaissance survey (Karale et al. 1991). Remote sensing methods facilitate mapping inaccessible areas by reducing the need for extensive time-consuming and costly field surveys.

The application of satellite remote sensing data products for small- and medium-scale soil mapping is widely accepted, but they have not been used for large-scale soil mapping due to coarse resolution of satellite data. With advances in terms of

Table 18.2 Web resources for development of land resource information

	Thematic layers	Scale	Website	Organization
1.	*Land use /land cover*			
(i).	Global coverage	30 m (resolution)	https://landcover.usgs.gov/glc/ http://glcf.umd.edu/data/	GLCF, Maryland, USA
(ii).	Indian region	1: 2,50,000 1: 50,000 1:10,000	http://bhuvan.nrsc.gov.in/gis/thematic/index.php	NRSC, ISRO, Hyderabad
(iii).	Forest cover	1:50,000-	http://bhuvan.nrsc.gov.in/	NRSC, ISRO, Hyderabad
2.	*Digital elevation models (DEMs)*			
(i).	Global: SRTM	Spatial resolution 90 m 30 m	http://glcf.umd.edu/data/srtm/	GLCF, Maryland, USA
(ii).	Global: ASTER	Spatial resolution 30 m	https://asterweb.jpl.nasa.gov/gdem.asp	JPL, NASA, USA
(iii).	Indian region: CartoDEM	Spatial resolution 30 m	http://bhuvan.nrsc.gov.in/data/download/index.php	NRSC, ISRO, Hyderabad
3.	*Soil map*			
(i).	Global: FAO	1:5000000	http://www.fao.org/soils-portal/soil-survey/soil-maps-and-databases/harmonized-world-soil-database-v12/en/	FAO
(ii).	Indian region	1:500,000	https://www.nbsslup.in/	ICAR-NBSS&LUP, Nagpur
(iii).	India: River Valley projects	1: 50,000	http://slusi.dacnet.nic.in/	Soil and Land Use Survey of India, New Delhi
(iv).	India: Selected block of the districts under IMSD projects	1:50,000		NRSC, ISRO, Hyderabad
4.	Geomorphology	1: 50,000	http://bhuvan.nrsc.gov.in/gis/thematic/index.php	NRSC, ISRO, Hyderabad
5.	Watershed/sub-watershed	1:50,000	http://bhuvan.nrsc.gov.in/gis/thematic/index.php	NRSC, ISRO, Hyderabad
6.	Land degradation	1:50,000	http://bhuvan.nrsc.gov.in/gis/thematic/index.php	NRSC, ISRO, Hyderabad
7.	Wasteland	1:50,000	http://bhuvan.nrsc.gov.in/gis/thematic/index.php	NRSC, ISRO, Hyderabad
8.	Climate data	Long term data	http://www.indiawaterportal.org/met_data/ http://www.mosdac.gov.in/	Ministry of Water Resources, New Delhi

Table 18.3 Typical scale for land resource inventory and required vis-a-vis Indian remote sensing sensors

Resource inventory level	Scale	Required spatial resolution	Remote sensing sensor/ satellite	Remarks
Exploratory	1: 1 M	100–1000 m	IRS WiFS, MODIS	Country level, LULC level I
Reconnaissance	1: 2,50,000	25–100 m	IRS LISS-II and LISS-III	State level, LULC level II
Semi-detailed	1: 50,000	5–25 m	IRS LISS-III	District level, LULC level III
Detailed	1: 20,000	2–5 m	IRS LISS-IV	Block level, LULC level IV
Very detailed	1: 10,000	1–2 m	Cartosat – PAN and LISS-IV + PAN merged	Village level

spatial, spectral, and radiometric resolutions of the sensors, studies have been initiated to characterize soils at large scale through physiography–land use–soil relationship (Srivastava and Saxena 2004). However, with the availability of high-resolution multispectral data, LISS-IV (5.8 m) from Resourcesat-1 and Resourcesat-2 is being used for large-scale soil mapping, i.e., up to 1:20,000 scale. Cartosat-1 and Cartosat-2 with spatial resolution of 2.5 m and 1 m, respectively, are available today for mapping soils at large scale of 1:10,000 or larger suited for micro-level planning. Standard soil surveys are normally carried out in different mapping levels, e.g., reconnaissance (1:2, 50,000 scale), semi-detailed (1: 50,000 scale), and detailed (1: 20,000 or larger), depending on the requirement of the area. Different satellite sensors and their use in land resource inventory are shown in Table 18.3.

18.7 Digital Soil Mapping

Digital soil mapping is the computer-assisted production of digital maps of soil type and soil properties. It typically implies use of a mathematical and statistical model that combines information from soil observations with information contained in correlated environmental variables and remote sensing data. The advent of GIS-based digital terrain modeling has facilitated soil–landscape analysis quantitatively (Mahmoudabadi et al. 2017). Several researchers developed quantitative concept for soil–landscape models using an integration of digital terrain analysis and statistical modeling methods to map predicted soil properties (McSweeney et al. 1994; Moore et al. 1993). Moreover, remotely sensed imagery is used as a data source supporting digital soil mapping (Ben-Dor et al. 2008; Slaymaker 2001).

Digital terrain analysis is used to analyze and model the land surface and relationships between the topography and geological, hydrological, biological, and anthropogenic components of landscape. However, by combining remotely sensed imagery, DEMs, and soil sample data using digital soil mapping (DSM) methods, a complete coverage can be produced, and the accuracy of the estimated soil properties can be improved (Grunwald 2009). Moore et al. (1993) and Gessler et al. (1995) showed strong correlation and predictive utility between terrain parameters with several soil properties. Soil and terrain parameters derived from remote sensing or soil proxies can be used as secondary variables to improve the interpolation of existing soil data (Kumar and Singh 2016). Soil spectroscopy has been recognized as a potentially effective and cost-efficient technology, but it is not yet routinely used in soil surveys.

Recent development in computer software enables the production of digital elevation models (DEMs) using remote sensing data. DEMs are used to generate a 3-D model of the landscape that enables precise delineation of land surfaces and geomorphological features and to establish a relationship between landscape elements to delineate soil–landscape units through visual interpretation. Information derived from DEMs, such as elevation, slope, and aspect maps, can be used with the remote sensing imageries to improve their capabilities for soil mapping. Several workers have investigated automated soil mapping using DTMs.

18.8 Terrain Information

Terrain information is needed for land resource management planning, environmental analysis, and monitoring the environmental impact of development. Terrain information has been increasingly derived from digital representation of topography, generally called the DEM (Moore et al. 1991; Martz and Garbrecht 1992). The most widely used sources of DEM data are SRTM derived from SAR and stereo-correlation of images; ASTER Global Digital Elevation Map (GDEM), created by stereo-correlation of ASTER imagery; and CartoDEM of India generated from IRS Cartosat-1 stereo pair of PAN data. A fine-resolution (2.5 m) DEM can be generated with the ALOS/PRISM, which is a panchromatic remote sensing instrument specially designed for stereo mapping (Earth Observation Research Center 2010). Terrain information such as slope, aspect, configuration of land surface, and geomorphic characteristics are derived from DEMs that helped in land resource appraisal, watershed management, and land degradation assessment. It has been widely used in watershed delineation, natural drainage network generation, and drainage density estimation. Watershed management requires geomorphologic parameters, viz., relative relief, shape factor, circulatory ratio, bifurcation ratio, drainage density, and hypsometric integral (HI), for watershed prioritization and implementation of soil and water conservation measures.

18.9 Inventory of Degraded Lands

Land degradation, defined here as "a reduction in the capability of the land to support a particular use," is considered to be one of the major problems facing the world (UNEP 1992). There are various types of land degradation, including soil erosion, salinization, soil contamination, loss of soil organic matter, decline in nutrient levels, acidification, and loss of soil structure. Land degradation processes such as salinization, alkalization, waterlogging, and soil erosion are a major cause to the development of degraded lands. It results in a decline in the quality of the land caused by human activities. Conventional techniques available for identifying and monitoring of these salt-affected soils are expensive and time-consuming and require intensive sampling to characterize spatial variability (Shepherd and Walsh 2002). In the recent past, remote sensing techniques have been established as most cost-effective tool for mapping of degraded lands. Availability of temporal remote sensing data in various spatial and spectral resolutions has provided edge in characterizing degraded lands. Satellite remote sensing and image processing techniques have enabled the detection, mapping, and monitoring of degraded lands in a timely and cost-effective manner.

18.10 Land Evaluation

Remote sensing data are used as a tool for generating the biophysical information, which helps to evolve the optimal land use plan for the sustainable development of an area. On the other hand, the use of GIS for handling a large database on soils and land is becoming increasingly common. Here, land evaluation methods commonly used to assess the potential of land employing biophysical information are discussed. These methods are grouped into qualitative and quantitative. These methods are essential for the proper management of land and optimal/potential land use planning (Kumar et al. 2002). A land resources planning is the process of evaluation of options and subsequent decision-making, which precedes implementation of a decision or plan. It is based on the socioeconomic conditions and expected developments of the population in and around a natural land unit. It involves making knowledgeable decisions about land use and the environment. Soil information is a vital component in the planning process, reflecting directly upon land use suitability. There are various land evaluation methods used to assess the potential and productivity of soil for agricultural purposes. Some of them are briefly described below:

- Land capability classification
- Soil and land irrigability classification
- Parametric methods of land evaluation
- Land productivity index (Storie 1976)
- Soil productivity index (Ricquier et al. 1970)
- FAO Framework of land evaluation (1976)

18.11 Land Capability Classification

Land capability is the inherent physical capacity of the land to sustain a range of land uses and management practices in the long term without degradation to soil, land, air, and water resources (Dent and Young 1981). It is a function of landscape features and processes and is influenced by terrain, soil, and climatic attributes and their interactions. It is based on an assessment of the biophysical characteristics of the land and the current technology that is available for the management of the land. It helps in providing guidance on the inputs and management requirements associated with different intensities of agricultural land use.

Land capability classification is a systematic classification of different kinds of land according to those properties, which determine the ability of the land to produce common cultivated crops and pasture plants virtually on permanent basis (Khybri 1979). The land is divided into eight capability classes, which are numbered in Roman numerals from I to VIII. These eight classes are grouped into two land use suitability groups, viz., (i) "land suited for cultivation and other uses" (Class I to Class IV) and (ii) "land not suited for cultivation, but suitable for other uses" (Class V to Class VIII). The land capability classes are based on the degree of erosion hazards and the intensity of limitations for use (Dhruva Narayana et al. 1990).

18.12 Soil and Land Irrigability Classification

The interpretation of soil and land conditions for irrigation is concerned primarily with predicting the behavior of soils under the greatly altered water regime brought about by irrigation. To carry out land irrigability classification of command areas, special interpretation and classification of soils for sustained use under irrigation are often required. The classification is based on effective soil depth, soil texture, soil permeability, coarse fragments, etc. Soil irrigability classes are assessed according to the suitability of soils for sustained use under irrigation, and they are grouped into different soil irrigability classes, whereas land irrigability classes are assessed according to the suitability of land for irrigation depending on physical factors like quality and quantity of irrigation water and socioeconomic factors like land development costs, provision of drainage facilities, and production costs of individual crops.

18.13 Parametric Methods of Land Evaluation

Semi-quantitative land evaluation methods such as parametric assessments are positioned halfway between qualitative and quantitative methods. Multiplying systems assign separate ratings to each one of several land characteristics or factors and then take the product of all factor ratings as the final rating index. These systems have the advantage in that any important productivity factor controls the rating. Another advantage is that the overall rating cannot be a negative number. A limitation of the system is that the overall final rating may be considerably lower than the ratings of each one of the individual factors.

1. *Land productivity index*: It is known as Storie Index Rating (SIR) and calculated by multiplying separate percent ratings for profile morphology (A), surface soil texture (B), slope angle (C), and modifying conditions such as soil depth, drainage, or alkalinity (X) and rainfall (Y) factors. The land productivity index can be stated as follows:

$$\text{Land productivity ndex} = A \times B \times C \times X \times Y \times \mathbf{100}$$

where

A = percentage rating for the general character of the soil profile
B = percentage rating for the texture of the surface horizon
C = percentage rating for the slope of the land
X = percentage rating for site conditions other than those covered by the factors A, B, and C (e.g., salinity, soil reaction, freedom from damaging winds)
Y = percentage rating for rainfall

2. *Soil productivity index*: The morphological, physicochemical properties of pedons were interpreted to assess their productivity using parametric approach described by Ricquier et al. (1970). The productivity index was calculated considering nine factors as determining soil productivity, viz., moisture (H), drainage (D), effective soil depth (P), texture/structure (T), base saturation (N), soluble salts (S), organic matter (O), nature of clay (A), and mineral reserves (M). Each factor was rated on a scale from 0 to 100. The actual factor-wise score was multiplied by each and expressed in percentage to derive the final index.

$$\text{Soil Productivity Index (SPI)} = H \times D \times P \times T \times SorN \times O \times A \times M$$

A case study was carried out to assess the potential of soil and land productivity indices (Fig. 18.1a and b) employing Ricquier et al. (1970) and Storie Index (1976) models, respectively. The study area falls between 72°27′30″ to 77°55′00″E longitude and 29°55′00″ to 30°17′30″N latitudes in Saharanpur district of Uttar Pradesh. The area lies below the foothill of southern slopes of Siwalik Hills with the most fertile alluvium in the form of old and recent flood plains. The soil map of the area was prepared by visual interpretation of satellite data (IRS-LISS-III Standard FCCs, March, 2003) in association with intensive field work for soil studies. Soils of the map units were characterized as deep to very deep in soil depth, well to moderately well drained, and sandy loam to silt loam/silt clay loam in soil texture.

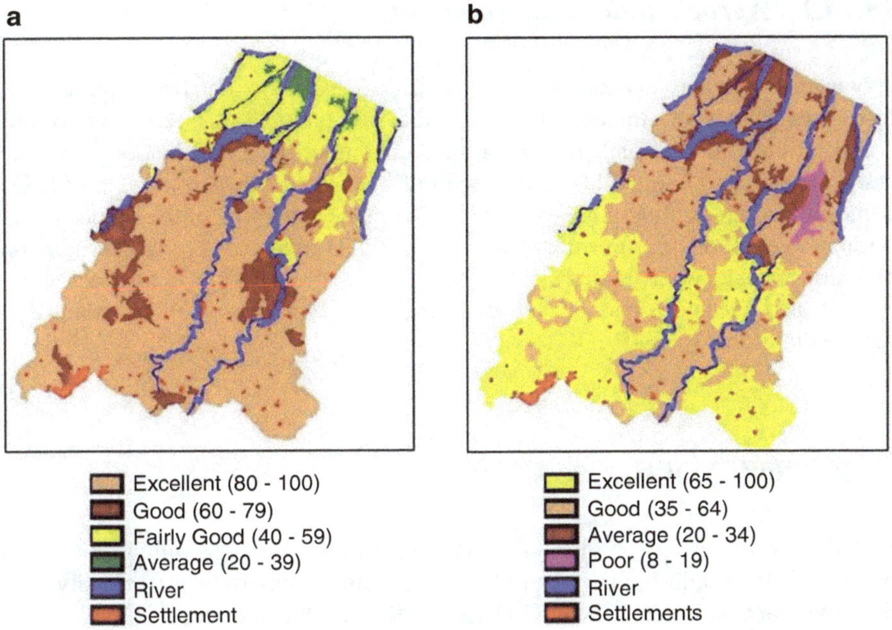

Fig. 18.1 (**a**) Land productivity index and (**b**) soil productivity index

18.14 FAO Framework of Land Evaluation

"A Framework for Land Evaluation" was proposed by FAO (1976) to assess the suitability of land for various crops. It provides standards, definitions, and a description of land qualities that can be formed from land characteristics. This assessment involves many factors that directly or indirectly control the ability of this part of land to host the land use under investigation. It sets the guidelines for physical and economical land evaluation. The land suitability method provides a widely used framework for assessing the physical land suitability for a specific use, based on expert knowledge. Suitability is expressed in qualitative terms: highly suitable (S1), moderately suitable (S2), marginally suitable (S3), and unsuitable with (N1) or without (N2) possibilities for land improvement. Within the FAO framework for land evaluation, the simple limitation method as described by Sys et al. (1991) uses the most limiting land characteristic for crop growth to determine the land suitability. FAO land suitability assessment measures how well the qualities of a land unit match the requirements of a particular form of land use. Suitability is assessed for each relevant use and each land unit identified in the study area.

18.14.1 Agroclimatic Crop Suitability

Advancement in land evaluation has focused on use of more detailed and quantitative elements to assess the agronomic potential of the biophysical resources. Availability of crop models and integration of their output with GIS helped in spatial assessment of crop suitability and preparing land use plan. Kumar et al. (2013) demonstrated the methodology to assess agroclimatic suitability of the soybean crop through integration of crop suitability based on FAO framework of land evaluation and biophysical (water limited) yield potential in the rainfed agroecosystem. Crop models also became a part in land evaluation and land use planning exercise at regional to global scale (FAO 1993).

18.15 Soil Quality Assessment

Soil quality is defined as "the capacity of a soil to function within ecosystem boundaries to sustain biological productivity, maintain environmental quality, and promote plant and animal health" (SSSA 1997). It shows direct relations to soil sustainability. Soil quality takes care of productivity with maintenance and sustenance of biodiversity, air quality, biomass production, nutrient cycle, and water quality. It is influenced by many factors known as indicators, viz., physical, chemical, and biological. These indicators are used to compute soil quality index (SQI). SQI is very important to understand the long-term effect of conservation practices of the landscape. It is used as the sensitivity risk for environment and in processes of land use planning. Andrews et al. (2004) developed a framework called Soil Management Assessment Framework (SMAF). This SMAF design follows three basic steps: (1) indicator selection, (2) indicator interpretation, and (3) integration into a SQ (soil quality) index value. SQI is a function of summation of weight multiplied by score for each and every parameter/indicator (Wu and Wang 2007):

$$SQI = \Sigma(\mathrm{Wi} \times \mathrm{Si})$$

where Wi is the weight assigned for each indicator and Si is the scoring done according to the importance of parameters.

18.16 Land Resource Monitoring

18.16.1 Land Cover Change Analysis

Land use/land cover (LULC) is a vital component in the interactions of the human activities with the environment understanding. Land use can be defined as the human activities toward the land such as agriculture, urban development, logging, grazing, and mining among many others. The application of remote sensing and GIS is capable to provide efficient methods for land use/land cover analysis and tools for land use planning and modeling. There are two main steps involved in this research, which are classification of the satellite data for LULC and the change detection analysis in the LULC types. Image registration, orthorectification, classification, and change detection using post-classification comparison were involved in the remote sensing data analysis. Radiometric and geometric errors are the most common types of errors encountered in remotely sensed imagery. Radiometric correction is necessary for detecting changes of surface reflectance (Jensen 1996). There are many techniques used such as post-classification comparison, conventional image differentiation, image ratio, image regression, and manual on-screen digitization of change, principal components analysis, and multi-date image classification.

18.16.2 Monitoring Land Degradation

The timely and early detection of degradation processes is necessary to prevent the continuing deterioration of land condition. Remote sensing data offers the synoptic view with temporal coverage of the large areas that helps in characterizing various kinds of degraded lands, viz., eroded lands, salt-affected soil, waterlogged area, decline in vegetal cover, etc. Land degradation processes develop significant changes in surface soil characteristics and conditions. These surface changes strongly affect reflectance characteristics in various spectral bandwidths. Satellite remote sensing data provide (1) multispectral observation; (2) multi-station observation to enable stereoscopy and different ground resolutions and scales of imagery; and (3) multi-date or multi-temporal observation of earth surface. These different aspects may contribute to object characterization, identification, and differentiation. Several researchers used remote sensing data in mapping and monitoring of salt-affected soils with either visual or computer remote sensing techniques or using combination of both methods. Metternicht and Zinck (2003) reviewed various sensors and approaches for identification and mapping of salt-affected areas. These studies mainly identified and mapped the salt-affected soils but did not characterize in terms of their severity.

Millennium Ecosystem Assessment (Safriel and Adeel 2005) defined land degradation as "the reduction in the capacity of the land to perform ecosystem goods, functions and services that support society and development" and desertification as the same process in arid and semiarid environments (collectively, the drylands). The terms desertification and degradation are interchangeably used. This definition considers the ability of land to support primary production as key ecosystem service which can be measured as net primary productivity (NPP). Thus, NPP at a site can potentially be viewed as land degradation (Wessels et al. 2012). Remote sensing data are widely used to characterize vegetation cover to assess vegetation degradation due to anthropogenic activities as well as land degradation. The assumption that land degradation reduces the NPP of a site is dependent upon the underlying mechanism. Land degradation processes result in an increase in bare ground area coupled with increased runoff and reduction of soil nutrients (Ravi et al. 2010). A reduction in vegetation cover leads to reduction in net primary production (NPP). Normalized difference vegetation index (NDVI) derived from remote sensing data is most widely used as proxy to estimate NPP (Wessels et al. 2006). Tucker et al. (1991) demonstrated the potential to apply time-series analysis on remote sensing data for monitoring and assessing ecosystem processes. Salient characteristics/indicators of land degradation provide unique opportunities for quantitative assessment using satellite data (De Jong et al. 2011).

Soil erosion due to water and wind is one of the most important processes of land degradation in the world. Remote sensing data assist erosion mapping through direct erosion detection or through analyzing erosion-controlling factors. Soil erosion is controlled by climate (rainfall erosivity), soil types (soil erodibility), topography (steepness and slope length factor), and vegetation cover (cover types and management practices). These controlling factors are integrated to map erosion employing erosion models. A large number of erosion models exist, which can be divided into empirical models and physically based models (Morgan 1995). A recent review of several current erosion models is provided by Merritt et al. (2003). Satellite imagery provides regional spatial data for several input parameters of erosion models. The Universal Soil Loss Equation (USLE) is the most widely used empirical model for assessing long-term averages of sheet and rill erosion. The USLE was modified as MUSLE (Smith et al. 1984) to predict sediment yield estimation at watershed/catchment level where it was revised as Revised Universal Soil Loss Equation (RUSLE) to adopt in various land uses/land covers and conditions and to apply at various spatial scales and region sizes in different environments worldwide.

Remote sensing data are extensively used in identifying various severities of soil erosion area. Lands suffering from rills and gully erosion have been mapped with visual interpretation techniques on satellite images of different sensors (Dwivedi et al. 1997; Vrieling 2006). Multi-temporal images allowed delineation of eroded areas and its increase (Fadul et al. 1999; Sujatha et al. 2000). Changes of surface conditions provide direct information on erosion occurrence. There are various methods available for analyzing change detection from remote sensing data (Coppin et al. 2004). Dhakal et al. (2002) used spectral image differencing, principal component analysis and spectral change vector analysis techniques in detecting area affected by flood and erosion using temporal remote sensing data.

Soil erosion was previously studied mainly considering and integrating erosion-controlling factor where soil and vegetation parameters were derived from remote sensing data. These had less accounted terrain and management practices. Now, the availability of good quality of DEMs from satellite data, such as stereo-optical imagery provided by IRS Cartosat-1 and Cartosat-2, SPOT, and ASTER or SAR imagery, has provided better assessment of soil erosion. These DEMs are increasingly being used in soil erosion modeling to derive spatial details of soil erosion. Availability of high-resolution remote sensing data had also facilitated in deriving land management practices at great details.

18.17 Microwave in Wetland Mapping/Soil Moisture

Microwave remote sensing can be used to estimate soil moisture on the basis of large contrast that exists between the dielectric constant values for dry and wet soils. Temporal SAR data are used in monitoring of water availability at soil root zone. Satellite remote sensing can be especially appropriate for wetland inventories and monitoring in developing countries, where funds are limited and little information is available on wetland areas (Henderson and Lewis 2008). Active microwave data from synthetic aperture radar (SAR) offers high potential not only for mapping and monitoring the extent of land submergence but also in identifying the wetland vegetation (Henderson and Lewis 2008). In addition, SAR data have been found to be extremely useful in delineating flood water boundaries beneath vegetation canopies (Kushwaha et al. 2000).

18.18 Land Resource Monitoring at Watershed Level

Watershed, a natural hydrologic entity, is conceptualized for optimal utilization of all soil and water resources in the catchment areas. Soil and water conservation measures are adopted on a watershed basis in order to facilitate sustainable development of natural resources. Land resource management at watershed level is important for restoring ecological services and ensuring food security as well as conserving soil and water against their degradation. Improper land management can lead to land degradation and a significant reduction in the productive and service (biodiversity niches, hydrology, carbon sequestration) functions of watersheds and landscapes. Temporal remote sensing data provides an excellent opportunity to monitor the land resources and to assess land cover changes in the watershed over a period of time. The satellite images were classified into various land use/land cover classes in the watershed using image classification techniques (Minaei and Kainz 2016). Multi-temporal remote sensing data are being used to assess the impact of watershed management programs (Shanwad et al. 2008). Vegetation cover serves as an integrated indicator of environmental factors of climate, soil, terrain as well as anthropogenic activities. Spectral vegetation indices derived from remote sensing data provide improved understanding of performance of land resources in the watershed. This understanding can be used to infer vegetation cover changes as well as ecosystem processes and services. Land use/land cover changes are as well considered as land degradation indicator (indicator of deforestation, encroachment of invasive species). Therefore, remote sensing-based monitoring of vegetation cover dynamics and vegetation parameters provides crucial information in monitoring land resources (Dubovyk 2017).

18.19 Conclusions

The spatial information technologies, viz., remote sensing, GIS, and GPS, have made it possible to characterize, assess, and monitor land resources most reliably, effectively, and in a cost-effective manner. It emerged as an effective tool in generating spatial information of natural resources. The planned and indiscriminate use of land has caused tremendous degradation of soils. Reliable inventory of degradaed soil and other resources need to be collected expeditiously. Development of new generation of high-spatial resolution, enhanced spectral coverage, revisit capabilities, and stereo viewing has opened new vistas in various issues of land resource management. GIS technology is bringing about rapid changes in the spatial analysis and management of natural resources. With increasing population pressure throughout the world and the need for increased agricultural production, there is a definite need for improved management of the agricultural resources. Therefore, it is necessary to obtain reliable data on not only the types of land resources but also the quality, quantity, and location of these resources. In the coming years, the field of remote sensing will change significantly with the projected increase in the number of satellites of high resolution in terms of spatial and temporal. Advancements in remote sensing technology will provide enormous opportunity to resource managers to monitor the state of land degradation as well as to assess potential of land resources to ascertain its optimal use to cater the need of growing population and insure food security.

References

Andrews SS, Karlen DL, Cambardella CA (2004) The soil management assessment framework. Soil Sci Soc Am J 68(6):1945–1962

Bandyopadhyay S, Jaiswal RK, Hegde VS, Jayaraman V (2009) Assessment of land suitability potentials for agriculture using a remote sensing and GIS based approach. Int J Remote Sens 30(4):879–895

Baumgardner MF, Silva LF, Biehl LL, Stoner ER (1985) Reflectance properties of soils. In: Brady N (ed) Advances in agronomy, vol 38. Academic, Orlando, pp 1–44

Bello OM, Aina YA (2014) Satellite remote sensing as a tool in disaster management and sustainable development planning: towards a synergistic approach. Procedia Soc Behav Sci 120:365–373

Ben-Dor E, Patkin K, Banin A, Karnieli A (2002) Mapping of several soil properties using DAIS-7915 hyperspectral scanner data: a case study over clayey soils in Israel. Int J Remote Sens 23(6):1043–1062

Ben-Dor E, Taylor RG, Hill J, Demattê JAM, Whiting ML, Chabrillat S, Sommer S (2008) Imaging spectrometry for soil applications. Adv Agron 97:321–392

Blaes X, Vanhalle L, Defourny P (2005) Efficiency of crop identification based on optical and SAR image time series. Remote Sens Environ 96:352–365

Clark RN, King TVV, Klejwa M, Swayze GA, Vergo N (1990) High spectral resolution reflectance spectroscopy of minerals. J Geophys Res 95(12):653–680

Coppin P, Jonckheere I, Nackaerts K, Muys B, Lambin E (2004) Digital change detection methods in ecosystem monitoring: a review. Int J Remote Sens 25(9):1565–1596

Das BS, Sarathjith MC, Santra P, Sahoo RN, Srivastava R, Routray A, Ray SS (2015) Hyperspectral remote sensing: opportunities, status and challenges for rapid soil assessment in India. Curr Sci 108(5):860–868

De Jong SM (1994) Applications of reflective remote sensing for land degradation studies in a Mediterranean environment. In: Netherlands geographical Studiesunknown: book. KNAG, Utrecht, 240 pp

De Jong R, de Bruin S, Schaepman M, Dent D (2011) Quantitative mapping of global land degradation using earth observations. Int J Remote Sens 32:6823–6853

Dent D, Young A (1981) Soil survey and land evaluation. George Allen & Unwin, London

Dhakal AS, Amada T, Aniya M, Sharma RR (2002) Detection of areas associated with flood and erosion caused by a heavy rainfall using multitemporal Landsat TM data. Photogramm Eng Remote Sens 68(3):233–239

Dhruva Narayana VV, Sastry G, Patnaik US (1990) Watershed management. Indian Council of Agricultural Research, New Delhi

Dubovyk O (2017) The role of remote sensing in land degradation assessments: opportunities and challenges. Eur J Remote Sens 50(1):601–613

Dwivedi RS (2001) Soil resources mapping: a remote sensing perspective. Remote Sens Rev 20(2):89–122

Dwivedi RS, Kumar AB, Tewari KN (1997) The utility of multi-sensor data for mapping eroded lands. Int J Remote Sens 18:2303–2318

Earth Observation Research Center, Japan Aerospace Exploration Agency (2010) ALOS user handbook. Available online at http://www.eorc.jaxa.jp/ALOS/en/doc/alos_userhb_en.pdf. Accessed on 15 July 2017

Fadul HM, Salih AA, Imadeldin AA, Inanga S (1999) Use of remote sensing to map gully erosion along Atbara river, Sudan. J Appl Earth Obs Geoinf 1(3–4):175–180

FAO (1976) A framework for land evaluation. Soils Bulletin, No. 32, Rome

FAO (1993) Guidelines for land-use planning. FAO development series 1. FAO, Rome

Farifteh J, Van der Meer F, Van der Meijde M, Atzberger CG (2008) Spectral characteristics of salt-affected soils: a laboratory experiment. Geoderma 145(3–4):196–206

Gessler PE, Moore ID, McKenzie NJ, Ryan PJ (1995) Soil-landscape modelling and spatial prediction of soil attributes. Int J Geogr Inf Sci 9(4):421–432

Grunwald S (2009) Multi-criteria characterization of recent digital soil mapping and modeling approaches. Geoderma 152:195–207

Guo H, Dou C, Zhang X, Han C, Yue X (2016) Earth observation from the manned low Earth orbit platforms. ISPRS J Photogramm Remote Sens 115:103–118

Henderson FM, Lewis AJ (2008) Radar detection of wetland ecosystems: a review. Int J Remote Sens 29(20):5809–5835

Hilwig FW (1975) Visual interpretation of Landsat imagery for reconnaissance soil survey of the Ganges river fan–southwest of Haridwar, India. Int Inst Aerial Surv Earth Sci (ITC) J 1:26

Hilwig FW, Karale RL (1973) Physiographic system and elements of photo-interpretation as applied to soil survey in Ganges plain. J Indian Soc Remote Sens 21:205–212

Jensen JR (1996) Introductory digital image processing: a remote sensing perspective, 2nd edn. Prentice Hall, Saddle River, 316 pp

Karale RL, Venkataratnam L, Sehgal JL, Saxena RK (1991) Soil mapping with IRS-1A data in areas of complex soilscape. Curr Sci 61(3 & 4)

Khybri ML (1979) Suggestions for land capability classification for the Himalayan Region. Indian J Soil Conserv 7(1):54–57

Kornelsen KC, Coulibaly P (2013) Advances in soil moisture retrieval from synthetic aperture radar and hydrological applications. J Hydrol 476:460–489

Kumar S (2017) Geospatial tools and techniques in land resource inventory. In: Reddy GPO, Patil NG and Chaturvedi A (eds) Sustainable management of land resources: an Indian perspective. CRC Press, Boca Raton, pp 273–300

Kumar S, Singh RP (2016) Spatial distribution of soil nutrients in a watershed of Himalayan landscape using terrain attributes and geostatistical methods. Environ Earth Sci 75:473

Kumar S, Dhaimodkar SB, Pande LM (2002) The assessment of potential land use change in irrigation command using remote sensing and GIS. J Indian Soc Remote Sens 30(3):157–166

Kumar S, Patel NR, Sarkar A, Dadhwal VK (2013) Geospatial approach in assessing agro-climatic suitability of soybean in Rainfed agro-ecosystem. J Indian Soc Remote Sens 41(3):609–618

Kushwaha SPS, Dwivedi RS, Rao BRM (2000) Evaluation of various digital image processing techniques for detection of coastal wetlands using ERS-1 SAR-data. Int J Remote Sens 21(3):565–579

Lillisand T, Kiefer R, Chipman J (2015) Remote sensing and image interpretation, 7th edn. Wiley, Hoboken

Lulla K, Nellis MD, Rundquist B (2012) Celebrating 40 years of Landsat program's Earth observation accomplishments. Geocarto Int 27(6):459–459

Mahmoudabadi E, Karimi A, Haghnia GH, Sepehr A (2017) Digital soil mapping using remote sensing indices, terrain attributes, and vegetation features in the rangelands of northeastern Iran. Environ Monit Assess 189(10):500

Manchanda ML, Kudrat M, Tiwari AK (2002) Soil survey and mapping using remote sensing. Trop Ecol 43(1):61–74

Martz LW, Garbrecht J (1992) Numerical definition of drainage network and subcatchment areas from digital elevation models. Comput Geosci 18:747–761

McSweeney KM, Gessler PE, Slater B, Hammer RD, Bell I, Peterson GW (1994) Towards a new framework for modelling the soil-landscape continuum. In: Factors of soil formation: a fifrieth anniversary retrospecrive, Special Publication 33. Soil Science Society of America, Madison, pp 127–145

Merritt WS, Letcher RA, Jakeman AJ (2003) A review of erosion and sediment transport models. Environ Model Softw 18(8–9):761–799

Metternicht GI, Zinck JA (2003) Remote sensing of soil salinity: potentials and constraints. Remote Sens Environ 85:1–20

Minaei M, Kainz W (2016) Watershed land cover/land use mapping using remote sensing and data mining in Gorganrood, Iran. Int J Geo-Inf 5:57–72

Moore ID, Grayson RB, Ladson AR (1991) Digital terrain modelling: a review of hydrological, geomorphological and biological applications. Hydrol Process 5(1):3–30

Moore ID, Gessler PE, Nielsen GA, Peterson GA (1993) Soil attribute prediction using terrain analysis. Soil Sci Soc Am J 57(2):443–452

Morgan RPC (1995) Soil erosion and conservation. Longman Group Limited, Essex, p 198

Mulder VL, de Bruin S, Schaepman ME, Mayr TR (2011) The use of remote sensing in soil and terrain mapping—a review. Geoderma 162(1):1–19

Price KP, Guo X, Stiles JM (2002) Comparison of Landsat TM and ERS-2 SAR data for discriminating among grassland types and treatments in eastern Kansas. Comput Electron Agric 37:157–171

Radhadevi PV, Solanki SS, Nagasubramanian V, Mahapatra A, Reddy DS, Jyothi MV, Sumanth K, Saibaba J, Varadan G (2010) New era of Cartosat satellites for large scale mapping. Photogramm Eng Remote Sens 76(9):1031–1040

Ravi S, Breshears DD, Huxman TE, D'Odorico P (2010) Land degradation in drylands: interactions among hydrologic–aeolian erosion and vegetation dynamics. Geomorphology 116:236–245

Ricquier J, Bramao DL, Cornet JP (1970) A new system of soil appraisal in terms of actual and potential productivity. Soil Resources Devision, and Conservation Service, FAO, Rome

Safriel U, Adeel Z (2005) Dryland systems. In: Hassan R, Scholes RJ, Ash N (eds) Ecosystems and human well-being: current state and trends, vol 1. Island Press, London, p 917

Shanwad UK, Patil VC, Gowda HH, Dasog GS (2008) Application of remote sensing technology for impact assessment of watershed development programme. J Indian Soc Remote Sens 36(4):375–386

Shepherd KD, Walsh MG (2002) Development of reflectance spectral libraries for characterization of soil properties. Soil Sci Soc Am J 66:988–998
Slaymaker O (2001) The role of remote sensing in geomorphology and terrain analysis in the Canadian Cordillera. Int J Appl Earth Obs Geoinf 3(1):11–17
Smith SJ, Williams JR, Menzel RG, Coleman GA (1984) Prediction of sediment yield from Southern Plains grasslands with the modified universal soil loss equation. J Range Manag 37:295–297
Soil Science Society of America (SSSA) (1997) Glossary of soil science terms 1996. Soil Science Society of America Inc, Madison, 139 p
Srivastava R, Saxena RK (2004) Technique of large-scale soil mapping in basaltic terrain using satellite remote sensing data. Int J Remote Sens 25(4):679–688
Storie RE (1976) Storie index soil ratings (revised). Special Publication Division of Agriculture Science, University of California, Berkley
Sujatha G, Dwivedi RS, Sreenivas K, Venkataratnam L (2000) Mapping and monitoring of degraded lands in part of Jaunpur district of Uttar Pradesh using temporal spaceborne multispectral data. Int J Remote Sens 21:519–531
Sys C, Van Ranst E, Debaveye J (1991) Land evaluation. Part I: Principles in land evaluation and crop production calculations. Agricultural Publication no. 7. General Administration for Development Cooperation. Brussels
Thenkabail PS (2015) Land resources monitoring, modeling, and mapping with remote sensing. CRC Press, Boca Raton, p 859
Toth C, Józ'ków G (2016) Remote sensing platforms and sensors: a survey. ISPRS J Photogramm Remote Sens 115:22–36
Tucker CJ, Dregne HE, Newcomb WW (1991) Expansion and contraction of the Sahara Desert from 1980 to 1990. Science 253:299–301
Ulaby FT, Moore RK, Fung AK (1986) Microwave remote sensing: active and passive, vol 3. Addison-Wesley Publishers, Reading
UNEP (1992) World atlas of desertification, 69 plates. UNEP, Nairobi & Edward Arnold, London, 69 pp
UNEP (1994) United Nations Convention to Combat Desertification in those countries experiencing drought and/or desertification, particularly in Africa. Text with Annexes, 71 p
Vrieling A (2006) Satellite remote sensing for water erosion assessment: a review. Catena 65:2–18
Weng Y, Gong P, Zhu Z (2008) Soil salt content estimation in the Yellow River delta with satellite hyperspectral data. Can J Remote Sens 34(3):259–270
Wessels K, Prince S, Zambatis N, MacFadyen S, Frost P, van Zyl D (2006) Relationship between herbaceous biomass and 1-km^2 advanced very high resolution radiometer (AVHRR) NDVI in Kruger National Park, South Africa. Int J Remote Sens 27:951–973
Wessels KJ, van den Bergh F, Scholes RJ (2012) Limits to delectability of land degradation by trend analysis of vegetation index data. Remote Sens Environ 125:10–22
Wu Q, Wang M (2007) A framework for risk assessment on soil erosion by water using an integrated and systematic approach. J Hydrol 337(1):11–21

Chapter 19
Remote Sensing and GIS in Agriculture and Forest Resource Monitoring

Jugal Kishore Mani and A. O. Varghese

Abstract Biological land resources are highly vulnerable to variations in climate, soil and topography of different regions, and all these factors need to be analysed on spatio-temporal basis for proper management and planning. The advanced techniques like remote sensing, global positioning system (GPS) and geographic information system (GIS) can be of great use for their assessment, management and monitoring. The availability of remotely sensed data from different sensors of various platforms with a wide range of spatio-temporal, radiometric and spectral resolutions made remote sensing as, perhaps, the best source of data for large-scale applications in agriculture and forest resource monitoring. The integrated use of remotely sensed data, GPS and GIS will enable natural resource managers and researchers to develop management plans for a variety of agriculture and forest resource management applications. In the present chapter, examples of remote sensing and GIS applications in agriculture and forest resource monitoring were reviewed.

Keywords Geographic information system · Global positioning system · Agriculture and forest resource monitoring · Remote sensing

19.1 Introduction

Land is a vital resource and is the basis of the existence of mankind, and management of this resource including its conservation and utilization is of crucial importance. The land resource is a base on which all life depends, and in most countries of the world, it is the life support system. In the recent past, with burgeoning populations and the national goals of seeking self-sufficiency in food and fibre

---

J. K. Mani (✉) · A. O. Varghese
Regional Remote Sensing Centre-Central, National Remote Sensing Centre, ISRO, Nagpur, India
e-mail: jugalmani@gmail.com

G. P. O. Reddy, S. K. Singh (eds.), *Geospatial Technologies in Land Resources Mapping, Monitoring and Management*, Geotechnologies and the Environment 21,
https://doi.org/10.1007/978-3-319-78711-4_19

production, the resource base is slowly dwindling. In many countries across the globe, land is subjected to varying degrees and forms of degradation due to the pressure of the growing population, increased demand for food, fodder and fuel wood and intensive industrial activity (FAO 2005). While natural systems often adapt to stress in a remarkable fashion, some relationships once destroyed can never be restored (Dill 1990). The main result is man-induced degradation of land resources through inadvertent, inappropriate or misuse of technological innovations. When degradation becomes a continuing process, yields decline, and the farmer is forced to eke a living on another piece of land, which in most instances may be a fragile ecosystem like steep lands or coastal swamps since much of the better arable land is already under cultivation. The system then becomes iterative to the detriment of all.

Satellite remote sensing technology due to its synoptic and repetitive coverage provides consistent and valuable information about the Earth's surface and has become important for cropping pattern analysis and monitoring crop acreage and production at different spatial scales from global to local scales. Satellite remote sensing coupled with conventional methods has been used to generate information on the current status and utilization potential of land resources. These databases have been analysed to identify local specific alternate land use management practices. In the last three decades, the applications of remote sensing in India have shown tremendous progress and established the potential use in natural resource management and monitoring, which has contributed significantly to national development in various sectors. Landsat Multispectral Scanner (MSS)/Thematic Mapper (TM) data have been used during the first decade, and the nationwide forest cover and wasteland mapping have been carried out. The successful launch of the Indian Remote Sensing (IRS) series of satellites and the establishment of ground segment for data reception, processing and archival facilities have further enhanced the remote sensing applications potential. The availability of high spectral/spatial resolution data like Resourcesat 1 and 2, Cartosat 1 and 2 and RISAT-1 opens new vistas in remote sensing applications, and it will be relevant at this juncture to take stock of the present and project the future scenario.

GIS have proved to be immensely helpful in the organization of the huge database generated through space technology (Trotter 1991). The utility of GIS in the analysis and modelling of integrated information is well established (Burrough 1989). GIS has been used in the development of digital databases, in the assessment of status and trends of resource utilization of the areas and to support and assess various resource management alternatives (Clark 1990). Spectacular developments in the field of GIS particularly in synthesizing various thematic information in conjunction with collateral data have not only made this technology effective and economical but also a tool to arrive at development strategies for sustainable land and water resource management.

19.2 Remote Sensing and GIS in Agriculture Resource Monitoring

Remote sensing technology has been applied in agriculture extensively since its early stage in the 1960s. Now several global and national operational systems of monitoring agriculture on remote sensing platforms are being operated. These systems provide consistent, timely and valuable information for agricultural production, management and policy-making. On the other hand, the demands arising from the applications in agricultural sectors have also enhanced the progress and innovation in remote sensing technology. The important applications of remote sensing in agriculture resource monitoring include crop identification and cropland mapping; crop growth monitoring and yield estimation/prediction; inversion of key biophysical, biochemical and environmental parameters; crop damage/disaster monitoring; precision farming; etc. In this chapter, crop mapping, yield prediction, soil moisture monitoring and crop phenology monitoring with remote sensing were discussed. The utilization of space-borne multispectral data for crop acreage and production estimation started in the 1970s with the launching of the Large Area Crop Inventory Experiment (LACIE) jointly by National Aeronautics and Space Administration (NASA), United States Department of Agriculture (USDA) and National Oceanic and Atmospheric Administration (NOAA) in 1974. In India, the satellite remote sensing is being used for the crop acreage and production estimation of agricultural crops. The methodology for crop acreage and production estimation using Landsat MSS/TM and IRS-1A/1B-Linear Imaging Self-Scanning Sensor (LISS-I&II) has been operationalized for major crops, namely, wheat, paddy, sorghum, soybean, groundnut and cotton in the monocropped areas. However, the use of remote sensing in plantation crops began in the 1970s. In 1970, Indian Space Research Organization (ISRO) carried out an innovative and promising experiment for Indian Agricultural Research Institute (IARI) in detecting coconut wilt disease in Kerala before it was visible on ground.

19.2.1 Crop Acreage Estimation and Monitoring

Monitoring, estimating and forecasting agricultural production are very important for the management of world/regional or local food demand and supply balance for social security. In India, reliable, accurate and timely information on the type of crops grown and their acreage, crop yield and crop growth conditions are of vital importance. India is one of the few countries which has a well-established system of collection of agricultural statistics, and detailed statistics of land utilization are continuously available since 1984. The estimate of the crop acreage at district level is obtained through enumeration, whereas the average yield is obtained through general crop estimation surveys based on crop cutting experiments conducted on a number of randomly selected fields in a sample at villages in the districts. This technique is more time consuming and less accurate.

The advent of remote sensing technology and its great potential in the field of agriculture have opened new vistas of improving the agricultural statistics systems all over the world. Space-borne remotely sensed satellite data has been widely used in the estimation of agricultural crops because of less time requirement and more accuracy (Panigrahy et al. 2002; Ray et al. 2005). Panigrahy et al. (2009) used multi-date optical data such as IRS AWiFS/WiFS for deriving the rabi and summer cropping patterns in India. Realizing the potential of remote sensing technology in generation of spatial information on natural resource of various scales in near-real time, it has become a prerequisite in all the scientific planning activities. Department of Space (DOS), Government of India, launched a major programme entitled 'Natural Resource Census (NR-Census)'; mapping of land use/land cover (LULC) on 1:250,000 scale has been envisaged under this, using multi-temporal IRS-P6 AWiFS data on an annual basis since 2004. A detailed LULC map of Maharashtra prepared under this programme is shown in Fig. 19.1. Figure 19.1 depicted that the percent area of agriculture and forest in Maharashtra during 2013–2014 is distributed under kharif (27.5%), rabi (3.0%), zaid (0.2%), double or triple crop (21.6%), fallow land (9.8%), plantation or orchard (1.4%), evergreen or semi-evergreen forest (2.2%), deciduous forest (11.7%) and degraded forest (1.1%).

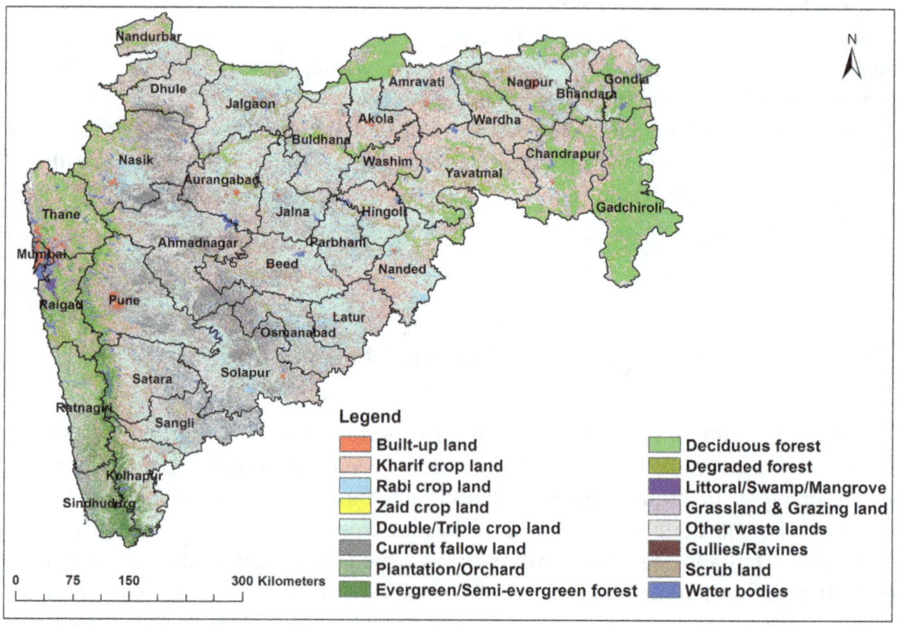

Fig. 19.1 Land use and land cover map of Maharashtra, India, during 2013–2014

19.2.2 Crop Yield Modelling and Forecasting

Reliable crop yield estimate is one of the most important components of crop production forecasting (Moulin et al. 1998; Gallego et al. 2010). In recent years, considerable attention has been given to the development of crop yield models, which could be derived with remote sensing inputs. The crop yield is dependent on many factors such as crop variety, availability of nutrients and water, protection against pests and diseases and weather factors. Plant shows the integrated effect of these factors, which results in the ultimate yield. The mathematical formulation of the relationship between various physically observable and yield-affecting parameters is crop yield modelling. Spectral response of a crop is the integrated manifestation of various crop growth factors. In a given time domain, the growth and decay of spectral response indicate the crop performance. Spectral data as such or through its relation with the biometrics parameters are being used for prediction of yield. The information on production of crops before the harvest is very vital to the national food policy planning and economy of the country. Remote sensing data have been hitherto used for crop production forecasting in the monocropped regions owing primarily to spatial resolution limitation and to spectral similarity of different crops at a particular growth stage. Geospatial technologies, in combination with modern tools like crop simulation modelling, have the capability to generate the requisite information for crop monitoring, which form a vital input for the decision-makers at different levels. Mani et al. (2017b) estimated the gross primary productivity of agricultural areas of Vidarbha region of Maharashtra using remote sensing-based model for understanding the most productive area in this region in terms of carbon accumulation.

Well-planned programmes for crop acreage and production forecasts like Crop Acreage and Production Estimation (CAPE) and Forecasting Agricultural Output Using Space, Agro-meteorology and Land-based Observations (FASAL) culminated in the establishment of a dedicated Mahalanobis National Crop Forecast Centre (MNCFC) by the Ministry of Agriculture (GoI), for crop inventorying and drought assessment. Crops covered under this are rice, wheat, cotton, mustard, sugarcane, etc. The use of satellite spectral data in crop yield estimation surveys was examined by Singh et al. (1992). Suitability analysis for orange orchards in Vidarbha region of Maharashtra, India, using remote sensing and GIS-based models is carried out by Mani et al. (2016) (Fig. 19.2).

19.2.3 Integrated Pest Management

Integrated pest management is an important component of sustainable agriculture. Satellite remote sensing can play an important role in surveying and identifying the inaccessible large desertic regions where locusts breed and lay eggs, thereby developing strategies for preventing their spread and effective control measures.

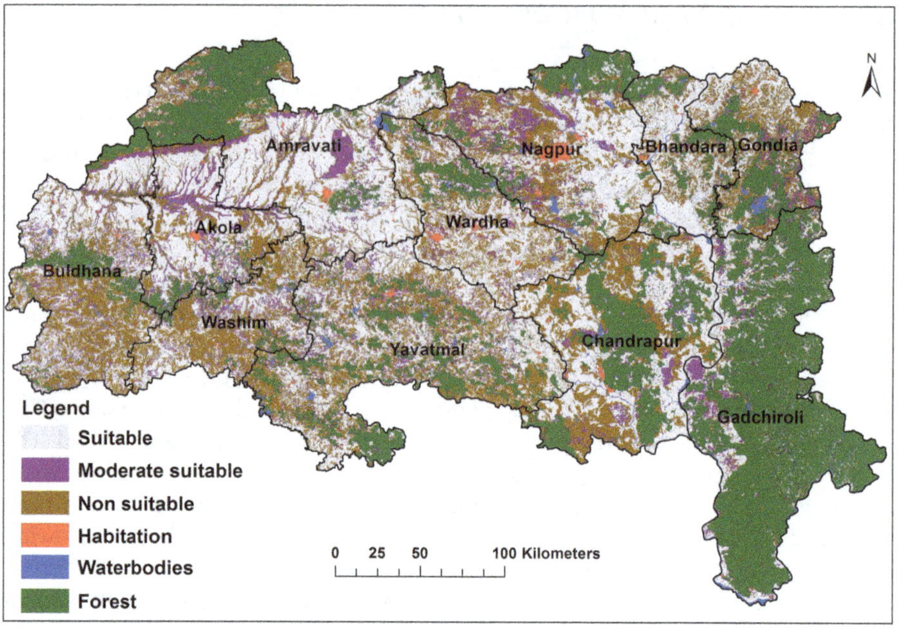

Fig. 19.2 Estimated suitable areas for orange orchards using deductive modelling approach in Vidarbha region, Maharashtra, India

Methodologies need to be perfected for the identification of locust breeding grounds based on vegetation or moisture status. Since the area is very large and the cycle of locust breeding, growth and its spread is very short, satellites having high temporal resolution and large area coverage like advanced very-high-resolution radiometer (AVHRR) may be suited well to identify and prioritize the areas to arrest and control desert menace. Research into vegetative spectral reflectance can help us gain a better understanding of the physical, physiological and chemical processes in plants due to pest and disease attack and to detect the resulting biotic stress. This has important implications to effective pest management. Pest surveillance programmes such as field scouting are often expensive, time consuming, laborious and prone to error. As remote sensing gives a synoptic view of the area in a non-destructive and non-invasive way, this technology could be effective and provide timely information on spatial variability of pest damage over a large area. Rani and Jyothi (2017) successfully studied the use of hyperspectral imagery in the physical and physiological processes in plants due to pest attack and thus detected the resulting biotic stress. The analysis of spectral reflectance pattern of cowpea infested with disease (cowpea rust) was carried out by NRSC, and the reflectance entered at 740, 660 and 1448 nm was found to give significant difference compared to normal crop (NRSC 2010).

19.2.4 Drought Assessment and Monitoring

The district-level drought assessment and monitoring using Normalized Difference Vegetation Index (NDVI) generated from NOAA-AVHRR data helps in taking timely preventive and corrective measures for combating drought. This is yet another area wherein remote sensing data has been used at operational level. Currently, the drought assessment is being done through visual interpretation of remote sensing data. Comprehensive drought assessment would also involve the integration of information on cropping zones, soil types, coherent rainfall zones, etc. Scheduling irrigation based on environmental parameters is useful for quick adoption of management practices suitable over large areas. Drought-monitoring indices which are derived from AVHRR and MODIS data are normally radiometric measures of vegetation condition and dynamics, exploiting the unique spectral signatures of canopy elements, particularly in the red and near-infrared (NIR) portions of the spectrum (Huete et al. 1997, 2002), and are sensitive to vegetation type, growth stage, canopy cover and structure (Thenkabail 2003).

Monitoring and assessment of drought conditions at different scales and timely dissemination of information constitute the most vital part of drought management system. Therefore, a sound, operationally feasible, objective and economically viable system for drought monitoring and decision support would enable efficient management of this hydrometeorological disaster. Along with a robust monitoring system, a mechanism for rainfall prediction and drought early warning brings out total solutions for drought management. In India, the National Agricultural Drought Assessment and Monitoring System (NADAMS) was initiated towards the end of 1986, with the participation of National Remote Sensing Agency, Dept. of Space, Government of India, as nodal agency for execution, with the support of India Meteorological Department (IMD) and various state departments of agriculture. Impact of agricultural drought and drought early warning can be assessed with multiple indices like Shortwave Angle Slope Index (SASI), Normalized Difference Water Index (NDWI), Normalized Difference Vegetation Index (NDVI) and Soil Moisture Index (NRSA 1990).

19.2.5 Soil Moisture Estimation

Soil moisture is a critical parameter in agriculture hydrology and meteorology. It is an important parameter for many weather forecasting and prediction models. Changes in soil moisture have a severe impact on agricultural productivity, forest and ecosystem health. So, regular measurements of soil moisture are essential for effective water resource management, drought forecast and management, understanding ecological processes, etc. The conventional soil moisture estimation methods are based on point observations, which do not permit extrapolation over large areas. The availability of space-borne sensors in infrared, thermal and microwave regions of the electromagnetic spectrum provides an opportunity to study the spatial variability of soil moisture. Moreover, remote sensing satellites cover the same area at regular intervals, which enables to monitor the soil moisture over a given period of time. The launch of Indian remote sensing satellite with microwave sensors, Radar Imaging Satellite 1 (RISAT-1), enables studying the soil moisture in top 15 cm of surface layer.

Many experts have established different methods for retrieving soil moisture based on the relationships between the soil moisture and satellite-derived land surface parameters. Using the visible infrared band, the soil moisture can be derived from the relationships between soil reflectance and different soil moistures or vegetation traits that occur under water stress. These methods can be divided into two categories: (1) the single spectral analysis method (Bower and Smith 1972 and Ishida et al. 1991) and (2) the vegetation index method (Kogan 1990, 1995). For thermal infrared wavebands, the parameters related to the soil thermal properties are mainly used to derive the soil moisture. This method includes two categories: (1) the thermal inertia method and (2) the temperature index method. Combining visible and thermal infrared remotely sensed data can provide more information for estimating soil moisture than the single one (Zhang and Zhao 2016). Spatial and temporal information-based methods have their own advantages and disadvantages. It is therefore important to determine how to combine these methods reasonably to obtain highly accurate soil moisture.

19.3 Remote Sensing and GIS in Forest Resource Monitoring

Forest is a vital organ of our ecosystem; it impacts human lives in several ways; despite of having huge importance, the world forest has been declining at an alarming rate. Being a renewable resource, forest cover can be regenerated through sustainable management. Hence, using remote sensing data and GIS techniques, a forest manager can generate information regarding forest cover, types of forest present within an area of interest, human encroachment extent into forest land/ protected areas, encroachment of desert-like conditions and so on (FSI 2017). This information is crucial for the development of forest management plans and in the process of decision-making to ensure that effective policies should be put in place to control and govern the manner in which forest resources can be utilized. The suitability and status of sites/forest area for a particular species of wildlife can also be assessed using remote sensing data using multi-criteria analysis (Varghese and Suryavanshi 2017). In the past, forest management inventories were primarily for timber management and focused on capturing area and volume by species. The use of remote sensing by forest managers has steadily increased, promoted in large part by better integration of imagery with GIS technology and databases, as well as implementations of the technology that best suit the information needs of forest managers. In the past decades, forest management responsibilities have broadened. As a result, inventory data requirements have expanded to include measures of non-harvest-related characteristics such as forest cover and type mapping, working plan preparations, monitoring of afforestation programmes, forest fire monitoring and risk mapping, biodiversity monitoring, species-habitat modelling, wildlife habitat evaluation, biomass and productivity assessments, etc.

19.3.1 Forest Cover Mapping and Monitoring

Satellite remote sensing data has been extensively used to map forests of the tropics, where up-to-date data about spatial distribution are absent or lacking. In India, the initial attempt at national level has been carried out on 1:250,000 scale using visual interpretation of false colour images. National Remote Sensing Agency (NRSA) for the first time studied 1:1 million scale images for 1972–1975 and 1980–1982 period, and forest cover was classified into three categories based on canopy closure. Closed (>40%), open (10–30%) and mangrove categories were identified (NRSC 2010). Subsequently, Forest Survey of India (FSI) also used a similar technique for the period of 1981–1983 for forest mapping on 1:250,000 scale. This experience has been extended in evolving the national programme to monitor vegetation cover. Presently, the forest cover at national level is biennially monitored by FSI using satellite remote sensing as one of the prime data sources. In FSI scheme, <10% canopy cover forests are mapped as scrub, while NRSA showed this as non-forest (FSI 2003). With the availability of improved spatial and temporal resolution data from Resourcesat 1 and 2, the application potential of satellite remote sensing has enhanced substantially. The more recent mapping scheme of FSI takes into account three densities, viz. open, dense and very dense forest (FSI 2009). The methodology adopted by FSI, i.e. hundred percent forest mapping each time, could be further improved by detecting only the changes in forest cover from base year each time. This will naturally minimize the workload of otherwise stupendous task. Data such as that from IRS LISS-IV with 5.8 m and IKONOS with 4 m spatial resolution provide still better opportunity to map the forests into five densities, viz. $\leq$ 20%, 20–40%, 40–60%, 60–80% and > 80%. The past experience with IRS PAN data showed better utility of this data in forest density mapping (compared to PAN and LISS-III merged data).

The revised forest-type classification of Champion and Seth (1968) is the most widely used classification system for India's forests. They classified forest into five major groups based on climatic factors. These major groups have been further divided into 16 type groups based on temperature and moisture contents. A few of these type groups have been further divided into several subgroups. Ultimately the type groups have been classified into 202 forest types and subtypes based on location-specific climate factors and vegetation formation. In the year 2002, Indian Institute of Remote Sensing (IIRS) produced biodiversity characterization maps of parts of the country like Andaman and Nicobar Islands, North Eastern region,

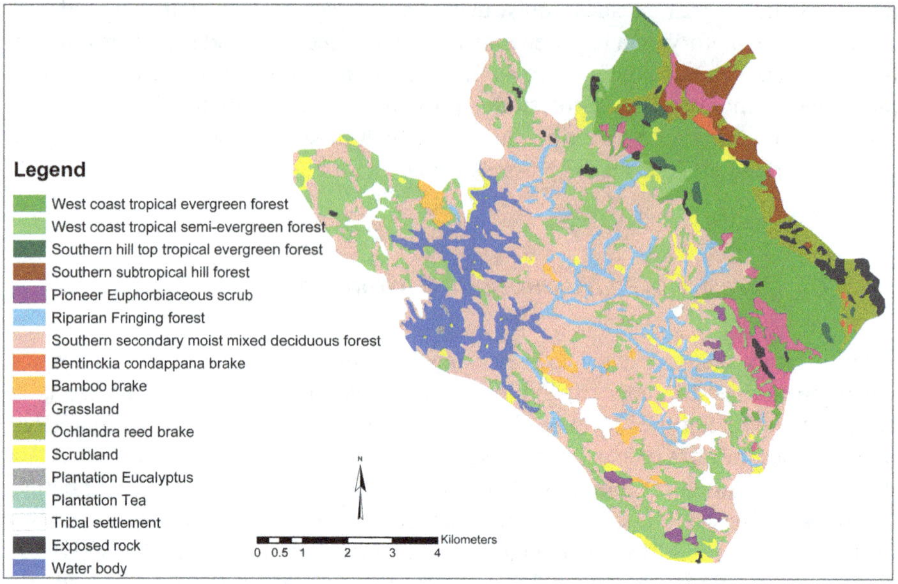

Fig. 19.3 Forest-type map of Peppara Wildlife Sanctuary prepared from LISS-IV data

Western Ghats, etc. through remote sensing technologies (Roy et al. 2012). Forest Survey of India brought out 'Atlas Forest Types of India' in the year 2011 containing forest-type maps according to Champion and Seth classification (1968) of the country (FSI 2011). ISRO prepared a seamless vegetation-type map of India (scale 1: 50,000) using medium-resolution IRS LISS-III images (Roy et al. 2015). The map was created using an on-screen visual interpretation technique and has an accuracy of 90%, as assessed using 15,565 ground control points. This vegetation-type map is the most comprehensive one developed in India so far. The digital map is now available through a web portal. With the advent of high-resolution (5.8 m) multispectral data of Linear Imaging Self-Scanning (LISS-IV) of Indian Remote Sensing Satellite (IRS) Resourcesat-1, the textural and spectral information can be better utilized in generating detailed forest-type classification (Varghese and Krishnamurthy 2006). A detailed forest type mapped using remote sensing and GIS is shown in Fig. 19.3.

19.3.2 Forest Working Plan Preparations

Indian forests are managed through the working plans, the instruments for scientific forest management, which are revised once in 10 years. More than 2/3 of the districts are covered under working plans. The conventional method of making inputs for working plan using ground survey is tedious and time consuming and is based on ocular estimation. Each division has hundreds of coupes due for working each year; hence the above task is assigned to many persons with varying degrees of experience and knowledge, which results in rather inaccurate and inconsistent treatment maps. For the delineation of working circles, felling series and coupes, the first requirement is the generation of treatment types. Each compartment for the purpose of annual coupe operations and site-specific silvicultural treatment has to be divided into various treatment types on the basis of capability of that land unit (Rao et al. 2006). Zonation of forest division into different working circles based on the preponderance and continuity of treatment types, generation of management map for coupe operations and site-specific treatment can be achieved quickly much more accuracy with the use of remote sensing and GIS (Rao et al. 2007). Utilization of RS and GIS inputs for working plan preparation can further enhance the measurement of tree height and stand volume (Varghese et al. 2011). However, tree height and stand volume could not be estimated directly from optical remote sensing data owing to poor correlation between these parameters with reflectance values. The analysis of radar data acquired at L band showed that sensitivity and correlation of radar backscatter with tree density and volume (Varghese et al. 2011). Research work is going on in this direction for standardizing methodology for density and stand volume mensuration for tropical forests of India by using microwave data (Varghese and Joshi 2015 and Varghese et al. 2016a, b).

19.3.3 Monitoring Afforestation/Deforestation Programmes

The major programmes launched in afforestation could be systematically assessed and monitored using temporal satellite data. So far, no systematic monitoring has been administered to evaluate the afforestation program. The improved spatial resolution data from IRS series of satellites could be used for monitoring changes due to afforestation/deforestation. Social forestry aims at raising plantations for the rural populations so as to meet the growing demand for timber, fuel wood, fodder, etc., thereby reducing the pressure on the notified forest area. One of the major lacunae in social forestry is the lack of information about the place or suitable sites where to implement the social forestry activities. Information about the length of roads, canals and railways along with the population pressure in and around it and the proximity to protected areas are crucial for the implementation of social forestry activities. With the advent of high-resolution satellite data of IRS Cartosat series, it is possible to generate the requisite digital database for the requirement of social forestry applications along with the other thematic information. Suitable sites for social forestry activities can be modelled in GIS for selecting linear and block plantations (Suryavanshi et al. 2015).

Though the information on forest plantations and their spatial distribution is available in the state plantation records, reliable maps are, however, lacking. This leads to erroneous estimates of their extent and spatial distribution. Hence, any change in the plantation areas could not be assessed or updated precisely. The satellite-derived information for selected plantation would provide better insight into their effective management. Assessment of forest structural attributes has major implications in the management of forestry by providing information of ecological and economic importance. The traditional methods of assessment involve collecting data in the field and are regarded as labour-intensive and expensive. In plantation forestry, field campaigns are generally time consuming and costly and may compromise profit maximization. The introduction of Light Detection and Ranging (LiDAR) remote sensing in forestry has shown promise to add value to the traditional field inventories mainly through large spatial coverages in a timely and cost-effective manner. LiDAR remote sensing is an advanced system capable of acquiring information in both the vertical and horizontal dimensions at relatively high resolutions. Numerous studies have established that these qualities of LiDAR data are suited to estimating forest structural attributes at acceptable high accuracies. The generic approach in most studies is to use LiDAR data in combination with field data. Such an approach still warrants a high cost of inventory. It is, therefore, useful to explore alternative methods that rely primarily on LiDAR data by reducing the necessity for field-derived information.

19.3.4 Forest Fires and Risk Mapping

Next to deforestation, the forest fire is the most important cause of damage to our forests, which destroys seedlings and regenerating trees. Monitoring and management of forest fires is very important in tropical countries like India, where 55 percent of the total forest cover is prone to fires annually causing adverse ecological, economic and social impacts (Menon et al. 1999). It is estimated that 25% of the carbon dioxide, a major 'greenhouse' gas, is released into the atmosphere due to forest biomass burning. Forest fire management aims at preventing the occurrence of these incidents. Planning such strategies and employing preventive measures will require fire-risk prioritization based on scientific data. The major factors affecting the spread of forest fire are the type and characteristics of the vegetation. Both overstorey and understorey are crucial as they represent the total fuel available for the fire to spread. Satellite-derived vegetation maps could be integrated with the topography, climate and ground-based observations in a GIS domain to model fire-risk areas (Prasanth et al. 2009). The active and dormant or scars of the forest fire are detectable using satellite data. Based on such study, a national-level strategy for forest fire management, in Indian Forest Fire Response and Assessment System (INFFRAS), has been developed at NRSC, which would help in building up natural regeneration, edaphic environment and soil moisture levels in the moist-deciduous and dry-deciduous forest areas of the country (NRSC 2006). The current structure of INFFRAS is available on the NRSC home page (www.nrsa.gov.in). Users need to query using the appropriate query option. The site is mainly about the information on satellite-based services for forest fire detection, assessment and mitigation. The fire images generated from Moderate Resolution Imaging Spectroradiometer (MODIS) and Defence Meteorological Satellite Program (DMSP)-Operational Linescan System (OLS) sensors over Indian region are provided in the form of JPGs and PDF file on the website (Kiran et al. 2006). The burnt area information in terms of area and maps will be useful and are the most required database for forest managers and the scientific community at different levels. Studies were carried out using Indian Remote Sensing Satellite (IRS) LISS-III data for assessing burnt area over different forest regions of India (Prasanth et al. 2009).

Fire risk is the chance of starting a fire as determined by the presence and activity of causative agents. Fire risk involves both ignition and spreading risk. In forests early detection of forest fires is essential for reduction of damage caused due to fire. Traditional methods of mapping forest fires do not meet the requirement of fast, accurate and repeated mapping. In such situation satellite data in combination with GIS plays a vital role in identifying, monitoring and mapping forest fire and recording the effect of fire on forest. They can be used effectively to combine different forest fire, causing factors and its effects together for the creation of the fire-risk zone map. The forest fire-risk areas mapped using remote sensing and GIS in Melghat Tiger Reserve, Maharashtra, are shown in Fig. 19.4 (Varghese and Suryavanshi 2017).

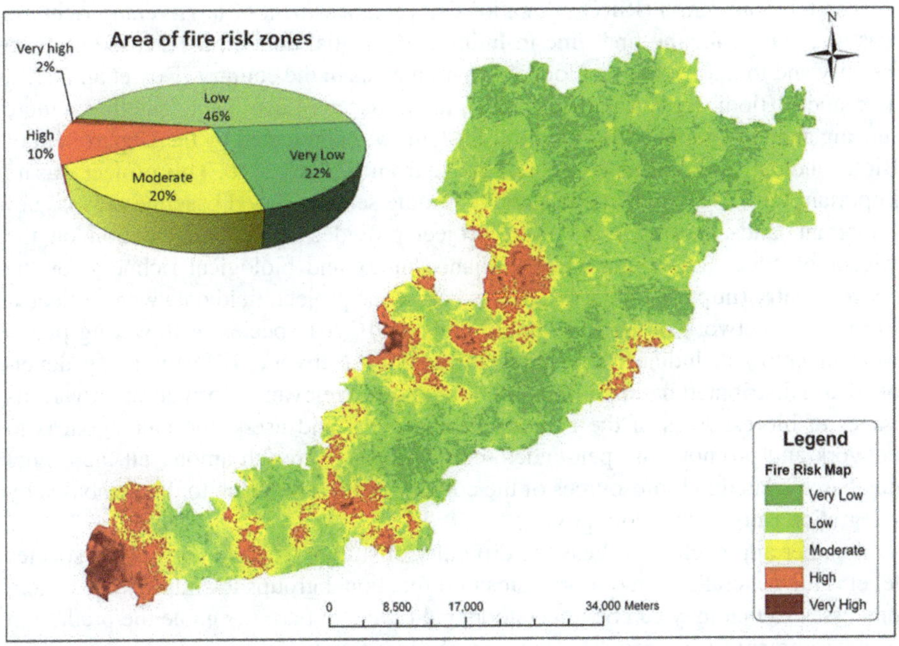

Fig. 19.4 Forest fire-risk mapping of Melghat Tiger Reserve using geoinformatics

19.3.5 Biodiversity Monitoring

Remote sensing and GIS have proven to be very effective tools to analyse biodiversity at various levels, viz. macro-level, meso-level and micro-level. Remote sensing provides spatial data, whereas GIS has made planner's and the manager's job too easy and has provided very useful insights to characterize biodiversity at meso- or macro-level monitoring (Menon and Varghese 2000). We will never be able to monitor extinctions or changes in species distribution directly with remote sensing. Biodiversity is commonly measured by 'species richness', which is the number of species in a site or habitat. Explaining patterns of biodiversity at the species level is difficult as it is usually the outcome of many contributing factors whose relative importance varies with spatial and temporal scale. Therefore, indicators, i.e. surrogate measures of other components of forest biodiversity, are used to monitor temporal and spatial changes in biodiversity. Thus, by analysing the patches, commonly known as patch characterization, the very nature of the landscape can be defined, and patch characterization is based on the analysis of various parameters like fragmentation, porosity, patchiness, interspersion, juxtaposition, etc. Biodiversity characterization at the landscape level using remote sensing and GIS is such kind of work for monitoring biodiversity of India conducted by Indian Space Research Organisation (ISRO). A national-level assessment of biodiversity richness was undertaken for the first time in India using spatial data on a 1:50,000 scale to identify and to map potential biodiversity-rich areas in the country (Roy et al. 2012). The study 'Biodiversity Characterization at Landscape Level using satellite remote sensing and Geographical Information System' was conducted by the Department of Biotechnology and the Department of Space from 1998 to 2010. This project was an important initiative to develop a satellite remote sensing-based baseline database of important landscapes of India. This project provided spatial information on the vegetation type, fragmentation, disturbance index and biological richness for the entire country (http://bis.iirs.gov.in/). As part of the project, field data were collected from a vast network of 16,500 sample plots, and 7761 species of flowering plants were inventoried. Indian Bioresource Information Network (IBIN) is being developed as a distributed national infrastructure to serve relevant information on various issues of bioresources of the country to a range of end users. Its major goal is to network and promote an open-ended, coevolutionary growth among all the digital databases related to bioresources of the country and to add value to the databases by integration (http://www.ibin.gov.in).

Another approach to address the difficulty of studying ecosystems at the species level is the generalization of organisms into functional groups to simplify ecosystem processes so that they can be understood and tested, in part, to enable the prediction of environmental changes. The concept of plant functional type (PFT) uses structural, physiological and/or phenological features to group species in response to environmental conditions or according to their impacts on ecosystems. 'Optical types' are PFTs that can be distinguished on the basis of optical properties detectable by remote sensing. Biodiversity has been successfully assessed through multiple traits evident in vegetation optical properties.

19.3.6 Species-Habitat Modelling and Suitability Assessment

A basic concept of biogeography is that every species has a set of requirements of resources and physical conditions where it can live and reproduce. These requirements define its 'ecological niche'. Although no species niche has been fully described, a lot is known about the general requirements of many taxa. Vertebrate habitat preference is generally better documented than those of plants and invertebrates. Species-habitat relationship models are the suitability ratings assigned by biologists to each habitat class, either as a binary (suitable and unsuitable), ordinal (high, medium, low) or ratio (an index score) values. Each species distribution can thus be mapped by mapping environmental factors. Richness can then be tallied from the model prediction of each species. In reality, a species' 'fundamental niche' corresponding to its suitable habitats has been often broader than the 'realized niche' where it actually occurs. In general, two modelling approaches are distinguished: inductive and deductive approaches (Corsi et al. 2000). The inductive approach is used to derive the ecological requirements of the species from localities in which the species occur. In inductive approaches, a habitat choice of an organism is observed, and the chosen habitat characteristics are extrapolated to wider areas (Skidmore 2002). In one such study, Varghese et al. (2010) mapped the realized and fundamental niches of threatened plant species of Peppara Wildlife Sanctuary using geoinformatics (Varghese et al. 2010). In this study, realized niches of the threatened plant species of Peppara Wildlife Sanctuary were gathered using GPS, and distribution and abundance of each species were measured by stratified random sampling techniques. The realized niches of each threatened species were gathered by GPS, and inductive habitat suitability assessment modelling has been done in GIS along with environmental, edaphic, topographic and phytosociological data for their fundamental niche. The study resulted in documenting 151 tree species belonging to 51 families with 8 threatened species. The distribution of each threatened species, their realized niches and its status, ecological amplitude among forest types and their geographical positions in the landscape were documented and mapped. This will lead to the exact locality information of the concerned taxa, their population status, edaphic and climatic characters of their habitat, their niche width, potential localities to be searched, knowledge of their habitat requirements, particularly the factors limiting their distributions, the factors leading to local and biological extinction and critical habitats. The species-habitat relationship and hot spots of threatened species mapped using remote sensing and GIS are shown in Figs. 19.5 and 19.6 (Varghese et al. 2010).

The deductive approach uses known species ecological requirements to extrapolate suitable areas from the environmental variable layers in the GIS database. Varghese et al. (2016a) used a deductive habitat suitability assessment to map the

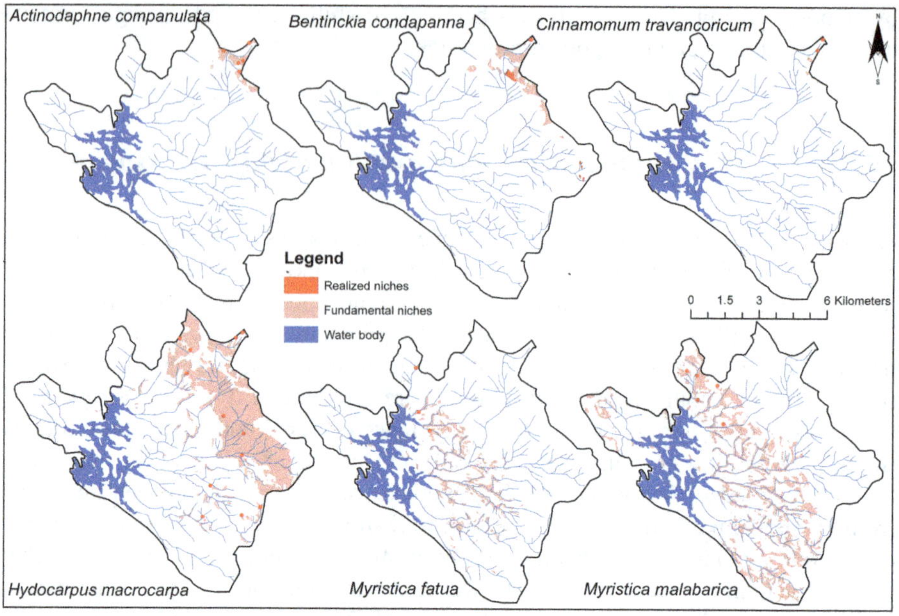

Fig. 19.5 Species-habitat relationship modelling using geoinformatics for the threatened species

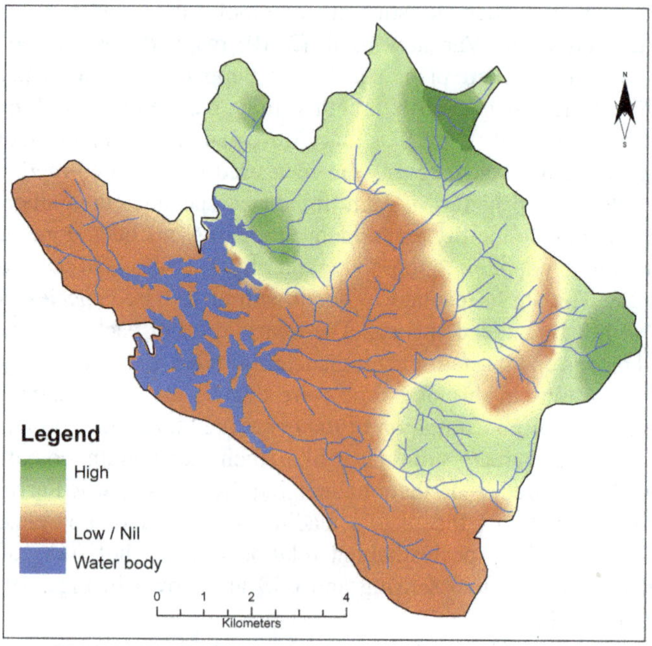

Fig. 19.6 Hot spots of threatened species mapped using geoinformatics

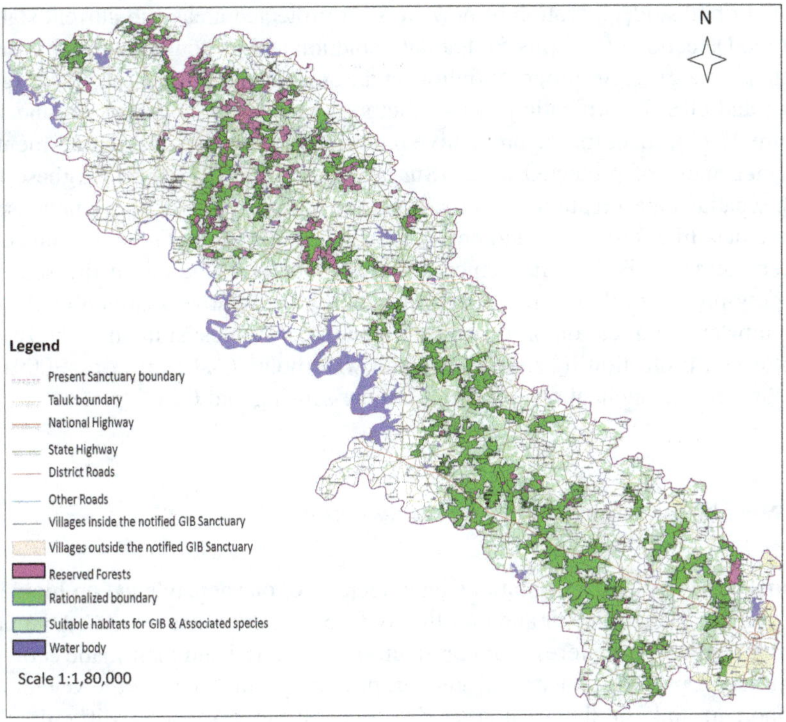

Fig. 19.7 Rationalization of GIB Sanctuary using habitat suitability assessment by remote sensing and GIS

habitat of Great Indian Bustard (GIB) in the GIB Sanctuary in order to rationalize the sanctuary boundary by using remote sensing and GIS. Overlay operation with intersect was used to combine geometries and attributes of all data layers generated for the study to make a composite feature layer. By using logical expressions based on the habitat requirement of GIB, the habitat suitability map was generated. The deductive approach uses known species ecological requirements to extrapolate suitable areas from the variable layers in the GIS database. Based on the criteria set for GIB, optimum set of topographic, edaphic and biological conditions were derived from the integrated layers. The habitat suitability assessment of a threatened and endemic species, Great Indian Bustard, mapped using remote sensing GIS is shown in Fig. 19.7.

19.3.7 *Wildlife Habitat Evaluation*

Remote sensing can be applied to wildlife habitat inventory, evaluation and wildlife census. The role of remote sensing has also been emphasized in quick appraisal of habitat attributes, identification of new sites for protected areas and current status of corridors. Detection of reasons for habitat depletion and its consequences has served as inputs for suggesting proper monitoring and management methods using remote sensing and GIS. Identification of risk factors, sensitivity categorization and infrastructure development for the preventive measures are the foremost requirements for the management of protected areas (Suryavanshi et al. 2013 and Varghese et al. 2012). Spatial data creation for infrastructure, terrain and multi-thematic spatial resource data like forest type and crown density; modelling in GIS for management plan requirements like fire frequency and fire-prone areas based on fire scar mapping; prey-predator relationship; mapping/plotting the relative spatial abundance of wild animals; identification of territories of major carnivores; suitability analysis for watchtowers, protection huts, salt licks and waterholes (Ashwini et al. 2009); etc. were some of the applications by using remote sensing and GIS.

19.3.8 Biomass/Productivity Assessment

As human and livestock populations have increased, our dependence on biomass in terms of food, fodder, timber and non-timber forest products has increased with time. Standing biomass in the forests and productivity of natural and man-made ecological systems are of primary concern. Moreover, primary productivity of the vegetation is an important link in the carbon cycle. Since harvest is often difficult, forest inventory-based biomass assessments are undertaken. A two-stage biomass inventory design employs forest type and density stratification on satellite imagery, followed by on-ground biomass assessment, adopting simple random or stratified random sampling design (Tushar et al. 2009; Juwarkar et al. 2011), and relates these two using regression to assess the biomass of the remaining areas. Another useful structural property of vegetation canopies is the Leaf Area Index (LAI). Since leaves form the chief organs responsible for photosynthesis, transpiration and respiration, it is, therefore, essential to account the total area covered by the leaf surface which in turn can be used to quantify vegetation-atmosphere interaction more precisely. Estimate of LAI provides meaningful interpretation of vegetation-atmosphere interactions because the photosynthetically active radiation (PAR) available through solar electromagnetic spectrum is utilized by the leaves to carry out photosynthesis. The proportion of PAR absorbed by the leaves is known as an absorbed photosynthetically active radiation (APAR). It has been found that APAR is directly related to Leaf Area Index (LAI). In the past three decades, broadband indices have been widely applied to estimate canopy LAI (Mani et al. 2017a). The broadband indices, usually constructed with near-infrared and red bands, use average spectral information over broad bandwidths, resulting in the loss of critical information available in specific narrow bands. Narrow-band hyperspectral remote sensing can be crucial for providing additional information over broadband in quantifying biophysical properties of vegetation and establishing an ecological fingerprinting of forest succession stages and biotic pressure (Hitendra and Varghese 2011). Predicting the gross primary productivity (GPP) of terrestrial ecosystems has been a major challenge in quantifying the global carbon cycle (Canadell et al. 2000). Among all the predictive methods, the light use efficiency (LUE) model may have the most potential to adequately address the spatial and temporal dynamics of GPP because of its theoretical basis and practicality (Running et al. 2000). The LUE model is built upon two fundamental assumptions (Running et al. 2004): (1) that ecosystem GPP is directly related to absorbed photosynthetically active radiation (APAR) through LUE, where LUE is defined as the amount of carbon produced per unit of APAR and (2) that realized LUE may be reduced below its theoretical potential value by environmental stresses such as low temperatures or water shortages (Landsberg 1986). At present, different remote sensing-based models are available to assess GPP and NPP (net primary productivity) like Carnegie-Ames-Stanford Approach (CASA) model (Potter et al. 1993), Global Production Efficiency Model (GLOPEM) (Prince and Goward 1995), 3-PG model (Physiological Principles in Predicting Growth) (Landsberg and Waring 1997), MODIS-GPP algorithms (Running et al. 1999, 2000) and Vegetation Production Model (VPM) (Xiao et al. 2004). The total carbon stock and potential for carbon sequestration mapped using remote sensing and GIS are shown in Figs. 19.8 and 19.9.

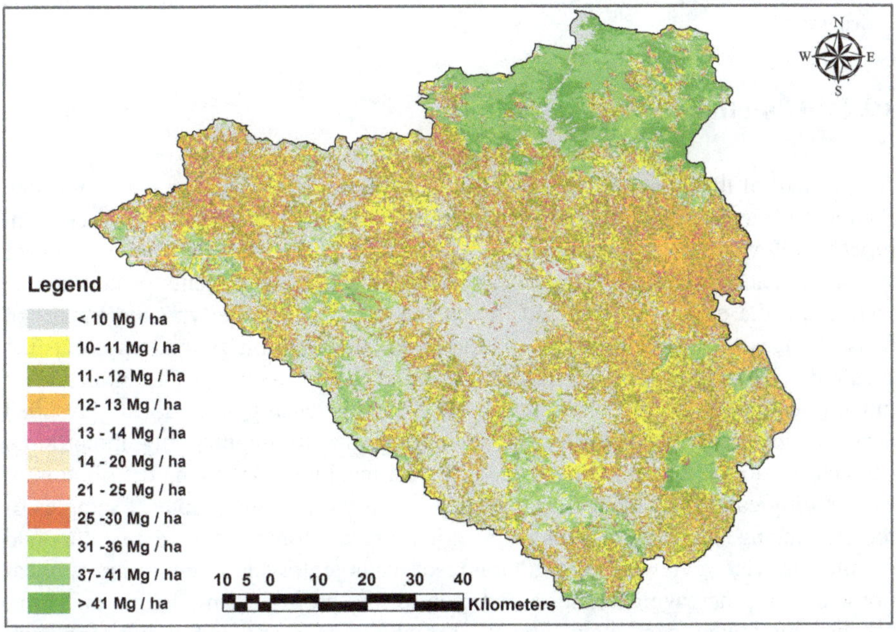

Fig. 19.8 Total carbon stock (phyto- and soilcarbon) of Nagpur district assessed using geoinformatics

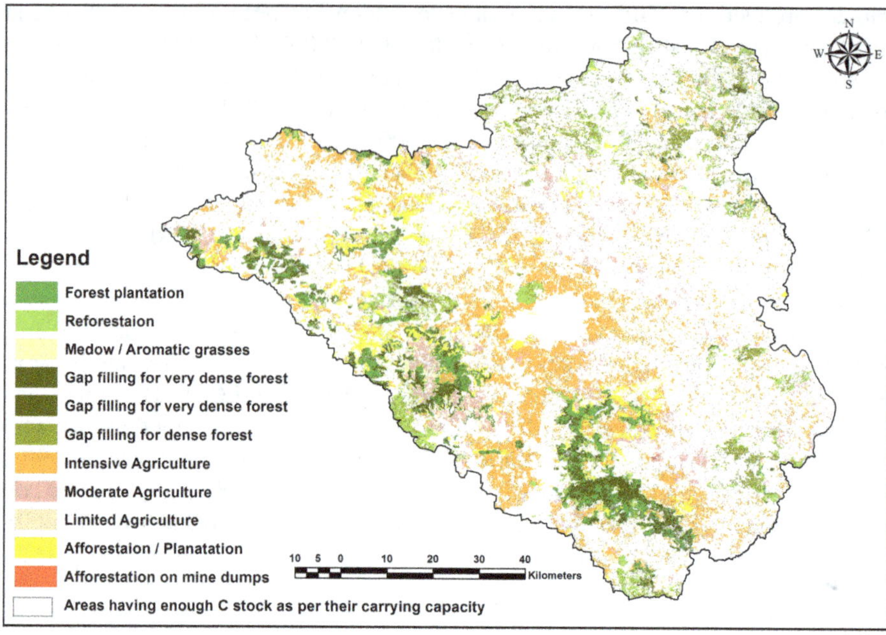

Fig. 19.9 Potential for carbon sequestration assessed by geoinformatics in Nagpur district, Maharashtra

19.4 Conclusions

Assessment of the availability and status of the biological resources is the first step for the sustainable resource development and management plan. The assessment process follows a continuum that involves determining the baseline data or levels of various phenomena, establishing the trends in these measurements or conditions, identifying the causes of rates and trends and determining the type and impact of consequences of rates and trends. An additional element, mitigation, represents the required follow-up on actions in terms of policies or directives. The induction of modern technologies of geospatial tools like remote sensing and GIS has provided very powerful methods of surveying, identifying, classifying, mapping, monitoring, characterization and tracking changes in the composition, extent and distribution of earth biological resources and its constituents. In the present chapter, examples of remote sensing and GIS applications in agriculture and forest resource inventory and monitoring were reviewed. Agricultural resource inventory and monitoring applications like crop acreage estimation and monitoring, crop yield modelling and forecasting, integrated pest management, drought assessment and monitoring, soil moisture estimation, etc. were explained with examples. Likewise forestry applications such as forest cover and type mapping, forest working plan preparations, monitoring of afforestation programmes, forest fire monitoring and risk mapping, biodiversity monitoring, species-habitat relationship modelling, wildlife habitat evaluation, biomass and productivity assessments, etc. were also discussed. Applications of remote sensing are numerous and expanding. With the availability of a variety of satellites, including Indian Remote Sensing (IRS) Satellite with increasing range and sensitivity of their sensors and the sophisticated image processing and interpretation facilities, the future of remote sensing looks very promising.

References

Ashwini SP, Prasanth DK, Varghese AO, Joshi AK (2009) Sensitivity categorization and management of forest protection huts using Geoinformatics. In: Proceedings of the ISRS symposium on advances in geo-spatial technologies with special emphasis on sustainable rainfed agriculture, 17–19, September 2009, Nagpur

Bower SA, Smith SJ (1972) Spectrophotometric determination of soil water content. Soil Sci Soc Am Proc 36:978–980

Burrough PA (1989) Principles of geographical information systems for land resources assessment. Clarendon, Oxford

Canadell JG, Mooney HA, Baldocchi DD, Berry JA, Ehleringer B, Field CB, Gower ST, Hollinger DY, Hunt JE, Jackson RB, Running SW, Shaver GR, Steffen W, Trumbore SE, Valentini R, Bond BY (2000) Carbon metabolism of the terrestrial biosphere: a multi-technique approach for improved understanding. Ecosystems 3:115–130

Champion HG, Seth SK (1968) A revised survey of forest types of India. Government of India Publication, New Delhi

Clark WF (1990) North Carolina Estuaries-A pilot study for managing multiple use in the state's public trust waters- Study report, 90–10, Raleigh, North Carolina

Corsi F, De Leeuw J, Skidmore AK (2000) Modeling species distribution with GIS. In: Boitani L, Fuller TK (eds) Research Techniques in animal ecology; controversies and consequences. Columbia University Press, pp 389–434. ISBN0231501390

Dill HN (1990) Conservation of natural resources, a few concepts. In: Agarwal VP, Rana SVS (eds) *Environmental and natural resources*. Society of Biosciences, Muzaffarnagar

FAO (2005) The state of food insecurity in the world 2005: Eradicating world hunger e key to achieving the millennium development goals. Food and Agriculture Organization, Rome

FSI (2003) India State of Forest Report, 2003, Forest Survey of India, Ministry of Environment and Forest, Dehradun

FSI (2009) India State of Forest Report, 2009, Forest Survey of India, Ministry of Environment and Forest, Dehradun

FSI (2011) India State of Forest Report, 2011, Forest Survey of India, Ministry of Environment and Forest, Dehradun

FSI (2017) India State of Forest Report, 2011, Forest Survey of India, Ministry of Environment and Forest, Dehradun

Gallego J, Carfagna E, Baruth B (2010) Accuracy, objectivity and efficiency of remote sensing for agricultural statistics. Agricultural survey methods. Wiley, West Sussex, pp 193–211

Hitendra P, Varghese AO (2011) Evaluating sensitivity of hyper spectral remote sensing satellite data based indices to in-situ measured biophysical characteristics of dry Deciduous forests. Report number: NRSC-RC-Mar-2010-TR-619, Regional Remote Sensing Center-Central, NRSC, Nagpur

Huete AR, Liu HQ, Batchily K, van Leeuwen W (1997) A comparison of vegetation indices over a global set of TM images for EOS-MODIS. Remote Sens Environ 59:440–451

Huete A, Didan K, Miura T, Rodriguez EP, Gao X, Ferreira LG (2002) Overview of the radiometric and biophysical performance of the MODIS vegetation indices. Remote Sens Environ 83:195–213

http://bis.iirs.gov.in/. Biodiversity information system, Indian Institute of Remote Sensing, Dehradun. Accessed on Mar 15, 2018

http://www.ibin.gov.in. Indian Bio-resource Information Network, Department of Biotechnology, Government of India. Accessed on Mar 16, 2018

Ishida T, Ando H, Fukuhara M (1991) Estimation of complex refractive index of soil particles and its dependence on soil chemical properties. Remote Sens Environ 38:173–182

Juwarkar AA, Varghese AO, Singh SK, Aher VV, Thawale PR (2011) Carbon sequestration potential in above ground biomass of natural reserve forest of Central India. Int J Agric Res Rev 1(2):80–86

Kiran CT, Badarinath KVS, Prasad VK, Murthy MSR (2006) Monitoring forest fires over the Indian region using DMSP-OLS nighttime satellite data. Remote Sens Environ 103:165–168

Kogan FN (1990) Remote sensing of weather impacts on vegetation in non-homogeneous areas. Int J Remote Sens 11:1405–1419

Kogan FN (1995) Application of vegetation index and brightness temperature for drought detection. Adv Space Res 15:91–100

Landsberg JJ (1986) Physiological ecology of forest production. Academic Press, London, pp 165–178

Landsberg JJ, Waring RH (1997) A generalised model of forest productivity using simplified concepts of radiation-use efficiency, carbon balance and partitioning. For Ecol Manage 95:209–228

Mani JK, Varghese AO, Kameshwara Rao SVC (2016) Suitability analysis for Orange orchards in Vidarbha region of Maharashtra using remote sensing and GIS based models, In: Proceedings of international conference on integrated land use planning for smart agriculture -an agenda for sustainable land management (ICILUPSA-2016). November 10–13, 2016, Nagpur

Mani JK, Varghese AO, Kameshwara Rao SVC (2017a) Estimation of leaf area index of Teak forest in Central India using satellite remote sensing, In: Proceedings of the 38th Asian Conference of Remote Sensing (ACRS-2017). October 23–27, 2017, New Delhi

Mani JK, Varghese AO, Joshi AK (2017b) Assessment of gross primary productivity in semi arid agricultural region of central India using temporal MODIS data. In: Reddy GPO, Patil NG, Chaturvedi A (eds) Sustainable management of land resources: an Indian Perspective. Apple Academic Press, USA, pp 365–386

Menon ARR, Varghese AO (2000) Biodiversity monitoring. In: Sivadasan M, Mohanan KV (eds) Biodiversity and ecology –concepts and facts. University of Calicut, Calicut, pp 102–116

Menon ARR, Varghese AO, Lowel KJM (1999) Impact of fire on moist deciduous forest ecosystem of southern tropical forests of India. In: Impacts of fire and human activities on forest ecosystems in the tropics Proceedings of the international symposium on Asian Tropical Forest Management III, pp 52–61

Moulin S, Bondeau A, Delecolle R (1998) Combining agricultural crop models and satellite observations: from field to regional scales. Int J Remote Sens 19:1021–1036

NRSA (1990) National agricultural drought assessment and monitoring system. NRSA, Hyderabad, p 124

NRSC (2006). Perspective of Geoinformatics in Forest Fire Management, Forestry and Ecology division, NRSC, Hyderabad

NRSC 2010 Remote sensing applications. In: Roy PS, Dwivedi RS, Vijayan D (eds). NRSC/ISRO, Hyderabad, p 377

Panigrahy S, Ray SS, Sood A, Patel LB, Sharma PK (2002) Cropping system analysis using remote sensing and GIS—Bathinda District, Punjab. Scientific Note RSAM/SAC/HORT/SR/01/02 Space Applications Centre, Ahmedabad

Panigrahy S, Ray SS, Manjunath KR, Pandey PS, Sharma SK, Sood A, Yadav M, Gupta PC, Kundu N, Parihar JS (2009) A spatial database of cropping system and its characteristics to aid to climate change impact assessment studies, In: Panigrahy S, Ray SS, Parihar JS (eds) The international archives of the photogrammetry, remote sensing and spatial information sciences' vol XXXVIII Part 8/W3, pp 150–155

Potter CB, Randerson JT, Field CB, Matson PA, Vitousek PM, Mooney HA, Klooster SA (1993) Terrestrial ecosystem production: a process model based on global satellite and surface data. Global Biogeochem 7:811–841

Prasanth DK, Ashwini SP, Varghese AO, Joshi AK (2009) Mapping of forest fire risk zones and identification of suitable sites for fire watchtowers using remote sensing and GIS, In Proc. of the ISRS Symposium on Advances in Geo-spatial technologies with special emphasis on sustainable rainfed Agriculture, 17–19, September 2009, Nagpur

Prince SD, Goward SN (1995) Global primary production: a remote sensing approach. J Biogeogr 22:815–835

Rani AS, Jyothi S (2017) A study on hyper spectral remote sensing pest management. Int J Recent Innovat Trend Comput Commun 5(6):497–503

Rao MK, Varghese AO, Krishna Murthy YVN (2006) Use of Geospatial database in Sustainable Forest Management, In: Proceedings of ISPRS international symposium on geospatial database for sustainable development, 27–30, September 2006, Goa, India, WG-IV-3-12

Rao MK, Varghese AO, Krishna Murthy YVN (2007) Remote sensing and GIS inputs for working plan preparation. Indian Forester 133(1a):65–76

Ray SS, Sood A, Panigrahy S, Parihar JS (2005) Derivation of indices using remote sensing data to evaluate cropping systems. J Indian Soc Remote Sens 33:475–481

Roy PS, Kushwaha SPS, Murthy MSR, Roy A, Kushwaha D, Reddy CS, Behera MD, Mathur VB, Padalia H, Saran S, Singh S, Jha CS, Porwal MC (2012) Biodiversity characterization at Landscape level, National Assessment 2012. IIRS, Dehradun

Roy PS et al (2015) New vegetation type map of India prepared using satellite remote sensing: comparison with global vegetation maps and utilities. Int J Appl Earth Obs Geoinf 39:142–159

Running SW, Nemani R, Glassy JM, Thornton PE (1999) MODIS daily photosynthesis (PSN) and annual net primary production (NPP) product (MOD17), algorithm theoretical basis document, version 3.0. http://modis.gsfc.nasa.gov/

Running SW, Thornton PE, Nemani RR, Glassy JM (2000) Global terrestrial gross and net primary productivity from the earth observing system. In: Sala O, Jackson R, Mooney H (eds) Methods in ecosystem science. Springer, New York, pp 44–57

Running SW, Ramakrisha R, Nemani FAH, Zhao M, Reeves M, Hashimoto H (2004) A continuous satellite-derived measure of global terrestrial primary production. Bioscience 54(6):547–560

Singh R, Goyal RC, Saha SK, Chhikara RR (1992) Use of satellite spectral data in crop yield estimation surveys. Int J Remote Sens 13:2583–2592

Skidmore AK (2002) Taxonomy of environmental models in the spatial sciences. In: Skidmore A (ed) Environmental modelling with GIS and remote sensing. Tylor & Francis, London, pp 8–24

Suryavanshi AS, Varghese AO, Patil V, Joshi AK. (2013) Wildlife trail tracking system – a wildlife monitoring system in GIS environment. ISRS & ISG, December 4–6, Visakhapatnam

Suryavanshi A, Varghese AO, Joshi AK., Sharma JR (2015) Social Forestry Information System (SOFIS): a standalone software for holistic planning of Social Forestry activities. In: Proceedings of "Open source geospatial tools in climate change research and natural resource management, (FOSS4G-India 2015)

Thenkabail PS (2003) Biophysical and yield information for precision farming from near-real-time and historical Landsat TM images. Int J Remote Sens 24:839–877

Trotter CM (1991) Remotely sensed data as an information source for geographical information systems in natural resource management: a review. Int J GIS 5(2):225–239

Tushar W, Varghese AO, Tawale P, Nair R, Singh SK, Juwarkar AA (2009) Assessment of the potential for carbon sequestration through afforestation, forestry and agriculture models using remote sensing and GIS. In: Proceedings of the ISRS Symposium on Advances in Geo-spatial technologies with special emphasis on sustainable rainfed Agriculture, 17–19, September 2009, Nagpur

Varghese AO, Joshi AK (2015) Polarimetric classification of C-band SAR data for forest density characterization. Curr Sci 108(1):100–106

Varghese AO, Krishna Murthy YVN (2006) Application of geoinformatic in conservation and management of rare and threatened plant species. Curr Sci 96(6):763–769

Varghese AO, Suryavanshi A (2017) Forest cover transformation analysis and Management of Melghat Tiger Reserve using remote sensing and GIS, RRSC-C/NRSC Technical Report, September 2017

Varghese AO, Joshi AK, Krishna Murthy YVN (2010) Mapping of realized and fundamental niches of threatened tree species using geoinformatics: a species level approach for sustaining biodiversity, special issue on biodiversity and landscape ecology. J Indian Soc Remote Sens 38(3):523–534

Varghese AO, Joshi AK, Krishna Murthy YVN (2011) Tropical dry deciduous forest stand variable estimation using SAR data. J Indian Soc Remote Sens 39(4):583–589

Varghese AO, Suryavamshi A, Rajashekhar SS, Joshi AK, Krishna Murthy YVN (2012) Decision support systems for production and protection forestry. NNRMS (B) 36:81–86

Varghese AO, Suryavanshi A, Joshi AK (2016a) Analysis of different polarimetric target decomposition methods in forest density classification using C band SAR data. Int J Remote Sens 37(3):694–709

Varghese AO, Sawarkar VB, Rao YLP, Joshi AK (2016b) Habitat suitability assessment of *Ardeotis nigriceps* (Vigors) in great Indian bustard sanctuary, Maharashtra (India) using remote sensing and GIS. J Indian Soc Remote Sens 44(1):49–57

Xiao XM, Zhang QY, Braswell B, Urbanski S, Boles S, Wofsy S, Moore B, Ojima D (2004) Modeling gross primary production of temperate deciduous broadleaf forest using satellite images and climate data. Remote Sens Environ 91:256–270

Zhang D, Zhao H (2016) Estimation of soil moisture from optical and thermal remote sensing: a review. Sensors 16:1308

Chapter 20
Remote Sensing and GIS in Mapping and Monitoring of Land Degradation

G. P. Obi Reddy, Nirmal Kumar, and S. K. Singh

Abstract The information on the extent and spatial distribution of various kinds of degraded lands is essential for strategic planning and development of degraded lands. Processes of land degradation can be broadly grouped into physical, chemical, and vegetal (biological) degradation. The physical processes include land degradation mainly due to water and wind erosion, compaction, crusting, and waterlogging. The chemical process includes salinization, alkalization, acidification, pollution, and nutrient depletion. The vegetal or biological processes on the other hand are reduction of organic matter content in the soils and degradation of vegetation. The use of remote sensing and geographic information system (GIS) techniques makes land degradation estimation and its spatial distribution feasible with reasonable costs and better accuracy in larger areas. The use of spaceborne multispectral data shown its potential in deriving information on the nature, extent, spatial distribution, and magnitude of various kinds of degraded lands. Assessment and monitoring of land degradation through remote sensing offer a series of advantages such as consistency of data, fairly near real-time reporting, and a source for having spatially explicit data. The integration of high-resolution remote sensing data and digital elevation models derived from satellites data like Cartosat-1 and Cartosat-2 and Light Detection and Ranging (LiDAR) with ground data has immense potential in assessment and monitoring of land degradation in local scales. In this chapter, application of remote sensing and GIS in assessment and mapping of physical, chemical, and vegetal degradation has been discussed. The study indicates that integrated remote sensing and GIS applications have immense potential in assessment, mapping and monitoring of land degradation with reasonable cost and better accuracy in larger areas that would otherwise require large inputs of human and material resources.

Keywords Remote sensing · Geographic information system · Land degradation · Physical degradation · Chemical degradation · Vegetal degradation

---

G. P. O. Reddy (✉) · N. Kumar · S. K. Singh
ICAR-National Bureau of Soil Survey & Land Use Planning, Nagpur, India

G. P. O. Reddy, S. K. Singh (eds.), *Geospatial Technologies in Land Resources Mapping, Monitoring and Management*, Geotechnologies and the Environment 21,
https://doi.org/10.1007/978-3-319-78711-4_20

20.1 Introduction

Land degradation remains a challenge in the twenty-first century for many developing countries across the globe because of its effect on the sustainability of agricultural production and impacts on livelihoods among the marginal and rural poor. Bai et al. (2008) reported that more than 20% of all cultivated areas, 30% of forests, and 10% of grasslands in the world undergo various categories of land degradation. About a quarter of world population is threatened by the effects of degradation (Eswaran et al. 2001), which affect nearly 84% of agricultural lands (FAO 2008a). Land degradation means reduction or loss of the biological or economic productivity and complexity of rainfed cropland, irrigated cropland, or range, pasture, forest, and woodlands resulting from land uses or from a process or combination of processes, including processes arising from human activities and habitation patterns. Land degradation is the temporary or permanent lowering of the productive capacity of land. UNEP define it as a rate of adverse changes in soil quality resulting in decline in productivity of lands (UNEP 1992). It thus covers the various forms of soil degradation, adverse human impacts on water resources, deforestation, and lowering of the productive capacity of rangelands. Land degradation includes soil erosion due to wind and/or water; deterioration of the physical, chemical and biological, or economic properties of soil; and long-term loss of natural vegetation. Land degradation is an obstacle to sustainable development due to its impact on the environment, food security, agroecosystem services, and people's livelihoods (UNCCD 2015). Metternicht (2006) considers that land degradation is the "reduction in the capability of the land" to produce benefits from a particular land use under a specified form of land management. In some cases, the cause for land degradation could be solely natural or purely human, but often both human and natural causes combine to land degradation. On-site effects of land degradation are the lowering of the productive capacity of the land, causing either reduced outputs (crop yields, livestock yields) or the need for increased inputs. Off-site effects of water erosion occur through changes in the water regime, including decline in river water quality and sedimentation of riverbeds and reservoirs.

There is a large variation in estimations of extent and rate of land degradation at global scale due to variation in definitions and methodologies adopted. It varied from 3.6 billion ha (Dregne and Chou 1994) to 1.9 billion ha (Oldeman 1994). Many methods have been applied to assess land degradation through different approaches, which use either qualitative or quantitative measures or both. GLASOD (Global Assessment of Soil Degradation) approach mapped the status of global soil degradation at a scale of 1:10 million by indicating type, extent, degree, rate, and main causes of degradation (Oldeman et al. 1991). LADA (Land Degradation Assessment in Drylands) developed an integrated assessment methodology for land degradation to understand the degradation processes at different scales (global, national, and local) by identifying the status and trends of land degradation, root causes, effects, and consequences (LADA 2009). In India, the earliest assessment of the area affected by the land degradation was made by the National Commission on Agriculture at 148 Mha (NCA 1976). The National Bureau of Soil Survey and Land Use Planning (NBSS&LUP) estimated an area of 187 Mha (Sehgal and Abrol 1994) under various categories of degraded lands by using GLASOD methodology (Oldeman 1988). Nearly 175 Mha hectares of land in India is subject to one or other kind of degradational process (Das 1985). About 150 Mha of land are suffering from different types of erosion, out of which 69 Mha are in severe deterioration phase (Anonymous 1976). The salt-affected soils and waterlogged areas are reported as 7 Mha and 6 Mha, respectively (Bali 1985). The National Wasteland Development Board (NWDB 1985) estimated an area of 123 Mha under wastelands. National Remote Sensing Centre (NRSC) (formerly known as NRSA) followed the remote sensing data-based assessment in preparation of wasteland maps with adequate field checks on 1:50,000 scale and reported that 63.85 Mha area in India is under various categories of wastelands (NRSA 2005). These estimations are due to the use of varying definitions of land degradation, data sources, classification systems, methodologies, and scales (Gautam and Narayan 1988). Maji et al. (2010) harmonized the land degradation and wasteland datasets of India by adopting systematic approach in GIS environment and reported that about 120.72 Mha area is suffering from various kinds of land degradation. It includes 82.57 Mha area affected by water erosion, 12.4 Mha by wind erosion, 6.74 Mha by salinity/alkalinity, and 17.94 Mha by soil acidity, and 1.07 Mha is under other complex problems (Table 20.1).

Remotely sensed data was effectively used in identifying and mapping of land degradation risks (Lu et al. 2007; Reddy et al. 2002). Satellite sensor data provide spatially continuous, replicable, and homogenous information on the condition,

Table 20.1 Extent of degraded and wastelands in India

Degradation type	Area (in Mha)	Open forest (<40% canopy) (Mha)
Water erosion	73.27	9.30
Wind erosion	12.40	–
Chemical degradation soils		
Exclusively salt-affected soils	5.44	–
Salt-affected and water-eroded soils	1.20	0.10
Exclusively acidic soils (pH < 5.5)	5.09	–
Acidic and water-eroded soils	5.72	7.13
Physical degradation		
Mining and industrial waste	0.19	–
Waterlogging	0.88	–
Total	104.19	16.53
Grand Total	120.72	

Source: Maji et al. (2010)

distribution, and dynamics of vegetation status and land degradation in a cost-effective manner and over large areas. Remote sensing applications are often considered as cost-effective and time-efficient procedures for the collection of data over large areas that would otherwise require a very large input of human and material resources. The integrated remote sensing and GIS technologies helps immensely to spatially analyze and improve the understanding of causative factors in land degradation assessment. Qualitative assessment, delineation, and mapping of eroded lands were attempted using Landsat, MSS/TM, SPOT-PLA/MLA, and IRS LISS-I/II data (Dwivedi et al. 1997a, b). Landsat MSS data have been used for predicting soil loss in the rangelands of Western Australia (Pickup and Chewings 1986). Landsat MSS/TM data have been used for mapping ravines (Karale et al. 1987; Singh and Dwivedi 1989). Raina et al. (1991) have used Landsat TM data to map the type, extent, and degree of degradation. NDVI derived from temporal satellite data has been widely used in studies of land degradation from the field scale to the global scale (Wessels et al. 2004; Singh et al. 2006). Many studies used Landsat TM and ETM+ data in assessment of soil erosion, soil salinity, and crusting in drylands (Metternicht et al. 2009; Vrieling et al. 2008; Vågen et al. 2013; Nawar et al. 2014). Pandey et al. (2013) have used spectral indices such as CI (Crust Index), NDSDI (Normalized Difference Sand Dune Index), and GSI (Topsoil Grain Size Index) and compared with NDVI (normalized difference vegetation index) to assess land degradation and sand encroachment in Western India.

20.2 Methods in Assessment of Land Degradation

There are many methods used to assess land degradation, viz., expert opinions, land users' opinions, field monitoring, observations and measurement, modeling, estimates of productivity changes, and remote sensing (Kapalanga 2008). The important methods like expert opinions, field monitoring, observations, and remote sensing have been discussed below in detail.

20.2.1 Expert Opinion

Early land degradation assessments were essentially based on expert opinion/judgment, as in the case of the GLASOD. The expert-based GLASOD (Oldeman et al. 1991) approach maps the status of global soil degradation at a scale of 1:10 million by indicating type, extent, degree, rate, and main causes of degradation based on responses to a questionnaire, which was sent to recognized experts in countries around the world. GLASOD survey provides basic data on distribution and intensity of erosional, chemical, and physical types of degradation at global scale (Bridges and Oldeman 1999). A total of 1965 Mha land of the world was found to be degraded and out of which water erosion affecting 1094 Mha of the land. Sehgal and Abrol

(1992) following the criteria and guidelines of the GLASOD methodology estimated an area of 187 Mha under various categories of degraded lands in India. Another example of an expert approach is the soil erosion risk map of Western Europe (De Ploey 1989). Kessler and Stroosnijder (2006) utilized historical data and farmers' knowledge to identify eroded lands and severity of erosion in the Bolivian mountain valleys based on indicators of soil, productivity, and vegetation cover loss. Soil degradation in South and Southeast Asia (ASSOD) is another approach in which the degree of soil degradation is expressed by degradation subtypes using qualitative terms such as impact on productivity (Van Lynden and Oldeman 1997). LADA considers both biophysical factors and socioeconomic driving forces for assessing the land degradation (FAO 2008b). Koohafkan et al. (2003) developed the guidelines for a methodological approach for assessing land degradation under LADA project to assess the causes, status, and impact of land degradation and possible responses. The approaches based on experts and users' opinion are subjective and qualitative (Bai et al. 2008) and have proven inconsistent and hardly reproducible (Sonneveld and Dent 2009).

20.2.2 *Field Monitoring and Observation*

The assessment of land degradation requires reliable analyses based on field monitoring and observations. It is thus necessary to accurately describe the different types of degradation and quantify the degree and extent of each type of degradation using relevant indicators for targeted applications through using geospatial technologies like satellite imaging, GIS, and global positioning system (GPS) at different scales. A systematic field survey is necessary to determine the extent of a given type of land degradation. During the field surveys, surveyor can pinpoint the areas in the field affected by the type of degradation, transferring the observations on a large-scale map, and then calculate the degraded area in GIS to determine its extent. The visual observations can be supplemented by legacy data and available high-resolution satellite images. GPS can be used in the field to accurately locate the observations. The collected field samples can be analyzed in the laboratory to determine various chemical properties of soils relevant to land degradation. These types of field surveys can also be supplemented by surveys of farmers/inhabitants to determine the cropping practices and history of the crops cultivated. Through such field surveys and observations, various subtypes of chemical and biological degradation can be assessed and mapped. In India, based on soil resource mapping on 1:250,000 scale, it was reported that strongly acidic soils (pH <4.5) and moderately acidic soils (pH 4.5–5.5) cover 1.9% and 7.4% of TGA of India, respectively (Maji et al. 2012). In the recent times, high-resolution satellite images are being widely used in mapping of land degradation through field surveys and observations.

20.2.3 Remote Sensing

Land degradation monitoring through remote sensing could be achieved through two approaches through the comparative analysis of independently produced classification for different dates and the simultaneous analysis of multi-temporal satellite data. Different change detection techniques include univariate image differencing, vegetation index differencing, image regression, image ratioing, principal component analysis, post-classification comparison, direct multi-date comparison, change vector analysis, and background subtraction which could be used to assess the land degradation. Many authors adopted various approaches such as visual interpretation, unsupervised and supervised classification, and remote sensing-derived indices for mapping of land degradation (Gupta et al. 1998; Saini et al. 1999; Jafari et al. 2008). Though researchers devised best techniques to derive the results, these techniques seem to yield different levels of results for different environmental features and applications. Image fusion techniques such as image sharpening, improvement of registration accuracy, creation of stereo data sets, feature enhancement, improved classification, temporal aspect for change detection, and overcoming data gaps due to clouds could improve and yield more information than a single sensor data can provide (Pohl and Van Genderen 1998). Metternicht and Zinck (1998) investigated synergistic use of JERS-1 and Landsat TM for mapping water-induced surface erosion features. Metternicht and Zinck (1997) found out the highest separability between erosion classes upon integration of seven bands of the Landsat TM and JERS-1 SAR with an overall classification accuracy of 87%. Dwivedi et al. (1997a, b) revealed that fusion of Landsat TM and SPOT MSS data provided an overall accuracy of 92% for erosion mapping.

20.3 Remote Sensing and GIS in Mapping and Monitoring of Land Degradation

Various processes of land degradation have been broadly grouped into physical, chemical, and vegetal degradation. Remote sensing and GIS technologies have immense potential in mapping and monitoring of land degradation with adequate field surveys.

20.3.1 Physical Degradation

Physical degradation covers land degradation due to water erosion, wind erosion, waterlogging, lowering of the water table, mining and quarrying, and urban and industrial waste. Global water erosion and wind erosion affect 1094 and 549 Mha, respectively (Lal 2003).

20.3.1.1 Water Erosion

The information on extent and spatial distribution of soil erosion is essential for formulating effective mitigation strategies and implementing appropriate conservation measures (Vrieling 2006; Panagos et al. 2015). Water erosion covers all forms of soil erosion by water, including sheet and rill and gully erosion. Soil erosion by water involves the processes of detachment and transportation by impact of raindrop and flowing water (Wischmeier and Smith 1978). Soil erosion is a natural process that removes soil particles and deposit as sediment in some other location. Out of total 1965 Mha of degraded lands in the world, 1094 Mha is under soil erosion due to water, and it accounts for 55% and causing up to a 17% reduction in crop productivity (Oldeman et al. 1990). Dhruvanarayana and Ram Babu (1983) reported that in Indian conditions about 16.4 t ha^{-1} $year.^{-1}$ of top soil, of which 29% is lost permanently into the sea, 10% gets deposited in the reservoirs reducing their capacity by 1–2% every year, and the remaining 61% gets displaced from one place to another. Mandal and Sharda (2011) reported that soil erosion caused by water is a major factor contributing to land degradation in India and many other countries, as it exceeds the natural soil formation rates.

Remote sensing and GIS techniques make soil erosion estimation and its spatial distribution feasible with reasonable costs and better accuracy in larger areas (Millward and Mersey 1999; Wang et al. 2003). Integrated remote sensing, GIS, and RUSLE provide the potential to estimate soil erosion loss on a cell-by-cell basis (Millward and Mersey 1999). Wang et al. (2003) demonstrated that integration of ground dataset, Thematic Mapper (TM), and digital elevation model (DEM) data through geostatistical methods provides significantly better results than using traditional methods in predicting soil erosion loss. Remote sensing data has immense potential to develop the cover management factor through land cover classifications (Reusing et al. 2000; Ma et al. 2003), whereas GIS tools could be effectively used to integrate the USLE factors in calculation of soil erosion (Bartsch et al. 2002; Wang et al. 2003). The utility of GIS capabilities increased when they coupled with empirical and predictive models in assessment of soil loss (Reddy et al. 2004, 2013, 2016; Srinivas et al. 2002). Wilson and Lorang (2000) reviewed the applications of GIS in estimating soil erosion, discussed the limitations of previous approaches, and identified that GIS provided tremendous potential for improving soil erosion estimation.

Many researchers used soil erosion models such as the Universal Soil Loss Equation (USLE) (Wischmeier and Smith 1978) and its subsequent Revised Universal Soil Loss Equation (RUSLE) (Renard et al. 1997) and GIS techniques to make soil erosion estimation and its spatial distribution (Wang et al. 2003; Fu et al. 2005). In practice, the Universal Soil Loss Equation (USLE) (Wischmeier and Smith 1978) and later the Revised Universal Soil Loss Equation (RUSLE) (Renard et al. 1997) has been the most widely used model in predicting soil erosion loss. The use of remote sensing and GIS techniques makes soil erosion estimation and its spatial distribution feasible with reasonable costs and better accuracy in larger areas

(Millward and Mersey 1999; Wang et al. 2003). The integrated use of GIS and erosion models, such as USLE/RUSLE, has been proved to be an effective approach for estimating the extent, magnitude, and spatial distribution of erosion (Mitasova et al. 1996; Molnar and Julien 1998; Millward and Mersey 1999; Fernandez et al. 2003). Availability of spatial databases in GIS, digital elevation models (DEMs), and temporal satellite imageries have immense potential to predict erosion potential on a cell by cell (Reusing et al. 2000).

In both models, the average soil erosion per year is computed from the product of six factors, namely, rainfall erosivity (R), soil erodibility (K), slope length (L), slope steepness (S), vegetation cover (C), and support practice factor (P). In the RUSLE, the mean annual soil loss is expressed as a function of six erosion factors:

$$A = R \times K \times L \times S \times C \times P \qquad (1)$$

where A is the computed amount of the average soil loss in tons per hectare per year, R the rainfall erosivity factor in megajoules per millimeter per hectare per hour per year, K the soil erodibility factor in tons per hour per megajoules per millimeter, L the slope length (meters), S the slope steepness (%), C the crop management factor, and P the erosion control practice factor. Factors C and P are dimensionless. R factor, in the USLE and RUSLE models, is an index of rainfall erosivity, which is the potential ability of the rain to cause erosion. The soil erodibility factor (K) represents both susceptibility of soil to erosion and the amount and rate of runoff, as measured under standard plot conditions. The cover factor (C) is an index which reflects, on the basis of the land use, the effect of cropping practices on the soil erosion rate. After computation of R, K, LS, C, and P maps as data layers, they can be multiplied in the GIS to assess spatial distribution of soil loss. Other than the USLE and RUSLE, other erosion models such as the Morgan and Finney method (Morgan et al. 1984), ANSWERS (Beasley et al. 1980), WEPP (NSERL 1995), and PCARES (Paningbatan 2001) were also used to predict soil erosion.

20.3.1.2 Wind Erosion

It refers to loss of soil by wind, occurring primarily in dry regions. Wind erosion process is often found to be one of the major causes of land degradation in arid and semiarid regions. In wind-induced land degradation mapping, sand dunes, wind streaks, paleo-aeolian features, desert pavements, sand encroachments, blowouts, and changes in the vegetation cover are indicators commonly to be considered. Wind erosion is controlled by several factors such as wind velocity, rainfall pattern, stability of the surface on which wind is acting upon, vegetation cover, and also socioeconomic condition of the region. Wind erosion not only impacts just the land but also the whole ecosystem and adversely affects socioeconomic conditions of the population. The focus of wind erosion has gradually been shifted from qualitative studies to semiquantitative and quantitative wind tunnel studies (Steffens et al. 2009). Over the past decades, the significance of the wind erosion problem is increased because of the changing agricultural practices (Riksen et al. 2003), and further increase can be expected due to the projected climate change (IPCC 2014). Therefore, it is important to precisely map the extent and spatial distribution of wind erosion.

Many authors used remote sensing techniques to monitor trends of land degradation as well as to identify and characterize sand dunes and their temporal dynamism (Chen et al. 1998; Tucker et al. 1994). Some of the other techniques applied for extracting data on wind erosion indicators are image transformation techniques (Carneiro and Zinck 1994), digital image classification using neural networks (Collado 2000), spectral mixture analysis (Collado 2000), and supervised maximum likelihood classification (Carneiro and Zinck 1994; del Valle et al. 2008). The satellite data based derived products subsequently used as input in GIS analysis to estimate sand mobilization rates and sand dune migration (del Valle et al. 2008). Image segmentation and object-oriented classifications of Terra-ASTER and textural details derived from RADARSAT were applied to discriminate desert pavements, active and stabilized dunes, and shrub encroachment (Blanco et al. 2009). Ajai et al. (2007, 2009) demonstrated applications of remote sensing data and GIS in land degradation and desertification mapping and reported about 105.48 Mha of India is undergoing the process of land degradation. The causative factors for desertification include overgrazing, cultivation on marginal lands and high slopes, non-sustainable land use practices, wrong agricultural management, mining, urbanization, and other activities that disturb the natural ecosystem. In addition to these factors, frequent droughts, extreme weather conditions, climate change, etc. are natural causes of land degradation and desertification. The socioeconomic condition of the local population also contributes to the land degradation/desertification process. The unstabilized longitudinal sand dunes in part of Jaisalmer district of Rajasthan mapped through analysis of sentinel 2 data (10 m) of February 10, 2016, is shown in Fig. 20.1.

20.3.1.3 Waterlogging

Waterlogging and subsequent salinization and/or alkalization are the major land degradation processes in irrigated agricultural lands of arid and semiarid regions. Waterlogging is the rise of the water table into the root zone of the soil profile, such that plant growth is adversely affected by deficiency of oxygen. Waterlogging lowers the land productivity through the rise in groundwater close to the soil surface. It also included surface ponding, where the water table rises above the surface. Waterlogging is linked with salinization, both being brought about by incorrect irrigation management. Waterlogging should be distinguished from naturally occurring poorly drained areas and also from the different problems of flooding. In the GLASOD estimate, waterlogging affects 4.6 Mha, largely in the irrigated areas of India and Pakistan. It is closely linked with salinization. In India, Ahmad and Kutcher (1992) monitored the progressive rise in the water table beneath the Indo-Gangetic plains since the commencement of large-scale irrigation schemes in the 1930s.

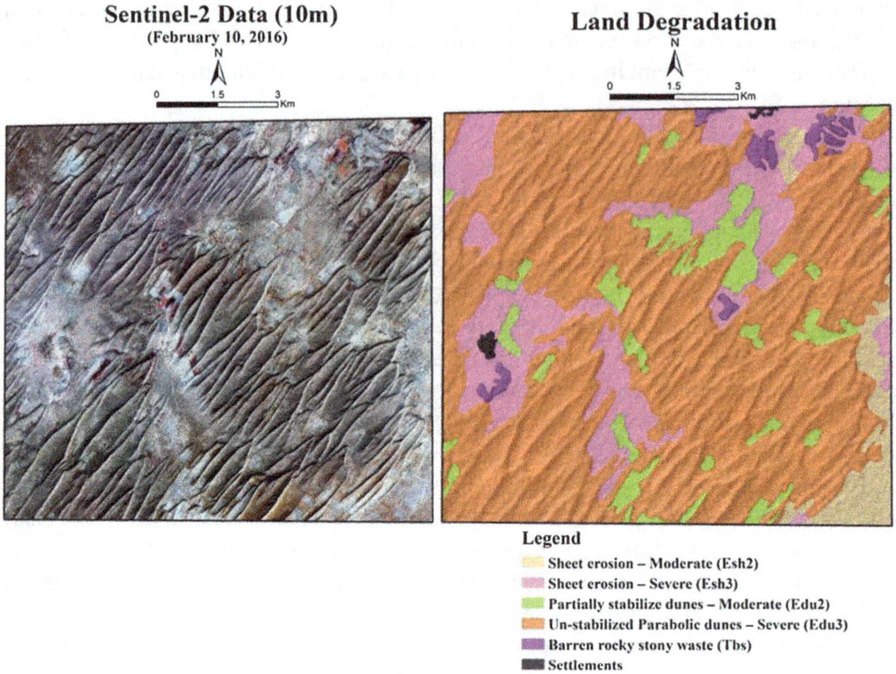

Fig. 20.1 Unstabilized longitudinal sand dunes in part of Jaisalmer district of Rajasthan mapped through analysis of sentinel 2 satellite data of February 10, 2016

Choubey (1997) used temporal IRS-IA-LISS-I, land use, and drainage data to delineate waterlogged areas and area sensitive to waterlogging in the Tawa command, and results were validated with water table data. He demonstrated that since the water table cannot be detected directly from satellite observations, the best integrative indicator can be the crop stress due to high-water table. Choubey (1998) made an attempt to identify waterlogged areas in Sriram Sagar command area by using remote sensing data. Barret and Curtis (1976) indicated that stream channel development and network, stream length, and the location of ponds and lakes can be mapped from Landsat-MSS data and it can be integrated in GIS to assess the waterlogging and drainage problems by identifying the drainage network and its characteristics in a basin besides the information on presence of high-water table, high morphology, soil color, plant stress, and drainage water collection in lower spots. The permanent waterlogged areas in the part of Sultanpur district of Uttar Pradesh, India, as appeared in Resourcesat-2 LISS-IV (5.8 m) of October 2016 are shown in Fig. 20.2.

Fig. 20.2 Permanent waterlogged areas in the part of Sultanpur district of Uttar Pradesh as appeared in Resourcesat-2 LISS-IV (5.8 m) of October 2016

20.3.1.4 Lowering of the Water Table

It is a self-explanatory form of land degradation, brought about through tube well pumping of groundwater for irrigation exceeding the natural recharge capacity. In areas of deep alluvial deposits and where the groundwater has not become saline, tube well irrigation has become widespread and has led to substantial increases in crop production. In many parts of Northwestern India, due to overextraction of groundwater, its level has been progressively lowered. Singh (1992) reported that in parts of the Punjab, the water table has fallen by between 0.5 and 4.0 m in the 8 year period (1978–1986) and is receding at 0.3–0.5 m per year. Joshi and Tyagi (1991a, b) indicated that in the Sudhar block of Ludhiana district, groundwater has fallen from 3 m to 11 m during 1965 and 1989 and in Haryana from 4.8 m to 7.7 m during 1974 and 1989. Rodell et al. (2009) used GRACE (Gravity Recovery and Climate Experiment) data to monitor groundwater storage changes for long term in high plains aquifer in the Central United States. Noomen (2007) also presented the groundwater monitoring techniques using both GRACE and ERS satellite images in continental or global scale. Suphan et al. (2004) also proposed a method for estimation of spatial variation of subsurface water level change caused by crop growth from Landsat TM data and its relationship with groundwater level in an irrigation project in Thailand.

20.3.1.5 Mining and Industrial Waste

A major problem across the globe is the loss of prime agricultural land for housing, industry, roads, and other nonagricultural purposes. As urban centers expand, good quality land is converted, usually permanently, to other uses. To compensate for this, farmers are forced on to poorer land – frequently steeply sloping ground with shallow, poor soils, which can quickly erode and lead to land degradation, flooding, siltation of dams and waterways, and an accompanying cycle of poverty. During the mining operations, removal of vegetation cover results in soil degradation due to accelerated water erosion, soil compaction, and soil crusting, which affects the land productivity. Mining activities also disturb large tract of land due to overburden dumps, which change the natural topography and drainage pattern of the area (Dhar et al. 1991). Remote sensing and GIS have been widely used in mapping land use/land cover changes and environmental degradation caused by mining activities. Remote sensing provides multi-temporal data, which gives valuable temporal information about the process and pattern of land use/land cover change, and it may be analyzed and mapped in GIS to find the impact of mining and industrial activities on land degradation.

20.3.2 Chemical Degradation

20.3.2.1 Salt Affected Soils

Salinization is defined as the presence of excessive salts on the top layer of the soil, resulting in deterioration of its chemical and physical properties. Soil salinization not only causes the destruction of land and plant resources and immense decline of agricultural productivity, but also threatens the ecosystem of the region. It often occurs in areas where soil's evaporation is very intense, and the water table is high and contains high dissolubility salt. Soil salinization is a serious issue, particularly in Argentina, Egypt, India, Iraq, Pakistan, Syria, and Iran (Rhoades 1990). Soil salinity is a prevalent environmental hazard in arid and semiarid regions of the world (Hillel 2000). Koohafkan and Stewart (2012) reported that saline soils covered 397 Mha of the total land area of the world. Ghassemi et al. (1995) estimated approximately one billion hectares of the earth's continental extent is affected soil salinity. Landsat data have been extensively used for separating different levels of soil salinity/sodicity in the United States (Wiersma and Horton 1976), India (Venkatratnam 1983), Iraq (Al Mahawili 1983), and Canada (Sommerfeldt et al. 1985). Most of the authors are able to distinguish only 2–3 classes (strong and medium) of salinity levels with errors between moderately saline and normal soils. Rao and Venkataratnam (1991) studied the spectral behavior of salt-affected soils of Indo-Gangetic alluvial plain and concluded that salt-affected soils as compared to normal cultivated soils showed relatively higher spectral response in visible and near-infrared regions. Further, strongly saline-sodic soils were found to have higher spectral response as compared to moderately saline-sodic soils. Joshi and Sahai (1993) compared the accuracy of TM, MSS, and SPOT and found TM to be the superior multispectral radiometer for soil salinity mapping. Metternicht and Zinck (1996) mapped salt- and sodium-affected surface by combining digital image classification with field observations of soil degradation features and laboratory determination.

Remote sensing technology, with its unique characteristics of systematic, synoptic, rapid, and repetitive coverage, has emerged as a cost-effective and time-efficient approach for studying and mapping salt-affected soils and other degraded lands in space and time domains (Navalgund et al. 2007; Metternicht and Zinck 2008). Johnston and Barson (1993) reported that the use of satellite data in discriminating the saline areas was the most successful approach during the peak vegetation growth. Goossens and Van Ranst (1996) demonstrated that the combination of remote sensing with GIS is very promising, especially for the monitoring of soil salinization. Goossens and Van Ranst (1998) reported that single image may be suitable for detecting severely salinized soils, but more gradations can be determined by using temporal images. Goossens and Van Ranst (1998) monitored and predicted soil salinity in the Nile Delta, Egypt, using GIS and remote sensing techniques. Khan et al. (2001) used IRS-LISS-II digital data and different remote sensing-derived indices such as salinity index (SI), normalized difference salinity index (NDSI), brightness index (BI), and normalized difference vegetation index (NDVI) for mapping salt-affected soils in Punjab, Pakistan. Koshal (2010) used wetness index (WI), soil brightness index (SBI), and soil-adjusted vegetation index (SAVI) for degraded land characterization and delineation with emphasis on salinity and sodicity problems.

In Indian conditions, Seghal et al. (1988) applied Landsat MSS data for mapping salt-affected soils in the frame of the reconnaissance soil map of India. Dwivedi (1992) used Landsat MSS and TM data for more detailed mapping and monitoring of the salt-affected soils in the Indo-Gangetic alluvial plains of India. Many authors reported that the delineation of saline soils using remote sensing data and GIS techniques has been proved efficient (Dwivedi 1992; Dwivedi and Sreenivas 1998; Rao et al. 1991; Sharma et al. 1988). Verma et al. (1994) demonstrated that the addition of the thermal band of Landsat TM to the visible NIR bands helped overcome spectral similarity issues with saline soils. Dwivedi et al. (2008) reported that fusion of IKONOS imagery with IRS-ID LISS-III sensor data significantly improves the overall accuracy in soil salinity mapping and detection. The first systematic mapping of salt-affected soils of the country has been carried out in 1996 with various project partners including NRSA, CSSRI, NBSS&LUP, all India soil and land use survey, state soil survey departments, and state/regional remote sensing application centers (NRSA 1996, 2008). To address the problem of diverging national estimates by remote sensing for arriving at an acceptable figure, CSSRI, NRSA, and NBSS&LUP held a series of consultations and have developed a GIS-based approach to reconcile the national estimates as 6.73 Mha (CSSRI 2007; Maji 2007). Maji et al. (2010) harmonized land degradation datasets of India and reported that exclusively salt affected soils and salt affected soils with water covers 5.44 and 1.30 Mha, respectively. The planners and decision-makers are using this information for planning reclamation programs. Salt-affected soils in the part of Sultanpur district of Uttar Pradesh as detected on Resourcesat-2 LISS-IV (5.8 m) on May 27, 2016, are shown in Fig. 20.3.

Fig. 20.3 Salt-affected soils in the part of Sultanpur district of Uttar Pradesh as detected on Resourcesat-2 LISS-IV (5.8 m) on May 27, 2016

20.3.2.2 Soil Fertility Decline

It is used as a short term to refer to what is more precisely described as deterioration in soil physical, chemical, and biological properties. While decline in fertility is indeed a major effect of erosion, the term is used here of cover effects of processes other than erosion. The main processes involved in soil fertility decline are lowering of soil organic matter, with associated decline in soil biological activity; degradation of soil physical properties, as brought about by reduced organic matter; and adverse changes in soil nutrient resources, including reduction in availability of the major nutrients, onset of micronutrient deficiencies, development of nutrient imbalances and buildup of toxicities, primarily acidification through incorrect fertilizer use.

GLASOD defines this form of degradation as "loss of nutrients and/or organic matter". The GLASOD assessment shows 65% of agricultural land in Bangladesh and 61% in Sri Lanka affected by this type of degradation. In Bangladesh, the average organic matter (presumably of topsoils) is said to have declined by 50%, from 2 to 1%, over the past 20 years (Bangladesh 1992). For the Indian state of Haryana, soil test reports over 15 years show a decrease in soil carbon (Chaudhary and Aneja 1991). Decreased organic matter leads to degradation of soil physical properties, including water holding capacity, reduced nutrient retention capacity, and longer release of nutrients, including micronutrients, from mineralization of organic matter. Negative soil nutrient balances have been reported for all three major nutrients in Bangladesh and Nepal, for phosphorus and potassium in Sri Lanka, and a large deficit for potassium in Pakistan (FAO 1986). Nutrient depletion has been reported for each of the 15 agroclimatic regions of India (Biswas and Tewatia 1991; Tandon 1992). For India, a deficiency between nutrient removal and addition of 60 kg/ha per year, or 9 Mt for the whole country, has been estimated (Tandon 1992). As per secondary and micronutrient deficiencies are concern, sulfur deficiency has been reported for India, Pakistan, and Sri Lanka and zinc deficiency for

India and Pakistan (FAO/RAPA 1992; Bowonder 1981; Chaudhary and Aneja 1991; Abrol 1990). For Bangladesh, 3.9 Mha are reported deficient in sulfur and 1.75 Mha in zinc, including areas of continuous swamp rice cultivation (Bangladesh 1992). Pakistan, because of its generally alkaline soils, is particularly liable to micronutrient deficiencies, which are being increasingly reported (Twyford 1994).

20.3.3 Vegetal Degradation

Vegetal degradation basically covers deforestation, overgrazing, and shifting cultivation.

20.3.3.1 Deforestation

Deforestation is one of causes of land degradation, firstly, when the land that is cleared is steeply sloping or has shallow or easily erodible soils and, secondly, when the clearance is not followed by good management. Deforestation and forest degradation lead to water erosion in steeply sloping humid environments. It is also a contributory cause of wind erosion, soil fertility decline, and salinization. The drivers and intensity of forest degradation vary by region (Kissinger et al. 2012), but the severity and impact of forest loss and degradation can be observed at all scales, from global climate change to declining economic value of forest resources and biodiversity and threatened local livelihoods. The impact of forest degradation varies from fine-scale structural changes in canopy cover and height (Franke et al. 2012; Hirschmugl et al. 2014), or subtle disruptions to ecosystem services, to broad-scale loss of biomass (Miettinen et al. 2014). These changes can occur over a range of spatial and temporal scales, which can be mapped and monitor through temporal satellite data. Many authors used satellite data to analyze the spatial and temporal patterns of deforestation and the identification of key variables related to deforestation and identify the driving forces behind changes to forest cover (Jha et al. 2000, Gautam et al. 2003; Panta et al. 2008).

20.3.3.2 Overgrazing

Overgrazing is the status of grazing of natural pastures as stocking intensifies above the livestock carrying capacity. Overgrazing adversely affects soil properties, which result in reduced infiltration, accelerated runoff, and soil erosion. Oldeman et al. (1991) reported that overgrazing is considered to be the major cause of soil degradation worldwide especially widespread in Australia and Africa, where it accounts for 80.6% and 49.2%, respectively, of all soil degradation (Warren and Khogali 1992). Degradation of vegetation cover and erosion leads to decline of soil organic matter and physical properties. Overgrazing especially in arid regions reduced infiltration and accelerated runoff and soil erosion. Results of several studies conducted in Argentina and India indicate that at the macro- and mesoscales, soil erosion can increase dramatically due to overgrazing, causing increases of 5–41 times over the control at the mesoscale and 3–18 times at the macroscale (Sharma, 1997).

20.3.3.3 Shifting Cultivation

In the past, shifting cultivation was a sustainable form of land use, at a time when low population densities allowed forest fallow periods of sufficient length to restore soil properties. Population increase and enforced shortening of fallow periods have led to it becoming non-sustainable. Shifting cultivation is found in the hill areas of Northeast India, where it is a cause of water erosion and soil fertility decline. In Northeast India, out of the total forest cover, 1.5 Mha is currently managed by shifting cultivation (Roy et al. 2012). Shifting cultivation in Northeast India not only degrades land productivity but also causes excess runoff and accelerates soil erosion in steep slope regions and deposit sediments on the riverbed in the adjoining basins and lowlands.

20.4 Management of Degraded Lands

GIS based reliable data on extent and spatial distribution on nature and degree of degraded lands; it includes soils, climate, vegetation, and topography which are needed to develop land management strategies and sound land use plans. This information can provide the background to the policies and strategies that are required by planners and policy makers to develop policies and programs in management of land degradation hazards. Subsidies, incentives, and taxes can all have an effect on what crops are grown, where, and whether or not the land is well managed. The best way to protect soil from erosion is through a dense cover of living or dead vegetation. Healthy, densely growing crops not only produce high yields but they also provide good ground cover and protection from erosion. Any conservation program should therefore promote good crop management. There is considerable potential for increasing the use of green manures in order to improve soil fertility and improve levels of organic matter in the soil.

Land degradation has been fairly high in Northeastern states of India like Nagaland, Sikkim, and Meghalaya and in some cases, it accounts for 50% of the total geographical areas. So development, reclamation, and management of degraded lands should be prioritized through proper land conservation programmes with an aim to encourage land users, at the level of the farm unit, to adopt land use systems and management practices that will lead to conservation. In order to impart essential knowledge and skills in conservation program, practical training on conservation program needs to be conducted. Overexploitation of groundwater has reached danger levels in Punjab, Haryana, and Tamil Nadu. For sustained agriculture and livelihood security in the future, rational planning and utilization of groundwater resources are essential. Salinized soils can be restored to productive use, although at a high cost, through salinity control and reclamation projects. In other cases, the land can only be restored by taking it out of productive use for some years, as in reclamation forestry. The cost of reclamation, or restoration to productive use, of degraded soils is invariably less than the cost of preventing degradation before it occurs.

20.5 Conclusions

In order to acquire more accurate data, it is necessary to define the type and degrees of land degradation in terms that offer practical means of observation, monitoring, and mapping. The study indicates that remote sensing technology, with its unique characteristics of systematic, synoptic, rapid, and repetitive coverage, has emerged as a cost-effective and time-efficient approach for studying and mapping land degradation in time and space domains. Remote sensing and GIS techniques have immense potential to map and monitor various types of physical degradations due to water erosion, wind erosion, waterlogging, mining and quarrying, and urban and industrial waste. The use of remote sensing and GIS techniques makes soil erosion estimation and its spatial distribution feasible with reasonable costs and better accuracy in larger areas. The temporal remote sensing data has immense potential to monitor and characterize sand dunes and their temporal dynamism. Integrated remote sensing and GIS could be effectively used to map and monitor the water-logged- and salt-affected soils. The spatiotemporal patterns of deforestation and identification of key driving forces behind changes in forest cover could be effectively mapped and monitored.

Reclamation and management of degraded lands should be a priority through proper land conservation program with the aim to adopt suitable land use systems and management practices. The information generated through integrated remote sensing and GIS on extent and spatial distribution of degraded lands could be effectively used in watershed management programmes for soil and water conservation, reclamation of salt-affected soils, afforestation towards sustainable management of land resources and improvement of the status of soil organic matter. In general, lightly degraded soils can be improved by crop rotation, minimum tillage techniques, and other on-farm practices. Moderately damaged land takes more resources to restore the land resources. More severely degraded soils could be used for afforestation and mechanical measures. In developing countries like India, the program of reclamation and management of degraded lands could be effectively linked with government-run employment guarantee schemes to provide employment to the landless and rural poor for effective management of land resources toward sustainable agriculture and livelihood security.

References

Abrol IP (1990) Problem soils in India. In: Problem soils of Asia and the Pacific. FAO/RAPA, Bangkok, pp 153–165

Ahmad M, Kutcher GP (1992) Irrigation planning with environmental considerations: a case study of Pakistan's Indus basin. World Bank Technical Paper 166. World Bank, Washington DC, 196 pp

Ajai RR, Arya AS, Dhinwa PS et al (2007) Desertification/land degradation atlas of India. Space Applications Centre, Ahmedabad. ISBN No. 978-81-909978-2-9

Ajai RR, Arya AS, Dhinwa PS, Pathan SK, Ganeshraj K (2009) Desertification/land degradation status mapping of India. Curr Sci 97(10):1478–1483

Al Mahawili SMH (1983) Satellite interpretation and laboratory spectral reflectance measurements of saline and gypsiferous soils of West Baghdad, Iraq. M.S. Thesis, Purdue University, West Lafayatte, Indian, USA

Anonymous (1976) Report of National Commission on Agriculture, part V, IX and abridged report, Ministry of Agriculture and Irrigation, Govt. of India, New Delhi

Bai ZG, Dent DL, Olsson L, Schaepman ME (2008) Proxy global assessment of land degradation. Soil Use Manag 24:223–234

Bali JS (1985) Problems of gullied and ravinous area -Policy, programmes and progress. In: Proceedings national seminar on soil conservation and watershed management, New Delhi, 17–18 September 1985

Bangladesh (1992) Land degradation. Paper presented to FAO 21st regional conference for Asia and the Pacific, New Delhi

Barrett EC, Curtis LF (1976) Introduction to environmental remote sensing. Chapman and Hall, London, p 336

Bartsch KP, van Miegroet H, Boettinger J, Dobrwolski JP (2002) Using empirical erosion models and GIS to determine erosion risk at Camp Williams. J Soil Water Conserv 57:29–37

Beasley DB, Huggins LF, Monke EJ (1980) ANSWERS: a model for watershed planning. Trans Am Soc Agric Eng 23(4):938–944

Biswas A, Tewatia RK (1991) Nutrient balance in agro-climatic regions of India – an overview. Fert News 36(6):13–18

Blanco PD, Metternicht GI, del Valle HF (2009) Improving the discrimination of vegetation and landforms patterns in sandy rangelands: a synergistic approach. Int J Remote Sens 30:2579–2605

Bowonder B (1981) The myth and reality of high yield varieties in Indian agriculture. Dev Chang 12(2):293–313

Bridges EM, Oldeman LR (1999) Global assessment of human-induced soil degradation. J Arid Soil and Rehabil 13(4):319–325

Carneiro FA, Zinck JA (1994) Mapping paleo-aeolian sand cover formations in the northern Amazon basin from TM images. ITC J 3:270–282

Chaudhary MK, Aneja DR (1991) Impact of green revolution on long-term sustainability of land and water resources in Haryana. Indian J Agric Econ 45:428–432

Chen Z, Elvidge CD, Groenveld DP (1998) Monitoring of seasonal dynamics of arid land vegetation using AVIRIS data. Remote Sens Environ 65:255–266

Choubey VK (1997) Detection and delineation of waterlogging by remote sensing techniques. J Indian Soc Remote Sens 25(2):123–135

Choubey VK (1998) Assessment of waterlogging in Sriram Sagar command area, India, by remote sensing. Water Resour Manag 12(5):343–357

Collado AD (2000) Spatio-temporal dynamics of dune patterns in semiarid Argentina: a neural network analysis. Edaphomatics Bull. no. 20. AICET-INTA, Buenos Aires, Argentina

CSSRI (2007) Annual report 2006–07. Central Soil Salinity Research Institute, Karnal, India

Das DC (1985) Problem of soil erosion and land degradation in India. Proceedings of the national seminar on the soil conservation and watershed management held on 17–18 September New Delhi

De Ploey J (1989) A soil erosion map for Western Europe. Catena Verlag

del Valle HF, Rostagno CM, Coronato FR, Bouza PJ, Blanco PD (2008) Sand dune activity in north-eastern Patagonia. J Arid Environ 72:411–422

Dhar BB, Jamal A, Ratan S (1991) Air pollution problem in an Indian open cast coal mining complex: a case study. Int J Surf Min Reclam Environ 5(2):83–88

Dhuruvanarayana VV, Ram Babu NP (1983) Estimation of soil erosion in India. J Irrig Drain Eng 109:419–434

Dregne HE, Chou NT (1994) Global desertification dimensions and costs. In: Dregne HE (ed) Degradation and restoration of arid lands. Texas Technical University, Lubbock

Dwivedi R (1992) Monitoring and the study of the effects of image scale on delineation of salt-affected soils in the Indo-Gangetic plains. Int J Remote Sens 13(8):1527–1536

Dwivedi RS, Sreenivas K (1998) Delineation of salt-affected soils and waterlogged areas in the indo-Gangetic plains using IRS-1C LISS-III data. Int J Remote Sens 19:2739–2751

Dwivedi RS, Kumar AB, Tewari AN (1997a) The utility of multi-sensor data for mapping eroded lands. Int J Remote Sens 18:2303–2318

Dwivedi RS, Ravi Sankar T, Venkataratnam L, Karale RL, Gawande SP, Rao KVS, Senchuandhary S, Bhaumik KR, Mukherjee KK (1997b) The inventory and monitoring of eroded lands using remote sensing data. Int J Remote Sens 18:107–119

Dwivedi RS, Kothapalli RV, Singh AN (2008) Generation of farm-level information on salt-affected soils using IKONOS-II multispectral data. In: Metternicht G, Zinck J (eds) Remote sensing of soil salinization: impact on land management. CRC Press, Boca Raton

Eswaran H, Lal R, Reich PF (2001) Land degradation: an overview. In: Bridges EM, Hannam ID, Oldeman LR, Pening de Vries FWT, Scherr SJ, Sompatpanit S (eds) Responses to land degradation. Proceedings of. 2nd international conference on land degradation and desertification, Khon Kaen, Thailand. Oxford Press, New Delhi

FAO (1986) Status report on plant nutrition in fertilizer programmes countries in Asia and Pacific region. AGL/MISC/86/7. FAO, Rome

FAO (2008a) Land degradation on the rise – one fourth of the world's population affected says new study. Report by Food and Agricultural Organization. http://www.fao.org/newsroom/en/news/2008/1000874/index.html

FAO (2008b) LADA project documents FAO, internet website: http://www.fao.org/ ag/agl/agll/lada/ladaprojectdoc.pdf. Accessed 10 July 2017

FAO/RAPA (1992) Environmental issues in land and water development. FAO/RAPA, Bangkok. 488 pp (Includes country papers on Bangladesh, India, Nepal, Pakistan and Sri Lanka)

Fernandez C, Wu JQ, McCool DK, Stockle CO (2003) Estimating water erosion and sediment yield with GIS, RUSLE, and SEDD. J Soil Water Conserv 58:128–136

Franke J, Navratil P, Keuck V, Peterson K, Siegert F (2012) Monitoring fire and selective logging activities in tropical peat swamp forests. IEEE J Select Top Appl Earth Observ Remote Sens 5(6):1811–1820

Fu BJ, Zhao WW, Chen LD, Zhang QJ, Lu YH, Gulinck H, Poesen J (2005) Assessment of soil erosion at large watershed scale using RUSLE and GIS: a case study in the loess plateau of China. Land Degrad Dev 16:73–85

Gautam NC, Narayan LRA (1988) Wastelands in India. Pink Publishing House, Mathura, p 96

Gautam AP, Webb EL, Shivakoti GP, Zoebisch MA (2003) Land use dynamics and landscape change pattern in a mountain watershed in Nepal. Agric Ecosyst Environ 99:83–96

Ghassemi F, Jakeman AJ, Nix HA (1995) Salinisation of land and water resources: human causes, extent, management and case studies. CABI, Wallingford

Goossens R, Van Ranst E (1996) The use of remote sensing and GIS to detect gypsiferous soils in the Ismailia Province (Egypt). In the international conference on soils with gypsum, Lleida, Spain, 15–21. Accessed 15 Oct 2017

Goossens R, Van Ranst E (1998) The use of remote sensing to map gypsiferous soils in the Ismailia Province (Egypt). Geoderma 87(1–2):47–56

Gupta SK, Ahmed H, Hussain M, Pandey AS, Singh S, Saini KM, Das SN (1998) Inventory of degraded lands of Palamau district, Bihar: a remote sensing approach. J Indian Soc Remote Sens 26:161–168

Hillel D (2000) Salinity management for sustainable irrigation: integrating science, environment, and economics. World Bank Publications, Washington, DC

Hirschmugl M, Steinegger M, Gallaun H, Schardt M (2014) Mapping forest degradation due to selective logging by means of time series analysis: case studies in central Africa. Remote Sens 6:756–775

IPCC (2014) Synthesis report. Contribution of working groups I, II and III to the fifth assessment report of the intergovernmental panel on climate change [Core Writing Team, Pachauri RK, Meyer LA (eds)]. IPCC, Geneva, Switzerland, 151 pp

Jafari R, Lewis MM, Ostendorf B (2008) An image-based diversity for assessing land degradation in an arid environment in South Australia. J Arid Environ 72:1282–1293

Jha CS, Dutt CBS, Bawa KS (2000) Deforestation and land use changes in Western Ghats, India. Curr Sci 79:231–238

Johnston RM, Barson MM (1993) Remote sensing of Australian wetlands: an evaluation of Landsat TM data for inventory and classification. Mar Freshw Res 44(2):235–252

Joshi MD, Sahai B (1993) Mapping of salt-affected land in Saurashtra coast using Landsat satellite data. Int J Remote Sens 14(10):1919–1929

Joshi PK, Tyagi NK (1991a) Sustainability of existing farming system in Punjab and Haryana – some issues in groundwater use. Indian J Agric Econ 46(3):412–421

Joshi PK, Tyagi NK (1991b) Sustainability of existing farming system in Punjab and Haryana some issues on groundwater use. Indian J Agric Econ 46:412–421

Kapalanga TS (2008) A review of land degradation assessment methods. Land restoration training programme Keldnaholt, 112 Reykjavík, Iceland, pp 17–68

Karale RL, Saini KM, Narula KK (1987) Mapping and monitoring ravines using remotely sensed data. J Soil Water Conserv (India) 32(1–2):75–82

Kessler CA, Stroosnijder L (2006) Land degradation assessment by farmers in Bolivian mountain valleys. Land Degrad Dev 17:235–248

Khan NM, Rastoskuev VV, Elrna VS, Yohei S (2001) Mapping salt affected soils using remote sensing indicators. In: 22nd Asian conference on remote sensing, Singapore, 5–9 November 2001

Kissinger G, Herold M, De Sy V (2012) Drivers of deforestation and forest degradation: a synthesis report for REDD+ policymakers. Lexeme Consult, Vancouver

Koofhafkan AP, Lantieri D, Nachtergaele F (2003) Land Degradation in Drylands (LADA): guidelines for a methodological approach. FAO, Rome

Koohafkan P, Stewart BA (2012) Water and cereals in drylands, the food and agriculture. Organization of the United Nations and Earthscan, Rome

Koshal AK (2010) Indices based salinity areas detection through remote sensing and GIS. In parts of south waste Punjab. In: 13th annual international conference and exhibition on geospatial information technology and applications, Gurgaon, India

LADA (2009) Guidelines for the identification, selection and description of nationally based indicators of land degradation and improvement, land degradation assessment in drylands. UNEP, Rome

Lal R (2003) Soil erosion and the global carbon budget. Environ Int 29:437–450

Lu D, Batistella M, Mausel P, Moran E (2007) Mapping and monitoring land degradation risks in the Western Brazilian Amazon using multitemporal Landsat TM/ETM+ images. Land Degrad Dev 18:41–54

Ma JW, Xue Y, Ma CF, Wang ZG (2003) A data fusion approach for soil erosion monitoring in the upper Yangtze river basin of China based on Universal Soil Loss Equation (USLE) model. Int J Remote Sens 24:4777–4789

Maji AK (2007) Assessment of degraded and wastelands of India. J Indian Soc Soil Sci 55:427–435

Maji AK, Reddy GPO, Sarkar D (2010) Degraded and wastelands of India, status and spatial distribution. ICAR and NAAS Publication, New Delhi, pp 1–158

Maji AK, Reddy GPO, Sarkar D (eds) (2012). Acid soils of India – their extent and spatial variability, NBSS Publ. No.145, NBSS&LUP, Nagpur, 147 p

Mandal D, Sharda VN (2011) Assessment of permissible soil loss in India employing a quantitative bio-physical model. Curr Sci 100(3):383–390

Metternicht G (2006) UN-Zambia-ESA regional workshop on the applications of GNSS in sub-Saharan Africa

Metternicht GI, Zinck JA (1996) Modelling salinity-sodicity classes for mapping salt affected top soils in the semi-arid valleys of Cochabamba (Bolivia). ITC J 11:125–135

Metternicht GI, Zinck JA (1997) Spatial discrimination of salt- and sodium-aff ected soil surfaces. Int J Remote Sens 18:2571–2586

Metternicht GI, Zinck JA (1998) Evaluating de information content of JERS-1 SAR and Landsat TM data for discrimination of soil erosion features. ISPRS J Photogramm Remote Sens 53:143–153

Metternicht GI, Zinck JA (eds) (2008) Remote sensing of soil salinization: impact on land management. CRC Press/Taylor & Francis, Boca Raton

Metternicht G, Zinck JA, Blanco PD, del Valle HF (2009) Remote sensing of land degradation: experiences from Latin America and the Caribbean. J Environ Qual 39:42–61

Miettinen J, Stibig H-J, Achard F (2014) Remote sensing of forest degradation in Southeast Asia-aiming for a regional view through 5–30 m satellite data. Glob Ecol Conserv 2:24–36

Millward AA, Mersey JE (1999) Adapting the RUSLE to model soil erosion potential in a mountainous tropical watershed. Catena 38:109–129

Mitasova H, Hofierka J, Zlocha M, Iverson LR (1996) Modelling topographic potential for erosion and deposition using GIS. Int J Geogr Inf Sci 10:629–641

Molnar DK, Julien PY (1998) Estimation of upland erosion using GIS. Comput Geosci 24:183–192

Morgan RPC, Morgan DDV, Finney HJ (1984) A predictive model for the assessment of soil erosion risk. J Agric Eng Res 30(1):245–253

National Commission on Agriculture (1976) Report of the National Commission on Agriculture: part V, IX and abridged. Ministry of Agriculture and Irrigation, New Delhi

Navalgund RR, Jayaraman V, Roy PS (2007) Remote sensing applications: an overview. Curr Sci 93:1747–1766

Nawar S, Buddenbaum H, Hill J, Kozak J (2014) Modeling and mapping of soil salinity with reflectance spectroscopy and Landsat data using two quantitative methods (PLSR and MARS). Remote Sens 6:10813–10834

Noomen M (2007) Groundwater monitoring using GRACE and ERS satellite image. In: Proceeding of the 28th Asian conference on remote sensing, Malaysia

NRSA. (1996). Mapping salt-affected soils of India on 1:250,000. NRSA, Hyderabad

NRSA (2005) Wasteland atlas of India. Ministry of Rural Development and NRSA Publ., NRSA, Hyderabad

NRSA (2008) National wide mapping of land degradation on 1:50,000 scale using multi-temporal satellite data. NRSA, Hyderabad

NSERL (1995) WEPP user summary. National Soil Erosion Research Laboratory, US Department of Agriculture. Accessed 10 September 2017

NWDB (1985) Description, classification, identification and monitoring of wastelands. National Wastelands Development Board, Ministry of Environment and Forests. Govt. of India, New Delhi

Oldeman LR (ed) (1988) Global Assessment of Soil Degradation (GLASOD). Guidelines for general assessment of status of human-induced soil degradation. ISRIC, Wageningen. Working paper and reprint no. 88 (4), pp 12

Oldeman LR (1994) The global extent of land degradation. In: Greenland DJ, Szabolcs I (eds) Land resilience and sustainable land use. CABI, Wallingford, pp 99–118

Oldeman RL, Hakkeling RTA, Sombroek WG (1990) World map of the status of human induced soil degradation. International soil reference and information Centre, Wageningen

Oldeman LR, Hakkeling RTA, Sombroek WG (1991) World map of the status of Human Induced soil degradation: an explanatory note. ISRIC (International Soil Reference and Information Center), Wageningen

Panagos P, Borrelli P, Poesen J, Ballabio C, Lugato E, Meusburger K, MOntanarella L, Alewell C (2015) The new assessment of soil loss by water erosion in Europe. Environ Sci Pol 54:438–447

Pandey PC, Rani M, Srivastava PK, Sharma LK, Nathawat MS (2013) Land degradation severity assessment with sand encroachment in an ecologically fragile arid environment: a geospatial perspective. Q Sci Connect 43:17

Paningbatan EP (2001) Geographic information system assisted dynamic modeling of soil erosion and hydrologic processes at a watershed scale. Philippine Agric Sci 84(4):388–393

Panta M, Kim K, Joshi C (2008) Temporal mapping of deforestation and forest degradation in Nepal: applications to forest conservation. For Ecol Manag 256:1587–1595

Pickup G, Chewings VH (1986) A grazing gradient approach to land degradation assessment in arid areas from remotely-sensed data. Int J Remote Sens 15:597–617

Pohl C, van Genderen JL (1998) Review article. Multisensor image fusion in remote sensing: concepts, methods and applications. Int J Remote Sens 19(5):823–854

Raina P, Joshi DC, Kolarkar AS (1991) Land degradation mapping by remote sensing in the arid region of India. Soil Use Manag 7(1):47–51

Rao BRM, Venkataratnam L (1991) Monitoring of salt-affected soils- a case study using aerial photographs, Salyut-7 space photographs and Landsat-TM data. Geocarto Int 6(1):5–11

Rao BRM, Dwivedi RS, Venkataratnam L, Ravishankar TSS, Bhargawa GP, Singh AN (1991) Mapping the magnitude of sodicity in part of the indo-Gangetic plains of Uttar Pradesh, northern India using Landsat-TM data. Int J Remote Sens 12(3):419–425

Reddy GPO, Maji AK, Srinivas CV, Velayutham M (2002) Geomorphological analysis for inventory of degraded lands in a river basin of basaltic terrain, using remote sensing data and geographical information systems. J Indian Soc Remote Sens 30(1&2):15–31

Reddy GPO, Maji AK, Chary GR, Srinivas CV, Tiwary P, Gajbhiye KS (2004) GIS and remote sensing applications in prioritization of river sub basins using morphometric and USLE parameters – a case study. Asian J Geoinform 4(4):35–49

Reddy GPO, Kurothe RS, Sena DR, Harindranath CS, Naidu LGK, Sarkar D, Sharda VN (2013). Soil erosion of Goa. NBSS Publ. 155. NBSS&LUP (ICAR), Nagpur, p 54

Reddy GPO, Kurothe RS, Sena DR, Harindranath CS, Niranjana KV, Naidu LGK, Singh SK, Sarkar D, Mishra PK, Sharda VN (2016) Assessment of soil erosion in tropical ecosystem of Goa, India using universal soil loss equation, geostatistics and GIS. Indian J Soil Conserv 44(1):1–7

Renard KG, Foster GR, Weesies GA et al (1997) Predicting soil erosion by water – a guide to conservation planning with the Revised Universal Soil Loss Equation (RUSLE), Agricultural Handbook No. 703, US Government Printing Office, Washington, DC

Reusing M, Schneider T, Ammer U (2000) Modeling soil erosion rates in the Ethiopian highlands by integration of high resolution MOMS-02/D2-stereo-data in a GIS. Int J Remote Sens 21:1885–1896

Rhoades JD (1990) Soil salinity-causes and controls. In: Goudie AS (ed) Techniques for desert reclamation. Wiley, Chichester, pp 109–134

Riksen M, Brouwer F, De Graaf J (2003) Soil conservation policy measures to control wind erosion in north-western Europe. Catena 52:309–326

Rodell M, Velicogna I, Famiglietti JS (2009) Satellite-based estimates of groundwater depletion in India. Nature 460:999–1002

Roy PS, Kushwaha SPS, Murthy MSR, Roy A, Kushwaha D, Reddy CS, Behera MD, Mathur VB, Padalia H, Saran S, Singh S, Jha CS, Porwal MC (2012) Biodiversity characterization at landscape level: national assessment 2012. Indian Institute of Remote Sensing, Dehradun, India, p 140

Saini KM, Deb TK, Mitra PP, Ghatol SG (1999) Assessment of degraded lands of Puruliya district, West Bengal using remotely sensed data. J Indian Soc Remote Sens 27:23–30

Seghal JL, Saxena RK, Verma KS (1988) Soil resource inventory of India using image interpretation techniques, remote sensing is a tool for soil scientists. In the 5th symposium of the working group remote sensing ISSS, Budapest, pp 17–31

Sehgal J, Abrol IP (1992) Land degradation statue: India. Desertification Bull 21:24–31

Sehgal J, Abrol IP (1994) Soil degradation in India: status and impact. Oxford/IBH, New Delhi

Sharma RC, Bhargawa GP (1988) Landsat imagery for mapping saline soils and wetlands in northwest India. Int J Remote Sens 9:69–84

Sharma KD, Walling DE, Probst JL (1997) Assessing the impact of overgrazing on soil erosion in arid regions at a range of spatial scales. Human impact on erosion and sedimentation. Proceedings of an international symposium of the fifth scientific assembly of the International Association of Hydrological Sci. (IAHS), Rabat, Morocco, pp 119–123

Singh B (1992) Groundwater resources and agricultural development strategy: Punjab experience. Indian J Agric Econ 47:105–113

Singh AN, Dwivedi RS (1989) Delineation of salt affected soils through digital analysis of Landsat-MSS data. Int J Remote Sens 10:83–92

Singh D, Meirelles MSP, Costa GA, Herlin I, Berroir JP, Silva EF (2006) Environmental degradation analysis using NOAA/AVHRR data. Adv Space Res 37(4):720–727

Sommerfeldt TG, Thompson MD, Pront NA (1985) Delineation and mapping of soil-salinity in southern Alberta from Landsat data. Can J Remote Sens 10(104–1):18

Sonneveld BGJS, Dent DL (2009) How good is GLASOD? J Environ Manag 90:274–283

Srinivas CV, Maji AK, Reddy GPO, Chary GR (2002) Assessment of soil erosion using remote sensing and GIS in Nagpur district, Maharashtra, for prioritization and delineation of conservation units. J Indian Soc Remote Sens 30(4):197–211

Steffens M, Koebl A, Giese M, Kogel-Knabner I (2009) Spatial variability of top soils and vegetation in a grazed steppe ecosystem in inner Mongolia (PR China). J Plant Nutr Soil Sci 172:78–90

Suphan S, Honda K, Gupta AD, Eiumnoh A, Chen X (2004) Estimation of subsurface water level change from satellite data. In: Proceeding of 25th Asia conference remote sensing, Thailand, 1, pp 512–516

Tandon HLS (1992) Assessment of soil nutrient depletion. Paper presented to FADINAP seminar, Fertilization and the environment, Chiang Mai, Thailand

Tucker CJ, Newcomb WW, Dregne HE (1994) AVHRR data sets for determination of desert spatial extent. Int J Remote Sens 15:3547–3565

Twyford I (1994) Fertilizer use and crop yields. Paper presented to 4th National Congress of the Soil Science Society of Pakistan, Islamabad. 1992

UNCCD (2015) Land matters for climate reducing the gap and approaching the target. Available from: http://www.unccd.int/Lists/SiteDocumentLibrary/Publications/2015Nov_Land_matters_For_Climate_ENG.pdf 2015

UNEP (1992) Desertification, land degradation [definitions]. Desertification Control Bull 21

Vågen TG, Winowiecki LA, Abegaz A, Hadgu KM (2013) Landsat-based approaches for mapping of land degradation prevalence and soil functional properties in Ethiopia. Remote Sens Environ 134:266–275

Van Lynden GWJ, Oldeman LR (1997) The assessment of the status of human-induced soil degradation in south and South East Asia. UNEP/FAO and ISRIC, Wageningen

Venkatratnam L (1983) Monitoring of soil salinity in indo-Gangetic plain of NW India using multi-date Landsat data. In: Proceedings of 17th international symposium on remote sensing of environment. Ann. Arbor, Michigan, USA 1, pp 369–377

Verma KS, Saxena RK, Barthwal AK, Deshmukh SN (1994) Remote sensing technique for mapping salt affected soils. Int J Remote Sens 15(9):1901–1914

Vrieling A (2006) Satellite remote sensing for water erosion assessment: a review. Catena 65:2–18

Vrieling A, de Jong SM, Sterk G, Rodrigues SC (2008) Timing of erosion and satellite data: a multi-resolution approach to soil erosion risk mapping. Int J Appl Earth Obs Geoinf 10:267–281

Wang G, Gertner G, Fang S, Anderson AB (2003) Mapping multiple variables for predicting soil loss by geostatistical methods with TM images and a slope map. Photogramm Eng Remote Sens 69:889–898

Warren A, Khogali M (1992) Assessment of desertification and drought in the Sudano-Sahelian region IW–/99/. United Nations Sudano-Sahelian Office

Wessels KJ, Prince SD, Frost PE, VanZyl D (2004) Assessing the effects of human-induced land degradation in the former homelands of northern South Africa with a 1 km AVHRR NDVI time-series. Remote Sens Environ 91:47–67

Wiersma JL, Horton M (1976) Remote sensing applications for detection of saline seep. Report no. OWRD-44, South Dakota State University, Brookings, South Dakota, USA

Wilson JP, Lorang MS (2000) Spatial models of soil erosion and GIS. In: Fotheringham AS, Wegener M (eds) Spatial models and GIS: new potential and new models. Taylor & Francis, Philadelphia, pp 83–108

Wischmeier WH, Smith DD (1978) Predicting rainfall erosion losses – a guide to conservation planning, USDA agricultural handbook no 587

Chapter 21
Monitoring of Spatiotemporal Dynamics of Rabi Rice Fallows in South Asia Using Remote Sensing

Murali Krishna Gumma, Prasad S. Thenkabail, Pardhasaradhi Teluguntla, and Anthony M. Whitbread

Abstract Cereals and grain legumes are the most important part of human diet and nutrition. The expansion of grain legumes with improved productivity to cater the growing population's nutritional security is of prime importance and need of the hour. Rice fallows are best niche areas with residual moisture to grow short-duration legumes, thereby achieving intensification. Identifying suitable areas for grain legumes and cereal grains is important in this region. In this context, the goal of this study was to map fallow lands followed by rainy season (*kharif*) rice cultivation or post-rainy (*rabi*) fallows in rice-growing environments between 2005 and 2015 using temporal moderate-resolution imaging spectroradiometer (MODIS) data applying spectral matching techniques. This study was conducted in South Asia where different rice ecosystems exist. MODIS 16 day normalized difference vegetation index (NDVI) at 250 m spatial resolution and season-wise-intensive ground survey data were used to map rice systems and the fallows thereafter (*rabi* fallows) in South Asia. The rice maps were validated with independent ground survey data and compared with available subnational-level statistics. Overall accuracy and kappa coefficient estimated for rice classes were 81.5% and 0.79%, respectively, with ground survey data. The derived physical rice area and irrigated areas were highly correlated with the subnational statistics with R^2 values of 94% at the district level for the years 2005–2006 and 2015–2016. Results clearly show that rice fallow areas increased from 2005 to 2015. The results show spatial distribution of rice fallows in South Asia, which are identified as target domains for sustainable intensification of short-duration grain legumes, fixing the soil nitrogen and increasing incomes of small-holder farmers.

Keywords Grain legumes · Ground survey data · MODIS 250 m · NDVI · Potential areas · Seasonal rice mapping · Rice fallows · Spectral matching techniques

---

M. K. Gumma (✉) · A. M. Whitbread
International Crops Research Institute for the Semi-Arid Tropics, Patancheru, Hyderabad, India
e-mail: M.Gumma@cgiar.org

P. S. Thenkabail · P. Teluguntla
U.S. Geological Survey (USGS), Western Geographic Science Center, Flagstaff, AZ, USA

G. P. O. Reddy, S. K. Singh (eds.), *Geospatial Technologies in Land Resources Mapping, Monitoring and Management*, Geotechnologies and the Environment 21,
https://doi.org/10.1007/978-3-319-78711-4_21

21.1 Introduction

Agriculture is the key to food security and nutrition for the growing populations of Asia and Africa improving incomes and employment. We need to produce at least 50% more to feed the projected 9.15 billion people by 2050 (Alexandratos and Bruinsma 2012). Agricultural and rural development are critical to the eradication of poverty and improvement livelihoods for sustainable and equitable growth. There is a need to increase cropping intensity (two to three crops within a crop year) where there is no scope for extensification in most of South Asia, which is known as food bowl of the world, with 60 million ha planted with rice. South Asia accounts for 40% of the world's harvested rice area (USDA 2010) and almost 25% of the world's population (FAO 2015). Rice is grown under irrigated and rainfed conditions in various ecosystems in South Asia. In most irrigated rice-grown areas, rice followed by rice, rice followed by wheat, rice followed by mustard, and rice followed by pulses are the regular practice. Major rice-growing area under rainfed conditions and followed by fallow is mainly located in the eastern part of India, Bangladesh, and Sri Lanka (Gumma et al. 2011b). Growing food demand has increased pressure to intensify existing croplands with increased use of irrigation technology, fertilizer, and mechanization (Garnett et al. 2013; Gray et al. 2014). Information on existing rice systems is important to intensify with short-duration grain legumes and dryland cereals.

Monitoring irrigated croplands, rainfed croplands, crop intensity, and land-use changes spatially is critical for planning agriculture development and sustainable food production. Spatial information is very important for prioritizing based on environmental and social factors and also where we could target short-duration crops (Gumma et al. 2014). Remote sensing is the ideal tool and provides an alternative, quick, and independent approach to estimate cropland extent seasonally and crop intensity over large areas (Gumma et al. 2011a; Gumma et al. 2016; Subbarao et al. 2001) and changes in croplands of a country (Badhwar 1984; Lobell et al. 2003; Thenkabail 2010; Thiruvengadachari and Sakthivadivel 1997). Many researchers worked on spatiotemporal analysis to map agriculture areas by irrigation source (Gumma et al. 2011c, d, 2015a; Knight et al. 2006; Thenkabail et al. 2005, 2007; Velpuri et al. 2009), specific crop-type mapping and temporal changes (Gumma et al. 2015b; Gumma et al. 2014), and crop intensity (Gumma et al. 2014; Sakamoto et al. 2005). Several studies have mapped rice areas with optical and synthetic-aperture radar (SAR) imagery. The number of studies have been conducted with MODIS time series data applying various methods. NDVI was widely used to map rice areas (Gumma et al. 2011b). Land surface water index (LSWI) and enhanced vegetation index (EVI) derived from temporal MODIS data were also used to map rice areas (Sakamoto et al. 2005; Shao et al. 2001). Spectral matching techniques were widely used to map irrigated areas, land use/land cover

(LULC), and rice crop mapping (Biradar et al. 2009; Gumma et al. 2015a; Thenkabail et al. 2007, 2009).

Given the above background, this paper mapped the fallow lands followed by rainy season (*kharif*) rice cultivation or post-rainy (*rabi*) fallows in rice-growing environments between 2000 and 2015. Accurate and up-to-date information on the spatial distribution of rice fallows is widely recognized as very important domain to target short-duration grain legumes. The aim of this research was to map rice fallows in South Asia as derived initially by mapping cropland extent by irrigation source and different rice systems using MODIS 16 day interval time series imagery for the years 2005–2006 to 2014–2015 applying spectral matching technique, phenological approaches, and intensive ground survey information.

21.2 Study Area

South Asia is located between 5°38′40″ and 36°54′30″ latitudes and 61°05′00″ and 97°14′15″ longitudes, with total geographical area of 477 Mha (Fig. 21.1). It has six agroecological zones: humid tropics, subhumid tropics, semiarid tropics, semiarid, subtropics, and arid (Choice 2009). South Asia is surrounded by Western Asia, Central Asia, East Asia, Southeast Asia, and the Indian Ocean. It includes six countries, Pakistan, India, Nepal, Bhutan, Bangladesh, and Sri Lanka. In South Asia, 80% of poor live in rural areas, and they mostly depend on agriculture for their livelihood (World Bank 2015). Nine major river basins were included in the study area: the Indus, Ganges, Brahmaputra, Narmada, Tapti, Godavari, Krishna, Kaveri, and Mahanadi. There are many major and minor irrigation projects in South Asia, covering a total command area of 133 Mha (Thenkabail et al. 2008). However, the ultimate potential is 139 m.ha, the increase being primarily due to upward revision in assessed potential of minor groundwater schemes and minor surface water schemes to 64 m.ha and 17 Mha, respectively. Rice is the major crop in this region, and it is grown two times a year in the entire region, but Bangladesh grows it three times a year (Table 21.1).

21.3 Data and Methods

The methodology for the identification of land-use changes and targeting of new technologies is shown in Fig. 21.2 and is described in the following sections.

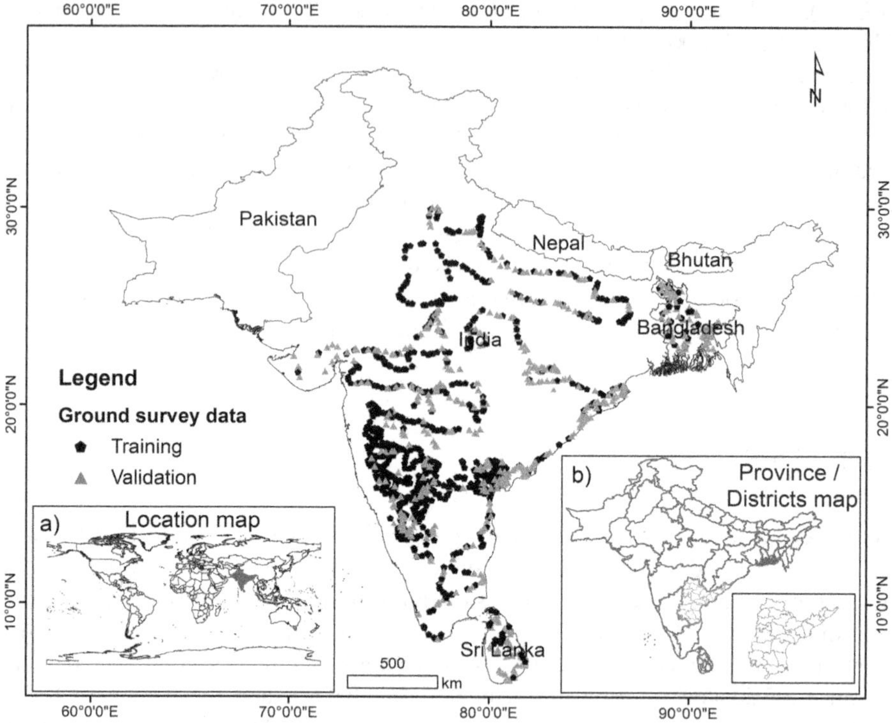

Fig. 21.1 Study area with countries and district boundaries. (**a**) Location map shown in the world and (**b**) administrative boundaries

Table 21.1 Country area and agricultural area in South Asia for 2010–2011

Country	Total geographical area ('000 ha)	Total gross planted area ('000 ha)	Net irrigated areas (NAS) ('000 ha)	Harvested area of rough rice (NAS) ('000 ha)
Bangladesh	14,804	15,002	6749	10,801
Bhutan	4365	121	27	26
India	3,45,623	1,84,443	63,601	44,712
Nepal	16,210	4208	1926	1560
Pakistan	89,167	22,817	19,270	2377
Sri Lanka	6453	2076	462	832
Total	4,76,622	2,28,668	92,035	60,308

Source: World Rice Statistics, FAO

21.3.1 Satellite Data and Processing

The present study used MOD13Q1.005 product, which provides 16 day composite images at 250 m spatial resolution. MOD13Q1.005 product includes vegetation indices, blue, red, and near-infrared (NIR) and mid-infrared (MIR) bands. Twelve

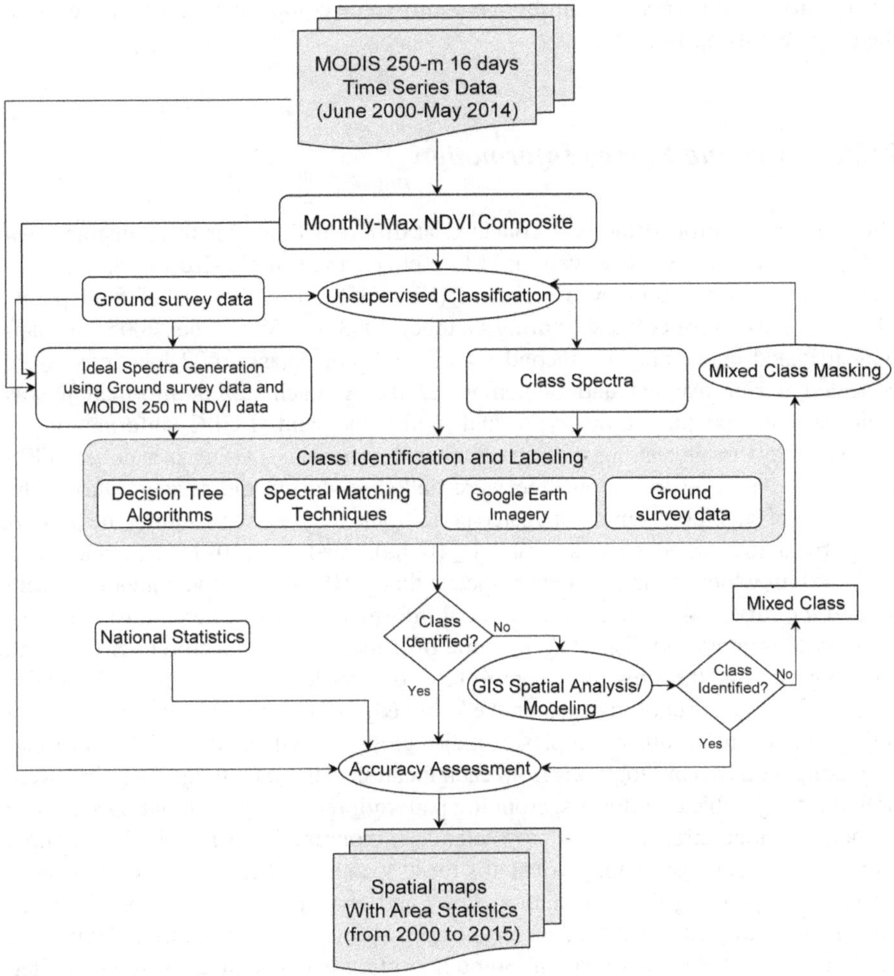

Fig. 21.2 Overview of the methodology for mapping rice fallows (rabi fallows) from 2000 to 2015 (each crop year)

tiles covering the South Asian region were downloaded from the Land Processes Distributed Active Archive Center (LP DAAC) (*https://lpdaac.usgs.gov*) (LPDAAC 2014). MODIS re-projection tool (MRT) was used to re-project and mosaic twelve tiles of study area and then stack them as single composite (Gumma et al. 2011b; Thenkabail et al. 2009). Altogether 23 images were stacked for the crop year from 2000 to 2015 (start from June to May).

$$\mathrm{NDVIMVC}_i = \mathrm{Max}(\mathrm{NDVI}_{i1}, \mathrm{NDVI}_{i2}) \tag{21.1}$$

where MVC_i is monthly maximum-value composite of ith month (e.g., "i" is Jan–Dec) and i_1 and i_2 are every 16 day composite in a month. The NDVI data was further processed to create monthly maximum-value composites (MVC) for each of the crop year using Eq. 21.1.

21.3.2 Ground Survey Information

Ground survey information was collected at different times for three distinct projects, which could be collectively used to increase the sample size for class identification as well as accuracy assessment (Fig. 21.1). The first set of field points (996 locations) were collected during October 2003 and September 2005 for mapping irrigated areas, and the second set of field-plot dataset (402 locations) were collected during August and September 2010. At each point, information was collected on existing crop type and soil type and LULC information at 250 m $\times$ 250 m plot along with geographical coordinates using a handheld GPS unit. Crop type and irrigation type were collected during the survey. During the collection of such points under irrigated conditions, the area surrounding the point is categorized into three classes, small ($\leq$10 ha), medium (10–15 ha), and large ($\geq$15 ha), in which "small" category means that $\leq$10 ha of irrigated area is present around the surveyed point. Also additional information was gathered through interviews with farmers and district agricultural officers to determine crop intensities and crop type during the previous year. A total of 830 locations covering the major cropland areas were chosen based on the knowledge of district agricultural extension officers to ensure adequate samples of major crops as well as other LULC information along with two photographs from each location. The remaining 568 points were only the geographic coordinates, cropping pattern/intensity by landholding size. The farmers provided information on crop calendars, cropping intensity (single or double crop), and percentage canopy cover for these locations. The interview included a question on planting dates, irrigation type, and cropping pattern. Ground survey information samples were based on local expert knowledge, distinct LULC type, and preliminary land-use classification. Some important irrigated areas were not visited due to road conditions and time constraints. Information was obtained in these areas from agriculture and irrigation departments. LULC names of class labels were assigned in the field using a labeling protocol (Gumma et al. 2014).

21.3.3 Ideal Spectra Signatures

Ideal spectra signatures were generated using 16 day NDVI time series composite and precise ground survey information, which was also used for class identification process (Fig. 21.3) (Gumma et al. 2016). Ideal spectral signatures were based on 303 ground survey information; these samples were grouped according to their unique categories and grouped major rice systems as shown in Fig. 21.3. Ideal spectra signatures were generated by considering crop intensity, crop type, and

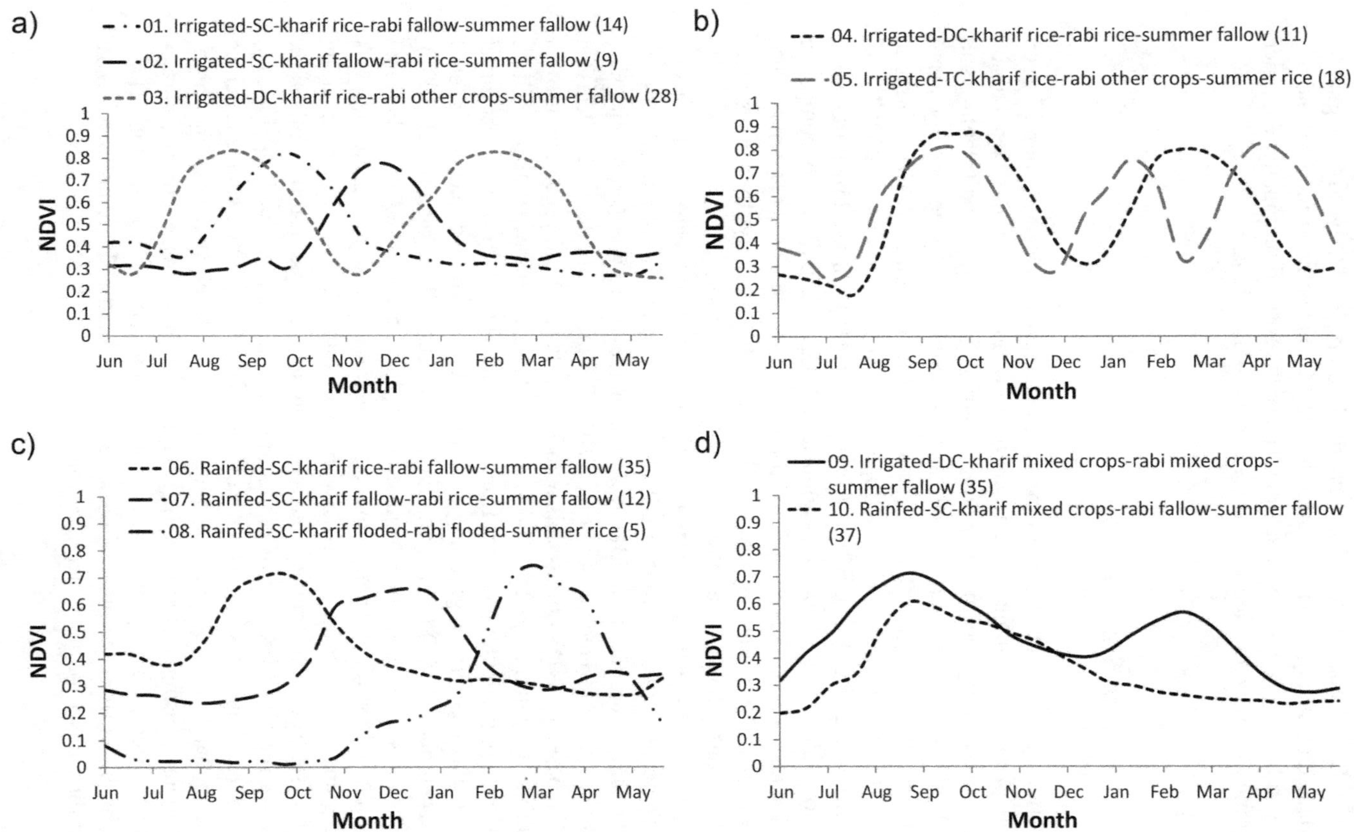

Fig. 21.3 Ideal spectra signatures. (**a**) Irrigated rice: (01) kharif (June–October) irrigated rice, rabi fallow; (02) rabi (November–February) irrigated rice, kharif fallow; (03) summer (March–May) irrigated rice, kharif and rabi fallow. (**b**) Irrigated rice: (04) kharif irrigated rice, rabi irrigated rice, summer fallow; (05) kharif irrigated rice, rabi fallow, summer rice. (**c**) Rainfed rice: (06) kharif rainfed rice, rabi, and summer fallow; (07) kharif fallow, rabi rice, summer fallow; (08) kharif and rabi fallow, summer rice; (09) kharif-mixed crops, rabi-mixed crops, summer fallow; (10) kharif-mixed crops, rabi fallow, summer fallow (Gumma et al. 2016)

cropping systems. Each signature was generated with group of similar samples. For example, Fig. 21.3a, class 1 “01. Irrigated single crop rice fallow (14)” signature defines irrigated as during kharif season followed rabi and summer fallows, and this was generated by 14 ground survey sample temporal information. Finally, eight rice spectra profiles were aggregated from the 132 sample locations.

21.3.4 Rice Classification

The generation of ideal spectra using time series imagery at selected ground survey points is the primary step in class identification. Ground survey points were chosen region-wise representing specific crop type, length of growing period (planting and harvesting), and irrigation source. Each ideal spectrum was generated by using 10–15 samples for each region (Gumma et al. 2011b; Gumma et al. 2014; Gumma et al. 2016; Thenkabail et al. 2007). Unsupervised classification using ISOCLASS cluster algorithm (ISODATA in ERDAS Imagine 2014™) followed by progressive generalization (Cihlar et al. 1998) was applied on MODIS 250 m 16 day composite NDVI for the crop year 2010–2011. The initial classification was set at a maximum of 100 iterations and a convergence threshold of 0.99. Initially 15 classes were extracted from unsupervised classification.

Class spectra were generated using ISOCLASS *k*-means classification using MODIS composite (Tou and Gonzalez 1975). The signatures were plotted for each LULC over time (Gumma et al. 2011b). Initial classes were grouped to ten using decision tree algorithm (Dheeravath et al. 2010; Gumma et al. 2011b, d). Decision tree rules are based on NDVI thresholds at different stages in the vegetation growth cycle, and this algorithm helps to identify similar classes. Rules were also based on field knowledge of particular locations from ground survey information. After initial grouping of classes, further grouping of similar classes was based on spectral matching techniques (SMT) (Homayouni and Roux 2003; Thenkabail et al. 2007). Temporal profiles of each group of classes were further compared with the ideal spectra in order to identify and name the class accurately. Additional verification was conducted using high-resolution imagery from Google Earth and GeoCover by overlaying district administrative boundaries in the Google Earth application. Mixed classes remained because of the large extent and diverse land use in small holdings. For resolving these mixed classes, we used various other sources such as irrigation command area boundaries, rainfall, district-level statistics, and high-resolution imagery using spatial modeling (Gumma et al. 2014). Some rare classes may not resolve conclusively even after using ground survey information. These classes were subset from the initial time series stack, and the above protocols were applied on this subset for class identification.

21.3.5 Rice Fallow Identification

The NDVI plots are ideal for understanding the changes within and between cropping seasons and between classes and exhibits the length of growing period. Temporal NDVI signature clearly elicits the planting time, peak growth, and harvesting stage in Fig. 21.3. Figure 21.3 also illustrates how the NDVI gradually goes up during the middle of July and reaches a peak (0.7) in October and gradually comes down to become a flat line indicating the cultivation of *kharif* rice followed by fallow.

NDVI time series plays a major role in class identification and determining crop growth stages season wise. Separation of rice-growing areas from other land-use/ land cover classes is based on annual average NDVI values and timing of the onset of "greenness." The annual NDVI of double-cropped and triple-cropped rice areas was higher than the other crops; meanwhile, the difference between irrigated and rainfed rice areas is also clearly seen in the study area.

The dates of vegetation transitions were determined using the NDVI time series and a double-logistic model of vegetation phenology (Biggs et al. 2006; Fischer 1994):

$$\mathrm{NDVI}_t = v_s + \frac{k}{1+\exp(-c(t-p))} - \frac{k+v_s-v_e}{1+\exp(-d(t-q))} \tag{21.2}$$

where v_s is the starting of rice-growing season, v_e is the ending of rice-growing season, k is an asymptotic maximum value of NDVI, c and d are the slopes of the NDVI time series at the inflection points, and p and q are the dates of the inflection points (Fig. 21.3). The starting of the time series was defined as the date of minimum (boro rice), which starts in February and harvested in May. In the present study, we used physical comparison of individual class signatures with ideal spectra of crop. All classes were quantified with NDVI signatures and cropping calendars of each crop classes.

21.3.6 Area Calculation and Accuracy Assessment

MOD13Q1.005 pixel covers 250 m a side, and its area is 6.25 hectares; this is larger than many different crop fields in South Asia. Due to large pixel area, there is a possibility of diverse crops except rice crops because this crop is grown in large areas (Tungabhadra, Nagarjuna Sagar, and delta areas). This study was classified at 250 m spatial resolution; it is necessary to calculate subpixel areas to get actual LULC areas. We used ground survey points to generate subpixel areas; however, ground information observations include a visual estimation of LULC percentages at 250 m × 250 m (each point) during ground survey. Considering the fall of ground survey points in each LULC class will help in estimating the subpixel area for each class based on average cropland area. This SPA crop fraction was applied to each land-use class and estimates net cropland areas and irrigated areas for study area with other LULC fractions in crop dominance classification.

Accuracy assessment was conducted on both resultant irrigated areas and rice maps (Congalton 2001; Jensen 2004), which generates an error matrix and accuracy map. Accuracy assessment was performed with 575 independent ground survey data. These data points were not used in class identification and labeling process.

21.3.7 Comparison with National Data

The final resultant irrigated subpixel area and subpixel rice areas generated from present study were compared against district-wise national statistics, such as irrigated area statistics adopted from Indiastat (www.indiastat.com) (INDIASTAT 2015). Rice statistics for India were obtained from the website of the Ministry of Agriculture's Directorate of Rice Development (http://dacnet.nic.in/rice/) and, for Bangladesh, Nepal, Pakistan, and Bhutan, from the national statistical departments. Irrigated area statistics were compared at province/state level (62 administrative units) and rice area compared at subdistrict level (812 administrative units).

21.4 Results and Discussion

21.4.1 Spatial Distribution of Croplands in South Asia

A typical agricultural landscape can be seen in the South Asian region (Fig. 21.4) for crop year 2015–2016. Most of the rainfed crops are spread over the Deccan Plateau, and irrigated cropping is prevalent mostly along the river courses. The rainfed cropped areas grown during the southwest monsoon are mostly highlands limited by water availability. Irrigated areas occur extensively and are concentrated in deltas and command areas of major river basins, such as Indo-Ganges, Godavari, Krishna, and Kaveri. Groundwater-irrigated areas are located across the study regions and mostly located in India. Figure 21.4 illustrates five irrigated classes, in which class 3 is irrigated by groundwater (GW) with double crop (DC), class 4 is irrigated by canal (surface water) with continuous crops, and classes 5 and 6 are irrigated by surface water (SW) with triple-cropped and single-cropped areas, respectively. Canal-irrigated double crops were located in headenders of command areas, and single-cropped areas were located at the tailenders. Rainfed areas were located outside of command areas and mainly located in the central part of India and uplands. The resultant class names (Fig. 21.4, Table 21.2) are based on dominance of particular land cover with in the irrigated areas. Table 21.2 shows area under different irrigated classes along with other LULC areas estimated with full pixel areas (FPA).

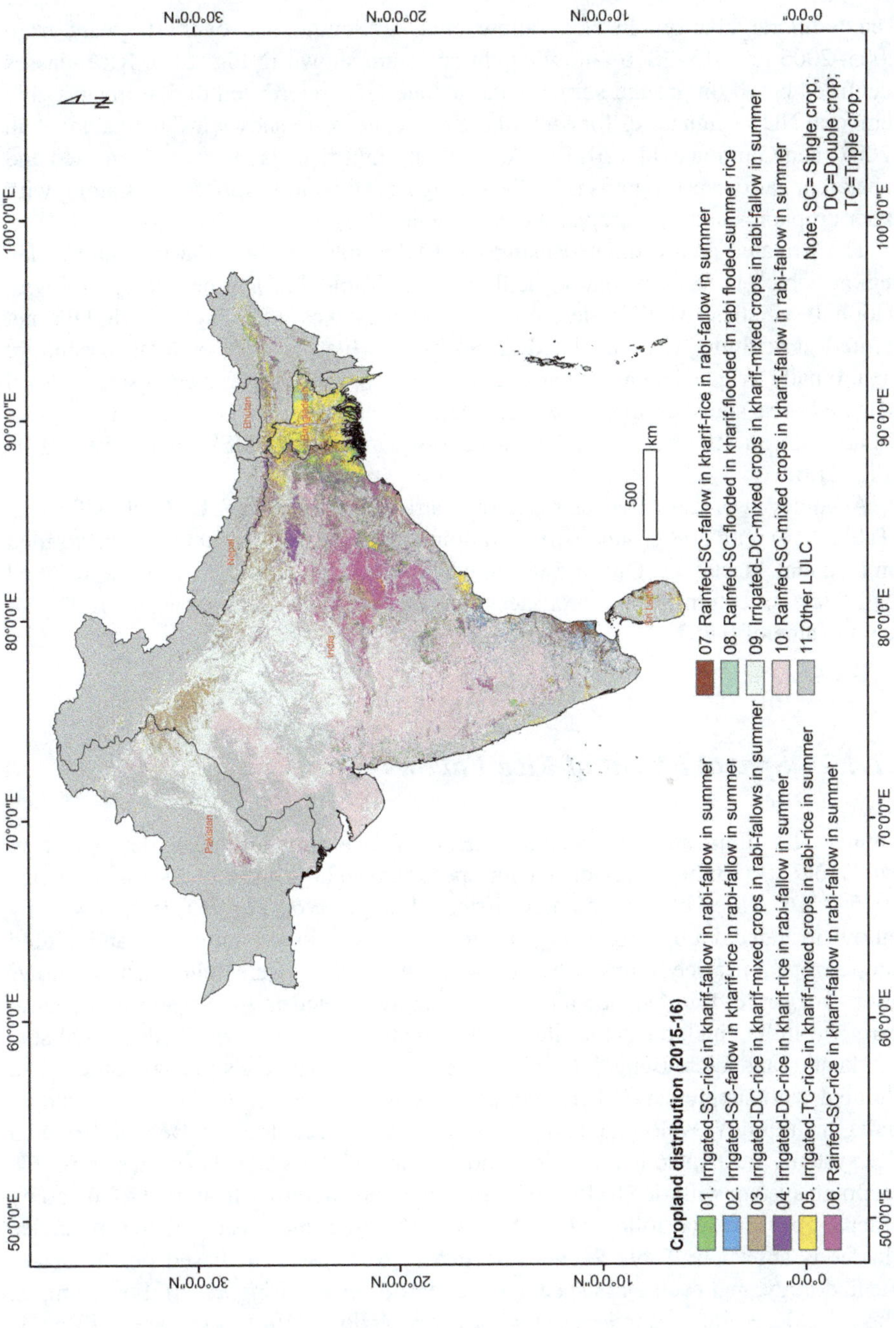

Fig. 21.4 Spatial distribution of land use/land cover in South Asia for the years 2015–2016

21.4.2 Spatiotemporal Distribution of Rice Fallows

The temporal changes in rice fallow area of South Asia over 10 years from 2005–2006 to 2015–2016 fallow extent areas are shown in Fig. 21.5. Rice classes identified based on ground survey data include GPS-referenced digital images, and temporal NDVI signatures for each of the rice classes are shown in Thenkabail et al. (2007) and Gumma et al. (2011b). Altogether, eight rice classes were identified and labeled in various ecoregions (Fig. 21.4). Figure 21.5 shows rabi fallows along with other croplands and non-cropland areas (Table 21.3).

The irrigated area estimated using the LULC map is 106 Mha in South Asian region, which is 58% of total agriculture area (Table 21.2). When compared with national statistics, MODIS-derived irrigated area was 15% higher. MODIS net rainfed agriculture was calculated as 84 Mha, which is 42% of total agriculture area. Finally irrigated area estimated using the map was compared against state-level irrigated area statistics available from national agencies. A good correlation was found with an R^2 of 0.93, and root-mean-square error (RMSE) was 859,000 ha (Fig. 21.6).

Accuracy was estimated using error matrices on resultant LULC classification (Table 21.4) with 568 ground surveyed points. Accuracy assessment was performed on resultant 11 classes. Out of 568 points, 463 locations matched with the resulted LULC classification. The overall accuracy of nine LULC classes was 81.5% and kappa value of 0.787.

21.4.3 Spatial Extent of Rice Fallows

Figure 21.4 illustrates the spatial extent of rice-growing areas derived from MOD13Q1.005 time series data with spectral matching techniques for the years 2015–2016. Three classes, namely, irrigated single crop *kharif* rice followed by fallow, irrigated double crop *kharif* rice followed by fallow summer rice, and rainfed single crop *kharif* rice followed by fallow summer fallow, were delineated as fallows after *kharif* rice. Rainfed rice areas were mainly located in the eastern and central parts of India, including the states of Chhattisgarh, Madhya Pradesh, Odisha, Jharkhand, and West Bengal. Significant parts in Sri Lanka also grow rainfed rice. Rainfed with supplemental irrigated areas are mainly located in the eastern part of India, including West Bengal and parts of Odisha adjacent to West Bengal. Irrigated rice systems are located in major irrigation command areas and also grown in double cropping system with rice followed by rice, rice followed by wheat, rice followed by other crops, and rice followed by fallows. The area under each of the three rice classes is shown in Table 21.3. Class names were assigned based on the above methodology, and each class specifies clearly rice crop during *kharif*. For example, class 1 "06 Rainfed single crop *kharif* rice fallows" full pixel area (FPA) is 12.17 Mha, which is dominated by rice along with other land cover areas, such as

Table 21.2 Rice systems in South Asia including other agriculture areas with irrigation source (crop year 2015–2016)

Class description	Full pixel area (FPA) (000′ ha)	% of total area (FPA)	Cropland fraction (%)			Actual cropland area (000′ ha)			Total gross cropland land area (000′ ha)
			Kharif	Rabi	Summer	Kharif	Rabi	Summer	
1. Irrigated SC rice in kharif fallow in rabi fallow in summer	6822	4.9	**96.1**	**8.1**	**3.2**	6556	553	218	7327
2. Irrigated SC fallow in kharif rice in rabi fallow in summer	2540	1.4	3.1	92.2	3.1	79	2342	79	2499
3. Irrigated DC rice in kharif-mixed crops in rabi fallows in summer	25,391	9.9	91	62.2	3.3	23,106	15,793	838	39,738
4. Irrigated DC rice in kharif rice in rabi fallow in summer	6823	1.3	96.3	86.6	3	6571	5909	205	12,685
5. Irrigated TC rice in kharif-mixed crops in rabi rice in summer	10,890	2.7	97.3	78.7	68.2	10,596	8570	7427	26,593
6. Rainfed SC rice in kharif fallow in rabi fallow in summer	12,171	6.8	**98.3**	**2.9**	**2.1**	11,965	353	256	12,573
7. Rainfed SC fallow in kharif rice in rabi fallow in summer	2983	0.4	3	93.6	3	89	2792	89	2971
8. Rainfed SC flooded in kharif flooded in rabi-flooded summer rice	543	0.3	3	3	91.9	16	16	499	532
9. Irrigated DC-mixed crops in kharif-mixed crops in rabi fallow in summer	73,490	35.0	86.7	83.2	3.5	63,715	61,143	2572	127,431
10. Rainfed SC-mixed crops in kharif fallow in rabi fallow in summer	77,375	37.3	77.7	18	0	60,120	13,927	0	74,048
Total croplands	219,029					182,814	111,400	12,183	306,397
Net cropland areas cultivated in South Asia, full pixel areas (FPA) = 219,029,000 ha									
Net cropland areas cultivated in South Asia, subpixel areas (SPAs) or actual areas during kharif season (June–October) = 182,814,000 ha									
Net cropland areas cultivated in South Asia, subpixel areas (SPAs) or actual areas during rabi season (November–February) = 111,400,000 ha									
Net cropland areas cultivated in South Asia, subpixel areas (SPAs) or actual areas during summer season (March–May) = 12,183,000 ha									

(continued)

Table 21.2 (continued)

Class description	Full pixel area (FPA) (000′ ha)	% of total area (FPA)	Cropland fraction (%)			Actual cropland area (000′ ha)			Total gross cropland land area (000′ ha)
			Kharif	Rabi	Summer	Kharif	Rabi	Summer	
Gross cropland areas cultivated in South Asia, subpixel areas (SPAs) or actual areas = 307,189,000 ha									
For classes 1 and 6 (Note: these two classes have rice during kharif and left fallow in rabi)									
Total net cultivated areas during kharif in South Asia, SPAs or actual areas of classes 1 and 6 = 18,520,789 ha									
Total net cultivated areas during rabi in South Asia, SPAs or actual areas of classes 1 and 6 = 905,579 ha									
Total uncultivated areas during rabi that were cultivated during kharif, SPAs or actual areas of classes 1 and 6 = 17,615,211 ha									

The table shows full pixel area (*FPA*), crop area fraction (*CAF*), and subpixel area (*SPA*) or actual area. SPA = FPA * CAF

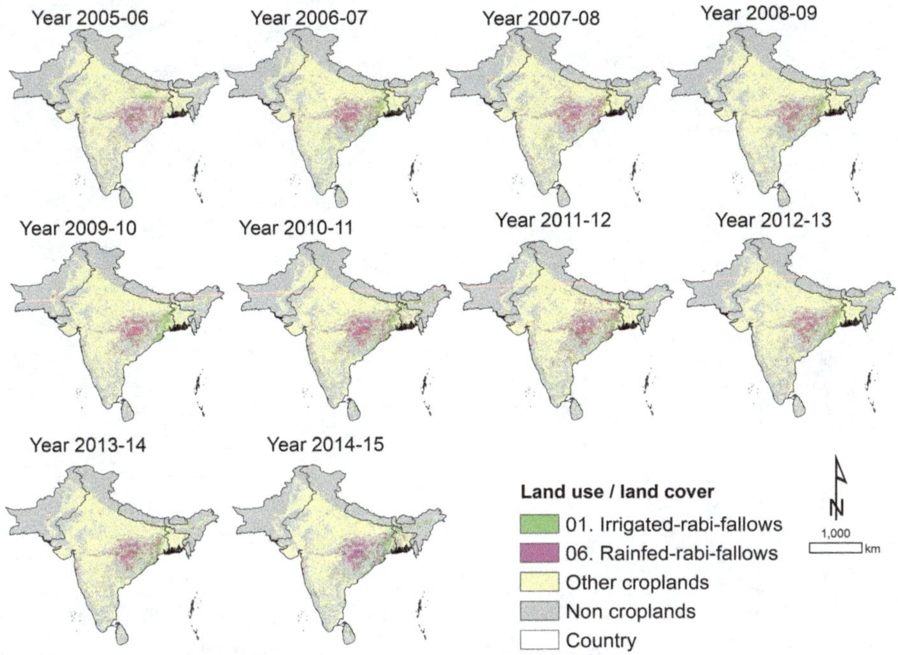

Fig. 21.5 Spatiotemporal distribution of rice fallows (rabi fallows)

Table 21.3 Net rice fallow in South Asia with irrigation source

Rice fallow (rabi fallow) classes	Rice fallows (000′ ha)	Percent of total rice fallows (%)
1. Irrigated SC rice in *kharif* fallow in *rabi* fallow in summer	6004	34
6. Rainfed SC rice in *kharif* fallow in *rabi* fallow in summer	11,612	66
Irrigated + rainfed rabi fallows + kharif rice: total	17, 615	100

trees, shrubs, grasses, water, fallows, and other crops. Rice fallows were mainly located in the central and eastern parts of India to a total of 17.6 Mha, which is 26% of total net rice-grown area in this region, which includes irrigated and rainfed areas. This means 26% of rice-growing area in South Asia is left fallow after growing rice in the rainy season. This elicits a large potential to intensify cropping in these lands based on the suitability. Subnation-wise rice fallows were shown in Table 21.5.

The rice estimates derived from the map generated in this study were compared against seasonal rice area statistics obtained from South Asian countries at the

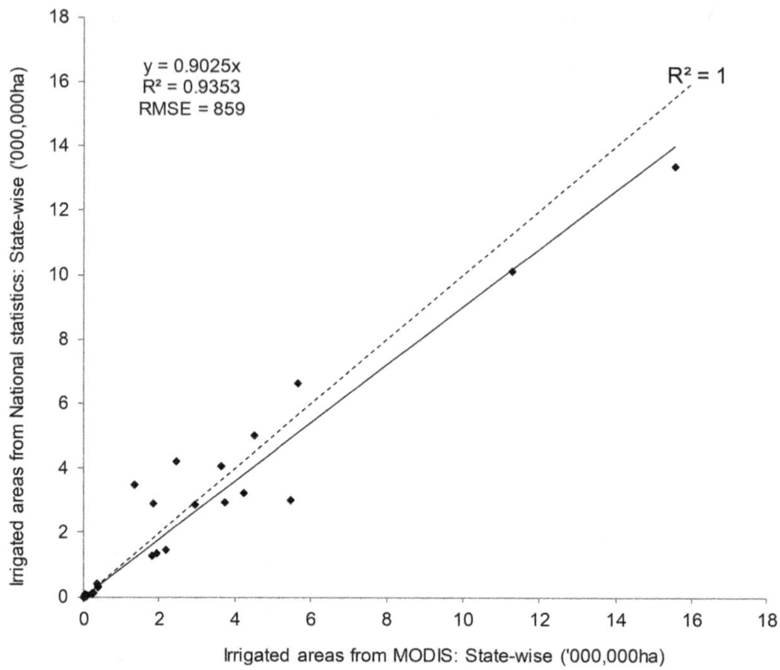

Fig. 21.6 The irrigated areas derived using MODIS 250 m imagery compared with agricultural census data for 2013–2014 (Administrative boundaries shown in Fig. 21.1b)

district level. Data was obtained for 840 administrative districts. The MODIS-derived rice areas were consistently nearer to the official estimates in all districts (Fig. 21.7). The R^2 value is 0.84 and RMSE was 40,079 ha (Fig. 21.7). The overestimation is obvious in 1:1 plot (Fig. 21.7). Present study clearly showed that statistical estimates can be generated at the district level from the thematic maps generated from satellite imagery.

21.4.4 Discussion on Methodology

The present research used MOD13Q1.005 temporal data to identify rice fallows with rice systems and irrigated areas across South Asia. MODIS captures imagery over the Earth on a daily basis, which makes it possible to get cloud-free data when available immediately after a rainfall event or cloudy day. The 16 day composites from the daily acquisitions bundle up to make a time series dataset over a crop year or a calendar year. This type of dataset provides temporal profiles of crop-growing locations to identify the start of season, peak growth stage, and harvest date during each season. The values of NDVI as function of time also help in identifying the type

Table 21.4 Accuracy assessment using error matrix (land-use/land cover classification)

Crop classification	I.	II.	III.	IV.	V.	VI.	VII	VIII	IX.	X.	XI.	Class totals	Reference totals	Classified totals	Users' accuracy (%)	Producers' accuracy (%)	Kappa
I. Irrigated SC rice in kharif fallow in rabi fallow in summer	21	0	0	1	1	0	0	0	0	0	0	24	46	37	91	88	0.909
II. Irrigated SC fallow in kharif rice in rabi fallow in summer	1	16	0	1	0	0	1	0	0	1	0	20	26	19	80	80	0.793
III. Irrigated DC rice in kharif-mixed crops in rabi fallows in summer	0	1	59	8	1	8	0	0	1	0	0	62	48	37	76	95	0.727
IV. Irrigated DC rice in kharif rice in rabi fallow in summer	0	0	1	20	2	3	0	0	1	0	0	37	55	49	74	54	0.723
V. Irrigated TC rice in kharif-mixed crops in rabi rice in summer	0	0	1	6	57	0	1	0	7	0	0	66	60	54	79	86	0.764
VI. Rainfed SC rice in kharif fal-low in rabi fallow in summer	1	1	0	0	4	68	0	0	3	1	0	92	57	43	87	74	0.847

(continued)

Table 21.4 (continued)

Crop classification	I.	II.	III.	IV.	V.	VI.	VII	VIII	IX.	X.	XI.	Class totals	Reference totals	Classified totals	Users' accuracy (%)	Producers' accuracy (%)	Kappa
VII. Rainfed SC fallow in kharif rice in rabi fallow in summer	0	1	0	0	0	0	29	0	1	1	0	34	10	5	91	85	0.9
VIII. Rainfed SC flooded in kharif flooded in rabi-flooded summer rice	0	0	0	0	0	0	0	5	0	0	0	5	5	5	100	100	1
IX. Irrigated DC-mixed crops in kharif-mixed crops in rabi fallow in summer	0	0	0	0	0	1	0	0	87	5	1	110	115	88	93	79	0.908
X. Rainfed SC-mixed crops in kharif fallow in rabi fallow in summer	1	0	1	1	1	11	2	0	10	76	1	86	109	105	73	88	0.683
XI. Other LULC	0	1	0	0	0	1	1	0	0	2	30	32	37	21	86	94	0.849
Column total	45	21	40	58	74	62	12	7	102	122	25	568	568	463			
Overall classification Accuracy = 82% overall Kappa statistics = 0.79																	

Table 21.5 MODIS rice areas with rice fallows and subnational statistics at state wise (2015–2016)

Country	State	MODIS-derived areas ('000 ha)			Subnational statistics ('000 ha)
		Total net rice area	Rice fallows (rainfed)	Rice fallows (irrigated)	
Bangladesh	Barisal	528	49	147	987
Bangladesh	Chittagong	1001	29	104	1093
Bangladesh	Dhaka	2103	21	60	1303
Bangladesh	Khulna	1011	29	69	904
Bangladesh	Rajshahi	2479	25	161	2015
Bangladesh	Sylhet	725	33	75	456
Bhutan	Bhutan	128	7	13	NA
India	Andhra Pradesh	4266	618	690	2922
India	Arunachal Pradesh	326	16	75	122
India	Assam	2150	165	365	2139
India	Bihar	4501	223	188	3232
India	Chandigarh	2	0	0	NA
India	Chhattisgarh	5839	4099	87	5820
India	Dadra and Nagar Haveli	8	2	4	10
India	Daman and Diu	0	0	0	2
India	Delhi	3	1	0	NA
India	Goa	52	2	4	31
India	Gujarat	757	139	164	728
India	Haryana	817	5	1	1243
India	Himachal Pradesh	119	3	6	NA
India	Jammu and Kashmir	200	2	10	212
India	Jharkhand	2522	921	770	2440
India	Karnataka	1509	342	233	NA
India	Kerala	246	3	4	162
India	Lakshadweep	0	0	0	NA
India	Madhya Pradesh	4193	2018	213	1522
India	Maharashtra	1500	377	278	1486
India	Manipur	151	29	42	NA
India	Meghalaya	342	13	45	95
India	Mizoram	2	0	0	NA
India	Nagaland	19	0	1	165
India	Orissa	4484	1653	1012	3933
India	Pondicherry	23	0	0	4
India	Punjab	2525	5	1	2517

(continued)

Table 21.5 (continued)

Country	State	MODIS-derived areas ('000 ha) Total net rice area	Rice fallows (rainfed)	Rice fallows (irrigated)	Subnational statistics ('000 ha)
India	Rajasthan	13	5	1	130
India	Sikkim	6	0	0	12
India	Tamil Nadu	2788	13	46	1906
India	Tripura	118	6	31	NA
India	Uttar Pradesh	5191	277	58	5632
India	Uttaranchal	212	4	9	270
India	West Bengal	5809	511	1138	NA
Nepal	Central	541	20	25	419
Nepal	East	561	17	42	419
Nepal	Far-Western	110	2	5	150
Nepal	Mid-Western	136	5	7	174
Nepal	West	354	7	8	335
Pakistan	Baluchistan	48	17	0	NA
Pakistan	F.C.T.	0	0	0	NA
Pakistan	Kashmir Pak	7	0	1	NA
Pakistan	Khyber	28	0	1	NA
Pakistan	Punjab Pak	1288	7	3	NA
Pakistan	Sind	374	51	4	NA
Pakistan	Tribal areas	7	0	0	NA
Sri Lanka	Central Sri	32	0	3	57
Sri Lanka	Eastern	209	1	9	89
Sri Lanka	North Central	213	1	20	78
Sri Lanka	North Western	32	0	1	25
Sri Lanka	Northern	107	1	2	265
Sri Lanka	Sabaragamuwa	17	0	0	34
Sri Lanka	Southern	138	1	1	33
Sri Lanka	Uva	119	1	3	91
Sri Lanka	Western	19	0	0	57

of crop in an ecoregion based on certain peak thresholds for that crop. This study applies spectral matching technique which is found to be ideal in mapping irrigated areas (Thenkabail et al. 2007) and mapping rice areas (Gumma et al. 2011a, b, c, d). Mapping spatial distribution of rice fallows using MODIS 250 m 16 day time series and ground survey information with spectral matching techniques is a significant new advancement in the use of this technique.

Some discrepancies were also found during the comparison of national statistics and MODIS-derived irrigated areas. The mismatch occurred in high-rainfall zones, where there was misclassification with irrigated areas due to similar growing

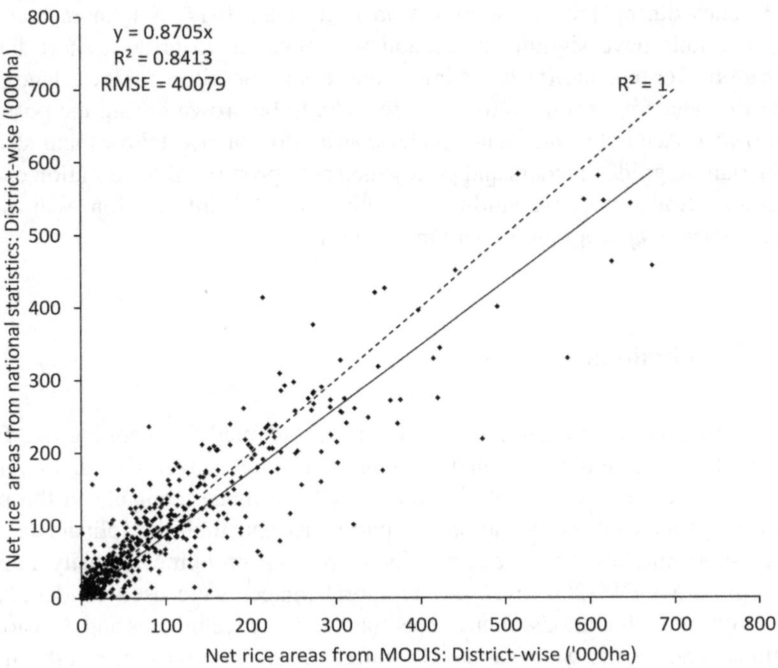

Fig. 21.7 The rice areas derived using MODIS 250 m compared with agricultural census data for 2010–2011 (Gumma et al. 2016) (Administrative boundaries in Fig. 21.1)

conditions during cropping season. Most of the areas were corrected using rainfall data and spatial modeling techniques.

Irrigated area fractions (the proportion of irrigated/rice area in a pixel) were assigned based on land-use proportions in each class to estimate the MODIS pixel area accurate to the real irrigated/rice area. Also, this method relies on ground survey information that is a truly representative sample of the fragmented rice systems. Higher-resolution imagery could be used to provide a more accurate estimate of pure classes but with a lot more mixed classes coming up. Results clearly show that present methods and MODIS time series data have many advantages such as capturing large-scale cropping pattern. But to minimizing errors, additional research will be attempted with multi-sensor images with advanced fusion techniques (Gumma et al. 2011c).

21.4.5 Discussion on Rice Fallows and Rice-Growing Areas

Mapping rice fallow is very important for the promotion of short-duration legumes for sustainable intensification and improving livelihoods. Rice fallow areas are high-rainfall zones during the monsoon season (Ali et al. 2014; Satyanarayana et al. 1997), and soils have significant residual soil moisture to grow a short-duration legume crop after monsoon rice. An important reason for fallowing these lands is the scarcity of water for a normal crop like rice, etc. to be grown during the post-rainy season (*rabi*). Accurate up-to-date spatial distribution of rice fallows and statistics are important to guide breeders and policymakers to promote short-duration crops in this region. Soil-water availability and phenological information will inform breeders to select appropriate variety in a region.

21.5 Conclusions

The present study demonstrated the methodology adopted for mapping rice fallow along with LULC including irrigated areas and rice systems over large areas. Also an estimated district-level extent of rice fallow will be made available in the public domain (maps.icrisat.org) for national departments and multidisciplinary teams to promote short-duration grain legumes for increasing cropping intensity in South Asian region. MODIS 250 m 16 day temporal images were used to identify rice fallows with suite of methods, such as spectral matching techniques and decision tree algorithms. The results were validated with extensive ground survey information. MODIS-derived rice areas at the district level were compared with national statistics. The R^2 value was 0.84, and root-mean-square error was 40,079 ha with 735 district administrative boundaries. MODIS-derived net rice cropped area is 56,940,000 ha, which is slightly higher than national statistics (55,139,300 ha). The total area under rice fallow in South Asia is 16,808,960 ha including in rainfed and irrigated environments. In rainfed environment, it is 14,380,250 ha, and in irrigated environment, it is 2,427,713 ha. Accuracy assessment was performed by error matrix with ground survey information, and overall accuracy is 75% with kappa coefficient of 0.644. Present methods and approaches are found to be ideal for mapping rice systems with rice fallows in over larger areas in South Asia. District-wise spatial extent of rice fallows and irrigated areas will guide breeders and social scientists to promote short-duration grain legumes. The research creates a broad contribution to the methods and products of the Group on Earth Observations (GEO) for monitoring agriculture areas, Agriculture and Water Societal Beneficial Areas (GEO Agriculture and Water SBAs), the GEO Global Agricultural Monitoring Initiative (GEO GLAM), the global cropland area database using Earth observation data, and studies pertaining to global croplands, their water use, and food security in the twenty-first century (https://powellcenter.usgs.gov/globalcroplandwater/). The information on these types of domains needs to be updated regularly to guide the decision-makers and agricultural scientists to plan for sustained food production for food and nutrition security.

Acknowledgments This research was supported by two CGIAR Research Programs: Grain Legumes and Dryland Cereals (GLDC) and (Water Land and Ecosystems (WLE)). The research was also supported by the global food security support analysis data at 30 m project (GFSAD30; http://geography.wr.usgs.gov/science/croplands/; https://croplands.org/) funded by the NASA MEaSUREs (Making Earth System Data Records for Use in Research Environments) funding obtained through NASA ROSES solicitation as well as by the Land Change Science (LCS), Land Remote Sensing (LRS), and Climate Land Use Change Mission Area Programs of the US Geological Survey (USGS). The authors would like to thank to the International Rice Research Institute (IRRI) for providing ground survey data and district-wise national statistics; Dr. Dheeravath Venkateshwarlu, Dr. Andrew Nelson, and Dr. Mitch Scull for supporting ground surveys in India; Dr. Saidul Islam for Bangladesh ground survey data and Dr. Nimal Desanayake for Sri Lanka ground survey data. Also thanks to Ms. Deepika Uppala for supporting data analysis.

References

Alexandratos N, Bruinsma J (2012) World agriculture towards 2030/2050: the 2012 revision. http://large.stanford.edu/courses/2014/ph240/yuan2/docs/ap106e.pdf. Accessed in 08 June 2015. ESA working paper 3

Ali M, Ghosh P, Hazra K (2014) Resource conservation technologies in rice fallow. Resource conservation technology in pulses, First edited by PK Ghosh, N Kumar, MS Venkatesh, KK Hazra, N Nadarajan, 01/2014: chapter 7: pp 83–88, Scientific publishers, ISBN: 978-81-7233-885-5

Badhwar GD (1984) Automatic corn-soybean classification using landsat MSS data. I. Near-harvest crop proportion estimation. Remote Sens Environ 14(1–3):15–29

Biggs TW, Thenkabail PS, Gumma MK, Scott CA, Parthasaradhi GR, Turral HN (2006) Irrigated area mapping in heterogeneous landscapes with MODIS time series, ground truth and census data, Krishna Basin, India. Int J Remote Sens 27(19):4245–4266

Biradar CM, Thenkabail PS, Noojipady P, Li Y, Dheeravath V, Turral H, Velpuri M, Gumma MK, Reddy GPO, Cai XL, Xiao X, Schull MA, Alankara RD, Gunasinghe S, Mohideen S (2009) A global map of rainfed cropland areas (GMRCA) at the end of last millennium using remote sensing. Int J Appl Earth Obs Geoinf 11(2):114–129

Choice H (2009) Agroecological zones. Available http://harvestchoice.org/production/biophysical/agroecology. Downloaded 01/05/ 2010

Cihlar J, Xiao Q, Chen J, Beaubien J, Fung K, Latifovic R (1998) Classification by progressive generalization: A new automated methodology for remote sensing multichannel data. Int J Remote Sens 19(14):2685–2704

Congalton RG (2001) Accuracy assessment and validation of remotely sensed and other spatial information. Int J Wildland Fire 10(4):321–328

Dheeravath V, Thenkabail PS, Chandrakantha G, Noojipady P, Reddy GPO, Biradar CM, Gumma MK, Velpuri M (2010) Irrigated areas of India derived using MODIS 500 m time series for the years 2001–2003. ISPRS J Photogramm Remote Sens 65(1):42–59

FAO (2015) FAOSTAT. http://faostat.fao.org/. Accessed on 02 June 2015

Fischer A (1994) A model for the seasonal variations of vegetation indices in coarse resolution data and its inversion to extract crop parameters. Remote Sens Environ 48(2):220–230

Garnett T, Appleby M, Balmford A, Bateman I, Benton T, Bloomer P, Burlingame B, Dawkins M, Dolan L, Fraser D (2013) Sustainable intensification in agriculture: premises and policies. Science 341(6141):33–34

Gray J, Friedl M, Frolking S, Ramankutty N, Nelson A, Gumma M (2014) Mapping Asian cropping intensity with MODIS. Sel Top Appl Earth Obs Remote Sens IEEE J PP 99:1–7

Gumma MK, Gauchan D, Nelson A, Pandey S, Rala A (2011a) Temporal changes in rice-growing area and their impact on livelihood over a decade: a case study of Nepal. Agric Ecosyst Environ 142(3–4):382–392

Gumma MK, Nelson A, Thenkabail PS, Singh AN (2011b) Mapping rice areas of South Asia using MODIS multitemporal data. J Appl Remote Sens 5:053547

Gumma MK, Thenkabail PS, Hideto F, Nelson A, Dheeravath V, Busia D, Rala A (2011c) Mapping irrigated areas of Ghana using fusion of 30 m and 250 m resolution remote-sensing data. Remote Sens 3(4):816–835

Gumma MK, Thenkabail PS, Muralikrishna IV, Velpuri MN, Gangadhararao PT, Dheeravath V, Biradar CM, Acharya Nalan S, Gaur A (2011d) Changes in agricultural cropland areas between a water-surplus year and a water-deficit year impacting food security, determined using MODIS 250 m time-series data and spectral matching techniques, in the Krishna River basin (India). Int JRemote Sens 32(12):3495–3520

Gumma MK, Thenkabail PS, Maunahan A, Islam S, Nelson A (2014) Mapping seasonal rice cropland extent and area in the high cropping intensity environment of Bangladesh using MODIS 500m data for the year 2010. ISPRS J Photogramm Remote Sens 91(5):98–113

Gumma MK, Kajisa K, Mohammed IA, Whitbread AM, Nelson A, Rala A, Palanisami K (2015a) Temporal change in land use by irrigation source in Tamil Nadu and management implications. Environ Monit Assess 187(1):1–17

Gumma MK, Mohanty S, Nelson A, Arnel R, Mohammed IA, Das SR (2015b) Remote sensing based change analysis of rice environments in Odisha, India. J Environ Manag 148:31–41

Gumma MK, Thenkabail PS, Teluguntla P, Rao MN, Mohammed IA, Whitbread AM (2016) Mapping rice-fallow cropland areas for short-season grain legumes intensification in South Asia using MODIS 250 m time-series data. Int J Digital Earth 9(10):981–1003

Homayouni S, Roux M (2003) Material mapping from hyperspectral images using spectral matching in urban area. In: Landgrebe P (ed) IEEE Workshop in honour of Prof. Landgrebe, Washington DC, USA, October 2003

INDIASTAT (2015) State-wise net area irrigated by source in India and state-wise irrigated area under crops in India. www.indiastat.com. Accessed in 30 May 2015

Jensen JR (2004) IIntroductory digital image processing: a remote sensing perspective, 3rd edn. Upper Saddle River, Prentice Hall, p 544

Knight JF, Lunetta RL, Ediriwickrema J, Khorram S (2006) Regional scale land-cover characterization using MODIS-NDVI 250 m multi-temporal imagery: a phenology based approach. GISci Remote Sens 43(1):1–23

Lobell DB, Asner GP, Ortiz-Monasterio JI, Benning TL (2003) Remote sensing of regional crop production in the Yaqui Valley, Mexico: estimates and uncertainties. Agric Ecosyst Environ 94(2):205–220

LPDAAC (2014) The land processes distributed active archive center – See more at: https://lpdaac.usgs.gov/#sthash.MxeroLVs.dpuf. Accessed on 15 April 2014

Sakamoto T, Yokozawa M, Toritani H, Shibayama M, Ishitsuka N, Ohno H (2005) A crop phenology detection method using time-series MODIS data. Remote Sens Environ 96(3–4):366–374

Satyanarayana A, Seenaiah P, Sudhakara Babu K, Prasada Rao M (1997) Extending pulses area and production in rice fallow. In: Asthana AN, Ali M (eds) Recent advances in pulses research. Indian Society of Pulses Research and Development, Indian Institute of Pulses Research, Kanpur, pp 569–580

Shao Y, Fan X, Liu H, Xiao J, Ross S, Brisco B, Brown R, Staples G (2001) Rice monitoring and production estimation using multitemporal RADARSAT. Remote Sens Environ 76(3):310–325

Subbarao G, KumarRao J, Kumar C, Johansen U, Irshad A, KrishnaRao L, Venkataratnam K, Hebbar K, Sai M, Harries D (2001) Spatial distribution and quantification of rice-fallows in south Asia: potential for legumes. ICRISAT, Patancheru, p 316 ICRISAT

Thenkabail PS (2010) Global croplands and their importance for water and food security in the twenty-first century: towards an ever green revolution that combines a second green revolution with a blue revolution. Remote Sens 2(9):2305–2312

Thenkabail PS, Schull M, Turral H (2005) Ganges and Indus river basin land use/land cover (LULC) and irrigated area mapping using continuous streams of MODIS data. Remote Sens Environ 95(3):317–341

Thenkabail P, GangadharaRao P, Biggs T, Gumma M, Turral H (2007) Spectral matching techniques to determine historical land-use/land-cover (LULC) and irrigated areas using time-series 0.1-degree AVHRR pathfinder datasets. Photogramm Eng Remote Sens 73(10):1029–1040

Thenkabail PS, Biradar CM, Noojipady P, Dheeravath V, Li YJ, Velpuri M, Reddy G, Cai X, Gumma M, Turral H (2008) A global irrigated area map (GIAM) using remote sensing at the end of the last millennium. International Water Management Institute, Colombo

Thenkabail P, Biradar C, Noojipady P, Dheeravath V, Li Y, Velpuri M, Gumma M, Reddy GPO, Turral H, Cai X, Vithanage J, Schull M, Dutta R (2009) Global irrigated area map (GIAM) derived from remote sensing, for the end of last millennium. Int J Remote Sens 30(14):3679–3733

Thiruvengadachari S, Sakthivadivel R (1997) Satellite remote sensing for assessment of irrigation system performance: a case study in India, Research report 9. International Irrigation Management Institute, Colombo

Tou JT, Gonzalez RC (1975) Pattern recognition principles. Addison-Wesley Publishing Company, Reading

USDA (2010) United States department of agriculture. foreign agricultural service. www.fas.usda.gov/psdonline/. Accessed on 21 May 2015. USDA

Velpuri NM, Thenkabail PS, Gumma MK, Biradar CB, Noojipady P, Dheeravath V, Yuanjie L (2009) Influence of resolution in irrigated area mapping and area estimations. Photogramm Eng Remote Sens 75(12):1383–1395

World Bank (2015) PovcalNet. http://iresearch.worldbank.org/PovcalNet/povcalNet.html. Accessed on 25 May 2015

Chapter 22
Land Resources Information System and NBSS BHOOMI Geoportal for Land Resource Monitoring and Management

G. P. Obi Reddy and S. K. Singh

Abstract In the study, Land Resources Information System (LRIS) of India has been designed and developed using geographic information system (GIS) to effectively store, update, manage and query land and allied resource database. Based on the standardized datasets generated under LRIS, NBSS BHOOMI Geoportal has been designed and developed to visualize, query and disseminate various theme-based Web Map Services (WMS) and applications for agricultural land use planning. The thematic information on major physiographic regions, sub-physiographic regions, agro-ecological regions, agro-ecological subregions, soil data of India on 1:1 m and 1:250,000 scale, soil loss status and land degradation status of India, 1:50,000 scale soil data for selected districts and 1:10,000 scale soil data for selected blocks of India has been processed and deployed in LRIS and Geoportal. The point database on climatic data, benchmark soils, soil fertility, soil loss grid points of India and soil series data for 18 states and 22 districts has been deployed in Geoportal. Digital elevation models of SRTM DEM (90 m and 30 m) and ASTER DEM (30 m), the thematic database on major crop growing areas, land management units and potential growing areas for 17 crops at India and state level have been processed and deployed in BHOOMI Geoportal. The developed LRIS and NBSS BHOOMI Geoportal have immense potential in sustainable management of land resources and develop smart agricultural land use plans in India.

Keywords Geographic information system (GIS) · Land resources information system · NBSS BHOOMI geoportal · Web map service (WMS)

G. P. O. Reddy (✉) · S. K. Singh
ICAR-National Bureau of Soil Survey & Land Use Planning, Nagpur, India

G. P. O. Reddy, S. K. Singh (eds.), *Geospatial Technologies in Land Resources Mapping, Monitoring and Management*, Geotechnologies and the Environment 21,
https://doi.org/10.1007/978-3-319-78711-4_22

22.1 Introduction

Decision-makers at various levels depend on reliable, up-to-date information on land resources in order to promote effective means of land use plans for sustainable agriculture. The accurate and timely availability of information on the extent and spatial distribution of land resources is of immense help for the planners and policymakers in decision-making processes for sustainable management of land resources and agricultural land use planning. Issues related to land resource management, land use and its planning are becoming increasingly complex, and land resource planners, policymakers, managers and farmers often lack the right information and right time to formulate and suggest judicious land use options and appropriate land management plans. The availability of site-specific information on land resources certainly guides the land resource planners, policymakers, managers and farmers to select appropriate land use, right technologies based on the potentiality and constraints of well-defined land management units. Earlier desktop-based applications were used extensively for generation and management of soil resources for generation of various thematic maps and land use planning (Maji et al. 2002a, 2004). The advance developments in GIS and computer technology in recent years allow a large amount of soils and associated data to be easily stored and retrieved (Reddy et al. 2014a, b). In order to utilize the land resources for sustained production, comprehensive information on land resources, more particularly on soils, is essential to provide systematic and location-specific information to the planners, decision-makers and developmental agencies. Soil maps generated through soil resource mapping illustrate the geographic distribution of landscape attributes, soil types and soil properties such as site and physical and chemical characteristics; such information has immense help in the planning process. Therefore, the knowledge of soils, in respect of their extent, distribution, characterization and use potential, is extremely important for optimized land resources utilization and management.

In recent times, soil information has become increasingly important in many agricultural applications (Maji et al. 2002b; Reddy 2012, 2016). With recent advances in geospatial technology, the ability for storing, manipulating, analysing and presenting spatial data has been greatly enhanced. It is now possible to handle vast amount of spatial data in digital form and information systems (Reddy et al. 2016). Development of GIS-based location-specific database plays a vital role in diverse planning, research and development activities of the present and future use of soil resources in agricultural land use planning in developing countries like India (Reddy et al. 2017a, b). GIS technology provides a powerful tool to present and analyse soil information (King et al. 1995) and integrates common database operations, such as query and statistical analysis with unique visualization offered through maps. Current agriculture systems use and/or combine different types of technology ranging from hardware deployments and remote sensing instruments and technology to distributed computing and web-based and mobile-based software applications (Geller 2016; Granell et al. 2016). In the present technology-driven landscape, geospatial technologies like remote sensing, GIS and web mapping technologies play a pivotal role in sustainable land resource management and agricultural land use planning (Reddy et al. 2017a).

In the study, GIS-based Land Resources Information System (LRIS) of India has been developed, and it contains, in addition to the soil information on various scales, point information on the characteristics of soil profiles, diagnostic horizons on the most important soil characteristics and derived products on land use planning. LRIS is able to store, search, process, synthesize and present the spatial soil information to connect them with the nonspatial information and to interpret and develop agricultural land use plans. Further, recent developments of geospatial technology have opened the new vistas to a new breed of spatial applications featuring real-time processing of large and varied data sources, integrated workflows and highly interactive map-based interfaces. With the available geospatial libraries and tools such as web mapping tools, geostatistical modelling software and geoprocessing capabilities can be combined to develop different evaluation and monitoring systems to different agricultural scenarios (Knörchen et al. 2015; Granell et al. 2016).

Implementation of Geoportal involves three distributed GIS components that present the geographic application or portal; web services publish geographic functionality as a web service; and data management software provides a managed relational environment for both raster and vector geographic content (Maguire and Longley 2005). Geoportal can be defined as an Internet or intranet entry point with the tools for retrieving metadata; searching, visualizing, downloading and disseminating of geoinformation; and providing the geoinformation-based services (Tait 2005; Fisher 2006; Reddy et al. 2017a). Based on the databases generated under LRIS, NBSS BHOOMI Geoportal has been designed and developed as web-based portal used to find and access geographic information and associated geographic services in a given domain. It is essentially a web-based system, which contains diversified thematic databases with metadata information about geographic data and service-related land resources. It was built using web infrastructures that include web servers and databases to provide applications and services. Geoportal also provides immense challenges in development of spatial data infrastructure and related services to deliver actionable and understandable information to meet the needs of stakeholders (Chen et al. 2016).

22.2 Land Resources Information System: Global Perspective

Globally, some of the countries have developed their own LRIS. Beaudette and O'Geen (2009) developed a framework to construct a web-based interface to the Natural Resources Conservation Service (NRCS), State Soil Geographic (STATSGO) Database and Soil Survey Geographic Database (SSURGO) to support sophisticated data storage, querying, map composition, data presentation and contextual links to related materials. Australian Soil Resource Information System (ASRIS) provides access to the best available soil and land resource information in a consistent format across the country; however, the level of details depends on the survey coverage in each region. ASRIS contains a set of spatial and temporal databases that maintain national soil and land information in a consistent and usable format (ASRIS 2011). The Canadian Soil Information Service (CanSIS) (http://sis.agr.gc.ca/cansis/) manages and provides access to soil and land resource information of Canada. It maintains the national repository of soil information such as soil data, maps, technical reports and standards and procedures through its National Soil Database (NSDB). Soil and Terrain Database (SOTER) programme was developed by FAO, ISRIC-World Soil Information and the United Nations Environmental Programme (UNEP) to create a soil and terrain digital database (Baumgardner 1986) with a global coverage of SOTER attribute data at a scale of 1:1 million. The SOTER concept is based on the relationship between the physiography, parent materials and soils within an area. The SOTER methodology was developed as a Land Resources Information System for the scale of 1:1 m (Van Engelen and Wen 1995). The SOTER methodology has been applied at a range of scales, from 1:50,000 to 1:5 m, using a similar standard database structure (ISRIC et al. 2003; ISSS et al. 1998).

Within the framework of the Global Earth Observation System of Systems (GEOSS), the e-SOTER (http://www.esoter.net) was developed by European countries to deliver a web-based regional pilot platform with data, methodology and applications, using remote sensing to validate, augment and extend existing data. Detailed digital elevation models (DEMs), advances in remote sensing and new analytical tools were extensively used in *e-SOTER* for landform analysis, parent material detection and soil pattern recognition – both to extend the legacy soil data and to build a framework for new data acquisition. ISRIC-World Inventory of Soil Emission Potential (WISE) is a comprehensive repository of global primary data on soil profiles. Under the WISE, ISRIC has consolidated select attribute data for over 10,250 soil profiles, with about 47,800 horizons, from 149 countries in the world. Profiles were selected from data holdings provided by the Natural Resources Conservation Service (USDA-NRCS), the Food and Agriculture Organization (FAO-SDB) and ISRIC (ISRIC-ISIS) (Batjes 2008). WISE data have been used for a wide range of applications, which includes the development of harmonized sets of derived soil properties of the main soil types of the world, gap filling in the primary SOTER database, global modelling of environmental change, analysis of global ecosystems, upscaling and downscaling of greenhouse gases, crop simulation and agroecological zoning (ISRIC-WISE 2008). The initiative of digital soil mapping (DSM) (www.globalsoilmap.net) by Working Group of the International Union of Soil Science (IUSS) aimed to assist decisions in a range of global issues like food production, climate change and land degradation. Many projects have been initiated aiming towards the compilation of new digital soil maps of the world. Under this initiative, a number of regional and national soil data collection programmes are on, like the Africa Soil Information Service (AFSIS), the European Soil Information System (EUSIS) and others. In recent times, the task of the Global Soil Partnership

(GSP) was initiated to build a partnership among the various soil data collection programmes in order to develop synergies and cost savings by avoiding duplication of efforts and provide a common soil data and information platform responding to the various user needs at global, regional, national and local scales.

Realizing the importance of digital soil resource databases in GIS, NBSS&LUP has developed Soil Information System on 1:1 m scale in GIS under Integrated National Agricultural Resource Information System (INARIS) of the National Agricultural Technology Project (NATP). The soil attribute database on soil mapping units, soil site and physical and chemical properties was compiled and codified, and extended legend was prepared as per the INARIS data structure (Maji et al. 2004). Attempts have been made to develop state-level Soil Information Systems on 1:250,000 scale for states like Arunachal Pradesh (Maji et al. 2001) and Nagaland (Maji et al. 2002c). In order to meet the demands of wider applications, digital soil resource databases of India in the form of Soil Information System have been designed and developed as a GIS-based spatial database management system with a set of files, viz. spatial and nonspatial to store and retrieve for systematic analysis and reporting (Reddy et al. 2014a, 2016).

22.3 Development of Land Resources Information System

Land Resources Information System of India consists of software and hardware systems, which support to arrange land information in logical order to store, analyse and interpret for data-intensive applications. The usefulness of LRIS depends on its ability to provide decision-makers with the right data at the right time. In order to meet the demands of wider applications, digital soil resource databases of India have been designed and developed in the form of LRIS in GIS-based spatial database management system with a set of files, viz. spatial and nonspatial to store and retrieve for systematic analysis and reporting. The LRIS provides a 'GIS-based standardized framework to develop land and allied resources database and computing tools for storing, managing, and analysing the resources to develop practical solutions'. GIS provides technological basis for development of LRIS and being able to combine disparate sources and types of geospatial data and provide the basis for integration of relevant information, such as remote sensing imageries. Typically, LRIS constitute a computerized geo-referenced database repository of land and allied resources, such as satellite imagery, meteorological observations and scanned legacy maps and images.

Advanced GIS techniques have been used to develop the seamless LRIS to integrate with other resource databases. Advance GIS capabilities have been used to efficiently capture, store, update, manipulate, analyse and display all forms of geographically referenced data. It allows mapping, modelling, querying, analysing and displaying large quantities of diverse data, all held together within a single database. The developed geocoded spatial data defines an object and relationships with other objects in two-dimensional space, also known as topological data, and stores in topological database. GIS links these two databases by manipulating a relationship between records of object location in the topological database and records of the object attribute in relational databases. GIS is also a potential tool for handling voluminous remotely sensed data and has the capability to support spatial statistical analysis (Balaselvakumar and Saravanan 2003) using the information generated in LRIS. Thus, LRIS platform provides the ability to search, discover, visualize and update geospatial soil resource database through a customized user interface. The design of GIS-based LRIS includes three major elements:

(a) Conceptual design: basically laying down the application requirements and specifying the end utilization of the database. The conceptual design is independent of hardware and software and could be a wish list of utilization goals.
(b) Logical design: it is basically the specification part of the database. This design sets out the logical structure of the database elements, schemas, protocols and work flows.
(c) Physical design: this pertains to the hardware and software requirements, consideration of file structure, memory and disk space, access and speed, etc.

Each stage is interrelated to the next stage of the design and impacts the organization in a major way. LRIS essentially consists of three main components, viz. the spatial database, containing all spatial soil maps, which shows the geographical extents to create thematic maps, the laboratory and soil profile database which contains the results of soil physical and chemical soil properties.

LRIS consists of spatial and attributes database and set of relationships that define the links among the databases. Geometrical database consists of geographic objects in an entity-oriented architecture and it includes three kinds of data, i.e. Polygons, representing delineations of map units, Lines, representing boundaries between delineations as well as linear features at the map scale and Points, represents point observations (sample sites) and point features. The attributes consist of information about the mapped entities. They are attributes of individual polygons, including their area and shape; attributes of polygon classes (map units), including the soil(s) found in polygons of the class and their spatial arrangement within the polygons; attributes of individual polygon boundary lines, including their length, accuracy and precision (implicit width); attributes of individual polylines, including their length and width; attributes of line classes, including the soil(s) or special feature(s) found along lines of the class; attributes of individual points, including the details of an observation; and attributes of point classes, including the soil(s) or special feature(s) found at points of the class.

The major steps involved in the deployment of LRIS include data organization to show the logical relation of data (e.g. a profile is made up of a sequence of horizons), a map unit that has several components of soil types, data storage to save data for later use, data retrieval to examine the saved data, data manipulation and transformation to derive new data from old (e.g. information about an entire map unit derived from information about its constituents) and data analysis to provide the solutions for the problems like land resource management, land evaluation, land use planning and soil erosion risk management, etc.

22.3.1 Assimilation and Processing of Spatial Databases

The common obstacle encountered in assimilation of spatial datasets is the diversity of data sources and their formats. These problems have been overcome by abstracting all possible data formats into a single generic model. Then, any data can be combined, visualized, overlaid and analysed together. The data catalogue is able to retrieve the data of which the metadata match the search criteria. The catalogue allows the user to find data relevant to application needs. The spatial soil resource database and the soil maps of different states generated at toposheet level (hard copies) on 1:250,000 scale under soil resource mapping (SRM) project carried out by NBSS&LUP have been used as input data in GIS for digitization and generation of state-wise soil information. In order to develop seamless digital soil resource resources of India, the state-wise soil resource polygon layers were brought on common platform in GIS with necessary geometric corrections during the edge matching and transformed to Lambert conformal conic (LCC) projection system with geographic extents in latitudes/longitudes. The topology was built for the seamless polygon layer of soils and other administrative unit maps.

Besides the soil resources data on 1:250,000 scale, soil resources data on 1:1 m scale, soil loss, climatic and soil profile point database, the raster databases on digital elevation models of SRTM DEM (90 m & 30 m) and ASTER DEM (30 m) of India were also used to develop LRIS. GIS-based LRIS provides platform to assimilate geo-referenced spatial and nonspatial database (Burrough and McDonnell 1998) with specific capabilities for developing spatially referenced data as well as perform a set of operations for working with the datasets.

22.3.2 Standardization of Geographic and Attribute Database

Standardization of geographic and attribute database and their structure starts with the data capture/data input techniques for a successful connection of geo-oriented databases and for building geographic information systems. The standardization of spatial data relates to various aspects of data preparation and management involving data formats, data dictionaries, data quality control, map layouts and many others. Although a comprehensive set of data standards are being followed in the generation soil and allied databases, number of standards and norms have adopted for enhancing the capabilities of data exchange and compatibility geographic layers and attribute databases. The uniform standards have been followed in generation of soil physico-chemical properties in the dBase IV environment and linked to the soil polygon layer in GIS. Attributions of various soil parameters at the state level were standardized in a seamless soil polygon layer of India for entry of various thematic databases to master soil layer. The standardized soil resource database contains information about the soil site and soil's morphological, physical and chemical properties.

22.3.3 Development of Schema for Spatial Databases

Schema can be defined as arrangement of tables and the relationships among them. The critical part of a database schema is actually to create a diagram that shows the relationships among the various tables in the database. Relationships have a property called cardinality that describes the type of relationship. The possibilities for relationships are one to one, one to many, many to one and many to many. Additionally, relationships may have the property of being required (mandatory) or optional. The schema has been developed for input parameters to define the data elements, sub-elements, short names, data type, unit of measure and maximum and minimum value of each parameter and brief description to effectively store, update, manage and query the geospatial soil database from LRIS. The standardized schema allows to store, analyse, visualize and query the soil resource and allied datasets across the administrative units and geographic regions in LRIS. Relational Database Management System (RDBMS) has been used to develop the spatial databases, and it provides many advantages to manage and work with large amount of data, such as the support for multiusers (concurrency), data consistency and integrity, persistent storage and query capabilities through a programming interface (Garcia-Molina et al. 2002).

22.3.4 Development of Metadata Standards

Metadata is nothing but 'data about the data', and it enables the users to integrate data from multiple sources and formats. Metadata is the term used to describe the summary information or characteristics of a set of data. Standardization and development of metadata standards are particularly beneficial for data management and integration. Metadata has been specified and assembled based on the existing, international OGC standards. In LRIS, common standards are adopted for soil data, which allow to integrate soil information with allied spatial information systems for wider applications in agricultural and land use planning. Metadata are structured, encoded data that describe characteristics of information-bearing entities to aid in the identification, discovery, assessment and management of the described entities. It explains the database structures of LRIS and its related databases. This includes the structure of each table, data element definitions, lists of codes and relationship information. The developed metadata schema with minimum fields for

LRIS provides a formal structure to identify the soil resource databases of a given state and obtain the details of soil information of that state. Broadly, metadata contains the following principal sections:

- Identification_Information: identifies the dataset and gives its geographic limits
- Data_Quality_Information: explains how the data were collected, sampling designs, analytical techniques, etc.
- Spatial_Data_Organization_Information: explains which spatial model was used to represent the data
- Spatial_Reference_Information: explains the coordinate system used for georeference
- Entity_and_Attribute_Information: explains the attributes (variables) in the database
- Distribution_Information: explains how to obtain the data, including online access, and any restrictions on the use of the dataset
- Metadata_Reference_Information: explains who is responsible for the metadata and which standard it follows

22.4 Design and Development of NBSS BHOOMI Geoportal

Realizing the importance of the availability of the accurate and reliable information on soil resources, ICAR-NBSS&LUP took a major initiative to develop a NBSS BHOOMI Geoportal as a geospatial data gateway for land resource management and promote sustainable agricultural development in the country. Geoportal can be defined as an Internet or intranet entry point with the tools for retrieving metadata, searching, visualizing, downloading, disseminating and in some cases the ordering of geographic information services. The capabilities of geospatial technologies have been extensively used in processing the soil geospatial database, developing protocols and procedures of LRIS and NBSS BHOOMI Geoportal and its applications. In order to meet the goal of development of Geoportal for improving access the digital soil data in the form of a Geoportal, the procedure is structured in four main parts. In the first part, a framework for the provision of spatial soil and soil related data has been defined. In the second part, the development of a schema for describing the spatial soil and related data and services take centre stage. In the third part, the focus was given on harmonization and semantic interoperability of selected soils and associated data sets. In this part, datasets have been systematically harmonized and developed LRIS. In the fourth part, the establishment of an integrated network, Geoportal helps to improve the access to data and metadata. In order to consider the needs of the target users of the Geoportal, user requirements have been analysed and considered in implementation.

NBSS BHOOMI Geoportal framework has the ability to systematically organize, search, discover, access, visualize and even update geospatial data and services via a customized user interface for internal and external users. The fundamental objective of the NBSS BHOOMI Geoportal is to provide a platform for referencing and accessing geospatial information that is distributed and made available using a variety of technologies. It also supports all principal metadata standards and electronic data communication standards. It also has capabilities that integrate data made available in a large variety of formats. GIS server technology has been used in development of Geoportal, and it provides a powerful tools to present and analyse soil information (King et al. 1995) and integrates common database operations, such as query and statistical analysis with unique visualization offered through maps. It is a web environment that provides the information and services related to land resources to the users and share geographic content. Further, Geoportal helps to maintain one master database in order to reduce data handling and data duplication, as well as to describe how desired products and databases should be dynamically developed 'on the fly' from the single database using an automated generalization procedure.

BHOOMI Geoportal organizes content and services, such as directories, search tools, support resources, data and applications. They provide capabilities to query metadata records for relevant data and services and then link directly to the online content services themselves. Data and data services were catalogued systematically according to a metadata standards and schema designated for Geoportal. This data cataloguing and maintenance reflect the type of data that was published using the Geoportal. Though data can be maintained and associated metadata can be created and published on a Geoportal by entities other than the portal's host organization, the metadata could be generated prior to the publication to ensure its completeness and conformity to established standards and schemas.

22.4.1 NBSS BHOOMI Geoportal Architecture

The core aim of NBSS BHOOMI Geoportal is to organize, catalogue, store, search and visualize all land and allied information and provide services. The provided services can be viewed and queried at different administrative levels: from national, regional, state and district levels. Within the Geoportal, different kinds of information shall be bundled, especially different kinds of data, textual documents, metadata and maps. In NBSS BHOOMI Geoportal, various spatial layers were deployed, and Web Mapping Services (WMS) were created using the ArcGIS server platform. Initially, the spatial layers deployed in the Geoportal as Web Map Services were created and saved as .mxd format in ArcMap and then shared as a map package. The map package was accessed through Geoportal and published to the server hosting the Web Mapping Services. HTML and JavaScript have been used to develop the frontend of the BHOOMI Geoportal. In order to enhance the speed of browsing of services, services have been grouped based on their nature and content and published as bundle of layers as a service. To access Geoportal services of BHOOMI, Arcgis

JavaScript API 3.22 was used to build full-featured 2D web applications with capabilities such as editing and support for all existing layer types. To enable the popup, Popup Widget provided by ArcGIS JavaScript API was used. Various user-friendly utility tools, like data query, extract and legend development were developed in the Geoportal. After the necessary details added, the Geoportal was configured to point to the public host. Web services are independent, self-describing, modular applications that can be published, located and dynamically invoked (Weerawarana et al. 2005). They are entirely based on Extensible Markup Language (XML) technologies and are discoverable based on their descriptions and conditions available based on the web service metadata. The web service architecture has a flexibility to allow a new service to be created in the existing web services. WMS are able to produce a map of a specific selection of data, answer queries about the content of the map and provide clients with a list of other derived maps it can produce.

22.4.2 Implementation of NBSS BHOOMI Geoportal

NBSS BHOOMI Geoportal provides the ability to securely search, discover, access, visualize and even update geospatial data via a customized web user interface. The main obstacle encountered for making databases and services widely available is the diversity of their formats. A Geoportal overcomes this problem by abstracting all possible data formats into a single generic model. This abstraction is performed by OGC/ISO standards-based web services, which can expose initially heterogeneous data, and then any data can be combined, visualized, overlaid and analysed together. The catalogue itself is a service that is able to retrieve the data of which the metadata match the search criteria. The catalogue allows the user to find data relevant to application needs. Implementation of Geoportal involves the construction of a database according to the specification of a logical schema. It includes the specification of an appropriate storage schema, security enforcement, external schema and so on. During implementation, the choice of available DBMSs, database tools and operating environment was also considered. There are additional tasks beyond simply creating a database schema and implementing the constraints – data must be entered into the tables, issues relating to the users and user processes need to be addressed and the management activities associated with wider aspects of corporate data management need to be supported. In practice, implementation of the logical schema in a given DBMS requires detailed knowledge of the specific features and facilities that the DBMS has to offer. In an ideal world, the first stage of implementation would involve matching the design requirements with the best available implementing tools and then using those tools for the implementation. In database terms, it involves choosing the products with DBMS and SQL variants suited to the database to be implemented.

22.4.3 Populating Geospatial Database

After a spatial database has created, there are two ways of populating the tables – either from existing data or through the use of the user applications developed for the database. For some tables, the existing data from another database or data files were used. The simplest approach to populate the database is to use the import and export facilities found in MySQL. Facilities to import and export data in various standard formats are usually available. Importing enables a file of data to be copied directly into a table. When data are held in a file format that is not appropriate for using the import function, then it is necessary to prepare an application programme that reads in the old data, transforms them as necessary and then inserts them into the database using SQL code specifically produced for that purpose. The transfer of large quantities of existing data into a database is referred to as a bulk load; it facilitates to transfer large quantities of data.

22.4.4 Publishing of Data Content

The publishing process of data content in NBSS BHOOMI Geoportal involves addition, deletion and modification of metadata content. Depending on the data content, volume and necessity of services, databases have been published as services through a web page interface or automated through a web service interface (metadata harvesting). The administration function is simply an extension of the publishing function with one additional capability: the review/approval of metadata content submitted for publishing on the portal web site. Geoportal requirements have been controlled through administrative privileges granted to certain administrators who can edit and validate the content to publish. Additionally, administrators are responsible for publishing content-level metadata.

22.5 Development of NBSS BHOOMI Geoportal Services

Web frameworks support the development of dynamic web applications by providing libraries, packages and templates into a single connected piece of software, thus alleviating the tasks of dealing with low-level details. The developed NBSS BHOOMI Geoportal framework allows to combine, publish and visualize services as WMS without having to deal with every single detail regarding projection and interactivity functions. BHOOMI Geoportal also offers basic graphical user interface (GUI) objects that can be extended and modified with additional functions and/or plugins. They usually offer an API (application programming interface) to allow an easier use and access to the libraries. The services and standards of Geoportal allow devices to communicate through services using a common protocol set by standards and therefore enabling the core aspects of the interoperability concept between devices. Web services of Geoportal are applications that offer functionality and that are accessible by other applications over the Web and through Internet protocol. They support direct and remote interactions from machine to machine, using standard-based interface (Alonso et al. 2004). Thus, they allow delivering remotely available functionality from a service provider to a service consumer. Web services use the concept of request-response mechanism: a client or service consumer sends a request to the server, and in return the server sends a response, containing the results asked for, to the client. In Geoportal, web services used for a wide range of tasks including the display and retrieval of data, processing operations and search functions. The concept of web framework and of the available options for web mapping and Geoportal allows the presentation of the geospatial data to the users in a simple and coherent manner. The implemented WMS in the NBSS BHOOMI provides the client with spatially referenced 2D maps in the form of an image, such as JPG and PNG, or SVG dynamically from geographic information that can be displayed in a browser. Figure 22.1 depicts the soil series point data as WMS of NBSS BHOOMI Geoportal through user interface.

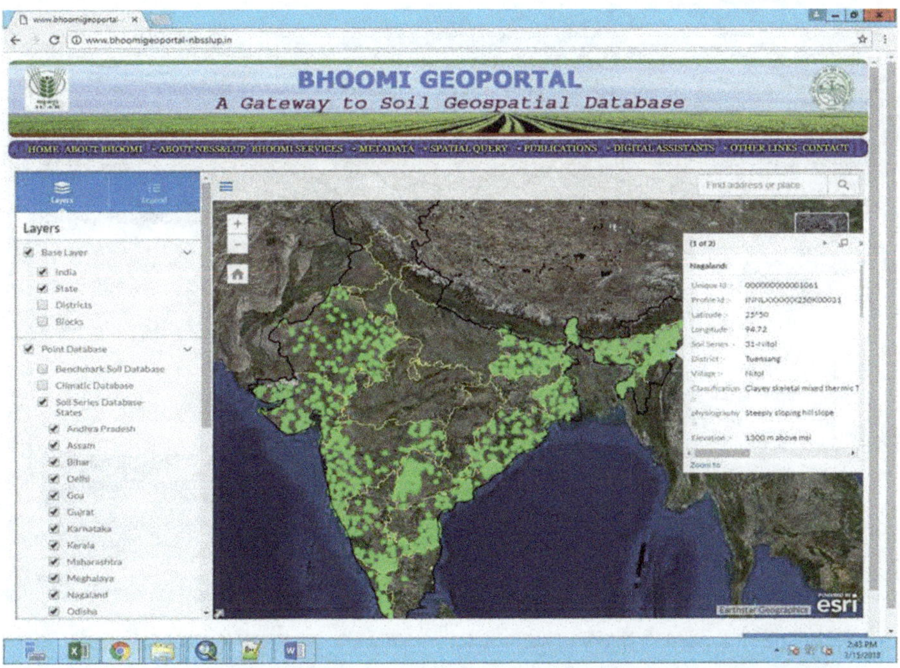

Fig. 22.1 Soil series point data service as WMS through NBSS BHOOMI Geoportal

22.5.1 Spatial Data Query

Spatial data queries of soil information can be divided as spatial, nonspatial and mixed. The languages like SQL and QBE could be used to query the structured data. In NBSS BHOOMI Geoportal, search functions are aggregations of building block tools, which are executed in sequential steps. The first step in many applications is to locate a place through one of several methods including a place name search using a gazetteer tool, an address search using a geocode tool or simply selecting a location from a list. For example, using a search tool, user can enter a place name, execute a search for that place and return a list of candidate locations, allowing a selection to be made. Once a place is identified, Geoportal applications then execute a second step in the search process; they search for a particular set of features or objects that are usually the focus of the Geoportal. This search could be for a particular location based on a neighbourhood name or for geographic web services with coverage of a particular location, and it allows both geographic attribute and content metadata for searching. The attribute data query of soils of 1:1 m service of NBSS BHOOMI Geoportal is shown in Fig. 22.2.

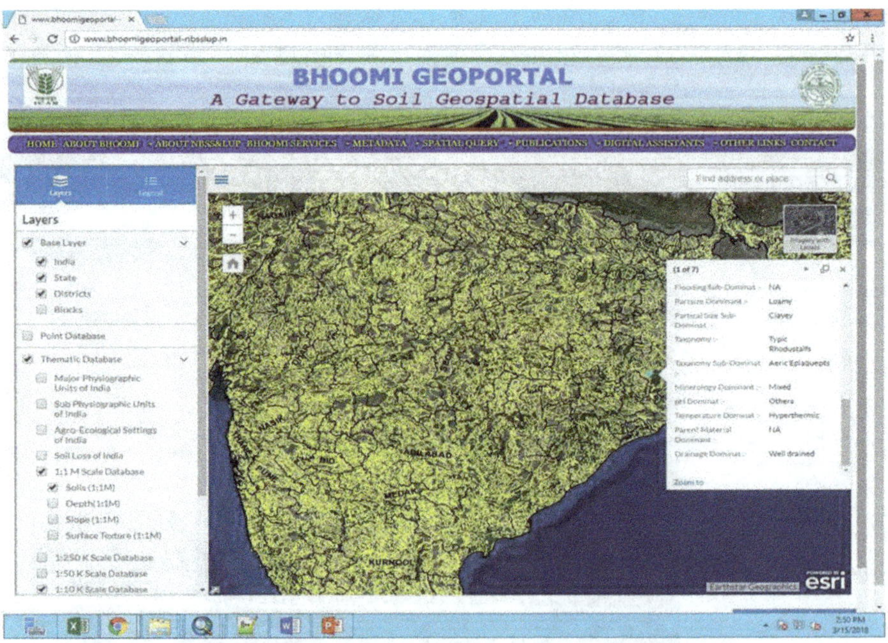

Fig. 22.2 The attribute data query of soils of 1:1 m service of NBSS BHOOMI Geoportal

22.5.2 Geographic Data Visualization

Visualization interfaces in Geoportals can be a very powerful tool for spatial data, which can aid the users with information visualization and data exploration. A thematic service is an example of information visualization, a specific type of map and attribute database that is designed to communicate information about a single topic or theme (Wade and Sommer 2006; Tyner 2010). BHOOMI Geoportal allows the user to fully visualize, explore and examine the published services and their content. Additional functions like pan, zoom and feature that identify capabilities aid the user in more thoroughly evaluating the published content. It also supports the ability to view multiple map services in a fused or single map service. The developed interface tools, Geoportal functionalities and deployed WMS in Geoportal provide the platform for visualization of services as individual or combined. The Catalogue Service component provides the underlying database, and it enables users to discover metadata records. BHOOMI Geoportal provides an easy-to-use interface for users to access and visualize resources to obtain the information for use in their own applications. Visualization of various land degradation classes through land degradation thematic service of NBSS BHOOMI Geoportal is shown in Fig. 22.3.

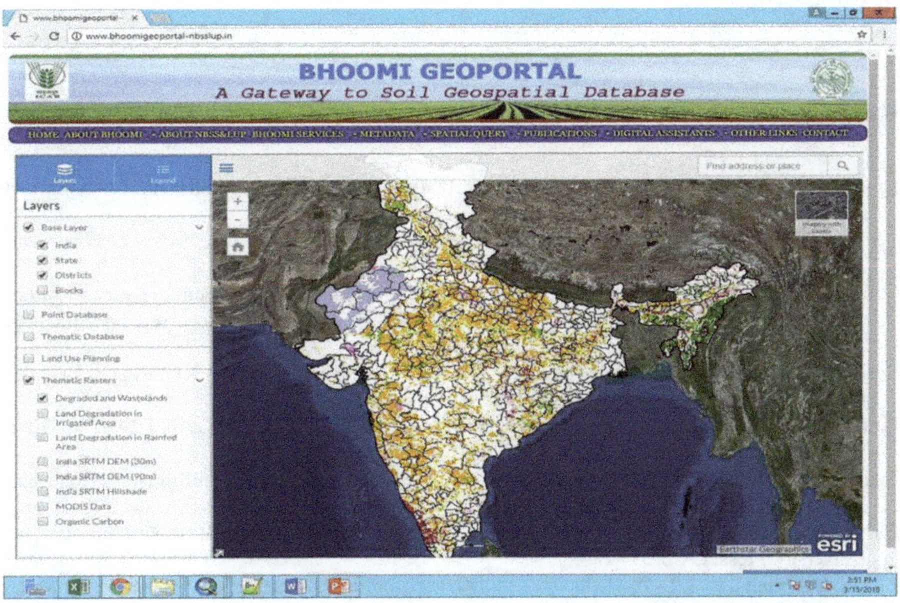

Fig. 22.3 Visualization of land degradation thematic service in NBSS BHOOMI Geoportal

22.6 Conclusions

The developed GIS-based LRIS of India enables to catalogue, store, manage, process and analyse large land resource databases in a digital environment. In this framework, identification of spatial land and allied databases and development of a schema for describing the databases are the crucial. NBSS BHOOMI Geoportals offer various capabilities for providing wide range of services, ranging from thematic service to attribute information that are usually required for decision-making purposes at various level. BHOOMI Geoportals also provide a platform to leverage geospatial technology, avoid the development of duplicate datasets, reduce data management costs and present cost-effective way of access to information and services. The designed and implemented overall service-oriented architecture of NBSS BHOOMI Geoportal and the ecosystem of tools help to deliver WMS to the planners, decision-makers and land users at different levels for various land resource management and land use planning applications.

Acknowledgements Part of the research work was carried out under institute-funded project. The authors are grateful to the past directors, heads of the regional centres, heads of divisions, scientists and technical officers of ICAR-NBSS&LUP, who have directly and indirectly contributed in the project. The contributions of Mr. Abhishek S. Chikore, SRF and Mr. Raghav A. Deoghare, YP-II are duly acknowledged.

References

Alonso G, Casati F, Kuno H, Machiraju V (2004) Web services: concepts, architectures and applications. Springer Berlin-Heidelberg, Berlin

ASRIS (2011) ASRIS-Australian soil resource information system. http://www.asris.csiro.au. Accessed on 19th Sept 2017

Balaselvakumar S, Saravanan S (2003) Remote sensing techniques for agriculture survey. http://www.gisdevelopment.net/application/agric/overview/agric00114.htm. Accessed on 25th Sept 2017

Batjes NH (2008) ISRIC-WISE harmonized global soil profile data set (Ver 3.1). Report 2008/02, ISRIC-world soil information, Wageningen (with data set)

Baumgardner MF (1986) World soils and terrain digital database at a scale of 1:1M (SOTER). Wageningen, ISSS

Beaudette D, O'Geen AT (2009) Soil-web: an online soil survey for California, Arizona and Nevada. Comput Geosci 35:2119–2128

Burrough PA, McDonnell RA (1998) Principles of geographical information systems. Oxford University Press, Oxford

Chen J, Dowman I, Li S, Li Z, Madden M, Mills J, Paparoditis N, Rottensteiner F, Sester M, Toth C, Triender J, Heipke C (2016) Information from imagery: ISPRS scientific vision and research agenda. ISPRS J Photogramm Remote Sens 115:3–21

Fisher F (2006) Report about recent trends in the GIS sector: INTERGEO 2006. GEOInformatics 9(7):6–8

Garcia-Molina H, Ullman JD, Widom J (2002) Database systems: the complete book (International edn.). Prenctice Hall, Upper Saddle River

Geller T (2016) Farm automation gets smarter. Commun ACM 59:18–19

Granell C, Havlik D, Schade S, Sabeur Z, Delaney C, Pielorz J, Usländer T, Mazzetti P, Schleidt K, Kobernus M, Havlik F, Bodsberg NR, Berre A, Mon JL (2016) Future internet technologies for environmental applications. Environ Model Softw 78:1–15

http://sis.agr.gc.ca/cansis/. Canadian soil information service. Accessed on 25 Oct 2017

http://www.esoter.net. Accessed on 21 Oct 2017

ISRIC, FAO and UNEP (2003) Soil and terrain database for Southern Africa (1:2 million scale), FAO land and water digital media series no. 25. FAO, Rome

ISRIC-WISE (2008) References and examples of WISE database applications. PDF ISRIC 2007. ISRIC world soil database

ISSS., ISRIC., & FAO (1998) World reference base for soil resources. Food and Agriculture Organization of the United Nations, Rome

King D, Burrill A, Daroussin J, Le Bas C, Tavernier R, Van Ranst E (1995) The EU soil geographic database. In: Agriculture; European land information systems for agro-environmental monitoring. European Commission Joint Research Centre, Ispra, pp 43–60

Knörchen A, Ketzler G, Schneider C (2015) Implementation of a near-real time cross-border web-mapping platform on airborne particulate matter (PM) concentration with open-source software. Comput Geosci 74:13–26

Maguire DJ, Longley PA (2005) The emergence of geoportals and their role in spatial data infrastructures. Comput Environ Urban Syst 29:3

Maji AK, Nayak DC, Krishna NDR, Srinivas CV, Kamble K, Reddy GPO, Velayutham M (2001) Soil information system of Arunachal Pradesh in a GIS environment for land use planning. Int J Appl Earth Obs Geoinf 3:69–77

Maji AK, Srinivas CV, Thamgadge DB, Reddy GPO, Velayutham M (2002a) Soil information of baster district in GIS for land use planning. In: Patil VC et al (eds) Proceedings of the first national conference on agro-informatics. NCAI, Dharwad, pp 30–33

Maji AK, Tamgadge DB, Reddy GPO, Gajbhiye KS (2002b). GIS based soil information system of Bilaspur district for land use planning, In: Navalgund RR et. al (eds) Proceedings of the ISPRS commission VII symposium on resources and environmental monitoring, Dec, 3–6, NRSA, Hyderabad, pp 746–749

Maji AK, Srinivas CV, Dubey PN, Reddy GPO, Kamble KH, Velayutham M (2002c) Soil resources information of Nagaland in GIS for land use planning in mountainous region. GIS India 11(4):13–16

Maji, AK, Reddy GPO, CCPI- Regional Centres (2004) NATP-(MM) project on integrated national agricultural resource information system (INARIS sub-project- soil resource database. Annu Rep 2004: 47–51

Reddy GPO (2012) Conceptual design of soil information system – a geoportal. In: Reddy GPO, Sarkar D (eds) Remote sensing and GIS in digital terrain analysis and soil-landscape modeling, NBSS & LUP Publication no. 152, pp 217–226

Reddy GPO (2016) Soil geoportal developmental and its applications. In: Singh SK, Bhatta RK, Chatterji S (eds) Land use planning for arresting land degradation, combating climate change and ensuring food security - a training manual, NBSS & LUP Publication no.171, pp 50–56

Reddy GPO, Mondal C, Srivastava R, Bhattacharyya T, Naidu LGK, Sidhu GS, Baruah U, Singh RS, Kumar N, Singh SK, Sarkar D (2014a) Development of Indian soil information system – a geoportal, NBSS&LUP, Nagpur, Project report, pp, 34

Reddy GPO, Singh SK, Mondal C, Srivastava R, Bhattacharyya T, Naidu LGK, Sidhu GS, Baruah U, Singh RS, Kumar N, Sarkar D (2014b) Development of district soil information system (DSIS) on 1:50,000 scale (50 districts). NBSS&LUP, Nagpur, Project Report, p 160

Reddy GPO, Sarkar D, Mandal C, Srivastava R, Bhattacharyya T, Naidu LGK, Sidhu GS, Baruah U, Singh SK, Singh RS, Nair KM, Sen TK, Chandran P, Sahoo AK, Srinivas S, Kumar N, Sapana C (2016) Digital soil resource database and information system. In: Dwivedi RS, Roy PS (eds) Geospatial technology for integrated natural resources management. Yes Dee Publishing Pvt Ltd, Chennai, pp 321–351

Reddy GPO, Singh SK, Patil NG, Chaturvedi A (2017a) Land resource inventory, mapping and management - an Indian perspective. In: Reddy GPO, Patil NG, Chaturvedi A (eds) Sustainable management of land resources-an Indian perspective. Apple Academic Press, Waretown, pp 1–30

Reddy GPO, Singh SK, Kumar N, Mondal C, Srivastava R, Maji AK, Sarkar D (2017b) Development of soil information system and prototype geoportal for resource management. In: Reddy GPO, Patil NG, Chaturvedi A (eds) Sustainable management of land resources - an Indian perspective. Apple Academic Press, Waretown, pp 411–442

Tait M (2005) Implementing geoportals: applications of distributed GIS. Comput Environ Urban Syst 29:33–47

Tyner JA (2010) Principles of map design. The Guilford Press, New York

Van Engelen VWP, Wen TT (eds) (1995) Global and National Soils and terrain digital databases (SOTER). Procedures manual (revised edition). ISSSUNEP- FAO-ISRIC, Wageningen

Wade T, Sommer S (2006) A to Z GIS. ESRI Press, Redlands

Weerawarana S, Curbera F, Leymann F, Storey T, Ferguson DF (2005) Web services platform architecture. Prentice Hall PTR, Upper Saddle River

www.globalsoilmap.net. Accessed on 25 Oct 2017

Part IV
Geospatial Technologies in Land Resources Management

Chapter 23
Land Use Planning: Basics and Approaches

T. K. Sen and S. Chatterji

Abstract Land use planning (LUP) is systematic assessment of land and water potential, alternatives for land use and economic and social conditions in order to select and adopt the best land use options. It is an iterative process. Biophysical and socio-economic data collected through standard surveys can be interpreted and evaluated for several purposes in land use planning like suitability for agriculture through technical classification of soils, hydrological groupings, etc. Land use planning can be applied at three broad levels: national, district and local. These are not necessarily sequential but correspond to the levels of government at which decisions about land use are taken. Land evaluation is an integral part of land use planning. A general evaluation, based on limitations of land characteristics, is land capability classification. The capability units represent similar proposals of use and management. The paper also discusses other major methods of land evaluation including mathematical method of land evaluation, namely, fuzzy modeling. Watershed-scale planning is gaining popularity among communities and agencies so that biological, physical and socio-economic components of the landscape system can be integrated into the land use planning framework.

Keywords Land capability classification · Land irrigability classification · Land use planning · Land evaluation

23.1 Introduction

The human population of India is estimated to increase to 1.6 billion in 2050 from the current level of 1.2 billion, and the food grain production would need to be raised to 581 Mt. from the present level of 241 Mt. (Ghosh 2013). This additional food requirement will have to be produced from the projected cultivable land of 142 Mha

T. K. Sen (✉) · S. Chatterji
ICAR-National Bureau of Soil Survey & Land Use Planning, Nagpur, India
e-mail: tksen2005@rediffmail.co

G. P. O. Reddy, S. K. Singh (eds.), *Geospatial Technologies in Land Resources Mapping, Monitoring and Management*, Geotechnologies and the Environment 21,
https://doi.org/10.1007/978-3-319-78711-4_23

(Kathpalia and Kapur 2011). However, the fact that per capita landholding will go down to an abysmally low 0.087 ha poses a grave concern. Also the price of land is likely to be almost 10 times higher than the present value in the land market. In such a scenario, the resource-poor farmers will no longer be interested to cultivate their lands and would much rather sell their land to corporate houses. Moreover, with the improvement in literacy, the younger generations will prefer to shift to manufacturing and service sector which provides much better employment opportunities and improved standards of living.

Another issue affecting the productivity is the rampant encroachment of prime agricultural lands from competing sectors. There will be fierce competition from other users like housing, industrial and health sectors, as well as other economic activities cardinal to the country's development. When prime lands are no longer available, there would be a corresponding pressure to destroy forests and fragile lands in an attempt to convert them to agricultural lands. All efforts need to be made to prevent the diversion of lands from agricultural to nonagricultural uses through negotiations and legislations. Before undertaking these efforts, we need to collect information on the extent and distribution of prime agricultural land available in the country, document it and reserve it solely for agriculture. There is, sadly enough, no information available at present on how much of prime agricultural land has been taken away for nonagricultural purposes, especially during the last two decades of economic growth.

This problem becomes even more severe in a market-driven, unplanned diversification, as well as urbanization of that land to non-sustainable development. Demand-driven or market-driven land use changes can severely impact natural resources of the country which may not be conspicuous immediately, but would cause long-term damage. Hence, judicious land use planning becomes imperative. Biophysical characteristics largely determine the land use patterns. Hence, we need soil and land resources information base for each parcel of land.

23.2 Land Use Planning

Land use planning is the systematic assessment of land and water potential, alternatives for land use and economic and social conditions in order to select and adopt the best land use options. Its purpose is to select and put into practice those land uses that will best meet the needs of the people while safeguarding resources for the future. The driving force in planning is the need for change, the need for improved management or the need for a quite different pattern of land use dictated by changing circumstances. LUP is an iterative process. Biophysical and socio-economic data collected through standard surveys can be interpreted and evaluated for several purposes like suitability for agriculture through technical classification of soils, hydrological groupings, suitability for sewage disposal, traffic ability, building construction, etc. Soil survey interpretation and land evaluation precede land use planning. The principles of LUP are by and large similar at all the levels be it regional or watershed/village level.

Land use planning is a decision-making process that facilitates the allocation of land to different uses that provide optimal and sustainable benefit. As land use is shaped by society-nature interaction, in land use planning, different components/ facets play a significant role in land use planning involving soil, water, climate, animal (ruminant/non-ruminant) and others, including forestry and the environment needed for survival of mankind. At times these components are moderated by human interference. Thus land use planning being a dynamic phenomenon is not guided by a single factor, but by a complex system working simultaneously, which largely affects the sustainability.

Goals Goals define what is meant by the "best" use of the land. They should be specified at the outset of a particular planning project. Goals may be grouped under the three headings of efficiency, equity and acceptability and sustainability.

Efficiency Land use must be economically viable, so one goal of development planning is to make efficient and productive use of the land. For any particular land use, certain areas are better suited than others. Efficiency is achieved by matching different land uses with the areas that will yield the greatest benefits at the least cost.

Equity and Acceptability Land use must also be socially acceptable. Goals include food security, employment and security of income in rural areas. Land improvements and redistribution of land may be undertaken to reduce inequality or, alternatively, to attack absolute poverty.

Approach Every land use planning project is different. Objectives and local circumstances are extremely varied, so each plan will require a different treatment. A sequence of ten steps has been found useful as a guide (FAO 1993). Each step represents a specific activity or set of activities, and their outputs provide information for subsequent steps.

Step 1. Establish goals and terms of reference. Ascertain the present situation; find out the needs of the people and of the government; decide on the land area to be covered; agree on the broad goals and specific objectives of the plan; settle the terms of reference for the plan.

Step 2. Organize the work. Decide what needs to be done; identify the activities needed and select the planning team; draw up a schedule of activities and outputs; ensure that everyone who may be affected by the plan, or will contribute to it, is consulted.

Step 3. Analyse the problems. Study the existing land use situation, including in the field; talk to the land users and find out their needs and views; identify the problems and analyse their causes; identify constraints to change.

Step 4. Identify opportunities for charge. Identify and draft a design for a range of land use types that might achieve the goals of the plan; present these options for public discussion.

Step 5. Evaluate land suitability. For each promising land use type, establish the land requirements, and match these with the properties of the land to establish physical land suitability.

Step 6. Appraise the alternatives: environmental, economic and social analysis. For each physically suitable combination of land use and land, assess the environmental, economic and social impacts, for the land users and for the community as a whole. List the consequences, favourable and unfavourable, of alternative courses of action.

Step 7. Choose the best option. Hold public and executive discussions of the viable options and their consequences. Based on these discussions and the above appraisal, decide which changes in land use should be made or worked towards.

Step 8. Prepare the land use plan. Make allocations or recommendations of the selected land uses for the chosen areas of land; make plans for appropriate land management; plan how the selected improvements are to be brought about and how the plan is to be put into practice; draw up policy guidelines, prepare a budget and draft any necessary legislation; involve decision-makers, sectoral agencies and land users.

*Step 9. Implement the plan***.** Either directly within the planning process or, more likely, as a separate development project, put the plan into action; the planning team should work in conjunction with the implementing agencies.

Step 10. Monitor and revise the plan. Monitor the progress of the plan towards its goals; modify or revise the plan in the light of experience. Steps 1 to 6 are, in fact, processes of *land evaluation***.** Figure 23.1 shows the steps involved in the iterative process of land use planning.

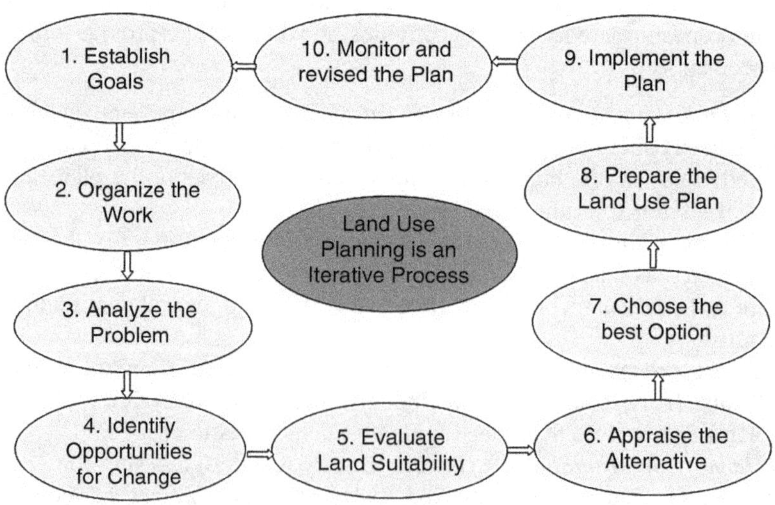

Fig. 23.1 Steps involved in the iterative process of land use planning

The objectives of land use planning can be summarized as follows:

- To quantify the agro-ecology factors that define the categories of land use and optimal cropping patterns
- To integrate the economic/market relationships of the input-output matrix that governs the acreage and production relations under the particular crop regime identified for the category of land use identified
- As a tool for benchmarking farmers' managerial index and their capacity to absorb capital and technology and charting out ways and means of enhancing it
- To identify the institutional infrastructure needs for promoting "brand equity" and usher in the commercial/industrial status to agriculture
- To help in reorienting the departmental perspective from present top-to-bottom departmental (compartmentalized) approach to a bottom-up management process with the participation of all stakeholders in agricultural development followed by ten steps

23.3 Planning at Different Levels

Land use planning can be applied at three broad levels: national, district and local. These are not necessarily sequential but correspond to the levels of government at which decisions about land use are taken. Different kinds of decision are taken at each level, where the methods of planning and kinds of plan also differ. However, at each level there is need for a land use strategy, policies that indicate planning priorities, projects that tackle these priorities and operational planning to get the work done. At village/watershed/farm level, the detailed information on physical factors at soil series phase level, existing climate and present land use coupled with socio-economic information help to plan at farm family level. At district level climate, present land use information and information on benchmark soils help to have isogrow zones for sectoral allocation of resources at district level. At state/regional level agro-ecozone, land systems approaches provide information for development and fund allocation to priority areas for planning at state/regional level. The greater the interaction between the three levels of planning, the better. The flow of information should be in both directions (in all the steps). At each successive level of planning, the degree of details needed increases, and so does the direct participation of the local people.

23.4 Perspective Land Use Planning

Conceptually, perspective land use planning involves altering the land area allocations over alternative uses through suitable technological and institutional devices such that (1) the supplies of the various commodities and service which follow from the stipulated land use pattern broadly conform to the projected demand for such item, (2) the uses are sustainable in the sense that the current uses by one group do not jeopardize the uses of another group or those of the future generation and (3) the biomass production, i.e. the streams of output which follow from the stipulated land use pattern (and even income and employment following from the streams of output), is maximized (Sarkar et al. 2009).

23.4.1 Farmers' Perspectives

For small farmers LUP is an effective utilization of land and water resources for crop production in order to minimize crop failures and risks and to sustain basic family needs. Quite differently, large farmers perceive LUP as an effective utilization of resources to maximize profit from the whole farm, based on the principle of comparative advantage. These perceptions introduce the farming systems dimension into LUP.

Thus, LUP means utilization of resources for sustainable production for professionals, whereas farmers expect family sufficiency and profit with or without sustained use of natural resources. Nevertheless, LUP aims to make the best use of limited resources by (1) assessing present and future needs and systematically evaluating the land's ability to supply them; (2) resolving conflicts between competing uses, between the present and future needs of individuals for production and between the needs of individuals and/or society; (3) identifying sustainable alternative uses and choosing those that best meet the needs; (4) planning to bring about needed changes; and (5) learning from mistakes (Sarkar et al. 2009).

23.5 Limitations of Current Land Use Planning Procedure

The main limitations of current land use planning procedure are as follows:

- Applying top-down approach.
- Socio-economic conditions are not considered.
- Participatory approach is not adopted to involve all stakeholders.
- Technology element is not considered.
- Relations in process of land use are not quantified.
- Multiple scales are not integrated in land use plans.
- Multidisciplinary work needs to be encouraged.
- Feasibility of a plan is low.

23.6 An Improved Approach to Integrated Land Use Planning: The Need

Conventional land use planning has frequently failed to produce a substantial improvement in land management or to satisfy the priority objectives of the land users. As a result, rural development programmes have had mixed success in meeting production and conservation aims. In calling for an integrated approach to the planning and management of land resources, the following specific needs are identified:

- Development of policies which will result in the best use and sustainable management of land
- Improvement and strengthening of planning management, monitoring and evaluation systems
- Strengthening of institutions and coordinating mechanisms
- Creation of mechanisms to facilitate the active involvement and participation of communities and people at local level

An improved approach is necessary to meet these needs. It must take into account the problems and must ensure:

- That programmes are holistic and comprehensive, so that all factors which are significant in relation to land resources development and environmental conservation are addressed and included. The planning process must include and consider all competing needs for land and also that in selecting the "best" use for a given area of land in terms of needs and objectives, all possible land use options must be considered, not only agricultural crops.
- That all activities and inputs are integrated and coordinated with each other. There must be built-in mechanisms to combine the efforts and inputs of all disciplines and groups.
- That all actions and programmes are based on partnership between governments and other institutions and people or stakeholders. This requires that the people should be fully informed and consulted and that mechanisms are established to ensure this.
- That the institutional structures need to be developed to carry out agreed proposals at all levels.

23.7 The Planning Method

Land use planning is based on the premise that land resources vary and that the particular properties and characteristics of any land area set the limits of possible land use options. A set of systematic technical procedures is needed to evaluate the

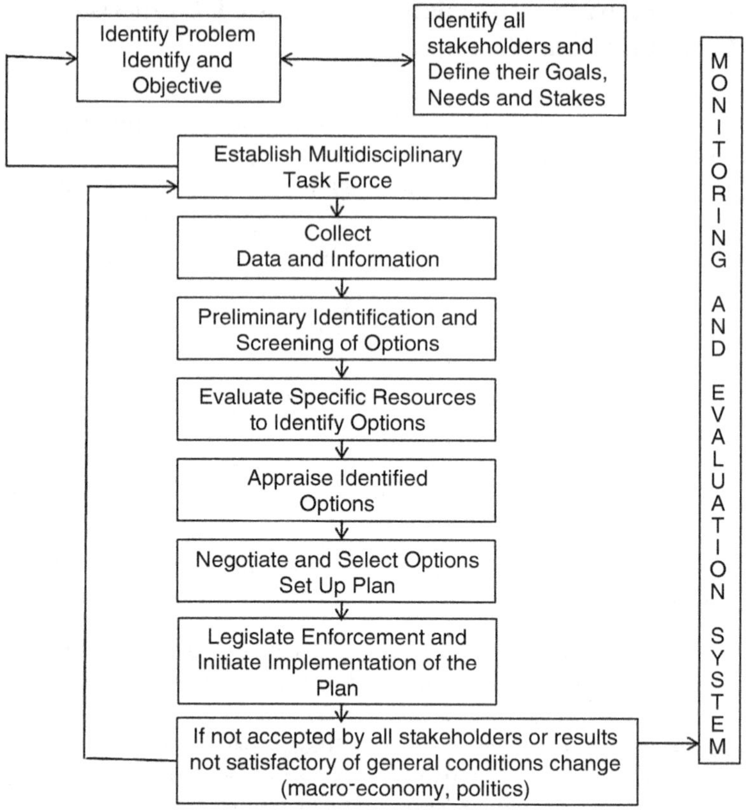

Fig. 23.2 The planning method. (Source: FAO-UNEP 1999)

resources and to guide the choice of those options which are sustainable and which satisfy the objectives of the land users. Markets, infrastructure and other external factors as well as personal preferences are also considered in reaching a final decision. The schematic planning method is furnished in Fig. 23.2.

23.8 Land Evaluation

Land evaluation is the assessment of land performance when used for specified purposes. As such it provides a rational basis for taking land use decisions based on analysis of relations between land use and land, giving estimates of required inputs and projected outputs. Land evaluation is an integral part of land use planning.

23.9 Major Land Evaluation Approaches

A general evaluation, based on limitations of land characteristics, is best illustrated in the United States Department of Agriculture (USDA) Capability Classification. The system of Riquier et al. (1970) is an example of a parametric approach for general evaluation, with however specific reference to arable land, pasture and forest and tree crops. The suitability for irrigation can be achieved through different methods. The system elaborated by the United States Bureau of Reclamation (USBR) and adapted to FAO standards illustrates a methodology based on limitations of land characteristics. The principle of the FAO classification, as presented in the framework for land evaluation, is important and commonly used across the world. A methodology has been suggested by Sys et al. (1991) to apply FAO classification system for evaluation of specific land utilization type. Although many systems have been employed to evaluate the quality of agricultural soils from time to time, the following are some of the important ones.

23.9.1 USDA Land Capability Classification

This method was established by the Soil Conservation Service of the USA according to the system proposed by Klingebiel and Montgomery (1966) and has been widely used throughout the world with numerous adaptations. It is a categorical system that uses qualitative criteria. The inclusion of a soil within a class is made in the inverse manner, that is, without directly analysing its capacity, but rather its degree of limitation with respect to a parameter according to a concrete use. Some factors that restrict soil use can be used to define the productive capacity (intrinsic, soil depth, texture, structure, permeability, rockiness, salinity, soil management; extrinsic, temperature and rainfall) and yield loss (slope of the terrain and degree of erosion). Five systems of permanent agricultural exploitation are considered: permanent soil cultivation, occasional soil cultivation, pasture, woods and natural reserves. This system seeks maximum production with minimum losses in potential. Three levels of classification were established: classes, subclasses and units. Also, eight classes with increasing limitations in use are defined from I to VIII. As a function of the permitted uses, four use groups can be distinguished: permanent soil cultivation (or any type of exploitation; Class I, suitable soils; Class II, good soils but with some limitations; Class III, soils acceptable but with severe limitations); occasional soil cultivation (pastures, woods or natural reserves; Class IV, not recommended for agricultural use for severe limitations and/or required careful management); no soil cultivation, only pastures (in forests or natural reserves; Classes V, VI and VII); and natural reserves (Class VIII).

Depending on the type of limitation, various subclasses of capacity are established: e, for erosion risks; w, for wetness and drainage; s, for rooting and tillage limitations resulting from shallowness, drought risk, stoniness or salinity; and c, for climatic limitations. The capability units represent similar proposals of use and management. This system has many advantages. The classes are defined with criteria that are very general, simple and easy to understand as well as adapt to very diverse regions, but it proves difficult to apply with objective criteria. All the evaluation characteristics that make up the agricultural capacity have identical weight. The same class, with only one parameter (the maximum limiting factor) that classifies the soil within a certain class, embraces highly different soils. This system provides a highly general classification of soil capacity, since it dispenses with many soil characteristics of undeniable interest, but has the advantage of not requiring a detailed knowledge of the soil. Its use proves quite subjective, though it adapts well to the experience of the evaluator. Its results materialize very well on a map, avoiding the erroneous evaluations that parametric methods can produce. The definitions of the different classes are given in Tables 23.1 and 23.2.

Table 23.1 Principles for the definitions of the arable land classes

Parameters	Class I	Class II	Class III	Class IV
Definition	Few limitations restrict their use	Moderate limitations	Severe limitations	Very severe limitations
Range of crops	All crops give optimal yields	Most crops give nearly optimal	Limited crops don't yield satisfactorily	Yield marginal
Slope erosion (e)	Level no or low erosion	Gentle slope, moderate Moderate susceptibility to wind or water erosion	Moderate steep slope. Wind and water erosion	Steep slope Very high wind and water erosion
Wetness (w) Flooding Drainage	Not subject to damaging overflow Well drained	Occasional overflow Moderate permeability limitation	Frequent overflow Water logging, very slow permeability	Frequent overflow Excessive water logging
Physical soil condition(s)	Hold water well, good workability Deep (+100 cm)	Unfavourable workability, less ideal depth (50–100 cm)	Low moisture holding, shallow depth (25–50 cm)	Low moisture <25 cm
Fertility	Well supplied with plant nutrients	Responsive to fertilizers	Low fertility	Low fertility
Salinity and alkalinity	No or slight	Slight to moderate, easy to correct	Moderate salinity sodium hazard	Severe salinity, sodium hazard
Management requirement	Ordinary	Careful management	Very careful	Very careful

Table 23.2 Principles for the definitions of the nonarable land classes

	Pastures		Forest	Recreation and wildlife
Parameters	Class V	Class VI	Class VII	Class VIII
Definition	Not suited to cultivation	Severe limitations	Very severe limitations	Unsuitable for any crop
Range of crops	Pastures	Pasture or range	Woodland	Recreation and wildlife
Slope and erosion (e)	Nearly level no erosion	Very steep severe erosion	Very steep severe erosion	Erosion hazard
Wetness (w) flooding	Frequent overflow	–	–	–
Drainage	Drainage feasible	–	Too wet soils	Too wet soils
Soils (s) conditions	Stony or rocky	Stoniness low moisture capability. Too shallow	Stoniness. Too shallow	Low moisture strong capacity stoniness. Too shallow
Fertility	–	–	–	–
Salinity and alkalinity	–	Severe salinity and sodium hazard	–	–
Management requirement	Pasture	Pasture	–	–

23.9.2 Productivity Index of Riquier et al. (1970)

The basic concept of this method is that agricultural-soil productivity, under optimal management conditions, depends on the intrinsic characteristics. This is a multiplicative parametric method to evaluate soil productivity, from a scheme similar to the Storie index. The concept of productivity is defined as the capacity to produce a certain quantity of harvest per hectare per year, expressed as a percentage of optimal productivity, which would provide a suitable soil in its first year of cultivation. The introduction of improvement practices leads to a potential productivity or potentiality. The quotient between the productivity and the potentiality is called the improvement coefficient.

The evaluation is made for three general types of use: agricultural crops, cultivation of shallow-rooted plants (pastures) and deep-rooted plants (fruit trees and forestation). The determining factors of soil depth are wetness, drainage, effective depth, texture/structure, base saturation of the adsorbent complex, soluble-salt concentration, organic matter, cation-exchange capacity/nature of the clay and mineral reserves. The parameters of the soil surface (e.g. slope, erosion, flood tendency or climate) are not considered.

The different parameters are evaluated as occurs in the Storie index, the evaluation factors represent different weights. Productivity is expressed as the product of all these factors expressed in percentages. Five productivity classes are defined: class P1 = excellent; class P2 = good, valid for all types of agricultural crops; class

P3 = medium, for marginal agricultural use, suitable for non-fruiting trees; class P4 = poor, for pasture or forestation or recreation; and class P5 = very poor or null, soils not adequate for any type of exploitation.

The improvement coefficient is the ratio between the productivity and the potentiality and represents a good index for evaluating the feasibility of these possible improvements. This is a quantitative method, precise (although the partial scoring of the different parameters is quite arbitrary), objective (the only objective is the mathematical method), simple and easy to calculate. The evaluations reflect the degree of suitability of the different evaluation parameters, so that it proves easy to determine the possible improvements for each soil. The evaluation parameters as well as the resulting assessment can be adapted to local conditions.

23.9.3 Soil Fertility Capability Classification (FCC)

This was proposed by Buol et al. (1975) and modified by Sánchez et al. (1982) to evaluate soil fertility. In this system, three levels or categories were established. The first, the type, was determined by the texture of the arable layer, or of the first 20 cm, if this is thinner. Its denomination and range are S, sandy (sandy and sandy loam); L, loams <35% clay (excluding sandy and sandy loam); C, clayey >35% clay; and O, organic >30% organic matter to 50 cm or more. The type of substrate is the second level and is used when there is a significant textural change in the first 50 cm of the soil. It is expressed with the same letters, adding ÒRÓ when a rock or a hard layer is found within this depth. The third level is comprised of the modifiers, which are the chemical and physical parameters that negatively influence soil fertility. These are numerous and are represented by lower-case letters. In the denomination of the soil class, the principle limitations for use are directly represented. For example, for an Orthic Solonchaks, the FCC class that represents it is LCds, which signifies that it is a soil susceptible to severe erosion (L), limited drainage (C), dry soil moisture regime (d) and salinity (s).

23.9.4 Land Irrigability Classification

A general system for evaluation for irrigation has been elaborated by the United States Bureau of Reclamation (USBR 1951). In this system (USBR 1953), soils are first categorized according to physical factors (topography, drainage and water quality) and socio-economic factors (development costs, etc.). Separation of land irrigability classes is made on specified limits of soil properties and other physical parameters. Land irrigability system can be used for selection of irrigable lands, estimation of water requirements and development costs and benefits from irrigation. Such information will help in land use planning decisions (AISLUSO 1970; Beek 1981). This system also provides six suitability classes for irrigation based on the soil and land characteristics and the payment capacity. The subclasses provided are based on deficiencies or problems with respect to topography (t), soil (s) and drainage (w).

- Soils are categorized based on their suitability for sustained use under irrigation.
- Physical factors (topography, drainage and water quantity).
- Socio-economic factors.

23.9.5 The FAO Framework for Land Evaluation (1976)

The FAO Framework for Land Evaluation (FAO 1976) and subsequent guidelines for rainfed agriculture (1983), forestry (1984), irrigated agriculture (1985) and extensive grazing (1991) is considered to be a standard reference system in land evaluation throughout the world (Dent and Young 1981; van Diepen et al. 1991) and has been applied both in developed and developing countries.

This framework is an approach, not a method. It is designed primarily to provide tools for the formulation of each concrete evaluation. The system is based on the following concepts: (1) The land is qualified, not only the soil. (2) Land suitability must be defined for a specific soil use (crop and management). (3) Land evaluation was to take into account both the physical conditions and economic ones. (4) The concept of land evaluation is essentially economic, social and political. (5) The evaluation requires a comparison between two or more alternative kinds of use. (6) The evaluation must propose a use that is sustainable. (7) A multidisciplinary approach is required (Purnell 1979; van Diepen et al. 1991).

In the scheme, four categories are recognized. The highest category is the order that it reflects, in broad features, whether a soil is suitable or not for a given use. Two orders are recognized:

S = Suitable. Land in which the benefits exceed the costs and sustained use does not incapacitate the soil over a sufficiently long period of time.

N = Not suitable. Land can be classified as not suitable for a certain use for diverse reasons. The use proposed may be deemed technically impractical, as in irrigation of abrupt rocky terrain, or that it causes serious environmental degradation, as in cultivation on steep slopes. Frequently, however, the reason is economic, in that the profit expected does not justify the cost required.

The second category is the class that reflects degrees of suitability within the order. These are numbered consecutively in Arabic numerals.

For the order S, three classes are considered:

S1 = Highly suitable. Without limitations for sustained use or minor limitations that do not affect productivity nor appreciably increase costs.

S2 = Moderately suitable. Moderately serious limitations that reduce profits or involve risks of degradation in the sustained use of the soil.

S3 = Marginally suitable. The limitations for the sustained use are serious, and the balance between the costs and benefits makes the use only marginally justifiable. Its use is normally justified on other than economic grounds.

In the order N, three classes are also recognized:

N1 = Not currently suitable. Land with limitations that could be eliminated by technical means or investment but that these changes are at present unfeasible
N2 = Permanently unsuitable. Serious limitations of generally a physical nature, which are assumed to be beyond solving over the long term
X = Land for conservation. Unsuitable for exploitation, being lands of special protection, due to their conservation, wildlife, of special scientific, ecological or social interest (e.g. parks, reserves or recreational zones)

The limits between the orders (S and N) and between the different classes (S1, S2, S3 and N1, N2) are established by the presence of limiting factors. One limiting factor is a characteristic of the soil that hampers its use, reduces productivity, increases costs and implies degradation risk or all of the above.

These limiting factors are used to define the third category of the system, which is the subclass. In the symbol of each subclass, the number of limitations involved should be kept to the minimum one letter or, rarely, two. The limitations proposed include t, slope; e, erosion risk; p, depth; s, salinity; d, drainage; c, bioclimatic deficiency; r, rockiness; and i, flood risk.

Finally, the fourth category is the unit that establishes the differences within the subclasses as a function of the desired use. All of the units within a subclass (S2rA, S2rM, etc.) have the same degree of suitability at the subclass level (S2) and analogous characteristics of limitation at the subclass level (r). The units differ from each other in their characteristics of production or in secondary aspects of their management demands. Their examination enables a detailed interpretation at the planning level of the exploitation. The units are distinguished by upper-case letters that are placed at the end. There is no limit at all for the number of units examined within a subclass. These defined are A, intensification in the agricultural use without need of great improvements; M, intensification in the agricultural use with need of major improvements (irrigation, etc.); P, use for pasture for livestock; and F, forestation.

A very important role in this framework is represented by the concept of land utilization type (LUT). It represents a thorough description of soil use for a proposed use, in terms of crop (type and rotation), soil management (working of the soil, additives and possible irrigation) and socio-economic framework (labour cost, market, distribution, expenses, profits and subsidies). To propose a certain LUT, the crop must meet certain requirements of the land that satisfy the LUT, and these are grouped under the concept of land use requirements (LUR). For these requirements to be fulfilled, the land must have certain features, known as land qualities (LQs), which are supported by particular characteristics of the land, called land characteristics (LCs). The LCs represent characteristics of the soils that can be examined and measured in the profile or in the laboratory, while the LQs are complex qualities that are not directly measurable but rather are estimated from a certain combination of LCs. The LQs are a direct result of the LCs. The LQs used for defining the LUT are numerous (although much less than the LCs) and as diverse as nutrient availability, workability, flooding hazards, resistance to degradation or accessibility of the terrain.

In addition, each of these LQs is based on a set of LCs, and thus, for example, the LQ (nutrient availability) is a combination of the following LCs: organic matter, nutrients (N, K, P), cation-exchange capacity, degree of saturation, pH, salinity, carbonates, depth and gravel. Meanwhile, the LQ (water availability) is based on the following LCs: texture, structure, drainage, available water and water reserve. In our opinion, to facilitate the elaboration of a certain LUT, it might be useful to use the support of LQ organized as soil qualities (SQ), climate qualities (CQ), topography qualities (TQ), management qualities (MQ) and socio-economic qualities (EQ), which are calculated on the basis of their corresponding characteristics SC, CC, TC, MC and EC.

Altogether, the FAO Framework represents a highly useful and flexible system which is easily adapted to local characteristics. The main drawbacks are (i) the conceptual confusion caused by using pre-existing terms and (ii) the poor distinction between the physical, social and economic characteristics.

An evaluation requires (1) the selection and description of the LUTs, (2) definition of the necessary LQs on the basis of the LCs, (3) definition of the suitable values of the LCs, (4) measurements of the LCs in the land units, (5) an economic and social study and (6) presentation of the final suitability evaluation and reflection on a map with the corresponding evaluation units. For the LUTs, LQs and LCs, five conditions are established according to suitability (S1, S2, S3, N1 and N2). As the differentiating feature, the greatest limitation alone may be used. A LUT would be defined as pertaining to class S2ab if it presented the LQa and LQb of level S2 and the LQc of level S1. Or a combination of limitations (number and intensity) can be taken. The previous LUT would be defined as belonging to class S3 for having two LQs with an S2 value. Also, a correction factor can be introduced according to the soil depth at which the limitations are found.

23.9.6 *Using Mathematical Approach (Fuzzy Modelling-Based Method)*

This method consists of three steps namely generation of membership values for the soil characteristics, determination of weights for the membership values and combination of weighted membership values to produce a composite soil index (CSI) (Burrough 1989 and Chatterji 2000):

a.

$$\sum_{A} \mu(Z) = \frac{1}{1 + a_i(z_i - c_i)} 2 \text{ for } 0 \le z \le £ \tag{23.1}$$

where A is the soil characteristic set, a is the dispersion index that determines the shape of the function, c (called the ideal point or standard index) is the value of the property z at the centre of the set and £ is the maximum value that z can take.

b. The joint membership function (JMF) for each pedon and for each parameter is computed using the convex combination rule, which is a linear weighted combination of membership values of each land characteristic Ai:

$$\underset{i=1}{\overset{N}{I}} = \sum w_i \mu \tag{23.2}$$

where I is the joint membership function and W_i are the weights of the membership value μ.

c. To ensure that weights sum up to unity, the rank ri of a land characteristic, A was converted to weight Wi using the equation:

$$w_i = \frac{r_i}{\sum_i^n = 1^{ri}} \tag{23.3}$$

Equation (23.2) shows that the choice of weights W_i is crucial in the determination of the overall land suitability index.

d. Simple ranking procedure was used in deriving weights. This ranking was based on literature (Sys 1985; NBSS&LUP 1994; Kadu et al. 2003; Naidu et al. 2006) which identified the relative importance of a particular parameter to the cultivation of cotton crop.
e. The composite soil index (CSI) of a SQ parameter was determined as the average of the aggregated JMF values of the parameters for a particular pedon which in concept and for all practical purposes holds the same implication as that by SQI.
f. The suitability classes for the crops were identified by placing the CSI values in a set of equally spaced classes on a 0–100 scale, with a 20-unit gradation. CSI lying between 100 and 80 comes under class I, CSI lying between 80 and 60 comes under class II, CSI lying between 60 and 40 comes under class III, CSI between 40 and 20 comes under class IV and class V has CSI values ranging from 20 to 0.
g. The CSI values were correlated with average yield for validation.

Modelling There are a large and increasing number of computer models relevant to different aspects of land use planning. Most models consist essentially of quantitative predictions based on input data, for example, the prediction of plant evapotranspiration from weather data or the prediction of net present return from data on inputs, production, costs and prices. Note that:

- Models are only as reliable as the data which are entered into them.
- Wherever possible, models should be calibrated for the planning area, its climate, soil types, etc.; data should be entered and the results compared with an independent measure, for instance, crop yield.

23.10 The Future of Land Use Planning

New ways of effective land use planning include information management through geographic information system (GIS), computer simulation and spatial-temporal data modelling on present land use, alternative scenarios and assessment of consequences. While zoning and regulation are the primary methods adopted by land use planners, public education often is a neglected area that is increasingly being recognized. Other methods that planners use include economic incentives, institutional reform and investment through multiagency cooperative projects.

Land use planning is becoming complex and multidisciplinary as planners face multiple problems that need to be addressed within a single planning framework. Such problems include nonpoint source pollution, water allocation, urbanization, ecosystem deterioration, global warming, poverty and unemployment, deforestation, desertification, farmland deterioration and low economic growth. Watershed-scale planning is gaining popularity among communities and agencies so that biological, physical and socio-economic components of the landscape system can be integrated into the planning framework.

23.11 Conclusions

Land use planning can be best described as a decision-making process that facilitates the allocation of land to different uses that provide optimal and sustainable benefit. As land use is shaped by society-nature interaction, different components/facets play a significant role in land use planning involving soil, water, climate, animal (ruminant/non-ruminant) and others, including forestry and the environment needed for survival of mankind. At times these components are moderated by human interferences. Thus land use planning being a dynamic phenomenon is not guided by a single factor, but by a complex system working simultaneously, which largely affects the sustainability. Different approaches adopted to land use planning are governed by the purpose of undertaking the planning process.

References

All India Soil and Land Use Survey Organization (AISLUSO) (1970) Soil survey manual. IARI, New Delhi, 123 pp

Beek KJ (1981) From soil survey interpretation to land evaluation. Parts 1 and 2. Soil Survey Land Eval 1:6–12, 23–30

Buol SW, Sanchez PA, Cate RB Jr, Granger MA (1975) Soil fertility capability classification. In: Bornemisza E, Alvarado A (eds) Soil management in tropical America. North Carolina State University, Raleigh, pp 126–141

Burrough PA (1989) Fuzzy mathematical methods for soil survey and land evaluation. J Soil Sci 40:477–492

Chatterji A (2000) Suitability determination of some benchmark soils of India for rice using fuzzy and Boolean logic: a comparative study. Journal of Indian Soceity of Soil Science 48:708–711
Dent D, Young A (1981) Soil survey and land evaluation. George Allen & Unwin, London
FAO (1976) A framework for land evaluation. FAO Soils Bulletin No. 32. Rome, FAO, 72 pp
FAO (1983) Guidelines: land evaluation for rainfed agriculture, FAO soils bulletin no. 52, Rome
FAO (1984) Land evaluation for forestry. Forestry paper 48. FAO, Rome
FAO (1985) Guidelines: land evaluation for irrigated agriculture. Soils Bulletin 55. Food and Agriculture Organization of the United Nations
FAO (1991) Guidelines: land evaluation for extensive grazing. Soils Bulletin 58. FAO, Rome
FAO (1993) Guidelines for land use planning, FAO developmental series-1. FAO, Rome, Italy, 96 pp
FAO-UNEP. (1999). The future of our land facing the challenge, FA. UNEP, Rome, pp1–88
Ghosh PK (2013) A paradigm shift in managing natural resources and food security: strategy to action. In: Proceedings of the 100th session of the Indian Science Congress organized by The Indian Science Congress Association, 3–7 January, 2013, Kolkata
Kadu PR, Vaidya PH, Balpande SS, Satyavathi PLA, Pal DK (2003) Use of hydraulic conductivity to evaluate the suitability of vertisol for deep rooted crops in semi-arid part of Central India. Soil Use Manag 19:208–216
Kathpalia GN, Kapur R (2011) Management of land and water for inclusive growth: India 2050, Alternative futures: NRM South Asia
Klingebiel AA, Montgomery PH (1966) Land capability classification. Agri. Book No. 210 Soil Conservation Series, U.S.D.A., Washington, DC
Naidu LGK, Ramamurthy V, Challa O, Hegde R, Krishnan P (2006) Manual soil-site suitability criteria for major crops. Technical Bulletin No. 129. NBSS&LUP, Nagpur (India), pp 118
NBSS&LUP (1994) Proceedings National Meet on soil site suitability criteria for different crops. NBSS&LUP, Nagpur. Publ. Feb 7–8, pp. 32
Purnell MF (1979) The FAO approach to land evaluation and its application to land classification for irrigation. World Soil Resource Report 50:4–8
Riquier J, Bramo D, Cornel JF (1970) A new system of soil appraisal in terms of actual and potential productivity. FAO-AGL-TESR/70/6, Rome
Sánchez PA, Couto W, Buol SW (1982) The fertility capability soil classification system: interpretation, applicability and modification. Geoderma 27(4):283–309
Sarkar D, Chaturvedi A, Mahapatra SK (2009) New vistas in perspective of land use planning. J Indian Soc Soil Sci 57(4):587–600
Sys C (1985) Land evaluation. State University of Ghent, The Netherlands
Sys C, Van Ranst E, Debaveye J (1991) Land evaluation, Part l, principles in land evaluation and crop production calculations. International Training Centre for Post-graduate Soil Scientists, University Ghent
USBR (1951). Bureau of Reclamation Manual, vol. V. Irrigated land use. Part 2. *Land classification*. Bureau of Reclamation, Dept. of Interior, Denver Federal Center, Denver, Col. 80225, USA
USBR (1953) Department of the Interior Bureau of Reclamation irrigated land use, part 2: Land classification. B.R. Manual. vol. 5. U.S. Government Printing Office, Washington
van Diepen CA, Van Keulen H, Wolf J, Berkhout JAA (1991) Land evaluation: from intuition to quantification. In: Stewart BA (ed) Advances in soil science. Springer, New York, pp 139–204

Chapter 24
Integrated Remote Sensing, GIS, and GPS Applications in Agricultural Land Use Planning

G. P. Obi Reddy, V. Ramamurthy, and S. K. Singh

Abstract Land use planning is the systematic assessment of land potential, alternatives for land use, and economic and social conditions in order to select and adopt the best land use options to achieve accelerated growth through judicious management of land and water resources. Inappropriate land management and land use practices and climatic aberrations like droughts and floods lead to degradation of soil, water, and vegetative cover, which affects the quality of land resources, land use patterns, and ecosystem functions. There is a felt need to develop detailed database on extent and nature of land resources, their potentials, and limitations for land resource-based scientific sustainable agricultural land use planning. Advance geospatial tools like remote sensing, geographic information system (GIS), and global positioning system (GPS) could be effectively used in land resource mapping and agricultural land use planning. The integration of spatial data and their combined analysis could be performed through GIS and simple database query systems to complex analysis and decision support systems for effective land resource management. The scope and potential applications of geospatial technologies in inventory, mapping and management of land resources, agricultural resources surveys, agro-ecological zonation, land degradation mapping and assessment, soil moisture estimation, soil fertility mapping and assessment, crop acreage estimation, crop yield forecasting, water resources planning and management, development of land resource information system, development of decision support systems, land evaluation, crop suitability evaluation, precision agriculture, characterization of land management units, development of land use options, and future perspectives for sustainable land use planning were briefly discussed in the chapter.

Keywords Agricultural land use planning · Agroecological zoning · Geographic information system · Remote sensing · Crop suitability evaluation

---

G. P. O. Reddy (✉) · S. K. Singh
ICAR-National Bureau of Soil Survey & Land Use Planning, Nagpur, India

V. Ramamurthy
ICAR-National Bureau of Soil Survey & Land Use Planning, Regional Centre, Bengaluru, India

G. P. O. Reddy, S. K. Singh (eds.), *Geospatial Technologies in Land Resources Mapping, Monitoring and Management*, Geotechnologies and the Environment 21,
https://doi.org/10.1007/978-3-319-78711-4_24

24.1 Introduction

India has over 17% of the world's population living on 2.6% of the world's geographical area, and per capita availability of agricultural land declined from 0.48 hectares in 1951 to 0.16 hectares in 2010. This posed challenges in protecting the prime agricultural lands and meeting the food security for growing population and protection of environment. About 60% of India's population depends on agriculture, and currently, India produces about 273 million tons of food grains in the year 2016–2017, and it is estimated that the demand for food grains shall rise to 377 million tons by 2050. Thus, there is a need to protect agricultural areas like prime agricultural lands, command areas of irrigation projects, and double-cropped areas for sustained food production and livelihood security of millions of people depending on agriculture and allied sectors (Reddy and Singh 2014).

Land use planning aimed to select those combinations of land and land use that will best suit the specified goals (FAO 1989). Land use planning is the process of evaluating land and alternative patterns of land use and other physical, social, and economic conditions for the purposes of selecting and adopting those kinds of land use best suited to achieve specified objectives. Its purpose is to select and put into practice those land uses that will best meet the needs of the people while safeguarding resources for the future. Land use plans can be implemented to increase the agriculture production by improving the productivity of crops, conservation of soil and water resources to improve input use efficiency, ensure livelihood security, generate employment, and last but not the least ensuring food security. Land use planning may be at national, regional, state, district, watershed, or village and also at farm level. The techniques and even the strategy of land use planning can be different at different levels. Land use policy is essentially an expression of the governments' perception of the direction to be taken on major issues related to land use and the proposed allocation of the national land resources over a fixed period of time.

The detailed spatial information on land resources is essential for many land use applications and environmental modeling (Burrough 1989; Maji et al. 2001; Reddy et al. 2001, 2017a, 2017b). Remote sensing data can be used to delineate distinct physiographic units besides deriving ancillary information about site characteristics, land use, slope, water resources, and cropping systems of the study area (Rao et al. 1996; Reddy and Maji 2003; Panigrahy et al. 2005). However, detailed information on rainfall, soil properties, and socioeconomics could be collected through field surveys to develop a comprehensive land use planning of the region. Hence, field survey data are indispensable for generating derived thematic information like crop suitability and cropping systems. In this direction, GIS capabilities, which includes overlaying different types of thematic maps of a particular area, integrating maps at different scales, development of customized models, etc., will be of immense help to develop derived thematic maps through integrating spatial and attribute databases. This in turn could be superimposed on a map showing length of growing seasons, thereby producing a land suitability map for a given crop. Establishing relationships between various attributes is another important aspect of GIS applications. Applications of GIS range from simple database query systems to complex analysis and decision support systems. GIS techniques are playing an increasing role in facilitating integration of multilayer spatial information with statistical attribute data to arrive at alternate developmental scenarios (Burrough 1986). Land evaluation using a scientific procedure is essential to assess the potential and constraints of a given land parcel for agricultural purposes (Rossiter 1996). Remote sensing data coupled with soil survey information can be integrated in GIS to assess crop suitability for various soil and biophysical conditions. The potential of the integrated remote sensing and GIS approach for quantitative land evaluation has been demonstrated earlier by several researchers (Beek et al. 1997; Merolla et al. 1994).

In order to facilitate these complex tasks, a perspective land use plan that would involve alterations in the land area allocation over alternative uses, through suitable technological and institutional devices, taking into account the potentials and limitations of each land management unit (LMU), is needed. Perspective land use plan ensures the supplies of various inputs/commodities and services, which follow from the stipulated land use pattern, broadly conform to the projected demands for such items.

In the past decades, continuous utilization of agriculture land regardless of its potential land suitability has caused much more ill effects than provide the resources (FAO 1976, 1983, 2007). In different 5-year plans of India, the necessity of developing land use plans at different levels has been increasingly felt and emphasized. Hence, proper land evaluation based agriculture land use planning is essential for sustainable agriculture land use planning. Cropland suitability analysis is a prerequisite to achieve optimum utilization of available land for agricultural production in a sustainable manner. In order to practice it, one has to grow the crops where they suit best and for which first and the foremost requirement is to carry out land suitability analysis (Sys 1985; Ahamed et al. 2000). The Government of India developed a draft on agricultural and land use policy with an objective to promote efficient management of land and water resources for sustainable growth of agricultural to increase food production to meet the needs of the growing population (DoLR 2013). There is growing interest in considering socioeconomic dimensions and peoples' participation in planning land use and management of land resources. Hence, an integrated approach in planning and management of land resources entails involvement of all stakeholders in the process of decision-making on the future of the land and identification and evaluation of all the biophysical and socioeconomic attributes of land units. In this chapter, an attempt has been made to discuss the scope and potential applications of integrated remote sensing and GIS in agricultural land use planning.

24.2 Land Use Scenario and Trends in India

The analysis of land use pattern in India since 1950–1951 to 2007–2008 shows that area under forests increased from 40.5 Mha in 1950–1951 to 69.6 Mha in 2007–2008 and it accounts for nearly 22.8% of the total reporting area. Area under nonagricultural uses increased from 9.4 Mha in 1950–1951 to 26.0 Mha in 2007–2008. In percentage terms, this increased from 3.3% in 1950–1951 to 8.5% in 2007–2008. According to the Bureau of Statistics, Govt. of India, the proportions of area under barren and unculturable wasteland (13.4–5.7%), lands under miscellaneous trees, groves, etc. (7.0–1.1%), and culturable wasteland and permanent fallow lands (8.1–4.3%) decreased significantly during the same time. The net area sown increased from 118.8 Mha in 1950–1951 (41.8%) to 140.8 Mha (46.1%) in 2007–2008. This was largely due to extension of cultivation in marginal lands including culturable waste and lands under miscellaneous trees, groves, etc. During the past several years, nearly 10 Mha of land have been kept permanently fallows other than current fallows. During the same time, cropping intensity increased from 111 in 1950–1951 to 132 in 2007–2008, mainly due to rise in the gross irrigated area from 22.6 Mha in 1950–1951 to 86.7 Mha in 2007–2008. At the national level, net sown area increased only marginally in the 1980s and remained constant at about 141 million hectares in the late years 2007–2008 (Indiastat 2011). At the same time, the area under nonagriculture uses is continuously rising primarily because of expansion of urban and industrial areas.

24.3 Land Use Planning: Issues and Constraints

In a developing country like India, land is not only an important factor of production but also the basic means of subsistence for majority of the people. Agriculture contributes less than 30% to India's gross domestic product but absorbs nearly 64% of the country's working population. About three-fourth of the total population draw their livelihood from agriculture. But there is evidence to indicate that the land sector cannot bear the burden of growing population, notwithstanding the untapped potentials for agricultural productivity growth in many regions. The major issues being faced by Indian agriculture in increasing the agricultural production are stagnating of net sown area, low agricultural productivity due to low soil fertility (Sarkar et al. 2012; Maji et al. 2012), lack of appropriate soil and water conservation measures in rainfed regions, depletion of groundwater table in different regions, increase of fallow lands over the years, land degradation due to salinity/alkalinity mainly in irrigated command, etc. Therefore, there is a felt need to develop detailed database on extent and nature of land resources, their potentials and limitations for both horizontal and vertical diversification of the cropping systems. This is particularly so because all lands and locations are not equally suitable for profitable, alternative farming, and hence, there is need for land resource-based scientific approach for sustainable agricultural land use planning.

24.4 Land Use Planning: Strategies and Approaches

At the national level, land use planning is concerned with national goals and the allocation of resources. National land use plan may cover land use policy to balance the competing demands for land among different sectors of the economy like food production, exports/imports, tourism, wildlife conservation, housing and public amenities, roads, and industry; national development plans, project identification and allocation of resources for development; coordination of sectoral agencies involved in land use; and legislation on such subjects like land tenure, forest clearance, and water rights. National goals are complex, while policy decisions, legislation, and fiscal measures affect many people and wide areas. Decision-makers cannot possibly be specialists in all facets of land use, so the planners' responsibility is to present the relevant information in terms that the decision-makers can both comprehend and act on.

District-level land use planning refers not necessarily to administrative districts but also to land areas that fall between national and local levels. Development projects are often at this level, where planning first comes to grips with the diversity of the land and its suitability to meet project goals. When planning is initiated nationally, national priorities have to be translated into local plans. Conflicts between national and local interests will have to be resolved. The kinds of issues tackled at this stage include the setting of development activities such as new settlements, forest plantations, and irrigation schemes; the need for improved infrastructure such as water supply, roads, and marketing facilities; and the development of management guidelines for improved kinds of land use on each type of land.

The local planning unit may be the village, a group of villages, or a small water catchment. At this level, it is easiest to fit the plan to the people, making use of local people's knowledge and contributions. Where planning is initiated at the district level, the program of work to implement changes in land use or management has to be carried out locally. Alternatively, this may be the first level of planning, with its priorities drawn up by the local people. Local-level planning is about getting things done on particular areas of land – what shall be done where and when and who will be responsible. It includes formal and informal institutions that affect the use and the transfer of assets to future generations to assure the quality of life in the long run. Institutional mechanism and policies in place help in succeeding the sustainable land resource management approaches and interventions. Attempts were made in development of soil resource-based land use planning at village (Ramamurthy et al. 2006; Ramamurthy and Sarkar 2009) and watershed level (Ramamurthy et al. 2015).

24.5 Integrated Remote Sensing and GIS in Agricultural Land Use Planning

Ever since independence, India's planners and policy makers have shown concern for efficient use of land, water, and other natural resources for accelerated as well as sustainable economic development. However, notwithstanding these concerns, it is often reported that the problems of land degradation and groundwater depletion have assumed serious proportions in many areas, which threaten not only the sustainability of agriculture but also the overall livelihood system of the people (Maji et al. 2010). The integration of spatial data and their combined analysis could be performed through GIS and simple database query systems to complex analysis and decision support systems for effective land resource management (Reddy et al. 2013a, b). The potential applications of integrated remote sensing and GIS technologies in management of land resource and agricultural land use planning have been briefly discussed below.

24.5.1 Agricultural Resource Inventory and Surveys

India is one of the few countries in the world which has a well-established system of collection of agricultural statistics and detailed statistics of land utilization since 1884. As early as 1919, the Board of Agriculture in India recommended a method for conducting crop cutting experiments for estimation of crop yield through a random selection of villages, fields, and plots. Sampling methods were adopted by Mahalanobis (1945), where crop cutting experiments were conducted for the yield estimation of wheat and gram in two districts as part of Bihar crop survey in 1943–1944. Subsequently, Sukhatme and Panse (1951) introduced a random sampling method for estimation of crop yield, where the main emphasis was given on the selection of plot for harvesting by a strict process of randomization in place of subjective selection. This technique has been adopted by the Ministry of Agriculture (MOA), Govt. of India, and has by and large been followed by the Central and State Departments of Agriculture for estimating the production of major crops in the country. With the advent of remote sensing technology, satellite data has been widely used for obtaining crop statistics. In India, systematic multi-crop and large-scale investigation on use of remote sensing data for crop statistics has been undertaken under Crop Acreage and Production Estimation (CAPE) project by the Department of Space at Space Application Center (SAC), Ahmedabad, since 1986.

24.5.2 Land Resource Inventory and Mapping

The knowledge of soils with respect to their extent, distribution, characterization, and use potential is extremely important for the optimized land utilization. The information on soil characteristics and various soil-based thematic maps are required on different scales to meet the requirements of agricultural land use planning at various levels. The conventional methods of gathering information on soils are expensive and time-consuming due to collection of large number of observations. However, recent advancements in the field of remote sensing, GIS, and GPS technologies augmented the terrain analysis (Reddy and Maji 2004; Gangopadhyay et al. 2015), geomorphological mapping (Reddy and Maji 2003), and soil resource mapping activities (Sarkar et al. 2006; Reddy et al. 2013a, b; Vishakha et al. 2013; Sahu et al. 2014, 2015). The remote sensing technology is found to be more efficient and economical than conventional survey (Srivastava and Saxena 2004; Reddy et al. 2013a, b). There are a wide variety of satellite sensors operating in the optical and microwave regions and providing varied spatial, spectral, and temporal resolutions data sets, which could be effectively utilized for varying intensity of soil resource inventory and mapping. Depending upon the scale of land resource inventory and mapping, satellite data like Landsat-MSS/TM, SPOT, IRS-P6 LISS-III and LISS-IV, Cartosat-1 and Cartosat-2, etc. could be effectively used to map the soils. The high resolution satellite data in conjunction with GIS could be effectively used in soil resource inventory and mapping at large scale to identify the problems and potentials of the soils and suggest management practices to develop appropriate and sustainable cropping systems to improve the livelihoods for farming community.

24.5.3 Agroecological Zoning and Mapping

FAO (1983) has given the concept of agroclimatic zones and agroecological regions. An agroclimatic zone is defined as the land unit which is uniform with respect to climate and length of growing period (LGP). This is climate-wise suitable for a certain range of crops and cultivars. An agroecological region (AER) is characterized by distinct ecological responses to macroclimate as expressed in terms of vegetation, soils, fauna, and aquatic systems. Therefore, an agroecological region is the land unit on the earth's surface carved out of agroclimatic region by superimposing climate on landforms and soils, which are the modifiers of climate and LGP. The thematic layers on physiographic regions, bioclimatic types, LGP, soil moisture, and temperature regimes could be generated and integrated in GIS for delineation of agroecological regions and agroecological subregions. The first agroecological region map of India was generated based on physiography, bio-climate, and soils (Krishnan and Singh 1968). Later, Murthy and Pandey (1978) attempted to prepare another agroecological region map of the country with the similar parameters. Subramaniam (1983) divided the country into 29 agroecological zones based on the dominant soil groups (FAO/UNESCO 1974); however, due consideration was not given to the two important parameters, namely, physiography and bio-climate. Sehgal et al. (1992) delineated twenty agroecological regions (AERs) in India based on the climatic conditions, length of growing period (LGP), landforms, and soils. The major advantage of LGP-based criteria is that it is a better indicator of moisture availability in soils than the total rainfall. It is further useful in delineating agroecologically comparable resource regions for generating and transferring agro-technology to enhance crop production.

The twenty AERs were further subdivided into sixty agroecological subregions (AESRs) (Velayutham et al. 1999) by narrowing down further the limits of LGP, physiography, soil depth, soil texture, and available water holding capacity. Further, keeping the changing climatic conditions and rainfall patterns with more precise data on soils and climate on finer resolution refined the twenty AERs (Mandal et al. 2016) for ensuring pragmatic regional-/country-level agricultural (crop) planning. The agroecological region concept characterizes tracts of land by quantified information on climate, soils, and other physical factors, which are used to predict the potential productivity for various crops according to their specific environmental and management needs. Agroecological zoning helps to delineate productivity efficiency zones, agro-technology transfer, and scientific land use planning.

24.5.4 Land Degradation Mapping and Assessment

India ranks very high among the developing countries in respect of both extent and severity of land degradation (Yadav 1996). In India, degraded lands, i.e., eroded lands, salt-affected soils, waterlogged area, ravine lands, area under shifting cultivation, and lands affected by toxic elements of industrial wastes, are either lying waste or are being partially utilized (Das 1985). From time to time, various agencies projected the nature and extent of degraded lands of the country (National Commission on Agriculture 1976; Bhumbla and Khare 1984; Govt. of India 1994; Sehgal and Abrol 1994). Sehgal and Abrol (1994) reported that about 187.7 Mha of the total geographical area of the country is suffering from various forms of soil degradation–water erosion and physical and chemical deterioration. GIS helps in integrated analysis of USLE parameters in quantification of potential and actual soil loss (Reddy et al. 2004; Reddy and Sarkar 2012). Reddy et al. (2016) assessed soil loss in Goa state of India through integrated analysis of rainfall erosivity (R), soil erodibility (K), slope length (LS), cover (C), and management (P) factors of USLE using Gaussian kriging model in GIS. According to the Ministry of Agriculture, Govt. of India, degraded land is of the order of 92 Mha, of which 73 Mha are arable land and 19 Mha are non-arable land. While there are ongoing efforts to improve the quality of these lands through various soil and water conservation and watershed development programs, there is also a growing threat of land degradation due to high incidence of groundwater depletion in low rainfall areas – and because of the poor drainage and water management as well as low utilization of available groundwater in high rainfall areas.

24.5.5 Soil Moisture Estimation and Assessment

Soil moisture is an essential variable in the earth system, as it influences the exchange of water, energy, and carbon fluxes between the land surface and the atmosphere. In the hydrological cycle, its distribution is of great importance due to its high spatial and temporal variability. Soil moisture has a strong influence on the relative distribution of water between various components of the hydrological cycle (Western et al. 2002). Due to heterogeneity of soil type, land use, and topography, soil moisture may change considerably in space and depth during any time retrieval (Haider et al. 2004). Accurate assessment of spatiotemporal patterns of soil moisture has wide applications in soil science, agriculture, hydrology, and climate sciences to understand various processes of hydrology and climate regimes. Soil moisture can be measured by instruments in the field (Romano 2014) or estimated by remote sensing techniques (Ochsner et al. 2013). Various theoretical and empirical models have been devised to retrieve soil moisture from active and passive remote sensing data (Ulaby et al. 1982; Shi et al. 1997; Wen et al. 2003; Das et al. 2008). Remote sensing of soil moisture from space is advantageous because of its spatial coverage and temporal continuity. Observations from both microwave and optical/thermal infrared sensors have been successfully used to retrieve surface soil moisture (Peng et al. 2015). In particular, soil moisture retrieval from microwave observations has been highly advanced for all-weather observations, and several global soil moisture products have been produced (Njoku et al. 2003; Wagner et al. 2007). Besides the estimation of soil moisture from microwave data, studies also exploit the feasibility of estimating soil moisture at high spatial resolution from optical/thermal satellite observations. Choker et al. (2017) reported that soil moisture could be effectively mapped using Sentinel-1 SAR data.

24.5.6 Soil Fertility Mapping and Management

In order to improve the crop productivity without environmental impact, there is the need for the utilization of novel procedures and techniques to optimize available soil nutrients. The modern geospatial technologies such as remote sensing, GIS, GPS, and information technology offer immense potential for soil fertility mapping and management. Collection of GPS-based soil samples assumed a greater importance in preparing thematic soil fertility maps for location-specific soil fertility management (Mishra et al. 2013). GPS helps to know precise latitude and longitude of the sample site, and it has great significance in agriculture for future monitoring of soil nutrient status of different locations/region. The databases generated through soil surveys/ soil fertility surveys could be analyzed in GIS to assess inherent status of soil fertility (Petersen et al. 1995). In soil fertility mapping and assessment of variability in soil properties, the sampling spacing, sampling method, and interpolation techniques used are important. McBratney et al. (2003) provided the comprehensive maps for physical, chemical, and biological soil properties by means of geostatistics, GIS, and remote sensing techniques. GIS and geostatistical tools have been widely used in producing soil fertility maps (Mueller et al. 2001; Mulla 1991; Webster and Mcbratney 1987). Geostatistical methods could be used effectively to spatially interpolate soil fertility data by using different spatial interpolation methods such as kriging (Bishop and McBratney 2001). GIS-based spatial interpolation methods estimate the values of a point source data at unsampled locations (Hengl et al. 2004). There are, however, many factors affecting the spatial variability of data, which should be taken into consideration when mapping soil fertility status, such as the number (Mueller et al. 2001) and time (Corstanje et al. 2007) of sampling, intensity of sampling (Röver and Kaiser 1999), land management (Cambardella et al. 1994), and different sampling schemes (Tan and Lal 2005). The soil fertility maps generated through GPS and GIS technologies help farmers, scientists, planners, and students in providing online soil test-based fertilizer recommendation, site-specific nutrient management, and monitoring of the soil health for intensive and sustainable crop production. Web-based fertilizer information system could be developed to recommend the location-specific amount and type of fertilizer and the timing of fertilizer application to enhance the soil health and crop productivity.

24.5.7 Crop Inventory and Acreage Estimation

The timely and reliable information on crop sown area become the most significant fraction in development of management strategies by the planners and administrators in the field of agriculture and allied sectors. The crop inventory and acreage statistics form a base for the numerous applications such as yield forecasting, managing the grain storage, procurement of price, making the strategies like import and export of commodities, requirement of fertilizers and pesticides, crop insurance, etc. (Anup et al. 2005). Remote sensing plays a fundamental role in providing precise and near real-time information on nature, extent, and spatial distribution of different crops on the earth. Remote sensing-based crop identification, discrimination, and mapping are gaining much importance and are mainly dependent upon the spectral signatures of each crop type, their canopy conditions, and crop-growing period. Remote sensing-based assessment of extent and spatial distribution of crop types has proven useful to a wide range of end users, including various government departments, policy makers, farmers, and scientists. It reduces the fieldwork to a considerable extent and helps in delineation of features boundaries more precisely than in conventional methods. Hence, it is a highly proven technology that is effective for mapping and characterizing different crops and vegetation types (Thilagam and Sivasamy 2013). The spectral reflection of a field will vary with respect to changes in the phenology, stage type, and crop health and thus can be measured and monitored by multispectral sensors (Singha et al. 2016). Moderate Resolution Imaging Spectroradiometer (MODIS) instrument provides a unique opportunity for monitoring agricultural systems over a large area (Townshend and Justice 2002). The advanced time series and multispectral coarse resolution remote sensing data have immense potential for national-scale croplands mapping on periodical basis (Wardlow et al. 2007; Thenkabail et al. 2009; Wardlow and Egbert 2008; Thenkabail et al. 2005). With increase in the spatial, spectral, and temporal resolutions of the sensors, the satellite data becomes an effective source for cropland mapping. Satellite images and other related information can be further processed, managed, and analyzed with the help of GIS. The locational information can be collected from ground in the form of ground truth database with the help of GPS tool that linked with satellite images in analysis process for cross checking and feature identification that results in better accuracy in crop identification, mapping, and acreage estimation. Optical remote sensing images like MODIS, AVHRR, SPOT-VGT, and AWiFS, which have relatively low spatial resolution range from 56 m to 1 km could be effectively used to monitor and map the cropland areas. At times, it would be difficult to obtain cloud-free images of crop-growing regions from optical remote sensing. The integration of optical with microwave remote sensing would be more beneficial for better accuracy in different crop mapping and acreage estimation.

24.5.8 Crop Yield Forecasting

Predicting yields prior to harvest is crucial in agriculture sector. The agricultural crop production of principal agricultural crops in India is usually estimated as a product of area under the crop and the average yield per unit area of the crop. The estimates of the crop acreage at a district level are obtained through complete enumeration, whereas, the average yield is obtained on the basis of crop cutting experiments conducted on a number of randomly selected fields in a sample of villages in the district. The technique developed during the 1940s has, by and large, been followed for estimating the production of major crops in the country. However, the traditional system of estimation of crop production has several problems, viz., lack of timely information, variation in statistical figures, accessibility, and quick retrieval of data (Reddy et al. 2009). Singh et al. (1992, 2002) used the satellite spectral data along with the survey data on crop yield from general crop yield estimation surveys to develop more efficient post-stratified estimators of crop yield at district level and also small area estimators of crop yield at tehsil level. Remote sensing has been used to forecast crop yields based primarily upon statistical–empirical relationships between yield and vegetation indices (Casa and Jones 2005). On a regional scale, crop yield estimation was carried out based on vegetation indices derived from AVHRR/NOAA satellite image data (Prasad et al. 2006). Galvao et al. (2009) studied the possibility of using satellite Hyperion hyperspectral images to estimate the yield of soybean obtaining a high correlation ($r = 0.74$) between vegetation indices and weight of harvested seed. Doraiswamy et al. (2004) also studied the possibility of using MODIS satellite data for forecasting yields using a calibrated form of the model developed by Li et al. (2008). Vegetation stages of development of plant are influenced by a variety of factors such as available soil moisture, air temperature, day length, and soil condition. These factors therefore also influence plant conditions and their productivity. Using various biophysical and plant morphological properties, yield model could be developed in GIS to estimate the crop yields.

24.5.9 Water Resources Planning and Management

Access to reliable groundwater sources plays an important role in food security in many cases as the access to reliable sources of water reduces the production risk. Yields in groundwater-irrigated areas are higher (often double) compared to those in the canal-irrigated areas (Shah 1993). Groundwater availability acts as a trigger to enable the farmers to invest in complementary inputs that, in combination, increase crop yields substantially. In the dry land semiarid tropics, an integrated watershed management resulted in increased groundwater availability that served as an entry point for increasing agricultural production and improving rural livelihoods (Wani et al. 2009). Integrated Watershed Management (IWM) is the strategy adopted to enhance the water use efficiency for sustainable development of dry land areas. The IWM strategy demonstrated that dry land areas with good quality soils could support double cropping, while the surplus rainwater could recharge the groundwater. Temporal data from remote sensing enables identification of groundwater aquifers and assessment of their changes along with the land use changes, whereas GIS enables integration of multi-thematic data. Analysis of surface and subsurface hydrological parameters through analysis of remotely sensed data in conjunction with field surveys provides a scope to identify hydro-geomorphological units. It can be carried out in GIS through database generation on drainage, landforms, lithology, land use/land cover, and lineament thematic parameters from data and their overlay analysis (Krishnamurthy and Srinivas 1995; Reddy et al. 1999). GIS can support in hydrologic modeling and development of water resource decision support systems (Wilson et al. 2000). The delineated hydro-geomorphological units can be analyzed to assess their groundwater potential (Reddy and Rao 1994). However, the necessary field verification and validation of the results with available collateral data are essential for accurate assessment of the groundwater prospects.

24.5.10 Development of Land Resource Information System

Accurate, up-to-date and spatially referenced land resource information is necessary for various agricultural applications. The convergence of geospatial technological remote sensing, GIS, GPS, and information technology provides an ideal framework for comprehensive development of land resource information system (Reddy et al. 2017a). GIS could be effectively used to integrate, store, edit, analyze, share, and display georeferenced land resources information. GIS plays essential roles in integrating a variety of data layers to derive meaningful information. The usefulness of an information system will depend on its ability to provide decision-makers with the right data at the right time in the proper manner. In recent years, considerable efforts have been made to develop GIS-based soil information systems, and some organizations have compiled and harmonized local and national soil information at regional to global scales. The development of GIS-based soil information system enables to catalogue, store, manage, process, and analyze voluminous soil resource databases in a digital environment. Soil maps and databases usually contain information on soil properties associated with the soil units described as being present in the polygons of the map, while soil profile databases contain information on the soil classification unit. Globally, many countries developed their own soil information systems; notably among them are Australian Soil Resource Information System (ASRIS) (ASRIS 2011), Canadian Soil Information Service (CanSIS) (http://sis.agr.gc.ca/cansis/), State Soil Geographic (STATSGO) (http://soils.usda.gov/survey/geography/ssurgo/description_statsgo2.html) and Soil Survey Geographic (SSURGO) (http://soils.usda.gov/survey/geography/ssurgo/ description.html) Database of NRCS, Soil Terrain Databases (SOTER) (Baumgardner 1986; FAO/AGL and ISSS 1986), eSOTER of Global Earth Observation System of Systems (GEOSS) (http://www.esoter.net), GlobalSoilMap.net (www.globalsoilmap.net), and World Inventory of Soil Emission Potential (WISE) of International Soil Reference and Information Centre (ISRIC) (ISRIC-WISE 2008). Maji et al. (2001, 2002) and Bhattacharyya et al. (2010) have developed state-level soil information systems for Nagaland and Tripura, respectively. Database on district-level planning has also been prepared for different districts of the country (Maji et al. 1998a, b). In order to meet the demands of wider land resource management and land use planning applications, digital soil resource databases of India in the form of soil information system have been designed and developed as a GIS-based spatial database management system (Reddy et al. 2014). The availability of digital land resource resources and allied databases in GIS framework has wide applications in the fields digital terrain analysis, soil–landscape modeling, assessment and monitoring of degraded lands, watershed hydrology, soil and water conservation, crop suitability evaluation, development of spatial decision support systems, etc.

24.5.11 Data Integration and Decision Support Systems

Decision support systems (DSS) are defined as computer-based information systems designed to support decision-makers interactively in thinking and taking decisions about relatively unstructured problems. Traditionally, DSS have three major components, a database, a model base, and a user interface. An extension of the DSS concepts Spatial Decision Support Systems (SDSS), which are the integration of DSS and GIS was initiated by Densham and Goodchild (1988). A significant capability of the SDSS is the ability to use spatial analysis and the display tools with the sectorial models that would form the model base of SDSS. The modeling capability of GIS allows the user of the SDSS to stimulate changes in objects and attributes. GIS has the capability to integrate geographical referenced data, together with local knowledge in relational databases to accurately display complex interactions of the system in simple formats. The database components of the SDSS can supply input data for the models. After the models are run, the resulting output can be returned to the database for later display through user interface, in tabular, chart, or map form. This GIS-based SDSS could be used in modeling, which can integrate with simulation techniques to solve complex nutrient management in various field and plantation crops.

24.5.12 Land Evaluation

Land evaluation using a scientific procedure is essential to assess the potential and constraints of a given land parcel for agricultural purposes (Rossiter 1996; Sonneveld et al. 2010). Agriculture land suitability assessment is defined as the process of assessment of land performance when used for alternative kinds of agriculture (He et al. 2011; Mu 2006; Prakash 2003). Remote sensing data coupled with soil survey information can be integrated in GIS to assess crop suitability for various soil and biophysical conditions. The use of land is not only determined by the user but also by the land capability. Land capability mapping is the basis for sustainable development by ensuring correct land use according to its capability for sustained economic production. The current land capability classification includes eight classes of land designated by Roman numerals I thru VIII. The first four classes are arable land suitable for cropland in which the limitations on their use and necessity of conservation measures and careful management increase from I thru IV. The criteria for placing a given area in a particular class involve the landscape location, slope of the field, depth, texture, and reaction of the soil. The remaining four classes, V thru VIII, are not to be used for cropland, but may have uses for pasture, range, woodland, grazing, wildlife, recreation, and esthetic purposes. Within the broad classes are subclasses which signify special limitations such as (e) erosion, (w) excess wetness, (s) problems in the rooting zone, and (c) climatic limitations. Within the subclasses are the capability units which give some prediction of expected agricultural yields and indicate treatment needs. The capability units are groupings of soils that have common responses to pasture and crop plants under similar systems of farming (Klingebiel and Montgomery 1961). The land capability is governed by the different land attributes such as the types of soil, its depth and texture, underlying geology, topography, hydrology, etc. In the past, land capability mapping for agriculture was carried out using traditional methods. There were no effective tools to help merge information from diverse sources and of different scales. The parameters like slope, soil depth, texture, and land use/land cover assessed by applying remote sensing and GIS techniques could be effectively used for land capability classification. Land evaluation of hill lands using soil survey information in GIS has been reported by Maji et al. (2000).

The main objective of land evaluation is to appraise the potential of land for alternative kinds of land use by a systematic comparison of the requirements of this land use with the resources offered by the land (Dent and Young 1981). In contemporary land evaluation exercises, the geometric and semantic definition of the land units and their performance attributes are typically stored in a geospatial database. Statistical and geospatial tools are then used to assess the land units and to present the results as suitability maps. Identifying and comparing present and potential land uses for different land types, making decisions on land uses types, and implementing development programs and activities on the basis of these decisions is an activity that is being implemented by governments, communities, and individuals to organize space. Sometimes it is rigid and formal like in urban and industrial planning with hard copy plans passed from top to bottom through a wide number of different administrations. Elsewhere, it is flexible and more informal when considering, for instance, annually renegotiated trekking routes between different nomadic tribes. These land use plans are orally agreed upon but are supported by very precise though fictitious drawings. A major breakthrough in land use planning methodology for rural areas dates back to the 1960s with the introduction of the land capability concept (Klingebiel and Montgomery 1961). This approach was developed to identify and classify areas to be allocated for different agrarian activities like irrigation, mechanized farming, grazing, and forestry.

24.5.13 Crop Site: Suitability Evaluation

Land use planning aims to encourage and assist land users in selecting options that increase their productivity, are sustainable, and meet the needs of society (FAO 1993). Land suitability analysis is a method of land evaluation, which measures the degree of appropriateness of land for a certain use. Land evaluation is a scientific procedure to assess the potential and constraints of a given land parcel for agricultural purposes (Rossiter 1996). Land suitability evaluation is an examination process of the degree of land suitability for a specific utilization type (Sys et al. 1991). Since land suitability analysis requires the use of different kinds of data and information (soil, climate, land use, topography, etc.), the GIS offers a flexible and powerful tool than conventional data processing systems, as it provides a means of taking large volumes of different kinds of data sets and combining the data sets into new data sets (Foote and Lynch 1996). Remote sensing data coupled with soil survey information can be integrated in GIS to assess crop suitability for various soil and biophysical conditions. The potential of the integrated approach in using GIS and remote sensing data for quantitative land evaluation has been demonstrated earlier by several researchers (Beek et al. 1997; Merolla et al. 1994; Maji et al. 2005). With the assistance of a GIS, Liu and Deng (2001) developed a land resource management system to evaluate land suitability. Matching of the land attributes with the specific crop growth requirements and definition of the preliminary suitability classes was worked out. Walke et al. (2012) made an attempt to evaluate the soil resources of the Ringnabodi watershed, Nagpur district of Maharashtra, Central India, in soil-site suitability evaluation for cotton using multi-criteria overlay analysis techniques in GIS. Naidu et al. (2006) developed soil suitability criteria for major crops of the country to take land evaluation an activity of land use planning. LUP has been developed at different levels like village (Ramamurthy et al. 2006 and Ramamurthy and Sarkar 2009), watershed, and district (Ramamurthy et al. 2015). As part of land use planning, potential areas for cereals, pulses, oil seeds (Ramamurthy et al. 2012 and Naidu et al. 2015), rubber, tea, and medicinal and aromatic crops (Ramamurthy and Singh 2015) have been identified.

24.5.14 Precision Agriculture

Precision agriculture is the art and science of utilizing advanced technologies for enhancing crop production while minimizing potential environmental pollution (Khosla and Shaver 2001). Bramley (2001) defined precision agriculture as the term that incorporates technologies that permit the improved management of agricultural production through the recognition that land productivity and input-output relations can vary even in small distances in the field. Bramley (2001) defined precision agriculture as the term that incorporates technologies that permit the improved management of agricultural production through the recognition that land productivity and input–output relations can vary even in small distances in the field. Precision agriculture is the art and science of utilizing advanced technologies for enhancing crop production while minimizing potential environmental pollution (Khosla and Shaver 2001). Precision agriculture concept was initiated for site-specific crop management as a combination of positioning system technology, variable rate technology, remote sensing, yield mapping, etc. to optimize the profitability and sustainability with a reduced environmental impact. Geospatial technologies could be effectively used in precision agriculture to improve productivity and resources use efficiency either through increased yields or reduced inputs and adverse environmental effects. GPS, GIS, remote sensing, variable rate technology (VRT), yield mapping, and advances in sensor and information technology have enabled the crop growers to visualize the entire field in a way that could help to manage the agricultural operations efficiently and improve overall productivity. Ground-based sensors or "on the go" sensors have developed rapidly in recent years. Such efforts provide soil organic matter, electrical conductivity, nitrate content, and compaction (Barnes et al. 2003) and when integrated with other data sources maximize the information for the farm manager. ICT may be an effective tool for facilitating data collection, validation, access, exploration, and communication, by allowing various stakeholders to assemble and share data pertaining to the interests of any given project. Accordingly, in the last few decades, many projects have applied ICT systems to numerous fields, including the environmental monitoring of agricultural production. In terms of geospatial science, technology, and monitoring, using ICT systems has several distinct advantages, as real-time sensors seamlessly track, store, and recover field data, which can be used to investigate a large range of different issues.

24.6 Characterization and Delineation of Land Management Units: A Case Study

Land management units (LMUs) are the homogenous areas with respect to climate and land attributes and require similar management for agricultural production. Land management and planning is crucial for present and future use of land and the sustainability of land resources. Site, physical, chemical characteristics of land resources collected through land resource inventory can be used to define land management units (LMUs) that aid in decision-making for managing land and communicating information between different research and application domains. LMUs reflect potential and actual condition of the pieces of land resources. It is expected that the same classes of LMUs reflect similar land potentials and respond to particular decisions in a similar fashion. As a case study, characterization and delineation of land management units has been carried out for Gajwel Mandal, Medak District of Telangana State, India. Integration of data sets on land resources has been carried out in GIS for delineation of land management units (LMUs). The factors considered for grouping are soil depth, texture, gravelliness, slope, and erosion (Ramamurhty et al. 2010, 2017). The process involved spatial integration of external land features, soils, agroecology, present land use, administrative divisions, and farming systems data. The input data sets and their processing are presented in schematic diagram (Fig. 24.1).

Land resource inventory of Gajwel Mandal was carried out at 1:10,000 scale. Before suggesting land use plan for the Gajwel Mandal, 16 soil phases were grouped into 5 LMUs (Table 24.1, Fig. 24.2).

Each LMU was evaluated for biophysical suitability of different crops and their economic viability. Considering potential and constraints of each LMU, best suitable and economically viable crops and cropping systems (short-term plan) and agri-silvi-horti-pasture systems (long-term plan) were suggested with nutrient plan and

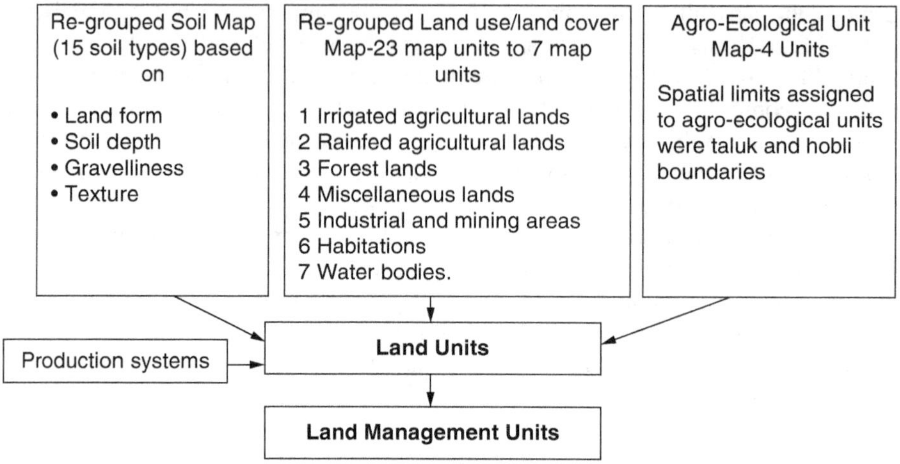

Fig. 24.1 Identification of land management units

Table 24.1 LMUs and their detailed description in Gajwel Mandal

LMU	Description
1	Shallow to very shallow, somewhat excessively drained, loamy skeletal soils with moderate and stoniness
2	Moderately deep to deep, well-drained to moderately well-drained, loamy sand soils with moderate erosion
3	Moderately deep to deep, moderately well-drained to well-drained, clayey soils with slight erosion
4	Very deep, well- to moderately well-drained, loamy soils with slight to moderate erosion
5	Very deep, well- to moderately well-drained, clayey soils with slight erosion

management plan (soil and water conservation measures). Crop wise specific management strategies were also provided to enhance the productivity of crops in the village (Table 24.2).

24.7 Future Challenges and Perspectives

The major challenges being faced in land resource management and land use planning are lack of adequate soil and water conservation measures especially in rainfed regions, depletion of groundwater table in different regions, land degradation due to salinity/alkalinity mainly in irrigated command areas, stagnating of net cultivable area, declining agricultural productivity due to low soil fertility, etc. Under these circumstances, a perspective land resource management and land use planning is needed that would involve alterations in the land area allocation over alternative uses, through suitable technological and institutional devices, taking into

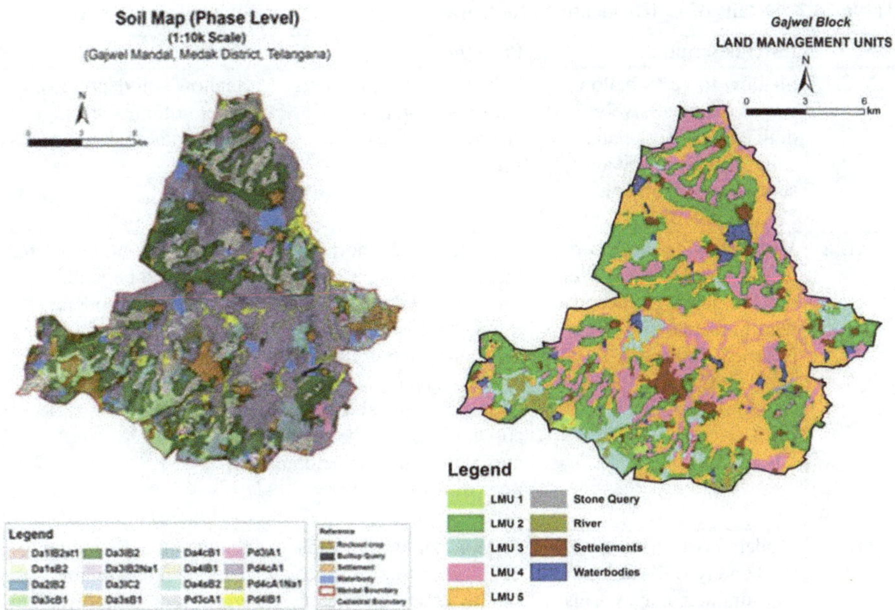

Fig. 24.2 Soil and LMU maps of Gajwel Mandal of Medak District of Telangana State

account the potentials and limitations of each land management unit (LMU). Some of the potentials and perspectives in land resource management and land use planning are summarized below.

- Advance geospatial tools like remote sensing, GIS, and GPS need to be effectively used in land resource inventory and mapping particularly in development of location-specific land resource database at block/watershed/village level for agricultural land use planning.
- Modern geospatial tools like remote sensing techniques, GPS, and GIS could be effectively used in periodically mapping and monitoring of croplands, degraded lands, and delineation of prime agricultural lands for scientific agricultural land use planning.
- A systematic appraisal of land resources and appropriate agricultural land use planning is necessary through adopting improved management practices, such as residue recycling, integrated use of inorganic fertilizers, and organic manures to enhance soil quality and production system.
- Develop and provide GIS-based soil fertility status to the planners, policy makers, and farmers for optimization of fertilizers and integrated nutrient management.
- Delineate land management units and crop-efficient zones based on the potentials and limitations of biophysical, resource availability and socioeconomic parameters to develop strategies for site-specific land use planning and land resource management on sustainable basis.

Table 24.2 Details of LMUs identified for Gajwel Mandal and their potential and constraints

LMU	LMU description	Potential	Constraints
LMU 1	Shallow to very shallow, somewhat excessively drained, loamy skeletal soils with moderate and stoniness (K1Ble2st1, K1Bse2, K2Ble2)	Suitable for small millets and short-duration pulses and silvi-horti-pasture system	Shallow soil depth, low water holding capacity, poor productivity of crops
LMU 2	Moderately deep to deep, well-drained to mod. Well-drained, loamy sand soils with moderate erosion (C3Ble2, C3Ble2Na1, C3Cle2, C3Bse1, Cv3Ale1)	Suitable for agri-horti-pasture system, maize, cotton, sunflower, red gram, jowar, maize/sunflower (short duration)-gram, millets, cowpea, green gram, black gram, and vegetables like coriander, cluster bean, pumpkin, paddy as aerobic rice cultivation with irrigation	Soil depth is limitation for deep-rooted perennial crops, low to moderate water holding capacity; protective irrigation is a must along with SWC measures
LMU 3	Moderately deep to deep, moderately well-drained to well-drained, clayey soils with slight erosion (C3Bce1, Cv3Ace1)	Suitable for agri-horti-pasture system, maize, cotton, sunflower, red gram, jowar, maize/sunflower (short duration)-gram, millets, cowpea, green gram, black gram, and vegetables like coriander, cluster bean, pumpkin, paddy as aerobic rice cultivation with irrigation	Soil depth is limitation for deep-rooted perennial crops, low to moderate to high water holding capacity; protective irrigation is a must along with SWC measures
LMU 4	Very deep, well- to moderately well-drained, loamy soils with slight to moderate erosion (C4Ble1, C4Bse2, Cv4Ble1).	Suitable to all crops	Erosion needs to be taken care
LMU 5	Very deep, well- to moderately well-drained, clayey soils with slight erosion (C4Bce1, Cv4Ace1, Cv4Ace1Na1)	Suitable to all crops	Erosion and water stagnation reduce the crop productivity

Source: Ramamurthy et al. (2010, 2017)

- Three-tier agricultural land use planning, i.e., block/village (1:10,000), district (1:50,000), and state (1:250,000), needs to be promoted to meet the demands of food grain requirements.
- Climate-resilient interventions need to be developed for sustainable land management in different agroecologies to enhance adaptive capacity to climate change and other climatic aberrations especially in rainfed agriculture.

- Sound agricultural land use policy is needed for region-specific scientific land use planning to improve farm productivity and profitability through conservation of land and water resources.
- There is a need for regular capacity building in the fields of land resource inventory, mapping and assessment of land degradation, and land use planning using emerging geospatial tools.

24.8 Conclusions

Remote sensing, GIS, and GPS technologies will continue to be very important in acquiring information on land resources and agricultural land use planning. Time series satellite imageries have unique ability to provide the actual synoptic view of large area as compared to conventional surveys, and processing and analysis of such aquired data are very fast in GIS. The integration of spatial and nonspatial data of land resources and their combined analysis could be performed through GIS for effective land resource management and land use planning. Remote sensing data coupled with soil survey information can be integrated in GIS to assess crop suitability for various biophysical conditions. The study shows that remote sensing, GIS, and GPS technologies have immense potential in land resource inventory, mapping and management, agroecological zonation, land degradation mapping and assessment, soil fertility mapping and assessment, development of land resource information system and decision support systems, crop suitability evaluation, and characterization of land management units for sustainable management of land resources in agricultural land use planning at various levels.

References

Ahamed TRN, Rao KG, Murthy JSR (2000) GIS based fuzzy membership model for crop land suitability analysis. Agric Syst 63:75–95

Anup KP, Chai L, Ramesh PS, Kafatos M (2005) Crop yield estimation model for Iowa using remote sensing and surface parameters. Int J Appl Earth Obs Geoinf 8:26–33

ASRIS (2011) ASRIS-Australian soil resource information system. http://www.asris.csiro.au. Accessed on 15th July, 2017

Barnes EM, Sudduth KA, Hummel JW, Lesch SM, Corwin DL, Yang C, Daughtry CST, Bausch WC (2003) Remote- and ground-based sensor techniques to map soil properties. Photogramm Eng Remote Sens 69(6):619–630

Baumgardner MF (1986) World soils and terrain digital database at a scale of 1, 1M (SOTER). Wageningen, ISSS

Beek KJ, De Bie K, Driessen P (1997) Land information and land evaluation for land use planning and sustainable land management. Land Chatham 1:27–44

Bhattacharyya T, Sarkar D, Pal DK, Mandal C, Baruah U, Telpande B, Vaidya PH (2010) Soil information system for resource management – Tripura as a case study. Curr Sci 99(9): 1208–1217

Bhumbla DR, Khare A (1984) Estimates of wastelands in India. Society for Promotion of Wasteland Development, New Delhi, 16p

Bishop TFA, McBratney AB (2001) A comparison of prediction methods for the creation of field-extent soil property maps. Geoderma 103:149–160

Bramley RGV (2001) Progress in the development of precision viticulture – variation in yield, quality and soil properties in contrasting Australian vineyards. In: Currie LD, Loganathan P (eds) Precision tools for improving land management. Occasional report No. 14. Fertilizer and Lime Research Centre, Massey University, Palmerston North

Burrough PA (1986) Principles of geographical information system for land resource assessment. Monograph on soil and land resource survey No. 12. Oxford University

Burrough PA (1989) Fuzzy mathematical methods for soil survey and land evaluation. J Soil Sci 40:477–492

Cambardella CA, Moorman TB, Novak JM, Parkin TB, Karlen DL, Turco RF, Konopka AE (1994) Field-scale variability of soil properties in Central Iowa. Soil Sci Soc Am J 58:1501–1511

Canadian Soil Information Service http://sis.agr.gc.ca/cansis/. Accessed on 29 July, 2017

Casa R, Jones HG (2005) LAI retrieval from multiangular image classification and inversion of a ray tracing model. Remote Sens Environ 98:414–428

Choker M, Baghdadi N, Zribi M, El Hajj M, Paloscia S, Verhoest N, Lievens H, Mattia F (2017) Evaluation of the Oh, Dubois and IEM backscatter models using a large dataset of SAR data and experimental soil measurements. Water 9:38

Corstanje R, Reddy KR, Prenger JP, Newman S, Ogram AV (2007) Soil microbial eco-physiological response to nutrient enrichment in a sub-tropical wetland. Ecol Indic 7(2):277–289

Das DC (1985) Problem of soil erosion and land degradation in India. In: Proceedings National Seminar on soil conservation and watershed management, New Delhi

Das NN, Mohanty BP, Cosh MH, Jackson TJ (2008) Modeling and assimilation of root zone soil moisture using remote sensing observations in walnut gulch watershed during SMEX04. Remote Sens Environ 112:415–429

Densham PJ, Goodchild MF (1988) Spatial decision support systems. A research agency. In: GIS/LIS' 88 proceeding value 2 Bethesda. American Congress on surveying and mapping, Maryland, November-December 1988

Dent D, Young A (1981) Soil survey and land evaluation. George Allen & Unwin, London

DoLR (2013) Draft National Land Utilisation Policy- framework for land use planning & management. Department of Land Resources, Ministry of Rural Development, Government of India, p 32

Doraiswamy PC, Hatfield JL, Jackson TJ, Akhmedov B, Prueger J, Stern A (2004) Crop condition and yield simulations using Landsat and MODIS. Remote Sens Environ 92:548–559

eSOTER of Global Earth Observation System of Systems (GEOSS) http://www.esoter.net. Accessed on 2 August, 2017

FAO (1976) A framework for land evaluation. Food and Agriculture Organization of the United Nations, Soils Bulletin 32. FAO, Rome

FAO (1983) Guidelines: land evaluation for rainfed agriculture. Food and Agriculture Organization of the United Nations, Soils Bulletin 52. Rome, Italy

FAO (1989) Guidelines: land evaluation extensive grazing, Soil Bulletin No, 55. Food and Agricultural Organization of the United Nations, Rome

FAO (1993) Guidelines for land use planning, FAO development series no.1, Rome

FAO (2007) Land evaluation towards a revised framework. Food and Agriculture Organization of the United Nations, Rome, Italy

FAO/AGL and ISSS (1986) Global soil and terrain database (WORLD-SOTER). The Food and Agricultural Organization of the United Nations. http://www.fao.org/ag/agl/agll/soter.stm

FAO-UNESCO (1974) Soil map of the world,. vol I – legend. Paris, 59 pp

Foote KE, Lynch M (1996) Geographic information systems as an integrating technology: context, concepts and definition, vol 2. University of Texas, Austin

Galvão LS, Roberts DA, Formaggio AR, Numata I, Breunig FM (2009) View angle effects on the discrimination of soybean varieties and on the relationships between vegetation indices and yield using off-nadir Hyperion data. Remote Sens Environ 113:846–856

Gangopadhyay SK, Reddy GPO, Mukhopadhyay S, Singh SK (2015) Characterization of landforms and soils in complex toposequence of Subarnarekha catchment, Chhotanagpur plateau using remote sensing and GIS. Agropedology 25(1):95–109

Government of India (1994) Indian agriculture in brief, 25th edn. Ministry of Agriculture, New Delhi, p 495

Haider SS, Said S, Kothyari UC, Arora MK (2004) Soil moisture estimation using ERS-2 SAR data: a case study in Solani river catchment. Hydrol Sci J 49(2):323–334

He Y, Yao Y, Chen Y, Ongaro L (2011) Regional land suitability assessment for tree crops using remote sensing and GIS. In: Computer Distributed Control and Intelligent Environmental Monitoring (CDCIEM) IEEE, Changsha, pp 354–363

Hengl T, Heuvelink GBM, Stein A (2004) A generic framework for spatial prediction of soil variables based on regression-kriging. Geoderma 120:75–93

Indiastat (2011) Land use statistics of India. www.indiastat.com. Accessed in 30 August 2017

ISRIC-WISE (2008) References and examples of WISE database applications. PDF ISRIC 2007. ISRIC World Soil Database

Khosla R, Shaver T (2001) Zoning in on nitrogen needs. Colorado State University. Agronomy Newsletter 21(1):24–26

Klingebiel AA, Montgomery PH (1961) Land-capability classification. USDA-SCS Agriculture Handbook 210. U.S. Government Print Office, Washington, DC

Krishnamurthy J, Srinivas G (1995) Role of geological and geomorphological factors in groundwater exploration: a study using IRS LISS data. Int J Remote Sens 16:2595–2618

Krishnan A, Singh M (1968) Soil climatic zones in relation to cropping patterns. In: Proceedings of the symposium on cropping patterns. Indian Council of Agricultural Research, New Delhi, pp 172–185

Li J, Liu Q, Liu Q, Tang Y, Xiao Q (2008) A patch spectral purification method to extract field patch average parameters from moderate resolution data. Int J Remote Sens 29:4993–5011

Liu YS, Deng XZ (2001) Structural patterns of land types and optimal allocation of land use in Qinling Mountainous. J Geogr Sci 11(1):99–109

Mahalanobis PC (1945) A report on the Bihar crop survey, 1943-44. Sankhya 7:29–118

Maji AK, Krishna NDR, Sehgal J, Velayutham M (1998a) Dissemination of soil information for district level planning using GIS. GIS India 7:21–25 Hyderabad

Maji AK, Krishna NDR, Challa O (1998b) Geographical information system in analysis and interpretation of soil resource data for land use planning. J Indian Soc Soil Sci 46:260–263

Maji AK, Nayak DC, Krishna NDR, Srinivas CV, Kamble K, Reddy GPO, Velayutham M (2001) Soil information system of Arunachal Pradesh in a GIS environment for land use planning. Int J Appl Earth Obs Geoinf 3:69–77

Maji AK, Srinivas CV, Dubey PN, Reddy GPO, Kamble KH, Velayutham M (2002) Soil resources information of Nagaland in GIS for land use planning in mountainous region. GIS India 11:13–16

Maji AK, Reddy GPO, Tamgadge DB, Gajbhiye KS (2005) Spatial modeling for crop suitability analysis using AGROMA GIS software. Asian J Geoinf 5(3):47–56

Maji AK, Reddy GPO, Sarkar D (2010) Degraded and wastelands of India, Status and spatial distribution. ICAR and NAAS Publication, pp 1–158

Maji AK, Reddy GPO, Sarkar D (eds) (2012) Acid soils of India – their extent and spatial variability. NBSS Publication No. 145, NBSS&LUP, Nagpur, 147p

Mandal DK, Mandal C, Singh SK (2016) Agro-Ecological Regions of India (revised). NBSS&LUP Publication No 170. ICAR-NBSS&LUP, Nagpur, India, pp 89

Global soil Map.net. www.globalsoilmap.net Accessed on 3 August, 2017

McBratney AB, Mendonca Santos ML, Minasny B (2003) On digital soil mapping. Geoderma 117:3–52

Merolla S, Armesto G, Calvanese G (1994) A GIS application for assessing agricultural land. ITC J 3:264–269

Mishra A, Das D, Saren S (2013) Preparation of GPS and GIS based soil fertility maps for Khurda district, Odisha. Indian Agric 57(1):11–20

Mu Y (2006) Developing a suitability index for residential land use: a case study in Dianchi drainage area. University of Waterloo, Canada

Mueller TG, Pierce F, Schabenberger O, Warncke D (2001) Map quality for sitespecific fertility management. Soil Sci Soc Am J 65(5):1547–1558

Mulla D (1991) Using geostatistics and GIS to manage spatial patterns in soil fertility. Wiley, Chichester. https://doi.org/10.1002/9780470515419.ch7

Murthy RS, Pandey S (1978) Delineations of Agro-Ecological Regions of India. Paper presented in Commission V, 11th Congress of ISSS, Edmonton, Canada, June 19–27

Naidu LGK, Ramamurthy V, Challa O., Hegde R, Krishanan P (2006) Manual soil-site suitability criteria for major crops. NBSS Publication No. 129, NBSS &LUP, Nagpur, 118 pp

Naidu LGK, Dharumarajan S, Lalitha M, Vasundhara R, Ramamurthy V, Reddy GPO, Singh RS, Tailor B, Singh R, Baruah U, Padmaiah M, Suresh G, Varaprasad KS (2015) Identification and delineation of potential castor growing areas in different agro-eco sub regions of India. J Oilseeds Res 32:39–48

National Commission on Agriculture (1976) Report, Ministry of Agriculture & Irrigation. Government of India, New Delhi

National Resource Conservation Service, USDA http://soils.usda.gov/survey/geography/ssurgo/description.html. Accessed on 21 July, 2017

National Resource Conservation Service, USDA. http://soils.usda.gov/survey/geography/ssurgo/description_statsgo2.html. Accessed on 21 July, 2017

Njoku EG, Jackson TJ, Lakshmi V, Chan TK, Nghiem SV (2003) Soil moisture retrieval from AMSR-E. IEEE Trans Geosci Remote Sens 41:215–229

Ochsner TE, Cosh MH, Cuenca RH, Dorigo WA, Draper CS, Hagimoto Y, Kerr YH, Njoku EG, Small EE, Zreda M (2013) State of the art in large-scale soil moisture monitoring. Soil Sci Soc Am J 77:1888–1919

Panigrahy S, Manjunath KR, Ray SS (2005) Deriving cropping system performance indices using remote sensing data and GIS. Int J Remote Sens 26(12):2595–2606

Peng J, Niesel J, Loew A (2015) Evaluation of soil moisture downscaling using a simple thermal-based proxy—the remedhus network (Spain) example. Hydrol Earth Syst Sci 19:4765–4782

Petersen GW, Bell GC, McSweeney K, Nielsen GA, Robert PC (1995) Geographical information system in agronomy. Adv Agron 55:67–111

Prakash TN (2003) Land suitability analysis for agricultural crops: a fuzzy multicriteria decision making approach, science in Geoinformatics. ITC, Netherlands, pp 6–13

Prasad AK, Chai L, Singh RP, Kafatos M (2006) Crop yield estimation model for Iowa using remote sensing and surface parameters. Int J Appl Earth Obs Geoinf 8:26–33

Ramamurthy V, Sarkar D (2009) Soil based agro technologies for livelihood improvement. NBSS Publication No. 144. NBSS&LUP, Nagpur, 44p

Ramamurthy V, Singh SK (2015) Land use planning for important medicinal and aromatic plants in Karnataka. NBSS Publication No. 1081, 22pp

Ramamurthy V, Gajbhiye KS, Lal S, Maji AK (2006) Land resource management through technology assessment and refinement. NBSS Publication No. 133. NBSS&LUP, Nagpur, 106p

Ramamurthy V, Nair KM, Ramesh Kumar SC, Srinivas S, Naidu LGK, Sarkar D (2010) Land management unit approach: step towards precision farming in XIX National symposium on "resource management approaches towards livelihood security" held at Bangalore from 02–04, December 2010

Ramamurthy V, Naidu LGK, Srinivas S, Reddy GPO, Sarkar D (2012) Delineation of potential areas for safflower production in Deccan Plateau. Journal of Oilseeds Research 29:226–233

Ramamurthy V, Naidu LGK, Nair KM, Ramesh Kumar SC, Srinivas S, Thayalan S, Sarkar D, Chaturvedi A, Singh SK (2015) District land use planning. NBSS Publication No. 169. Mysore, Karnataka, 95pp

Ramamurthy V, Nair KM, Ramesh Kumar SC, Srinivas S, LGK N, Sarkar D, Singh SK (2017) Delineation of land management units in Tropical South Deccan Plateau of India. Agropedology 26(02):105–116

Rao DP, Gautam NC, Nagaraja R, Ram Mohan P (1996) IRS-IC application in land use mapping and planning. Curr Sci 70(7):575–578

Reddy GPO, Maji AK (2003) Delineation and characterization of geomorphological features in a part of lower Maharashtra Metamorphic Plateau, using IRS-ID LISS-III data. J Indian Soc Remote Sens 31(4):241–250

Reddy GPO, Maji AK (2004) Characterization of biophysical land units using remote sensing and GIS. J Indian Soc Remote Sens 32(2):159–165

Reddy GPO, Rao MS (1994) Hydro-geology and hydro-geomorphological conditions of Anantapur district (AP), India. Indian Geogr J 69(2):128–135

Reddy GPO, Sarkar D (2012) Assessment of soil loss for prioritization of sub watersheds-a remote sensing and GIS approach, NBSS Publication No. 137. NBSS&LUP, Nagpur, pp 54

Reddy GPO, Singh SK (2014) Achieveing food security through land management. Geospatial Today:14–20

Reddy GPO, Chandramouli K, Srivastav SK, Maji AK, Srinivas C (1999) Evaluation of groundwater potential zones using remote sensing data – a case study. J Indian Soc Remote Sens 28(1):19–32

Reddy GPO, Maji AK, Srinivas CV, Thayalan S, Velayutham M (2001) Landscape ecological planning in a basaltic terrain, Central India, using remote sensing and GIS techniques. J Indian Soc Remote Sens 29:3–16

Reddy GPO, Maji AK, Chary GR, Srinivas CV, Tiwary P, Gajbhiye KS (2004) GIS and remote sensing applications in prioritization of river sub basins using morphometric and USLE parameters - a case study. Asian J Geoinform 4(4):35–49

Reddy GPO, Dheeravath V, Thenkabail PS, Chandrakantha G, Biradar CM, Noojipady P, Velpuri M, Maji AK (2009) Irrigated areas of India derived from satellite sensors and national statistics – a way forward from GIAM experience. In: Thenkabail PS, Lyon JG, Turral H, Biradar CM (eds) Remote sensing of global croplands for food security, Taylor and Francis Series in Remote Sensing Applications. CRC Press, London, pp 139–176

Reddy GPO, Nagaraju MSS, Ramteke IK, Sarkar D (2013a) Terrain characterization for soil resource mapping in part of semi-tract of Central India using high resolution satellite data and GIS. J Indian Soc Remote Sens 41:331–343

Reddy GPO, Sarkar D, Prasad J, Ramamurthy V (2013b) Geospatial modeling in assessment of biophysical resources for sustainable land resource management. Trop Ecol 54(2):227–238

Reddy GPO, Mondal C, Srivastava R, Bhattacharyya T, Naidu LGK, Sidhu GS, Baruah U, Singh RS, Kumar N, Singh SK, Sarkar D (2014). Development of Indian soil information system- a Geoportal. Project Report, pp 34

Reddy GPO, Kurothe RS, Sena DR, Harindranath CS, Niranjana KV, Naidu LGK, Singh SK, Sarkar D, Mishra PK, Sharda VN (2016) Assessment of soil erosion in tropical ecosystem of Goa, India using universal soil loss equation, geostatistics and GIS. Indian J Soil Conserv 44(1):1–7

Reddy GPO, Patil NG, Chaturvedi A (2017a) Sustainable management of land resources-an Indian perspective. Apple Academic Press, Oakville, p 769

Reddy GPO, Singh SK, Kumar N, Mondal C, Srivastava R, Maji AK, Sarkar D (2017b) Development of soil information system and prototype GeoPortal for resource management. In: Reddy GPO, Patil NG, Chaturvedi A (eds) Sustainable management of land resources - an Indian perspective. Apple Academic Press, New York, pp 411–442

Romano N (2014) Soil moisture at local scale: measurements and simulations. J Hydrol 516:6–20

Rossiter GD (1996) A theoretical frame work for land evaluation. Goederma 72:165–190

Röver M, Kaiser EA (1999) Spatial heterogeneity within the plough layer: low and moderate variability of soil properties. Soil Biol Biochem 31(2):175–187

Sahu N, Reddy GPO, Kumar N, Nagaraju MSS, Srivastava R, Singh SK (2014) Characterization of landforms and land use/land cover in basaltic terrain using IRS-P6 LISS-IV and Cartosat-1 DEM data: a case study. Agropedology 24(2):166–178

Sahu N, Reddy GPO, Kumar N, Nagaraju MSS (2015) High resolution remote sensing, GPS and GIS in soil resource mapping and characterization- a review. Agric Rev 36(1):14–25

Sarkar D, Gangopadhyay SK, Sahoo AK (2006) Soil resource appraisal towards land use planning using satellite remote sensing and GIS- a case study in Patloinala microwatershed, District Puruliya, West Bengal. J Indian Soc Remote Sens 34:245–260

Sarkar D, Reddy GPO, Naidu LGK (2012) Genesis and spatial distribution of acid soils in India towards sustainable crop production. In: Sudhir S et al (eds) Acid soils of India- distribution, properties and management for sustainable crop production, pp 1–14

Sehgal JL, Abrol IP (1994). Soil degradation in India: status and impact. Oxford and IBH Publishing, New Delhi, 80p

Sehgal JL, Mandal DK, Mandal C, Vadivelu S (1992) Agro-ecological regions of India, 2nd edn. Technical Bulletin. NBSS Publication No. 24. Nagpur, India

Shah T (1993) Groundwater markets and irrigation development: political economy and practical policy. Oxford University Press, Bombay

Shi J, Wang A, Hsu Y, Neill PEO', Engman ET (1997) Estimation of bare surface soil moisture and surface roughness parameter using L-band SAR image data. IEEE Trans Geosci Remote Sens 35(5):1254–1266

Singh R, Goyal RC, Saha SK, Chhikara RS (1992) Use of satellite spectral data in crop yield estimation surveys. Int J Remote Sens 13(14):2583–2592

Singh R, Semwal DP, Rai A, Chhikara RS (2002) Small area estimation of crop yield using remote sensing satellite data. Int J Remote Sens 23(1):49–56

Singha M, Bingfang W, Zhang M (2016) An object-based paddy rice classification using multi-spectral data and crop phenology in Assam, Northeast India. Remote Sens 8:1–20

Sonneveld MPW, Hack-ten Broeke MJD, van Diepen CA, Boogaard HL (2010) Thirty years of systematic land evaluation in the Netherlands. Geoderma 156:84–92

Srivastava R, Saxena RK (2004) Techniques of large scale soil mapping in basaltic terrain using satellite remote sensing data. Int J Remote Sens 25(4):679–688

Subramaniam AR (1983) Agro-ecological zones of India. Arch Met Geoph Biocl Ser Bull 32:329–333

Sukhatme PV, Panse VG (1951) Crop surveys in India - II. J Indian Soc Agric Stat 2:95–168

Sys C (1985) Land evaluation. Algemeen Bestuur vande Ontwikkelingss. International Training Centre for Post-Graduate Soil Scientists. State University of Ghent, Ghent

Sys C, Van Ranst E, Debaveye J (1991) Land evaluation. Part I: principles in land evaluation and crop production calculations. General Administration for Development Cooperation. Agricultural Publication No. 7

Tan Z, Lal R (2005) Carbon sequestration potential estimates with changes in land use and tillage practice in Ohio, USA. Agric Ecosyst Environ 111(1–4):140–152

Thenkabail PS, Schull M, Turral H (2005) Ganges and Indus river basin land use/land cover (LULC) and irrigated area mapping using continuous streams of MODIS data. Remote Sens Environ 95:317–341

Thenkabail PS, Dheeravath V, Biradar CM, Reddy GPO, Noojipady P, Gurappa C, Velpuri M, Gumma M, Li Y (2009) Irrigated area maps and statistics of India using remote sensing and National Statistics. Remote Sens 1(2):50–67

Thilagam VK, Sivasamy R (2013) Role of remote sensing and GIS in land resource inventory- a review. Agric Rev 34:223–229

Townshend JRG, Justice CO (2002) Towards operational monitoring of terrestrial systems by moderate-resolution remote sensing. Remote Sens Environ 83:351–359

Ulaby F, Moore R, Fung A (1982) Microwave remote sensing: active and passive, vol 3. Artech House, Norwood

Velayutham M, Mandal DK, Mandal C, Sehgal JL (1999) Agro-ecological subregions of India for planning and development. NBSS Publication No. 35. NBSS&LUP, Nagpur, India, 372 p

Vishakha TD, Reddy GPO, Maji AK, Ramteke IK (2013) Characterization of landforms and soils in complex geological formations – a remote sensing and GIS approach. J Indian Soc Remote Sens 41(1):91–104

Wagner W, Blöschl G, Pampaloni P, Calvet JC, Bizzarri B, Wigneron JP, Kerr Y (2007) Operational readiness of microwave remote sensing of soil moisture for hydrologic applications. Hydrol Res 38:1–20

Walke N, Reddy GPO, Maji AK, Thayalan S (2012) GIS-based multi-criteria overlay analysis in soil-suitability evaluation for cotton (Gossypium spp.): a case study in the black soil region of Central India. Comput Geosci 41:108–118

Wani SP, Sreedevi TK, Rockström J, Ramakrishna YS (2009) Rain-fed agriculture – past trend and future prospects. In: Wani SP, Rockström J, Oweis T (eds) Rain-fed agriculture: unlocking the potential, Comprehensive assessment of water management in agriculture series. CAB International, Wallingford, pp 1–35

Wardlow BD, Egbert SL (2008) Large-area crop mapping using time series MODIS 250 m NDVI data: an assessment for the US Central Great Plains. Remote Sens Environ 112(3):1096–1116

Wardlow BD, Egbert SL, Kastens JH (2007) Analysis of time-series MODIS 250 m vegetation index data for crop classification in the US Central Great Plains. Remote Sens Environ 108(3):290–310

Webster R, Mcbratney AB (1987) Mapping soil fertility at brooms barn by simple kriging. J Sci Food Agric 38(2):97–115

Wen J, Su Z, Ma Y (2003) Determination of land surface temperature and soil moisture from tropical rainfall measuring mission/microwave imager remote sensing data. J Geophys Res 108(D2):805–826

Western AW, Grayson RB, Bloschl G (2002) Scaling of soil moisture: a hydrologic perspective. Annu Rev Earth Planet Sci 30:149–180

Wilson JP, Helena M, Dawn JW (2000) Water resource applications of geographic information systems. URISA J 12(2):61–79

Yadav JSP (1996) Land degradation and its effects on soil productivity, sustainability and environment. Fourth U.P Bali Memorial Lecturer, Soil Conservation Society of India, New Delhi

Chapter 25
Trends in Land Resource Management and Land Use Planning

V. Ramamurthy

Abstract Land is finite, pressure on land resources increasing, and competition and conflicts also rising among the stakeholders. In order to avoid competition and land degradation, proper land use planning and land resource management approaches needs to be followed. Agenda 21, the Programme of Action for Sustainable Development at UNCED conference, recognizes the need to allocate land for sustainable uses and promotes integrated planning and management of land resources. Different types of conflicts arise during land management like inter micro-micro, intra micro-micro, and micro-macro conflicts. The components of a conflict management plan will vary with each situation. Since the 1950s, different policies and programs were formulated and implemented by government on land resource management. Land use planning is a tool to implement land resource management in relation to land evaluation. Modern approaches of LUP are always focused on one or more specific objectives and are closely linked to the concept of efficiency, equity, and sustainability. To achieve these objectives, participatory and integrated approaches are to be applied to manage land resource on sustainable basis.

Keywords Land · Land resources · Land use planning · Land management · Participatory land use planning

25.1 Introduction

Land is one of the most essential natural resource for the survival of mankind, and it is the platform on which human activities take place. Food and Agricultural Organization (FAO) defined land and land resources as "Land and Land resource refers to a delineable area of the earth's terrestrial surface, encompassing all attributes of the biosphere immediately above or below this surface, including those of the near-surface, climate, the soil and terrain forms, the surface hydrology, the near surface

V. Ramamurthy (✉)
ICAR-National Bureau of Soil Survey & Land Use Planning, Regional Centre, Bengaluru, India

G. P. O. Reddy, S. K. Singh (eds.), *Geospatial Technologies in Land Resources Mapping, Monitoring and Management*, Geotechnologies and the Environment 21,
https://doi.org/10.1007/978-3-319-78711-4_25

sedimentary layers and associated ground water and geo-hydrological reserve, the plant and animal population, the human settlement pattern and physical results of past and present human activity" (FAO 1997). This highlights that land includes all material components needed for human activity. Management of these resources including its conservation and utilization is of crucial importance. In many countries across the globe, land is subjected to varying degrees and forms of degradation due to the pressure of the growing population; increased demand for food, fodder, and fuel wood; and intensive industrial activity.

One out of every three people on earth is in some way affected by land degradation. Latest estimates indicate that nearly 2 billion ha of land worldwide – an area twice the size of China – are already seriously degraded, some irreversibly (FAO 2011). This includes large areas of cropland, grassland, and woodland and forest areas whose degradation reduces productivity, disrupts vital ecosystem functions, negatively affects biodiversity and water resources, and increases vulnerability to climate change (FAO 2011).

Land is scarce in India, even though the country has a land area of about 328 million hectares, which is the seventh largest land area among the countries of the world and is equal to only 2.4% of the world's geographical area, supporting approximately 16% of the world's human population and 20% of the world's livestock population (DES 2016). India is burdened with a population of 1210 million as per the 2011 census, which grew from 345 million in 1947 with a growth rate of 1.76 in the last decade. Population density has increased from 117 per sq.km in 1951 to 368 in 2011. The increasing human and animal population has reduced the per capita availability of land over the decades. The per capita availability of land has declined from 0.89 ha in 1951 to 0.37 ha in 1991 and is projected to slide down to 0.20 ha in 2035 and less than 0.07 ha in 2050. As far as agricultural land is concerned, the per capita availability of land has declined from 0.48 ha in 1951 to 0.16 ha in 1991, and it is likely to decline further to 0.08 ha in 2035.

In India, land is a source of livelihood for 60% of the population through agriculture and related activities. Population growth and the consequent demand for land, water, and biological resources have put tremendous pressure on land. Competition for land among the different uses is becoming acute, and conflicts arising out of this competition are more frequent and more complex. Land is crucial for all developmental activities, for natural resources, for ecosystem services, and for food security. As a consequence of various developmental endeavors, present arable land is shrinking because of diversion of prime agricultural lands to other nonagricultural uses, and agriculture-related activities are being taken up on marginal lands. Due to these activities, ecological balances are being impaired by soil erosion, siltation of dams, shortage of groundwater, land and water pollution, water logging, etc. Agenda 21 recognizes the need to allocate land for sustainable uses and promote the integrated planning and management of land resources.

25.2 Status of Land Resources

India with a total geographical area of 329 Mha has the seventh largest land area (occupying 2.4% of the earth's surface) and is the second largest populated (population density of 368 persons per sq.km) country in the world. The area available for agriculture, forestry, pasture, and other biomass production in India is 262 Mha, and the net sown area is 140 Mha (Table 25.1). The remaining area is not suitable for agriculture due to inaccessibility of the terrain or harsh nature of the climate.

The degradation of land resources is taking place at an alarming rate, and not all the cultivated lands at present are in its prime productivity. A number of surveys and assessments have been carried out by different agencies over the past several decades on the extent and type of land degradation occurring in the country.

However, recently harmonized datasets of land degradation and wastelands of India shows that about 120.72 Mha area is under different categories of degraded and wastelands in India (Maji et al. 2010) (Table 25.2).

A close look at the present health of the soil and water resources of India reveals the need of comprehensive land use policy. About 120.72 Mha are threatened by various types of degradations like salinity, alkalinity, ravine and gully erosion areas, areas under ravages of shifting cultivation, and desertification. The highest proportion of degradation is caused by soil erosion (9.86%). There are also specific problems of land degradation due to opencast mining.

Currently, India produces about 257 million tons of food grains (2014–2015), whereas by the end of the decade, it is estimated that the demand for food shall rise to 307 million tons. Further, land use planning (LUP) has so far not been comprehensive and adequate, particularly to deal the competitive demands by various sectors. Thus, there is a need to protect agricultural areas that are essential for food security including the prime agricultural lands, command areas, double-cropped land, and other lands that are essential for livelihood of rural population. Thus, conserving

Table 25.1 Land use in India (Mha) (2012–2013)

S.No.	Land use	Area (in Mha)	% Area to TGA
1	Total geographical area	328.72	
2	Forests	70.00	22.89
3	Area under nonagricultural uses	26.45	8.06
4	Barren and uncultural land	17.28	5.60
5	Permanent pastures and grazing lands	10.24	3.35
6	Misc. tree crops and groves	3.16	1.03
7	Culturable wasteland	12.58	4.11
8	Old fallow lands	11.00	3.60
9	Current fallows	15.28	5.00
10	Net sown area	139.93	45.76

Source: Directorate of Economics & Statistics, Ministry of Agriculture

Table 25.2 Extent of degraded and wastelands in India

Degradation type	Area (in million hectares)	Open forest (<40% canopy) (m. ha)
Water erosion	73.27	9.30
Wind erosion	12.40	–
Chemical degradation soils		
Exclusively salt-affected soils	5.44	
Salt-affected and water-eroded soils	1.20	0.10
Exclusively acidic soils (pH < 5.5)	5.09	–
Acidic and water-eroded soils	5.72	7.13
Physical degradation		
Mining and industrial waste	0.19	
Water logging	0.88	
Total	**104.19**	**16.53**
Grand total area (arable land and open forest)	**120.72**	

Source: Maji et al. (2010). Values in parenthesis are % of TGA (329 Mha) of India

prime agricultural lands by proper land use planning and sustainable management of land resources at different levels assumes mammoth importance. Agenda 21, the Programme of Action for Sustainable Development at UNCED conference in June 1992 in Rio de Janeiro, recognizes the need to allocate land for sustainable uses and promotes integrated planning and management of land resources. FAO (1993) complained that there is currently very little awareness of the importance of planning for land resource management (LRM) at any level; very little progress has been made in developing a relationship between government policy and land use decisions; and very few countries have an effective institutional structure for land resource development and conservation.

25.3 Land Resource Management and Land Use Planning

Land resource management is the process of managing the use and development of land resources. Land use planning means different things to different people. Most practitioners would put it as the assessment of land and water potential, alternatives for land use, and economic and social conditions in order to select and adopt the best land use options (FAO 1993). Dent (1993) described it as an exercise of foresight in making appropriate, informal decisions regarding the use of land. So, the purpose of LUP is to provide the information and foresight to support decision-making. It should also resolve the conflicts of interest between the disparate groups of people who have a hand in the use of the land and, also, the interest of the generations that will inherit it.

25.4 Regional Issues/Conflicts in Land Management

The word "conflict" carries negative connotations (Warner 2000). Different types of conflicts can be categorized in terms of whether they occur at the micro-micro or micro-macro levels (Table 25.3), i.e., among community groups or between community groups and outside government and among private or civil society organizations (Warner 2000).

Conflicts arising from poor enforcement of natural resource management regulations include (Umesh Babu and Nautiyal 2003):

- Private companies avoiding compliance and sanctions by threatening to withdraw their investment or by manipulating courts
- A general lack of understanding of environmental laws and regulations by industries and government agencies and the general population
- Noncompliance arising from unrealistic requirements for pollution control technology and poor implementation of environmental impact mitigation plans
- Failure of the courts to enforce regulations because of prolonged legal processes, with the outcome often unsupported by one or more parties
- Perverse incentive structures promoted by conventional cost-benefit analysis

Table 25.3 Types of conflicts arising in land management

Intra micro-micro conflicts	Inter micro-micro conflicts	Micro-macro conflicts
Disputes over land and resource ownership, e.g., between private and communal land owners Disputes over land boundaries between individuals or groups Latent family and relationship disputes Disputes due to natural resource projects being captured by elites and/or those who happen to own resources of a higher quality Breaking of CPR constitutional or operational rules, such as protection agreements for grazing areas, fish net sizes, forests, or misappropriation of funds, etc. Disputes over unfair distribution of work and profits	Conflict between landowners and resource users Conflict between indigenous CPR groups and more recent settlers Disputes generated by jealousy related to growing wealth disparities Lack of cooperation between different community groups Disputes over renewal arrangements for leased land Internal land ownership disputes ignited by the speculative activities of commercial companies Resentment built up due to lack of representation in village committees	Contradictory natural resource needs and values, e.g., between wildlife habitat protection and local livelihood security Cultural conflicts between community groups and outsiders Disputes over project management between community groups and outside project sponsors Disputes caused by political influence (national, provincial, or local) Disputes arising from differences between the aspirations of community groups and expectations of NGOs or commercial companies Off-site environmental impacts affecting unintended third parties

Source: (Warner 2000)

The conflict management plan describes the overall strategy for managing the conflict, combined with the proposed process of consensus-building and an initial set of conflict mitigation or prevention options. The components of a conflict management plan will vary with each situation.

25.5 Government Policies on Land Conservation/ Management

The constitution of India enables the central government and the states to enact laws for the preservation and conservation of natural resources. Article 39(b) and (c) of the Directive Principles of State Policy lays down that it is the duty of the state and the center to develop natural resources for common good. There is a constitutional provision for the involvement and participation of the people at local level for participatory planning and decision-making. The Eleventh Schedule (Article 243-G) of the constitution lists matters pertaining to land improvement, implementation of land reforms, land consolidation, soil conservation, and watershed development and management under the authority and responsibility of Panchayats. Table 25.4 presents the important policies formulated, programs implemented, and the institutional framework adopted in India.

Sustainable land management (SLM) is crucial to minimizing land degradation, rehabilitating degraded areas, and ensuring the optimal use of land resources for the benefit of present and future generations. SLM is based on four common principles:

- Land-user-driven and participatory approaches
- Integrated use of natural resources at ecosystem and farming systems levels
- Multilevel and multistakeholder involvement
- Targeted policy and institutional support, including development of incentive mechanisms for SLM adoption and income generation at the local level

Its application requires collaboration and partnership at all levels – land users, technical experts, and policy makers – to ensure that the causes of the degradation and corrective measures are properly identified and that the policy and regulatory environment enables the adoption of the most appropriate management measures.

25.6 Strategies for Sustainable Land Management

- High priorities to site-specific programs such as watersheds, river valleys, arid areas, and wastelands.
- Public policies on land use and the impact of subsequent land use on natural resources.

Table 25.4 Policies, acts, and programs by government on land resource management

Year	Initiatives	Features
1950s	National land reform policy	Abolition of intermediary tenures Tenancy reforms Celling on agricultural holdings and redistribution of surplus land Updating and maintenance of land records Consolidation of land holdings Distribution of government wasteland
1972–1973	Drought Prone Areas Programme (DPAP)	Minimize adverse effects of droughts on the productivity of land, water, and human resource Promote overall economic development and improve the socioeconomic condition of poor and disadvantaged sections inhabiting the program areas Capacity building and empowerment of village community ensuring participation of Panchayat Raj institutions and NGOs in program implementation at grassroot level and transfer of funds as well as decision-making power to the local people Since 1995–1996, a watershed development-based approach has been adopted
1977–1978	Desert Development Programme (DDP)	Mitigate adverse effects of desertification and adverse climatic conditions on crops and human and livestock population Restoration of ecological balance by harnessing, conserving, and developing natural resources, i.e., land, water, and vegetative cover, and raise land productivity Capacity building and empowerment of village community and ensuring participation of Panchayat Raj institutions and NGOs
1985	National Land Use and Conservation Board	Formulate a national policy and perspective plan for conservation, management, and development of land resources of the country Review of the progress of implementation of ongoing schemes and programs connected with conservation and development of land resources and soils Take measures to restrict the conversion of good agricultural land to nonagricultural uses Coordinate the work of State Land Use Boards
1985	National Wastelands Development Board (NWDB)	Formulate perspective plan and programs for the management and development of wastelands in the country Identify the wastelands in the country, review the progress of implementation of programs and schemes for the development of wasteland Create a reliable database and documentation center on related aspects of wasteland development

(continued)

Table 25.4 (continued)

Year	Initiatives	Features
1988	National Land Use Policy	To install an efficient and effective administrative structure for prescribing and regulating land by all concerned and revitalize the land use boards in this respect Prevent further deterioration of land resources, restore the productivity of degraded lands Allocate land for different uses based upon land capability, land productivity, and national production goals Complete the inventory of land resources based on prescribed land use
1989–1990	Integrated Wastelands Development Project (IWDP)	Adopt soil and moisture conservation measures such as terracing, bunding, trenching, vegetative barriers, etc. Encourage natural regeneration, enhance people's participation in wasteland development programs at all stages resulting in equitable sharing of benefits Employment generation, poverty alleviation, community empowerment, and development of human and other economic resources of the village Training, extension, and creation of awareness among the participants
1992	Constitution (Seventy third Amendment) act, 1992	Gives land-related subject to the Panchayat Raj institutions (local self-governments) at the village, block, and district levels to ensure participatory planning, decision-making, and monitoring of programs by the local self-governments
1992	Constitution (Seventy-Fourth Amendment) Act, 1992	Regulation of land use and urban planning were brought under the functional domain of urban self-governing bodies
1999	Department of Land Resources	Coordination of land administration in India Formulation of integrated land resource management policies Implementation of land-based development programs
2013	Department of Land resources	Draft national land reforms policy Draft national land utilization policy

- Coordinate the activities of all line departments and adopt an integrated approach.
- Expansion and intensification of irrigated agriculture.
- Address weaknesses in land use policies as well as options that are available to deal with natural resource management and conservation issues.
- Establish the horizontal linkages between various agencies that are involved in land resource management.
- Involve the stakeholders from the planning stage onward and address socioeconomic and poverty issues in land development programs.

- The government should take the lead role in capacity building at the grassroot level by planning, implementing, and monitoring integrated land resource management programs.
- Intensification of high-quality rainfed lands.

25.7 Land Use Planning

Soil survey and land evaluation are the means of helping farmers, foresters, engineers, planners, developmental agencies, and other users to make wise decisions about land use and management. Various national and state research institutes including State Agricultural Universities (SAUs) are involved in the development of agriculture in different production systems. In spite of all these efforts, there exist a number of bottlenecks for sustainable land resource management or agriculture, important ones being noninvolvement of farmers in the planning process and policy making as well as lack of coordinated multidisciplinary approach toward proper land use planning.

Considerable emphasis was given on carrying out different types of land resource surveys. In fact, the main aim of national soil surveys was to produce maps and related inventories for land use planning and management issues. Despite innovative research on pedology, land capability assessment, and soil survey interpretations during the last few decades, this information was of little use in making and translating land use plans. A *customer-oriented approach* has become increasingly common whereby soil survey institutes to varying extent have had to finance their activities through contract funding. One benefit from this change has been the move toward the integration of land resource surveys and land evaluation with the processes of land use planning and management. Gradually soil surveyors have found themselves in public inquiries when arguments exist over land quality in connection with proposed developments, which will result in the permanent loss of land from agriculture. The gradual adoption of government schemes designed to maintain or enhance environmental quality has necessitated the need to monitor the effects of such policies. Resource monitoring schemes increasingly became important as a means of determining the success of particular policies designed to encourage particular land management strategies.

The 1980s was a decade of dramatic change in the evaluation of land resources, a trend that further accelerated in the 1990s. The technological developments since the mid-1980s in geographic information systems (GIS) have had considerable impact on the methods of input, processing, and output of spatially referenced data. This period was distinguished by the increased application of computer technology and a greater emphasis on shaping the results from projects to particular circumstances and planning needs. Such developments were built upon the major methodological advance of the framework for land evaluation (FAO 1976) for land use planning.

Despite continued efforts to provide soil information and partly because of the highly technical terminology created by different approaches and formats, agronomists, environmentalists, or other potential users have used the information to a limited extent. Farmers, who are the main end users, have hardly benefited from soil surveys, and the soil information that they do use includes very simple soil fertility tests. The main reasons expressed by nonspecialists for not using soil data are that the scales of the maps are not convenient, that the information provided is too complex, and that the definitions of the soil units are not practical enough. The question of scale is a major limitation, as small-scale maps can be used for general planning as compared to large-scale maps. Agronomists and farmers are mostly looking for micro variations at and between the farmer's fields. The environmentalists prefer at watershed level, and economists/planners may prefer broad regional level. The soil scientists, however, are more at ease at the regional or country level. Understandably, there has been limited information and output at microlevel.

25.8 Land Use Planning: Approaches

"Conventional or top-down" and "participatory or bottom-up" are the two basic approaches to land use planning. Land use planning addresses the present and projected land utilization patterns under agricultural and nonagricultural sectors. It attempts to strike a balance between agricultural and nonagricultural sectors as per the potential of land and demand of the growing population in an area. The soil resource, climatic, land use information and the database of the land management units need appropriate planning for sustainable development at state/regional, district, and watershed/village levels

The state-/regional-level planning is primarily concerned with the priority allocation of the resources between the competing demands of different sectors, which may form the keys for sustainable LUP. The land systems, land facets (Challa and Dalrymple 1993), agro-eco regions and subregions (Velayutham et al. 1999), and land management units (Ramamurthy et al. 2017) provide information on the potentials and problems of an area. This enables a planner or policy maker to allocate the land resources under different sectors of agricultural and nonagricultural uses. A perspective land use plan includes a set of decisions about the ways and means to bring out the desired land use patterns.

In the participatory or bottom-up approach, individual farm holding and its management through participation of individual farmer as stakeholder in the developmental planning is considered as prime criterion. In this approach the planning starts at farm level, i.e., at microlevel, and proceeds toward national level, i.e., at macro-level with generalization at each higher level.

25.8.1 Land Use Planning at National/Regional Level

At the national level, planning is concerned with national goals and the allocation of resources. In many cases, national land use planning does not involve the actual allocation of land for different uses, but the establishment of priorities for district-level projects. A national land use plan should have the following:

- A strong land use policy: to balance the competing demands for land among different sectors of the economy, food production, export crops, tourism, wildlife conservation, housing and public amenities, roads, industry, etc.
- National development plans and budget: project identification and the allocation of resources for development
- Coordination of sectoral agencies involved in land use
- Legislation on such subjects as land tenure, forest clearance, and water rights

National goals are complex, while policy decisions, legislation, and fiscal measures affect many people and wide areas. Decision-makers cannot possibly be specialists in all facets of land use, so the planners' responsibility is to present the relevant information in terms that the decision-makers can both comprehend and act on.

Regional level refers not necessarily to administrative state boundaries but also land areas that fall between states and district levels. Planning often at this level considers climate and land areas that are suitable for particular crop to identify commodity regions at national level. Commodity region planning requires 1:250,000 scale soil resource information which is available and climate and socio-economic data like commodity-specific market network, processing industries, and storage network details. This planning not only enhances the productivity of particular commodity but also resource use efficiency and reduces cost of production. Specific commodity region planning helps district/state decision-makers to take timely decisions on pooling commodity-specific seed and other inputs to particular region; thereby farming community can get quality input timely with customized commodity-specific land use plans/scientific management knowledge. Also it helps in forecasting of particular commodity production and gives input in policy formulation at national level. Besides agricultural planning, this approach can be used to identify or delineate industrial zones, eco-sensitive zones, etc.

25.8.1.1 Case Study: Potential Soybean-Growing Areas

Potential soybean growing areas in India are delineated as per the conceptual model presented in Fig. 25.1. The analysis shows that Vidarbha region of Maharashtra and Malwa region of Madhya Pradesh are highly potential for cultivation of soybean (Fig. 25.2). Parts of Marthawad, Khandesh, eastern Maharashtra, Mysore plateau of Karnataka, and eastern plains of Rajasthan are moderately potential for soybean. Parts of northern Vidarbha and central Madhya Pradesh are marginally potential for

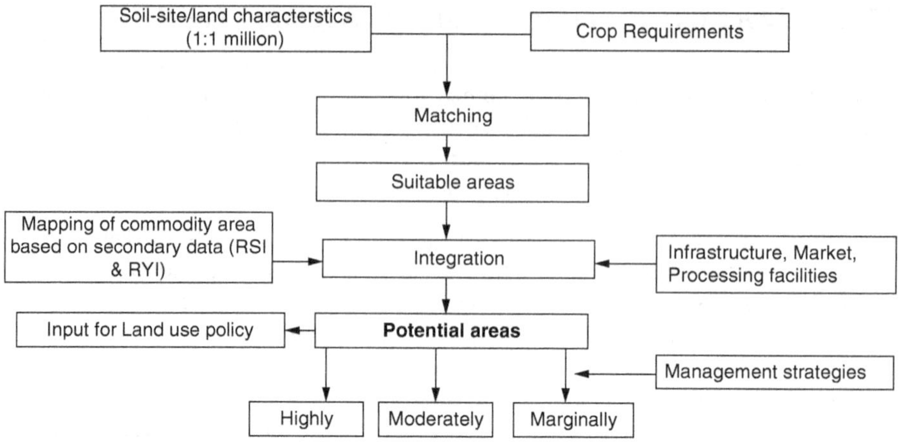

Fig. 25.1 Methodology for identification of potential areas .(Ramamurthy et al. 2017)

soybean. There is a need to substitute high-yielding varieties/hybrids with latest agro-technologies to maintain productivity and soil health on sustainable basis in highly potential areas, whereas in moderate potential areas, focus should be on to enhance the productivity by transfer of technology on high-yielding varieties, integrated nutrient and plant protection management, timely availability of inputs, capacity building, and establishing market and processing units. In marginally potential areas, priority should be given for area expansion by conducting frontline demonstrations (FLDs), farmer participatory research (FPR), establishment of markets and processing units.

This approach helps in availability of agro-inputs and information on crop production at the right time at the right place and leads to reduction in cost of production and enhancement of productivity and profitability. Establishment of markets nearby production points minimizes postharvest losses and helps in establishing value-added infrastructure that could lead to doubling of profitability of farming community.

25.8.2 Land Use Planning at District Level

Development projects are often at this level, where planning first comes to grips with the diversity of the land and its suitability to meet project goals. When planning is initiated nationally, national priorities have to be translated into local plans. Conflicts between national and local interests will have to be resolved. The kinds of issues tackled at this stage include developments such as new settlements, forest plantations, and irrigation schemes; delineation of land management zones; identification of prime agricultural lands and marginal lands for nonagricultural purposes; the need

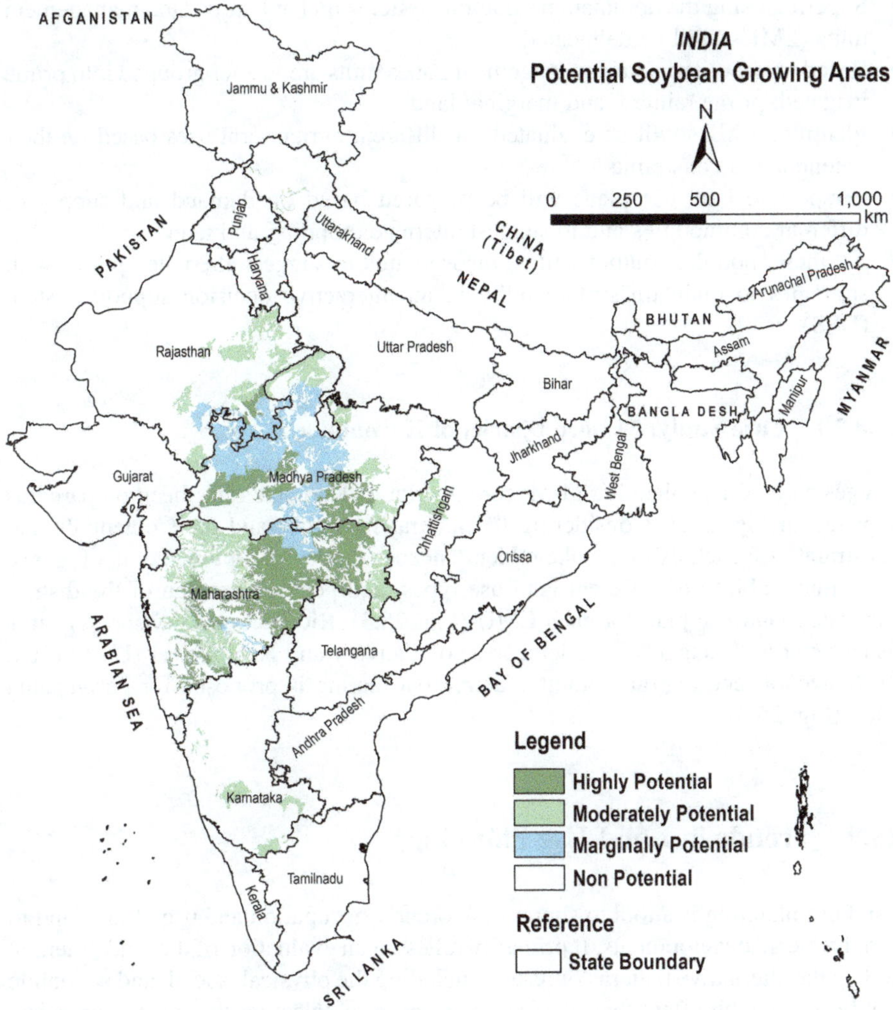

Fig. 25.2 Potential soybean growing areas in India

for improved infrastructure such as water supply, roads, and marketing facilities; and development of management guidelines for improved kinds of land use on each type of land.

District-level land use plan includes the following steps. They are:

- Grouping of soil map (1:250,000 or 1:50,000) and land use land cover maps into meaningful number of units.
- Grouped soil and land use maps will be spatially integrated in GIS environment along with agro-ecological zones/units map to derive land units.

- Superimposing the dominant production systems on land units, land management units (LMUs) will be delineated.
- Based on set criteria, land management zones/units are further grouped into prime irrigated, prime rainfed, and marginal lands.
- Identified LMUs will be evaluated for different agricultural uses based on their potential and constraints.
- Perspective land use plans will be prepared based on demand and supply of different commodities and to suggest alternate cropping and land use.
- All these modules' output will be incorporated to suggest alternate options with strategies to maintain sustainability in an interactive decision support system (DSS).

25.8.2.1 Case Study: Mysore District of Karnataka

Suggested land use planning of Mysore district of Karnataka has been prepared as depicted in Fig. 25.3. Considering the natural resources and their potentials and constraints of each LMU, stakeholders' needs, output of optimization of goals, economic viability of different land use types, and perspective plan of the district suggested land use plan for each LMU (Fig. 25.3). Rice-rice + dairy/sheep/goat + fishery rearing is suggested in delta areas of Cauvery and Kabini river (LMU 1 and 2). Maize/tobacco-ragi/groundnut + dairy/goat rearing is proposed for Periyapatna taluk (Fig. 25.3).

25.9 Trends in Land Use Planning

Land use planning is a tool to support the orderly occupation and use of land and to avoid adverse developments. It primarily relies on an evaluation of the land potential and on the alternative patterns of its use, including the physical, social, and economic conditions, which affect that use for the purpose of selecting the most appropriate use. Its main objective is to select and choose the options, which meet the best the needs, and to draw up the policy for its sustained use.

LUP does not stand on its own, but constitutes an intermediate step between land evaluation and land management. Land evaluation primarily identifies and rates the land potential and recommends alternative use scenarios; LUP focuses more on effective choices, with the basic information at hand among the options provided by the land evaluation process. Implementation of scenarios is mainly achieved by land use management techniques. Hence, while land evaluation and land use management operate mainly on technical criteria, land use planning deals with decision-making, and this is mainly a political action.

Conceptual road to LUP can be traced from the 1960s onward.

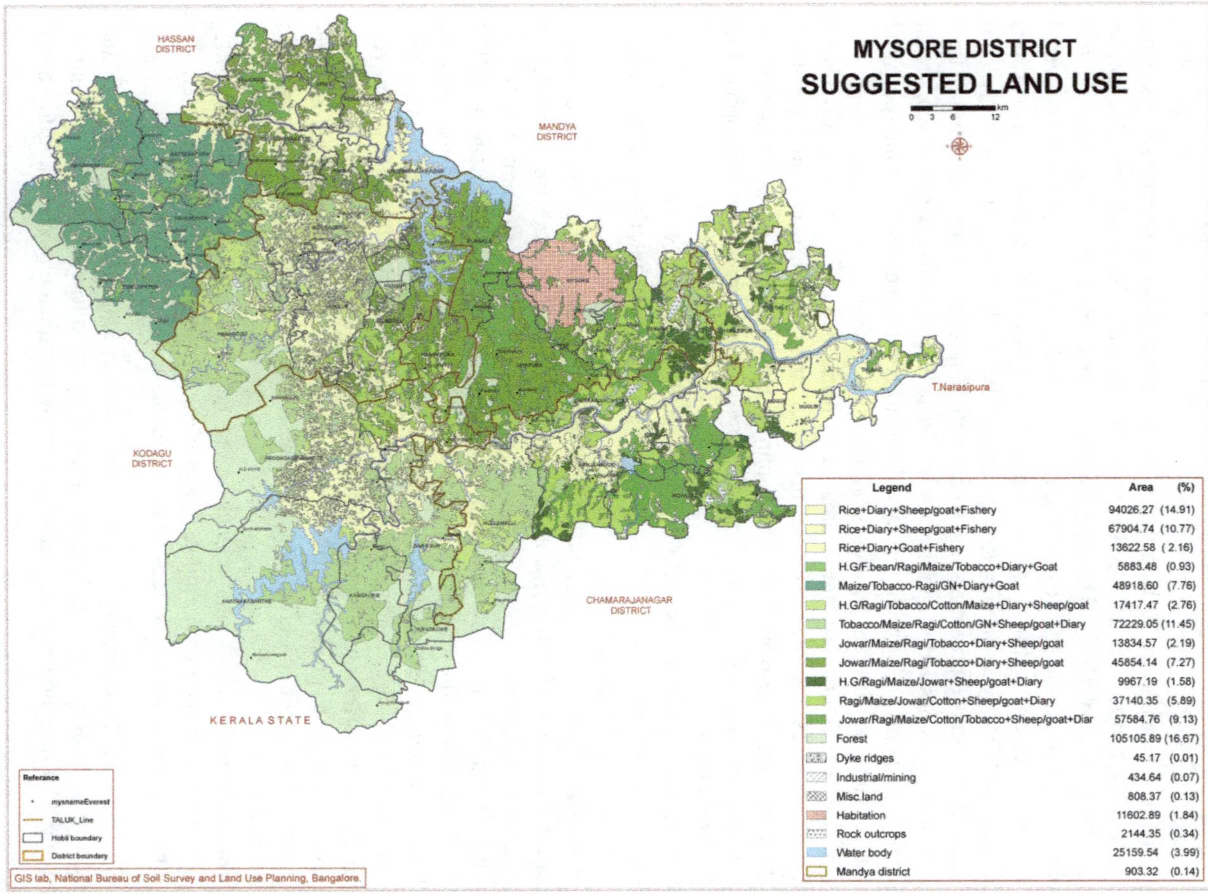

Fig. 25.3 Suggested land use planning at district level for Mysore district

- Identifying and comparing present and potential land uses for different land types, making decisions on land use types, and implementing development programs and activities on the basis of these decisions is an activity that is being implemented by governments, communities, and individuals to organize space before the 1960s.
- A major breakthrough in land use planning methodology for rural areas dates back to the 1960s with the introduction of the land capability concept (Klingebiel and Montgomery 1961). This approach was developed to identify and classify areas to be allocated for different agrarian activities like irrigation, mechanized farming, grazing, and forestry. This initial work was followed in the late 1960s and early 1970s by an attempt to add value to the results of soil surveys in order to determine land use potentials and to promote land use patterns according to land potential. This has led to the framework for land evaluation (FAO 1976), which provides a solid and unique technical basis to compare the present and potential performance of different land uses on different lands.
- The UN Conference on the Human Environment, known as the Stockholm Conference (1972), where, for the first time, due concern was expressed about the shrinking of the world's natural resources. Hence, the need was formulated for strong environmental action – oriented approach to sustainable development of land resources.
- The publication of the World Conservation Strategy (IUCN 1980), where governments were urged to use their natural resources for promoting human welfare while respecting the carrying capacity of ecosystem.
- Brundtland Commission emphasized that it is the present generation's responsibility to safeguard future generation's options and opportunities for development by protecting the planet's environment and natural resources (Anon. 1989).
- The United Nations Conference on Environment and Development (UNCED 1992) known as the Earth Summit of Rio de Janeiro emphasized the need for both economic development and a respect for the environment. The output of this conference includes a set of global conventions on climatic change and biodiversity, a set of principles for governments and people (in the Earth Charter), an action program to promote sustainability (Agenda 21), and institutional arrangements to implement programs and awareness creation among policy makers at the highest level.
- The Millennium Development Declaration and the Johannesburg Meeting (2002) where a major additional emphasis was given to more efficient land, water, and other natural resources besides poverty eradication.

Over time, LUP has undergone a number of conceptual changes. In the 1980s and early 1990s, LUP was generally considered a direct follow-up of land evaluation. Planning of future land use was mainly done by technical experts, using top-down approach, wherein results were poor. As a result, more participative approach was introduced based on the following considerations (Verheye et al. 1997).

- LUP should not be a too technical approach.
- LUP should avoid a too-top-down approach.
- LUP is a multidisciplinary exercise and should avoid being conceived in a too sectoral context.

Sustainable land use requires that all of the many relevant factors linked to production incentives and benefits. In this context, land use planning has become a modern tool for optimizing the use of land and for providing an optimal sustainable benefit to its users. It has become a systematic way of addressing the problems of land scarcity and land competition at different levels. For a land use plan to be operational and successful, three conditions must be met:

- There must be a need for change in land use, or there must be an action to prevent some unwanted change, acceptable by the people involved.
- There must be a political will and ability to put the proposed land use plan into effect.
- Stakeholders must have the feeling that they "own" the plan.

25.10 Modern Land Use Planning

Initially, LUP was mainly focused on agricultural development, and often referred LUP is a tool for rural development. Now, with the wide array of land uses involved, ranging from agriculture to urban, industrial, residential, and environmental uses, LUP deals with all types of uses and demands local action-oriented planning (participatory land use planning).

Planning at the local level, normally a community or a village, deals with the identification of concrete actions and activities within a global framework, called the strategic plan, to promote local development. It is a demand- and opportunity-driven exercise where realistic development options are identified mainly but not solely by the community itself. Indigenous communities usually have more experience of local conditions than outsiders when managing natural resources like pastures, forests, water, and wildlife. Technical institutions and development organizations often have experience and knowledge of the outside world that is unavailable to those at the local level. Bringing these different sources of information and knowledge together opens more perspectives for sustainable development.

25.10.1 Participatory Land Use Planning (PLUP)

PLUP approach was evolved to induce local-level involvement in sustainable land management. It can be defined as "an iterative process based on dialogue among all parties involved, and aiming to reach decisions on sustainable forms of land use in rural areas. It also includes the initiation of, and support to appropriate implementation measures" (GTZ 1995). In the Indian context, it is essentially a collective consensus on the use of resources. PLUP is based on the assumption that development is a process brought about "from bottom" and based on self-help and collective responsibility. It is an interdisciplinary task and is intended to improve the participant's capacity to prioritize, plan, and act. This requires free access to information for all participants in the process.

25.10.1.1 Case Study: LUP for Kokarda Village

Land resource inventory-based land use planning was implemented in Kokarda village of Kalmeshwar tehsil of Nagpur from 2000 to 2005 under the Institute Village Linkage Programme. The mean annual rainfall of the area is 976 mm, which is received mostly from southwest monsoon, from second fortnight of June to October. About 96% of the total rainfall occurs during June to September in 67 rainy days with length of growing period workout to be about 150–170 days. Delayed onset, early withdrawal, and prolonged dry spells are the common characteristics of the monsoon. These aberrations have a severe impact on the crop production and productivity.

The soils are shallow to moderately deep and deep, well drained/moderately well drained, with gravely clay loam/sandy clay loam soils. Shallow soils occur on hill slopes and deep soils in valley bottom. Shallow soils are under *kharif* monocrops. Deep soils are cultivated for *kharif* and *rabi* crops.

Before implementation of soil-based land use plans, about 26% of the area was under sorghum, 22% under cotton, 17% under soybean, and 13% of the area under orange orchards. The small and marginal farmers were mostly the wage earners. The cattle and goat rearing formed an integral part of their livelihood. Dairying was supplementary enterprise to most of the farmers.

After LRI, participatory approach (Ramamurthy and Sarkar 2009) was followed to prepare and implement land use planning of the village (Fig. 25.4). At the end of the project, impact assessment was carried out and found that socially acceptable, technically feasible, and economically viable technologies are by and large considered sustainable.

Due to village-level land use planning (VLUP), area under cultivation has increased in the villages. The uncultivable land (12%) and current fallows (34%) were brought under cultivation by way of fodder production, afforestation, and

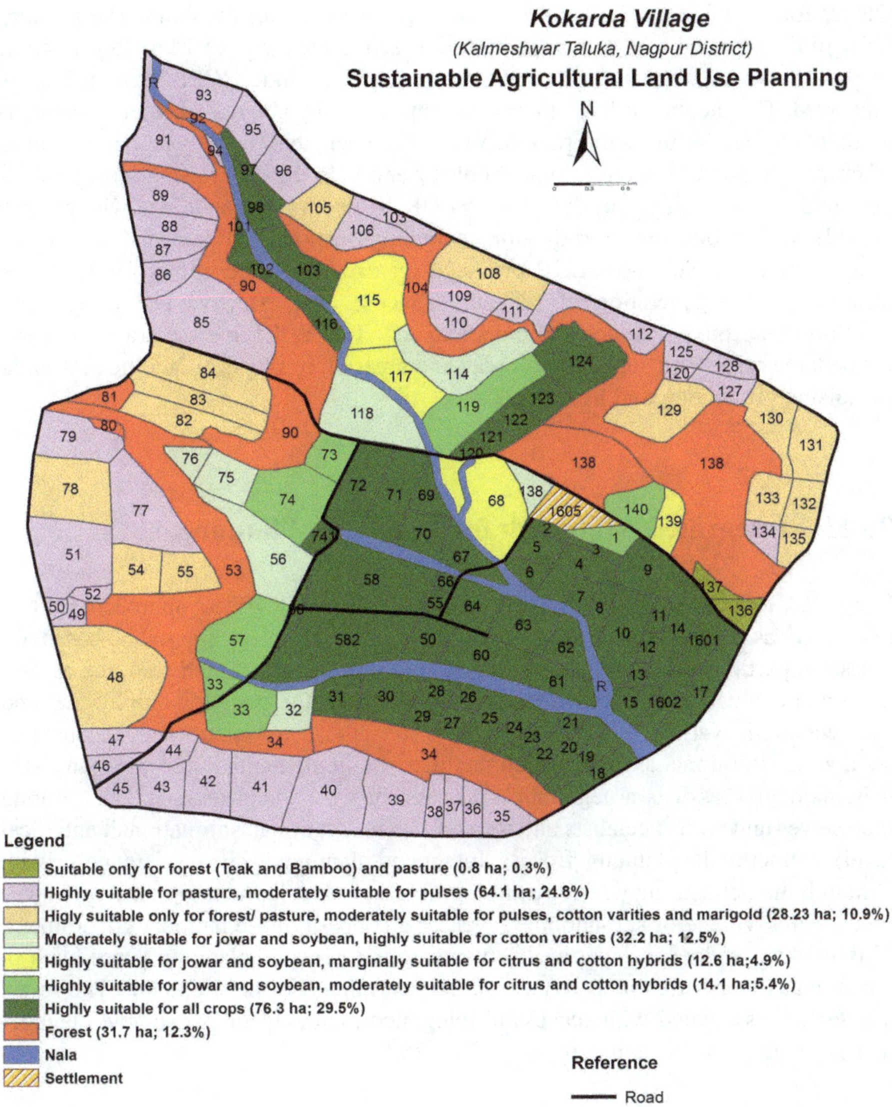

Fig. 25.4 Land use plan map of Kokarda village

agroforestry. There was 5% increase in net sown area. Changes in land use and cropping pattern were significant due to VLUP. Before the implementation of the VLUP, maximum area was under cotton hybrids and sorghum. After 5 years, cotton hybrids have shifted to deep soils and varieties to shallow soils. Similarly, sorghum and citrus has been shifted to soybean. This indicates that productivity level and market forces influence the land use rather than soil suitability alone. The productivity of dryland and irrigated crops has increased in the range of 14 to 48% in grain crops and 400% in fodder crops. The Crop Productivity Index (CPI) of rainfed crops improved. The income of farmers has been increased by about 32% after implementation of the VLUP. In a short period, VLUP has been able to motivate large number of farmers to adopt modern farm technology and help them in raising farm production and consequently income. The VLUP helped to generate knowledge and provide skills about new production technology among adopted and also other farmers of the area. The VLUP provided an excellent mechanism for feedback information for generation of refined, low-cost, and effective technology and develop interpretive base for soil units mapped. The VLUP has generated demand for better services and supplies of inputs and creation of physical facilities essential for raising production and livelihood.

25.11 Integrated Approach for Land Use Planning

Integrated approach combines both interactive and integrated approaches. It is integrated as it combines elements of both the bottom-up approach, based on grassroot participation, and traditionally top-down aspects of land resource assessment and evaluation of options; it takes into account the complex biophysical and socioeconomic variables, which determine the land use system; and it considers legal and institutional aspects, which facilitate the implementation of the plan. Also it is interactive as it is a negotiation process, in which land users interact among themselves and with specialists and different levels (national, sub-national, and local level) interact in the planning process. Integrated planning is always demand-driven, although the demand may result from a problem or development opportunity either perceived at village or sub-national level, or a concern of the national government. This marks a welcome departure from previous top-down planning procedures in which plans were often prepared as routine instruments of development. The seven key factors associated with successful integrated planning for sustainable management of land resources are depicted in Fig. 25.5.

25.11.1 Case Study: Gajwel Mandal of Telangana State

Land resource inventory of Gajwel Mandal was carried out at 1:10,000 scale. Integrated approach (Fig. 25.6) was adopted for suggesting land use plan of the

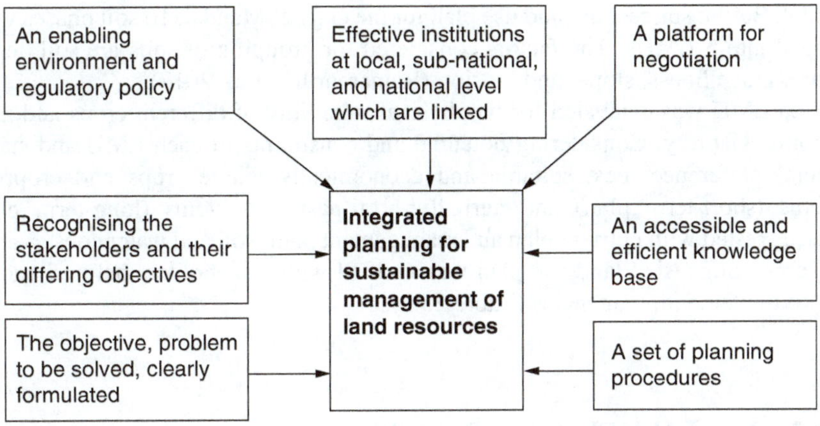

Fig. 25.5 Key factors associated with sucessful implementation of integrated planning for sustainable resource management

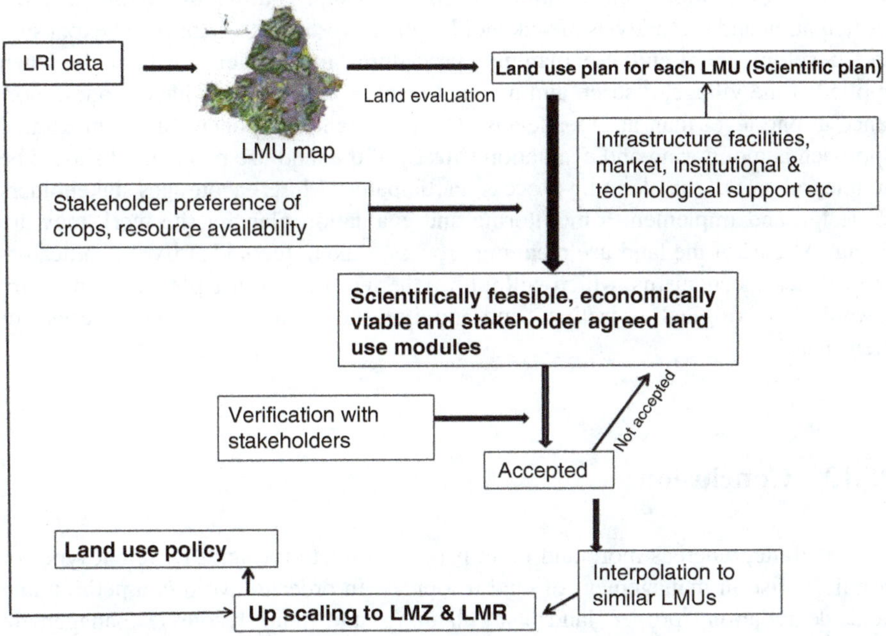

Fig. 25.6 Methodology adopted for Mandal-level land use planning

Mandal. Before suggesting land use plan for the Gajwel Mandal, 16 soil phases were grouped into 5 LMUs. The factors considered for grouping of soils are soil depth, texture, gravelliness, slope, and erosion (Ramamurthy et al. 2017).

Each LMU was evaluated for biophysical suitability of different crops and their economic viability. Considering potential and constraints of each LMU and stakeholders' preference, best suitable and economically viable crops and cropping systems (short-term plan) and agri-silvi-horti-pasture systems (long-term plan) were suggested with nutrient plan and management plan (soil and water conservation measures). Suggested land use plan was verified with stakeholders before handing over to implementing agencies (Table 25.5).

25.12 Land Use Planning Procedures

Adoption of an integrated approach in planning and sustainable management of land resources calls for a critical look at planning procedures. Clearly, some of the technical methods used in conventional land use planning remain valid components (e.g., land evaluation), but certain aspects, particularly those involving people's participation and the analysis of stakeholder objectives, require significant expansion and development. Land use planning procedures may differ substantially when applied at the village, district, and national level. Some elements have more importance at one level than at other levels. The important component of the integrated approach is monitoring and evaluation (M&E) of the land use plan, and it should be an integral part of the planning process. Participatory M&E encourages stakeholders to design and implement a monitoring and evaluation plan for the work they are doing. As part of the land use planning process, stakeholders identify the indicators or feedback mechanisms, which will inform them if the land use plan is taking them toward their original objectives and monitor the indicators of the success of their plan.

25.13 Conclusions

Land is finite, whereas more and more people compete for land. Different types of conflicts arise in management of land resources. In order to avoid competition and land degradation, proper land use planning and land resource management approaches needs to be followed. Land use planning and land resource management approaches should be in holistic manner. Modern approaches are always focused on one or more specific objectives and are closely linked to the concept of efficiency, equity, and sustainability. To achieve these objectives, participatory and integrated approaches are to be applied to manage land resource at different levels. These approaches also provide input to prepare land use policy of the country.

Table 25.5 Suggested land use plan for Gajwel Mandal

LMU	Crops and cropping systems	Agri-horti-silvi-pasture systems	Nutrient plan	Contingency plan
Shallow to very shallow, somewhat excessively drained, loamy skeletal soils with moderate stoniness (K1Ble2st1, K1Bse2, K2Ble2)	Small millets, jowar, horse gram, green gram and black gram, cotton, and maize	Silvi-horti-pasture system (*C. ciliaris*, stylo-grass mixture, custard apple, amla, ber, and suitable MPTs like *Dalbergia sissoo*, *Faidherbia albida*, *A. auriculiformis*, neem). To conserve soil and water, open intermittent contour trenches at 3–7 m interval with 0.3 m base width and 0.3 m height with 1:1 side slope, afforestation with MPTs on the berm side of trench and stone terracing at an interval of 30 m	Apply recommended dose of NPK with 5 kg borax/ha	*Early-season drought (delayed onset)* *Delay by 2 weeks (up to the end of June)* Prefer medium-duration dual-purpose millet varieties and YMV-resistant varieties of pulses *Delay by 4 weeks (July second to fourth week)* Prefer short-duration dual-purpose millet varieties and YMV-resistant varieties of pulses *Delay by 8 weeks* Prefer fodder-type cereals and pulses. High-density mixed cropping of fodder maize, cowpea, and horse gram *Normal onset followed by 15–20-day dry spell after sowing leading to poor germination/crop stand, etc.* If population is sparse, resowing of short-duration cultivars and spray systemic

(continued)

Table 25.5 (continued)

LMU	Crops and cropping systems	Agri-horti-silvi-pasture systems	Nutrient plan	Contingency plan
				insecticides to control sucking insects *Mid-season drought (long dry spell, consecutive 2 weeks rainless (<2.5 mm) period) at vegetative stage* Irrigate alternate row to protect the crop if water is available, and topdress urea after supplemental irrigation to boost the growth *At reproductive stage* – lifesaving irrigation if water is available
Moderately deep to deep, well-drained to mod. well-drained, loamy sand soils with moderate erosion (C3Ble2, C3Ble2Na1, C3Cle2, C3Bse1, Cv3Ale1)	Maize, cotton, sunflower, red gram, jowar, maize/sunflower (short duration)-gram, millets, paddy as aerobic rice cultivation, cowpea, green gram, black gram, and vegetables like coriander, cluster bean, pumpkin	Agri-horti-pasture system (*C. ciliaris*, stylo, *Brachiaria* on bunds, mango, guava, custard apple, amla, jamun, ber, and suitable MPTs like *Dalbergia sissoo, Faidherbia albida, A. auriculiformis*, neem) with SWC measures like stone bunds on the boundaries across major slope and water harvesting farm ponds needs to be created	Apply recommended dose of NPK and zinc with 5 kg borax/ha follow INM approach. Phosphorus should be applied as SSP. Boundary planting of *Gliricidia* helps in improving organic source	*Early-season drought (delayed onset)* *Delay by 2 weeks (up to the end of June)* Sow primed seeds of cotton and maize. Use higher seed rate for jowar. Prefer medium-duration dual-purpose millet/jowar varieties and YMV-resistant varieties of pulses *Delay by 4 weeks (July second to fourth*

(continued)

Table 25.5 (continued)

LMU	Crops and cropping systems	Agri-horti-silvi-pasture systems	Nutrient plan	Contingency plan
				week) Sow primed seeds of cotton and maize. Use higher seed rate for jowar. Prefer short-duration dual-purpose millet/jowar varieties and YMV-resistant varieties of pulses *Delay by 8 weeks* Prefer fodder-type cereals and pulses. High-density mixed cropping of fodder maize/jowar, cowpea, and horse gram *Normal onset followed by 15–20-day dry spell after sowing leading to poor germination/crop stand, etc.* If population is sparse, resowing of short-duration cultivars and spray systemic insecticides to control sucking insects *Mid-season drought (long dry spell, consecutive 2 weeks rainless (<2.5 mm) period) at vegetative stage*

(continued)

Table 25.5 (continued)

LMU	Crops and cropping systems	Agri-horti-silvi-pasture systems	Nutrient plan	Contingency plan
				Irrigate alternate row to protect the crop if water is available, and topdress urea after supplemental irrigation to boost the growth *At reproductive stage* – lifesaving irrigation if water is available
Moderately deep to deep, moderately well-drained to well-drained, clayey soils with slight erosion (C3Bce1, Cv3Ace1)	Cotton, maize, sunflower, red gram, jowar, maize/sunflower (short duration)-gram, millets, paddy as aerobic rice cultivation, cowpea, green gram, black gram, and vegetables like coriander, cluster bean, pumpkin, tomato, chili	Silvi-horti-pasture system (*C. ciliaris*, stylograss mixture, custard apple, amla, jamun, ber, and suitable MPTs like *Dalbergia sissoo, Faidherbia albida, A. auriculiformis*, neem) with SWC measures like stone bunds on the boundaries across major slope and water harvesting farm ponds needs to be created	Apply recommended dose of NPK with 5 kg borax/ha follow INM approach. Boundary planting of *Gliricidia* helps in improving organic source	*Early-season drought (delayed onset)* *Delay by 2 weeks (up to the end of June)* Sow primed seeds of cotton and maize. Use higher seed rate for jowar. Prefer medium-duration dual-purpose millet/jowar varieties and YMV-resistant varieties of pulses *Delay by 4 weeks (July second to fourth week)* Sow primed seeds of cotton and maize. Use higher seed rate for jowar. Prefer short-duration dual-purpose millet/jowar varieties and YMV-resistant

(continued)

Table 25.5 (continued)

LMU	Crops and cropping systems	Agri-horti-silvi-pasture systems	Nutrient plan	Contingency plan
				varieties of pulses *Delay by 8 weeks* Prefer fodder-type cereals and pulses. High-density mixed cropping of fodder maize/ jowar, cowpea, and horse gram *Normal onset followed by 15–20-day dry spell after sowing leading to poor germination/crop stand, etc.* If population is sparse, resowing of short-duration cultivars and spray systemic insecticides to control sucking insects *Mid-season drought (long dry spell, consecutive 2 weeks rainless (<2.5 mm) period) at vegetative stage* Irrigate alternate row to protect the crop if water is available, and topdress urea after supplemental irrigation to boost the growth *At reproductive*

(continued)

Table 25.5 (continued)

LMU	Crops and cropping systems	Agri-horti-silvi-pasture systems	Nutrient plan	Contingency plan
				stage – lifesaving irrigation if water is available
Very deep, well- to moderately well-drained, loamy soils with slight to moderate erosion (C4Ble1, C4Bse2, Cv4Ble1)	Maize, cotton, sunflower, red gram, jowar, maize/sunflower (short duration)-gram, paddy, cowpea, green gram, black gram, and vegetables like beans, chili, watermelon, muskmelon, tomato, coriander, cluster bean, pumpkin, tomato, chili	Banana, mango, guava with drip. In elevated uplands grow millets and pulses like cowpea and green and black gram, and agri-horti-pasture system (*C. ciliaris*, stylo, *Brachiaria* on bunds, mango, guava, custard apple, amla, jamun, ber, and suitable MPTs like *Dalbergia sissoo, Faidherbia albida, A. auriculiformis*, neem) with SWC measures like graded bunds, stone bunds on the boundaries across major slope, vegetative bunds within the field boundaries, and water harvesting farm ponds needs to be created	Apply recommended dose of NK and boron, use PSP with 25% reduced recommended dose of P and follow INM approach. Phosphorus should be applied as SSP. Boundary planting of *Gliricidia* helps in improving organic source	*Non-release of water in canals under delayed onset of monsoon in catchment* Grow sunflower and red gram instead of rice, and open dead furrow to conserve in situ soil moisture *Lack of inflows into tanks due to insufficient/ delayed onset of monsoon* Grow rice-sunflower/maize instead of rice-rice *Insufficient groundwater recharge due to low rainfall* Go for low water requirement crops like sunflower, red gram, gram, safflower, etc. instead of rice. Irrigate in critical stages to save the crop *Continuous high rainfall in a short span leading to water logging (or) high-speed winds in a short span*

(continued)

Table 25.5 (continued)

LMU	Crops and cropping systems	Agri-horti-silvi-pasture systems	Nutrient plan	Contingency plan
				Drain out excess water, spray 1% KNO3 or urea 2% solution 2–3 times, topdressing of booster dose of 12 kg MOP +30 kg urea per acre as soon as possible
Very deep, well- to moderately well-drained, clayey soils with slight erosion (C4Bce1, Cv4Ace1, Cv4Ace1Na1)	Cotton, maize, paddy, sunflower, red gram, jowar, maize/sunflower (short duration)-gram, cotton+red gram, millets, cowpea, green gram, black gram, and vegetables like coriander, cluster bean, pumpkin, tomato, chili,	Banana, mango, guava with drip. In rainfed areas grow millets and pulses like cowpea and green and black gram, and agri-horti-pasture system (*C. ciliaris*, stylo, *Brachiaria* on bunds, mango, guava, custard apple, amla, jamun, ber, and suitable MPTs like *Dalbergia sissoo, Faidherbia albida, A. auriculiformis*, neem) with SWC measures like graded bunds, stone bunds on the boundaries across major slope, vegetative bunds within the field boundaries, and water harvesting farm ponds needs to be created	Apply recommended dose of NPK and zinc with 5 kg borax/ha and follow INM approach. Apply P through SSP. Boundary planting of *Gliricidia* helps in improving organic source	*Irrigated areas: non-release of water in canals under delayed onset of monsoon in catchment* Grow sunflower and red gram instead of rice, and open dead furrow to conserve in situ soil moisture *Lack of inflows into tanks due to insufficient/delayed onset of monsoon* Grow rice-sunflower/maize instead of rice-rice *Insufficient groundwater recharge due to low rainfall* Go for low water requirement crops like sunflower, red gram, gram, safflower, etc. instead of rice. Irrigate in critical stages to save the crop

(continued)

Table 25.5 (continued)

LMU	Crops and cropping systems	Agri-horti-silvi-pasture systems	Nutrient plan	Contingency plan
				Continuous high rainfall in a short span leading to water logging (or) high-speed winds in a short span Drain out excess water, spray 1% KNO3 or urea 2% solution 2–3 times, topdressing of booster dose of 12 kg MOP +30 kg urea per acre as soon as possible *Rainfed areas: delay by 2 weeks (up to the end of June)* Sow primed seeds of cotton and maize. Use higher seed rate for jowar. Prefer medium-duration dual-purpose millet/jowar varieties and YMV-resistant varieties of pulses *Delay by 4 weeks (July second to fourth week)* Sow primed seeds of cotton and maize. Use higher seed rate for jowar. Prefer short-duration dual-purpose millet/jowar

(continued)

Table 25.5 (continued)

LMU	Crops and cropping systems	Agri-horti-silvi-pasture systems	Nutrient plan	Contingency plan
				varieties and YMV-resistant varieties of pulses *Delay by 6 weeks* Go for wilt-resistant medium-duration varieties of red gram and castor with closer spacing instead of maize, green gram, jowar, and long-duration red gram *Delay by 8 weeks* Go for wilt-resistant medium-duration varieties of red gram and castor with closer spacing instead of maize, green gram, jowar, cotton, and long-duration red gram *Normal onset followed by 15–20-day dry spell after sowing leading to poor germination/crop stand, etc.* If population is sparse, resowing of short-duration cultivars and spray systemic

(continued)

Table 25.5 (continued)

LMU	Crops and cropping systems	Agri-horti-silvi-pasture systems	Nutrient plan	Contingency plan
				insecticides to control sucking insects *Mid-season drought (long dry spell, consecutive 2 weeks rainless (<2.5 mm) period) at vegetative stage* Spray of systemic insecticides of control sucking pests. Foliar spray 2% urea with receipt of rains after cessation of drought, apply booster dose of topdress with nitrogen *At reproductive stage* – foliar spray with 2% urea

References

Anon (1989) Our common future. The World Commission on Environment and Development, Oxford University Press

Challa O, Dalrymple JB (1993). Land evaluation for land use planning, eds. Dent DL, Deshpande SB. NBSS Publication No. 42. Nagpur, India, 167 p

Dent DL (1993) What do we mean by land use planning. In: Land evaluation for land use planning. NBSS Publication No.42

DES (2016) State of Indian agriculture 2015-16. Ministry of Agriculture & Farmers Welfare Department of Agriculture, Cooperation & Farmers Welfare, Govt. of India

FAO (1976) A framework for land evaluation. FAO Soils Bulletin 32

FAO (1993) Guidelines for land use planning, FAO development series no.1, Rome

FAO (1997) Land quality indicators and their use in sustainable agriculture and rural development. FAO Land and Water Bulletin 5, 212p

FAO (2011) The state of the World's land and water resources for food and agriculture. FAO, Rome

GTZ. (1995) Landnutzungs planung. Strategien, Instumente, methoden, Arbeitsgruppe LNP, Eschborn

IUCN (1980) World conservation strategy: living resource conservation for sustainable development. IUCN-UNEP-WWF, Gland

Klingebiel A, Montgomery PH (1961) Land capability classification. USDA Agriculture

Maji AK, Reddy GPO, Sarkar D (2010) Degraded and wastelands of India, status and spatial distribution. ICAR and NAAS Publication, pp 1–158

Ramamurthy V, Sarkar D (2009) Participatory land use planning improved employment opportunity. ICAR News 15(1):4–5

Ramamurthy V, Nair KM, Ramesh Kumar SC, Srinivas S, LGK N, Sarkar D, Singh SK (2017) Delineation of land management units in Tropical South Deccan Plateau of India. *Agropedology* 26(2):105–116

Umesh Babu MS, Nautiyal S (2003) Historical issues and perspectives of land resource Management in India: a review. Working paper 309, The Institute for Social and Economic Change, Bangalore

UNCED (1992) The Rio declaration on environment and development. Rio de Janeiro, Brazil

Velayutham M, Mandal C, Mandal DK, Sehgal J (1999) Agro-ecological sub-regions of India for planning and development. Bulletin no. 35. NBSS Publication, Nagpur, India

Verheye W, Brinkman R, Sims D (1997) Elements of a different approach to land development issues. The Land 1(2):143–152

Warner M (2000) Conflict management in community-based natural resource projects: experiences from Fiji and Papua New Guinea. Working paper 135. Overseas Development Institute

Chapter 26
Soil-Based Land Use Planning: Impacts on Livelihood Security

V. Ramamurthy

Abstract Soil is a vital natural resource. To ensure optimum agricultural production, it is imperative to know the basic facts of the soils, their potential and limitations and management strategies needed to overcome the limitations. To improve the productive capacity of soils and further degradation of soil resources, there is a need to develop soil-based technologies and their assessment through participatory approach for quick and wider adoption. Identification of biophysical, socio-economic setup and need of the stakeholders will help to design sustainable, acceptable and economically viable land use plans. In this chapter, impact of soil-based agrotechnology transfer in two agroecological subregions of India has been reported. These plans are based on participatory and integrated farming system approach, where soil-specific agrotechnologies improved the productivity of soils and also the livelihood of farming community.

Keywords Soil based · Land use planning · Participatory · Farming systems · Livelihood improvement

26.1 Introduction

More than 60% of population of India derives its livelihood and environmental securities directly from natural resources like soil, water, vegetation and livestock and village enterprises. Increased demographic pressure and different users like agriculture, urban and rural development, industrial development, mining and transportation have reduced the per capita cultivated land from 0.48 ha in 1951 to 0.16 ha in 1991, and it is likely to decline further to 0.08 ha by 2035. Livelihood needs of rural communities are to be realized from increased productivity of agriculture, animal husbandry and fisheries by integrated farming system approach without degrading the quality of natural resources (Ramamurthy and Sarkar 2009). Science

V. Ramamurthy (✉)
ICAR-National Bureau of Soil Survey & Land Use Planning, Regional Centre, Bengaluru, India

G. P. O. Reddy, S. K. Singh (eds.), *Geospatial Technologies in Land Resources Mapping, Monitoring and Management*, Geotechnologies and the Environment 21,
https://doi.org/10.1007/978-3-319-78711-4_26

and technology play a crucial role in augmenting production and bringing prosperity among the rural poor. Transfer of technologies and their widespread adoption by the farmers are the key components to maximize the production, and increasing household income, thereby improving the livelihood of rural people.

The most important factors, which govern agriculture and livelihood, are natural resources like soil, water and climate and socio-economic conditions of an area. Among the natural resources, importance of soil is increasing due to shrinking of land holding size and its various kinds of degradation. Many authors reported some typical and common mechanisms, which establish vulnerability of smallholder farmers in developing countries that are characterized by increasing pressures on natural resources, soil degradation and breakdown of traditional coping mechanisms, which in turn posed barriers to alternative livelihoods (Geist and Lambin 2004; Safriel et al. 2005; Safriel and Adeel 2008). To ensure optimum agricultural production, it is imperative to know the basic facts of the soils, their potential and limitations and management strategies needed to overcome the limitations. In order to improve the productive capacity of soils, there is a need to develop soil-based land use plans.

The land use planning process covers all steps extending from the collection of data and information through its processing, analysis, discussion, evaluation, identification and transfer of right agrotechnology at the right time and right place. Naidu et al. (2014) characterized and delineated the prime, moderate prime and marginal lands of Andhra Pradesh based on important soil and climatic parameters to develop or frame strong land use legislations to restrict use of prime lands for nonagricultural purpose. The study suggested that pressure on marginal lands has to be reduced by evaluating the capability for a present land use and accordingly suitable alternate land use can be suggested. Ramamurthy et al. (2015) have developed an integrated land use plan for enhancing tribal livelihood in H.D. Kote taluk of Mysore district. The study reported that livelihood of tribals is mainly based on agriculture and allied activities rather than cultivation of crops alone. Therefore, it is essential to enhance the income level of small land holders and landless agricultural labourers through additional livelihood opportunities, i.e. by diversifying farm and nonfarm occupations (Walingo et al. 2009). Costales et al. (2007) reported security of livelihood, income, food and nutrition, etc. of the rural poor through mixed crop-livestock systems. Land use planning may be at national, regional, state, district, watershed or village and also at farm level. Soil-based land use plans have been implemented in different parts of India, and their impact has been summarized in this chapter.

26.2 Soil-Based Land Use Planning (SLUP) in Eastern Maharashtra Plateau

The National Bureau of Soil Survey and Land Use Planning (NBSS&LUP) implemented soil-based land use plans from 1979 to 1982 under lab-to-land programme (Mondha and Sukali villages of Nagpur) and from 2000 to 2005 under

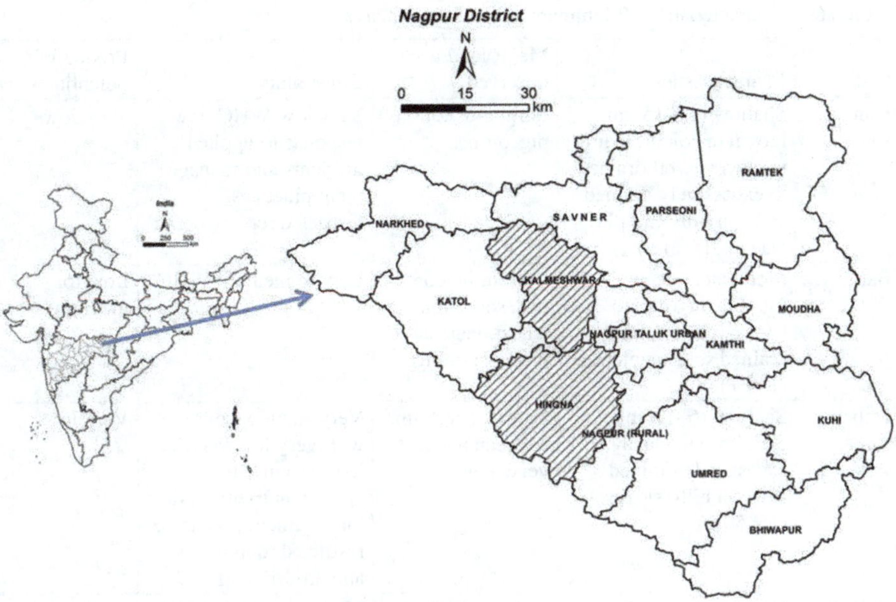

Fig. 26.1 Location map of Kalmeshwar and Hingna tehsils of Nagpur district covered under Eastern Maharashtra Plateau

Institute-Village Linkage Programme (IVLP) in Kokarda, Kaniyadol, and NATP-MM-LUP in Panubali of Kalmeshwar tehsil of Nagpur district (Fig. 26.1). The village Mondha is located in Hingna tehsil of Nagpur district, and Sukali village is 20 km to the south-west of Nagpur city. The climate of study area is tropical dry subhumid with well-expressed summer. The mean annual rainfall of the area is 976 mm received mostly from south-west monsoon, from second fortnight of June to October. About 96% of the total rainfall occurred during June to September in 67 rainy days. The length of growing period of the study area is about 150–170 days. Delayed onset, early withdrawal and prolonged dry spells are the common characteristics of the monsoon. These aberrations have a severe impact on the crop production and productivity.

26.2.1 Soil Resources

The soils are shallow to moderately deep and deep, well drained/moderately well drained, with gravely clay loam/sandy clay loam soils (Table 26.1). Shallow soils occur on hill slopes and deep soils in valley bottom. Shallow soils are under *kharif* monocrops. Deep soils are cultivated to *kharif* and *rabi* crops.

Table 26.1 Characteristics of dominant soils of project area

Soil	Characteristics	Major land use observed	Constraints	Productivity potential
Dongari soil	Shallow (11–15 cm), brown in colour, loamy to clayey, well-drained to excessively drained soils on hills and plateau (1–15%)	Sorghum, cotton, pigeon pea	Very low WHC, low response to applied nutrients and management practices, restricted rooting depth	Very low
Bardi soil	Moderately deep (36–60 cm), dark brown, clayey, well-drained soils on plateau (1–8%)	Cotton, pigeon pea, sorghum, wheat, vegetables and orange	Low to medium WHC	Low to medium
Pathari soil	Shallow (6–17 cm), dark brown, clayey, excessively drained soils on hilly slopes (15–50%)	Cotton, sorghum, soybean and vegetables	Very shallow soils with very low WHC, low response to applied nutrients and management practices, restricted rooting depth and eroded soils	Very low
Halki kanhar	Deep clayey (80–97 cm), very dark brown and well drained to moderate permeability on Piedmont (1–8%)	Cotton, sorghum, pigeon pea, wheat, gram, vegetables, orange	Medium to high WHC, narrow workable time and insufficient drainage	Medium
Bhari kanhar	Very deep clayey (150 cm), dark brown to very dark brown in colour, moderately to well drained, on Piedmont (1–3%)	Cotton, sorghum, pigeon pea, wheat, gram, vegetables, orange, sugarcane	Medium to high WHC, narrow workable time and insufficient drainage	Medium

26.2.2 *Farming Systems*

Rainfed agroecosystem has been divided into five production systems, viz. sorghum, cotton, soybean, orange and livestock systems. Before implementation of soil-based land use plans, about 26% of the area is under sorghum, 22% under cotton, 17% under soybean and 13% of the area under orange orchards. The small and marginal farmers are mostly the wage earners. The cattle and goat rearing form an integral part of their livelihood. Dairying is supplementary enterprise to most of the farmers (Table 26.2).

On the basis of land use types, five distinct types of farming situations were identified in the village.

Table 26.2 Major production problem related to farming system situation

Micro-farming situation	Problem	Related causes of low yields
Sorghum and cotton cultivated in medium deep to deep soil on summit (rainfed)	Poor yield of sorghum	Late sowing Use of local varieties Unbalanced nutrient supply
	Low and unstable cotton yield	Unbalanced nutrient supply Low plant population of varieties due to higher spacing
Sorghum, cotton and soybean cultivated under rainfed in shallow to medium deep soils on escarpment	Low and unstable cotton yield	Unbalanced nutrient supply Use of hybrids
	Low yield of sorghum	Unbalanced nutrient supply, use of local varieties Late sowing
	Poor yield of soybean	Low plant population Unbalanced nutrient supply
Sorghum, cotton, soybean, groundnut and pigeon pea cultivated under rainfed in shallow to medium deep soils on escarpment	Low and unstable yield of sorghum	Unbalanced nutrient supply Use of local varieties Late sowing
	Low and unstable yield of cotton	Unbalanced nutrient supply Low plant population of varieties due to higher spacing
	Low yield of soybean	Low plant population Unbalanced nutrient supply
	Low yield of groundnut	Low nutrient status of soils White ant (termite) incidence, wild boar (pig) menace
	Low yield of pigeon pea	Poor WHC of eroded soils low yield of local varieties, poor nutrient status of soils, water erosion, pod borer
Sorghum, cotton, soybean, cotton + pigeon pea under rainfed; wheat, gram and orange under rainfed and irrigated conditions cultivated in medium to deep soils of plateau	Low and unstable yield of Sorghum	Unbalanced nutrient supply Use of local varieties Late sowing
	Low yield of cotton	Unbalanced nutrient supply Low plant population of varieties due to higher spacing
	Low yield of soybean	Low plant population Unbalanced nutrition
	Poor yield of cotton + pigeon pea (intercrop)	Use of local varieties, delayed sowing Incidence of *Helicoverpa* and pod fly Wilting Poor drainage
	Unstable yield of wheat	Delayed sowing in rotation with cotton, use of local variety (long duration) Unbalanced nutrition Untimely rains at the time of harvesting

(continued)

Table 26.2 (continued)

Micro-farming situation	Problem	Related causes of low yields
	Low yield of gram	Use of local varieties Imbalanced use of fertilizers Incidence of pod borer
	Low fruit yield from orange orchards	Incidence of 'Kolshi' Irregular bearing, gummosis High water holding capacity of deep soils – wilt, flower drop, termites and stem borer attack
Shallow soils supporting with bushes, shrubs, shallow-rooting trees on escarpment associated with rock out crops	Poor carrying capacity of animals	Eroded soils with thin top layer of soil Very low moisture and nutrient retention capacity Indiscriminate grazing and cutting of plants for wood
Shallow soils with surface stoniness but cultivated to rainfed sorghum, cotton + pigeon pea and soybean	Poor yield of sorghum	Unbalanced nutrient supply Use of local varieties Late sowing
	Low productivity of cotton + pigeon pea (intercrop)	Unbalanced nutrient supply Low plant population of varieties due to higher spacing Incidence of *Helicoverpa*
	Low yield of soybean	Unbalanced nutrition Use of local varieties, delayed sowing
Livestock Cattle and buffaloes	Poor productivity of milch animals and mortality	Poor milk yield from milch animals Fodder scarcity Poor genetic potential of native animal breeds Malnutrition Diseases
Goats	Poor productivity of goats	Susceptibility to diseases Malnutrition Fodder scarcity Poor genetic potential of native breeds Lack of knowledge about animal care Incidence of diseases

- Rainfed moderately deep to deep soils on summit having undulating terrain with sorghum and cotton as major crops
- Rainfed shallow to moderately deep soils on escarpment cultivated to sorghum, cotton, soybean, groundnut and pigeon pea

- Rainfed and irrigated moderately deep to deep soils of plateau dominated by sorghum, cotton, soybean, cotton + pigeon pea, wheat, gram and orange orchards. Inadequate natural drains
- Rainfed soils with bushes, shrubs, shallow-rooting trees on escarpment having shallow soils and rock out crops used predominantly for grazing
- Rainfed shallow soils with high stoniness and rolling slope cultivated for sorghum, cotton + pigeon pea and soybean

26.3 Suitability of Crops to Different Soils

The land suitability for defined uses and the impact of the uses on environment are determined by land conditions and land qualities. Each plant species require specific soil-site conditions for its optimum growth. The adaptability of crops in one or the other area is the interaction between existing edaphic conditions and fitness of the cultivar under these conditions. Soil-crop suitability studies provide information on choice of crops to be grown on best suited soil unit for maximizing crop production per unit of land, labour and inputs (Naidu et al. 2006). However, farmers grow different crops without considering soil suitability.

Perception of the farmers towards soil suitable to crops varies. Most of the small holders preferred effective utilization of available soils for crop production to sustain their family needs. Quite differently, educated and large farmers preferred that effective utilization of soils as per the potential to grow crops in order to get maximum profit from the farm.

Focused PRA was carried out to assess the farmers' perception about soil suitability and how they are implementing their perception in actual use (Ramamurthy et al. 2000). More than 60% of the farmers opined that the soil-site characteristics (soil depth, slope per cent and stoniness) corroborated with the scientific suitability framework. However, in practice only 60% of cotton; 50% of sorghum, soybean and groundnut; and 20% of the orange are being grown in suitable land (Fig. 26.2).

It is observed that all the crops recorded higher yield in deep soils followed by medium deep and shallow soils under standard management practice (Table 26.3). But, in practice this is not possible to put only deep soils for crop production, and many stakeholders especially small and marginal farmers have mostly shallow and medium deep soils. Therefore, it is necessary to identify suitable management strategies and agrotechnologies specific to soil to improve the productivity of crops in sustainable manner.

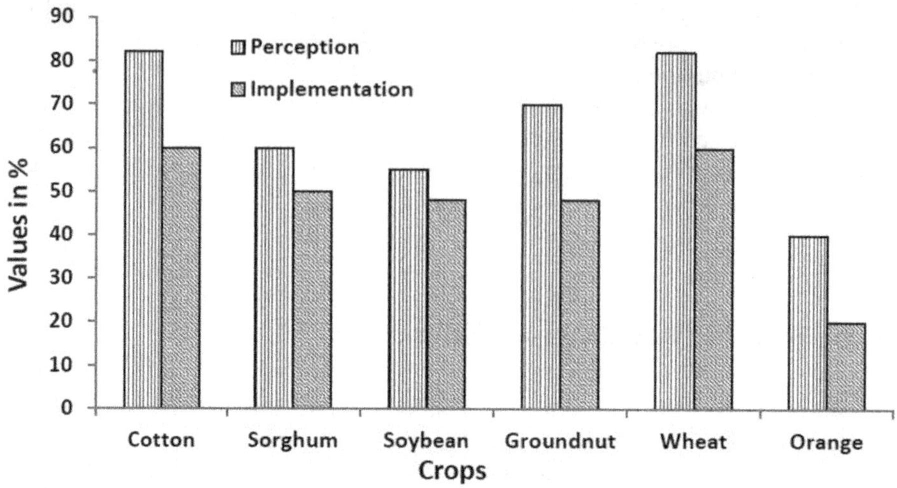

Fig. 26.2 Perception and implementation of scientific soil-site suitability in the watershed

Table 26.3 Performance of different crops on different soils

Soil depth (cm)	Sorghum grain yield (q/ha)	Seed cotton yield (q/ha)	Soybean grain yield (q/ha)
Shallow (<25)	17.1	03.3	07.9
Medium deep (25–50)	21.8	06.5	11.6
Deep (>50)	32.1	13.4	14.1

26.4 Sustainable Cropping Systems Identified for Different Soils

Improvement in productivity of cropping systems assumes significance in farming system to utilize all the resources in maximum possible way to achieve sustainability (Table 26.4). In this context most economical and sustainable crops and cropping systems were identified to different soils (Ramamurthy and Sarkar 2009).

26.5 Soil-Based Agrotechnologies Identified for Sustainability

Farmers grow different crops without considering soil suitability. More than 60% of the farmers' opined that the soil-site characteristics (soil depth, slope per cent and stoniness) corroborated with the crop production. But, in practice only about 50% of

Table 26.4 Identified cropping systems for different soils

Soil type	Technological interventions	Crops	B:C ratio	Situation
Shallow soils (<25 cm)	Farmer's practice	Cotton (hybrid)	0.41	Rainfed
		Sorghum	0.69	Rainfed
	IMP (suggested combinations)	Cotton (varieties)	0.94	Rainfed
		Soybean-wheat (Lok-1)	0.82	Rainfed
		Soybean-gram	0.57	Rainfed
		Sorghum-coriander	0.76	Irrigated
		Mulberry + urd/mung bean-gram	2.50	Irrigated
		Mulberry + marigold	2.00	Irrigated
Medium deep soils (25–50 cm)	Farmer's practice	Cotton (Hyb)	1.79	Rainfed
		Sorghum	1.21	Rainfed
		Soybean-gram	0.57	Rainfed
		Soybean-wheat (Lok-1)	0.82	Irrigated
	IMP (suggested combinations)	Sorghum-gram (Vijay)	0.95	Rainfed
		Sorghum-wheat	0.62	Irrigated
		Cotton-wheat	1.27	Irrigated
		Soybean-gram	1.88	Rainfed
		Soybean-wheat (Lok-1)	1.19	Irrigated
		Sorghum-fodder maize (A.tall)	1.62	Irrigated
Deep soils (>50 cm)	Farmer's practice	Cotton (Hyb)	2.40	Rainfed
		Soybean-wheat	1.19	Irrigated
		Sorghum-wheat	0.62	Irrigated
	IMP (suggested combinations)	Cotton-gram (Vijay)	3.06	Irrigated
		Cotton-wheat	1.74	Irrigated
		Soybean-wheat (HD-2189)	1.88	Irrigated
		Sorghum-wheat	0.64	Irrigated
		Soybean-fodder maize (A.tall)	2.23	Irrigated
		Soybean-gram (Vijay)	2.72	Rainfed
		Sorghum-gram (Vijay)	1.32	Rainfed

sorghum, 60% of cotton and 50% of soybean are being grown in suitable land. Hence, land evaluation (assessment of land performance) for growing sorghum, cotton and soybean was carried out, and it was observed that soils with more than 50 cm depth associated with more than 50% clay were found to be more suitable for higher production of sorghum, cotton, soybean and orange. But, in practice this is not possible to put only deep soils for crop production, and many stakeholders especially small and marginal farmers have mostly shallow and medium deep

Table 26.5 Appropriate agrotechnologies identified for different soils (Ramamurthy and Sarkar 2009)

Agrotechnologies	Shallow soils (Dongari and Pathari soil)	Medium deep soils (Bardi soil)	Deep soils (Halki and Bhari kanhar)
Sorghum-BJH-117 as an alternative to the existing CSH-9	*	**	***
Sorghum-CSV-15 dual purpose (grain and green fodder) variety	***	**	*
Sowing of sorghum between 25 June and 10 July, maintenance of optimum plant population through higher seed rate (10 kg/ha) and recommended dose of NPK (80:40:40 kg/ha)	***	***	***
Seed priming – soaking seeds in water for 10 h, surface drying and sowing for delayed sowing of sorghum	***	***	*
Opening of broad bed and furrow (BBF) at 30 days after sowing after every eight lines of sorghum	***	**	***
Cotton varieties with recommended dose of fertilizers (50:25:0 kg NPK/ha), closer spacing 60 × 60 or 90 × 45 cm spacing and inoculation of seed with *Azotobacter*	***	**	*
Bt cotton hybrid (MECH-184) with recommended dose of fertilizers (100:50:50 kg NPK/ha) and seed inoculation with *Azotobacter*	***	***	**
Cotton hybrids with recommended dose of fertilizers (100:50:50 kg NPK/ha) and seed inoculation with *Azotobacter*	*	**	***
Opening of ridge and furrows and INM technique for cotton hybrids	***	**	***
Drilling of *Rhizobium*-inoculated seed coupled with FYM + phosphate-solubilizing bacteria (PSB) mixture in 1:1 ratio along with 75% of RDF (22.5:56.25 kg NP/ha) and maintenance of optimum seed rate for soybean	***	***	**
Liquid *Rhizobium* seed treatment instead of carrier-based *Rhizobium* inoculation	**	***	**
Application of 5 kg of vermicompost per plant along with 20 kg FYM in the month of November and May (during basin preparation) for increased number of fruits per tree, fruit weight and fruit yield in orange	–	***	***
Growing of marigold as a pure crop or as an intercrop in cotton as a contingent crop plan (when the onset of *kharif* rains are delayed by 3–4 weeks)	***	***	**

(continued)

Table 26.5 (continued)

Agrotechnologies	Shallow soils (Dongari and Pathari soil)	Medium deep soils (Bardi soil)	Deep soils (Halki and Bhari kanhar)
Improved forages and grasslands profitable	***	–	–
Growing of fodder maize (African tall) with pre-monsoon rains to ease the fodder problem during July–August. Fodder maize-coriander is the profitable cropping system for fodder and cash resources	–	**	***
Fodder maize as intercrop in orange orchards with or without cowpea for fodder	–	**	***
Introduction of pasture legumes like *S. hamata* to improve the quality of native grasses and productivity of native grasslands	***	–	–
Soybean-coriander, soybean-gram and sorghum-gram cropping systems for rainfed conditions	**	***	***
Soybean-wheat, mulberry + urd/mung-gram and mulberry + marigold for irrigated/ protective irrigated conditions	***	–	–
Bt cotton-wheat/gram, sorghum-fodder maize, soybean-wheat and sorghum-wheat	*	***	***

*Note: * less appropriate; ** appropriate; *** most appropriate; – not verified*

soils. Therefore, it is necessary to identify suitable management strategies and agrotechnologies specific to soil to improve the productivity of crops in sustainable manner (Table 26.5).

26.6 Sustainable Farming Systems

Integration of these soil-based agrotechnologies with other enterprises would give sustainability and profitability of farming system. Integrated farming system modules were implemented in the project area with different categories of farmers and identified the sustainable farming systems.

- *Case 1*: Introduction of Giriraja breed as backyard poultry and Osmanabadi buck with local breed of goat (1 + 10) increased the family income of land less farmers by 60% besides providing employment throughout the year to one of the family members.
- *Case 2*: Small farmers could earn 79% higher income by adopting soil-based agrotechnologies in shallow soils integrated with backyard poultry, goat rearing,

mushroom cultivation and honeybee keeping. This integrated farming system (IFS) also generated 450 additional man days of self-employment.

- *Case 3*: Medium farmer cultivating shallow, medium deep and deep soils can generate 60% and more additional income by integration of soil-specific crop management practices along with dairying, sericulture and vermicompost. This system also provided 600 man days of additional self-employment.
- *Case 4*: Large farmers owning deep soils can get 70% higher income mostly from agriculture by adopting site-specific management practices. Integration of dairying, goat rearing, honeybee keeping and vermicompost. Vermicompost serves as an input supply activity to their main activity of agriculture than monetary point of view

26.7 Soil-Based Land Use Planning (SLUP) in Southern Transition Zone of Central Karnataka Plateau

Integrated land use planning exercise was initiated in 15 tribal hamlets of H.D.Kote taluk of Mysore district, Karnataka. The annual rainfall of H.D. Kote ranges from 611.7 to 1053.9 mm and average rainfall is around 832 mm. This area is having 55 rainy days and the length of growing period is 150–180 days. Heavy rains during the monsoons cause heavy runoff, which results in massive soil erosion. Tribes practise subsistence agriculture and depend mainly on minor forest produces for their livelihood. Agriculture is predominantly rainfed and monocropped. Most of the tribal farmers are growing ragi, field bean, cotton, horse gram and cowpea. Only 54% of the families are having land and land holdings range from 0.4 to 1 ha.

26.7.1 Soils and Their Constraints

Soils are shallow to very deep (25 to >150), gravelly and non-gravelly sandy clay loam to clay in texture. The colour is dark red to dark reddish brown. The slope is generally 1 to 5%. These soils are developed from granite. Most of the cultivators are applying neither organic manures nor fertilizers (imbalanced) in cotton cultivation resulting in multi-nutritional deficiencies. The major soil-related constraints identified are undulating land terrain with moderate to steeply sloping landform, moderately deep to shallow soils with low water holding capacity, moderate to severe erosion and gravelliness/stoniness.

Table 26.6 Constraints, crops suitable and mutually agreed land use plan to each LMU

LMU	Existing land use	Soils and constraints	Crop suitability	Mutually agreed LUP
LMU 1	Cotton	Shallow, well-drained, gravelly red loam soils with low WHC and low productivity	Cotton-N; finger millet-S3; maize-S3; horse gram/green gram-S1; field bean-S2; pigeon pea-N; cowpea-S2	Cotton (0.5 ac), finger millet + field bean (1 ac), sesame-finger millet (1 ac), vermicomposting
LMU 2	Cotton, finger millet	Medium deep, well-drained, sandy clay loam soils with medium WHC and medium productivity	Cotton-S3; finger millet-S2; maize-S2; pigeon pea-S2; green gram/cowpea/black gram-S1	Maize-banana (1 ac); cotton-ragi (1 ac); vegetables (0.5 ac) + dairy (1 milch cow)
LMU 3	Cotton	Deep, well-drained, sandy clay loam soils with high water retention and low to medium productivity	Cotton-S2; finger millet-S1; maize-S1; pigeon pea-S1; green gram/cowpea/black gram-S1; mango-S2, amla-S1; banana-S1	Maize, chilli, cotton, finger millet, banana (each 0.5 ac), dairy, vermicomposting
LMU 4	Cotton	Deep to very deep, moderately well-drained, clayey soils with shade and low to medium productivity	Cotton-S1; finger millet-S1; maize-S2; pigeon pea-S1; green gram/cowpea/black gram-S1; mango-S2, amla-S1; banana-S1	Vegetables (1.5 ac); coffee + pepper + yam + drumstick (0.7 ac); turmeric and zinger (0.3); backyard poultry and vermicomposting

26.7.2 Identification of Crops Suitable to Different Soils

Land resource inventory of adopted tribal hamlets was carried out at 1:5 K scale. Based on soil depth, gravelliness and texture, soils were grouped into different land management units (LMUs). The LMUs identified and their details of the study area are presented in Table 26.6. Each LMU was evaluated for their suitability to different crops, and farmer preference and his/her resources were considered while suggesting crops and cropping system along with other enterprises in a farming system approach. Farming system analysis was carried out in each LMU through participatory approach to identify the constraints and potentials of different components and also to prioritize the farming system components.

26.7.3 Performance of Soil-Based Integrated Land Use Plan

Shallow, Well-Drained, Gravelly Red Loam Soils (LMU 1) Suggested integrated land use plan (cotton (0.5 ac), finger millet + field bean (1 ac), sesame-finger millet (1 ac), vermicomposting) for shallow gravelly red loam soils was validated in

Table 26.7 Economic analysis of suggested and farmers practice land uses

Land use	Net returns (Rs)	BCR
Cotton (farmers practice) (1 ha)	5750	0.57
Cotton (0.5 acre)	1800	0.75
Finger millet + field bean (1 ac)	6800	1.70
Sesame-field bean (1 ac)	8600	2.15
Net income from crop components	17,200	
Vermicompost	8000	
Net returns from 1 ha integrated land use plan	**25,200**	

Table 26.8 Economic analysis of suggested and farmers practice land uses

Land use	Net returns (Rs)	BCR
Cotton (FP) (1 ha)	11,000	1.10
Ragi (FP) (1 ha)	2000	0.25
Maize (1.2 ac)	6000	1.00
Ragi (1.2 ac)	5600	1.40
Cotton (1.2 ac)	9200	1.92
Banana (1.2 ac)	100,000	2.50
Net income from crop components	120,800	
Dairy (fodder in 0.10 ac)	18,000	1.00
Net returns from 1 ha ILUP of maize-banana; cotton-ragi + dairy (1 milch cow)	**138,800**	

farmers field comparing with farmers practice (cotton). The suggested different components found more economical than farmers practice (Table 26.7).

Medium Deep, Well-Drained, Sandy Clay Loam Soils (LMU-2) Cotton and finger millet are the major land use followed by farmers in this LMU. Suggested integrated land use plan (maize-banana (1.20 ac); cotton-ragi (1.20 ac) + dairy (1 milch cow with 0.10 acre fodder)) was validated in farmers field by comparing farmers practice (cotton and finger millet). Suggested integrated land use plan recorded higher net returns and BCR than farmers practice (Table 26.8).

Deep, Well-Drained, Sandy Clay Loam Soils (LMU-3) Integrated land use plan (maize, chilli, ragi, cotton, banana each in 0.25 acre with dairy and vermicompost) was compared with farmers' major land use (cotton) on deep, well-drained sandy clay loam soils. Banana is found to be the most profitable one with highest BCR, and all other crop components are found more economical than farmers practice (Table 26.9).

Deep to Very Deep, Well-Drained to Moderately Well-Drained, Clayey Soils (LMU-4) Existing crop of this area is mostly paddy in valleys though tribals cultivate vegetables like yam, tapioca, tomato, cabbage, cotton and finger millet in forest fringe areas for own consumption and very little is sold to others. Soils are moderately to highly suitable for all the crops. Suggested land use plan includes

Table 26.9 Economic analysis of suggested and farmers practice land uses

Land use	Net returns (Rs)	BCR
Cotton (FP) (1 ha)	18,000	1.80
Maize (0.5 ac)	5000	1.67
Chilli (0.5 ac)	6200	1.24
Ragi (0.5 ac)	3000	1.50
Cotton (0.5 ac)	4600	1.92
Banana (0.5 ac)	60,000	3.00
Net income from crop components	78,800	
Dairy	18,000	1.00
Vermicompost	11,000	

Table 26.10 Annual net return obtained from different crops/cropping systems

Crops/cropping system	Net returns (Rs/year)
Field bean + tomato + beans + bhendi + veg. Cowpea (1 acre)	30,000
Bitter gourd (0.5 acre)	5000
Coffee+ pepper+ yam+ turmeric +drumstick (1 acre)	54,000
Net income from crop components	*89,000*
Vermicompost	3000
Backyard poultry (Giriraja)	2000
Net returns (Rs/ha/year)	*94,000*

cotton, field bean, vegetable cowpea, beans, radish, greens, cabbage, tomato, bitter gourd, yam, turmeric and ginger. Multi-storied system comprising of pepper, coffee, yam, drumstick and tapioca comes up very well in these areas, and there is good demand for these crops. Besides these crops, goat farming and backyard poultry are more remunerative enterprises. Integrated land use plan consisting of vegetable, spices, vermicompost and poultry was evaluated in 1 ha area, and net return obtained is presented in Table 26.10.

Maximum net return obtained was from vegetable followed by spices. Furthermore, farmer used most of the vermicompost produced to his own field. However, by selling worms he could earn an amount of Rs. 3000 per year. Similarly, from backyard poultry by selling eggs and three young birds, he got net return of Rs.2000.

26.8 Impact of Soil-Based Land Use Plans

- Area under cultivation has increased due to soil-based land use planning in the villages. The uncultivable land (12%) and current fallows (34%) were brought under cultivation by way of fodder production, afforestation and agroforestry. There was 5% increase in net sown area.
- Changes in land use and cropping pattern were significant due to SLUP. Before the implementation of the plans, maximum area was under cotton hybrids and

sorghum. After 5 years, cotton hybrids have shifted to deep soils and varieties to shallow soils. Similarly, sorghum and citrus have been shifted to soybean. This indicates that productivity level and market forces influence the land use rather than soil suitability alone.
- The productivity of dryland and irrigated crop has increased in the range of 14–48% in grain crops and 400% in fodder crops.
- The Crop Productivity Index of rainfed crops is higher than irrigated crops.
- The income of farmers was increased by about 32% after implementation of the SLUP.
- In a short period, SLUP has been able to motivate large number of farmers to adopt modern farm technology and help them in raising farm production and consequently income.
- The SLUP helped in generating knowledge and providing skills about new production technology among adopted and also other farmers of the region.
- The programme provided an excellent mechanism for feedback information for generation of refined, low-cost and effective technology and developed interpretive base for soil units mapped.
- The SLUP has generated demand for better services and supplies of inputs and creation of physical facilities essential for raising production and livelihood.

26.9 Conclusions

Soils are heterogeneity in nature and land use plans can't be implemented uniformly. Soil-based land use plans help in enhancing the productivity, profitability and livelihood of farming communities than existing land use. Crop performance and their input requirement vary with soils and site characteristics; thereby cost of production could be minimized by way of improving resource use efficiency. SLUPs implemented in central and southern plateau of India indicated that crop diversification and cropping intensity in both rainfed and irrigated increased by utilizing available land area and area under fallowing decreased. SLUP has generated demand for better services and supplies of inputs and creation of physical facilities due to increase income. Soil-based integrated farming systems are sustainable livelihood and economically viable systems for farming community.

References

Costales A, Delgado C, Catelo MA, Lapar ML, Tiongco M, Ehui S, Bautista AZ (2007) Scale and access issues affecting smallholder hog producers in an expanding peri-urban market: Southern Luzon, Phillippines. Research report 151, International Food Policy Research Institute, Washington, DC, USA

Geist HJ, Lambin EF (2004) Dynamic causal patterns of desertification. Bio Science 54(9):817–829

Naidu LGK, Ramamurthy V, Challa O, Hegde R, Krishnan P (2006) Manual on soil-site suitability criteria for major crops. NBSS publication. 129, Nagpur

Naidu LGK, Dharumarajan S, Lalitha M, Srinivas S, Ramamurthy V, Singh SK (2014) Agropedology 24(02):253–261

Ramamurthy V, Sarkar D (2009) Soil Based Agro technologies for Livelihood Improvement. NBSS Publ. No. 144, NBSS&LUP, Nagpur, 44p

Ramamurthy V, Patil NG Bankar WV, Gajbhiye KS, Venugopalan MV (2000) Land use planning: farmers' perception and priorities – a case study. Paper submitted for National Symposium on perspectives and policies for land use planning, held at NBSS&LUP from 27th Dec 2000

Ramamurthy V, Singh SK, Ramesh Kumar SC, Nair KM, Patil NG, Angadi S (2015) Integrated land use planning for enhanced tribal livelihood in H.D. Kote Taluk of Mysore district Karnataka, NBSS&LUP, Nagpur 34p

Safriel U, Adeel Z (2008) Development paths of drylands: thresholds and sustainability. Sustain Sci 3:117–123

Safriel U, Adeel Z, Niemeijer D, Puigdefabres J, White R, Lal R, Winslow M, Ziedler J, Prince S, Archer E, King C (2005) Dryland systems. In: Hassan R, Scholes R, Ash N (eds) Ecosystems and human well-being: current state and trends, vol 1. Island Press, Washington, DC, pp 623–662

Walingo MK, Liwenga ET, Kangalawe RYM, Madulu NF, Kabumbuli R (2009) Perceived impact of land use changes and livelihood diversification strategies of communities in the Lake Victoria Basin of Kenya. J Agric Biotechnol Sustain Dev 1(3):69–78

Chapter 27
Geospatial Technologies in Integrated Watershed Management

G. P. Obi Reddy

Abstract Watershed is a geo-hydrological unit area that drains to a common point and is considered as an appropriate physical unit for natural resources evaluation, planning, and management. Watershed management implies the rational utilization of land and water resources for optimum production with minimum hazard to natural environment. Watershed prioritization is one of the most important aspects of planning for implementations of its development and management programs. Geospatial tools can be effectively used in various aspects of integrated watershed management, which includes terrain analysis, land resource inventory, assessment of soil erosion, watershed prioritization, assessment of land capability and irrigability, land use planning, and identify the critical areas for treatment within the watershed for planning and implementation of watershed plans. High-resolution remote sensing, geographic information system (GIS), global positioning system (GPS), simulation modeling, and information and communication technologies (ICT) opened new opportunities to develop intelligent watershed management information systems. Synergy of remote sensing, GIS, and Web-based technologies allow to develop and access dynamic geospatial watershed information without burdening the users with complicated and expensive software. Participatory monitoring and evaluation must be an integral part of integrated watershed management.

Keywords Geographic information system · Remote sensing · Watershed prioritization · Watershed management

27.1 Introduction

Watershed-based development has been the prime strategy for rainfed regions of India since the 1980s to conserve natural resources, enhance agricultural productivity, and improve rural livelihoods. Although soil and water conservation was initially

G. P. O. Reddy (✉)
ICAR-National Bureau of Soil Survey & Land Use Planning, Nagpur, India

G. P. O. Reddy, S. K. Singh (eds.), *Geospatial Technologies in Land Resources Mapping, Monitoring and Management*, Geotechnologies and the Environment 21,
https://doi.org/10.1007/978-3-319-78711-4_27

the primary objective of watershed program that saw large public investments since inception, its focus later shifted to people's participation, equity, and livelihood security, particularly from the mid-1990s. The basic goal of watershed management is to reduce rural poverty and improve livelihood security, while protecting the environment and enhancing sustainability of natural resources (Reddy et al. 2006). Watershed management is a landscape-based strategy that aims to implement improved natural resource management systems for improving livelihoods and promoting beneficial conservation, sustainable use, and management of natural resources. Watershed management is the implementation of management systems to ensure the preservation, conservation, and sustainable use of land resources. The development of watershed management is recognized as a prerequisite for the sustainable management of land resources and the improvement of upland inhabitants' living conditions. In the context of changing climate, there is need to adopt an integrated approach in land management considering its ecological potentials and limitations (Reddy et al. 2008). Instead of focusing exclusively on biophysical processes that improve resource conditions, integrated watershed management includes multiple crop- and livestock-based income strategies that support and diversify livelihood opportunities for the poor and create synergies between targeted technologies, policies, and institutions to improve productivity, resource use sustainability, and market access (Kerr 2001). It is generally accepted that sustainable use and management of land resources will only be achieved by adopting a system of improved land, water, and vegetation management and use based on an integrated approach to land resources development with the direct involvement and participation of the people. Ramesh Kumar et al. (2002) evaluated the impact of watershed development in different agroclimatic zones of Karnataka, Southern India.

27.2 Watershed-Based Approach

Watershed is a geo-hydrological unit area that drains to a common point and is considered as an appropriate physical unit for natural resources evaluation, planning, and management. Watershed management implies the rational utilization of land and water resources for optimum production with minimum hazard to natural environment (Bhat 1989). In fact, watershed management integrates various aspects of climate, soils, agriculture, forestry, hydrology, ecology, and other resources to provide guidelines for choosing acceptable management alternatives within the specific social and economic context. Integrated watershed management through people's participation has become widely accepted as the approach that ensures sound sustainable natural resources management and a better agriculture economy for upland inhabitants as well as the people living in downstream areas. The concept of watershed management involves accurate mapping and timely monitoring of natural resources using available latest geospatial technologies (Reddy et al. 2013; Reddy 2014). The common guidelines for watershed development projects emphasized to use remote sensing, GIS, and modeling techniques to bring a paradigm shift in preparing detail project reports (DPRs) for implementation of the watershed development programs (NRAA 2008).

27.3 Integrated Watershed Management: Strategies and Approaches

Integrated watershed management is an effective means for the appraisal, conservation, and development of land and water resources. As an interdisciplinary approach, it integrates the biophysical, technological, and socioeconomic aspects of the development. Remote sensing and GIS technologies have immense potential in inventory, characterization, and management of land resources particularly in terrain characterization, soil resource inventory (Reddy et al. 2013), land use/land cover, groundwater assessment, soil loss assessment, land capability classification (Reddy et al. 2000), soil-site suitability evaluation (Walke et al. 2012), watershed prioritization (Reddy et al. 2004a, b, c; Reddy and Sarkar 2012), participatory planning, and knowledge dissemination. In broader sense, watershed management is basically to maintain the equilibrium between elements of the natural ecosystem or vegetation, land or water on the one hand and human activities on the other hand.

27.4 Role of Geospatial Technologies in Integrated Watershed Management

The capabilities of remote sensing and GIS in conjunction with field surveys have immense potential in resources inventory, mapping, geospatial database generation, and analysis in ecological perspective and applications that are explicitly spatial and temporal in nature (Smith and Blackwell 1980; Trotter 1991; Maji and Reddy 2004; Reddy et al. 2009). Satellite data with synoptic view, repetitive coverage, and multispectral view provide accurate information about the present land use/land cover to prepare integrated plans for optimal utilization of natural resources (Anderson et al. 1976; Clevers et al. 1999). Field surveys data on terrain, soils, and different land use systems coupled with the satellite data-based derived inputs in a GIS framework enable us to understand the interrelationships among the climatic, edaphic, agronomic, and economic factors, in order to develop conservation-effective and sustainable production systems. In watershed-based studies, GIS and remote sensing are being used in inventory of biophysical resources, geospatial database generation, analysis of spatiotemporal changes, and overlay of different thematic databases to identify the potentials and limitations of the resources in formulation of site-specific action plans for sustainable management of land resources and land use systems.

A variety of new-generation high-resolution remote sensing satellites like IRS-P6, Cartosat-1, Cartosat-2, QuickBird, and IKONOS are available to the Earth scientists for generation of spatiotemporal database on natural resources for various applications in watershed development. High spatial and temporal resolution satellite data could be effectively used for watershed management and monitoring activities. The innovative techniques are also successfully used for inventory and preparation of detailed thematic databases, watershed development plans, and continuous monitoring of the natural resources in rainfed areas. GIS, high-resolution remote sensing, simulation modeling, and ICT open up new opportunities to develop intelligent watershed management information systems. Synergy of remote sensing, GIS, and Web-based technologies have capabilities to develop dynamic watershed-based geospatial information and dissemination to the users. Geospatial tools can be effectively used in terrain analysis, land resource inventory, estimating soil erosion, land capability classification, land use planning, and identification of critical areas, which needs treatment within the watershed area through implementation of watershed plans. Decision-making process at watershed level requires the integration of knowledge, data, simulation models, and expert judgment to solve practical problems and provide a scientific basis in implementation of plans (NRC 1999).

27.4.1 Morphometric Analysis of Watersheds

The surface drainage characteristics of various watersheds and sub-watersheds in different parts of the globe have been studied earlier using conventional methods (Horton 1945; Strahler 1952, 1957, 1964; Morisawa 1959; Leopold and Miller 1956). Evaluation of morphometric parameters necessitates analysis of drainage parameters, i.e., ordering of the various streams, measurement of area and perimeter of basin, length of drainage channels, drainage density, drainage frequency, bifurcation ratio, texture ratio, and circulatory ratio (Kumar et al. 2000; Sahu et al. 2016, 2017). Quantitative description of basin morphometry involves measurement of linear and areal features, gradient of channel network, and contributing ground slopes of the drainage basin. Such type of analysis is of immense help in understanding the drainage morphometry and its interactive impacts on landforms and their characteristics (Reddy et al. 2004a, Reddy 2012).

Remote sensing and GIS techniques are being effectively used in determining the quantitative description of the basin geometry (Biswas et al. 1999; Wilson and Gallant 2000; Reddy et al. 2002a, b, c, 2004a, b, c). The high-resolution remotely sensed data coupled with topographical data analysis procedures have made satellite data-based morphometric analysis highly effective tool to understand and manage natural resources. The analysis of satellite data in conjunction with drainage morphometry improves the capabilities in delineation of geological and landform units.

It also helps in identification of drainage channels, which are altered by natural forces or human-induced activities. The generated data geo-coded resource database at river basin level in the core of GIS provides an excellent means of storing, retrieving, manipulating, and analyzing spatial information in watershed-scale applications. It also provides a powerful mechanism not only to upgrade and monitor morphometric parameters but also to permit the spatial analysis of other resources data (Jain et al. 1995).

The morphometric parameters at watershed level like higher drainage density, stream frequency, lower form factor, circulatory ratio, and elongation ratio indicates severe soil erosion, whereas, low drainage density, stream frequency, texture ratio, moderate form factor, and circulatory ratio are favorable factors for high infiltration and low surface runoff. The quantitative evaluation of USLE parameters using GIS and remotely sensed data is of immense help in prioritization of sub-basins. The analysis of morphometric parameters of landscape helps to assess and evaluate erosion risk, groundwater potential, and watershed characterization and develop soil and water conservation strategies.

27.4.2 Terrain Characterization and Landform Mapping

Terrain characterization is a process through which fractal nature of terrain and its biophysical processes are quantified or attributed into thematic layers. Terrain consists of the physiography, lithology, morphometry, soil geography, and to some extent land cover (Meijerink 1988). Remote sensing and GIS help in characterization and delineating the distinct units of terrain in the watershed. The resultant terrain units presented in the form of map could be effectively used in inventory of various biophysical resources at watershed level. The availability of high-resolution digital elevation models (DEM) at 2.5 m resolution from Cartosat-1 is an ideal data set for precise terrain analysis and topographic characterization in terms of the nature of relief aspects, patterns of lineaments and surface slope, topographic profiles and their visualization, correlation between geology and topography, hypsometric attributes, and, finally, the hierarchy of terrain subunits in the watershed area. The delineated landform units delineated from analysis of different terrain parameters formed as base for soil resource inventory and mapping (Speight 1990) and planning of land use systems in the watershed. Semiautomated landform mapping can be carried out to obtain landform information from digital data like DEM and satellite imagery using object-based image analysis (OBIA) by considering geometric and contextual information (Camargo et al. 2011). Further, terrain characterization and landform analysis of a terrain have immense help to develop landscape-soil relationship for precise soil resource mapping, land degradation mapping, and other environmental applications at water level (Reddy et al. 1999a, b, 2002a, b, c).

27.4.3 *Soil Resource Inventory*

Soil resource inventory and mapping is the process of classifying soil types and other soil properties in watershed area and geo-encoding such information. Soil survey provides an accurate and scientific inventory of different soils, their kind and nature, and extent of distribution so that one can make prediction about their characters and potentialities (Manchanda et al. 2002). Several authors have reported that the satellite remote sensing and GIS are proved as promising tools in large-scale soil resource mapping (Maji et al. 2001, 2004; Srivastava and Saxena 2004; Sarkar et al. 2006; Reddy et al. 2013). The surface features reflected on satellite image provide enough information to accurately delineate the boundaries, which is accomplished effectively through systematic interpretation of satellite imageries (Velmurugan and Carlos 2009).

27.4.4 *Land Use/Land Cover Analysis*

The knowledge of both land use and land cover within the watershed area is more important for monitoring and improving ecosystems, landscape, hydrology, land planning, land management activities, habitats, and ecological studies for sustainable planning (Rao et al. 1996). Application of remotely sensed data made possible to study the changes in land use/land cover in less time, at low cost, and with better accuracy (Kachhwala 1985), and GIS provided suitable platform for data analysis, update, and retrieval (Chilar 2000). Various methods have been proposed and used to detect changes in land use/land cover using remote sensing data such as image differencing, vegetation index differencing, selective principal components analysis, direct multi-date classification, univariate image differencing, image rationing, change vector analysis, and post-classification (Mas 1999).

27.4.5 *Assessment of Soil Loss*

Soil erosion is a major concern in landscape management and conservation planning (FAO 1985). Geospatial technologies play a critical in the generation of spatial data layers and their integration to estimate soil loss at watershed level by adopting the suitable models like Universal Soil Loss Equation (Wischmeier and Smith 1978). GIS integrated with image analysis systems (IAS) and database management systems (DBMS) is a vital tool in the generation of spatial data layers and their integration to estimate soil loss of a given area. These techniques have been used extensively in quantification of USLE parameters to estimate soil erosion (Jain and Kothari 2000; Reddy et al. 2004a, b, c, 2016). The factors of rainfall erosivity (R), soil erodibility (K), slope length (LS), cover (C), and management (P) are the components of USLE and can be computed in GIS using rainfall data, series-wise soil information, slope generated from contours, and land use/land cover interpreted from remotely sensed data and field data, respectively. GIS has immense potential in integrated analysis of USLE parameters in quantification of potential and actual soil loss (Reddy et al. 2004a, b, c and Srinivas et al. 2002; Das et al. 2013; Sahoo, et al. 2014, 2014a; Gangopadhyay et al. 2014; Bandyopadhyay et al. 2015). Following multi-criteria overlay of several critical parameters, the units for soil conservation can be delineated to suggest suitable agronomic and mechanical conservation measures based on the soil and site characteristics in each of the delineated units.

27.4.6 Assessment of Groundwater Resources

Access to reliable groundwater sources plays an important role in food security in many cases as the access to reliable sources of water reduces the production risk. Yields in groundwater-irrigated areas are higher (often double) compared to those in the canal-irrigated areas (Shah 1993). Groundwater availability acts as a trigger to enable the farmers to invest in complementary inputs that, in combination, increase crop yields substantially. In the dry land semiarid tropics, an integrated watershed management resulted in increased groundwater availability that served as an entry point for increasing agricultural production and improving rural livelihoods (Wani et al. 2003). Integrated watershed management (IWM) is the strategy adopted to enhance the water use efficiency for sustainable development of dry land areas. The IWM strategy demonstrated that dry land areas with good quality soils could support double cropping, while the surplus rainwater could recharge the groundwater. Temporal data from remote sensing enables identification of groundwater aquifers and assessment of their changes along with the land use changes, whereas, GIS enables integration of multi-thematic data. Analysis of surface and subsurface hydrological parameters through analysis of remotely sensed data in conjunction with field surveys provides a scope to identify hydro-geomorphological units (Reddy et al. 1999a, b). It can be carried out in GIS through database generation on drainage, landforms, lithology, land use/land cover, and lineament thematic parameters from data and their overlay analysis (Krishnamurthy and Srinivas 1995; Krishnamurthy et al. 1996). The delineated hydro-geomorphological units can be analyzed to assess their groundwater potential. However, the necessary field verification and validation of the results with available collateral data are essential for accurate assessment of the groundwater prospects.

27.4.7 Assessment of Land Capability

The use of land is not only determined by the user but also by the land capability. Land capability mapping is the basis for sustainable development by ensuring correct land use according to its capability for sustained crop production (Reddy et al. 2000). The land capability is governed by the different land attributes such as the types of soil, its depth and texture, underlying geology, topography, hydrology, etc. The capability units are groupings of soils that have common responses to pasture and crop plants under similar systems of farming (Klingebiel and Montgomery 1961). In eightfold LCC the first four classes are arable land, which are suitable for cropland in which the limitations on their use and necessity of conservation measures and careful management increase from I to IV. The criteria for placing a given area in a particular class involve the landscape location, slope of the field, depth, texture, and reaction of the soil. The remaining four classes, from V to VIII, are not to be used for cropland, but may have uses for pasture, range, woodland, grazing, wildlife, recreation, and esthetic purposes. Within the broad classes, there are subclasses which signify special limitations such as (e) erosion, (w) excess wetness, (s) problems in the rooting zone, and (c) climatic limitations. Within the subclasses there are capability units, which give some prediction of expected agricultural yields and indicate treatment needs. The parameters like slope, soil depth, texture, and land use/land cover assessed by applying remote sensing and GIS techniques could be effectively used for land capability classification at watershed level.

Soils of Borgaon Manju watershed, Akola district, Maharashtra state, have been evaluated for land capability and land irrigability classification. Soils of Borgaon Manju watershed were grouped under three land capability classes and four land capability subclasses of IIs, IIIs, IVs, and IVes. Soils of Bonderkhed, Dhotardi, Warudi, Apoti Khurd, Kumbhari-1, Kumbhari-2, and Sanglud Buzurg were grouped under class IIs with minor soil limitation. Soils of Washimba and Shamabad have been classified as class IIIs with moderate soil limitation. Soils of Borgaon Manju-2 have been grouped under IVs, whereas the soils of Gorva, Borgaon Manju-1, and Sivani have been grouped under class IVes with severe limitations of soil depth and erosion. The data indicates that the land capability subclasses IIs, IIs-IIIs, IIIs-IIs, IVes-IIIs, and IVes-IVs cover 43.4, 10.9, 3.9, 8.6, and 26.8%, respectively (Fig. 27.1).

27.4.8 Assessment of Land Irrigability

Land irrigability classification is an interpretative grouping based on soil and land characteristics, which indicate relative suitability of land for irrigation as well as predicted behavior of soils under irrigation. The factors like soil depth, texture, moisture holding capacity, soil surface cover, soil salinity and alkalinity hazards, drainage, soil erosion, groundwater table, and landform (relief and slope) decide the irrigability class of soil unit. Soils of Borgaon Manju watershed, Akola district, Maharashtra, have been evaluated for land irrigability according to their limitation for sustained use under irrigation (AISLUS 1971). Lands suitable for irrigation are grouped under classes I to IV according to their limitations. Lands not suitable for irrigation are grouped under classes V and VI. Land irrigability classes have

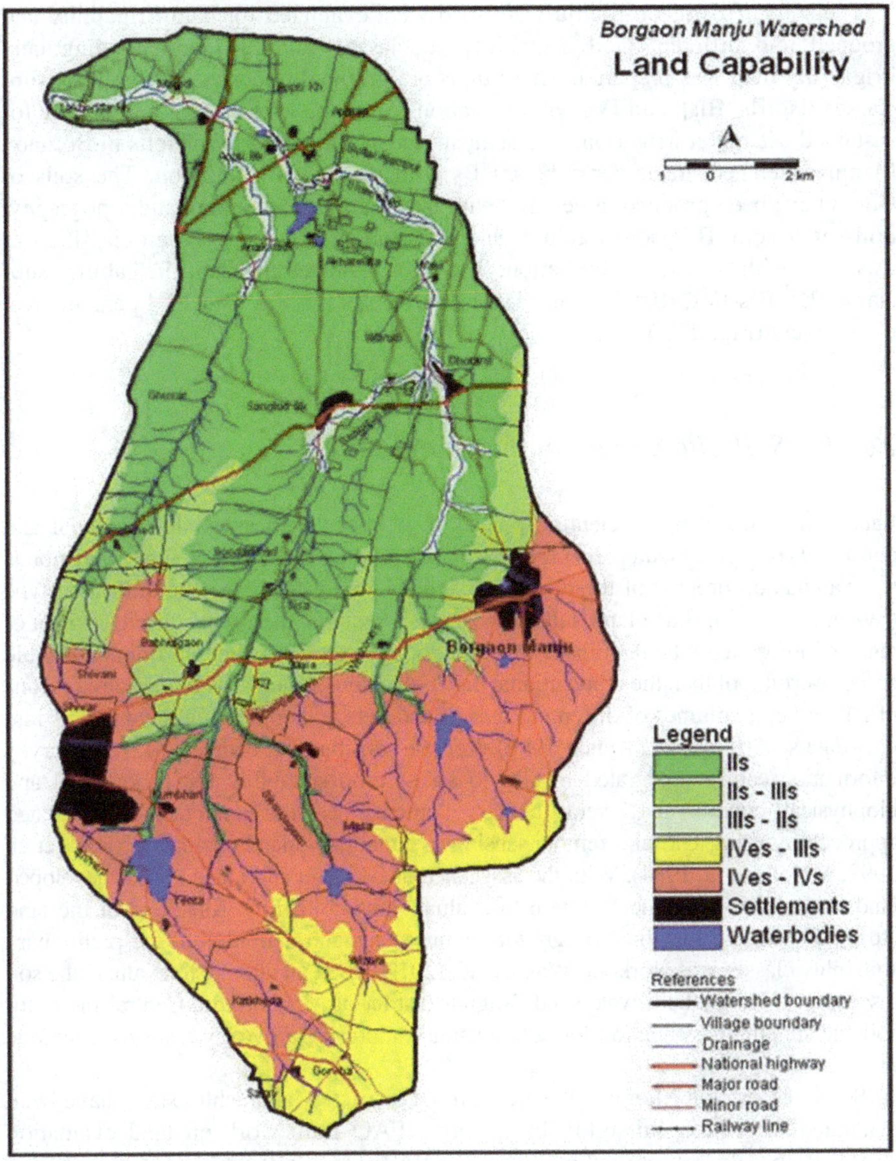

Fig. 27.1 Land capability of Borgaon Manju watershed, Akola district, Maharashtra

subclasses to indicate their dominant limitations for irrigation purposes. Three subclasses are denoted by "s" for soil limitations such as heavy clay or sandy texture, soil depth, and gravel/stones, "d" for drainage problems, and "t" for limitations of topography.

The soils of Borgaon Manju watershed were evaluated for land irrigability and grouped into different land irrigability subclasses, and the corresponding land irrigability map was prepared. These units are grouped under land irrigability subclasses IIs, IIIs, IIIst, and IVs with limitations of depth, texture, and topography for sustained use under irrigation. The analysis shows that majority of soils in Borgaon Manju watershed are under subclass IIs with minor soil limitation. The soils of Gorva have been grouped under IIIst with major limitations of soil and topography. Soils of Vizora, Borgaon Manju-1, and Borgaon Manju-2 have been classified as class IVs with major soil limitation. The data indicates that land irrigability subclasses IIs, IIIs-IVs, IIIst-IVs, and IVs cover 58.2, 11.5, 8.6, and 15.4 percent area, respectively (Fig. 27.2).

27.4.9 Soil-Site Suitability Evaluation

Land evaluation using a scientific procedure is essential to assess the potential and constraints of a given land parcel for agricultural purposes. Land suitability evaluation is an examination process of the degree of land suitability for a specific utilization type (Sys et al. 1991b). Since land suitability analysis requires the use of different kinds of data and information (soil, climate, land use, topography, etc.), the GIS offers a flexible and powerful tool than the conventional data processing systems, as it provides a means of taking large volumes of different kinds of data sets and combining the data sets into new data sets (Foote and Lynch 1996). Remote sensing data coupled with soil survey information can be integrated in GIS to assess crop suitability for various soil and biophysical conditions. Several authors demonstrated the potential of integrated approach by using GIS and remote sensing in quantitative land evaluation (Beek et al. 1997; Merolla et al. 1994). With the assistance of GIS, Liu and Deng (2001) developed land resources management system to evaluate land suitability. Matching of the land attributes with the specific crop growth requirements and definition of the preliminary suitability classes was workout. Walke et al. (2012) made an attempt to evaluate the soil resources of Ringnabodi watershed, Nagpur district of Maharashtra, Central India, for soil-site suitability evaluation for cotton using multi-criteria overlay analysis technique in GIS.

Soils of Borgaon Manju watershed, Akola district, Maharashtra state, have been evaluated for cotton suitability by adopting FAO framework on land evaluation (FAO, 1976, 1983), modified by Sys et al. (1991a) and NBSS&LUP (1994). As per the soil suitability criteria for cotton, soils of Bonderkhed, Dhotardi, Warudi, Apoti Khurd, Kumbhari-1, Kumbhari-2, Shamabad, and Sanglud Buzurg are moderately suitable (S2) with moderate limitations of texture, calcium carbonate, and low organic carbon content (Fig. 27.3). Soils of Sivani and Washimba are marginally suitable (S3) with severe limitations of depth. Soils of Gorva, Vizora, Borgaon Manju-1, and Borgaon Manju-2 are not suitable (N) with very severe limitation of depth and erosion. The extent of area at soil series association level under S2, S2-S3, S3-N, and N is 47.3, 10.9, 11.5, and 24.0 percent, respectively.

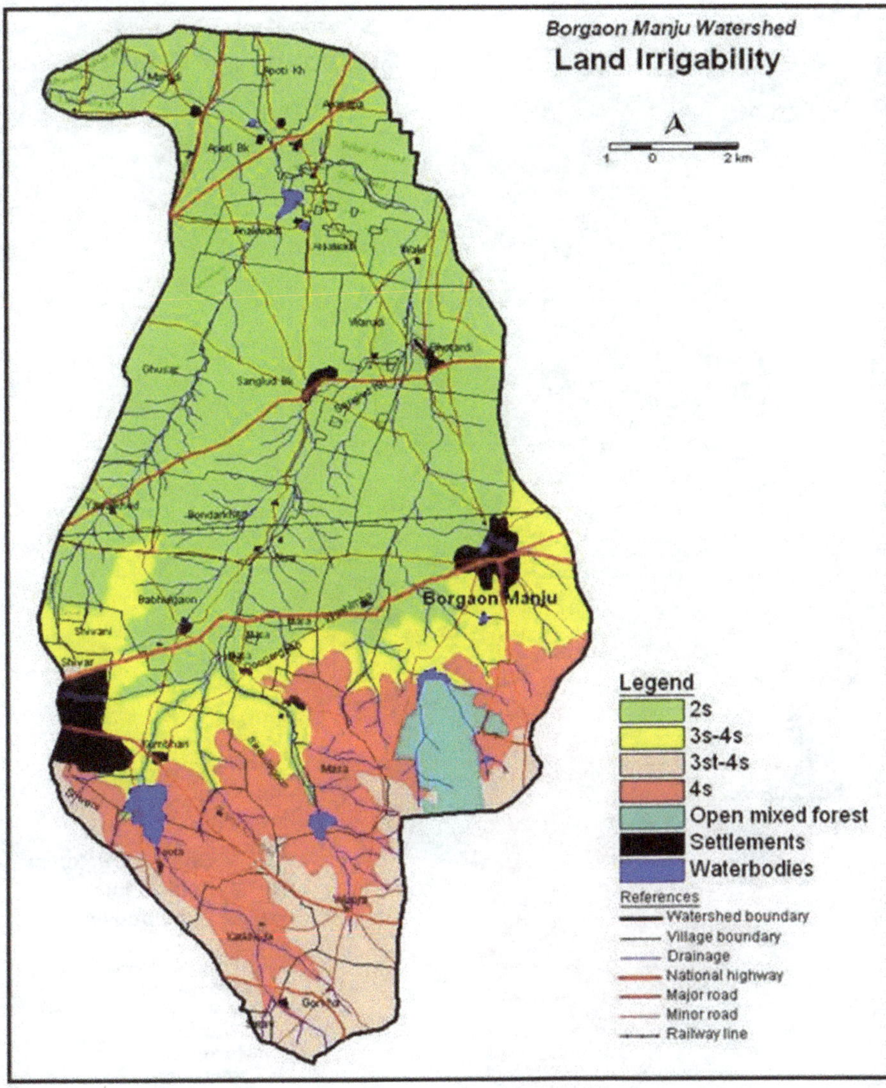

Fig. 27.2 Land irrigability of Borgaon Manju watershed, Akola district, Maharashtra

27.4.10 Watershed Prioritization for Soil and Water Conservation

Watershed prioritization is one of the most important aspects of planning for implementations of its development and management programs more particularly in arid and semiarid regions (Reddy et al. 2004a). Watershed prioritization on the basis of soil erosion has become inevitable component of watershed management in

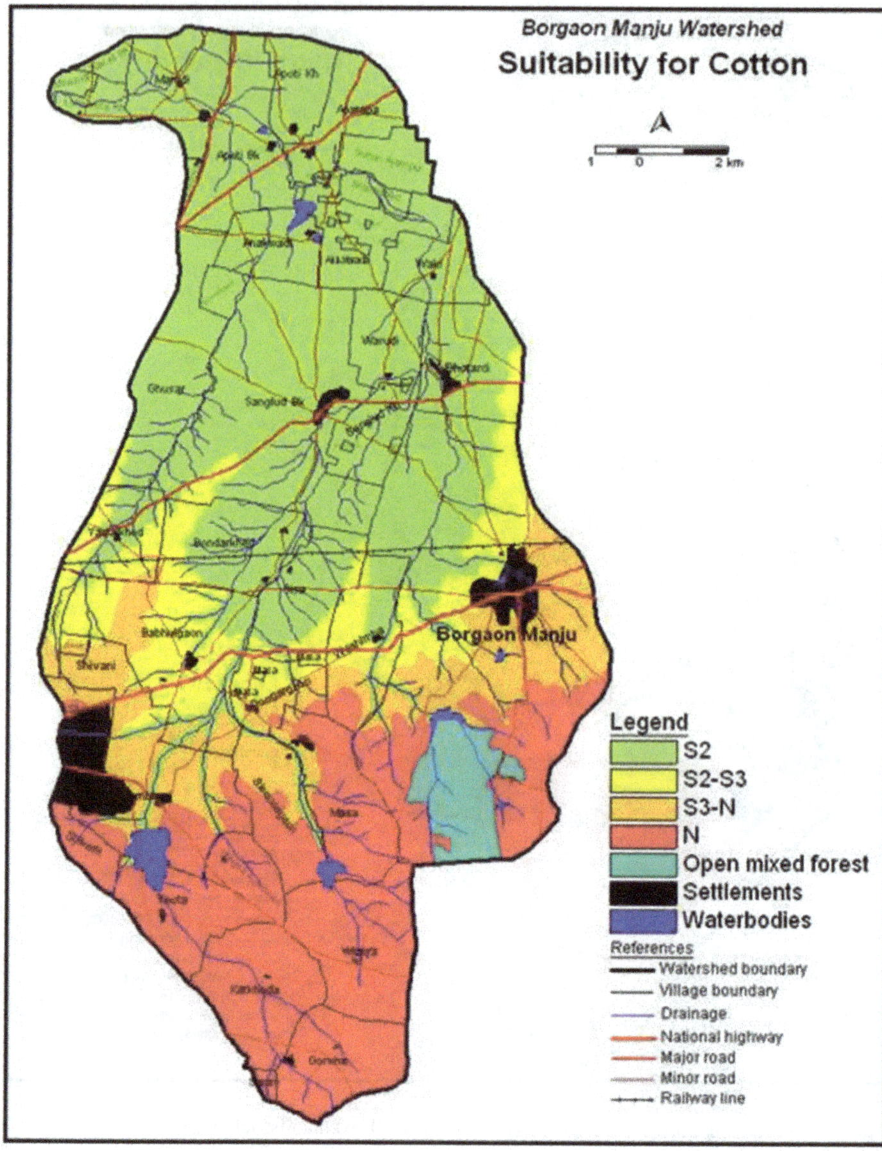

Fig. 27.3 Soil-site suitability for cotton and sorghum in Borgaon Manju watershed

order to conserve this precious natural resource. The major factors contributing to the soil erosion include the land use, soils, slope, climate, and land management practices. Biswas et al. (1999) reported that morphometric analysis is a significant tool for prioritization of sub-watersheds even without considering the soil map. Remote sensing and GIS techniques have been used extensively in quantification of USLE (R, K, LS, C, and P) parameters in assessment of soil loss (Pandey et al. 2007; Yoshino and Ishioka 2005). Reddy et al. (2002a, 2004a) reported that computation of morphometric parameters and USLE parameters at watershed level in GIS is found very useful in evaluation of erosion characteristics of watersheds for their prioritization.

27.4.11 Participatory Approaches in Watershed Management

Participatory approaches and community watershed management plans have been widely used, with varying success, to reconcile the overlay of human activity on naturally defined watersheds. The participatory processes succeeded where there were common purposes that could interest all or most of the population in the watersheds, where communities could see the economic benefits and were empowered. Participation does not, however, guarantee specific outcomes, and it is not a neutral concept: it involves shifts in decision-making power between the state and local communities and also between different segments of the local community. Participatory processes, therefore, have to be designed for the specific development and distributional outcomes intended. Participatory approaches impose a demanding set of requirements. Participatory monitoring and evaluation must be an integral part of participatory watershed management. In watershed management, ICT-based platforms could be effectively used for learning and synthesizing watershed development experiences in different agroecological regions of the country so that the evolving experience can be captured, validated, and made available for dissemination to a wider community.

27.4.12 Spatial Decision Support Systems in Watershed Development

The goal of the watershed-level spatial decision support system (SDSS) is to provide an integrated watershed-scale application that can be used to educate stakeholder groups and assist in the decision-making process. A user-friendly SDSS helps various stakeholder groups develop, understand, and evaluate alternative watershed management strategies. ICT, GIS, and SDSS provide a great opportunity for sharing information and applications with decision-makers. However, limitations in availability, architectures, bandwidth, and security present challenges for using this medium. Advances in communication of information through Internet GIS, simulation models, and integration of these technologies in SDSS provide opportunities for improving the transfer of information and knowledge from watershed scientists and managers to decision-makers. Integration of hydrological models with GIS provides an important platform for evaluating and assessing hydrologic systems, and managers can rely on this technology to support decision-making. Further, the Internet offers an efficient medium for sharing information among decision-makers and stakeholders at watershed level. Use of ICT to create and deploy watershed-based educational tools provides stakeholders with additional resources to incorporate into the bottom-up decision-making process.

27.5 Conclusions

High-resolution satellite data, GPS, and GIS can be effectively used in integrated watershed management particularly in terrain analysis, land resource inventory, estimating soil erosion, land capability classification, land use planning, and identify the critical areas, which needs treatment within the watershed for preparing and implementation of watershed plans. Watershed prioritization become an important component for planning and implementations of various development and management programs as well as conservation of this precious natural resource. Watershed management must be focused on rational utilization of available land and water resources for optimum production with minimum hazard to natural environment. Participatory monitoring and evaluation must be an integral part of watershed planning, management, and evaluation. The developments in high-resolution remote sensing, GIS, and ICT technologies have given new insights and dimension in addressing variety of land resources planning, development, and management activities at watershed level for sustainable land resource management, land use planning, and maintain ecological balance.

References

AISLUS (1971) All India soil and land use survey, soil survey manual. Indian Agricultural Research Institute (IARI) Publ, New Delhi

Anderson JR, Hardy EE, Roach JT, Witmer RE (1976) A land use and land cover classification system for use with remote sensor data. Geological Survey Professional Paper 964, U. S. Govt. Printing Office, Washington

Bandyopadhyay S, Reddy GPO, Dutta D, Reza SK, Dutta DP, Baruah U, Sah KD, Singh SK (2015) Assessment of Erosion susceptibility zones in Diring-Thanglong watershed north Eastern Hill regions of Assam using GIS techniques. Agropedology 25(2):169–180

Beek KJ, De Bie K, Driessen P (1997) Land information and land evaluation for land use planning and sustainable land management. Land Chatham 1:27–44

Bhat LS (1989) Status report on land system analysis for evaluation of resources at micro level. Discussion paper for the fifth review meeting for NRDMS project, DST, Govt. of India

Biswas S, Sudhakar S, Desai VR (1999) Prioritization of sub watersheds based on morphometric analysis and drainage basin: a remote sensing and GIS approach. J Indian Soc Remote Sens 27(3):155–166

Camargo F, Almeida C, Florenzano T, Heipke C, Feitosa R, Costa G (2011) ASTER/Terra imagery and a multilevel semantic network for semi-automated classification of landforms in a subtropical area. Photogramm Eng Remote Sens 77(6):619–629

Chilar J (2000) Land cover mapping of large areas from satellites: status and research priorities. Int J Remote Sens 21(67):1093–1114

Clevers JGPW, Muecher CA, Popov VP, Vandysheva NM, Vassilenko GI (1999). Agriculture land cover monitoring over Russia using remote sensing. In: Proceedings of the 18th EARSeL Symposium ITC, Enschede, The Netherlands, A.A. Balkema, Publishers Rotterdam

Das TH, Dutta D, Reddy GPO, Singh SK, Sarkar D, Dhyani BL, Mishra PK (2013) Soil erosion of Sikkim, by NBSS Publ. 164, NBSS&LUP, Nagpur, p 42

FAO (1976) A framework for land evaluation. Soils bulletin 32. FAO, Rome, 72 p

FAO (1983) Guidelines: land evaluation for Rainfed agriculture. Soils bulletin No. 52. FAO, Rome, 237 p

FAO (1985) Watershed development with special reference to soil and water conservation, Soil bulletin 44. Food and Agricultural Organization, Rome

Foote KE, Lynch M (1996) Geographic information systems as an integrating technology: context, concepts and definition, vol 2. University of Texas, Austin

Gangopadhyay SK, Reddy GPO, Sarkar D (2014) Erosion risk mapping in parts of semi-arid tropics of India using remote sensing and GIS. Indian J Soil Conserv 42(1):29–38

Horton RE (1945) Erosional development of streams and their drainage basins: hydrophysical approach to quantitative morphology. Bull Geol Soc Am 56:275–370

Jain MK, Kothari UC (2000) Estimation of soil erosion and sediment yield using GIS. Hydrol Sci J 45(5):771–786

Jain SK, Seth SM, Nema RK (1995) Morphometric analysis of Sei Dam catchment using GIS. In: Proceedings of National symposium on Hydrology, Jaipur, India, pp 5–7

Kachhwala TS (1985) Temporal monitoring of forest land for change detection and forest cover mapping through satellite remote sensing. In: Proceedings of the 6th Asian conference on remote sensing, Hyderabad, pp 77–83

Kerr J (2001) Watershed project performance in India: conservation, productivity, and equity. Am J Agric Econ 83(5):1223–1230

Klingebiel AA, Montgomery PH (1961) Land capability classification. Agriculture handbook No. 210. Soil Conservation Service, U.S. Department of Agriculture, Washington, DC, pp 1–3

Krishnamurthy J, Srinivas G (1995) Role of geological and geomorphological factors in ground-water exploration: a study using IRS-LISS-II data. Int J Remote Sens 16:2595–2618

Krishnamurthy J, Kumar NV, Jayaraman V, Manivel M (1996) An approach to demarcate ground water potential zones through remote sensing and a geographical information system. Int J Remote Sens 17(10):1867–1884

Kumar R, Kumar S, Lohani AK, Nema RK, Singh RD (2000) Evaluation of geomorphological characteristics of a catchment using GIS. GIS India 9(3):13–17

Leopold LB, Miller JP (1956) Ephemeral streams: hydraulic factors and their relation to the drainage network. U.S. Geological Survey Prof Paper 282-A

Liu YS, Deng XZ (2001) Structural patterns of land types and optimal allocation of land use in Qinling Mountainous. J Geogr Sci 11(1):99–109

Maji AK, Reddy GPO (2004) Geoinformatics for natural resources and environmental management at watershed level for sustainable development. In: Pandey BW (ed) Natural resource management. Mittal publications, New Delhi, pp 143–163

Maji AK, Nayak DC, Krishna NDR, Srinivas CV, Kamble K, Reddy GPO, Velayutham M (2001) Soil information system of Arunachal Pradesh in a GIS environment for land use planning. Int J Appl Earth Obs Geoinf 3(1):69–77

Maji AK, Reddy GPO, Thayalan S, Walke NJ (2004) Characterization and classification of landforms and soils over basaltic terrain in sub-humid tropics of Central India. J Indian Soc Soil Sci 53(2):154–162

Manchanda ML, Kudrat M, Tiwari AK (2002) Soil survey and mapping using remote sensing. Trop Ecol 43(1):61–74

Mas JF (1999) Monitoring land cover changes: a comparison of change detection techniques. Int J Remote Sens 20(1):139–152

Meijerink AMJ (1988) Data acquisition and data capture through terrain mapping units. ITC J 1:23–44

Merolla S, Armesto G, Calvanese G (1994) A GIS application for assessing agricultural land. ITC J 3:264–269

Morisawa ME (1959) Relation of morphometric properties to runoff in the Little Mill Creek, Ohio, drainage basin. Technical report 17. Columbia University, Department of Geology, ONR, New York

NBSS&LUP (1994) Soil site suitability criteria for different crops. In: Proceedings of national meet on soil site suitability criteria for different crops Feb, 7–8, 31p

NRAA (2008) Common guidelines for watershed development projects. National Rainfed Area Authority Government of India, New Delhi, p 60

NRC (1999) New strategies for America's watersheds. National Research Council, National Academy Press, Washington, DC

Pandey A, Chowdary VM, Mal BC (2007) Identification of critical erosion prone areas in the small agricultural watershed using USLE, GIS and remote sensing. J Water Resour Manage 21:729–746

Ramesh Kumar SC, Krishnan P, Velayutham M, Gajbhiye KS (2002) Economic land evaluation for sustainable land management of watersheds in different agroclimatic zones of Karnataka. NBSS technical report No. 582, vol. 2 (Database), Nagpur

Rao DP, Gautam NC, Nagaraja R, Ram Mohan P (1996) IRS-IC application in land use mapping and planning. Curr Sci 70(7):575–581

Reddy GPO (2012) GIS and remote sensing applications in watershed hydrology. In: Reddy GPO, Sarkar D (eds) Remote sensing and GIS in digital terrain analysis and soil-landscape modeling. NBSS&LUP Publ. No. 152, pp 159–173

Reddy GPO (2014) Innovative and cost effective spatial positioning-scope of geospatial technologies in large scale land resource inventory. GIS India 23(4):4–5

Reddy GPO, Sarkar D (2012) Assessment of soil loss for prioritization of subwatersheds – a remote sensing and GIS approach. NBSS Publ. No. 137, NBSS&LUP, Nagpur, p 55

Reddy GPO, Chandramouli K, Srivastav SK, Maji AK, Srinivas C (1999a) Evaluation of groundwater potential zones using remote sensing data - a case study. J Indian Soc Remote Sens 28(1):19–32

Reddy GPO, Shekinah DE, Maurya UK, Thayalan S, Prasad J, Ray SK, Bhasker BP (1999b) Landscape-soil relationship in part of Bazargaon plateau, Maharashtra. Geogr Rev India 61(3):280–291

Reddy GPO, Rao MS, Maji AK (2000) Land capability of Narayanappa Kunta micro watershed, Anantapur district (A.P.). Geogr Rev 62(2):126–134

Reddy GPO, Maji AK, Srinivas CV, Kamble KH, Velayutham M (2002a) GIS based basin morphometric information system for terrain and resources analysis. In: Patil VC et al (eds) Proceedings of the first national conference on agro-informatics. NCAI, Dharwad, pp 34–41

Reddy GPO, Maji AK, Srinivas CV, Velayutham M (2002b) Geomorphological analysis for inventory of degraded lands in a river basin of basaltic terrain, using remote sensing data and geographical information systems. J Indian Soc Remote Sens 30(1–2):15–31

Reddy GPO, Maji AK, Chary GR, Srinivas CV, Tiwary P, Gajbhiye, KS (2002c) Geo-spatial database for prioritization and evaluation of conservation strategies for landscape management in a river basin, R.R. In: Navalgund et al (ed) Proceedings of ISPRS commission VII. Symposium on resources and environmental monitoring, Hyderabad, India, vol. 1, pp 654–659

Reddy GPO, Maji AK, Chary GR, Srinivas CV, Tiwary P, Gajbhiye KS (2004a) GIS and remote sensing applications in prioritization of river sub basins using morphometric and USLE parameters - a case study. Asian J Geoinf 4(4):35–49

Reddy GPO, Maji AK, Gajbhiye KS (2004b) Drainage morphometry and its influence on landform characteristics in basaltic terrain – a remote sensing and GIS approach. Int J Appl Earth Observ Geoinf 6:1–16

Reddy GPO, Rao MS, Maji AK (2004c) Delineation and prioritization of macro watersheds in semi-arid Anantapur district, Andhara Pradesh. Geogr Rev India 66(1):72–83

Reddy GPO, Maji AK, Prasad J, Ramamurthy V, Ramteke IK (2006) Geo-spatial resource database for carrying capacity based resource planning - remote sensing and GIS approach. In: Patil VC, Ninomiya S (eds) Information technology in agriculture, the fifth international conference of the Asian Federation for Information Technology in agriculture. Macmillan, India, pp 181–201

Reddy GPO, Maji, AK, Nagaraju MSS, Thayalan S, Ramamurthy V (2008) Ecological evaluation of land resources and land use systems for sustainable development at watershed level in different agro-ecological zones of Vidarbha region, Maharashtra using Remote sensing and GIS techniques. Project Report, NBSS & LUP, Nagpur, 270p

Reddy GPO, Maji AK, Gajbhiye KS (2009) Role of Geoinformatics in micro-level planning for natural resources and environmental management. In: Nathawat MS, Pandey AC (eds) Geoinformatics for decentralized planning and governance. Rawat Publications, Jaipur, pp 159–172

Reddy GPO, Nagaraju MSS, Ramteke IK, Sarkar D (2013) Terrain characterization for soil resource mapping in part of semi-tract of Central India using high resolution satellite data and GIS. J Indian Soc Remote Sens 41(2):331–343

Reddy GPO, Kurothe RS, Sena DR, Harindranath CS, Niranjana KV, Naidu LGK, Singh SK, Sarkar D, Mishra PK, Sharda VN (2016) Assessment of soil erosion in tropical ecosystem of Goa, India using universal soil loss equation, geostatistics and GIS. Indian J Soil Conserv 44(1):1–7

Sahoo AK, Das K, Das AL, Reddy GPO, Singh SK, Sarkar D, Mishra PK (2014) Soil erosion of Andaman & Nicobar Islands, NBSS Publ. 165. NBSS&LUP (ICAR), Nagpur, p 30

Sahoo AK, Sarkar D, Singh SK, Reddy GPO, Dhyani BL, Mishra PK, Sharda VN (2014a) Soil erosion in Jharkhand, NBSS Publ. 159. NBSS&LUP (ICAR), Nagpur, 39 p

Sahu N, Reddy GPO, Kumar N, Nagaraju MSS, Srivastava R, Singh SK (2016) Morphometric analysis in basaltic terrain of Central India using GIS techniques - a case study. Appl Water Sci 7(5):2493–2499

Sahu N, Reddy GPO, Kumar N, Nagaraju MSS, Srivastava R, Singh SK (2017) Morphometri analysis using GIS techniques- a case study from basaltic terrain of Cerntral India. In: Reddy GPO, Patil NG, Chaturvedi A (eds) Sustainable management of land resources-an Indian perspective. Apple Academic Press, Oakville, Canada, pp 301–326

Sarkar D, Gangopadhyay SK, Sahoo AK (2006) Soil resource appraisal towards land use planning using satellite remote sensing and GIS- a case study in Patloinala microwatershed, District Puruliya, West Bengal. J Indian Soc Remote Sens 34(3):245–260

Shah T (1993) Groundwater markets and irrigation development: political economy and practical policy. Oxford University Press, Mumbai

Smith AY, Blackwell RJ (1980) Development of an information database for watershed monitoring. Photogramm Eng Remote Sens 46:1027–1038

Speight JG (1990) Australian soil and land survey: field handbook, 2nd edn. Inkata Press, Melbourne

Srinivas CV, Maji AK, Reddy GPO, Chary GR (2002) Assessment of soil Erosion using remote sensing and GIS in Nagpur District, Maharashtra, for prioritization and delineation of conservation units. J Indian Soc Remote Sens 30(4):197–211

Srivastava R, Saxena RK (2004) Techniques of large-scale soil mapping in basaltic terrain using satellite remote sensing data. Int J Remote Sens 25(4):679–688

Strahler AN (1952) Dynamic basis of geomorphology. Bull Geol Soc Am 63:923–938

Strahler AN (1957) Quantitative analysis of watershed geomorphology. Trans Am Geophys Union 38:913–920

Strahler AN (1964) Quantitative geomorphology of basins and channel networks. In: Te Chow V (ed) Handbook of applied hydrology. McGraw Hill, Newyork

Sys C, van Ranst E, Debaveye J (1991a) Land evaluation, Part-II. ITC Ghent, Belgium 247 pp

Sys C, Van Ranst E, Debvay J (1991b) Land evaluation, Part: III. General Adminstration for Development Coopration Agriculture, Brussels
Trotter CM (1991) Remotely-sensed data as an information source for geographical information systems in natural resource management a review. Int J Geogr Inf Syst 5:225–239
Velmurugan A, Carlos GG (2009) Soil resource assessment and mapping using remote sensing and GIS. J Indian Soc Remote Sens 37(3):511–525
Walke N, Reddy GPO, Maji AK, Thayalan S (2012) GIS-based multi-criteria overlay analysis in soil-suitability evaluation for cotton (*Gossypium* spp.): a case study in the black soil region of Central India. Comput Geosci 41:108–118
Wani SP, Pathak P, Sreedevi TK, Singh HP, Singh P (2003). Efficient management of rainwater for increased crop productivity and groundwater recharge in Asia. In: Kijne W, Barker R, Molden D (eds) Water productivity in agriculture: limits and opportunities for improvement, pp 199–215
Wilson JP, Gallant JC (2000) Terrain analysis – principles and applications. Wiley, New York, 479 pp
Wischmeier WH, Smith DD (1978) Predicting rainfall erosion losses – a guide to conservation planning. USDA Agricultural Handbook No. 587
Yoshino K, Ishioka Y (2005) Guidelines for soil conservation towards integrated basin management for sustainable development: a new approach based on the assessment of soil loss risk using remote sensing and GIS. Water Environ Res 3:235–247

Chapter 28
Geospatial Applications in Water Resource Management with Special Reference to Climate Change

Y. B. Katpatal

Abstract Global climate change has stressed the water availability scenario across the globe. As far as India is concerned, still about 60% of agriculture is rainfed, and global climate change has huge impact on its sustainability. Accurate information on the probable changes in monsoon patterns and availability of water resource is a must for sustainable rainfed agriculture. Proper monitoring and timely anticipation of spatial impacts also assume greater importance in water resource management. Drastic changes in the global environment have compelled us to increase the awareness amongst masses, especially related to and affected by water availability to convert our efforts into higher rate of success. Environmental management could be achieved only through monitoring of the changes in environment and making everybody aware of it. Geoinformatics is the potential technology for generating baseline data for monitoring of environmental parameters pertaining to several environmental changes. Remote sensing technology with various satellites collecting information from the space at different spectral, spatial and temporal resolutions is being widely used for extracting information related to many baseline parameters like vegetation, crops, forests, water resources, urban changes, rain water harvesting, etc. Geographic information system (GIS) has the capability to generate spatial digital data for projecting the parametric information at various levels of environmental monitoring. There is need to exploit the full potential of geoinformatics for timely monitoring and management of the water resources. In today's changing global climate, visualization of water resource parameters and management of water resources could be effectively achieved through application of geoinformatics. The chapter discusses applications of geoinformatics in various aspects of water resources monitoring, especially with reference to climate change.

Keywords Climate change · Geoinformatics · Water resource · Water management

Y. B. Katpatal (✉)
Visvesvaraya National Institute of Technology, Nagpur, India

G. P. O. Reddy, S. K. Singh (eds.), *Geospatial Technologies in Land Resources Mapping, Monitoring and Management*, Geotechnologies and the Environment 21,
https://doi.org/10.1007/978-3-319-78711-4_28

28.1 Introduction

Recent studies have provided sufficient evidences that the El Niño-Southern Oscillation (ENSO) is the dominant climate phenomenon affecting extreme weather conditions worldwide (Cai et al. 2015). The ENSO is not only affecting the climate, but its impacts have been observed to affect the water resources availability, and it is found responsible for disasters in several parts of the world (Banholzer et al. 2014). The ultimate impacts of change in the status of water resources availability are seen on the agricultural sector where monitoring and forecasting of the climate-induced impacts on crops and yields have become important for achieving the goals of production for several governments, especially developing and underdeveloped nations (Iizumi et al. 2014). Since the concept of global climate change and its impacts, especially on water resource management, have been accepted (IPCC 2013), there is eminent need for resorting to techniques which provide information in less time, repetitively so that such problems at global level could be tackled.

It has been well realized now that natural as well as man-managed systems, static or dynamic, call for a separate technology, which can not only provide the information at the baseline but also provide solutions through analysis of the background information, which would eventually bridge the gap between what exists on the ground and the implementation practices; a phenomenon seems to block the development of resources. Various developed countries have adopted geoinformatics, and India has also started generation of the baseline data and the applications based on these data. Several government organizations and the private institutes together have started adopting and using geoinformatics, which also promises as a major source of employment.

Depleting resources and the increasing complexities in the static and dynamic resources have compelled managers to adopt the spatial technology for time- and cost-effective management. Geoinformatics is an interdisciplinary field having potential of synergistic modelling and analysis for dealing with geospatial data and phenomena. The geoinformatics technology can be applied in various fields to solve problems and address issues related to geospatial data and phenomena. Geoinformatics encompasses geospatial data collection, geospatial information analysis and modelling and geospatial information systems development, implementation and processing.

Geoinformatics technologies mainly include remote sensing, geographic information systems (GIS) and global positioning system (GPS). Today, numerous applications are benefiting from geoinformatics techniques and tools, and with advances in geoinformatics and other technologies, such as mobile computing and wireless networks, the emergence of many new applications is expected.

Geoinformatics provides potential to generate the spatial models covering various aspects of environmental management. The most important aspects are rapid urbanization and environment, development in rural and urban areas, water resources and their relation to changing environment, environmental impacts due to anthropogenic activities, carrying capacity of a region, environmental disasters, any individual and his requirements from environment, pollution status in urban and rural areas, etc. Such spatial modules can be easily generated using geoinformatics, and pictorial representation of environmental parameters has highly appreciated, better understood, infused seriousness and would sensitize the young minds. Overall, it would be more effective in our endeavour of awareness and public participation for environmental management.

28.2 Climate Change and Facts of Climate Change: IPCC 2013

The detailed report (IPCC 2013) lists various facts related to climate change and its impacts which could be summarized as below.

- Averaged over the midlatitude land areas of the Northern Hemisphere, precipitation has increased since 1901.
- Over the period 1992–2011, the Greenland and Antarctic ice sheets have been losing mass.
- Over the period 1901–2010, global mean sea level rose by 0.19 (0.17 to 0.21 m).
- Anthropogenic greenhouse gas emissions have increased since the pre-industrial era, driven largely by economic and population growth.
- Total anthropogenic GHG emissions have continued to increase over 1970 to 2010 with larger absolute increases between 2000 and 2010.
- Changes in many extreme weather and climate events have been observed since about 1950.
- The globally averaged combined land and ocean surface temperature data as calculated by a linear trend show a warming of 0.85 (0.65 to 1.06 °C) over the period 1880–2012.

Taken as a whole, the range of published evidence indicates that the net damage costs of climate change are likely to be significant and to increase over time. The most important aspect of climate change study is 'monitoring' of environmental parameters and generating information on their status at any time which requires faster techniques, and the remote sensing and GIS have this capability.

The changes that are occurring due to global climate change which must be understood and recorded are:

1. Temperatures will continue to rise further affecting the water resources availability in warm countries where impacts of the temperature change would have to be recorded.
2. Crop growing season will lengthen due to changing rainfall patterns due to El Niño.
3. Changes in precipitation patterns will ultimately present a complex situation where farmers would require to change their agricultural practices or even will have to resort to genetically altered varieties.

4. More droughts and heat waves would directly affect the water resources storage and groundwater availability required for purposes other than agriculture.
5. Hurricanes will become stronger and more intense causing damages to infrastructure and would also impact the surface water management.
6. Sea level rise will cause the sea water incursions in coastal areas turning the freshwater into saline.
7. More melting at the glaciers by the mid-century would change the discharges in several rivers causing floods and also affect the sustainable supply to perennial rivers in several countries.

The climate forcings hence affect the shape of the climate, and these forcing mechanisms can be either internal or external (IPCC 2013). The internal forcings include meridional overturning circulation and ocean variability, while the external forcings include either natural (changes in solar irradiance) or anthropogenic (increased emission of greenhouse gases).

El Niño and La Niña have been identified as part of the El Niño-Southern Oscillation (ENSO) phases, which affect the global climate system. El Niño events are associated with a warming of the central and eastern tropical Pacific, while La Niña events are the reverse, with a sustained cooling of these same areas (NCDC 2009; NOAA 2006). It is also a known fact today that these phases generate due periodic fluctuation in sea surface temperature measured over Darwin, Australia and Tahiti. The El Niño and La Niña are two opposite phases of this ENSO cycle where El Niño is the warm phase while La Niña is the cold phase. The impact of El Niño on Indian southwest monsoon is associated with lower than normal rainfall and opposite for La Niña (Maity and Kumar 2006). The El Niño strengthens northeast monsoon opposite for La Niña (Zubair and Ropelewski 2006; Ropelewski and Halpert 1987). In this phase coastal areas of Tamil Nadu, India, receive more rainfall by NE monsoon.

28.3 Remote Sensing and GIS Applications

Remote sensing has been providing vital information for monitoring the change in the resources status due the impacts of the global climate change. There are sensors generating the data for specific requirements. Based on the need to understand the climate change and its impacts, especially on water resources availability, the following application objectives may be identified.

- To understand El Nino and La Nina phenomenon and its influence on Indian monsoon.
- To estimate the impacts of ENSO on climate change indicators.
- To identify the impact of regional climate variability on change in the global climate parameters like water resources availability, vegetation, sea surface temperatures, soil moisture, etc.

- Climate varies also in time: from season to season, year to year and decade to decade or on much longer time scales, such as the Ice Ages. Weather, climate change and variability operate on different time scales.
- To identify problems generated through global climate change and present records in support as evidences through tying up the impacts to the global climate changes.

The analysis of rainfall anomalies in the Indian subcontinent (Fig. 28.1) clearly shows that most of the El Nino years are related to droughts, while most of the La Nina years are associated with floods. The management of water resources must start with the understanding of the frequencies of these two phases, which would result into either surplus or deficiency in water availability, controlled by the tropical monsoon. The strategies and planning of managing the shortage or the excess water during these phases will have to be made in spatial context after recording the varying impacts in several river basins of India. The knowledge on the spatial patterns of the impacts of ENSO through several case studies conducted using satellite data and GIS will result into:

- Modelling for future
- Better understanding of problems
- Present status of the problems
- Best alternatives through data analysis

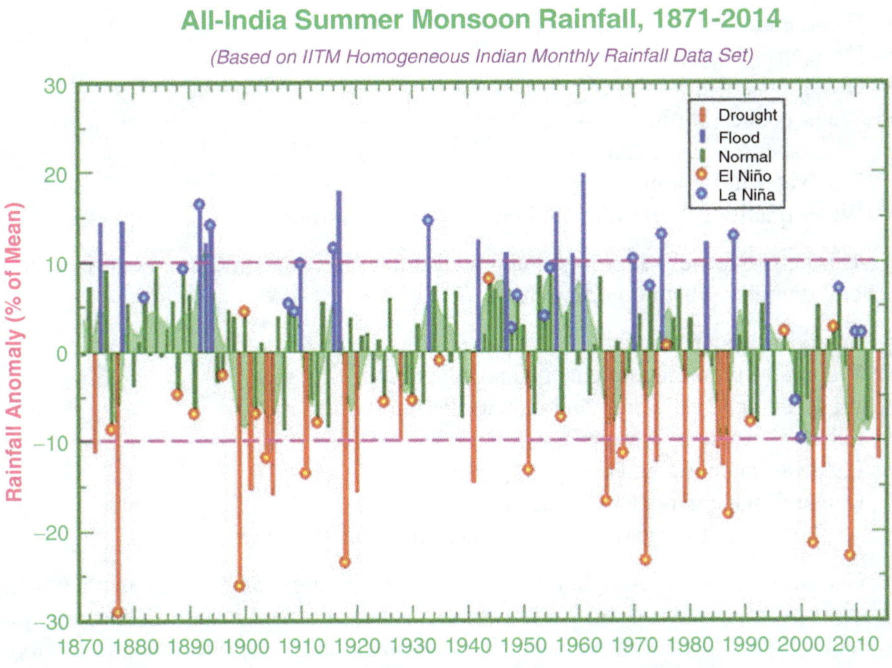

Fig. 28.1 Rainfall anomalies show relationship of droughts and floods to ENSO events (Source: http://www.tropmet.res.in/~kolli/MOL/Monsoon/Historical/air.html)

28.4 Water Resource Management Using Remote Sensing and GIS

Utilizable water resource is the quantum of water which may be withdrawn from its place of natural occurrence. India receives annual precipitation of about 4000 km^3, including snowfall. Out of this, monsoon rainfall is of the order of 3000 km^3. The annual potential natural groundwater recharge from rainfall in India is about 342.43 km^3, which is 8.56% of total annual rainfall of the country. The total utilizable water resources of the country are assessed as 1086 km^3 (Kumar et al. 2005). Most of the precipitation in India is received from southwest monsoon. In country like India, which has huge demand for surface as well as groundwater demand, if there is slightest change in the pattern or quantity of the precipitation, there is impact on the water resources availability and its utilization.

Traditionally, India has been an agriculture-based economy. Hence, development of irrigation to increase agricultural production (Selvaraju 2003) for making the country self-sustained and for poverty alleviation has been of crucial importance for the planners. Climate change has further disturbed and affected the normal orientation of self-sustenance programmes and plans in water sector. Remote sensing and GIS can be successfully utilized in its simplest form for following water resources problems either generated or further modified in scale and pattern due to the global climate change.

- Flood management
- Drought management
- Groundwater management
- Water conservation
- Watershed management
- Rainwater harvesting
- Water quality conservation and environment restoration

This can be achieved through integrated studies, which could be categorized into several domains which could be listed as below.

- Interbasin water transfer
- Vulnerability to climate change and adaptation strategies
- Water demand reduction and management measures
- Legal restrictions on water use
- Land use planning and cropping pattern
- Demand management for urban areas
- Applications of decision support system in water resources

Geoinformatics has brought the science and technology into management domain through its application right from characterization and monitoring till the routines governing management of resources in any known area of application. As a whole the technological developments in geoinformatics have led to many areas like spatial modelling, geo-visualization, knowledge discovery, data mining, geo-computation

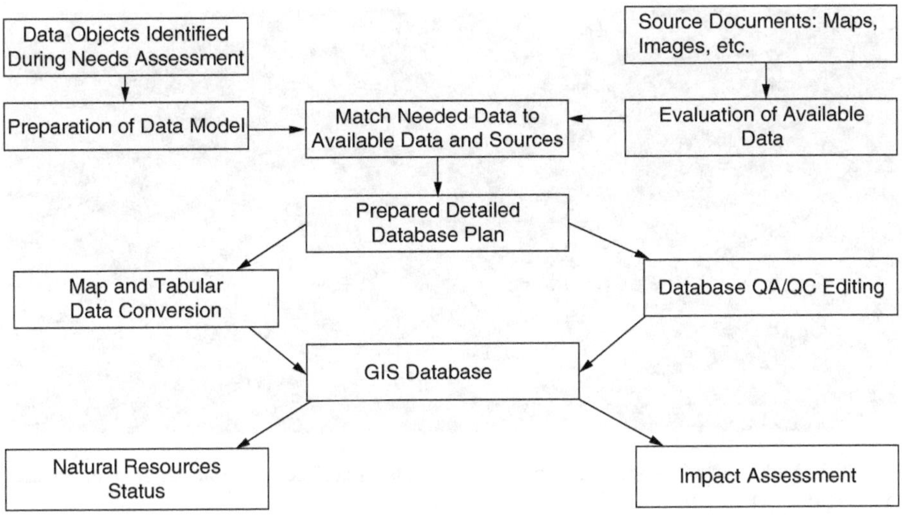

Fig. 28.2 General methodology for resource management using geoinformatics

and spatial information management. Large numbers of earth-observing satellite are continuously collecting the data on earth features, and their pictorial representation provides immensely valuable information.

Space-borne multispectral data have been used to generate baseline information on various natural resources, namely, soils, forest cover, surface water, groundwater and land use/land cover, and subsequent integration of such information with slope and socioeconomic data in a GIS to generate locale-specific prescription for sustainable development of land and water resources development on a watershed basis. The general methodology of the application involving remote sensing and GIS along with other data sets generated from other sources is indicated in Fig. 28.2.

It is important to understand the applicability of remote sensing and GIS in generating information related to micro-aspects of water resource management. Certain case studies reflecting these application areas are discussed below.

28.4.1 Surface Water Resource Inventory

The most required and basic potential of satellite observation for water resources is monitoring of surface water availability in reservoirs using time series satellite data. A simple but most powerful tool where moderate spatial resolution satellite data like IRS LISS-III may also be useful for such monitoring. Simple comparison of the water spreads within the reservoirs may be useful in determining the change (Fig. 28.3).

Patil and Katpatal (2008) have shown that multispectral data may be used even to quantify various parameters including turbidity within the surface reservoirs. The study utilized the multispectral IRS LISS-III data to demonstrate the change in the

Fig. 28.3 Multispectral images showing water spreads in surface reservoirs in Nagpur District (Source: Bhuvan)

turbidity zones in various reservoirs in Chandrapur District of Maharashtra (Fig. 28.4) during 1989 and 2007.

28.4.2 Surface Topography Mapping

Several workers have used the elevation data to generate the digital elevation model (DEM) from the Cartosat satellite for constructing the surface topography, which may be used further to analyse the surface runoff or surface drainage. The topographic information from remote sensing has been useful as it may be obtained temporally in time series and changes even in the topography caused by natural or anthropogenic causes may be understood. The natural slopes, especially in urban areas, get compromised due to heavy infrastructure development and in turn affect the surface runoff. Many flash floods have occurred in recent times due to incessant rain which may also be due to the result of global climate change. Here, the impacts are identified and mitigated using slope directions (aspect) which help in identifying the direction of surface runoff (Fig. 28.5).

28.4.3 Surface Runoff and Infiltration Studies

The surface runoff can be estimated using GIS, which again utilizes land use/land cover classes extracted through satellite data. The SCS curve number method (Mishra et al. 2005; Rishma et al. 2015) is widely used to estimate runoff depth (Fig. 28.6a). Similarly, infiltration index generated from ϕ index method (Mane and

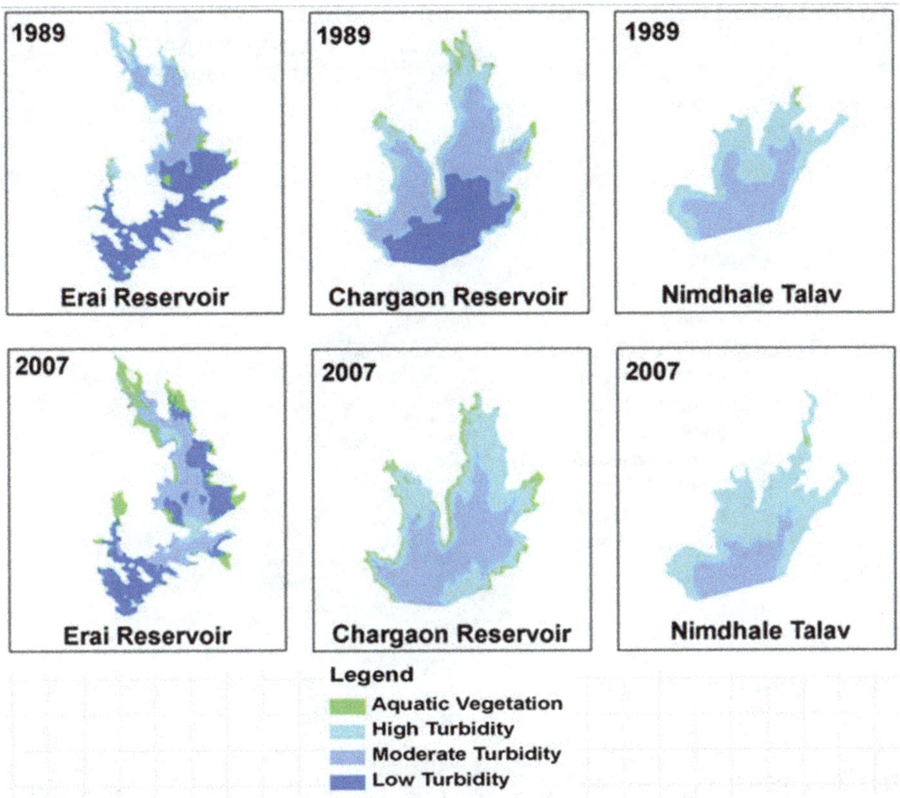

Fig. 28.4 Images showing change in the turbidity zones in reservoirs between 1989 and 2007

Katpatal 2010) can be spatially manifested to understand the control of lithological parameters on infiltration characteristics within the watersheds (Fig. 28.6b).

28.4.4 Groundwater Monitoring and Assessment

Information on groundwater availability is important in Indian scenario as most of the agricultural activity in India is either rainfed or irrigated with the groundwater. Groundwater monitoring data can be visualized in GIS for better understanding, and also, groundwater potential and recharge maps can be generated using GIS. The information on seasonal variations in groundwater levels is a prerequisite for planning, and it may be correlated with the climate change (Katpatal and Dube 2010; Katpatal et al. 2014). Time series spatial analysis of groundwater table (Fig. 28.7) is very helpful in understanding the groundwater fluctuation including recharge.

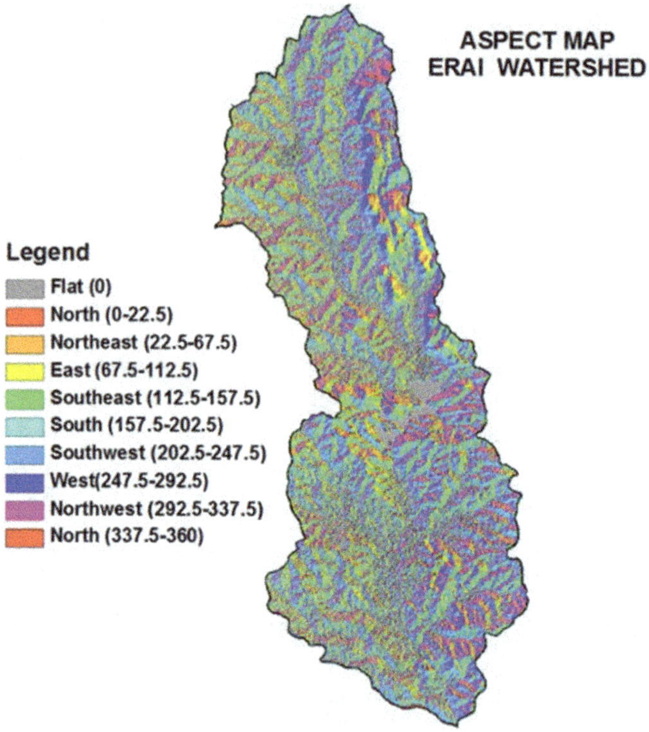

Fig. 28.5 Aspect map generated from Cartosat data for Erai watershed, Chandrapur District, Maharashtra, India

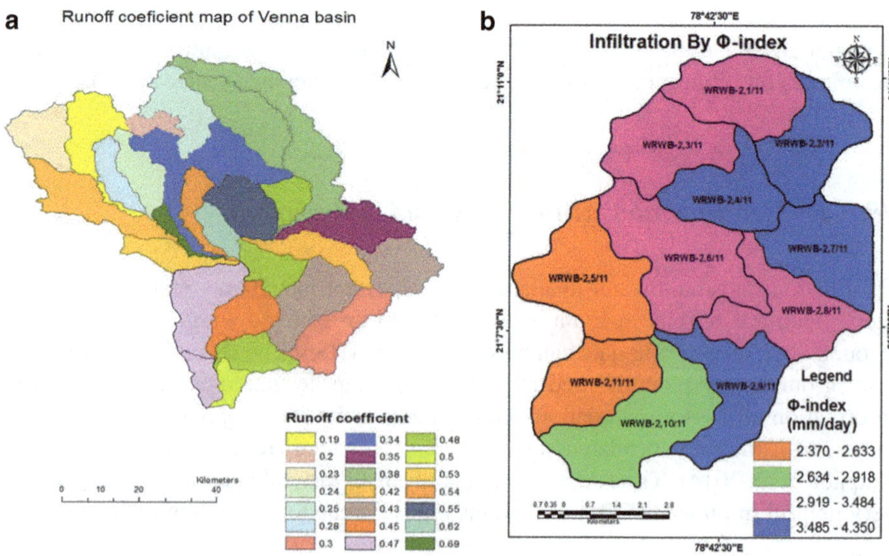

Fig. 28.6 (**a**) Runoff coefficients generated using SCS curve number method; (**b**) infiltration index generated from ϕ index method

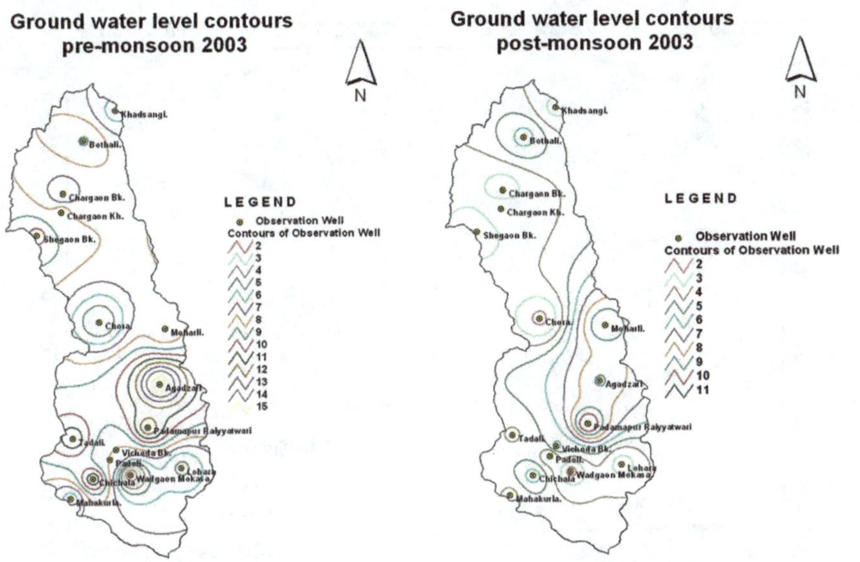

Fig. 28.7 The groundwater level variation during pre- and post-monsoon periods illustrated using interpolation methods in GIS

GIS is useful even at the micro-watershed level to analyse the impacts of small conservation structures on the groundwater development in a village. The performance-based analysis at such a small scale has been shown to be useful by Bhusari et al. (2016). It has been shown in Fig. 28.8 that observation wells recording increase in water levels within the catchment of a conservation structures can be monitored and the area of influence may be known.

28.4.5 Impacts of ENSO on Water Resources and Vegetation

Remote sensing data has been found very appropriate for identifying the impacts of ENSO on the water resources availability and its impacts on the agricultural practices in turn controlling the vegetation indices within the command and non-command areas. The Vegetation Index has been estimated by several workers using MODIS data (MODIS 2013), and the time series analysis has been conducted to show the variation of the status of vegetation. Similarly, Normalized Difference Vegetation Index (NDVI)-based studies have been conducted (Bothale and Katpatal 2014; Bothale and Katpatal 2015; Bing et al. 2014; Erasmi et al. 2009; Guo et al. 2014) to study various aspects of impacts of ENSO on agricultural practices, crop productivity, phenological changes, crop periods and yields, etc. The NDVI variability during ENSO and normal years (Rishma and Katpatal 2016) in watersheds makes the impact of ENSO evident (Fig. 28.9). GRACE data (Rodell et al. 2007) which offers total terrestrial water storage on a very coarse resolution has also been used for groundwater and surface water studies (Fig. 28.10).

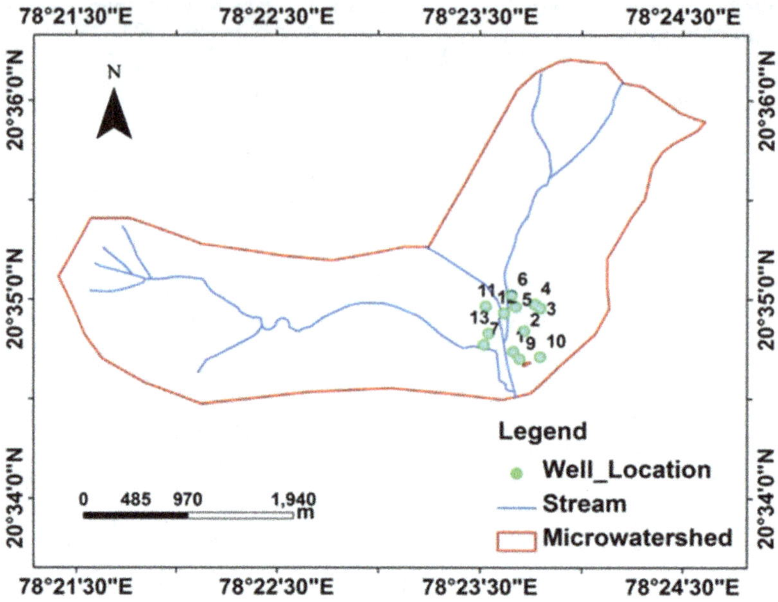

Fig. 28.8 The groundwater observation wells falling within the catchment of a conservation structure showing increase in water level in Wardha District, India

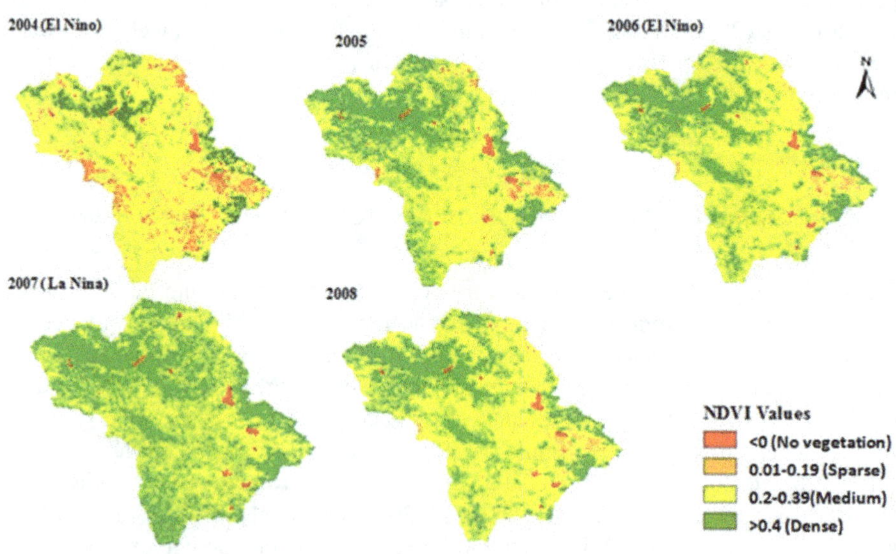

Fig. 28.9 NDVI variability during ENSO and normal years in different watersheds of Venna Basin, India

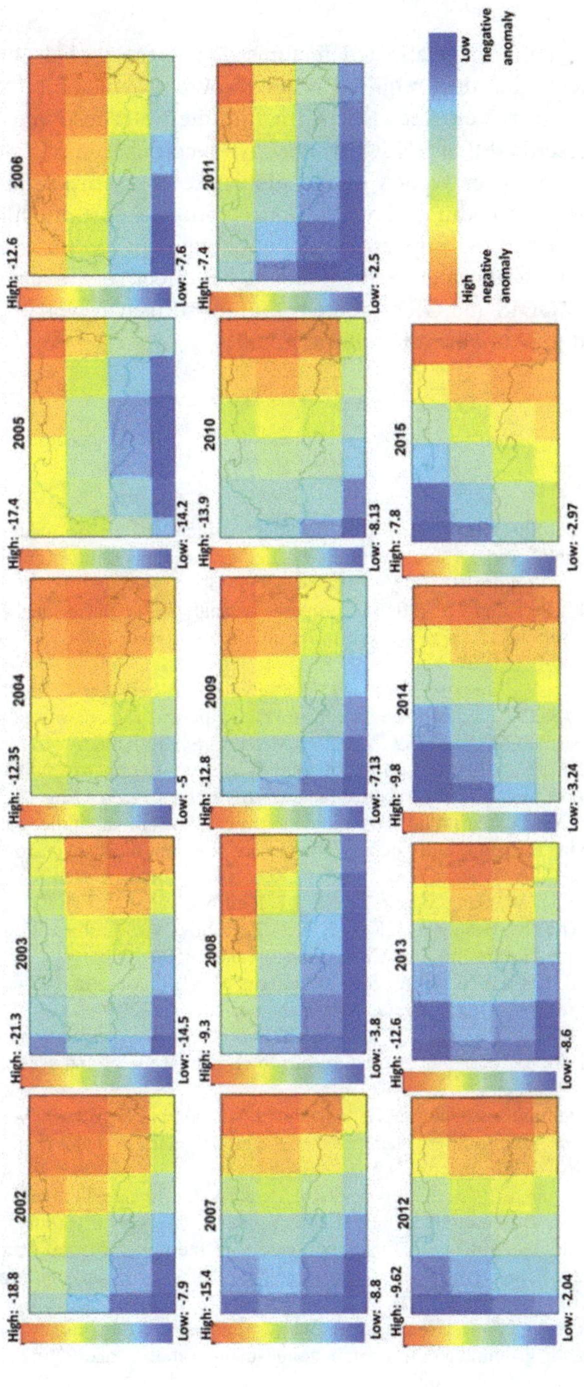

Fig. 28.10 Terrestrial water storage (cm) from GRACE satellite for the period 2002–2015 during pre-monsoon (May)

28.5 Conclusions

Water resources are the most affected features due to the global climate change which in turn affects the most required water resources availability scenario. The global climate change has closest association with the water resources availability, and it has been described that ENSO events have recordable impacts on the water resources situation at several parts of the globe. Remote sensing and GIS have immense potential in providing time series information in cost-effective and time-efficient manner, so that the impacts of global climate change on water resources availability and the water management could be effectively executed. Remote sensing data like MODIS and GRACE are useful for surface as well as groundwater investigations and to study the agricultural impacts.

References

Banholzer S, James K, Simon D (2014) The impact of climate change on natural disasters. In: Zommers Z, Singh A (eds) Reducing disaster: early warning systems for climate change. Springer. doi: https://doi.org/10.1007/978-94-017-8598-3_2

Bhusari V, Katpatal YB, Kundal P (2016) Comparative analysis for performance evaluation of reverse gradient artificial recharge system in basaltic terrain. Hydrogeol J 25:689. https://doi.org/10.1007/s10040-016-1499-3

Bing G, Yi Z, Shi-xin W, He-ping T (2014) The relationship between normalized difference vegetation index (NDVI) and climate factors in the semi-arid region: a case study in Yalu Tsangpo river basin of Qinghai-Tibet Plateau. J Meteorol Sci 11(4):926–940

Bothale RV, Katpatal YB (2014) Response of rainfall and vegetation to ENSO events during 2001–2011 in Upper Wardha Watershed, Maharashtra, India. J Hydrol Eng 19(3):583–592

Bothale RV, Katpatal YB (2015) Trends and anomalies in extreme climate indices and influence of El Nino and La Nina over Pranhita catchment in Godavari Basin, India. J Hydrol Eng 05015023 (12):1–12

Cai W, Agus S, Guojian W, Sang-Wook Y, An S-I, Kim MC, Mat C, Eric G, Fei-Fei J, Jong-Seong K, Matthieu L, McPhaden MJ, Ken T, Axel T, Gabriel V, Masahiro W, Lixin W (2015) ENSO and greenhouse warming. Nat Clim Change 5:849–859

Erasmi S, Propastin P, Kappas M (2009) Spatial patterns of NDVI variations over Indonesia and their relationship to ENSO warm events during the period 1982–2006

Guo L, Wu S, Zhao D, Yin Y, Leng G, Zhang Q (2014) NDVI-based vegetation change in inner Mongolia from 1982 to 2006 and its relationship to climate at the biome scale. Adv Meteorol 2014:1–12

Iizumi T, Luo JJ, Challinor AJ, Sakurai G, Yokozawa M, Sakuma H, Brown ME, Yamagata T (2014) Impacts of El Niño southern oscillation on the global yields of major crops. Nat Commun 5. https://doi.org/10.1038/ncomms4712

IPCC (2013) Climate change 2013: the physical science basis. In: Stocker TF, Qin D, Plattner G-K, Tignor M, Allen SK, Boschung J, Nauels A, Xia Y, Bex V, Midgley PM (eds) Contribution of working group I to the fifth assessment report of the Intergovernmental Panel on Climate Change. Cambridge University Press, Cambridge/New York, 1535 pp, doi:https://doi.org/10.1017/CBO9781107415324

Katpatal YB, Dube YA (2010) Comparative overlay analysis through analytical hierarchical process to delineate groundwater potential zones using satellite data. Int J Earth Sci Eng 3(5):638–653

Katpatal YB, Pophare AM, Lamsoge BR (2014) A groundwater restoration model for over exploited basaltic aquifer and Bazada formation in India. Environ Earth Sci 72(11):4413–4425

Kumar R, Singh RD, Sharma KD (2005) Water resources of India. Curr Sci 89:794–811

Maity R, Kumar DN (2006) Bayesian dynamic modelling for monthly Indian summer monsoon rainfall using El Niño–Southern Oscillation (ENSO) and Equatorial Indian Ocean Oscillation (EQUINOO). J Geophys Res 111(D7):1–12

Mane VP, Katpatal YB (2010) Study of lithological control over infiltration through water-level fluctuation and F-index in micro-watersheds. In: Proceedings of ninth International Conference on Hydro-Science and Engineering (ICHE 2010), IIT Madras, Chennai, India, 2–5, August 2010

Mishra SK, Jain MK, Singh VP (2005) Catchment area based evaluation of the AMC dependent SCS-CN-inspired rainfall-runoff models. Hydrol Process 19(14):2701–2718

MODIS (2013) Moderate resolution imaging spectroradiometer. http://modis.gsfc.nasa.gov/

NCDC (2009) El Niño/Southern Oscillation (ENSO) June 2009. (National Climatic Data Center), National Oceanic and Atmospheric Administration. https://www.ncdc.noaa.gov

NOAA (2006) Tropical Atmosphere Ocean Project. (National Oceanic, and Atmospheric Administration). http://www.pmel.noaa.gov/tao/El Niño/el-nino-story.html

Patil SA, Katpatal YB (2008) Impact assessment of coal mines in Erai watershed of Chandrapur district using Geoinformatics. J Environ Sci Eng 50(4):271–276

Rishma C, Katpatal YB (2016) Variability in rainfall and vegetation density as a response to ENSO events: a case study in Venna river basin of Central India. J Agrometeorol 18(2):300–305

Rishma C, Katpatal YB, Jasima P (2015) Assessment of Enso impacts on rainfall and runoff of Venna River basin, Maharashtra using spatial approach. Discovery 39(178):100–106

Rodell M, Chen J, Kato H, Famiglietti J, Nigro J, Wilson C (2007) Estimating groundwater storage changes in the Mississippi River basin (USA) using GRACE. Hydrogeol J 15:159–166

Ropelewski CF, Halpert MS (1987) Global and regional scale precipitation patterns associated with the El Nino/southern oscillation. Mon Weather Rev 115:1606–1626

Selvaraju R (2003) Impact of El Niño–southern oscillation on Indian foodgrain production. Int J Climatol 23:187–206

Zubair L, Ropelewski CF (2006) The strengthening relationship between ENSO and northeast monsoon rainfall over Sri Lanka and southern India. J Clim Am Meteorol Soc 19:1567–1575

Chapter 29
Remote Sensing and GIS Applications in Micro Level Planning

G. Sreenivasan and Y. V. N. Krishna Murthy

Abstract Micro level planning (MLP), sometimes also called as grassroots level planning or cadastral level planning, is a developmental planning exercise, which requires detailed spatial database on presently available natural resources in the area in terms of geology, geomorphology, soil resources, groundwater and surface water resources, rivers/streams, and land use/land cover. Though abundant data is available, not much of it is properly organized and analyzed, which has resulted in limited outcome and wastage of available resources. By making this resource information available to the people at grassroots level in easy and user-friendly manner, planning process could be catalyzed at micro level. Geospatial technologies, i.e., satellite remote sensing, global positioning system (GPS), information technology (IT), and geographic information system (GIS), have the capability in generation and integration of datasets derived from various sources. In recent decades, satellite remote sensing data has improved continuously in terms of the spectral, spatial, and temporal resolutions. This has led to revolutionary change in mapping in two ways – one, in terms of spatial extent, and, two, in terms of information content that can be mapped. Information derived from remote sensing data can be used in conjunction with other field data for prioritization and characterization of areas for micro level planning enabling us to prepare village cadastral level development plans. Further, remote sensing and GIS technology is also used in monitoring and evaluation of the implemented developmental plans mainly to assess the overall improvement in the health of the micro-watershed or *gram panchayat* (village) and to evaluate the long-term effects and impact of the developmental activities taken up.

Keywords Micro level planning · Spatial planning · Georeferenced cadastral maps · Cadastral level planning · Gram panchayat planning · Remote sensing · GIS

G. Sreenivasan (✉)
Regional Remote Sensing Centre-North, National Remote Sensing Centre (ISRO), New Delhi, India

Y. V. N. Krishna Murthy
National Remote Sensing Centre (ISRO), Hyderabad, India

G. P. O. Reddy, S. K. Singh (eds.), *Geospatial Technologies in Land Resources Mapping, Monitoring and Management*, Geotechnologies and the Environment 21,
https://doi.org/10.1007/978-3-319-78711-4_29

29.1 Introduction

The growing pressure of the ever-increasing population on land for food, fuel, fiber, and fodder has resulted in competing and conflicting demands on finite natural resources (Navalgund and Jayanthi 2004; FAO 2011). Micro level planning (MLP), sometimes also called as grassroots level planning or village cadastral level planning, is a developmental planning exercise to optimally utilize the natural resources of a small area and make them sustainable. Developmental planning with integrated approach has been accepted the world over for optimal management and better utilization of natural resources to improve living conditions of the people (PAGE 2016; Wang et al. 2016).

Micro level planning (MLP) is a bottom-up planning approach, as the planning starts at the smallest administrative unit, i.e., the village or the smallest natural entity, i.e., the micro-watershed (Ramamohana Rao and Suneetha 2015). The planning begins by understanding the limitations of an area, computing the surplus and shortfalls of the resources, and understanding the present and future demands of the people in the area. This requires detailed spatial database on presently available natural resources in the area in terms of geology, geomorphology, soil resources, groundwater and surface water resources, rivers/streams, and land use/land cover; it includes agriculture, forest, and wastelands. The advancement in remote sensing technology provided us the best tool for mapping of the natural resources (Melesse et al. 2007). Remote sensing technology initiated about six decades back, and since then the satellite remote sensing data has improved continuously in terms of the spectral, spatial, and temporal resolutions. This has led to revolutionary change in mapping in two ways – one, in terms of spatial extent, and, two, in terms of information content that can be mapped. Information derived from remote sensing data can be used in conjunction with other field data for prioritization and characterization of areas for micro level planning and development of management strategies for optimum utilization of land and water resources. This enabled us to address mapping distinct units at village/cadastral level thus making it possible for micro level planning.

29.2 Micro Level Planning (MLP)

Micro level planning aims at inclusive, participatory, and coordinated approach for local area development to ensure that each village (*gram panchayat*) or local body or micro-watershed is considered as a planning unit. The major concern expressed by planners all over the world is that the plans for development are generated for a large area, i.e., at the district level ignoring the locale-specific problems. If a large area is considered for planning, i.e., at district level, planning is at 1:50,000 scale or

1:25,000 scale, which cannot address the micro level problems. Hence, the resultant plan-outputs do not effectively address the local issues. These limitations are overcome by micro level planning. It is opined that "macro level planning at the state or national level manifests broadly the measures to be taken, whereas, the micro level planning provides detailed action plans. Although the macro level planning can set the overall goals and directions, it should be prepared and implemented by integration of micro level planning inputs. The upward integration of micro level planning, necessary to generate macro-level, would ensure sustainable development and provide us the tools to ensure balanced development" (Kale 1992).

For any micro level planning approach, effective utilization of natural resources and their management are essential, and the need is to catalyze the planning process at grassroots level. Reliable and timely information on resources is prerequisite for the development of a plan. Most of the information is not available in a form that can be depicted spatially as various layers. This information is mostly available in tabular format and spread out with many departments. Micro level planning requires information to be provided in the form of digital spatial layers, so that the decisions or planning exercises pertaining to developmental activities can be taken up in a scientific and timely manner for the benefit of the target group.

Micro level planning is synonymous with "decentralized planning" which comprises of what different micro planning units within a district can achieve by envisioning collectively, operating their budgets, exercising their skills, and leveraging their initiatives. Typically, in an ideal decentralized district planning exercise, each planning unit, namely, village (*panchayat*), municipalities, and line departments, would prepare the plan for execution after consultations with the people. These plans are called Gram Panchayat Development Plans (GPDP), and the planning approach is called spatial planning. Spatial planning needs to be part of the district planning process at all levels, and in this regard, India has rich experience in using space technology for supporting decentralized planning and benefiting the grassroots level. Comprehensive database comprising information on natural resources, socioeconomic data, infrastructure, and other collateral information is a prime requisite for proper and optimal planning, implementation, and impact assessment.

Spatial information – images and maps – forms the foundation and basis for the most planning and implementation of developmental activities, infrastructure development, disaster management support, environmental monitoring, and natural resources management. Even common citizens require maps and spatial information for their localized decision-making. Though abundant data is available these days, not much of it is properly organized and analyzed, which has resulted in limited outcome and wastage of available resources. This has become hindrance in utilizing resource information in most productive and timely manner. By making this resource information available to the people at grassroots level in easy and user-friendly manner, planning process can be catalyzed due to inherent benefits provided by this system. Satellite remote sensing, global positioning system (GPS), information technology (IT), and geographic information system (GIS) have the capability of generating, integrating, and analyzing spatial and nonspatial data from various sources to arrive at meaningful decisions for micro level planning.

29.3 Remote Sensing Tools and Techniques

Remote sensing is defined as the technology of acquiring information about the earth's surface and atmosphere from a distance without being in physical contact with them, using sensors onboard aircraft (airborne) or satellites and space shuttles (spaceborne) platforms. Remote sensing employs electromagnetic radiation (EMR) and relies on the interaction of electromagnetic energy with the material on the earth's surface. It refers to the sensing of EMR, which is reflected, scattered, or emitted from the material. The interaction of EMR with earth's surface features generates reflected and/or emitted signals, which are referred to as spectral response patterns or spectral signatures. The spectral response patterns play a central role in detection, identification, and analysis of earth's surface material. The regions of the EMR used for remote sensing and their principal applications are described in Table 29.1.

Remote sensing can be classified into two broad types based on the wavelength region of EMR that is used for sensing as optical and microwave. Optical remote sensing technique detects solar radiation in the visible, near-, middle-, and thermal-infrared wavelength regions, reflected, scattered, or emitted from the earth material, which is captured by the sensors on the airborne or space platforms. Microwave remote sensing either detects solar emitted energy or the scattered energy (in case of active remote sensing) in the microwave region which is returning from the terrain. The microwave remote sensing has inherent advantages over optical remote sensing,

Table 29.1 Regions of EMR used for remote sensing and their principal applications

Regions of EMR	Wavelength (μm)	Principal application
Blue	0.45–0.52	Coastal morphology and sedimentation study, soil and vegetation differentiation, coniferous and deciduous vegetation discrimination
Green	0.52–0.60	Vigor assessment, rock and soil discrimination, turbidity and bathymetry studies
Red	0.63–0.69	Plant species differentiation
Near infrared	0.76–0.90	Vegetation vigor, biomass, delineation of water features, land-forms/geomorphic studies
Mid-infrared	1.55–1.75	Vegetation moisture content, soil moisture content, snow and cloud differentiation
Mid-infrared	2.08–2.35	Differentiation of geological materials and soils
Thermal IR	3.0–5.0	For hot targets, viz., fires and volcanoes
Thermal IR	10.4–12.5	Thermal sensing, vegetation discrimination, volcanic studies

mainly in providing all weather remote sensing and information of the shallow subsurface due to penetration capability of the microwave energy.

The basic principle of remote sensing is dependent on the fact that different features on the earth's surface, such as soil, rocks, vegetation, built-up, barren land, water, snow, etc., reflect visible and infrared light in different ways. Interpretation of remote sensing optical data is carried out using this principle. The variation in the reflectance of the different features at different wavelengths is recorded in the form of spectral signature of that feature. The spectral signature has four important characteristics by which an object can be identified on the satellite image. These are spectral variations, spatial variations, temporal variations, and polarization variations. The change in the reflectance or emitted energy of an object as a function of wavelength is called spectral variation. Changes in the reflectance or emitted energy of an object as a function of its shape, size, and texture of the object are called spatial variation. Temporal variation is the seasonal or diurnal changes in reflectance or emitted energy of an object, and polarization variations are the changes in the polarization of the radiation reflected or emitted by an object (Navalgund et al. 2007).

The remote sensing data received from the satellite sensors is digitally processed in high-end computers by using various standard and customized algorithms for the purposes of image correction, image enhancement, and feature extraction. This process in general is called as digital image processing. The relevant information on the natural resources is extracted and analyzed from the satellite images either through semiautomatic algorithms or through visual interpretation of the enhanced images on the computer screen using GIS software. Further, GIS is also used for integrating and analyzing geospatial information derived from satellite images and nonspatial information collected from other sources to finally make planning decisions.

29.4 Geographic Information System (GIS) Tools and Techniques

GIS is a system which facilitates the storage and intelligent use of geographic data about all natural resources and socioeconomic aspects. The main components of GIS are the computer hardware and software to collect, store, update, manipulate, analyze, display, and process the geographic data. GIS encompass tools that allow for the processing of spatial data and its integration with nonspatial data to generate information for making decisions about various activities on the earth. GIS is a powerful tool to assist decision-making for a particular activity, and further GIS can assist in the process of monitoring the effect of that activity in the surrounding environment. Therefore, the generation of geographic database is essential for addressing various aspects for planning, implementation, and long-term monitoring of any activity. By using GIS tools and techniques, one can maximize the efficiency of planning and decision-making, thus cutting the overall cost and manpower involvement in the planning process. GIS also can efficiently manage the data through the use of database management systems (DBMS).

Four fundamental geoprocessing operations can be done in GIS, viz., overlay operations, neighborhood characterization operations, reclassification operations, and difference and connecting measurements. These operations help in integrated analysis of large number of spatial layers in the process of planning and decision-making. The purpose of any integration analysis would be to identify the relationship that can be extracted and understood from the integrated data – which could be either useful on its own or provide a beginning to more investigation. An important outcome of integration analysis is the function of pointing in the right direction for making the right decision (Openshaw 1990).

29.5 Advantages of Remote Sensing and GIS in Micro Level Planning

Remote sensing (RS) plays a significant role in providing geospatial information and also in determining, enhancing, and monitoring the overall capacity of the earth. Remote sensing and GIS techniques together can be applied effectively to generate data and information for sustainable development of natural resources through micro level planning.

Satellite or airborne remote sensing can provide valuable information with regard to mapping the various hydrological aspects, geology, soil types and properties, land use/land cover, terrain features in the form of DEMs, and location of the aquifers. Especially, Indian Remote Sensing program is directed toward the realization of operational capabilities in space and ground segments of the technology for regular monitoring of earth resources and environment.

The advantages of remote sensing data are its synoptic viewing capability, multispectral imaging capability, and repetitive coverage. These inherent advantages of remote sensing techniques help in generating the necessary parametric information on natural resources of an area that is essential for micro level planning. Generating similar information using conventional techniques is very difficult, time-consuming, manpower intensive, and also expensive. Remote sensing also is the best tool in achieving successful and accurate mapping of features like geological structures and geomorphology which are very important parameters for planning of water and land resources development and management strategies. The repetitive coverage capability of the satellite remote sensing sensors helps in acquiring the satellite data of the same place more than once, and these in turn are immensely useful in making analysis on the changes in the natural resources of the planning area over the time, the dimension of the changes, and the overall direction of the changes. The change detection studies using the satellite remote sensing data are valuable for time-to-time monitoring of micro-watershed health, evaluation of the implementation of developmental plans, and working out alternate strategies on a holistic basis for better development of the natural resources of a micro-watershed.

GIS is a tool, which allows synergism of spatial data obtained from the remote sensing and nonspatial databases from other sources in the most efficient manner. GIS has the ability to handle multiple layers of information in the spatial domain and allow for the integration and modeling of these parameters to arrive at inputs for decision-making. The modeling and integration will allow for visualization of information in newer perspectives and therefore is quiet useful in micro level planning applications. The advanced tools like network and proximity analysis are of immense use in finding out suitable solutions for societal problems related to demand and supply planning.

29.6 Steps in Micro Level Planning

Micro level planning through geospatial techniques can be achieved through a series of steps. The preparatory work includes procuring the required satellite data of the right spectral, spatial, and temporal resolution, its preprocessing, and enhancement. This is followed by bringing the existing administrative maps (such as village cadastral maps) to the image frame followed by preparing the thematic layers from satellite data interpretation and analysis. Next step is the integrated analysis of the thematic information and the field data and preparation of action plans for implementation. Finally, the process of implementation needs to be monitored, and the impact of the micro level planning needs to be evaluated.

Selecting the right satellite data is very important for any planning purpose. Satellite data of various spatial resolutions are now available from different sensors onboard the various satellites in orbit. Broadly, the selection of the required spatial resolution satellite data is directly linked with the scale of mapping and planning (Table 29.2). In case of micro level planning, the mapping scales will be of the order of 1:10,000 to 1:2000. So, the satellite data with spatial resolutions of 5.0–0.5 m or better will be ideal in carrying out micro level planning. We have various satellites launched by India and other space agencies giving this type of satellite data. The Resourcesat LISS-IV and the Cartosat series of satellites of ISRO provide the required data for large-scale mapping and micro level planning.

The preprocessing of the satellite data consists of radiometric correction, geometric correction, and georeferencing. Radiometric correction includes conversion of pixel digital value to apparent (at-sensor) radiance, subtraction of the atmospheric contribution, topographic normalization, and sensor calibration. The geometric correction and georeferencing involve transformation of a remotely sensed image so that it has the scale and projection properties of a given map projection. This includes removal of the systematic and nonsystematic distortions and also relief correction. By this process, the satellite images get tagged with the ground coordinate system, and one-to-one link is established between earth terrain and satellite data. Further, the satellite data is enhanced by various digital image processing

Table 29.2 Satellite data spatial resolution versus scale of mapping of natural resources

Satellites	Sensor	Spatial resolution (in meters)	Possible scales of resource mapping
IRS-1A/IRS-1B	LISS-I	72.5	1:250,000–1:150,000
IRS-1A/IRS-1B	LISS-II	36.25	1:100,000
LANDSAT-1,2,3,4,5	MSS	60	1:250,000–1:150,000
IRS-1C/IRS-1D, RESOURCESAT-1/2	LISS-III	23.5	1:50,000
LANDSAT-4 & 5	TM	30 (thermal: 120)	1:50,000 (visible, NIR)
LANDSAT-7	ETM+	30 (thermal: 60)	1:50,000 (visible, NIR)
LANDSAT-8	OLI	30	1:50,000
LANDSAT-8	TIRS	100	1:250,000
IRS-1C/IRS-1D	PAN	5.8	1:25,000–1:12,000
RESOURCESAT-1/2	AWiFS	56	1:250,000
RESOURCESAT-1/2	LISS-IV	5.8	1:25,000–1:12,000
SENTINEL-2	MSI	10–20	1:50,000–1:20,000
SPOT-7	PAN	1.5	1:10,000–1:3000
SPOT-7	MS	6.0	1:25,000–1:12,000
CARTOSAT-1	PAN	2.5	1:10,000–1:5000
KOMPSAT-3A	PAN	0.55	1:5000–1:2000
KOMPSAT-3A	MS	2.2	1:10,000–1:5000
QUICKBIRD	PAN	0.65	1:5000–1:2000
QUICKBIRD	MS	2.62	1:10,000–1:5000
GEOEYE-1	PAN	0.46	1:10,000–1:2000
GEOEYE-1	MS	1.84	1:10,000 to 1:4000
WORLDVIEW-4	PAN	0.31	1:5000–1:2000
WORLDVIEW-4	MS	1.24	1:10,000–1:3000
PLEIADES-1B	PAN	0.5	1:5000–1:2000
PLEIADES-1B	MS	2.0	1:10,000–1:5000
CARTOSAT-2 SERIES	PAN +MS	0.65	1:5000–1:2000

techniques, which are specific to the type of information we are trying to extract from the satellite data. These include contrast enhancement, edge enhancement, spatial filtering, and image transforms (such as principal component analysis).

29.6.1 *Georeferencing of Village Cadastral Maps*

Micro level developmental planning for any gram panchayat or micro-watershed has to consider the people's requirement and imbibe local wisdom in the action plans. Therefore, integration of the information available at the parcel level with spatial information base is a prerequisite (Krishna Murthy et al. 2000). Hence, the micro level planning exercise requires accurate, up-to-date, and standardized cadastral maps in digital environment and proper geodetic control framework for linking the high-resolution satellite data and the thematic information derived from them to each parcel of land at the grassroots level. Further, for pinpointing the location for a specific activity, action plans on the cadastral scales (1:8000–1:4000) with survey numbers and landowner details (private, govt., etc.) are needed for effective implementation (Krishna Murthy et al. 1996).

In most cases, the cadastral maps do not have geodetic coordinates, limiting the use of cadastral maps in conjunction with GIS. Therefore georeferencing of these maps is essential. Georeferencing is the process of defining the position of geographical objects relative to a standard reference grid (Bernstein 1983). This is achieved by assigning map coordinates to image data for making the data amenable to GIS analysis and to facilitate viewing in conjunction with maps. Once a cadastral map is georeferenced, it can be linked to the large-scale thematic maps to finally undertake the micro level planning at village cadastral level. The georeferencing process enables deriving geodetic coordinates for each map and in turn for each vertex of the cadastral map. Georeferencing facilitates linking of cadastral maps to the natural resources databases at various levels (Adiga and Krishna Murthy 2000; Gopal Rao et al. 2000).

The georeferencing of village cadastral maps using high-resolution satellite images can be achieved by establishing mathematical relationships between the addresses of pixels in an image and the corresponding coordinates on the ground through cadastral maps. These relationships can be used to correct the image geometry irrespective of the analyst's knowledge of the source and types of distortion (Bernstein 1983; Krishna Murthy et al. 2000; Srinivasa Rao et al. 2003). For georeferencing the cadastral map, the real-world coordinates for sufficient number of ground control points (GCPs) are required. The real-world coordinates of the GCPs are obtained through primary sources or secondary sources. The primary source consists of three modes, viz., ground control survey, topographical maps, and coordinates obtained from GPS. The secondary sources consist of aerial images or high-resolution satellite images. Aerial images contain maximum information and provide excellent control for registering the cadastral maps. Nevertheless, very high-resolution space images available today have a resolution comparable to the aerial images and are good substitute to aerial images for georeferencing of cadastral maps.

It is very expensive and time-consuming to obtain real-world coordinates of seven to ten GCPs from ground control survey, and hence this source is not used in operational mode for georeferencing of the cadastral maps. The topographical maps, at 1:50,000 or 1:25,000 scale, can be used for deriving the real-world coordinates of the GCPs, but the lack of compatibility between the scales of topographical maps and the cadastral maps may influence the error budget in the locational accuracy of the georeferenced cadastral maps. The other primary source of deriving real-world coordinates directly is GPS. The GPS coordinates can be used to georeference the high-resolution satellite image (HRSI). These HRSI will then be best reference data for georeferencing the village cadastral maps. An example of a cadastral map georeferenced with Cartosat-1 HRSI is shown in Fig. 29.1.

Fig. 29.1 Georeferenced village cadastral map of part of Chandur Bazar Taluk, Amaravati District, Maharashtra, overlaid on the natural color composite high-resolution satellite merged product of Cartosat-1 (PAN) and Resourcesat-1 LISS-IV (multispectral). The yellow line is the boundary between Masod & Tonglapur villages, and the grey lines are the parcel boundaries with the parcel numbers

29.6.2 Characterizing the Planning Area and Thematic Maps Generation

The response of a particular area to different hydrological processes and its behavior depends upon various physiographic, hydrogeological, and geomorphological parameters. These parameters are area specific and unique and provide an idea about its natural resource status, potentials, and limitations. When we consider large areas of the earth's surface, there are inherent heterogeneities with respect to the soils, geology, terrain slope, land use/land cover, etc. But, in case of micro level planning, the planning unit under reference is very small (a village or a micro-watershed), and therefore, we have better scope of characterizing it more precisely by overcoming these heterogeneities to a large extent. This is made possible by large-scale mapping (1:10 –1:2 K), which is achieved due to availability of very high-resolution satellite remote sensing data (2.5–0.5 m spatial resolution). Characterization of the planning area involves measurement of parameters that influence the characteristic behavior of the land parcel under consideration, whereas analysis aims at the critical study of these parameters to arrive at conclusions on micro-watershed response and behavior.

The development of any planning unit requires the proper understanding of the hydrologic process active therein. This is possible when the planning unit considered conforms to a smallest natural hydrological unit, i.e., a micro-watershed. Some of the important characteristics of a micro-watershed include the size, shape, physiography, slope, geology, soils, hydrology, land use, vegetation density, drainage pattern, and climate. These factors form inputs to the hydrological design, and therefore their timely and accurate computation becomes a prerequisite for micro level planning and generation of land and water resources action plans for sustainable management.

The characterization of the planning area also involves the inventory, mapping, and analysis of the existing natural resources therein using the remote sensing data. Apart from this there is a variety of other nonspatial information to be derived and integrated with the spatial layers in GIS for further modeling to arrive at the sustainable developmental plans for land and water resources. The different geospatial information required for micro level planning and its source is given in Table 29.3 (Sreenivasan et al. 2010).

Drainage and Water Body Delineation Drainage delineation can be done using manual methods using existing maps and high-resolution satellite imageries or semiautomated methods using a digital elevation model (DEM). The drainage updated using high-resolution satellite images can give more detailed drainage map on scales of 1:10 K or better (Fig. 29.2). It is well known that drainage is dynamic in nature and changes its course and nature over time. Therefore, the drainage maps updated from the latest satellite data will show latest information on present course of streams. The satellite updated maps also will be depicting many minor 1st order streams, which are not mapped in the published topomaps of 1:50 K or 1:25 K. In the manual method, the drainage lines are extracted from existing topomaps and updated by overlaying on the latest high-resolution georeferenced satellite imagery through the process of heads-up digitization in GIS environment. Water bodies can be easily mapped directly using the latest satellite data.

In the semiautomated method of drainage delineation, a hydrologically corrected depressionless DEM is used as an input. DEM can be generated using several methods. Indian satellites, like Cartosat-1 and their predecessors Cartosat series, are highly useful for generating DEMs by processing the stereo pairs. DEMs can also be generated by SAR interferometry technique. Delineation of drainage using DEM initially starts with the generation of a flow direction grid using the conditioned DEM as input; it creates a raster flow direction from each cell to its steepest downslope neighbor. Then, "flow accumulation" is calculated, which is a grid of accumulated flow to each cell, by accumulating the weight for all cells that flow into

Table 29.3 Geospatial information requirement for micro level planning

Parameter/theme	Source	Attribute	Relevance
Geomorphology	Satellite imagery	Landforms	Geomorphic process, slope, runoff and ground-water occurrence, recharge
Geology/lithology	Satellite imagery, published geological maps	Rock type	Hydrogeologic characteristic, recharge
Geological structures	Satellite imagery	Joints/fractures/faults	Groundwater occurrence and recharge
Land use/land cover	Satellite temporal data (three season data)	Present land use, wastelands, surface water	Runoff and sedimentation, suitable locations for water harvesting
Soils and derivative maps	Satellite imagery, NBSS published maps and field survey	Soil texture, depth, land capability, soil moisture	Land production potential, runoff, erosion, recharge, irrigation potential
Flow status of streams	Satellite temporal data	Dry/perennial	Suitable sites for water harvesting
Drainage network/water bodies	High-resolution satellite imagery and published topomaps	Stream order	Run-off & sedimentation, suitable sites for water harvesting
DEM/slope/topography/contour information	Satellite stereo data	Elevation and slope	Delineation of watershed, drainage, runoff and water impounding, planning land resources and water harvesting sites
Watershed	To be extracted from high-resolution satellite imagery taking reference of the drainage and terrain information	Area, geometric form, shape	Hydrological planning unit
Village/cadastral maps	Land records/census/tehsil/block maps	Parcel boundary, parcel number	Land ownership
Base map details	High-resolution satellite imagery and published topomaps	Infrastructure type (road, rail, canal, village/city, etc.)	Helps in connectivity during planning and implementation

each downslope cell. The output of "FLOWACCUMULATION" thus represents the amount of rain that would flow through each cell leading to the formation of the drainage network.

Slope The slope and related information is essential for delineating watershed boundary, for planning soil and water conservation measures, and for deriving the land resources action plan. Conventionally, slope map is prepared using the elevation information from topomaps. The elevation information is used to generate a

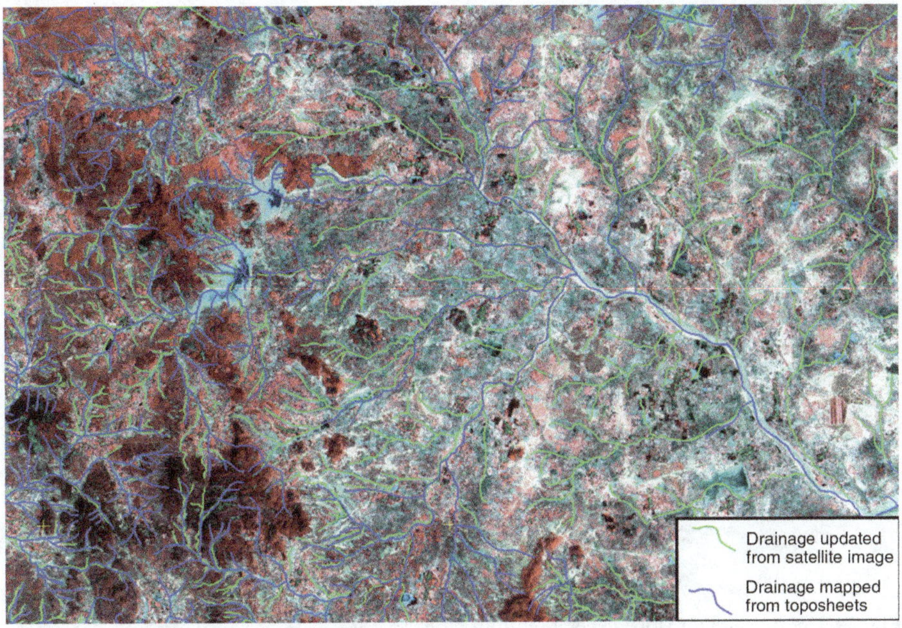

Fig. 29.2 Updating drainage map from high-resolution satellite images

DEM and is further used to generate slope map using the ArcGIS tools. Advances in the satellite technology have created new tools for generation of DEM from satellite stereo pairs. The latest Indian remote sensing satellite Cartosat-1 and its predecessors the Cartosat series satellite have the stereo imaging capabilities to generate the DEM. The satellite stereo pair-derived DEM is used for generating slope map using the procedure shown in Fig. 29.3 (NRSC 2010a). The DEM can also be draped with the ortho-satellite image to generate a digital terrain model (DTM) (Fig. 29.4). DEM is also used to generate derivatives such as aspect and elevation contours using the Satellite Photogrammetric Software Packages.

Soil Resource Inventory and Mapping Soil is an important segment of our ecosystem, as it serves as an anchorage for plants and source for the nutrients. Soil is the medium and fundamental raw material for plant growth. Soils are vital natural resources on whose proper use depend the life-supporting system of an area and the socioeconomic development of its people. Soil erosion has now become one of the major environmental problems and a major constraint for agricultural production. To abate the soil erosion and for sustained utilization of soil resources, it is necessary to take up suitable conservative measures, for which it is necessary to know the nature, characteristics, and extent of distribution of different soils, their qualities, potentials, and limitations.

Remote sensing data has been effectively used for mapping the soils and also for assessing their characteristics by correlating with the results of the limited ground

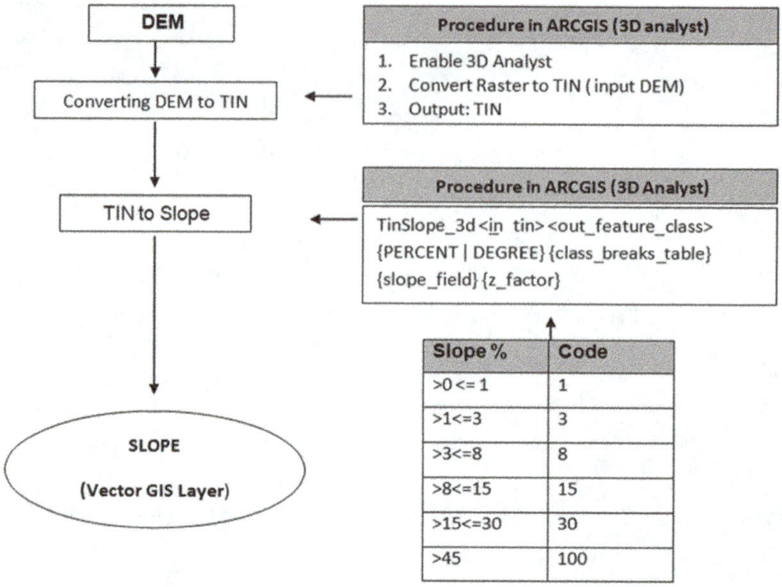

Fig. 29.3 Procedure for slope generation from DEM (NRSC 2010a)

Fig. 29.4 Digital terrain model of part of Amravati District, Maharashtra

profiling and sampling (Manchanda et al. 2002). The use of remote sensing data enables to map the soil boundaries accurately at relatively low cost and time. Soil map showing soil texture, depth, soil erodibility, major limitations, and land capability classes is prepared mainly by using the relationship between physiography and soils. The major physiographic units like hills and ridges, rolling lands, undulating lands, gently sloping lands, plains, and valleys are further subdivided based on the tonal variations on satellite image. Soil profiling is carried out in each of the unique soil physiographic units depending on the distribution and extent of these units. The soil profiles are used for analyzing the morphological and chemical properties of the soils for final classification of the soil physiographic units to soil units with attached attribute information such as soil texture, soil depth, land capability, etc.

The land resources development planning requires the soil parameters such as land capability and land irrigability. These two parameters are the soil derivatives based on the soil characteristics such as texture, depth, and erodibility. The generation of the water conservation strategies requires mainly two aspects of soils to be considered, viz., the physical properties and the land capability. From the point of physical properties, the water harvesting structures for storage purpose should be in areas with soils of low permeability like clayey soils. For groundwater recharge structures, the soils should be light and highly permeable, like sandy soils. The immediate downstream area or the command area of the water harvesting structure should have soils of good land capability and of moderate to good thickness that can support good agriculture.

Land Use/Land Cover Analysis

Sustainable land use forms an overall planning framework, while sound land use planning concepts, together with the adoption of appropriate land use practices, provide key guidelines for land and water resources development and management. A watershed, which is under stress due to the various social factors as well as due to limitations of the natural resources, has the need to go for alternative sustainable land use taking into consideration the present land use/land cover, soils, slope, land capability, surface/groundwater availability, and other socioeconomic constraints. Digital image processing of the remote sensing data facilitates quick and cost-effective land use/land cover classification using maximum likelihood algorithm and the standard classification schemes (Mather 1999; Navalgund et al. 2007). A multi-temporal (*kharif*, *rabi*, and summer seasons) remote sensing dataset is used first for deriving the seasonal land use, which further on refinement and aggregation will give the land use/land cover map for an area.

Geology and Geological Structures

The knowledge of the rock types and the geological structures is very important for taking up the watershed development activities, especially the water conservation measures and groundwater recharge. The nature of subsurface soil and rock strata controls the seepage or percolation of water. Compact igneous/metamorphic rocks, which have undergone very little of weathering and devoid of geological structures like joints/fractures/faults, act as non-permeable base. These conditions are favorable for storage of water, and therefore sites satisfying these conditions are ideal for water harvesting structures for storage purpose. In contrast, the water harvesting structures for recharge purpose should be located on a more porous and permeable soil/rock base with presence of geological structures like fractures/joints/lineaments, which allow percolation of water through them. The interpretation of the rock types is done from the multispectral satellite imageries using the image interpretation elements and other terrain information such as drainage pattern, etc., using the reference of the published quadrangle geological maps of GSI for assigning the stratigraphy to the interpreted units. Geological structures like joints/fractures and faults and in general the lineaments are best interpreted from the digitally enhanced satellite imageries using the drainage anomalies, image tonal anomalies, vegetation anomalies, geomorphic anomalies, and lithological anomalies (Gupta 2003).

Geomorphology
The geomorphology comprising of the study of geomorphic units and landforms is one of the important criteria for generation of the water resources development plan for a watershed (Sreenivasan et al. 2010). Landform exercises control on the distribution of various natural resources on the earth's surface. They give indirect information about the secondary porosity and permeability created in the parent rock due to the different processes which have acted on the landscape. The study of landform in terms of its morphometry facilitates understanding of the recharge conditions in an area. Landforms also give broad information about the slope of the landscape as the slope is also an expression of morphology.

The landforms based on their origin are divided into different types, viz., denudational, fluvial, aeolian, marine, structural, etc. Each landform, therefore, is a result of a set of natural processes and is characterized by unique geology, intensity of weathering and thickness of overburden/fill, nature of fill material, and general slope. Therefore, depending on the landform in an area, the suitability of a water harvesting structure can be decided. Landforms like pediment/buried pediment and pediplains and gently sloping valleys are suitable for "water-spread area" (WSA) of a water harvesting structure. The "immediate downstream" areas ideally should have buried pediplains and valley fills with good fill, which support good aquifer and also good agriculture. In aeolian terrain, the interdunal depressions form good zones for locating rainwater harvesting structure. The mapping of the landforms is carried out by visual interpretation of satellite image false color composites with various band combinations (Gupta 2003). Image transforms like PCA will bring out the contrast between landforms in a better way. Digital color composites of PCA band with original data bands will be useful in enhancing the landforms. The landforms are interpreted from the enhanced satellite data by heads-up digitization by displaying the satellite images in computer workstations in GIS environment. Standard classification schemes are used to classify each landform into a specific class based on its origin and morphology (NRSC 2010b).

Groundwater Potential Assessment
Groundwater distribution in the subsurface is subject to wide spatiotemporal variations. The controlling factors for the groundwater occurrence and movement are the nature of the subsurface litho-units, the amount and spatial variation of the

weathering, and the structural fabric mainly dictated by the joints/fractures (lineaments). The remote sensing data with sufficient ground truth information provides information on the hydrogeological conditions, geological structures, landforms, and recharge conditions (Sunitha et al. 2016), which define the groundwater regime of an area. This information when integrated in the GIS and suitably analyzed along with field well inventory data will facilitate the generation of groundwater prospect maps, showing the probable regions with different groundwater prospects. These maps give the probable yield and depth ranges and the type of wells suitable in an area. Apart from the thematic maps described in this section, many other baseline thematic information such as canals, roads and railway line, habitations (villages/towns/cities), etc. can also be mapped using the high-resolution satellite data. This baseline information is very important in the implementation of the micro level development plans in the field.

29.7 Micro Level Land and Water Resources Development Planning

Action plans for sustainable natural resources development essentially deal with area-specific and locale-specific measures. Micro-watersheds and village cadastral boundaries are shown on the integrated action plan maps to facilitate easier location in the field for implementation of action plan. The maps help to plan sustainable agricultural practices and soil and water conservation measures depending upon the land conditions and needs (Sreenivasan et al. 2010).

29.7.1 Land Resources Development Action Plan

The overall objective of Land Resources Development Plan (LRDP) is to facilitate allocation of land to the uses that provide the greatest sustainable benefits and to promote the transition to a sustainable and integrated management of land resources. Much of the land is already settled and used for some purpose, and the aim is to move toward uses that provide greater benefits but which are sustainable in the long term, while protecting essential natural ecosystems and biodiversity values. In doing this, geo-environmental, social, and economic issues should be taken into consideration, and the rights of individuals and categories, such as indigenous people and women, should be kept in mind.

LRDP is derived through integration of thematic maps such as slope, soil, geomorphology, drainage, and surface and groundwater prospect map to know the potential of each homogeneous portion of land in terms of natural resources utilization and is compared with the existing land use map. Depending upon the natural resources availability, requirements of the people, and government priorities, alternate land use practice is suggested to optimally use the available resources for overall sustainable development. All the thematic maps mentioned above are prepared by interpretation of high-resolution satellite images. The integration of the various thematic maps and attribute data and further manipulation/analysis for identifying alternatives for land and water development are carried out using the GIS. The thematic information which are digitally encoded to form digital spatial layers corresponding to soils and their derivatives, land use/land cover, litho-units, geomorphology, and geological structures are feature coded and integrated to arrive at the "composite mapping units" (CMU). The composite mapping units are three-dimensional landscape units homogeneous in respect of characteristics and qualities of land, water, and vegetation and separated from other dissimilar units by distinct boundaries. The CMU characteristics imply physical parameters of the component resources of a biophysical domain, whereas qualities are suggestive of their potential for specific uses under the defined sets of conditions.

Before suggesting the alternate resources development plans, surplus and shortfalls are computed at the present level of demand and supply considering the present land use, present availability of the surface/groundwater, agricultural productivity, socioeconomic problems, and any other local-specific problems. At the same time, the future demands of the population of the watershed are considered, which help us in identification and focusing the developmental plans toward the thrust areas for development.

Land capability classes based on the soil derivatives give optimum land use potential of a parcel, whereas the land irrigation classes help to assess the suitability of land for irrigated agriculture. These two, when considered with other factors such as the slope/landforms and water potential, enable predictions of optimal land use of a parcel, which if different from the present land use requires a revision.

To arrive at precise prescription, a multidisciplinary team of scientists who are involved in preparation of thematic maps undertakes an extensive and quick visit to the study area. A number of spot observations are made by the team covering almost all types of landform, soil, slope, groundwater potential and quality, rainfall and climate zones, present land use, etc. Each spot is marked on the high-resolution georeferenced satellite images overlaid with village cadastral maps on scales greater than 1:10,000. At each spot, the land characteristics (parameters) as mapped in the respective theme maps are recorded along with the present land use. A few pertinent site conditions with respect to soil moisture, native natural vegetation, and any other aspects, which are not reflected in the thematic maps, are also recorded. The existing cropping pattern and irrigation practice are also noted. Upon noting these details, a short deliberation is made on the optimality of the present land use especially keeping in view the sustainable production and quality of ecosystem. If the present land use is considered suboptimal, then a few possible options for such a site are discussed with an aim to achieve optimality within the overall framework of sustainability of production. Unless and otherwise the present land use is beyond the threshold limit of some land parameters, a drastically different option is not recommended since such a change will not meet with high level of acceptability. For example, if a land unit which is ideally suitable for horticulture or fodder and

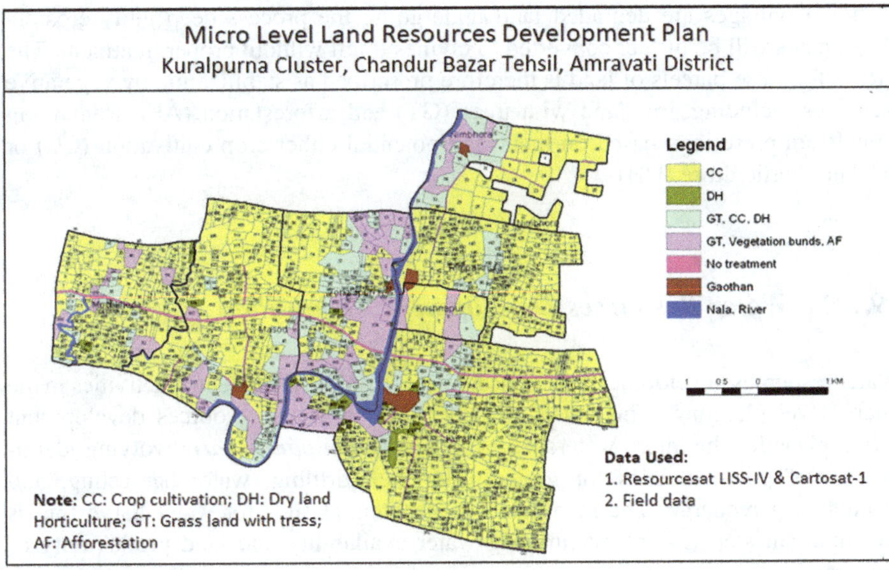

Fig. 29.5 Example of LRDP developed for a cluster of villages under the NABARD watershed development program

fuelwood plantation is presently under agricultural practice, then a modest change such as agro-horticulture or agroforestry is recommended. However, for a similar site, if the slope is very steep, then it becomes the limiting factor. Hence, agricultural practice is ruled out, and an altogether new land use practice like silvopasture or fodder and fuelwood plantation is recommended. The threshold limit of a particular parameter vis-à-vis its consideration for a particular recommended land use practice varies from area to area.

Futuristic interventions on the availability and exploitation of groundwater and water management by micro-irrigation techniques and other site improvements through soil and water conservation measures are also considered while preparing the LRDP. Availability of improved varieties of crops, trees, shrubs, and grasses and advantages of interdependency of agriculture, livestock, and other practices as in case of integrated farming system that have been made available through contemporary research are also taken into consideration.

An example of micro level LRDP generated for a cluster of villages (Kuralpurna Cluster) under the NABARD holistic watershed development program is shown in Fig. 29.5.

The action plan depicts the type of development that can be taken up in each parcel. The prescription is based on the potential of the existing resources like soil potential, slope, water availability, and the present land use of each parcel. All these layers have been integrated in GIS and overlaid with georeferenced cadastral map of the cluster of villages under consideration. The stretches along the river in this cluster of villages are degraded land undergoing the process deep gully erosion. These lands will be further converted to ravines if left without proper treatment. The LRDP for these parcels of land is therefore prescribed as stabilization by vegetative measures including grassland with trees (GT) and afforestation (AF). Remaining parcels are prescribed based on their land potential either crop cultivation (CC) or dry land horticulture (DH).

29.7.2 Water Resources Development Action Plan

Water resources development and conservation is one of the essential activities in the micro level planning. This caters to generation of water resources development action plans for the micro-watershed or village (*gram panchayat*) involving identification of potential sites for groundwater well drilling, water harvesting, and groundwater recharge. The identification of potential sites for water harvesting is an important step toward maximizing water availability and land productivity of any area.

The water resources development action plan mainly suggests locale-specific groundwater development plan and water conservation measures to be taken up in the area under planning. The development measures include the suggestions on the type of groundwater abstraction structures suitable in an area based on the nature of the subsurface formation, the landforms, and the groundwater dynamics in an area. With the availability of high-resolution satellite data, it is now possible to map village-level geomorphology, minor fractures/lineaments, and detailed terrain information from high-resolution DEM. Integrating all these parameters in GIS will provide us with the village-level groundwater prospect site map (Fig. 29.6). This map will narrow down to two or three zones of good groundwater potential within the village. The exact point for drilling a well within these zones is identified based on ground geophysical vertical electrical soundings.

The water conservation measures include the area-specific suggestions about the type of water harvesting and groundwater recharge structures to be taken up under the limited constraints of the various natural resources criteria and the socioeconomic feasibility.

The main objectives of surface water harvesting are storing excessive runoff for future use, augmenting percolation/recharge to aquifers, checking of floods, and checking of land erosion. Water harvesting measures must be implemented on a watershed basis for effectiveness and for better management. Appropriate location of water harvesting structures is important for effectiveness of these structures. There are a number of parameters, which need to be considered for selecting sites for a water harvesting structure. The essential parameters, which control the location and type of the water harvesting structures, are nature of subsurface soil and rock strata and their thickness, extent of weathering and lateral variations, geological structures, depth to bedrock, topography/slope, geomorphology, drainage characteristics, and

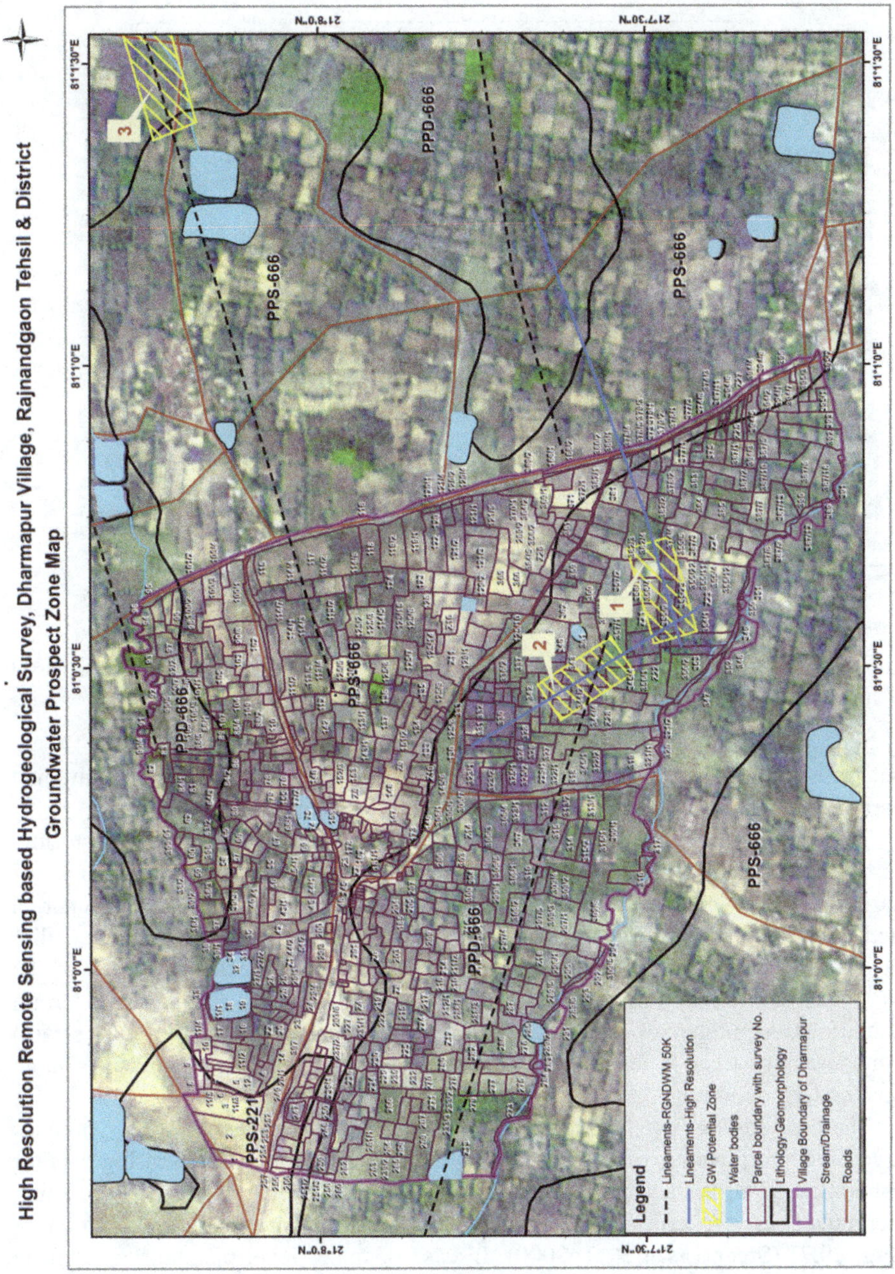

Fig. 29.6 Micro level groundwater prospect zone map

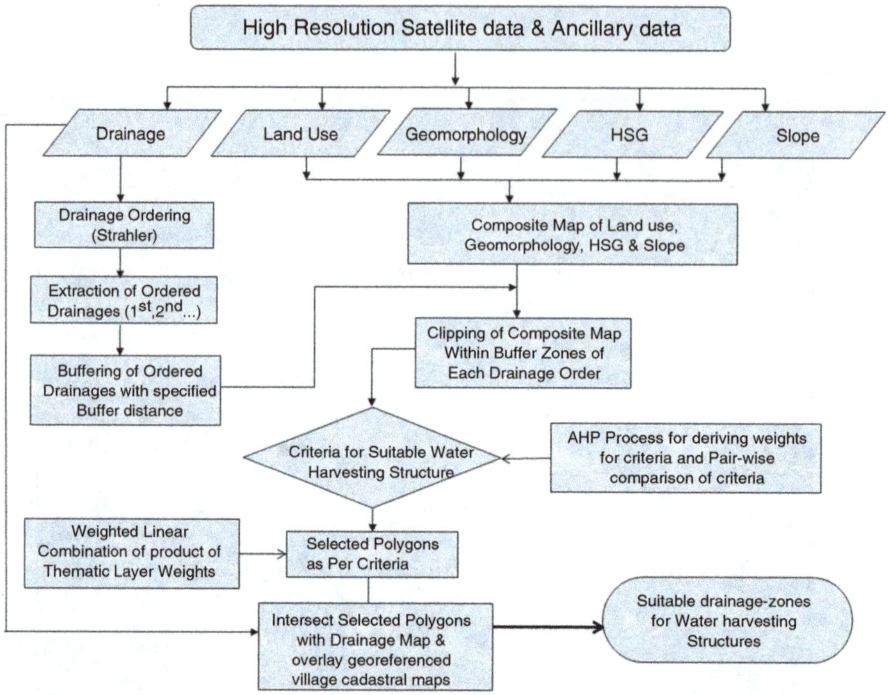

Fig. 29.7 Methodology for identification of suitable sites for water harvesting structures (modified after Sreenivasan et al. 2000)

land use pattern in an area. The information on many of the abovementioned parameters can be easily derived from the satellite remote sensing images.

There are four different methods of water harvesting, viz., obstruction, storing, spreading, and injection. The method to be adopted depends on the purpose of harvesting, i.e., whether the surface water harvesting is for storage or for recharge. It also depends on the site conditions like subsurface geology and soils, permeability of the subsurface material, the stream/drainage conditions, catchment area of the structure, etc. Depending on the type of water harvesting method and terrain conditions, a number of water harvesting structures are prevalent, such as farm ponds, minor irrigation tanks, check dams, nala bunds, percolation tanks, recharge pits, underground bhandaras, recharge wells, etc.

The methodology for locating sites for water harvesting involves a set of steps involving interpretation of information from high-resolution remote sensing data for generation of all the necessary thematic layers and further uses the GIS modeling techniques as per criteria requirement for specific water harvesting structure (Fig. 29.7) (Sreenivasan et al. 2000). In case of micro level planning, apart from considering the micro-watershed as the planning unit, it is also essential to consider the beneficiary villages and the people of the village. Therefore, the most important aspect that is considered here is the overlay of the georeferenced village cadastral

Table 29.4 Sample criteria for locating water harvesting structures (modified after Sreenivasan et al. 2010)

Water harvesting structure	Drainage order	Hydrological soil group	Land use	Slope	Geomorphology
Check dam	2	C, D	Wastelands/permanent fallows	< 10%	Pediplain/pediments/valley fill
Nala bund	1	C, D	Wastelands	< 15%	Pediplain/pediment
Percolation tank	> 3	B, C, D	Wastelands	< 5%	Pediplain/alluvial plain/valley fill
Kolhapur-type weir	> 3	C, D	Wastelands/permanent fallows	< 5%	Pediments, shallow Pediplains
Underground bhandara	2, 3	A, B	Wastelands/permanent fallows	< 5%	Pediplain/valley fill

maps onto the high-resolution satellite images. This facilitates the identification of exact location of the structure at parcel level, and therefore the group of beneficiaries also can be identified.

Most of the water harvesting structures are constructed along streams, as the rainfall runoff that is flowing in the streams is harvested. Therefore, the first step in GIS is the extraction of buffer area along the streams of the different orders. The selection of the order of the stream depends on the type of water harvesting structure that is planned (Table 29.4). All the thematic layers are then clipped with the drainage buffer layer. Drainages of lower orders, like first order, usually are not selected, as these drains will not be able to give required volume of water either for storage or recharge. These first-order drainages will be ideal for locating soil conservation structures like "nala bunds."

Criteria have been developed for locating various water harvesting structures based on the spatial parameters, viz., drainage order, geomorphology, slope, hydrological soil groups, land use, etc. (NRSA 1995; Oweis et al. 1998, 2012). The sample criteria are given in Table 29.4 (Sreenivasan et al. 2010). The criteria may include many other parameters like lithology, groundwater levels, etc., as per their availability for planning area. The criteria are flexible and could be altered based on the geographical setup and terrain conditions to optimize the process of selection of best zones for locating the water harvesting structures.

Selection of suitable locations for the water harvesting structures is carried out using multi-criteria decision analysis (MCDA) through analytical hierarchical process (AHP) for deriving weight for individual criteria and pairwise comparison of criteria. Initially, GIS is used to combine the thematic spatial layers representing the criteria selected for this analysis. Once all the thematic spatial layers are extracted, MCDA is applied within a GIS environment. AHP is used to identify the weights for each criterion or to determine the relative weight of each criterion (Tsiko and Haile 2011). The relative weights between criteria are determined by applying pairwise comparison matrices (PCMs) and assigning the weights to the thematic layers. The PCMs involve comparing all the possible pairs of criteria in order to determine which of all the criteria is of a higher priority. Saaty (1990) suggests a scale from 1 to 9 for PCM elements. The value of 1 on this scale indicates that all the criteria are equally important, whereas a value of 9 on this scale indicates that the criterion under consideration is extremely important compared to the other criteria. The assessment of the relative importance of each criterion to the other is usually done by domain experts' knowledge and judgment. PCM includes a consistency check where judgment errors are identified and a consistency ratio is calculated (Al-shabeeb 2016). The assessment of the consistency through pairwise comparisons to assign the consistency ratio (CR) involves calculating eigenvalues and eigenvectors of the square preference matrix. Finally the potential sites for water harvesting will be identified by weighted linear combination of the product of the thematic layer weights and the rating of the individual criterion within each thematic layer. Final selection of sites is done by overlaying the drainage layer on the final cumulative weighted layer (Fig. 29.8).

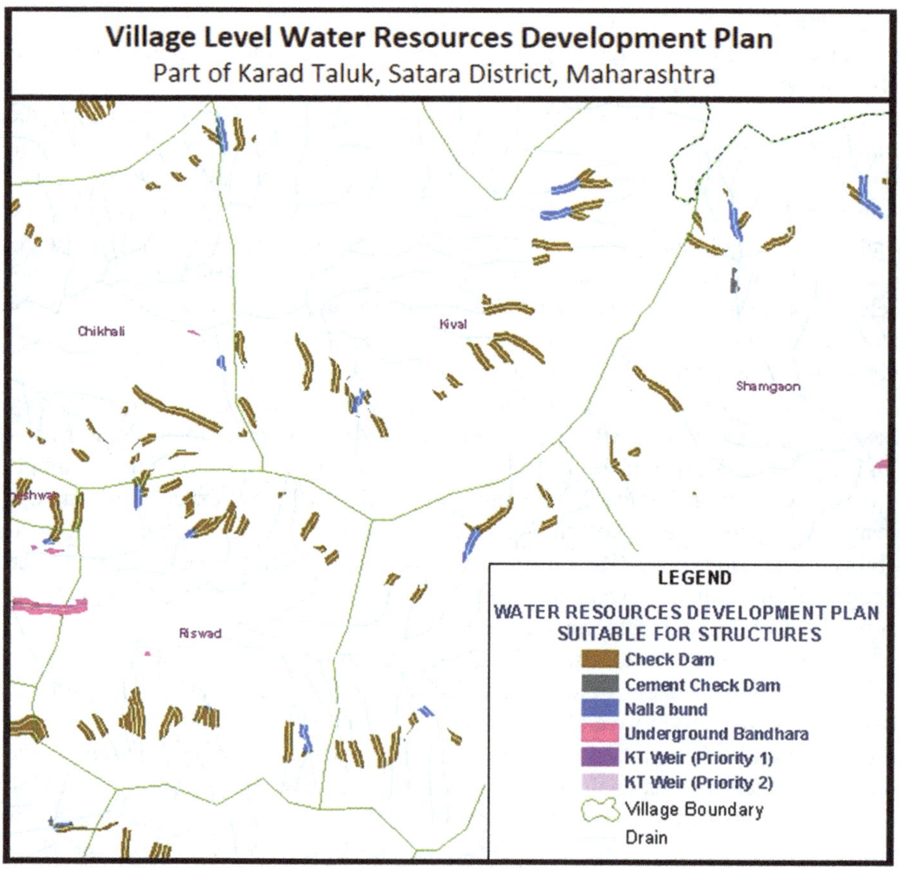

Fig. 29.8 Example of a village-level water resources development plan showing suitable zones for various water harvesting structures

29.8 Monitoring and Evaluation of the Micro Level Planning

Monitoring and evaluation of the impact of developmental plans taken up in the process of micro level planning is an important component in the management of natural resources. Monitoring is done during the process of implementation of the developmental plans. This is mainly to assess the overall improvement in the health of the micro-watershed/village and to evaluate the long-term effects and impact of the developmental activities taken up. Monitoring refers to the continuous assessment of project activities in the context of implementation schedules, the use and allocation of the project inputs by targeted beneficiaries, and the measurement of progress against the stated goals. It helps in the mid-course correction and to identify the factors responsible for success or failure of the planning process (NRSA 1995). Evaluation refers to periodic assessment of the relevance, performance, efficiency, and impact of the project in the context of its objectives. Evaluation of the micro level planning is carried out during the process of implementation as well as after the implementation of the plans, and this may continue for few years after implementation of the plans.

The monitoring and evaluation of watersheds mainly involves the establishment of the baseline information, tracking and documenting the physical progress and achievements vis-à-vis the set goals for the micro-watershed/village under consideration, developing a system for comparative analysis of project components, providing a mechanism for evaluating the need for project achievements, and executing studies to evaluate project performance (Dutta et al. 2002).

The impact evaluation is carried out at three levels, (i) at the study area level where action plans are implemented, (ii) in the surrounding buffer areas of the study area to account for impact due to the natural conditions, and (iii) at the parcel level where an individual farmer is the beneficiary (Shanmuganathan 1998).

The inherent capability of the satellites to visit the same area again and again within a scheduled period and the availability of high-resolution, multispectral datasets which can give synoptic and unbiased view of any area have facilitated the monitoring of the changes in a micro-watershed over a long time period. Satellite remote sensing technology is the most cost-effective tool for long-term monitoring and evaluation of the implemented developmental plans. This will further help us in evaluating the interventions and provide feedback for mid-course corrections if necessary.

There are many parameters by which the impact of developmental planning in any area can be monitored and evaluated. The following are important spatial parameters that can be analyzed through the satellite remote sensing and GIS technologies.

- Water bodies – number and extent: Can be mapped directly from the high-resolution satellite images. Indicates the increase or decrease of total water availability in the planning area, which has a direct impact on the agriculture, drinking water for human and livestock.

- Cropped area: Changes in areal extent, cropping pattern, extent of irrigated area (to infer groundwater status).
- Cropping pattern: Change in the cropping pattern to a sustainable nature. Switchover from marginal croplands to agro-horticulture, agroforestry, and increase in green cover are positive indicators of development.
- Productivity: Can be assessed using the satellite-based vegetation indices like Normalized Difference Vegetation Index (NDVI). Increase in productivity indicates positive impact of the developmental plan implementation.
- Overall increase in biomass: Can be assessed through NDVI. Indicates increase in green cover, and total biomass, thus storing more per capita carbon in the planning area. It can also provide support of livestock by increase in area under fodder cultivation.
- Irrigated cropped area: In rain fed agriculture areas, it can be mapped from the summer season satellite data. Usually, increase in cropped area in the summer season indicates groundwater-based irrigation. This indicates improvement in the groundwater status in the planning area.
- Moisture content in the soils: Sustenance of soil moisture for longer periods after monsoon season is a positive indicator of development. This can lead to increase in the *rabi* cropland, thus increasing the overall agricultural produce in the planning area.
- Forest density and vigor: Can be monitored through the NDVI. Indicates overall increase in the vegetation canopy and vigor of vegetation.
- Change in the overall land use pattern: Changes in areal extent of wastelands is one of the important parameters that can be monitored through the satellite remote sensing. Reclaiming wastelands and bringing them to productive use for agriculture or setup of solar power plants or other industries based on their potential can be planned during the micro level planning process.

Remote Sensing-Based Change Analysis

The assessment of changes with respect to many parameters mentioned above can be done using the satellite remote sensing time series datasets. The change analysis is carried out by selecting the pretreatment and posttreatment period satellite images. To start with, both image data are radiometrically corrected to remove the atmospheric interference effects. Then the land use/cover is classified using the standard digital image processing algorithms like maximum likelihood supervised classification techniques. More emphasis is given on the categories of cropland, forest and other natural vegetation, plantations, water bodies and streams, and wastelands. Vegetation indices such as Normalized Difference Vegetation Index (NDVI) are derived to assess the extent and vigor of vegetation in the area. NDVI is calculated using the red and near-infrared (NIR) band multispectral satellite data using the algorithm: NDVI = NIR-Red/NIR + Red. The comparison of pre-implementation and post-implementation NDVI maps for the planning area will give us the direct indications of the two important parameters, viz., increase in the extent and vigor of vegetation. These two parameters have direct relation with many other parameters like increase in the number and/or extent of surface water bodies and improvement in the groundwater status in the watershed and increase in the overall agricultural and natural vegetation status. The land use/cover and the NDVI maps together give the required information for geospatial assessment of the impact of the implementation of the development plans under the micro level planning activity. The geospatial assessment can be further integrated with socioeconomic impact indicators to make a complete evaluation of the micro level planning.

29.9 Conclusions

Geospatial tools and techniques involving combined use of remote sensing, GIS, and GPS technologies play an important role in generating the vital spatial parameters on natural resources. Further, these tools and techniques facilitate the integration and analysis of the spatial parameters with non-spatial database to generate application-specific developmental plans for micro level planning to achieve sustainable development of natural resources, their management, and conservation. The latest advancements in remote sensing technology in terms of high-resolution imaging with better spatial and radiometric resolution are immensely useful for deriving the geo-environmental indicators that can be integrated with collateral data and social indicators as inputs for preparation of developmental plans at micro-watershed and village (*gram panchayat*) levels. Synoptic view, high-resolution, multispectral, repetitive coverage of the satellite remote sensing data offers appropriate method for quick, unbiased mapping and monitoring of natural resources both in space and time domain. Timely and accurate information on spatial distribution of land use, soil, vegetation density, forest, geology and geological structures, drainage and surface water resources, etc. can be generated, collated, and integrated with socio-economic information. Apart from this, remote sensing data in conjunction with collateral data helps in generation of baseline information such as delineation of watershed boundaries, characterization and prioritization of micro-watersheds for development, identifying the erosion-prone areas, etc. HRSI also are important secondary source for georeferencing of village cadastral maps. The integration of these maps with the large-scale thematic database and the development action plans is essential for micro level planning.

GIS is a powerful tool for integration of natural resource information with socioeconomic data for assessment of the existing infrastructure to arrive at developmental schemes and the generation of locale-specific action plans for land and water resources at grassroots level. Since the sustainable development of natural resources is based on maintaining a fragile balance between productivity functions and conservation practices through monitoring and identification of problem areas and combined analysis of a multitude of parameters, remote sensing and GIS make the task easy, thus reducing the timeline for planning and implementation of the micro level plans for land and water resources development. Integrating space-based Earth observation (EO) products and services with multi-institutional framework and people's participation in decision-making process relevant to society is the principle for future direction in management of natural resources through micro level planning approach.

Acknowledgments The authors thank the teams of scientists from Regional Remote Sensing Centre, Nagpur, who have contributed in carrying out the different remote sensing application projects, the examples of which have been quoted in this work.

References

Adiga S, Krishna Murthy YVN (2000) Integrated sustainable development of land and water resources using space technology inputs. Space Forum 5:179–202

Al-shabeeb AR (2016) The use of AHP within GIS in selecting potential sites for water harvesting sites in the Azraq Basin, Jordan. J Geogr Inf Syst 8:73–88

Bernstein R (1983) Manual of remote sensing, vol. 1. American Society of Photogrammetry

Dutta D, Sharma JR, Adiga S (2002) Watershed characterization, prioritization, development planning and monitoring – remote sensing approach. ISRO-NNRMS-TR-103-2002. Indian Space Research Organisation, Bangalore

FAO (2011) The state of the world's land and water resources for food and agriculture (SOLAW) – managing systems at risk. Food and Agriculture Organization of the United Nations, Rome and Earthscan, London

Gopal Rao K, Nisar Ahamed TR, Murthy JSR (2000) GIS-based fuzzy membership model for crop-land suitability analysis. J Agric Syst 63(2):75–95

Gupta RP (2003) Geological applications. In: Remote sensing geology. Springer, Berlin/Heidelberg, pp 429–592

Kale P (1992) Sustainable development-Critical issues. J Indian Soc Remote Sens 20(4):183–184

Krishna Murthy YVN, Srinivasa Rao S, Srinivasan DS, Radhakrishnan K (1996) Land information systems for rural development. In: IX INCA conference, New Delhi

Krishna Murthy YVN, Srinivasa Rao S, Srinivasan DS, Adiga S (2000).A land information system (LIS) for rural development. Geomatics-2000, Pune

Manchanda ML, Kudrat M, Tiwari AK (2002) Soil survey and mapping using remote sensing. Trop Ecol 43(1):61–74

Mather PM (1999) Land cover classification revisited. In: Atkinson PM, Tate NJ (eds) Advances in remote sensing and GIS analysis. John Wiley, Chichester, pp 7–16

Melesse AM, Weng QS, Thenkabail P, Senay GB (2007) Remote sensing sensors and applications in environmental resources mapping and modelling. Sensors (Basel) 7(12):3209–3241

National Remote Sensing Agency (NRSA) (1995) Integrated mission for sustainable development. Technical guidelines for field survey and mapping, Hyderabad, p 127

National Remote Sensing Centre (NRSC) (2010a) Preparation of geospatial layers using high resolution (Carosat-1 PAN + LISS-IV mx) Ortho rectified satellite imagery. Technical document: manual of space based information for decentralized planning (SIS-DP). National Remote Sensing Centre, Hyderabad, p 237

National Remote Sensing Centre (NRSC) (2010b) Manual for national geomorphological and lineament mapping. Technical manual, NRSC-RS&GISAA-ERG-G&GD-FEB' 10-TR149. NRSC, Hyderabad, p 146

Navalgund RR, Jayanthi SC (2004) Role of earth observations for sustainable development: emerging trends. In: Proceedings of XXth ISPRS congress, technical commission VII, July 12–23 2004, Istanbul, Turkey. ISPRS Archives – Volume XXXV, Part B7

Navalgund RR, Jayaraman V, Roy PS (2007) Remote sensing applications – an overview. Curr Sci 93(12):1747–1766

Openshaw S (1990) Spatial analysis and GIS: a review of progress and possibilities. In: Scholten HJ, Stillwell JCH (eds) Geographic information systems for urban and regional planning. Kluwer, Dordrecht, pp 156–163

Oweis T, Oberle A, Prinz D (1998) Determination of potential sites and methods for water harvesting in Central Syria. Adv GeoEcol 31:83–88

Oweis T, Prinz D, Hachum AY (2012) Rainwater harvesting for agriculture in the dry areas. CRC Press, London, p 262

PAGE (2016) Integrated planning & sustainable development: challenges and opportunities. Partnership for Action on Green Economy (PAGE), UNDP

Ramamohana Rao P, Suneetha P (2015) Micro level planning and rural development through remote sensing & GIS. LAP Lambert Academic Publishing, p 200

Saaty TL (1990) How to make a decision: the analytic hierarchy process. Eur J Oper Res 48:9–26

Shanmuganathan (1998) Cost-benefit analysis of remote sensing applications. IIM-Lucknow

Sreenivasan G, Gupta AK, Sharma JR (2000) Selection of sites for water harvesting and ground water recharge structures using remote sensing and GIS approach. In: National Seminar on water conservation and drought proofing of arid areas and its people. Social Policy Research Institute, Jaipur, Rajasthan

Sreenivasan G, Joshi AK, Krishna Murthy YVN (2010) Role of remote sensing and GIS in integrated watershed development. In: National symposium on "The application of nuclear and space technologies for the benefit of farmers". Consortium of Indian Farmers Associations (CIFA), Raipur

Srinivasa Rao S, Krishna Murthy YVN, Joshi AK, Shantanu B, Das SN, Pandit DS (2003) Computerisation and geo-referencing of cadastral maps in Chhattisgarh State. Technical document. Regional Remote Sensing Service, ISRO, Nagpur

Sunitha V, Abdullah Khan J, Ramakrishna Reddy A (2016) Evaluation of groundwater resource potential using GIS and remote sensing application. Int J Eng Res Appl 6(1):116–122

Tsiko RG, Haile TS (2011) Integrating geographical information systems, fuzzy logic and analytical hierarchy process in modelling optimum sites for locating water reservoirs – a case study of the Debub District in Eritrea. J Water 3(1):254–290

Wang G, Mang S, Cai H, Liu S, Zhang Z, Wang L, Innes JL (2016) Integrated watershed management: evolution, development and emerging trends. J For Res 27(5):967–994

Index

G. P. O. Reddy, S. K. Singh (eds.), *Geospatial Technologies in Land Resources Mapping, Monitoring and Management*, Geotechnologies and the Environment 21,
https://doi.org/10.1007/978-3-319-78711-4

CPSIA information can be obtained
at www.ICGtesting.com
Printed in the USA
LVHW06*1746260918
591446LV00002B/2/P

9 783319 787107